Innovative Materials for Industrial Applications:

Synthesis, Characterization and Evaluation

Assia Mabrouk
Ibn Zohr University, Morocco

Ahmed Bachar
Ibn Zohr University, Morocco

Seitkhan Azat
Satbayev University, Kazakhstan

Rachid Amrousse
University of Chouaib Doukkali, Morocco

IGI Global
Scientific Publishing
Publishing Tomorrow's Research Today

Published in the United States of America by
IGI Global Scientific Publishing
701 East Chocolate Avenue
Hershey, PA, 17033, USA
Tel: 717-533-8845
Fax: 717-533-8661
E-mail: cust@igi-global.com
Website: https://www.igi-global.com

Library of Congress Cataloging-in-Publication Data

CIP DATA PENDING

ISBN13: 9798369375051
EISBN13: 9798369375075

Vice President of Editorial: Melissa Wagner
Managing Editor of Acquisitions: Mikaela Felty
Managing Editor of Book Development: Jocelynn Hessler
Production Manager: Mike Brehm
Cover Design: Phillip Shickler

British Cataloguing in Publication Data
A Cataloguing in Publication record for this book is available from the British Library.

Table of Contents

Preface xiii

Acknowledgment xix

Chapter 1
Cellular Ceramics Used as Catalytic Supports for Heterogeneous Catalyst Synthesis 1
Youssef Hairch, University of Chouaib Doukkali, Morocco
Ahmed E. S. Nosseir, Institute of Mechanical Intelligence Scuola Superiore Sant'Anna, Pisa, Italy
Meiram Atamanov, Kazakh National Women's Teacher Training University, Almaty, Kazakhstan
Rachid Amrousse, University of Chouaib Doukkali, Morocco

Chapter 2
Model Matrices for Nuclear Waste Storage: Structure, Characterization, and Mechanical Properties 39
Assia Mabrouk, Science and Technology Research Laboratory, The Higher School of Education and Training, Ibn Zohr University, Agadir, Morocco
Ahmed Bachar, Laboratory of Process Engineering, Faculty of Applied Science Ait Melloul, Ibn Zohr University, Agadir, Morocco
Aurélien Canizarès, CNRS, CEMHTI UPR 3079, Université d'Orléans, France
Yann Vaills, CNRS, CEMHTI UPR 3079, Université d'Orléans, France
Stuart Hampshire, The Bernal Institute, University of Limerick, Ireland

Chapter 3
Slow Sand Filtration for Water Treatment 75
Yelriza Nurakimkyzy Yeszhan, Satbayev University, Kazakhstan
Shynggyskhan Sultakhan, Satbayev University, Kazakhstan
Erzhan Kuldeyev, Satbayev University, Kazakhstan
Bostandyk Khalkhabay, Laboratory of Engineering Profile, Satbayev University, Kazakhstan
Qin Xu, Yangzhou University, China
Seitkhan Azat, Satbayev University, Kazakhstan

Chapter 4
New Generation of Glass Materials for Biomedical Applications: Properties, Structure, Bioactivity, and Viability 97
Ahmed Bachar, Laboratory of Process Engineering, Faculty of Applied Science Ait Melloul, Ibn Zohr University, Agadir, Morocco
Assia Mabrouk, Science and Technology Research Laboratory, The Higher School of Education and Training, Ibn Zohr University, Agadir, Morocco
Cyrille Mercier, CERAMATHS, University Polytechnique Hauts-de- France, France
Claudine Follet, CERAMATHS, University Polytechnique Hauts-de- France, France
Franck Bouchart, CERAMATHS, University Polytechnique Hauts-de- France, France
Arnaud Tricoteaux, CERAMATHS, University Polytechnique Hauts-de- France, France
Stuart Hampshire, The Bernal Institute, University of Limerick, Ireland

Chapter 5
Next-Gen Energy Storage Devices: Carbon-Based Materials for Flexible Supercapacitors 135
Krishna Kumar, Guru Gobind Singh Indraprastha University, India
Uplabdhi Tyagi, Guru Gobind Singh Indraprastha University, India
Saurav Kumar Maity, Guru Gobind Singh Indraprastha University, India
Gulshan Kumar, Guru Gobind Singh Indraprastha University, India

Chapter 6
Ammonium Dinitramide (ADN) Decomposition as Green Propellant: Overview of Synthesized Catalysts 169
Zakaria Harimech, Chouaib Doukkali University, Morocco
Mohammed Salah, Chouaib Doukkali University, Morocco
Rachid Amrousse, Chouaib Doukkali University, Morocco

Chapter 7
Molecular Interaction of Lactams With Mild Steel in Hydrochloric Acid Environment: Corrosion Inhibition Efficiency and Surface Adsorption Mechanisms 195
Nadia Faska, Faculty of Applied Sciences Ait Melloul, Ibn Zohr University, Morocco
Soukayna Maitouf, Ibn Zohr University, Morocco
Brahim Orayech, Powerco Battery, Spain

Chapter 8
Sol-Gel Synthesis of a New Composition of Bioactive Glass 225
Halima El Bouami, CERAMATHS, Université Polytechnique Hauts-de-France, France & Laboratory of Process Engineering, Faculty of Applied Science Ait Melloul, Ibn Zohr University, Agadir, Morocco
Assia Mabrouk, Science and Technology Research Laboratory, The Higher School of Education and Training, Ibn Zohr University, Agadir, Morocco
Cyrille Mercier, CERAMATHS, Université Polytechnique Hauts-de-France, France
Claudine Follet, CERAMATHS, Université Polytechnique Hauts-de-France, France
Ahmed Bachar, Laboratory of Process Engineering, Faculty of Applied Science Ait Melloul, Ibn Zohr University, Agadir, Morocco

Chapter 9
Electrochemical Performance of Biomass-Derived AC and CNTs-Based Supercapacitors 247
Meiram Atamanov, Kazakh National Women's Teacher Training University, Kazakhstan
Tolganay Atamanova, Al-Farabi Kazakh National University, Kazakhstan
Azamat Taurbekov, Al-Farabi Kazakh National University, Kazakhstan
Zulkhair Mansurov, Al-Farabi Kazakh National University, Kazakhstan

Chapter 10
Engineering the Invisible: Synthesis, Characterization, and Applications of Nanoporous Materials 271
Krishnappa Madhu Kumar, Sir M. Visvesvaraya Institute of Technology, Bengaluru, India
Manjappa Kiran Kumar, Sir M. Visvesvaraya Institute of Technology, Bengaluru, India
Nagaraj Sasi Kumar, Sir M. Visvesvaraya Institute of Technology, Bengaluru, India
Lakshman N. Sampath Kumar, Sir M. Visvesvaraya Institute of Technology, Bengaluru, India

Chapter 11
Synthesis and Characterization of Activated Carbon From Biomass 329
Makhabbat Kunarbekova, Satbayev University, Kazakhstan
Rosa Busquets, Kingston University, UK
Inabat Sapargali, Satbayev University, Kazakhstan
Laura Seimukhanova, Satbayev University, Kazakhstan
Ulan Zhantikeyev, Satbayev University, Kazakhstan
Kenes Kudaibergenov, Satbayev University, Kazakhstan
Seitkhan Azat, Satbayev University, Kazakhstan

Chapter 12
Bio-Piezoelectric Ceramic Coatings for Bone Tissue Engineering Applications 349
John Henao, National Council of Humanities, Science, and Technology (CONAHCYT), Mexico & CIATEQ A.C., Queretaro, Mexico
Astrid Giraldo Betancur, Center for Research and Advanced Studies of the National Polytechnic Institute, Mexico
Adriana Gallegos, National Council of Humanities, Science, and Technology (CONAHCYT), Mexico & InnovaBienestar de Mexico, San Luis Potosí, Mexico
Andrea Yamile Resendiz Mancilla, Center for Research and Advanced Studies of the National Polytechnic Institute, Mexico
Carlos Poblano Salas, CIATEQ, Queretaro, Mexico

Conclusion 393

Compilation of References 395

About the Contributors 479

Index 485

Detailed Table of Contents

Preface xiii

Acknowledgment xix

Chapter 1
Cellular Ceramics Used as Catalytic Supports for Heterogeneous Catalyst Synthesis 1
Youssef Hairch, University of Chouaib Doukkali, Morocco
Ahmed E. S. Nosseir, Institute of Mechanical Intelligence Scuola Superiore Sant'Anna, Pisa, Italy
Meiram Atamanov, Kazakh National Women's Teacher Training University, Almaty, Kazakhstan
Rachid Amrousse, University of Chouaib Doukkali, Morocco

Cellular ceramics have become essential in various industrial applications. This chapter explores the types, shapes, and characteristics of cellular ceramics, and their potential uses in space propulsion and catalysis. There are different varieties, including foams, honeycombs, and reticulated structures, classified by their composition, structure, and production processes. These materials are favored for their high porosity, low density, excellent mechanical strength, and thermal stability. Their large surface area and enhanced mass transfer capabilities make them effective as catalyst supports in processes like hydrocarbon reforming, oxidation, and hydrogenation, where they immobilize catalytically active species to boost activity and selectivity. In space propulsion, cellular ceramics are used in thruster components due to their ability to withstand high temperatures and maintain structural integrity under harsh conditions.

Chapter 2
Model Matrices for Nuclear Waste Storage: Structure, Characterization, and Mechanical Properties 39
Assia Mabrouk, Science and Technology Research Laboratory, The Higher School of Education and Training, Ibn Zohr University, Agadir, Morocco
Ahmed Bachar, Laboratory of Process Engineering, Faculty of Applied Science Ait Melloul, Ibn Zohr University, Agadir, Morocco
Aurélien Canizarès, CNRS, CEMHTI UPR 3079, Université d'Orléans, France
Yann Vaills, CNRS, CEMHTI UPR 3079, Université d'Orléans, France
Stuart Hampshire, The Bernal Institute, University of Limerick, Ireland

Glasses enriched with rare earth elements hold significant promise for nuclear waste storage, especially for immobilizing high-level waste (HLW) containing minor actinides. This chapter delves into the structural intricacies of peraluminous lanthanum-rich sodium aluminoborosilicate glasses, where lanthanum serves as a surrogate for lanthanides and minor actinides typical of HLW. Given the imperative to accommodate higher concentrations of HLW, the aluminum/boron substitution emerges as a pivotal factor influencing the glass matrix's properties. Employing a combination of infrared (IR), nuclear magnetic resonance (NMR), and Brillouin spectroscopy, we unravel the intricate interplay between composition and structure. Our investigation sheds light on the effects of aluminum/boron substitution on phase separation phenomena and structural rearrangements within the glass network. This in-depth analysis provides critical insights into the design and optimization of novel glass matrices tailored for advanced nuclear waste immobilization strategies.

Chapter 3
Slow Sand Filtration for Water Treatment 75
Yelriza Nurakimkyzy Yeszhan, Satbayev University, Kazakhstan
Shynggyskhan Sultakhan, Satbayev University, Kazakhstan
Erzhan Kuldeyev, Satbayev University, Kazakhstan
Bostandyk Khalkhabay, Laboratory of Engineering Profile, Satbayev University, Kazakhstan
Qin Xu, Yangzhou University, China
Seitkhan Azat, Satbayev University, Kazakhstan

Slow sand filtration (SSF) is one of the traditional methods of water treatment yet gaining attention as a promising method. The rising demand is owing to its simplicity, environmental friendliness and effectiveness, which can reach 90-95%. This chapter reviews the principles and mechanisms underlying SSF, particularly addressing the physical, biological, and chemical processes that contribute to its effectiveness. Furthermore, the chapter delves into the design and construction of SSF systems, with a particular emphasis on the critical components and factors that facilitate the attainment of optimal performance. The efficacy of the method in enhancing water quality is assessed through a variety of analytical methods, including the removal of turbidity, the reduction of pathogens, and the removal of organic matter. The chapter also encompasses the environmental impact assessment of SSF to guarantee its long-term sustainability.

Chapter 4
New Generation of Glass Materials for Biomedical Applications: Properties, Structure, Bioactivity, and Viability 97
Ahmed Bachar, Laboratory of Process Engineering, Faculty of Applied Science Ait Melloul, Ibn Zohr University, Agadir, Morocco
Assia Mabrouk, Science and Technology Research Laboratory, The Higher School of Education and Training, Ibn Zohr University, Agadir, Morocco
Cyrille Mercier, CERAMATHS, University Polytechnique Hauts-de- France, France
Claudine Follet, CERAMATHS, University Polytechnique Hauts-de- France, France
Franck Bouchart, CERAMATHS, University Polytechnique Hauts-de- France, France
Arnaud Tricoteaux, CERAMATHS, University Polytechnique Hauts-de- France, France
Stuart Hampshire, The Bernal Institute, University of Limerick, Ireland

The limited strength of bioglasses has confined their application to non-load-bearing contexts, thus necessitating enhancements in their mechanical properties. A viable approach to surmount this hurdle involves integrating nitrogen into the silicate network of the glass. This outlines the effect of nitrogen addition in bioglasses of the system SiO2-Na2O-CaO. The purpose is to determine the effects of nitrogen addition on the physical and mechanical properties and the structure of oxynitride bioglasses based on the system Na2O-CaO-SiO2-Si3N4. Properties were all observed to increase linearly with nitrogen content. These increases are consistent with N in the glass structure in 3-fold coordination with silicon and extra cross-linking of the glass network. These increases are consistent with the incorporation of N into the glass structure in three-fold coordination with silicon with result in extra cross-linking of the glass network.

Chapter 5
Next-Gen Energy Storage Devices: Carbon-Based Materials for Flexible Supercapacitors 135
Krishna Kumar, Guru Gobind Singh Indraprastha University, India
Uplabdhi Tyagi, Guru Gobind Singh Indraprastha University, India
Saurav Kumar Maity, Guru Gobind Singh Indraprastha University, India
Gulshan Kumar, Guru Gobind Singh Indraprastha University, India

The rapid growth of portable electronics and an increased need for sustainable energy sources have grown interest in developing next-generation energy storage technologies. This chapter deals with the development and optimization of supercapacitors synthesized with carbon materials such as graphene, carbon nanotubes, activated carbon, and carbon aerogels. In this regard, the recent progress in the synthesis of carbon-based materials is reviewed relevance to capacitance, conductivity, and mechanical stability. Moreover, this chapter gives overall coverage of issues relating to the production and fabrication of carbon-based flexible supercapacitors. Additionally, the applicability of these carbon-based flexible electrodes in wearable electronics, flexible displays, and next-generation smart devices is explained in detail. The chapter also discussed the general overview of the present status of carbon-based flexible supercapacitors in next-generation energy storage devices, in addition to the future perspectives.

Thermal decomposition of eco-friendly propellants such us ammonium dinitramide (ADN) aims to replace hydrazine in satellite systems. ADN, with formula [NH4]+[N(NO2)2]−, is a promising high-performance rocket propellant. It decomposes cleanly, producing gases such as NH3, H2O, NO, N2O, NO2, HONO, and HNO3, making it an attractive alternative to ammonium perchlorate (AP) and hydrazine. This chapter reviews catalyst systems for ADN decomposition, focusing on efficiency and thermal stability. Various catalysts, including metal oxides, transition metal complexes, and nanomaterials, enhance ADN decomposition. Iron and copper oxides lower decomposition temperatures, crucial for energy-efficient propellant compositions. Ruthenium and palladium complexes support homogeneous catalysis. Nanomaterials with high specific surface areas and distinct electronic activity improve ADN decomposition. Alloying carbon nanotubes with metals or using noble metal nanoparticles enhances decomposition rates at lower temperatures while maintaining thermal stability.

The inhibition effect of some lactams (Pyrrolidin-2-one, δ-valerolactam, and ε-caprolactam) on the corrosion behaviour of mild steel in 1M Hydrochloric acid solution was studied by weight loss and electrochemical techniques. The results demonstrated that both δ-valerolactam, and ε-caprolactam significantly inhibit corrosion. Specifically, δ-valerolactam achieved an inhibition efficiency of 85.2%, while ε-caprolactam exhibited a higher inhibition efficiency of 91.5%. The thermodynamic parameters governing the adsorption process such as adsorption heat, adsorption entropy, and adsorption free energy were determined and discussed. The adsorption of lactam compounds on the mild steel surface in 1M HCl follows the Langmuir adsorption isotherm model.

Bioactive glasses hold immense promise for tissue engineering and bone regeneration due to their ability to bond with living tissues and stimulate bone growth. This chapter explores the sol-gel method, a versatile technique offering advantages like lower processing temperatures and superior compositional control, for synthesizing a new bioactive glass composition. We discuss the fundamental principles of sol-gel chemistry and precursor selection for achieving the desired elements in the final glass. The chapter details the design rationale behind the new compositions, targeting the effect of therapeutic ions promotion specific antibacterial and angiogenesis properties and enhancement in bioactivity and osteoblast production. Additionally, the chapter offers a brief overview of in vitro bioactivity assessment methods for evaluating the glass's interaction with physiological fluids. Finally, we discuss the potential applications of the newly developed bioactive glass and propose future research directions for further optimization and exploration of its functionalities.

This study investigates the electrochemical performance of activated carbon (AC) derived from two biomass sources, rice husk (RH) and walnut shell (WS), utilizing two activation methods: chemical activation with potassium hydroxide (KOH) and physical activation with carbon dioxide (CO2). The aim is to evaluate the potential of these bio-derived carbon materials in energy storage applications, particularly for supercapacitors. The electrochemical behavior of the activated carbon electrodes was assessed using cyclic voltammetry (CV) and galvanostatic charge-discharge (GCD) techniques, along with electrochemical impedance spectroscopy (EIS), to evaluate the specific capacitance, energy density, charge transfer resistance, and ion diffusion properties.

Chapter 10
Engineering the Invisible: Synthesis, Characterization, and Applications of Nanoporous Materials 271
Krishnappa Madhu Kumar, Sir M. Visvesvaraya Institute of Technology, Bengaluru, India
Manjappa Kiran Kumar, Sir M. Visvesvaraya Institute of Technology, Bengaluru, India
Nagaraj Sasi Kumar, Sir M. Visvesvaraya Institute of Technology, Bengaluru, India
Lakshman N. Sampath Kumar, Sir M. Visvesvaraya Institute of Technology, Bengaluru, India

Nanoporous materials are a major research focus due to their unique structures and diverse applications in catalysis, drug delivery, and environmental remediation. This chapter reviews the synthesis, characterization, and practical applications of nanoporous materials. It examines various synthesis methods, detailing their principles, advantages, and limitations. Key characterization techniques such as scanning electron microscopy (SEM), X-ray diffraction (XRD), transmission electron microscopy (TEM), and atomic force microscopy (AFM) are discussed for assessing important properties like surface area and pore size. The chapter also highlights the performance of nanoporous materials in enhancing reaction rates in catalysis, improving drug delivery systems, and aiding in pollutant degradation. Ideally, this chapter contributes to providing researchers and practitioners with the knowledge necessary to advance the further development and application of nanoporous materials in various technological and industrial areas.

Chapter 11
Synthesis and Characterization of Activated Carbon From Biomass 329
Makhabbat Kunarbekova, Satbayev University, Kazakhstan
Rosa Busquets, Kingston University, UK
Inabat Sapargali, Satbayev University, Kazakhstan
Laura Seimukhanova, Satbayev University, Kazakhstan
Ulan Zhantikeyev, Satbayev University, Kazakhstan
Kenes Kudaibergenov, Satbayev University, Kazakhstan
Seitkhan Azat, Satbayev University, Kazakhstan

This chapter discusses current practices and new developments in the preparation of activated carbons for the adsorption of pollutants. There is widespread pollution such as pharmaceuticals and heavy metals in global surface water that needs to be mitigated and carbon sorbents made for it can be part of the solution. Preparing carbons for such challenge requires tailoring carbon's structure and surface chemistry to maximise interaction with low concentrations (part per billion level) of a variety of pollutants. The preparation of effective carbon sorbents constitutes a technical challenge. This chapter explores treatments and analytical approaches for the preparation and development of modified carbons to remove water pollutants. Effective carbons will help to control global contamination problems but should not be the source of more pollution: carbon dioxide emissions during the production and maintenance of carbon sorbents are a concern.

Chapter 12
Bio-Piezoelectric Ceramic Coatings for Bone Tissue Engineering Applications 349
John Henao, National Council of Humanities, Science, and Technology (CONAHCYT), Mexico & CIATEQ A.C., Queretaro, Mexico
Astrid Giraldo Betancur, Center for Research and Advanced Studies of the National Polytechnic Institute, Mexico
Adriana Gallegos, National Council of Humanities, Science, and Technology (CONAHCYT), Mexico & InnovaBienestar de Mexico, San Luis Potosí, Mexico
Andrea Yamile Resendiz Mancilla, Center for Research and Advanced Studies of the National Polytechnic Institute, Mexico
Carlos Poblano Salas, CIATEQ, Queretaro, Mexico

This chapter is focused on describing the latest advances related to the development of bio-piezoelectric coatings for bone regeneration applications. It starts with the description of the main concepts about bioelectrical phenomena in the human body and its role in the regeneration of bone. It explains concepts such as dielectric and electrical responses that includes piezoelectricity, pyroelectricity, and ferroelectricity and how the human bone can present these types of phenomena. The chapter also includes the definition of bio-ceramic materials and bioactive coatings, and a summary of the main bio-ceramic coatings employed nowadays for applications in bone tissue regeneration. Also, this chapter includes a review of the latest advances in the development of bio-piezoelectric coatings for bone tissue engineering and future perspectives on this topic. Overall, this chapter is focused on reviewing comprehensively the electrical response of natural tissues and the relevance of bio-piezoelectric ceramics for bone tissue regeneration.

Conclusion 393

Compilation of References 395

About the Contributors 479

Index 485

Preface

The increasing demand for creative materials in miscellaneous technical uses has incited significant progresses in material skill, manufacturing, and allure. The quest for resolutions to the progressing challenges confronted by businesses such as strength, preservation of natural resources, biomedical maneuvers, and aerospace has compelled researchers to survey new boundaries in material combination, description, and evaluation. The composite aim search out expand materials that are more effective, tenable, and compliant to the changeful needs of industrial processes.

This refined capacity, *Innovative Materials for Industrial Applications: Synthesis, Characterization and Evaluation*, presents an inclusive accumulation of 12 chapters written by masters engaged. The book covers an off-course range of contemporary materials and their useful requests across diversified rules. From catalysts and bioactive potteries to state-of-the-art nanomaterials and smart fabrics for tangible and energy requests, this book climaxes the increasing significance of interdisciplinary research in forceful novelty.

The book is arranged into having a theme sections that indicate key extents of research and mechanics progress in material science:

1. Catalysis and environmental applications: This portion involves topics in the way that natural potteries as catalysts, slow soil filtration for water situation, and creative materials for wastewater situation. It showcases fabrics that play a crucial function in converting mechanical processes and enhancing tangible sustainability.
2. Biomedical and biocompatible materials: From biography-piezoelectric pottery coatings for bone fabric design to generation after baby boom glass matters for biomedical uses, this division emphasizes the meaningful part of matters in improving healthcare and numbering healing sciences.
3. Advanced materials for aerospace and nuclear applications: The discussion on matters for environmental pollutant depository, as well as creative catalytic fabrics for space force, indicates the detracting role matters skill plays in aerospace and strength applications. The book further delves into the devices of disintegration inhibition, surveying in what way or manner leading materials can safeguard detracting foundation in these urgent environments.

Each division presents an all-encompassing investigation of the material's combining forms, characterization methods, and the efficient requests that have the potential to transform energies. The book offers not only a deep understanding of the matters themselves but further detracting intuitions into their efficiency in legitimate-world uses.

In the circumstances of an more and more pertain and tenable all-encompassing saving, this capacity specifies a timely citation for scientists, engineers, and experts the one are revere be a guest of the prominence of novelty in material skill. We hope that this book serves as two together a source of stimulus and a proficient guide for those complicated in the design and incident of fabrics that will shape the businesses of the future.

We would like to express our straightforward appreciation to the donating authors for their difficult labor, dedication, and knowledge that have fashioned this book likely. It is our assumption that the information held inside these pages will spark new plans, collaborations, and breakthroughs, forceful further progresses in material skill for mechanical requests.

CHAPTER OVERVIEW

Chapter 1: Cellular Ceramics used as Catalytic Supports for Heterogeneous Catalyst Synthesis

Authored by Youssef Hairch and Rachid Amrousse from University of Chouaib Doukkali, Morocco, Ahmed E.S. Nosseir from Institute of Mechanical Intelligence Scuola Superiore Sant'Anna, Italy, and Meiram Atamanov from Kazakh National Women's Teacher Training University, Kazakhstan.

This chapter discusses cellular ceramics, highlighting their types, shapes, and characteristics, and their applications in space propulsion and catalysis. Varieties include foams, honeycombs, and reticulated structures, noted for their high porosity, low density, and mechanical strength. Their large surface area enhances mass transfer, making them effective catalyst supports in various chemical processes. In space propulsion, they are utilized for thruster components due to their thermal stability and structural integrity under extreme conditions. Overall, cellular ceramics are crucial for advancing industrial applications.

Chapter 2: Model Matrices for Nuclear Waste Storage: Structure, Characterization and Mechanical Properties

Authored by Assia Mabrouk and Ahmed Bachar from Ibn Zohr University, Morocco, Aurélien Canizarès and Yann Vaills from Orleans University, France, and Stuart Hampshire from University of Limerick, Ireland.

This chapter explores rare earth element-enriched glasses for nuclear waste storage, particularly for immobilizing high-level waste (HLW) with minor actinides. It focuses on lanthanum-rich sodium aluminoborosilicate glasses, using lanthanum as a surrogate for lanthanides found in HLW. The study highlights the significance of aluminum/boron substitution in shaping the glass matrix's properties to accommodate higher HLW concentrations. Utilizing infrared, nuclear magnetic resonance, and Brillouin spectroscopy, the research examines how this substitution affects phase separation and structural changes within the glass network. The findings offer valuable insights for designing optimized glass matrices for effective nuclear waste immobilization.

Chapter 3: Slow Sand Filtration for Water Treatment

Authored by Yelriza Nurakimkyzy Yeszhan, Shynggyskhan Sultakhan, Erzhan Kuldeyev, Bostandyk Khalkhabay, and Seitkhan Azat from Satbayev University, Kazakhstan, and Qin Xu from Yangzhou University, China

This chapter reviews slow sand filtration (SSF), a traditional yet increasingly recognized method for water treatment, known for its simplicity, environmental benefits, and effectiveness, achieving up to 90-95% removal efficiency. It examines the principles and mechanisms behind SSF, focusing on the physical, biological, and chemical processes that enhance its performance. The chapter also discusses the design and construction of SSF systems, highlighting key components and factors for optimal operation. Various analytical methods are used to assess the method's efficacy in improving water quality by removing turbidity, pathogens, and organic matter. Additionally, it includes an environmental impact assessment to ensure the sustainability of SSF.

Chapter 4: New Generation of Glass Materials for Biomedical Applications: Properties, Structure, Bioactivity and Viability

Authored by Ahmed Bachar and Assia Mabrouk from Ibn Zohr University, Morocco, Cyrille Mercier, Claudine Follet, Franck Bouchart and Arnaud Tricoteaux from University Polytechnique Hauts-de-France, France, and Stuart Hampshire from University of Limerick, Ireland.

This chapter addresses the limited strength of bioglasses, which has restricted their use to non-load-bearing applications, and explores a solution by integrating nitrogen into their silicate network. It examines the effects of nitrogen addition on the physical and mechanical properties of oxynitride bioglasses within the SiO_2-Na_2O-CaO system. Findings indicate that the properties improve linearly with increased nitrogen content, attributed to nitrogen's three-fold coordination with silicon, leading to enhanced cross-linking in the glass network. This incorporation of nitrogen enhances the structural integrity and overall performance of the bioglasses, potentially expanding their applications.

Chapter 5: Next-Gen Energy Storage Devices: Carbon-Based Materials for Flexible Supercapacitors

Authored by Krishna Kumar, Uplabdhi Tyagi, Saurav Kumar Maity and Gulshan Kumar from Guru Gobind Singh Indraprastha University, India.

This chapter discusses the growing interest in developing next-generation energy storage technologies, particularly supercapacitors made from carbon materials like graphene, carbon nanotubes, activated carbon, and carbon aerogels. It reviews recent advancements in synthesizing these carbon-based materials, focusing on their capacitance, conductivity, and mechanical stability. The chapter also addresses challenges in producing and fabricating flexible carbon-based supercapacitors. Furthermore, it explores the applications of these flexible electrodes in wearable electronics, flexible displays, and smart devices. Finally, it provides an overview of the current status of carbon-based flexible supercapacitors and offers insights into future developments in energy storage.

Chapter 6: Ammonium Dinitramide (ADN) Decomposition as Green Propellant: Overview of Synthesized Catalysts

Authored by Zakaria Harimech, Mohammed Salah and Rachid Amrousse from University of Chouaib Doukkali, Morocco.

This chapter explores the thermal decomposition of eco-friendly propellants, specifically ammonium dinitramide (ADN), which is being considered as a replacement for hydrazine in satellite systems. ADN, with the formula $[NH_4]^+[N(NO_2)_2]^-$, is a high-performance rocket propellant that decomposes cleanly, producing gases like NH_3, H_2O, and NO, making it a viable alternative to ammonium perchlorate (AP) and hydrazine. The chapter reviews various catalyst systems that enhance ADN decomposition, focusing on their efficiency and thermal stability. It discusses catalysts such as metal oxides and transition metal complexes that lower decomposition temperatures, along with nanomaterials that improve decomposition rates. Additionally, it highlights how alloying carbon nanotubes with metals or utilizing noble metal nanoparticles can enhance decomposition at lower temperatures while maintaining stability.

Chapter 7: Molecular Interaction of Lactams with Mild Steel in Hydrochloric Acid Environment: Corrosion Inhibition Efficiency and Surface Adsorption Mechanisms

Authored by Nadia Faska and Soukayna Maitouf from Ibn Zohr University, Morocco, and Brahim Orayech, Powerco Battery Spain S.A., Spain.

This chapter investigates the corrosion inhibition effects of several lactams Azetidin-2-one, pyrrolidin-2-one, δ-valerolactam, and ε-caprolactam on mild steel in 1M hydrochloric acid using weight loss and electrochemical techniques. The findings reveal that δ-valerolactam significantly inhibits corrosion with an efficiency of 85.2%, while ε-caprolactam shows even greater efficiency at 91.5%. The chapter discusses thermodynamic parameters related to the adsorption process, including adsorption heat, entropy, and free energy. It concludes that the adsorption of these lactam compounds on mild steel surfaces in 1M HCl conforms to the Langmuir adsorption isotherm model.

Chapter 8: Sol-Gel Synthesis of a New Composition of Bioactive Glass

Authored by Halima El Bouami, Cyrille Mercier and Claudine Follet from University Polytechnique Hauts-de- France, France, Assia Mabrouk and Ahmed Bachar from Ibn Zohr University, Morocco.

This chapter highlights the potential of bioactive glasses in tissue engineering and bone regeneration, focusing on their ability to bond with living tissues and stimulate bone growth. It explores the sol-gel method, which allows for lower processing temperatures and better compositional control, to synthesize a new bioactive glass composition. The discussion includes the fundamental principles of sol-gel chemistry and the selection of precursors to achieve desired elements in the final glass. The chapter also examines the design rationale behind the new compositions, emphasizing therapeutic ions that promote antibacterial properties, angiogenesis, and enhance bioactivity and osteoblast production. Additionally, it reviews in vitro assessment methods for evaluating the glass's interaction with physiological fluids and outlines potential applications and future research directions for optimizing its functionalities.

Chapter 9: Electrochemical Performance of Biomass-Derived AC and CNTs based Supercapacitors

Authored by Meiram Atamanov from Kazakh National Women's Teacher Training University, Kazakhstan, Tolganay Atamanova, Azamat Taurbekov and Zulkhair Mansurov from Institute of Combustion Problems, Kazakhstan.

This study investigates the electrochemical performance of activated carbon (AC) produced from rice husk (RH) and walnut shell (WS) using two activation methods: chemical activation with potassium hydroxide (KOH) and physical activation with carbon dioxide (CO_2). The research aims to evaluate the potential of these bio-derived carbon materials for energy storage applications, specifically in supercapacitors. The electrochemical behavior of the activated carbon electrodes was analyzed through cyclic voltammetry (CV), galvanostatic charge-discharge (GCD) techniques, and electrochemical impedance spectroscopy (EIS). These methods were employed to assess specific capacitance, energy density, charge transfer resistance, and ion diffusion properties of the electrodes.

Chapter 10: Engineering the Invisible: Synthesis, Characterization and Applications of Nanoporous Materials

Authored by Krishnappa Madhu Kumar, Nagaraj Sasi Kumar, Manjappa Kiran Kumar and Lakshman N. Sampath Kumar from Sir M. Visvesvaraya Institute of Technology, India.

This chapter reviews nanoporous materials, emphasizing their unique structures and wide-ranging applications in catalysis, drug delivery, and environmental remediation. It covers various synthesis methods, outlining their principles, advantages, and limitations. Key characterization techniques, including scanning electron microscopy (SEM), X-ray diffraction (XRD), transmission electron microscopy (TEM), and atomic force microscopy (AFM), are discussed for evaluating properties such as surface area and pore size. The chapter also highlights the effectiveness of nanoporous materials in enhancing reaction rates for catalysis, improving drug delivery systems, and facilitating pollutant degradation. Ultimately, it aims to equip researchers and practitioners with essential knowledge for advancing the development and application of nanoporous materials across diverse technological and industrial fields.

Chapter 11: Synthesis and Characterization of Activated Carbon from Biomass

Authored by Makhabbat Kunarbekova, Inabat Sapargali, Laura Seimukhanova, Ulan Zhantikeyev, Kenes Kudaibergenov and Seitkhan Azat from Satbayev University, Kazakhstan, and Rosa Busquets from Kingston University, United Kingdom.

This chapter reviews methods for preparing activated carbons to adsorb pollutants like pharmaceuticals and heavy metals in surface water. Effective carbons must be tailored to enhance interaction with low concentrations of pollutants. It discusses the technical challenges in creating optimized carbon sorbents, including surface modifications. The chapter also highlights the need for environmentally responsible carbon production, mindful of CO_2 emissions. Ultimately, the goal is to address global contamination without creating further environmental harm.

Chapter 12: Bio-Piezoelectric Ceramic Coatings for Bone Tissue Engineering Applications

Authored by John Henao from CONAHCYT-CIATEQ A.C., Mexico, Astrid Giraldo Betancur and Andrea Yamilé Reséndiz from Cinvestav Unidad Queretaro, Mexico, Adriana Gallegos from CONAHCYT-Innovabienestar de Mexico, Mexico, and Carlos Poblano Salas from CIATEQ A.C., Mexico.

This chapter explores recent advances in bio-piezoelectric coatings for bone regeneration applications. It begins by discussing the bioelectrical phenomena in the human body and their role in bone healing, covering concepts like dielectric responses, piezoelectricity, pyroelectricity, and ferroelectricity, and how these apply to human bone. The chapter defines bio-ceramic materials and bioactive coatings, summarizing the primary bio-ceramic coatings currently used in bone tissue regeneration. Additionally, it reviews the latest developments in bio-piezoelectric coatings for tissue engineering and presents future perspectives on the topic. Overall, the chapter provides a comprehensive overview of the electrical responses of natural tissues and the significance of bio-piezoelectric ceramics in promoting bone regeneration.

Acknowledgment

We would like to extend our heartfelt gratitude to all the authors and contributors who have devoted their time and expertise to this book, ***"Innovative Materials for Industrial Applications: Preparation, Characterization, and Evaluation"***. Their collective efforts have been indispensable in shaping this volume, which brings together research and insights into the latest developments in material sciences.

Our special thanks go to the esteemed authors from various countries, including India, China, Kazakhstan, Spain, France, Mexico, and Morocco, whose dedication to advancing this field through innovative research has been crucial to the successful completion of this project. The diversity in their backgrounds has enriched the content, making this book a truly global collaboration.

We are equally grateful to the peer reviewers for their meticulous evaluation of the chapters. Their valuable feedback and constructive criticism ensured that the highest quality and rigor were maintained throughout the publication process. Without their commitment and expertise, this book would not have achieved the same level of excellence.

A very special thanks goes to Mrs. Nina Eddinger, Book Development Editor from IGI Global, for her unwavering support and guidance throughout the development of this project. Her dedication to ensuring the smooth progression of this publication has been instrumental, and we are deeply appreciative of her efforts.

We also wish to express our great thanks to ENSUS - UM6P - OCP Project Grant and AAP - PROGRES Research Grant for their generous support, which has been fundamental to the realization of this work.

Finally, we wish to thank IGI Global for providing the platform to disseminate this important work to the global academic and professional community.

Thank you all for your remarkable contributions.

Chapter 1
Cellular Ceramics Used as Catalytic Supports for Heterogeneous Catalyst Synthesis

Youssef Hairch
University of Chouaib Doukkali, Morocco

Ahmed E. S. Nosseir
https://orcid.org/0000-0001-6259-0956
Institute of Mechanical Intelligence Scuola Superiore Sant'Anna, Pisa, Italy

Meiram Atamanov
Kazakh National Women's Teacher Training University, Almaty, Kazakhstan

Rachid Amrousse
University of Chouaib Doukkali, Morocco

ABSTRACT

Cellular ceramics have become essential in various industrial applications. This chapter explores the types, shapes, and characteristics of cellular ceramics, and their potential uses in space propulsion and catalysis. There are different varieties, including foams, honeycombs, and reticulated structures, classified by their composition, structure, and production processes. These materials are favored for their high porosity, low density, excellent mechanical strength, and thermal stability. Their large surface area and enhanced mass transfer capabilities make them effective as catalyst supports in processes like hydrocarbon reforming, oxidation, and hydrogenation, where they immobilize catalytically active species to boost activity and selectivity. In space propulsion, cellular ceramics are used in thruster components due to their ability to withstand high temperatures and maintain structural integrity under harsh conditions.

DOI: 10.4018/979-8-3693-7505-1.ch001

1. INTRODUCTION

Cellular ceramics, including honeycombs, foams, and 3D structures, are crucial in the field of catalytic supports due to their unique structural and material properties. These ceramics are characterized by their high surface area, excellent thermal stability, and exceptional mechanical strength, making them ideal for a wide range of catalytic applications. Honeycomb structures, with their orderly and interconnected channels, offer minimal resistance to fluid flow, thereby enhancing the efficiency of catalytic reactions. This architecture is particularly beneficial in automotive catalytic converters, where the reduction of toxic emissions is achieved through efficient catalysis. Foam ceramics, with their highly porous and interconnected networks, provide an even greater surface area for catalytic reactions. This structure facilitates the dispersion of catalytic materials, allowing for more active sites and improving reaction rates. These foams are extensively used in applications such as air and water purification, where rapid and thorough catalytic processes are essential. The advent of 3D-printed cellular ceramics has further revolutionized catalytic supports. The precision of 3D printing allows for the creation of complex geometries tailored to specific catalytic processes, optimizing the interaction between reactants and catalysts. This customization leads to enhanced performance in chemical synthesis, energy storage, and environmental remediation. Therefore, the importance of cellular ceramics as catalytic supports lies in their ability to enhance reaction efficiency, durability, and adaptability across a multitude of industrial applications. Their unique structures not only maximize catalytic activity but also contribute to the sustainability and effectiveness of catalytic processes in various fields.

Cellular ceramics are essential for a variety of technological applications due to their stability in harsh environments, excellent high-temperature properties, and superior thermomechanical characteristics (Gianella et al., 2012; Xu et al., 2020; Lou et al., 2020). Specifically, cellular ceramics offer lightweight load-bearing capabilities and high specific surface area because of their open-cell porous nature (Al-Ketan et al., 2017). Today, cellular ceramics are widely used in various engineering fields, such as catalysis supports, concentrated solar energy, thermal protection or thermal storage, heat exchangers, radiant burners, nuclear fusion, gas streams, and biomedical implants (Papetti et al., 2018; Santoliquido et al., 2017; Feng et al., 2020).

However, due to their complex shapes and open-cell structures, manufacturing cellular ceramics at a low cost is challenging. Traditional processing methods, such as direct foaming (Du et al., 2019), freeze casting (Frank et al., 2017), and gelcasting (Deng et al., 2017), are unable to produce cellular ceramics with the desired intricate structures (Stochero et al., 2020). Recently, additive manufacturing, commonly known as three-dimensional printing (3D printing), has been increasingly used to create complex cellular ceramics (Vlasea et al., 2015; Tabard et al., 2021; Zhang et al., 2022).

In this chapter, we explore various forms of cellular ceramics, including honeycombs, foams, and 3D structures, along with their compositions, such as carbon, cordierite, alumina, silicon carbide, and mullite. Additionally, we examine the mechanical, thermal, and acoustic properties of these materials. Finally, we illustrate the diverse industrial applications of cellular ceramics in areas such as thermal insulation, catalysis and chemical processing, energy applications, environmental solutions, acoustic applications, the aerospace and automotive industries, and biomedical fields.

2. DIFFERENT FORMS OF CELLULAR CERAMICS

Cellular ceramics are a diverse group of advanced materials distinguished by their intricate cellular or porous structures. Engineered to harness a broad spectrum of properties, these materials find applicability across a multitude of industries, ranging from aerospace to healthcare. The classification of cellular ceramics is primarily based on their structural configuration, composition, and the methods used in their fabrication. Common forms include foams, honeycombs, and lattice structures, each offering unique advantages tailored to specific needs. For instance, ceramic foams are known for their high surface area and lightweight nature, making them ideal for thermal insulation and catalyst supports. Honeycomb structures, with their exceptional mechanical strength and thermal stability, are widely used in filtration systems and automotive applications. Meanwhile, lattice structures provide customized solutions for biomedical implants due to their biocompatibility and ability to mimic natural bone architecture. The versatility and innovation inherent in cellular ceramics lie in their ability to combine low density with exceptional thermal and chemical stability. This combination makes them indispensable in environments requiring high-performance materials that can withstand extreme conditions. Applications extend to areas such as thermal insulation in high-temperature settings, catalyst supports in chemical processing, filtration systems for clean energy solutions, and biomedical implants that require precise structural and functional integration with human tissue. As research and development in this field continue to advance, the potential applications of cellular ceramics are poised to expand even further, driving progress in technology and industry.

Cellular materials are generally divided into two main categories: honeycombs and foams as shown in Figure 1. Honeycombs consist of cells arranged in a two-dimensional array, while foams are made up of a three-dimensional array of hollow polygons. Ceramic honeycombs have gained significant use in catalytic converters, heat exchangers, and combustors. Foams are further classified based on whether the individual cells have solid faces. If the solid is only present at the cell edges, the material is known as open cell (or reticulated) and is permeable (Liu et al., 2014).

Figure 1. Different forms of cellular ceramics: honeycombs, foams and 3D printed

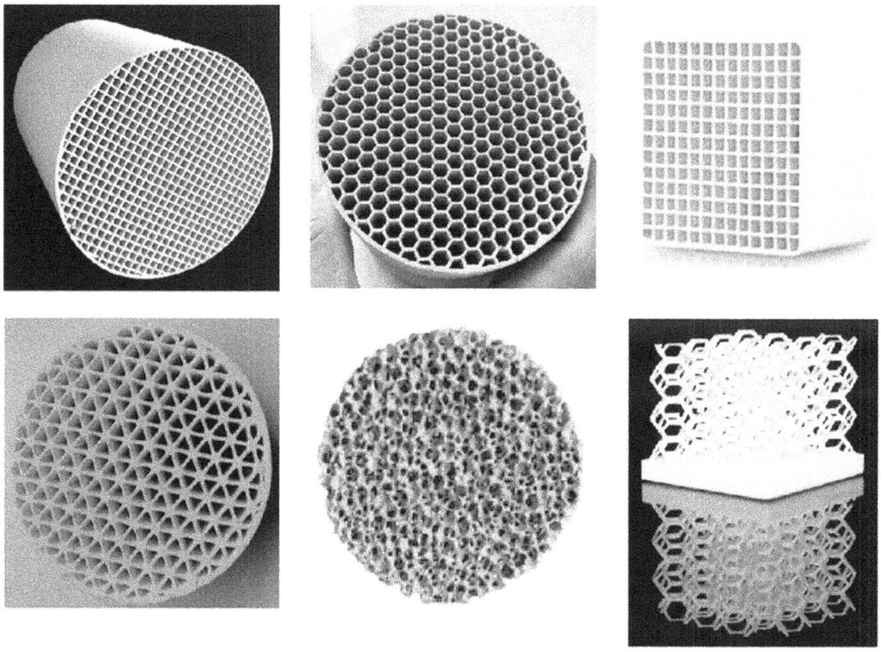

Ceramic honeycombs are typically created through extrusion, followed by a conventional binder burn-out and sintering process. The die used in extrusion can be designed to produce a wide range of cell geometric structures, such as square, triangular, hexagonal, and more. This process requires moldable raw materials, whose rheological properties can be adjusted using water-soluble additives. Achieving the right balance in rheology is crucial to maintain the stability of the cellular structure while minimizing excessive die wear (Liu et al., 2014).

Herein, we presented several cellular ceramic forms:

2.1 Honeycombs

The cellular structure of a honeycomb (Figure 2) is inspired by nature, providing the largest possible surface area with the smallest possible volume. This design is ideal for flow-related tasks, as the very thin walls maintain low pressure loss while ensuring full contact of gas molecules with the surface. By selecting specific ceramic materials, the surface characteristics can be widely varied. This adaptability makes honeycomb ceramics suitable for applications such as filters, catalyst carriers, and heat storage media. Honeycomb ceramics are used as carrier structures for the catalytic purification of combustion gases from gasoline engines, for filtering soot particles from diesel engine exhaust...etc. They are also employed in treating industrial exhaust flows that contain various organic and inorganic contaminants. Additionally, honeycomb ceramics find applications in heat exchangers and space propulsion.

Figure 2. Honeycombs with square and triangular channels fabricated by Fraunhofer Institute for Ceramic Technologies and Systems IKTS (Ravkina, 2024)

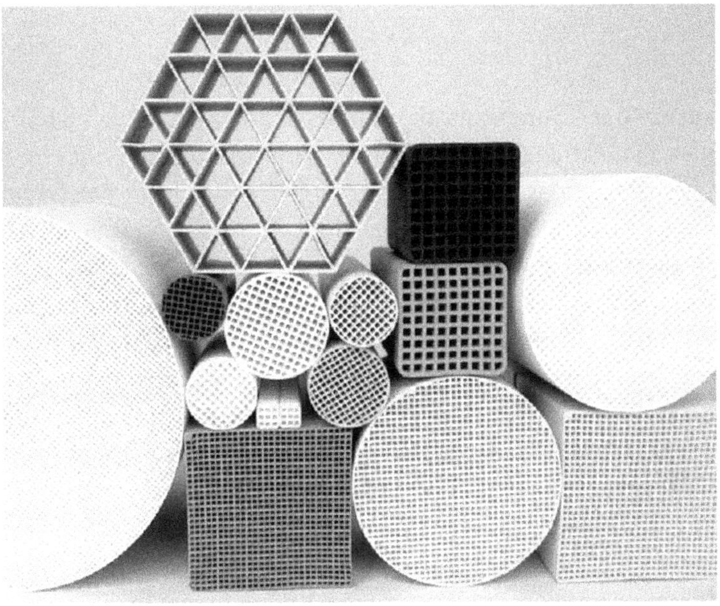

2.2 Foams

2.2.1 Preparation

The struts of cellular foams made by copying polyurethane foam have big, hollow gaps running along them. The absence of the polymeric precursor is the cause of these characteristics. After conducting extensive research on these materials, Green et al. (Hagiwara and Green, 1987; Brezny et al., 1989; Brezny and Green, 1989; Van Voorhees and Green, 1991) came to the conclusion that, although removing strut cracks and other significant flaws may increase strength, the large triangular holes in the struts do not always weaken the strut unless the apex is close to the outside of the strut or connected to a crack. They also discovered that, for the mechanical behavior of cellular materials, the majority of the theoretical correlations established by Gibson and Ashby (Gibson and Ashby, 1988) apply well to these reticulated ceramics, with any discrepancies being attributable to microstructural flaws. In order to understand the flow characteristics of reticulated ceramics made by reproducing polyurethane foam, Gauckler and Waeber (Gauckler and Waeber, 1985) employed stereological methods to analyze the microstructure of the materials. This was done with an eye on metal infiltration applications. They examined the structure, which was made up of spherical polyhedra with a nominal diameter joined by windows or openings, using fundamental stereological criteria. Together with the total internal cell surface area per unit volume and the separation between two pore centers, the total porosity of the reticulated body was identified. Anisotropy was evident in the structure; the pores were spherical in the plane of the filter plate but bigger and longer in the perpendicular direction. Pore size and window size were the two most crucial characteristics in relation to the permeability of the ceramics. Both the relationship between the internal surface area per unit volume and the size of the window and the relationship between the pore and window sizes were linear (Binner, 2006).

2.2.2 Structure

Even in nations that use SI measurements, the number of pores per linear inch (PPI value) is a feature of cellular foam supports that is frequently mentioned. Pore sizes and densities of ceramic foams made by replication techniques as previously mentioned commonly vary from 5 to 65 ppi (2 to 25 pores per cm) and 5% to 30% of the theoretical density, respectively. According to Binner (2006), these foams are typically 10–100 cm wide and 1-10 cm thick.

Ceramic foams, which can be either closed-cell or open-cell, have a cellular structure with a high-volume proportion of gas-filled pores, as seen in Figure 3. These foams have a fascinating mix of characteristics, such as low weight, low heat capacity, strong chemical resistance, high mechanical strength, and low thermal conductivity. These characteristics enable a wide range of applications, including: metal melt filtration, catalyst carriers, thermal insulation, ion-exchange filtration, heat sinks and heat exchangers, noise reduction, medical technology...etc. Fineway Ceramic foams are available in various pore sizes and are used in ceramic filters for a range of common metals, such as iron, stainless steel, and aluminum. To meet the highest standards and provide superior performance, Fineway Ceramic foam filters are also ideal for use in air and water treatment applications (http://finewayceramics.com/ceramic-foams/).

Figure 3. Foams with random distributed pores fabricated by FineWay Ceramics (http://finewayceramics .com/ceramic-foams/).

Moreover, the process of foaming ceramic slurries involves dispersing gas in the form of bubbles into a ceramic suspension. There are two main ways to accomplish this: 1) creating gas on the spot, or 2) utilizing an aerosol propellant or mechanical foaming to incorporate an external gas. In order to stabilize the gas bubbles inside the slurry, a surfactant is typically added to lower the surface tension at the gas–liquid interfaces. But bubble structure alterations result from the weakening of their surrounding lamellae, therefore this stabilization is only temporary. Since changes in the foam structure prior to solidification alter the ultimate cell size distribution, wall thickness, and microstructure of the solid foams, an extra mechanism is therefore required for long-term stabilization. These elements have a major impact on characteristics like strength and permeability. A closed-cell foam, for example, forms if the films surrounding the bubbles hold together until solidification. Partial rupture of these films produces open-celled foams; severe film rupture can cause foam collapse.

2.3 Silicon Carbide (SiC) Honeycombs and Foams

The straight cells and random pores of silicon carbide (SiC) were shown in the Figure 4.

Figure 4. Cellular ceramics on silicon carbide (SiC): (a) honeycombs with straight channels and (b) foams with random pores

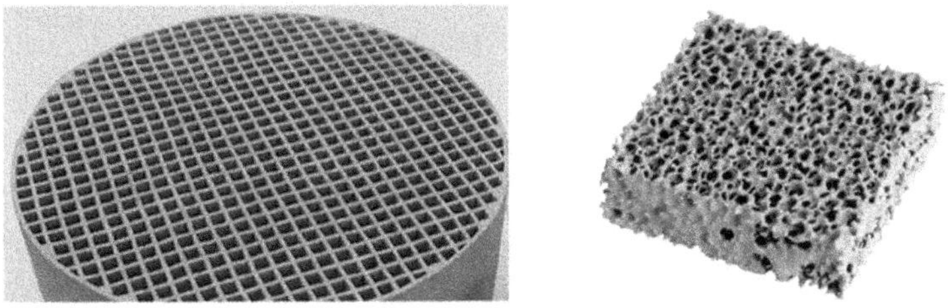

(a) (b)

A low-density permeable material with a wide range of applications is silicon carbide (SiC) honeycomb (Figure 4a). This honeycomb is distinguished by its extremely high porosity, with void spaces making up between 75 and 95 percent of the volume on average. Applications for metallic honeycomb are numerous and include lightweight optics, energy absorption, flow diffusion, and heat exchangers. Ceramic honeycomb is frequently used as a substrate for catalysts that need a lot of internal surface area, as well as for adsorption of environmental pollutants, filtration of molten metal alloys, and thermal and acoustic insulation. The silicon carbide honeycomb's geometric structure minimizes the amount of material utilized, resulting in reduced weight and cost. Strong to weight ratio: the honeycomb pattern has a high value. In most cases, silicon carbide honeycomb is readily available right away in most volumes. There is more technical, scientific, and safety (MSDS) data available. SiC honeycomb's strength and porosity make it useful in a wide range of scientific and engineering applications. Similar to diamond, which is a pure carbon compound, carbide compounds are frequently very hard, refractory, and resistant to heat, wear, and corrosion, which makes them great options for coatings for instruments like drills (https://www.americanelements.com/silicon-carbide-honeycomb-409-21-2).

In contrast, SiC foam ceramic filter (Figure 4b) uses imported polyurethane foam sponge from Germany as a carrier. Silicon carbide is primarily squeezed into the foam sponge, and after drying and baking at a high temperature, the polyurethane thermally decomposes to leave foam ceramic products, such as SiC foam ceramic filter. With its three-dimensional mesh interconnection structure of foam sponge, the SiC foam ceramic filter can effectively improve the purity of metal liquid, reduce waste rate, and reduce mechanical processing loss, all while improving production efficiency and cutting costs. This is achieved by making metal liquid reflect rectification, mechanical screening, filtration cake, adsorption, and other filtration mechanisms. SiC foam ceramic filter features excellent size accuracy, superior strength, superior thermal stability, and bigger surface area. Its primary applications include the filtration and purification of ductile cast iron, ash cast iron, aluminum, copper, and other alloy metal liquids; large inclusion removal in metal liquids; smell magazine adsorption; reduction of gas and harmful element content in metal liquids; enhancement of metal matrix organization; and mechanical property improvement of castings. (https://www.ikts.fraunhofer.de/en/departments/environmental_process_engineering/nanoporous_membranes/functional_carrier_systems_layers/honeycomb_ceramics.html).

2.4 Carbon Foams

Carbon (C) Foam (Figure 5) is a low density, porous elemental carbon substance that has several uses due to its unusual electrical properties and status as the stiffest and strongest known fiber. High purity, extremely high purity, coated, suspended, and dispersed types of carbon foam are available. The AE Nanofluid production group also offers them as a dispersion. In general, suspended nanoparticles in solution made with surface charge or surfactant technology are referred to as nanofluids. Technical advice is also given for coating choosing and nanofluid dispersion. Nanorods, nanowhiskers, nanohorns, nanopyramids, and various nanocomposites are examples of further nanostructures. Chemically attached polymers enable surface functionalized nanoparticles to be absorbed preferentially at the surface contact (https://www.americanelements.com/silicon-carbide-honeycomb-409-21-2).

Figure 5. Foams on carbon with random distributed pores

2.5 3D Structures

Researchers (Maurath and Willenbacher, 2017) printed cellular log-pile and honeycomb structures using capillary suspension-based inks containing AKP-50 and CT3000SG particles (Figure 6).

Figure 6. (a) Top view of a log-pile structure printed with a 610 μm nozzle. (b) Side view on the two log-pile structures (Maurath and Willenbacher, 2017)

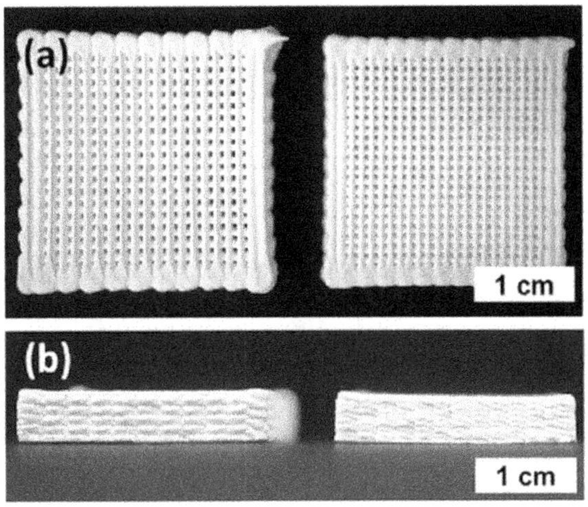

Log-pile structures were used to evaluate ink behavior, exhibiting small spanning elements typical in ink development for testing filament deliquescence and stability in Direct Ink Writing (DIW). Capillary suspension inks successfully extruded thin filaments (200-610 μm nozzles) without deliquescence issues, unlike pure suspensions. Printed pores in the xy-plane ranged from 490 to 570 μm, with excellent accuracy using all nozzle diameters tested. However, pores in the xz- and yz-planes varied in quality based on nozzle diameter, with larger nozzles performing better due to reduced material overflow.

Hexagonal honeycomb structures were also printed for lightweight construction, mimicking materials like wood. These structures had high aspect ratios (wall height to width ratios of 4-11) with wall thicknesses of 0.7-1.5 mm and heights of 6-8 mm, showing no deformation from deliquescence.

All printed green bodies underwent drying, thermal debinding, and sintering with uniform shrinkage (10-26%) and no visible defects or deformations. Filaments protruding from structures were observed due to printing dynamics aimed at preventing nozzle clogging and ensuring high-quality prints.

3. DIFFERENT PROPERTIES OF CELLULAR CERAMICS

Cellular ceramics, characterized by their highly porous structure, offer a multitude of advantageous properties that make them highly desirable in various applications, particularly in catalysis. These materials excel in mechanical, thermal, acoustic, and permeability aspects, which collectively enhance their performance and versatility as catalyst carriers. Mechanically, cellular ceramics boast impressive strength and durability despite their porous nature, ensuring they can withstand harsh operational conditions and maintain structural integrity over extended periods. Their thermal properties are equally notable, with high thermal stability and excellent heat resistance, allowing them to perform effectively in high-temperature environments without degradation. Acoustically, cellular ceramics exhibit sound-absorbing qualities, making them useful in applications where noise reduction is beneficial. Additionally, their permeability is a critical feature, facilitating efficient gas and liquid flow through the material. This attribute is particularly advantageous in catalytic processes, where the uniform distribution of reactants and products is essential for optimal performance. These combined properties render cellular ceramics exceptionally well-suited as catalyst carriers. They provide a large surface area for catalytic reactions, ensure effective mass transfer, and enhance the overall efficiency and longevity of the catalytic process. Consequently, cellular ceramics are employed in a wide range of catalytic applications, from environmental remediation to chemical synthesis, underscoring their significance and versatility in modern industrial processes.

3.1 Mechanical Properties

Because of their distinct porosity architectures and range of mechanical attributes, cellular ceramics are ideal for a number of applications, especially those requiring mechanical strength, lightweight, and thermal insulation. Among these characteristics are: (i) High strength-to-weight ratio; that is, cellular ceramics frequently have a sizable compressive strength in spite of their lightweight design. This makes them perfect for structural applications, such those in the automobile and aerospace sectors, where reducing weight is essential. (ii) Thermal shock resistance: cellular ceramics' porous nature prevents them from cracking in the face of sudden temperature fluctuations, which makes them beneficial for high-temperature activities like furnace linings and thermal barriers. (iii) Energy absorption: Cellular ceramics offer superior shock absorption and protection since they can absorb a large amount of energy upon impact. This makes them suitable for use in protective gear and impact-resistant materials, (iv) High-temperature stability indicates that cellular ceramics maintain their mechanical integrity at high temperatures, which is essential for applications in environments with extreme heat, such as in gas turbines and heat exchangers, (v) Resistance to creep and deformation; i.e. these materials exhibit good resistance to creep (the tendency to deform under mechanical stress over time), maintaining their shape and structural integrity under prolonged loading conditions, (vi) Fracture toughness; although ceramics are typically brittle, the cellular structure can help arrest crack propagation, providing a degree of toughness that can be beneficial in various mechanical applications, and (vii) Customizable characteristics refer to the ability to modify the porosity, pore size, and material composition of cellular ceramics in order to modify their mechanical properties. This customization allows for the optimization of specific properties for particular applications.

3.1.1 Modeling the Porosity Dependence of Mechanical Properties

- **Earlier models:** Four prominent models for assessing porosity effects on material properties are discussed, with one older model focused on elastic moduli, specifically Young's modulus (E). This model, derived empirically and analytically, uses the expression:

$$\frac{E}{E_0} = \left(\frac{\rho}{\rho_0}\right)^n = \left(1 - P\right)^n$$

where P is porosity and n an empirically determined exponent. Despite its frequent use, this model has significant limitations: it neglects microstructural effects and doesn't account for the critical porosity value (PC) beyond which a material can't sustain stress. Normalizing porosity values by PC has shown limited improvement. Despite its flaws, this model remains relevant for its insights into porosity effects.

- **Gibson-Ashby Models:** The Gibson and Ashby (G–A) models focus on the porous microstructure, idealizing the structure of tubular and box cells for honeycombs and foams, respectively (Figure 7). These models scale mechanical behavior with cell-structure parameters, fitting data to determine proportionality constants. Although they point out commonalities across cellular structures made of polymers, metals, and ceramics, they have trouble connecting features at high porosity (P) to those at low P. Despite this, G-A models and other approaches may help understand transitions in properties across different porosity levels, although cross-correlation of these models has received limited attention.

Figure 7. (A) The unit pores in this model, which depicts an open-cell foam structure, frequently have the shape of a cubic (or parallelepiped, but it isn't depicted here). In G-A-type investigations, these models are frequently employed to represent the anisotropy typical of numerous foam formations. The model highlights packing variations and associated cell-to-cell joining that must be taken into account in both kinds of models, demonstrating a methodical departure from the straightforward cubic packing of the cells. As a decent approximation of random packing in MSA models, these models (B) provide basic cubic packing of uniform spherical pores (or spherical particles) in a cubic cell structure (Rice, 1996)

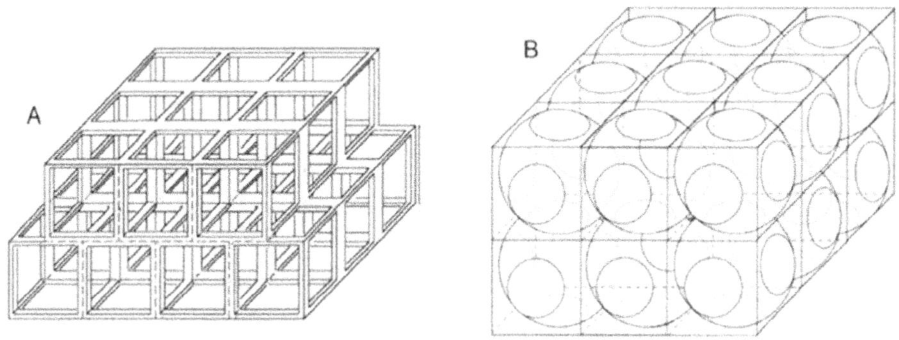

- **Minimum Solid Area (MSA) Models:** address a wide range of pore structures from low to high porosity, complementing the cellular models of Gibson and Ashby by focusing on similar microstructures and their evolution. These models assume a uniform packing of identical cells, each with a pore or a spherical particle at the center, to simulate porosity changes. Mechanical properties are scaled with the ratio of the MSA of a cell to its cross-sectional area. The output is typically a plot of relative MSA versus porosity, showing three stages of property decrease, with exponential relations approximating the initial linear decrease. This approach has shown good agreement with experimental data, validating its use.
- **Computer Models:** the fourth method of modeling utilizes computers, with promising extensions of MSA models and the more established finite-element method (FEM). FEM can be conducted in 2D or more rigorously in 3D. Studies by Agarwal et al. showed reasonable 3D results for spherical pores in a glass matrix, consistent with MSA models. Roberts and Garboczi's FEM analyses of ceramics with pores demonstrated results consistent with the equation:

$$\frac{E}{E_0} = \left(\frac{\rho}{\rho_0}\right)^n = \left(1 - \frac{P}{P_c}\right)^n$$

though with some limitations near critical porosity values (PC). This program has been used for 2D analyses of real pore structures, yielding useful results but with caution needed due to the differences between 2D and 3D modeling.

3.1.2 Porosity Effects on Mechanical Properties

- **Honeycombs:** show anisotropic mechanical properties. For honeycombs with straight tubular pores aligned parallel to applied stress, the following formula yields the Young's modulus (E):

$$\frac{E}{E_0} = 1 - P$$

establishing n = 1 as the porosity dependency upper bound. The characteristics are lower and reliant on the form and stacking of the pores when stressed perpendicular to aligned pores. While the properties of circular cross-sections are isotropic in planes normal to the pores, different shapes, such as square or triangular, might exhibit different degrees of anisotropy. Tensile strength is influenced by pore alignment and cell wall strength, and strength is often more dependent on porosity than E. Cell wall strength has a linear relationship with crushing strengths, however porosity has a stronger influence. Particular research has been done on shock failure and thermal stress in automotive applications.

- **Foams:** an extensive overview of experimental and modeling evaluations for various foam structures, focusing on their mechanical properties and porosity dependence. Here are the key points summarized:

For open-cell foams:

Young's Modulus (E): follows a relationship $\frac{E}{E_0} = 1 - P$, indicating a quadratic dependence on porosity P.

- Shear Modulus (G): similarly follows $\frac{G}{G_0} = \left(1 - P\right)^2$
- Poisson's ratio (ν): approximately constant (~ 0.3) regardless of porosity.
- Additional Mechanical Properties: Hardness, compressive strength, tensile strength, and fracture toughness all scale with $(1 - P)^{1.5}$.

For closed-cell foams:

- • Included parameters such as j (fraction of solid material in cell edges) in the model. 1 − f (portion of solid material in cell walls).
- Similar behavior to open-cell foams but with different structural considerations.

Specific Examples:

- Glass foams: show agreement with models predicting properties based on spherical bubble packing.
- SiO_2 foams from sol-gel processing: transition in pore structure from cubic packings at low porosity to spherical packings at higher porosity.
- Alumina foams: variation in mechanical properties attributed to microstructure of cell struts.

3.2 Thermal Properties

Cellular ceramics have exceptional thermal characteristics due to their distinctive microstructures. The large surface area-to-volume ratio of these materials improves their capacity for thermal insulation. Cellular ceramics' porous nature effectively controls heat transport, which makes them perfect for applications needing thermal management. Their low thermal conductivity also makes effective thermal insulation possible in a variety of settings by reducing heat loss. Because of these qualities, cellular ceramics are extremely beneficial in sectors such as energy, automotive, and aerospace where insulation and thermal stability are critical for effectiveness.

Herein, different thermal properties of cellular ceramics were illustrated

- **Thermal conductivity:** Fourier's law defines heat transport through a medium via thermal conductivity k [W m^{-1} K^{-1}], stating heat flux is proportional to k and temperature gradient ∇T. In non-stationary cases, transient heat flux is governed by the heat equation involving thermal diffusivity:

$$\alpha = \frac{k}{\rho c_p}$$

For cellular ceramics, which include solid frameworks filled with fluid, effective thermal conductivity k_{eff} accounts for various heat transfer mechanisms like conduction, convection, and radiation, simplified for computational purposes despite the material's complex structure and properties.

- **Specific heat capacity:** the specific heat capacity c_p [J kg^{-1} K^{-1}] quantifies a material's ability to store heat energy. For cellular ceramics, comprising solid s and fluid f, c_p is derived from their respective weight fractions m_f/m_{cc} and m_s/m_{cc}, and specific heat capacities c_p^F and c_p^S:

$$c_p^{cc} = c_p^F \cdot \frac{m_f}{m_{cc}} + c_p^S \cdot \frac{m_s}{m_{cc}}$$

Typically, the gas mass fraction is minimal, making c_p of the cellular ceramic close to that of the solid component s. For solids with known compositions, c_p^S can be estimated using Dulong and Petit's rule, which relates to Boltzmann's constant k, atomic mass $l_{AM,}$ and degrees of freedom f.

- **Thermal shock resistance:** this parameter is crucial for ceramics used in high-temperature applications, as it measures the material's ability to withstand sudden temperature changes without fracturing. For cellular ceramics, which typically have low fracture toughness, thermal shock resistance is influenced by factors such as strength r_C, elastic modulus E, thermal expansion α\alphaα, and thermal conductivity k. Experimental studies show that larger cell sizes enhance thermal shock resistance, while density has a weaker effect. Two parameters, R1 and R2, characterize thermal shock behavior: R1 estimates the maximum allowable temperature difference for rapid superficial shocks, while R2 considers long-term effects including heat flux into the material.

- **Volumetric convective heat transfer:** cellular ceramics find application in heat exchangers, solar receivers, and air heaters where effective convective heat transfer between solids and fluids is crucial. The convective heat transfer coefficient a [W m^{-2} K^{-1}] describes this capability, influenced by the specific surface area Av [m^2 m^{-3}] and temperature difference $\Delta T = T_S - T_F$ between the solid (ceramic) and fluid temperatures T_S and T_F. Experimental methods, like those by Younis and Viskanta, measure a by analyzing temperature waves in air streams through porous samples, accounting for phase shifts and attenuation. Results show significant enhancement in heat transfer with increasing fluid velocity and smaller cell sizes in the ceramic foams studied.

3.3 Acoustic Properties

The acoustic characteristics of cellular ceramics are distinct due to their porous structure and material composition. Cellular ceramics' interconnecting pores function as sound-absorbing chambers, efficiently attenuating sound waves over a broad frequency range. Cellular ceramics are appropriate for applications needing acoustic insulation and noise reduction because of their capacity to absorb sound. These materials are used in the construction, automotive, and industrial machinery industries to reduce noise levels and enhance acoustic comfort. Furthermore, cellular ceramics' adaptability enables customized designs to maximize acoustic performance in particular settings, highlighting their usefulness in both commercial and industrial acoustic applications. We can mention that the research group of De mello Innocentini et al. (De mello Innocentini et al., 2005) previously studied a number of acoustic parameters, including (i) *Acoustic impedance and admittance,* where the real part describes the acoustic resistance and is related to the speed at which the acoustic waves propagate, and (ii) *Acoustic wavenumber k,* which is defined by:

$$k = \frac{\omega}{c}$$

where ω is the angular frequency and c is the propagation speed. The wavenumber describes the propagation of the sound waves. If the propagation speed c and hence k is allowed to be complex, then the waves will decay or grow with distance, (iii) *Reflection Coefficient, Transmission Coefficient, and Transmission Loss* which are defined respectively as:

$$R_{refl} = \frac{R}{I}$$

R_{refl} is an intricate number. This definition's waves can be related to any fixed point.

$$T_{trans} = \frac{T}{I}$$

Although the waves can be referenced to any position, they are typically related to the material's center, much like the reflection coefficient, which is similarly thought of as a function of frequency. The definition of transmission loss (T_{loss}) is:

$$T_{loss} = 20\log(\frac{T}{I})$$

and has a decibel (dB) measurement.

Lastly, (iv) *Absorption coefficient:* even though the transmission and reflection coefficients provide useful information, it should be emphasized that very modest transmission or reflection can be obtained without damping any acoustic energy at all. Therefore, measuring the amount of absorbed acoustic energy is also beneficial. This is usually done using the absorption coefficient D, which indicates the percentage of incident acoustic energy that is absorbed.

3.4 Permeability

Cellular ceramics are characterized by their permeability, which allows fluids or gases to flow through their interconnected pore structure. This permeability is essential in applications such as filtration, catalysis, and thermal insulation, where the material's ability to allow fluid or gas passage while maintaining structural integrity is crucial.

Figure 8. Experimental apparatuses for assessing the permeability of ceramic cells. (a) Test of water flow. (b) (De Mello Innocentini et al., 2005) conducted an air-flow test

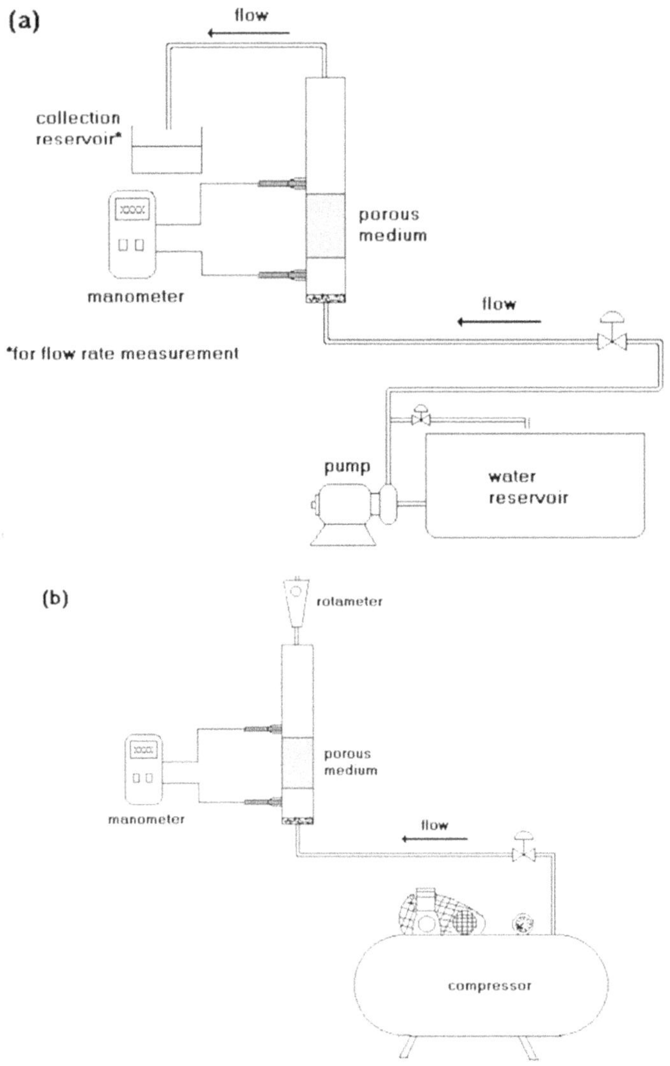

Experiments involving the manipulation of a fluid to pass through a stationary porous material are used to evaluate permeability parameters experimentally. The sample is usually tightly sealed between two chambers, say in the case of a disk or cylinder with thickness L and exposed area A. Absolute fluid pressures are monitored and recorded at the sample's intake (Pi) and outflow (Po) during the test while the flow rate (Q) or fluid velocity is changed $v_s = \frac{Q}{A}$. When applying Forchheimer's equation, the gathered data is fitted to a parabolic model of the form y = ax + bx^2 using the least-squares approach. Here, y represents either $P_i - P_o$ for liquids or $\frac{P_i^2 - P_0^2}{2\rho L}$ for gases and vapors, and x denotes the velocity v_s.

The velocity of gas flow needs to be modified to account for the sample's real average pressure and temperature. Then, using the fitted constants a and b, the permeability parameters k_1 and k_2 are calculated as $k_1 = \frac{L}{a}$ and $k_2 = \frac{\rho}{b}$. The literature (Moreira et al., 2004) describes a number of apparatuses for assessing permeability; the best configuration is primarily dependent on the sample's properties. Two typical settings for ceramic foams are shown in Figure 8. Liquids are recommended as the testing fluid for extremely porous materials in order to maximize the inertial term's $\frac{\rho v_s^2}{k_2}$ contribution, thereby ensuring reliable fitting of k_2 values. To ensure consistent data analysis, samples from different batches should have their permeability parameters fitted within a comparable pressure-velocity range (Innocentini et al., 1999). In cases where the sample is anisotropic, attention should be paid to the flow direction, and ideally, the experiment should be conducted with the sample oriented in its regular usage direction (De Mello Innocentini et al., 2005).

4. PREPARATION AND CHARACTERIZATION OF CATALYSTS BASED ON CELLULAR CERAMICS

4.1 Preparation

Cellular ceramics, also known as honeycomb ceramics and/or foam ceramics, are a class of materials with a cellular structure composed of straight cells and/or interconnected pores; respectively. These materials are typically made from oxides or from non-oxides. The unique structure of cellular ceramics provides several advantages: high surface area, thermal stability, mechanical strength and mass, and heat transfer. In general, the cellular ceramics were washed in acidic medium to remove all impurities such as water steam and CO_2 (Amariei et al., 2010; Amrousse et al., 2010) then the calcined supports were washcoated with slurries or suspensions in order to increase their surface areas. Usually, alumina was used as second support distributed inside pores and on the walls. Then, the impregnation is used as technique to incorporate active phases into porous structures of cellular ceramics. The calcination and reduction over appropriate gases was a crucial step to activate the cellular ceramic catalysts.

4.2 Preparation

Characterization of cellular ceramics is crucial for understanding and optimizing their properties and performance in various applications. Techniques such as scanning electron microscopy (SEM), transmission electron microscopy (TEM), X-ray diffraction (XRD), and Brunauer-Emmett-Teller (BET) analysis provide comprehensive insights into the microstructure, crystallography, and surface area of these materials. In fact:

SEM is indispensable for examining the surface morphology and microstructural features of cellular ceramics. It provides high-resolution images that reveal the pore size, shape, and distribution, which are critical for determining the material's mechanical strength and thermal insulation properties.

TEM offers even higher resolution imaging, allowing for the observation of the internal structure at the atomic or molecular level. This is particularly important for understanding the finer details of the ceramic's composition and the nature of grain boundaries and defects, which can significantly impact its thermal and mechanical behavior.

XRD is essential for phase identification and crystallographic analysis. It helps in determining the crystalline phases present in the ceramic and their relative abundances. This information is vital for correlating the material's composition with its properties and for ensuring the quality and consistency of the ceramic during manufacturing processes.

BET analysis is used to measure the specific surface area of cellular ceramics. This parameter is crucial for applications involving catalysis, adsorption, and surface reactions. A high specific surface area indicates a greater potential for interaction with gases or liquids, enhancing the material's functionality in these applications.

5. DIFFERENT CATALYTIC REACTIONS USING CELLULAR CERAMICS AS HETEROGENEOUS CATALYSTS

Cellular ceramics, with their unique open porous structures, have emerged as highly effective catalyst carriers. Unlike traditional catalyst forms such as powders, extruded shapes, or grains, cellular ceramics facilitate a laminar flow of liquids and gases. This distinct feature not only enhances the stability and longevity of the catalysts but also plays a crucial role in preventing the sintering of catalytic particles. The interconnected porosity of cellular ceramics ensures uniform distribution and optimal interaction of reactants, thereby maintaining catalytic efficiency and reducing the likelihood of deactivation over time. This makes cellular ceramics an ideal choice for applications demanding high-performance and durable catalytic processes. In this section, several catalytic reactions using cellular ceramics as catalysts were provided:

5.1 Hydrogen Production

Palma et al. (Palma et al., 2018) studied the decomposition of S_2H over coated honeycomb catalysts to produce hydrogen (H_2). Preliminary tests at 1000 and 1100°C evaluated the effect of washcoat loading on catalytic performance as shown in Figure 9. At 1000°C, increasing washcoat loading improved H_2S conversion but slightly reduced H_2 yield, with the best SO_2 selectivity observed at 30 wt% washcoat content. At 1100°C, similar trends in H_2S conversion and H_2 yield were noted, with higher SO_2 selectivity for the 30 wt% washcoat sample. The 30 wt% washcoat loading was chosen for further study due to its ability to maintain low SO_2 selectivity below 0.4% at both temperatures. The goal was to enhance selectivity towards sulfur and H_2 while minimizing SO_2 emissions.

Figure 9. Effect of washcoat loading at T = 1000°C (left) and T = 1100°C (right) on H_2S conversion (a, b), H_2 yield (c, d), and SO_2 selectivity (e, f) (Palma et al., 2018)

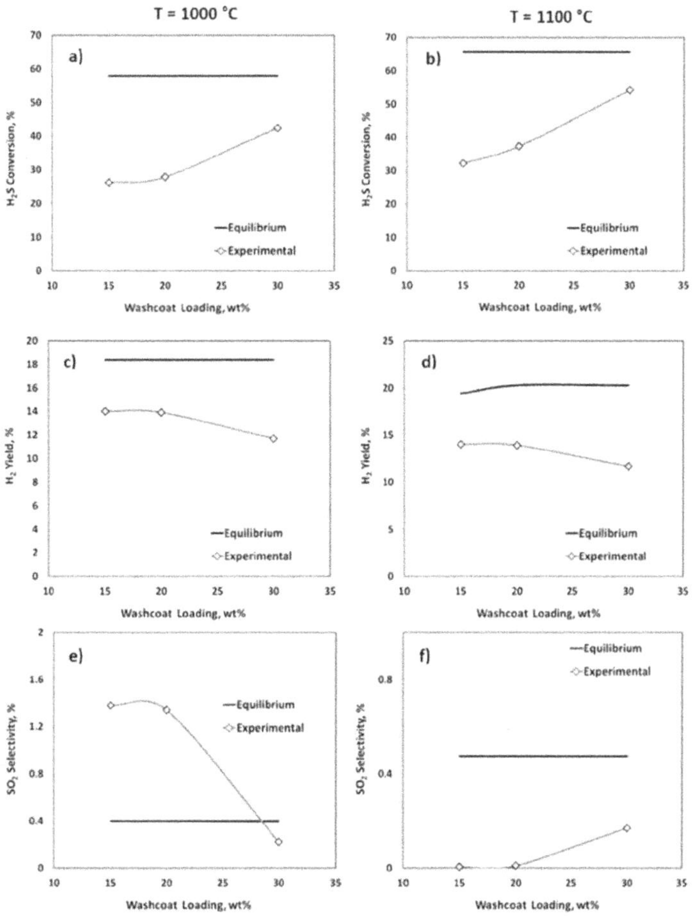

The vaporeforming of methanol is considered as another source for H_2 production. The loading capacity of catalyst washcoats was evaluated through weight change after ultrasonic vibration, showing less than 20% weight loss, indicating good mechanical stability. Catalytic performance tests revealed that In/Ce-1,4-benzenedicarboxylic acid (H_2BDC, 99%) (denoted BDC)-CZIC achieved the highest methanol conversion rates and hydrogen yields at 300°C, with a 100% conversion rate and a hydrogen yield of 0.336 mol/h. At higher methanol flow rates, the conversion rate decreased but hydrogen production increased. In/Ce-BDC-CZIC also demonstrated lower CO generation and maintained high stability with 93.9% methanol conversion after 30 hours. The optimal conditions for maximum energy efficiency were 300°C and a methanol flow rate of 0.112 mol/h (Liao et al., 2023).

Megapores with spherical-like cells connected through windows and high porosities make up catalyst supports in the form of ceramic foams to be used toward hydrogen production through the dry reforming of methane (CH_4) using a fixed-bed reactor was evaluated under certain operating conditions (Yeetsorn et al., 2022). In fact, this research group selected ceramics named CF-10NAM for its cost-effectiveness and efficiency in catalysis, especially in methane conversion processes. CF-10NAM is 10% nickel oxide on ceramic foam made from alumina magnesium support. The catalytic activity of 100CF-10NAM for dry reforming of methane (DRM) at 620°C was tested, showing stable CH_4 conversion after 50 minutes. The 100CF-10NAM/PU_5 °C/min catalyst had the highest CH4 conversion at 74%, followed by 100CF-10NAM/PU_2 °C/min and 100CF-10NAM/PU_0.5 °C/min with 63% and 56% conversions, respectively. H_2 yields exceeded theoretical values due to side reactions like methane decomposition and CO disproportionation, leading to coke deposition within the temperature range of 553–674°C.

Hydrogen production measured the photocatalytic activity of samples, revealing that pure CdS had low hydrogen-generation properties. When combined with H-Co_3O_4, the H-Co_3O_4@CdS honeycomb's performance increased dramatically, producing 16.3 mmol h^{-1} g^{-1}, which is 7.3 times higher than pure CdS. This boost is attributed to enhanced visible-light absorption and a high apparent quantum yield (AQY) of 32%. Stability tests showed that despite a decrease after four cycles, the activity remained superior to other CoO_x/CdS catalysts. Additional tests across different spectra confirmed H-Co_3O_4@CdS -4's highest H_2 production rates, demonstrating the synergistic effect between CdS and Co_3O_4. Structural stability and photoelectric response were maintained post-reaction. Photoelectrochemical measurements showed H-Co_3O_4@CdS-4's strong photocurrent density and low electron migration resistance, enhancing charge carrier generation, separation, and migration, thus boosting H_2 evolution rates (Zhang et al., 2021).

5.2 Conversion of Volatile Organic Compounds (VOCs)

The adsorption of volatile organic compounds (VOCs) by the catalyst is crucial for their complete oxidation, as it prolongs the VOCs' presence in the plasma discharge region. The adsorption capacity depends on factors like surface roughness, specific surface area, pore volume, pore size, and pore size distribution. A catalyst with pores larger than VOC molecules allows VOCs to be adsorbed internally, while smaller pores do not. With a pore size suitable for VOC accommodation, the catalyst readily adsorbs VOCs. For a flow rate of 60 L/min, with a toluene concentration of 15 ppm and 3.0% water content, the catalyst based on palladium (Pd) and deposited on monolith support reached saturation after 100 minutes, indicating an adsorption capacity of 280 µmol/g (Mokter Hossain et al., 2022).

Moreover, Figure 10 illustrates the degradation of methyl ethyl ketone (MEK) as VOCs in a plasma catalytic reactor using different discharge gases, showing that MEK removal efficiency decreases in the order: O_2 > air > N_2 in the presence of Pd-Mn/γ-Al_2O_3/honeycomb catalyst.

Figure 10. MEK removal efficiency and CO_2 selectivity for: (a-b) gas phase plasma; and (c-d) Pd-Mn/γ-Al_2O_3/honeycomb catalysts combined with plasma (flow rate: 5 L/min, water vapor: 3%, MEK inlet: 100 ppm, SIE: 180 J/L) (Nguyen et al., 2023)

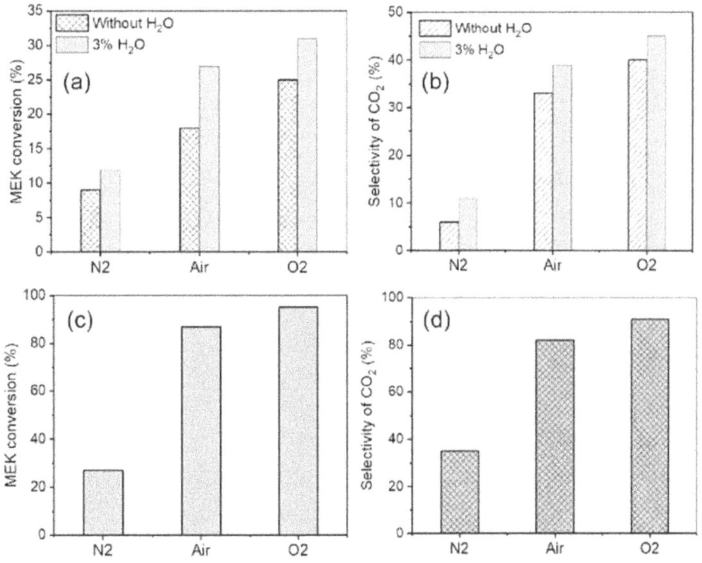

Oxygen content significantly enhances MEK removal efficiency by increasing oxygen radicals in the plasma process. In air plasma, various reactive species like oxygen atoms and hydroxyl radicals are generated, which aid MEK oxidation, albeit less efficiently than with pure oxygen due to nitrogen quenching effects. Nitrogen plasma, utilizing excited N_2 states and electrons, shows less efficient MEK removal compared to air or oxygen plasma. Introducing 3% H_2O into the gas stream enhances MEK oxidation, as water vapor boosts plasma activity and facilitates radical formation. The results in Figure 10(c-d) indicate that MEK oxidation rates increase with higher oxygen content, leading to CO_2 selectivity due to complete oxidation. Comparison in Table 1 highlights that the honeycomb catalytic discharge (HCP) reactor configuration used here achieves 87% MEK removal efficiency and 5.1 g kW^{-1} h^{-1} energy efficiency, slightly surpassing DBD plasma reactors in both metrics.

5.3 Carbon Dioxide (CO_2) Capture

The CO_2 capture is an engineering catalytic process to reduce its environmental impact. The study performed by (Zhu et al., 2022) evaluates various photocatalysts in a CO_2-water vapor system, focusing on CO production under Xenon lamp illumination. g-C_3N_4 (CN) foam/LDH heterojunction monolith (NCF) exhibit higher CO production rates compared to pure NiFe-LDH and CN foam, with NCF-3 achieving a peak yield of 275.3 μmol g^{-1} in 5 hours. The NiFe-LDH/CN heterojunction significantly enhances CO production rates, with NCF-3 achieving 55.1 μmol g^{-1} h^{-1}, about 10 and 6 times higher than pure NiFe-LDH and CN foam, respectively. The study emphasizes the importance of optimizing heterostructure design for efficient CO_2 photoreduction, highlighting NCF-3's enhanced activity due to superior CO_2 capture, light absorption, and charge separation properties.

Darunte et al. (Darunte et al., 2017) studied the CO_2 capture using monolith-supported amine-functionalized Mg_2(dobpdc) adsorbents. In fact, this catalyst showed significant potential for CO_2 capture from dilute gas mixtures.

Moreover, Regufe et al. (Regufe et al., 2019) focused on developing a tailored adsorbent for CO_2 capture using 3D printing technology. Honeycomb monoliths composed of 70% zeolite 13X and 30% activated carbon were manufactured via a paste mixture characterized by high viscosity suitable for 3D printing. Rheometer tests indicated a zero-shear viscosity of 15701 Pa s, confirming the paste's printability. Textural and mechanical characterization revealed high CO_2 adsorption capacity and mechanical stability. Heating experiments using Joule effect demonstrated efficient temperature control up to 377 K with a power consumption of 3.25 W in 180 seconds, validating the material's suitability for Electric Swing Adsorption (ESA) processes in CO_2 capture applications.

5.4 Water Treatment

Udayakumar et al. (Udayakumar et al., 2021) reviewed the effectiveness of foamed materials in separating oils and solvents from contaminated wastewater. It covers polymer and metal foams, detailing their historical context, oil absorption mechanisms, and surface wettability characteristics. Foams' micro/nano structured pores allow oils to permeate while repelling water due to higher surface energy. Effective oil absorption requires a hydrophobic surface with enhanced roughness, influenced by water contact angle. Notably, Ni foam coated with polytetrafluoroethylene (PTFE) demonstrated 99% oil removal from oily wastewater, while copper foam, known for its mechanical strength and durability, achieved 95-98% oil/solvent removal. Emerging foams like expanded polyethylene (EPE), polystyrene (PS), and Polyvinylidene fluoride foam (PVDF) show promising potential for enhanced oil-water separation performance. Environmental considerations and factors affecting foam performance conclude the review, emphasizing sustainable oil and solvent recovery strategies.

Gatica et al. (Gatica et al., 2013) evaluated activated clay monoliths for their potential to adsorb methylene blue (MB) from aqueous solutions. At low MB concentrations (10 mg l^{-1}), adsorption primarily occurred on the external surface area and larger macropores, with similar capacities observed across all samples except the 0M monolith. At higher concentrations (100 mg l^{-1}), differences in adsorption kinetics were notable among the activated monoliths, influenced by variations in coal milling procedures and resulting macropore structures. Despite lower maximum adsorption capacities compared to conventional adsorbents, the clay monoliths offer advantages such as simpler preparation processes and acceptable mechanical properties suitable for wastewater purification applications. Another research work performed by Shi et al. (Shi et al., 2019) highlighted that honeycomb ceramics efficient light absorption and energy utilization for solar-driven water evaporation. It introduces a novel consideration: the condensation of volatile organic compounds (VOCs) in distilled water within solar still devices. $CuFeMnO_4$ composite is proposed as a solution for VOCs removal, leveraging its high efficiency in the photo-Fenton reaction. Unlike homogeneous iron salts, this composite immobilizes iron oxides, preventing iron sludge formation and enabling material reusability through simple flushing. Integrating photo-Fenton chemistry with solar-driven evaporation expands the potential applications of both processes synergistically.

6. APPLICATION OF CELLULAR CERAMICS AS CATALYST SUPPORTS IN SPACE PROPULSION

Space propulsion systems operate under extreme thermal and mechanical conditions, and cellular ceramics are renowned for their high thermal stability and mechanical strength, making them ideal for withstanding the sever conditions encountered during space propulsion operations. Their ability to maintain structural integrity at high temperatures and high pressures ensures consistent catalytic performance against the deactivation of granular surfaces (Amrousse et al., 2013). The interconnected porosity of cellular ceramics allows for a high surface area-to-volume ratio, promoting optimal interaction between reactants and the catalytic surface, ensuring maximum catalytic efficiency, which is vital for the effective propulsion and maneuvering of spacecraft (Remissa et al., 2023; Souagh et al., 2022; Amrousse et al., 2015; Amrousse et al., 2024; Amrousse et al., 2012; Harimech et, 2023; Harimech et al., 2023b; Remissa et al.; 2022; Mansurov et al., 2020; Atamanov et al.; 2018; Remissa et al., 2024; Nosseir et al., 2024; Chai et al., 2024; Kumari et al., 2024; Ercan et al., 2024).

Moreover, by providing a stable and supportive environment, cellular ceramics help mitigate deactivation mechanisms such as fouling and thermal degradation. The prevention of such phenomena is critical in maintaining the operational readiness of propulsion systems, especially during long-term missions where maintenance opportunities are minimal (Krejci et al., 2012; Yoo et al., 2024; Amrousse et al., 2010; Amariei et al., 2010; Wu et al., 2015; An et al., 2010; Amrousse et al., 2011 and Micoli et al., 2013).

6.1 Catalyst Preparation for Space Propulsion Applications

Honeycomb catalysts were extensively used for thruster firing tests; especially for the decomposition of propellants; HAN thruster (Amrousse et al., 2017; Amrousse, 2021; Amrousse, 2022) and H_2O_2 thruster (Krejci et al., 2012).

Figure 11. Final Ir-based catalyst and monolith on cordierite 400 channel per square inch (cpsi) (Amrousse et al., 2014)

Wet impregnation of coated monoliths was used to create the monolith-based catalysts (Figure 11), following the instructions in (Amrousse et al., 2010). The procedure started with the mixing of wet nitric acid, Disperal boehmite, and urea to create a colloidal suspension at room temperature. Nitric acid was introduced to a beaker containing urea, which was added as white pellets and quickly dispersed. Then, in order to create a porous layer and improve the metallic active phase's dispersion, boehmite was added. The covered cellular ceramics were then heated in an airtight environment. For Ir-based catalysts, the active phase was deposited into the coating layer's porosity by impregnation with an active phase precursor, namely hexachloroiridic acid (H_2IrCl_6). Because the porous layer was only a few tens of micrometers thick, moist impregnation was the easiest method to use. This required mechanically stirring an aqueous solution of the metal salt precursor while submerging the coated cellular ceramics for an entire night. The impregnated cellular ceramic support was thoroughly dried and the excess solution was removed before it was subjected to heat treatment. In order to bring the metal precursor into its metallic state, the last heat treatment entailed H_2-reduction and O_2-calcination (Amrousse et al., 2014).

6.2 Catalyst Characterization for Space Propulsion Applications

Figure 12 shows SEM micrographs of grain and monolith catalysts. The grain catalyst and channel of the monolith are seen in this illustration from a perpendicular angle.

Figure 12. An SEM study of a monolith catalyst and a grain catalyst in cross-section (Amrousse et al. 2014)

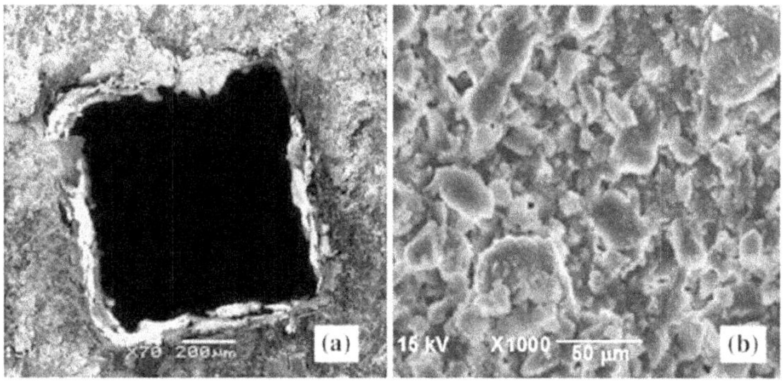

The sol-gel approach was utilized to deposit γ-alumina for the monolith catalyst, and the wet impregnation method was employed to apply the Ir active phase onto the coated monolith. These materials were integrated into the walls of the channel and the monolith pores. Additional information is obtained by EDX analysis of a particle, as described in our earlier work mentioned in (Amrousse et al., 2012). On the other hand, Figure 12b shows the shape of the porous grain catalyst. The alumina's micropores contained iridium particles, resulting in a heterogeneous porous structure. The active phase was distributed evenly and dispersed uniformly thanks to the alumina's substantial specific surface area.

Furthermore, TEM pictures of monolith catalysts both before and after the breakdown of liquid monopropellant based on HAN are shown in Figure 13. The TEM results showed that iridium particles, with an average particle size of 2.4 nm, were uniformly distributed over the washcoating layer of the monolith catalyst before the catalysts were evaluated in a liquid propellant thruster.

Figure 13. TEM images of monolith catalysts before and after firing tests (Amrousse et al. 2014)

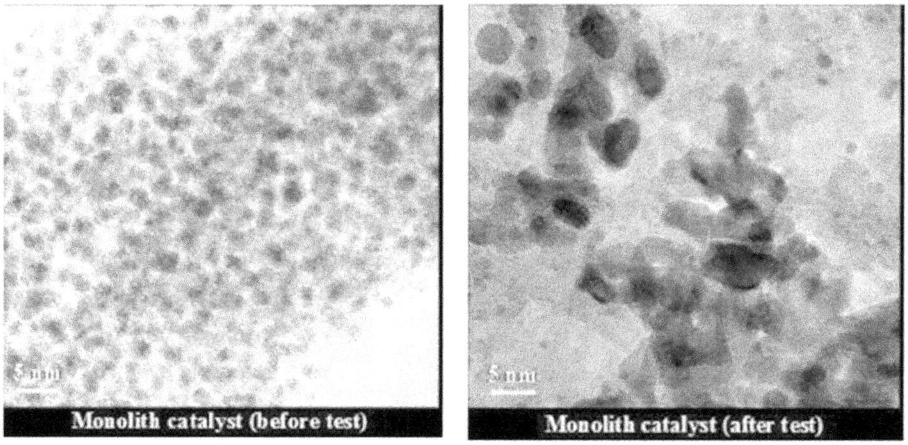

Following the catalytic breakdown experiments, a marginal rise in particle sizes was noted, suggesting that the strong connection between this metal and alumina was causing a partial resistance to sintering of iridium supported on alumina. Chemical research showed that following the catalytic breakdown of the HAN-based liquid monopropellant, the iridium level in the monolith catalyst dropped from 30% to 27.5%. The catalysts' mass loss can be the cause of this decrease in iridium concentration.

6.3 Thruster Catalytic Firing Tests

Figure 14 illustrates the thermal breakdown of HAN-based green monopropellant utilizing a 20 N-class thruster. The performance of the catalyst in dissolving the HAN-based liquid monopropellant was assessed using a monopropellant thruster. Nitrogen (N_2) from a gas cylinder was used to purge the upper portion of the HAN-based monopropellant tank before testing. To keep the catalytic decomposition reaction's HAN feed rate almost constant, the tank's N_2 pressure was regulated between 1 and 1.3 MPa. Before every test, the catalyst was put into the thruster and heated to 210°C. When the electromagnetic valve opened, N_2 pressure helped to pump highly concentrated HAN monopropellant into the thruster. Water was also added to the monopropellant to stabilize it, albeit it was observed that this could change the active metals' oxidation state. When HAN came into touch with the catalyst, it underwent catalytic breakdown (Amrousse et al., 2024b; Amrousse et al., 2014).

Figure 14. Test catalyst bed: (1) monolithic catalyst and (2) grainy catalysts (Amrousse et al. 2014)

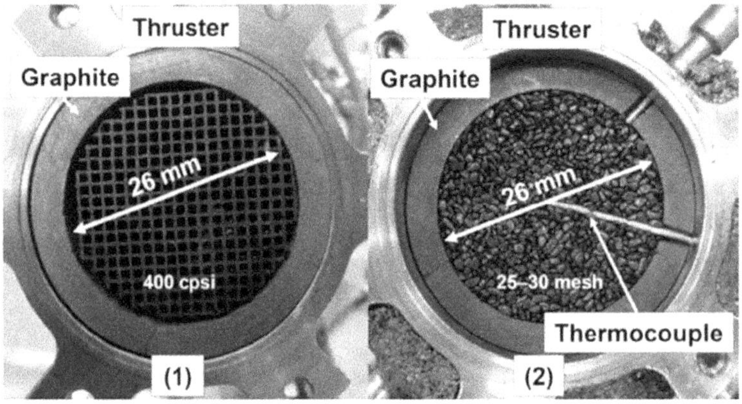

The decomposition data of HAN-based liquid monopropellant utilizing preheated monolith and grain catalysts, respectively, are shown in Figures 15a and 15b.

Figure 15. Using both a monolith and a grain catalyst, catalytic breakdown studies were carried out for HAN-based liquid monopropellant in a thruster. The estimated mass flow rates are shown in the insets.

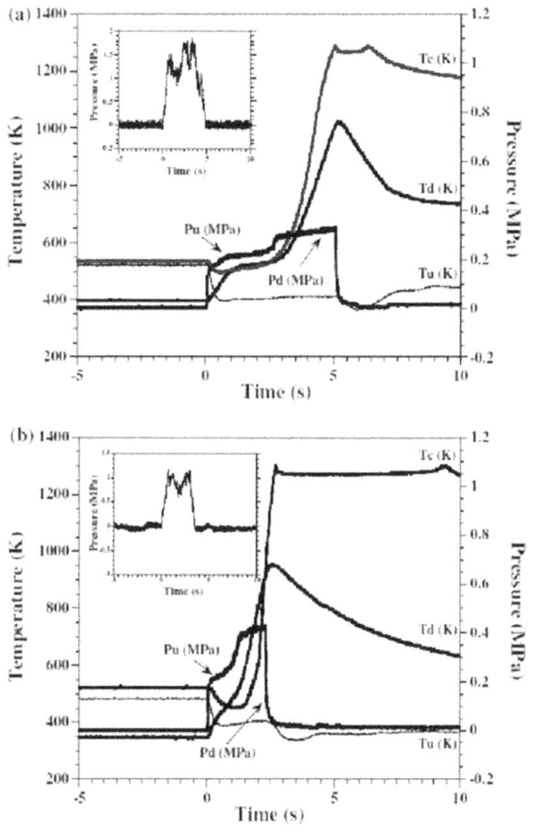

Injection pressures ranged from 0.5 to 1.4 MPa, with higher feed pressures resulting in increased liquid monopropellant flow rates, as depicted in the insets of Figure 15. The pressure drop across the monolith catalyst beds was negligible, whereas it was higher for the grain catalyst. Prior to startup, nitrogen gas was purged at a high flow rate, and continuous injection during reactions facilitated the removal of product gases. Gas evolution was observed during monolith catalyst testing. Results indicated that upstream temperatures for both catalyst types decreased upon injection due to the initial room temperature of the liquid monopropellant entering the preheated catalytic chamber. Under identical circumstances, the downstream temperatures of the product gases differed between grain and monolith catalysts. High liquid monopropellant flow rates did not significantly affect Td temperatures or pressure variations in the monolith catalyst bed. The monolith catalyst bed's Td temperature rose toward the end of the 5-second injection period, which was explained by improved heat transfer, a larger specific external catalyst surface area for mass transfer, and a longer solid-liquid contact (50 mm) than the grain catalyst (2 mm). Less pressure decrease was observed during the early induction phase of monolith catalyst testing than with the grain catalyst; this difference may have resulted from the grain catalyst's higher heat capacity and better heat transmission capabilities. Because there are more free channels in monolith catalysts, their overall surface area per reaction volume is smaller, which could explain why they are less reactive. On the other hand, dead volume in the catalytic chamber was decreased by the porous structure of grain catalysts, which offered a larger surface area per unit reaction volume. This structural difference improved reactant and product mixing efficiency during decomposition, favoring the grain catalyst. Furthermore, inefficient heat and mass transfer in monolith catalyst beds, characterized by channel presence, contrasted with the complex flow paths facilitated by grain catalysts. Thus, despite efforts to optimize reaction volumes, the reactivity of monolithic catalysts did not match that of grain catalysts, highlighting the latter's advantages in efficient heat and mass transfer (Amrousse et al., 2014).

The pressure profile throughout the SHP163 burning test is shown in Figure 16(a). Tests on honeycomb structures showed that, in comparison to the S405 catalyst, the pressure gradients were steeper with Ir honeycomb catalysts at 400 and 600 cpsi.

Figure 16. Ir-CuO honeycomb catalyst is contrasted with S405 catalyst in (a) and Ir honeycomb catalyst is contrasted with S405 catalyst in (b) (Amrousse et al., 2017)

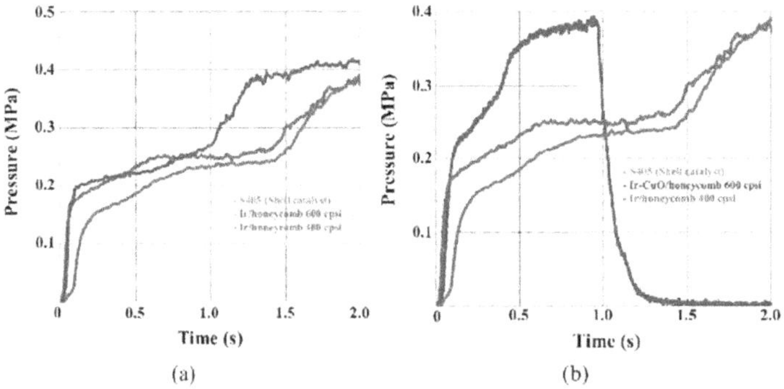

In particular, the 30-weight percent Ir honeycombs achieved up to 3.5 MPa/s, whereas the pressure slope of the S405 catalyst was roughly 1.5 MPa/s. Furthermore, secondary pressure gradients were visible on all three curves as a result of subsequent interactions between gas phase products and the catalysts. Notably, for the Ir/honeycomb 600 cpsi catalyst, the second pressure slope started early (1 sec after SHP163 injection), whereas for the S405 catalyst and honeycomb 400 cpsi, it started later. This discrepancy can be explained by the 600 cpsi honeycomb's smaller channel widths, which give gas products more contact or a longer residence period, enabling quicker interactions with the solid catalyst surface. A new 600 cpsi honeycomb catalyst based on Ir and CuO active phases was created to alleviate the secondary pressure slope delay. Ir and Cu precursors were successively impregnated into this catalyst, depositing 20 weight percent and 10 weight percent of each active phase, respectively. The Ir-CuO biphasic catalyst has two benefits over Ir monophasic catalysts, as illustrated in Figure 16(b): It accomplishes two goals: (i) it costs less and achieves the same pressure slope value as other Ir honeycomb catalysts; and (ii) it speeds up the second pressure slope by around 1-1.5 seconds, achieving a maximum pressure of 0.4 MPa in less time. Based on the accelerated and faster second slope, this catalyst might be useful for gas phase HNO_3 breakdown. Furthermore, the preheated catalyst bed temperature was adjusted between 210 and 350 °C because it is widely known that HNO_3 breakdown is sluggish and requires a high activation energy (Amrousse et al., 2017).

In Table 1, we have summarized the characteristics of existing cellular ceramics, including their compositions, forms, structures, advantages, and industrial applications:

Cellular ceramics, including honeycombs with straight cells, foams with random pores, and 3D printed structures, are developed using various compositions such as cordierite, mullite, alumina, silicon carbide, Zirconium oxide (ZrO_2), and carbon. These materials offer several advantages, including high-temperature resistance, low density, and high surface area. Due to these properties, cellular ceramics find applications in industries such as automotive (e.g., catalytic converters), aerospace (e.g., thermal protection systems), and environmental engineering (e.g., filtration and insulation).

Table 1. Use of different cellular ceramics for several industrial applications

Cellular ceramics	Different compositions	Different forms	Advantages	Different industrial applications
- Honeycombs (Monoliths) - Foams - 3D structures	- Cordierite ($2MgO{\cdot}2Al_2O_3{\cdot}5SiO_2$) - Mullite ($3Al_2O_3{\cdot}SiO_2$) - Alumina ($Al_2O_3$) - Silicon carbide (SiC) - Zirconium oxide (ZrO_2) - Mullite-Zirconium oxide ($3Al_2O_3{\cdot}SiO_2{\cdot}ZrO_2$) - Carbone (C)	- Honeycombs: Defined by number of cells per square inch (CPSI), from 100 to 800 CPSI. Cells usually are square, triangular, circular and hexagonal. - Foams: Defined by number of pores per inch (PPI), from 30 to 100 PPI.	- High surface area: catalysis as converters and filtration because of the porous structure. - Lightweight: structural applications and thermal insulation. - Thermal stability: high-temperature applications and thermal shock resistance. - Mechanical properties: high strength-to-weight Ratio and energy absorption. - Chemical resistance: corrosion resistance and inertness. - Acoustic properties. - Environmental benefits: recyclability, sustainability.	- Thermal insulation: furnaces and kilns, building materials. - Catalysis and chemical processing: catalyst support, gas and liquid filtration. - Energy applications: fuel cells and battery components, hydrogen production. - Environmental applications: pollution control (CO_2 capture, NOx and VOCs conversion) and water purification. - Aerospace and automotive industries: lightweight structural components, catalyst for thruster firing tests and thermal protection system. - Biomedical applications: abrasives and cutting tools, brake pads and clutches. - Acoustic applications: sound absorption.

7. FUTURE OUTPUTS AND RECOMMENDATIONS

In the future, basic potteries are likely to enhance even more necessary in differing industrial requests on account of their singular characteristics. Here are some potential future outputs and approvals for their use:

Hydrogenation Catalysts

Cellular potteries comprise excellent supports for hydrogenation incentives across miscellaneous industries, containing petrochemicals, pharmaceuticals, and foodstuff processing. Future research commit devote effort to something optimizing the pore structure and surface allure of these potteries to enhance catalytic project and discrimination for particular hydrogenation reactions. Moreover, progresses in 3D publication technologies commit allow the fabrication of well complex basic ceramic forms tailor-made to particular hydrogenation processes, leading to enhanced effectiveness and cost-effectiveness. By fine-bringing into harmony the material features and leveraging innovative production methods, the performance and business-related animation of hydrogenation catalysts maybe considerably improved.

Space Propulsion Systems

Cellular potteries provide inconsequential yet healthy fabrics ideal for use in space force systems, containing solid rocket propellants and warm guardianship components for spaceship. Future developments can involve mixing natural ceramics into leading propulsion sciences, such as ion thrusters and cosmic sails, to reinforce performance and endurance in harsh scope environments. Research works keep also devote effort to something designing novel basic ceramic architectures worthy enduring extreme temperature gradients and fallout exposure confronted during room responsibilities. By leveraging the unique characteristics of cellular potteries, space force wholes can achieve revised efficiency, dependability, and longevity, donating to the happiness of long-duration scope missions and investigation endeavors.

Environmental Remediation

Cellular potteries hold important potential for environmental remediation uses, containing catalytic converters for reducing issuances from automotive impoverish systems and filters for eliminating contaminants from industrial wastewater streams. Future research can survey the use of tailored basic stoneware composites with improved adsorption and catalytic characteristics to efficiently capture and dissolve differing environmental contaminators. Additionally, progresses in material processing methods keep enable the lie of economical and tenable cellular stoneware-located solutions to address material challenges on a worldwide scale. By optimizing these materials and leveraging creative result methods, basic potteries could play a critical part in mitigating contamination and advancing environmental sustainability.

Energy Storage and Conversion

Cellular potteries play a critical role in strength depository and conversion electronics, containing fuel cells, batteries, and thermoelectric maneuvers. Future novelties grant permission involve evolving novel basic ceramic architectures accompanying revamped porosity and conductivity to increase the efficiency and endurance of these devices. Furthermore, research works commit focus on leveraging the extreme surface region and thermal security of basic potteries to enhance the adeptness of strength conversion processes, to a degree hydrogen fuel result and solar power harvesting. The future of basic potteries in industrial uses looks hopeful, with continuous research and novelty anticipated to unlock new convenience for leveraging these fabrics in diverse fields, grazing from catalysis and force to environmental sustainability and strength science. By optimizing material features and exploring new lie methods, cellular potteries can considerably impact the efficiency and sustainability of miscellaneous strength structures.

8. CONCLUSIONS

In summary, the book chapter addresses that cellular ceramics stand out as highly versatile catalyst carriers due to their diverse forms, such as honeycombs, foams, and intricate 3D structures. These materials can be composed of various compounds, including silicon carbide (SiC), carbon, cordierite, mullite, alumina, and zirconium oxide, each imparting unique beneficial properties. The mechanical strength, thermal stability, acoustic properties, and permeability of these ceramics make them exceptionally well-suited for demanding catalytic applications.

Their utility spans a wide range of catalytic reactions. In hydrogen production, cellular ceramics provide the robustness needed for efficient and sustained performance. For water treatment processes, their structural integrity ensures durability and effectiveness in removing contaminants. Additionally, these materials play a crucial role in environmental applications, such as carbon dioxide (CO_2) capture and the conversion of toxic gases and volatile organic compounds (VOCs), helping to avoid pollution and improve air quality. Notably, the use of cellular ceramics extends to advanced technological applications, including space propulsion. Their role in thruster firing tests of green propellants such as hydroxylammonium nitrate (HAN), ammonium dinitramide (ADN), and hydrogen peroxide (H_2O_2) showcases their ability to withstand extreme conditions while maintaining optimal performance. This adaptability highlights the tremendous potential of cellular ceramics to improve the sustainability and efficiency of a range of catalytic technologies and processes. In conclusion, this book chapter showed that cellular ceramics are essential for a variety of catalytic applications due to their distinct forms, compositions, and qualities. Their importance in the field of catalysis will be cemented by their continuing growth and integration, which promise to push technological boundaries and tackle important environmental issues.

REFERENCES

https://www.ikts.fraunhofer.de/en/departments/environmental_process_engineering/nanoporous _membranes/functional_carrier_systems_layers/honeycomb_ceramics.html access date 7 July, 2024.

https://www.rsref.com/steel-industry/foam-ceramic-filter.html?network=g&keyword=silicon%20 carbide%20foam&matchtype=p&creative=593261174063&device=c&16559093407=16559093407 &target=&placement=&gclid=CjwKCAjwnK60BhA9EiwAmpHZw1EUMV-FtL6kHwjPTjo3Abg AimyC9VUjm6szERKcGZoIJ647zffjnhoC9iQQAvD_BwE&gad_source=1 Access date: July 7, 2024

http://finewayceramics.com/ceramic-foams/ Access_date: July 7, 2024 https://www.americanelements .com/silicon-carbide-honeycomb-409-21-2 Access date: July 7, 2024.

Al-Ketan, O., Soliman, A., AlQubaisi, A. M., & Al-Rub, R. K. A. (2017). Nature-Inspired Lightweight Cellular Co-Continuous Composites with Architected Periodic Gyroidal Structures. *Advanced Engineering Materials*, 20(2), 1700549. Advance online publication. DOI: 10.1002/adem.201700549

Amariei, D., Amrousse, R., Batonneau, Y., Kappenstein, C., & Cartoixa, B. (2010). Monolithic catalysts for the decomposition of energetic compounds. In *Studies in surface science and catalysis* (pp. 35–42). DOI: 10.1016/S0167-2991(10)75005-9

Amrousse, R. (2021) Thermal decomposition of HAN green propellant. In the Proceedings of the 72nd International Astronautical Congress, Dubai, UAE.

Amrousse, R. (2022) Survey on the green propulsion systems: from the lab-scale to the pilot scale-up: HAN is a good Example. In the Proceedings of the 73rd International Astronautical Congress, Paris, France.

Amrousse, R., Augustin, C., Farhat, K., Batonneau, Y., & Kappenstein, C. J. (2011). Catalytic decomposition of H_2O_2 using FeCrAl metallic foam-based catalysts. *International Journal of Energetic Materials and Chemical Propulsion*, 10(4), 337–349. DOI: 10.1615/IntJEnergeticMaterialsChemProp.2012005202

Amrousse, R., Batonneau, Y., Kappenstein, C., Théron, M., & Bravais, P. (2010). Preparation of monolithic catalysts for space propulsion applications. In *Studies in surface science and catalysis* (pp. 755–758). DOI: 10.1016/S0167-2991(10)75153-3

Amrousse, R., Elidrissi, A. N., Bachar, A., Mabrouk, A., Toshtay, K., & Azat, S. (2024). Nanosized catalytic particles for the decomposition of green propellants as substitute for hydrazine. In Advances in chemical and materials engineering book series (pp. 195–217). DOI: 10.4018/979-8-3693-3268-9.ch009

Amrousse, R., Elidrissi, A. N., Nosseir, A. E. S., Toshtay, K., Atamanov, M. K., & Azat, S. (2024b). Thermal and Catalytic Decomposition of Hydroxylammonium Nitrate (HAN)-Based Propellant. In *Space technology library* (pp. 33–60). DOI: 10.1007/978-3-031-62574-9_2

Amrousse, R., Hori, K., Fetimi, W., & Farhat, K. (2012). HAN and ADN as liquid ionic monopropellants: Thermal and catalytic decomposition processes. *Applied Catalysis B: Environmental*, 127, 121–128. DOI: 10.1016/j.apcatb.2012.08.009

Amrousse, R., Katsumi, T., Azuma, N., & Hori, K. (2017). Hydroxylammonium nitrate (HAN)-based green propellant as alternative energy resource for potential hydrazine substitution: From lab scale to pilot plant scale-up. *Combustion and Flame*, 176, 334–348. DOI: 10.1016/j.combustflame.2016.11.011

Amrousse, R., Katsumi, T., Azuma, N., & Hori, K. (2017). Hydroxylammonium nitrate (HAN)-based green propellant as alternative energy resource for potential hydrazine substitution: From lab scale to pilot plant scale-up. *Combustion and Flame*, 176, 334–348. DOI: 10.1016/j.combustflame.2016.11.011

Amrousse, R., Katsumi, T., Bachar, A., Brahmi, R., Bensitel, M., & Hori, K. (2013). Chemical engineering study for hydroxylammonium nitrate monopropellant decomposition over monolith and grain metal-based catalysts. *Reaction Kinetics, Mechanisms and Catalysis*, 111(1), 71–88. DOI: 10.1007/s11144-013-0626-6

Amrousse, R., Katsumi, T., Itouyama, N., Azuma, N., Kagawa, H., Hatai, K., Ikeda, H., & Hori, K. (2015). New HAN-based mixtures for reaction control system and low toxic spacecraft propulsion subsystem: Thermal decomposition and possible thruster applications. *Combustion and Flame*, 162(6), 2686–2692. DOI: 10.1016/j.combustflame.2015.03.026

Amrousse, R., Katsumi, T., Niboshi, Y., Azuma, N., Bachar, A., & Hori, K. (2013). Performance and deactivation of Ir-based catalyst during hydroxylammonium nitrate catalytic decomposition. *Applied Catalysis A, General*, 452, 64–68. DOI: 10.1016/j.apcata.2012.11.038

Amrousse, R., Keav, S., Batonneau, Y., Kappenstein, C. J., Theron, M., & Bravais, P. (2011). Catalytic ignition of cold hydrogen/oxygen mixtures for space propulsion applications. *International Journal of Energetic Materials and Chemical Propulsion*, 10(3), 217–230. DOI: 10.1615/IntJEnergeticMaterialsChemProp.2012004877

An, S., Lee, J., Kappenstein, C., & Kwon, S. (2010). Comparison of catalyst support between monolith and pellet in hydrogen peroxide thrusters. *Journal of Propulsion and Power*, 26(3), 439–445. DOI: 10.2514/1.46075

Atamanov, M. K., Amrousse, R., Hori, K., Kolesnikov, B. Y., & Mansurov, Z. A. (2018). Influence of activated carbon on the thermal decomposition of hydroxylammonium nitrate. *Combustion, Explosion, and Shock Waves*, 54(3), 316–324. DOI: 10.1134/S0010508218030085

Binner, J. (2006). Ceramic Foams In *Cellular Ceramics: Structure, Manufacturing, Properties and Applications. Ceramics Foams,* Scheffler, M., Colombo, P. (Eds.), 31–56. DOI: 10.1002/3527606696.ch2a

Brezny, R., & Green, D. J. (1989). Fracture behavior of Open-Cell ceramics. *Journal of the American Ceramic Society*, 72(7), 1145–1152. DOI: 10.1111/j.1151-2916.1989.tb09698.x

Brezny, R., Green, D. J., & Dam, C. Q. (1989). Evaluation of strut strength in Open-Cell ceramics. *Journal of the American Ceramic Society*, 72(6), 885–889. DOI: 10.1111/j.1151-2916.1989.tb06239.x

Chai, W. S., Liu, L., Sun, X., Li, X., & Lu, Y. Y. (2024). An Overview of Green Propellants and Propulsion System Applications: Merits and Demerits. In *Space technology library* (pp. 3–32). DOI: 10.1007/978-3-031-62574-9_1

Darunte, L. A., Terada, Y., Murdock, C. R., Walton, K. S., Sholl, D. S., & Jones, C. W. (2017). Monolith-Supported Amine-Functionalized Mg_2(dobpdc) Adsorbents for CO_2 Capture. *ACS Applied Materials & Interfaces*, 9(20), 17042–17050. DOI: 10.1021/acsami.7b02035 PMID: 28440615

Deng, X., Ran, S., Han, L., Zhang, H., Ge, S., & Zhang, S. (2017). Foam-gelcasting preparation of high-strength self-reinforced porous mullite ceramics. *Journal of the European Ceramic Society*, 37(13), 4059–4066. DOI: 10.1016/j.jeurceramsoc.2017.05.009

Du, Z., Yao, D., Xia, Y., Zuo, K., Yin, J., Liang, H., & Zeng, Y. (2019). The high porosity silicon nitride foams prepared by the direct foaming method. *Ceramics International*, 45(2), 2124–2130. DOI: 10.1016/j.ceramint.2018.10.118

Dupère, I. D. J., Lu, T. J., & Dowling, A. P. (2005). *Acoustic Properties*. 381–399. DOI: 10.1002/3527606696.ch4e

Ercan, K. E., Yurtseven, M. A., & Yilmaz, C. (2024). Development of Mono and Bipropellant Systems for Green Propulsion Applications. In *Space technology library* (pp. 249–280). DOI: 10.1007/978-3-031-62574-9_9

Fend, T., Trimis, D., Pitz-Paal, R., Hoffschmidt, B., & Reutter, O. (2005). *Thermal Properties*. 342–360. DOI: 10.1002/3527606696.ch4c

Feng, C., Zhang, K., He, R., Ding, G., Xia, M., Jin, X., & Xie, C. (2020). Additive manufacturing of hydroxyapatite bioceramic scaffolds: Dispersion, digital light processing, sintering, mechanical properties, and biocompatibility. *Journal of Advanced Ceramics*, 9(3), 360–373. DOI: 10.1007/s40145-020-0375-8

Frank, M. B., Naleway, S. E., Haroush, T., Liu, C., Siu, S. H., Ng, J., Torres, I., Ismail, A., Karandikar, K., Porter, M. M., Graeve, O. A., & McKittrick, J. (2017). Stiff, porous scaffolds from magnetized alumina particles aligned by magnetic freeze casting. *Materials Science and Engineering C*, 77, 484–492. DOI: 10.1016/j.msec.2017.03.246 PMID: 28532056

Gatica, J. M., Gómez, D. M., Harti, S., & Vidal, H. (2013). Clay honeycomb monoliths for water purification: Modulating methylene blue adsorption through controlled activation via natural coal templating. *Applied Surface Science*, 277, 242–248. DOI: 10.1016/j.apsusc.2013.04.034

Gauckler, L. J., & Waeber, M. M. (1985) in Light Metals 1985, Proc. 114th Ann. Meet. Metal. Soc. AIME, 1261–1283.

Gianella, S., Gaia, D., & Ortona, A. (2012). High temperature applications of SIðSIC Cellular ceramics. *Advanced Engineering Materials*, 14(12), 1074–1081. DOI: 10.1002/adem.201200012

Gibson, L. J., & Ashby, M. F. (1988). *Cellular Solids: Structure and Properties*. Pergamon.

Hagiwara, H., & Green, D. J. (1987). Elastic behavior of Open-Cell alumina. *Journal of the American Ceramic Society*, 70(11), 811–815. DOI: 10.1111/j.1151-2916.1987.tb05632.x

Harimech, Z., Hairch, Y., Atamanov, M., Toshtay, K., Azat, S., Souhair, N., & Amrousse, R. (2023). Carbon nanotube iridium-cupric oxide supported catalysts for decomposition of ammonium dinitramide in the liquid phase. *International Journal of Energetic Materials and Chemical Propulsion*, 22(3), 13–18. DOI: 10.1615/IntJEnergeticMaterialsChemProp.2023047555

Harimech, Z., Toshtay, K., Atamanov, M., Azat, S., & Amrousse, R. (2023). Thermal decomposition of Ammonium dinitramide (ADN) as green energy source for space propulsion. *Aerospace (Basel, Switzerland)*, 10(10), 832. DOI: 10.3390/aerospace10100832

Hossain, M. M., Mok, Y. S., Nguyen, V. T., Sosiawati, T., Lee, B., Kim, Y. J., Lee, J. H., & Heo, I. (2022). Plasma-catalytic oxidation of volatile organic compounds with honeycomb catalyst for industrial application. *Process Safety and Environmental Protection/Transactions of the Institution of Chemical Engineers. Part B, Process Safety and Environmental Protection/Chemical Engineering Research and Design. Chemical Engineering Research & Design*, 177, 406–417. DOI: 10.1016/j.cherd.2021.11.010

Innocentini, M. D. M., Salvini, V. R., Pandolfelli, V. C., & Coury, J. R. (1999). Assessment of Forchheimer's Equation to Predict the Permeability of Ceramic Foams. *Journal of the American Ceramic Society*, 82(7), 1945–1948. DOI: 10.1111/j.1151-2916.1999.tb02024.x

Inocentini, M. D., Pardo, A. R., & Pandolfelli, V. C. (2002). Permeability. *Journal of the American Ceramic Society*, 85(6), 1517–1521.

Krejci, D., Woschnak, A., Scharlemann, C., & Ponweiser, K. (2012). Structural impact of honeycomb catalysts on hydrogen peroxide decomposition for micro propulsion. *Process Safety and Environmental Protection/Transactions of the Institution of Chemical Engineers. Part B, Process Safety and Environmental Protection/Chemical Engineering Research and Design. Chemical Engineering Research & Design*, 90(12), 2302–2315. DOI: 10.1016/j.cherd.2012.05.015

Kumari, S., Agnihotri, R., & Oommen, C. (2024). Cerium Oxide-Based Robust Catalyst for Hydroxylammonium Nitrate Monopropellant Thruster. In *Space technology library* (pp. 187–215). DOI: 10.1007/978-3-031-62574-9_7

Liao, M., Yi, X., Dai, Z., Qin, H., Guo, W., & Xiao, H. (2023). Application of metal-BDC-derived catalyst on cordierite honeycomb ceramic support in a microreactor for hydrogen production. *Ceramics International*, 49(17), 29082–29093. DOI: 10.1016/j.ceramint.2023.06.184

Liu, P., & Chen, G. (2014). Fabricating Porous Ceramics. In *Elsevier eBooks* (pp. 221–302). DOI: 10.1016/B978-0-12-407788-1.00005-8

Lou, Y., Wang, F., Li, Z., Zou, Z., Fan, G., Wang, X., Lei, W., & Lu, W. (2020). Fabrication of high-performance $MgTiO_3$–$CaTiO_3$ microwave ceramics through a stereolithography-based 3D printing. *Ceramics International*, 46(10), 16979–16986. DOI: 10.1016/j.ceramint.2020.03.282

Mansurov, Z. A., Amrousse, R., Hori, K., & Atamanov, M. K. (2020). Combustion/Decomposition behavior of HAN under the effects of nanoporous activated carbon. In Springer eBooks (pp. 211–230). DOI: 10.1007/978-981-15-4831-4_8

Maurath, J., & Willenbacher, N. (2017). 3D printing of open-porous cellular ceramics with high specific strength. *Journal of the European Ceramic Society*, 37(15), 4833–4842. DOI: 10.1016/j.jeurceramsoc.2017.06.001

Micoli, L., Bagnasco, G., Turco, M., Trifuoggi, M., Sorge, A. R., Fanelli, E., Pernice, P., & Aronne, A. (2013). Vapour phase H_2O_2 decomposition on Mn based monolithic catalysts synthesized by innovative procedures. *Applied Catalysis B: Environmental*, 140–141, 516–522. DOI: 10.1016/j.apcatb.2013.04.072

Moreira, E., Innocentini, M., & Coury, J. (2004). Permeability of ceramic foams to compressible and incompressible flow. *Journal of the European Ceramic Society*, 24(10–11), 3209–3218. DOI: 10.1016/j.jeurceramsoc.2003.11.014

Nguyen, V. T., Dinh, D. K., Lan, N. M., Trinh, Q. H., Hossain, M. M., Dao, V. D., & Mok, Y. S. (2023). Critical role of reactive species in the degradation of VOC in a plasma honeycomb catalyst reactor. *Chemical Engineering Science*, 276, 118830. DOI: 10.1016/j.ces.2023.118830

Nosseir, A. E. S., Cervone, A., Amrousse, R., Pasini, A., Igarashi, S., & Matsuura, Y. (2024). Green Monopropellants: State-of-the-Art. In *Space technology library* (pp. 95–134). DOI: 10.1007/978-3-031-62574-9_4

Palma, V., Barba, D., Vaiano, V., Colozzi, M., Palo, E., Barbato, L., Cortese, S., & Miccio, M. (2018). Honeycomb structured catalysts for H_2 production via H_2S oxidative decomposition. *Catalysts*, 8(11), 488. DOI: 10.3390/catal8110488

Papetti, V., Eggenschwiler, P. D., Della Torre, A., Lucci, F., Ortona, A., & Montenegro, G. (2018). Additive Manufactured open cell polyhedral structures as substrates for automotive catalysts. *International Journal of Heat and Mass Transfer. International Journal of Heat and Mass Transfer*, 126, 1035–1047. DOI: 10.1016/j.ijheatmasstransfer.2018.06.061

Regufe, M. J., Ferreira, A. F., Loureiro, J. M., Rodrigues, A., & Ribeiro, A. M. (2019). Electrical conductive 3D-printed monolith adsorbent for CO_2 capture. *Microporous and Mesoporous Materials*, 278, 403–413. DOI: 10.1016/j.micromeso.2019.01.009

Remissa, I., Baragh, F., Mabrouk, A., Bachar, A., & Amrousse, R. (2024). Low-Cost Catalysts for Hydrogen Peroxide (H_2O_2) Thermal Decomposition. In *Space technology library* (pp. 61–94). DOI: 10.1007/978-3-031-62574-9_3

Remissa, I., Jabri, H., Hairch, Y., Toshtay, K., Atamanov, M., Azat, S., & Amrousse, R. (2023). Propulsion Systems, Propellants, Green Propulsion Subsystems and their Applications: A Review. *Eurasian Chemico-technological Journal*, 25(1), 3–19. DOI: 10.18321/ectj1491

Remissa, I., Souagh, A., Hairch, Y., Sahib-Eddine, A., Atamanov, M., & Amrousse, R. (2022). Thermal decomposition behaviors of hydrogen peroxide over free noble metal-synthesized solid catalysts. *International Journal of Energetic Materials and Chemical Propulsion*, 21(4), 17–29. DOI: 10.1615/IntJEnergeticMaterialsChemProp.2022043338

Rice ,R. (2005). *Mechanical Properties*. 289–312. DOI: 10.1002/3527606696.ch4a

Rice, R. W. (1996). Evaluation and extension of physical property-porosity models based on minimum solid area. *Journal of Materials Science*, 31(1), 102–118. DOI: 10.1007/BF00355133

Santoliquido, O., Bianchi, G., Eggenschwiler, P. D., & Ortona, A. (2017). Additive manufacturing of periodic ceramic substrates for automotive catalyst supports. *International Journal of Applied Ceramic Technology*, 14(6), 1164–1173. DOI: 10.1111/ijac.12745

Shi, L., Shi, Y., Zhuo, S., Zhang, C., Aldrees, Y., Aleid, S., & Wang, P. (2019). Multi-functional 3D honeycomb ceramic plate for clean water production by heterogeneous photo-Fenton reaction and solar-driven water evaporation. *Nano Energy*, 60, 222–230. DOI: 10.1016/j.nanoen.2019.03.039

Souagh, A., Remissa, I., Atamanov, M., Alaoui, H. E., & Amrousse, R. (2022). Comparative study of the thermal decomposition of hydroxylammonium nitrate green energetic compound: Combination between experimental and DFT calculation. *International Journal of Energetic Materials and Chemical Propulsion*, 21(4), 31–38. DOI: 10.1615/IntJEnergeticMaterialsChemProp.2022044056

Stochero, N., De Moraes, E., Moreira, A., Fernandes, C., Innocentini, M., & De Oliveira, A. N. (2020). Ceramic shell foams produced by direct foaming and gelcasting of proteins: Permeability and microstructural characterization by X-ray microtomography. *Journal of the European Ceramic Society*, 40(12), 4224–4231. DOI: 10.1016/j.jeurceramsoc.2020.05.036

Tabard, L., Garnier, V., Prud'homme, E., Courtial, E., Meille, S., Adrien, J., Jorand, Y., & Gremillard, L. (2021). Robocasting of highly porous ceramics scaffolds with hierarchized porosity. *Additive Manufacturing*, 38, 101776. DOI: 10.1016/j.addma.2020.101776

Udayakumar, K. V., Gore, P. M., & Kandasubramanian, B. (2021). Foamed materials for oil-water separation. *Chemical Engineering Journal Advances*, 5, 100076. DOI: 10.1016/j.ceja.2020.100076

Van Voorhees, E. J., & Green, D. J. (1991). Failure behavior of Cellular-Core ceramic sandwich composites. *Journal of the American Ceramic Society*, 74(11), 2747–2752. DOI: 10.1111/j.1151-2916.1991.tb06838.x

Vlasea, M., Pilliar, R., & Toyserkani, E. (2015). Control of structural and mechanical properties in bioceramic bone substitutes via additive manufacturing layer stacking orientation. *Additive Manufacturing*, 6, 30–38. DOI: 10.1016/j.addma.2015.03.001

Wu, C., Wang, X., Zhou, X., Yang, T., & Zhang, T. (2015). Supported MnOx /SRO-Al_2O_3 high-cell density honeycomb ceramic monolith catalyst for high-concentration hydrogen peroxide decomposition. *International Journal of Energetic Materials and Chemical Propulsion*, 14(5), 421–436. DOI: 10.1615/IntJEnergeticMaterialsChemProp.2015011176

Xu, T., Cheng, S., Jin, L., Zhang, K., & Zeng, T. (2020). High-temperature flexural strength of SiC ceramics prepared by additive manufacturing. *International Journal of Applied Ceramic Technology*, 17(2), 438–448. DOI: 10.1111/ijac.13454

Yeetsorn, R., Tungkamani, S., & Maiket, Y. (2022). Fabrication of a ceramic foam catalyst using polymer foam scrap via the replica technique for dry reforming. *ACS Omega*, 7(5), 4202–4213. DOI: 10.1021/acsomega.1c05841 PMID: 35155913

Yoo, D., Kim, M., Oh, S. K., Hwang, S., Kim, S., Kim, W., Kwon, Y., Jo, Y., & Jeon, J. (2024). Synthesis of Hydroxylammonium Nitrate and Its Decomposition over Metal Oxide/Honeycomb Catalysts. *Catalysts*, 14(2), 116. DOI: 10.3390/catal14020116

Zhang, C., Liu, B., Li, W., Liu, X., Wang, K., Deng, Y., Guo, Z., & Lv, Z. (2021). A well-designed honeycomb Co_3O_4@CdS photocatalyst derived from cobalt foam for high-efficiency visible-light H_2 evolution. *Journal of Materials Chemistry. A, Materials for Energy and Sustainability*, 9(19), 11665–11673. DOI: 10.1039/D0TA11433B

Zhang, X., Zhang, K., Zhang, L., Wang, W., Li, Y., & He, R. (2022). Additive manufacturing of cellular ceramic structures: From structure to structure–function integration. *Materials & Design*, 215, 110470. DOI: 10.1016/j.matdes.2022.110470

Zhu, B., Xu, Q., Bao, X., Yin, H., Qin, Y., & Shen, X. (2022). Highly selective CO_2 capture and photoreduction over porous carbon nitride foams/LDH monolith. *Chemical Engineering Journal*, 429, 132284. DOI: 10.1016/j.cej.2021.132284

Chapter 2
Model Matrices for Nuclear Waste Storage:
Structure, Characterization, and Mechanical Properties

Assia Mabrouk
https://orcid.org/0000-0001-6399-0644
Science and Technology Research Laboratory, The Higher School of Education and Training, Ibn Zohr University, Agadir, Morocco

Ahmed Bachar
Laboratory of Process Engineering, Faculty of Applied Science Ait Melloul, Ibn Zohr University, Agadir, Morocco

Aurélien Canizarès
CNRS, CEMHTI UPR 3079, Université d'Orléans, France

Yann Vaills
CNRS, CEMHTI UPR 3079, Université d'Orléans, France

Stuart Hampshire
https://orcid.org/0000-0002-8993-2570
The Bernal Institute, University of Limerick, Ireland

ABSTRACT

Glasses enriched with rare earth elements hold significant promise for nuclear waste storage, especially for immobilizing high-level waste (HLW) containing minor actinides. This chapter delves into the structural intricacies of peraluminous lanthanum-rich sodium aluminoborosilicate glasses, where lanthanum serves as a surrogate for lanthanides and minor actinides typical of HLW. Given the imperative to accommodate higher concentrations of HLW, the aluminum/boron substitution emerges as a pivotal factor influencing the glass matrix's properties. Employing a combination of infrared (IR), nuclear magnetic resonance (NMR), and Brillouin spectroscopy, we unravel the intricate interplay between composition and structure. Our investigation sheds light on the effects of aluminum/boron substitution on phase separation phenomena and structural rearrangements within the glass network. This in-depth analysis

DOI: 10.4018/979-8-3693-7505-1.ch002

provides critical insights into the design and optimization of novel glass matrices tailored for advanced nuclear waste immobilization strategies.

1. INTRODUCTION

In the 1970s, France embarked on a path towards energy independence by predominantly generating its electricity through nuclear power, spurred by two oil shocks. Now over a century since the nuclear era began in 1896 with the discovery of natural radioactivity by Henri Becquerel, followed by Pierre and Marie Curie's identification of the radioactive elements polonium (^{210}Po) and radium (^{226}Ra), advancements in nuclear technology have accelerated. Like any industry, nuclear power generates waste. In this context, managing radioactive waste remains a critical societal challenge, both today and in the future. For the confinement of these wastes, the industrial solution adopted over the past three decades has been vitrification. This process involves indiscriminately incorporating Fission Products and Minor Actinides into molten glass. Thereby, the fission product solution is permanently immobilized within the glass matrix, confining the radioactivity of activated elements. The primary advantage of glasses lies in their amorphous nature.

Unlike crystalline materials that have well-defined structures, glass is uniquely capable of incorporating the entire spectrum of elements present in fission product solutions into its disordered structure through chemical bonding with its constituent elements. Thus, radionuclides become integral to the glass structure, ensuring not mere encapsulation but true confinement of all elements down to the atomic scale. Various matrix types have been considered for confinement purposes, including borosilicate, aluminosilicate, and aluminoborosilicate. The latter, with high rare earth content, is particularly intriguing and serves as a model for nuclear waste containment glasses. Rich in TR_2O_3 (where TR = La), these glasses simulate the spectrum of lanthanides and minor actinides found in nuclear waste (Quintas et al., 2007; Angeli et al., 2013; Morin et al., 2016).

In light of the findings presented in this chapter, there is significant potential for these model matrices to influence future trends in nuclear waste management. As the industry increasingly focuses on sustainable practices and improved safety measures, the development of advanced glass matrices could lead to enhanced containment solutions that minimize environmental impact. Furthermore, ongoing research into the structural properties and stability of these glasses may pave the way for innovative vitrification techniques, setting new standards for waste management practices globally. To ensure the information is up-to-date, we also incorporate recent studies, such as Luet al. (2024), which explores advancements in glass formulations, and Lu et al. (2024b), which discusses the impact of new vitrification techniques on waste management. Other relevant studies include also Lu et al. (2024c), which investigate the long-term stability of various glass matrices, and Zhou et al. (2024), highlighting innovative approaches to incorporate higher waste loadings into glass compositions.Typically, these glasses contain network formers such as Al_2O_3, B_2O_3, SiO_2, and modifier oxides. Structural studies often manipulate the proportions and types of modifier oxides to understand their influence on glass properties (Morin et al., 2016). Structural investigations of aluminosilicate (Schaller et al., 1998; Florian et al., 2007 Stevensson et al., 2013), borosilicate (Quintas et al., 2007), aluminoborate (Deters et al., 2009), and aluminoborosilicate glasses (Morin et al., 2016) containing lanthanum and yttrium have been extensively pursued. The structure of lanthanum aluminoborosodosilicate glasses is inherently complex due to the presence of at least four oxides. As such, correlations between structure and properties in these glasses, and their relationship with composition, remain unclear (Zheng et al., 2012). Over recent decades, the development of numerous characterization methods (NMR, infrared, etc.) has provided crucial tools for unraveling this structural complexity.

This chapter focuses on aluminoborosodosilicate glasses within the peraluminous lanthanum series of composition: "$55SiO_2-(25-x)Al_2O_3-xB_2O_3-15Na_2O-5La_2O_3$ ($0 \leq x \leq 12.5$)". Lanthanum is intended to simulate the range of lanthanides and minor actinides present in nuclear wastes. Studies are ongoing to formulate new vitreous matrices with a fission product content exceeding the current 18.5% mass limit of R7T7 glass (Guo et al., 2023b), aiming for excellent long-term behavior. Thus far, structural studies of lanthanum-rich aluminoborosodosilicate glasses have been limited. Accordingly, this chapter focuses on investigating the influence of boron on glass structure using various characterization techniques. An experimental strategy in three stages was employed: Glass elaboration, Microstructural studies (XRD, DSC, TEM...) Characterization of short and medium-range order: NMR spectroscopy (^{29}Si, ^{27}Al, ^{23}Na, ^{11}B) for short-range structural properties, Infrared spectroscopy for intermediate-range structural properties, and Brillouin spectroscopy for macroscopic mechanical properties (long-range).

1.1 Glasses and Glass Transition

Definition of Glass

Glass, once considered a simple material known and used by humans for millennia, has become an essential element in numerous fields such as optics, construction materials and nuclear waste storage (Wondraczek et al., 2010; Ueda et al., 2011; Ye et al., 2011; Da et al., 2011). Scientifically, the term 'glass' generally refers to a non-crystalline solid that exhibits the phenomenon of glass transition; the corresponding physical state is termed the vitreous state (Wondraczek et al., 2007). According to this definition, the classical method of obtaining glass involves rapidly cooling a liquid so that crystallization does not have time to occur. A gradual solidification process occurs as the liquid's viscosity steadily increases with decreasing temperature. Partial crystallization can occur when the liquid is cooled below its melting temperature T_m, but rapid enough cooling can prevent crystallization entirely. At temperatures just below Tm, the system remains liquid, existing in a metastable state known as supercooled liquid. Further lowering the temperature significantly increases its viscosity, resulting in a macroscopically solid state: the system solidifies, thermodynamically out of equilibrium in a state known as the vitreous state. Over time, the system approaches thermodynamic equilibrium through relaxation phenomena, which are crucial for assessing the durability of glass.

Conditions for Glass Formation

Numerous studies have been conducted and models proposed to explain the formation or non-formation of glasses. Historically, structural approaches were the first to be developed and gave rise to various glass formation criteria. In the 1930s, Zachariasen proposed a random network theory for the structure of glass based on Goldschmidt's perspective (Zachariasen., 1932). The network hypothesis, supported by X-ray diffraction experiments by Warren, represented a significant advancement in understanding glass structure. Zachariasen and Warren formulated simple rules that oxides A_mO_n must satisfy to be capable of vitrification: (a) The coordination number of oxygen atoms around atom A, known as its coordination number, must be low (3 or 4), forming triangles or tetrahedra. (b) No oxygen atom can be bonded to more than two cations A. (c) Polyhedra with coordination numbers of 3 or 4 (triangles or tetrahedra) must contact each other only at their vertices, not through their edges or faces. (d) At least three vertices of one polyhedron must connect with neighboring polyhedra to establish a three-dimensional network of polyhedral (Zachariasen.,1932). The 'continuous random network' model developed by Zachariasen and Warren (Zachariasen.,1932) assumes a homogeneous and random distribution of network-modifying ions within the glass. Thus, the unordered linkage of SiO_4 tetrahedra is complemented by a perfectly random distribution of various Q_n units (where n denotes the number of bridging oxygens). More recently, with advanced characterization techniques such as EXAFS (Mastelaro et al., 2018), neutron or X-ray scattering (Heinz et al., 2016), and molecular dynamics (Zheng et al., 2014), it has been demonstrated that modifying ions are actually distributed inhomogeneously within the glass network. In the 'modified random network' model proposed by Greaves (1997), the glass consists of regions rich in network-forming elements separated by domains rich in network-modifying ions and non-bridging oxygens.

Classification of Glasses

According to Zachariasen, in oxide glasses, different cations can be classified into three categories: network-forming elements, network-modifying elements, and intermediate elements. (a) Network-forming elements such as Si, B, P, Ge, As, and Be have the capability to form the vitreous network structure of the glass. (b) Network-modifying elements such as Na, K, Ca, and Ba depolymerize the network of forming oxides and break bonds between oxygen and forming cations. These elements are primarily alkali or alkaline-earth metals. (c) Intermediate elements, depending on the composition, can act either as formers or modifiers. Aluminum is an example of such an intermediate element.

The Glass Transition

Glass is a solid characterized by the absence of long-range order. Of crystalline and amorphous silica, showing that short-range order is very similar in both structural modes, with only long-range order differing (Neuville et al., 2022). To understand the origin of such differences, we must delve into the formation of an amorphous solid. When examining a specific volume (or enthalpy) versus temperature diagram (Jiang et al., 2014), it becomes apparent that below a certain temperature, the most stable state of a material is the crystalline phase. However, there are instances where the liquid-to-crystal transformation does not occur upon cooling through the crystallization temperature. This phenomenon can be understood by considering that crystallization requires the formation of nuclei, where there is competition between the surface energy of the crystallized nucleus, which increases the system's free energy, and the volume energy of the nucleus, which decreases it. The growth of crystalline nuclei can only proceed if they reach a critical size. Under certain conditions, these nuclei may fail to reach this critical size, primarily due to excessively rapid cooling rates relative to atomic mobility. The system then finds itself in a metastable state known as a supercooled liquid. From a thermodynamic perspective, this indicates that the system adopts a structure where the free enthalpy extends toward lower temperatures from the liquid state. Below a certain temperature, the system cannot maintain this metastable state. This leads to what is known as the glass transition, where the system's free enthalpy gradually deviates from the supercooled state to follow a path parallel to the free enthalpy of the crystal. This transition does not occur at a fixed temperature but at a temperature that increases with faster cooling rates. The glass transition is thus not a classical transition in the thermodynamic sense because the systems studied are out of equilibrium.

Furthermore, the glass transition depends on thermal history, specifically on the quenching rate, unlike, for example, a melting temperature.

Figure 1. Shows a DSC (Differential Scanning Calorimetry) thermogram obtained from a glass prepared in the context of this study (V2X2.5). The glass transition is observed as an extended phenomenon spanning a range of temperatures. Assigning a specific value to the glass transition temperature involves a somewhat arbitrary approach and is primarily useful for making comparisons between different glassy systems. For our purposes, the glass transition temperature, T_g, will be defined as the average of the temperatures at the end and the beginning of the glass transition region as indicated in Figure 1. Furthermore, when reporting the value of T_g, it is essential to specify the conditions under which it was measured, as this value depends on the observation time and the rate of temperature change (dT/dt) as programmed in the experiment.

Figure 1. DSC thermogram obtained from the V2X2.5 glass. The glass transition is observed as an extended phenomenon covering a range of temperatures

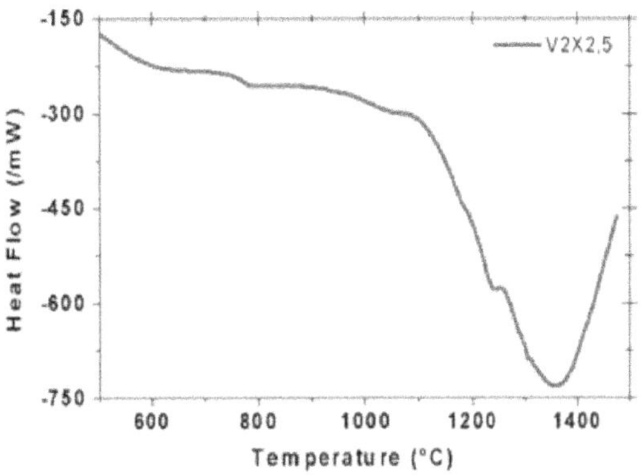

Structural Relaxation

The structure of glass is akin to that of a frozen liquid, representing a non-equilibrium state. The structure of the glass tends to evolve towards a crystalline configuration, which corresponds to the minimum energy state, thus achieving equilibrium. This physical phenomenon, where a macroscopic system disturbed from equilibrium returns to a thermodynamic equilibrium state, is referred to as relaxation. This phenomenon is more pronounced the further the glassy state is from equilibrium; in glasses, relaxation is associated with the temporal evolution of molecular rearrangements.

$T > T_g$: The system evolves rapidly towards equilibrium, and the characteristic times for this evolution (relaxation times) are generally short relative to the experimental time. Only an equilibrium system is detected.

$T < T_g$: The system is in a non-equilibrium state and evolves slowly, with the relaxation time becoming significantly longer compared to the experimental time. The glass is then considered as a liquid whose configurational arrangement can no longer achieve the minimum free energy.

Structural Relaxation in Oxide Glasses

Glasses composed of three, four, or even five oxides, such as those in our chapter, are relatively underexplored compared to binary glasses like SiO_2-K_2O (29 and 33 mol% K_2O). (Malfait et al., 2008) identified, through Raman spectroscopy, a single short-range dynamic at two temperatures close to T_g for the SiO_2-K_2O system. In contrast, (Naji et al., 2013) demonstrated the presence of two relaxation times in the SiO_2-Na_2O system. This aspect will be discussed in detail in the section dedicated to the results of Brillouin light scattering spectroscopy.

Scale Notion

The structure of glasses can be studied at different levels (Massiot et al., 2013). Diffraction (Saoût et al., 2008) and spectroscopy methods (Raman, NMR, etc.) have revealed short-range order within the first coordination sphere, which, in most cases, is similar to that of the corresponding crystalline phases. However, glasses lack long-range order and do not exhibit the periodic atomic arrangement characteristic of crystals. Therefore, it is necessary to probe different scales to understand their structure.

Short-Range Scale (Local Scale)

This scale reveals the fundamental structure. It often involves a polyhedron (e.g., SiO_4 tetrahedra in silicate glasses). The parameters describing this polyhedron include interatomic bonds, bond lengths, and intrapolyhedral angles. Characteristic distances range from a few angstroms to nanometers. Structural information at this scale is generally provided by Raman (Saoût et al., 2008), Infrared (De Sousa Meneses et al., 2006; De Sousa Meneses et al., 2006a), and NMR (Massiot et al., 2013) spectroscopies.

Intermediate-Range Scale

This scale focuses on the connectivity of structural units. Descriptive parameters include the angles between polyhedra and torsion angles. Unlike the short-range scale, where local order can still be discussed, the intermediate range already shows disorder through angular distributions. Characteristic distances are on the order of nanometers. Structural information at this scale can still be obtained through Raman, Infrared, X-ray, and neutron diffraction. Angular distributions between interatomic bonds are typically determined by NMR (Kiyono et al., 2012).

Long-Range Scale

Characterized by density fluctuations corresponding to inhomogeneities over distances of tens to hundreds of nanometers. These fluctuations impact material properties, which can be probed using Brillouin light scattering (Champagnon et al., 2000; Parc et al., 2001; Reibstein et al., 2011).

1.2 Structure of Aluminoborosilicate Glasses

Among oxide glasses, silica glass (SiO_2), being a network-forming oxide, can form a glass on its own by definition. This network is composed of SiO_4 tetrahedra linked together at their vertices, forming rings typically containing $3 < n < 8$ tetrahedral units. In the vitreous state, silicon adopts a tetrahedral coordination with the Si-O distance generally ranging between 1.60 and 1.63 Å.

The introduction of Na_2O, a network-modifying oxide, disrupts Si-O-Si linkages by introducing additional oxygens into the glass network. The extra oxygen will then complete the bonding of one of the silicon atoms, forming a $Si\text{-}O^-$ pair. Each negative charge is balanced by a nearby Na^+, with sodium serving as a charge compensator. Oxygens bonded to a single atom are referred to as Non-Bridging Oxygens (NBOs), while those involved in bridging are called Bridging Oxygens (BOs). This bond-breaking mechanism induces depolymerization of the silica and lowers the liquidus temperature of the oxide mixture. When the alkali oxide content is low (20% Na_2O), phase separation occurs within the glass, resulting in regions that are richer in silica and others that are richer in sodium oxide.

The addition of aluminum can alter the structure of glass in a more complex manner, as it can act either as a network former or a network modifier. Its unique characteristics enable it to substantially modify the structure and properties of the network depending on the glass composition. Thus, it does not strictly adhere to the definition of a network former. Several studies (Neuville et al., 2004; Neuville et al., 2004a) have been conducted on glasses from the ternary systems SiO_2-M_2O-Al_2O_3 and SiO_2-RO-Al_2O_3 (where M represents alkali metals or R represents alkaline earth metals). For certain compositional ranges, the structure remains poorly understood and the results are subject to various interpretations. Generally, in these glasses:

When $[M_2O] / [Al_2O_3] > 1$: The glasses exhibit an excess of modifiers relative to aluminum and are termed peralkaline. Al^{3+} ions tend to adopt an sp^3 hybridization, resulting in a coordination number of 4 (in terms of oxygen atoms) and are present as $[AlO_4]^-$ entities, with charge compensation provided by M^+ cations. In this case, Al^{3+} functions as a network former. The excess alkali relative to aluminum generates Non-Bridging Oxygens (NBOs) in the glass by creating $Si\text{-}O^-$ linkages. As the $[M_2O] / [Al_2O_3]$ ratio decreases, the number of Non-Bridging Oxygens in the vitreous network decreases, leading to increased polymerization of the glass.

When $[M_2O] / [Al_2O_3] = 1$: All modifier ions are involved in charge compensation for the $[AlO_4]^-$ tetrahedra, eliminating the presence of Non-Bridging Oxygens in the vitreous network. This represents a charge compensation line. Glasses in this composition range are fully polymerized (Lee et al., 2003; Neuville et al., 2004).

When $[M_2O] / [Al_2O_3] < 1$: The glasses have a deficiency of modifiers relative to aluminum and are termed peraluminous. Al^{3+} ions preferentially adopt a coordination number of 4 to form $[AlO_4]^-$ units, while the remaining aluminum incorporates into the glass network as $[AlO_5]^{2-}$ and $[AlO_6]^{3-}$ units. The $[AlO_6]^{3-}$ units each consist of three Bridging Oxygens and three Non-Bridging Oxygens. In this case, aluminum in a coordination number of 6 acts as a network modifier, creating Non-Bridging Oxygens.

1.2.1 Aluminoborosilicate Glasses Without Rare Earths

The composition of SiO_2 - B_2O_3 - M_2O - Al_2O_3 glasses is more complex than that of the SiO_2 - B_2O_3 - M_2O system due to the presence of alumina in the glass. Indeed, depending on whether the glass is peralkaline ($[M_2O] > [Al_2O_3]$) or peraluminous ($[M_2O] < [Al_2O_3]$), aluminum can be found as a network former, a modifier, or both. Additionally, the addition of Al_2O_3 to borosilicate glasses leads to significant modifications in the silicon and boron environments. Various studies using ^{11}B NMR (Wang et al., 1999; Yiannopoulos et al., 2001) have shown that the addition of Al_2O_3 decreases the number of BO_4 tetrahedra (commonly referred to as N_4 entities) in favor of BO_3 units. Modifier ions are thus more attracted to the aluminum environment rather than the boron environment. Consequently, Al^{3+} ions preferentially form AlO_4 tetrahedra, with their negative charge balanced by M^+ modifier ions. In peraluminous glasses where there is a charge compensator deficit relative to alumina ($[M_2O] < [Al_2O_3]$), the proportion of BO_4 (N_4) entities decreases. This preference of modifiers for the aluminum environment over the boron environment is often explained by the fact that boron atoms can adopt stable coordination numbers of 3 or 4 in glasses, unlike aluminum atoms which require charge compensation. Furthermore, the addition of Al_2O_3 decreases the number of non-bridging oxygens at the center of SiO_4 tetrahedra, leading to re-polymerization of the glass network. For "mixed" glasses ($[M_2O] = [Al_2O_3]$) or peraluminous glasses ($[M_2O] < [Al_2O_3]$), studies using ^{29}Si NMR have shown very few non-bridging oxygens in the glass. The chemical shift of the NMR signal, around -91 ppm, indicates a high proportion of Si-O-Al bonds in the glass network. Additionally, the nearly constant shape of the ^{27}Al NMR signals confirms that Al^{3+} ions are predominantly in 4-fold coordination in these glasses, regardless of their composition. However, the presence of a shoulder on all signals at lower chemical shifts indicates that there are only two types of tetrahedral sites for Al^{3+} ions. While the presence of Al^{3+} ions in 6-fold coordination cannot be excluded, the shoulder is more associated with Al(IV) ions bonded to boron atoms, while the main signal is associated with Al(IV) ions bonded to silicon atoms.

1.2.2 Aluminoborosilicate Glasses With Rare Earths

Until now, the structure of rare-earth-rich aluminoborosilicate glasses has been relatively understudied. A description has been provided (Barrow et al., 2020;Li et al.,2001), as well as by the works of (Quintas et al., 2009; Gasnier et al., 2014) and ((Mabrouk et al., 2021; Mabrouk et al., 2024). for new vitreous matrices used in waste immobilization. These studies focused on the effects of peralkaline and peraluminous chemical compositions, using the composition variable:

$$R' = \frac{[Na_2O] - [Al_2O_3]}{[B_2O_3]}$$

Peralkaline glasses, where $R' = \frac{[Na_2O] - [Al_2O_3]}{[B_2O_3]} < 0{,}5$, starting from a zero rare earth concentration, exhibit nanometric-scale heterogeneity. Similar to soda-borosilicate glasses, SiO_2 - B_2O_3 - M_2O - Al_2O_3 glasses are composed of two distinct zones: alkaline borate zones and silicate zones, where aluminum atoms substitute silicon atoms. Introducing a rare-earth element into this system increases the size of these heterogeneities. Beyond the solubility limit of the rare-earth element, macroscopic phase separation occurs, observed by TEM (Li et al., 2000) as dark nodules enriched in boron and rare-earth elements on a lighter continuous background enriched in silicon. In these glasses, rare-earth elements preferentially occupy borate zones, where competition arises between Na^+ and TR^{3+} ions, with the latter having a stronger field strength. The rare-earth element then adopts a structure resembling rare-earth metaborate $1BO_4$: 1TR: $2BO_3$, where the rare-earth element is surrounded by one BO_4 tetrahedron and two BO_3 triangles. When this double-chain structure becomes too bulky, it separates from the mixture, resulting in macroscopic phase separation, leading to a semi-opaque whitish appearance in the glass. Na^+ ions, excluded from borate zones, localize within silicate zones where they act as non-bridging oxygens in SiO_4 tetrahedra. Peralkaline glasses with $R' = \frac{[Na_2O] - [Al_2O_3]}{[B_2O_3]} > 0{,}5$: remain homogeneous in the absence of rare-earth elements. The addition of a rare-earth element causes phase separation between borate and silicate zones. A metaborate structure forms with all free boron atoms in the glass, and the excess rare-earth element relative to boron ($[TR_2O_3] - 1/3\ [B_2O_3]$) incorporates into silicate zones. With increasing $[TR_2O_3]$, TR – O – TR clusters form (beyond 30% by mass) in silicate zones, followed by the crystallization of $NaTR_9(SiO_4)_6O_2$ apatite beyond the rare-earth solubility limit in the system Peraluminous glasses without rare-earth elements exhibit a charge compensator deficit relative to alumina, leading to demixing of borate and silicate networks. In "mixed" glasses ($[Na_2O] = [Al_2O_3]$), the addition of rare-earth elements (Li et al., 2000b) increases the size of heterogeneous domains, resulting in macroscopic phase separation between a discontinuous phase rich in B and TR and a continuous silicate phase beyond the rare-earth solubility limit. For peraluminous glasses, the addition of rare-earth elements (Li et al., 2000b) induces the formation of a rare-earth metaborate structure. In these glasses, where not all AlO_4^- ions are compensated by the limited Na^+ ions, boron atoms preferentially adopt BO_3 structures (which do not require electrical compensation unlike BO_4^-), leaving rare-earth elements free to compensate AlO_4^- tetrahedra.

1.3 Phase Separation in Glasses

Phase separation in glasses has often proven problematic for glassmakers. It refers to the uncontrolled formation of phase separation within the glass during melting, quenching, or any other secondary processing operation. This decomposition alters the homogeneity and physicochemical properties of the glass in its liquid state and complicates shaping operations. Consequently, it leads to a deterioration in glass quality. However, the development of heterogeneous materials containing phases of different natures and morphologies can be seen as an advantage in the field of glass ceramics. The wide variety of heterogeneous materials opens up a broad range of industrial applications due to the wide range of compositions and physical properties they offer (Lee et al., 2003; Qian et al., 2004; Wheaton et al., 2007).

2. EXPERIMENTAL METHODS

2.1 Glass Synthesis

The first step in studying a glass material begins with selecting its composition, as it dictates the chemical and physical properties of the material. In our case, we have focused on studying LaNaSiAlBO-type matrices, which are rich in rare earth elements. These glasses are known as model glasses for the storage of fission products and minor actinides, appreciated for their excellent chemical durability, good thermal stability, and mechanical strength (Mabrouk et al., 2019; Mabrouk et al., 2020; Mabrouk et al., 2021). Below are some notable properties of the oxides composing this glass: **Silica (SiO_2)**: Acts as a network former and is the fundamental component of this type of glass. Silica has the interesting property of reducing the thermal expansion coefficient and enhancing the mechanical durability of the glass, although it raises the glass melting temperature. **Alumina (Al_2O_3)**: Used to improve mechanical strength. In this glass system, a significant portion of aluminum substitutes for silicon and acts as a network former. Its importance lies in its tendency to facilitate the incorporation of rare earth elements into the glass matrix. **Sodium oxide (Na_2O)**: Structurally, the addition of Na^+ ions break the bridging Si-O-Si bonds and creates non-bridging oxygen ions. This depolymerizes the network. In the presence of aluminum, sodium preferably acts as a charge compensator for $[AlO_4]^-$ tetrahedra. Studies show that the presence of Na reduces the glass transition temperature, thereby saving energy during the glass manufacturing process. **Boron oxide (B_2O_3)**: Another network-forming element with anomalous properties. Unlike other glass formers, such as Na_2O in silicate glasses, which exhibit monotonic variations, borate glasses often show specific concentration-dependent anomalies, known as the "boron anomaly. **Lanthanum oxide (La_2O_3)**: Used to model the behavior of actinides (nuclear fuel wastes). The Lanthanide family, in particular, is frequently used for this purpose. These components are carefully chosen to impart specific properties to the glass matrix, making it suitable for its intended application in the storage and containment of nuclear waste products.

2.2 Composition Analysis

Glasses with compositions $55\ SiO_2 - (25\text{-}x)\ Al_2O_3 - xB_2O_3 - 15Na_2O - 5La_2O_3$ ($0 \leq x \leq 12.5$) were prepared using a simple and rapid solid-state method. They were produced by mixing oxides; sodium carbonate (Na_2CO_3), silica (SiO_2), alumina (Al_2O_3), lanthanum oxide (La_2O_3), and boric acid (H_3BO_3). Table 1 shows the selected compositions of the glasses in molar percentage.

Table 1. Composition of the glasses in our study in mol%

Composition (% en mol)					
Glass	**SiO_2**	**Al_2O_3**	**B_2O_3**	**Na_2O**	**La_2O_3**
V1X0	55	25	0	15	5
V2X2,5	55	22,5	2,5	15	5
V3X5	55	20	5	15	5
V4X7,5	55	17,5	7,5	15	5
V5X10	55	15	10	15	5
V6X12,5	55	12,5	12,5	15	5

The starting materials are finely and thoroughly ground using a mortar to achieve a homogeneous mixture. The synthesis of the glasses is conducted in air using a platinum crucible according to the following reaction:

$$55SiO_2 + (25\text{-}x)\ Al_2O_3 + 2xH_3BO_3 + 15Na_2CO_3 + 5Al_2O_3\ \ 55SiO_2\text{-}(25\text{-}x)\ Al_2O_3\text{-}xB_2O_3\text{-}15Na_2O\text{-}5La_2O_3 + 15CO_2 + 3xH_2O$$

The reaction takes place in a platinum crucible placed in a furnace. Prior to this, dehydration is carried ut at 250°C for 1 hour to remove water from boric acid, followed by decarbonation at 900°C for 2 hours to decompose sodium carbonate. To achieve melting, each glass mixture is heated to a temperature ranging between 1300°C and 1450°C. The temperature required decreases with increasing B_2O_3 content in the composition.

Figure 2. Glasses obtained after quenching (a): V1X0 (b): V2X2.5 (c): V3X5 (d): V4X7.5 (e): V5X10 (f): V6X12.5

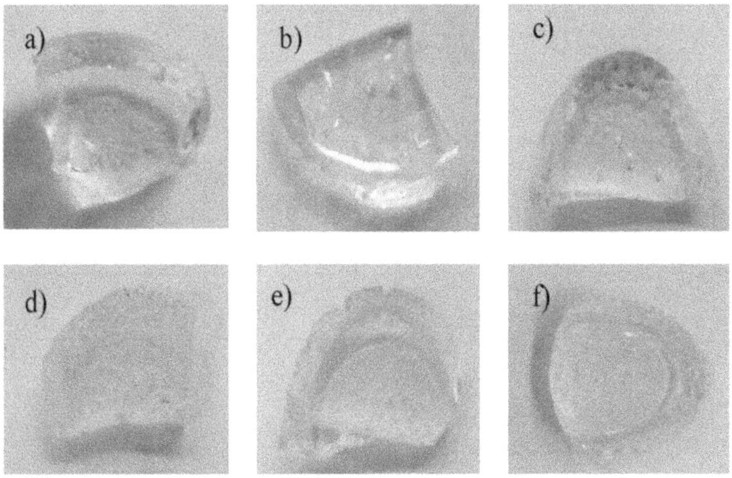

On the photographs in Figure 2. It can be observed that no visible heterogeneity is detectable in our glasses. It can be inferred that if any heterogeneity exists, its size is less than ≈100 μm.

Chemical Composition Analysis

For the chemical analysis using Scanning Electron Microscopy coupled with Energy Dispersive Spectroscopy (SEM-EDS) to be quantitative, the sample must be flat, polished, and coated with a metal layer. To ensure accuracy, each analysis is performed at ten different points on each sample and then averaged. The uncertainty in the chemical composition is thus estimated to be 0.1 mol%. Boron is known to be a volatile element during high-temperature glass melting, making its quantification in the finished glass the most challenging aspect of the fabrication process. As shown in Table 2, chemical analyses indicate that the compositions of the produced glasses are very close to the desired compositions presented in Table 1. This confirms retrospectively that the glass processing temperature was appropriately chosen according to our experimental protocol.

Table 2. The chemical compositions in molar percentages calculated from the atomic percentages obtained for each element by EDS/SEM for the V1X0 glass and by ICP/AES for the other glasses in the series

Composition (% en mol)						
Glass		**SiO_2**	**Al_2O_3**	**B_2O_3**	**Na_2O**	**La_2O_3**
V1X0	**Nominal** **MEB-EDS**	55 55	25 25	0 0	15 15	5 5
V2X2.5	**Nominal** **ICP-AES**	55 56.8	22.5 22.0	2.5 2.0	15 14.3	5 4.9
V3X5	**Nominal** **ICP-AES**	55 56	20 19.8	5 4.6	15 14.6	5 5
V4X7.5	**Nominal** **ICP-AES**	55 56.5	17,5 17.2	7.5 6.5	15 14.8	5 5
V5X10	**Nominal** **ICP-AES**	55 56.57	15 15	10 9.84	15 14.5	5 4.09
V6X12.5	**Nominal** **ICP-AES**	55 56.67	12,5 12.5	12.5 11.09	15 14.99	5 4.75

2.3 X-Ray Diffraction Measurements

By X-ray diffraction, we initially sought to confirm that the diffractograms of our samples exhibit only a continuous background, thereby ensuring the absence of crystalline phases on both nano and macroscopic scales. Below are the diffractograms of our glass series as shown in Figure 3.

Figure 3. Diffractogram of the glass series showing the presence of some peaks of SiO_2-Quartz and Cristobalite

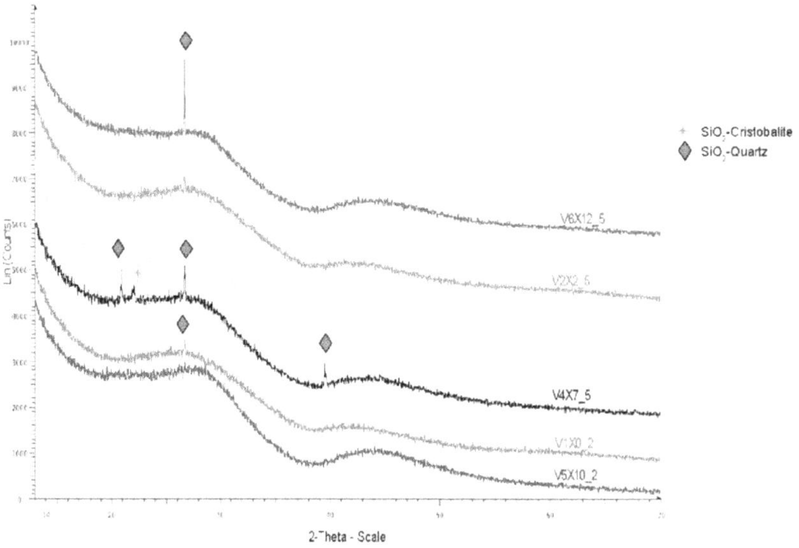

Figure 3. Shows the X-ray diffraction pattern of the glass series, clearly indicating the amorphous state of our samples with the presence of some crystallization peaks. These signatures were then compared to the diffractograms of crystalline phases from reference materials in the Powder Diffraction File (PDF) database of JCPDS-ICDD (Joint Committee on Powder Diffraction Standards and the International Center for Diffraction Data) and identified. These peaks correspond to SiO_2-quartz and cristobalite phases, which could be attributed to slight reactivity issues during melting or contamination of the samples during grinding in a glass mortar.

2.4 Transmission Electron Microscopy

The samples from the series were analyzed using Transmission Electron Microscopy (TEM) to verify their nanoscale homogeneity. The corresponding images are shown in Figure 4. The TEM images reveal phase separation in three glasses from our series, characterized by a distinct dark phase and a lighter phase. This phase separation becomes more pronounced with increasing boron content in the compositions.

Figure 4. TEM images corresponding to samples from the entire series

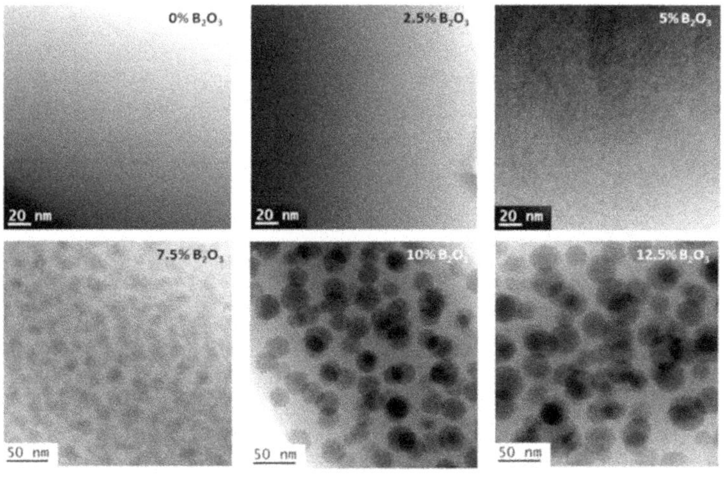

The glasses with 0%, 2.5%, and 5% B_2O_3, which are homogeneous and transparent, do not exhibit phase separation at the nanoscale, as they appear homogeneous under TEM. However, the glasses with 7.5%, 10%, and 12.5% B_2O_3, despite appearing homogeneous to the naked eye, show significant glass-glass phase separation. They display dark spherical inclusions of approximately 30 nm size embedded in a clear continuous background (Figure 4).

2.5 Differential Scanning Calorimetry Analysis

The prepared glasses were characterized using Differential Scanning Calorimetry (DSC). DSC measurements were conducted for all samples using approximately 0.3 g of powder deposited in a platinum crucible. Each sample was referenced against a second platinum crucible containing an equal mass of alumina, both under an argon atmosphere. The DSC technique measures the amount of heat required to raise the temperature of the reference crucible to match that of the sample crucible.

Figure 5. Thermograms of the entire series of glasses

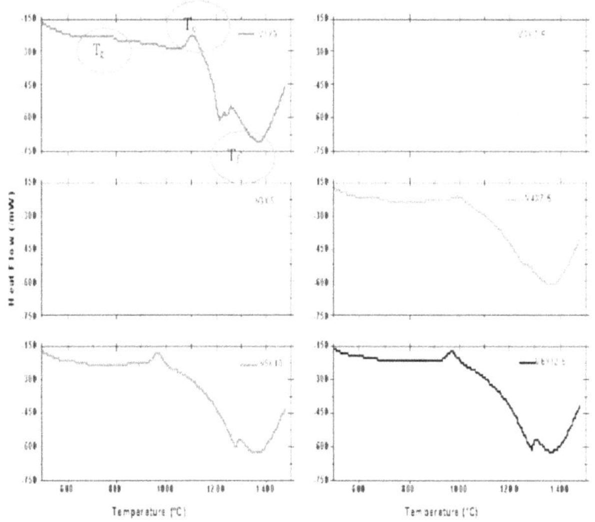

Table 3. Glass transition temperatures, crystallization temperatures, and melting temperatures (T_g, T_c, T_f) of the prepared glasses

	X= (%B_2O_3)	T_g(°C)	T_c(°C)	$T(_{1f}$ (°C))	T_f(°C)	Hr= $(T_c-T_g)/(T_f-T_c)^2$
V1X0	0	807	1057	1181	1365	0.00263
V2X2.5	2.5	772	-	1204	1359	-
V3X5	5	723	-	1194	1354	-
V4X7.5	7.5	707	943	1195	1365	0.00132
V5X10	10	677	929	1227	1358	0.00136
V6X12.5	12.5	671	930	1231	1362	0.00138

Figure 6. Variation of T_g as a function of B_2O_3 content

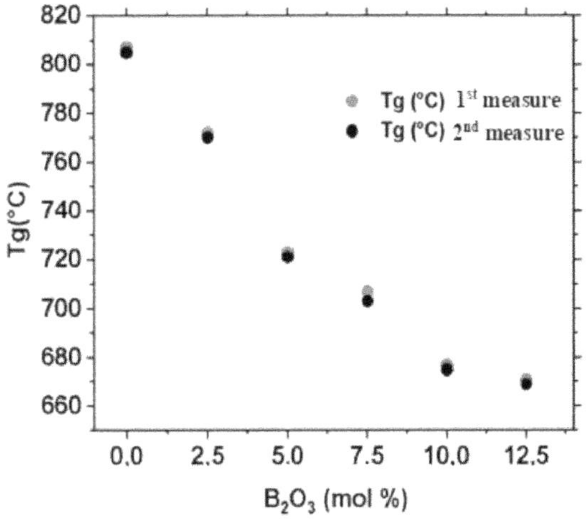

Increasing the boron content within the glass network causes a decrease in the glass transition temperature (T_g) across the entire glass series (Figure 6). This type of trend has been previously observed during the substitution of SiO_2 by B_2O_3 in soda-lime borosilicate glasses (Shamzhy et al., 2015; Liu et al., 2019). This decrease is attributed to a reduction in the interatomic bonding strengths within the network. Interplanar Si-O-B bonds with BO_3 entities are weaker than Si-O-Al bonds, and substituting aluminum with boron lowers the glass transition temperature. It is noted that the glasses exhibit crystallization peaks, with their temperatures clearly above T_g and below the melting temperature (T_f). These crystallization peaks shift to lower temperatures as the proportion of boron increases within the glass network. The production of chemically and physically stable glasses is crucial for technological applications. For instance, if glasses are not stable, it becomes challenging to use them for the storage of radioactive waste as their devitrification poses long-term hazards. Several criteria for thermal stability, based on temperatures measured by DSC, are proposed. One such criterion is the Hrüby parameter, defined as follows: $H_r = (T_c-T_g)/(T_f-T_c)^2$, which assesses glass stability based on measurements of glass transition, crystallization, and melting temperatures. A higher Hrüby parameter indicates greater glass stability. As the boron content increases, the Hrüby parameter increases in compositions containing boron (Table 3). It is observed that the thermograms of the entire series show an initial melting event, which could be attributed to microscopic heterogeneity, possibly involving clusters that melt initially. Transmission electron microscopy (TEM) studies have been conducted to confirm the presence of these heterogeneous zones within this series.

3. CHARACTERIZATION OF SHORT AND MEDIUM-RANGE ORDER BY NMR AND INFRARED SPECTROSCOPY

3.1. Structural study of Aluminoborosilicate Glasses by NMR Spectroscopy

Solid-state NMR spectroscopy has been extensively used in the study of silicate, aluminosilicate, and borosilicate glass structures (Morin et al., 2014; Morin et al., 2016) because it enables probing the local environment of silicon, aluminum, boron, and sodium atoms (Quintas et al., 2007; Quintas et al., 2009; Mabrouk et al., 2020). Recent instrumental advancements, including high-field spectrometers (B_0=20T) and fast magic-angle spinning (30 kHz - 70 kHz), now allow for spectra with very high resolution. This expanded chemical shift range provides distinct signals for BO_3 and BO_4^- entities, facilitating their identification. We investigated the environments of all atoms in our systems using ^{11}B, ^{29}Si, ^{27}Al, and ^{23}Na NMR to gain a comprehensive understanding of the complex structure and the role of each element.

^{11}B NMR

Through MAS (Magic Angle Spinning) ^{11}B NMR spectroscopy, we determined that boron exists predominantly as BO_3 units, with chemical shifts typically ranging between 5 and 20 ppm for BO_3 units and around 0 ppm for BO_4 units (Lee et al., 2003; Morin et al., 2014; Mabrouk et al., 2020). The spectra (Figure 7) indicate a network composition primarily composed of BO_3 units, with a smaller proportion of BO_4 units. A deconvoluted spectrum of glass V2X2.5, processed using Dmfit software (Massiot et al., 2001), is shown in Figure 8. Revealing shifts towards less negative chemical shifts for BO_3 entities.

Figure 7. ^{11}B-MAS NMR spectra at 750 MHz of our glasses

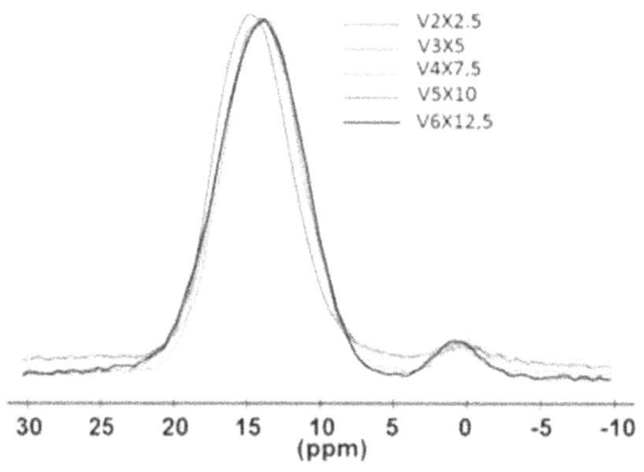

Figure 8. Experimental [11]B -MAS spectrum at 750 MHz (in blue) and simulated spectrum (in red) of glass V2X2.5

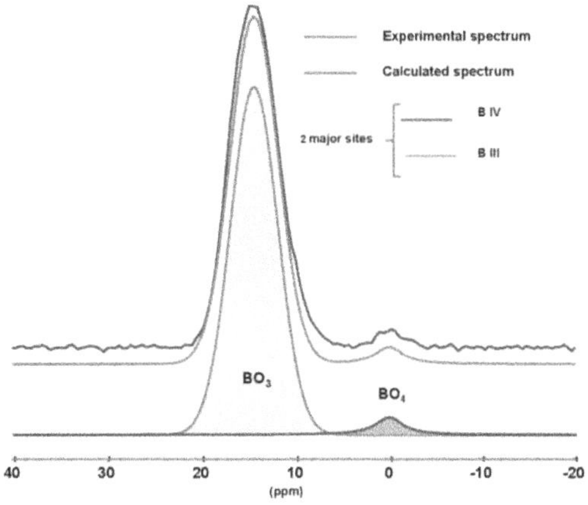

The ^{11}B NMR spectra of the glasses, shown in Figure 7. Were deconvoluted using contributions from BO_3 and BO_4 units. By integrating the signals associated with these different units, Table 4 shows the chemical shifts and relative contents of these units.

Table 4. NMR parameters corresponding to BO_3 and BO_4^- entities obtained from the simulation

%B_2O_3	**%BO_3**	**%BO_4**	**$\Delta\delta_{iso}$**	**BO_3 ppm**	**BO_4 ppm**	**$\Delta\delta_{iso}$**
2.5	96	4	±0.5	14,6	0,6	±0.5
5	96	4		14,1	0,2	
7.5	96	4		13,8	0	
10	96	4		14,1	0,6	
12.5	96	4		14	0,6	

The chemical shift of both BO_3 and BO_4^- species remains nearly constant, indicating that the nature of the entities does not seem to change. Boron is predominantly tricoordinated. We observe a shift in the chemical shift of the BO_3 entity in glass V2X2.5 compared to other glasses in the series. According to the literature (Kroeker et al., 2001; Donald et al., 2011b; Mabrouk et al.,2024), the low chemical shift value of the BO_3 units visible in the spectrum of glass V2X2.5 may be due to the contribution of BO_3 units containing non-bridging oxygens in their local environment (metaborates BO_3^-) (Kaneko et al., 2017). With increasing boron content, the evolution of the chemical shift value of both BO_3 and BO_4 species becomes almost constant. Indeed, in peraluminous glasses SiO_2-B_2O_3-Al_2O_3-Na_2O-CaO-Nd_2O_3 (Gasnier et al., 2014) with only calcium and sodium present (0Nd_2O_3), the BO_4 content is very low (5% of total boron), explaining that sodium and calcium are preferentially present in the aluminum environment, as described in many aluminoborosilicate systems (Angeli et al., 2006; Quintas et al., 2009). With increasing rare earth content, the proportion of BO_4^- increases, indicating that at least some of the rare earth ions can compensate for the charge of BO_4^- units. This is confirmed by the fact that when the rare earth content is constant, the proportion of BO_4^- entities also remains constant (Kiyono et al., 2012). This observation demonstrates that BO_4^- entities are not very sensitive to the nature of the charge-compensating cation. When all modifier ions are substituted with rare earth, the BO_4^- content increases, highlighting the charge compensating role of rare earth towards BO_4^- entities. This is consistent with our results showing a low and constant boron content since the amount of lanthanum does not vary throughout the series.

^{27}Al NMR

Figure 9. ^{27}Al-MAS NMR spectra of our glasses at 750 MHz

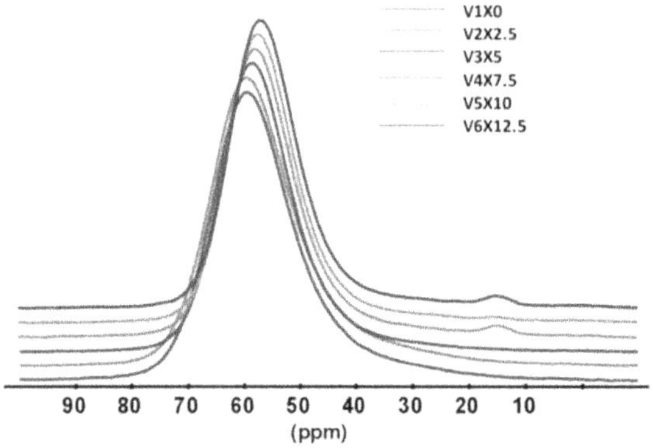

Figure 10. Experimental ^{27}Al MAS NMR spectrum (in blue) and simulated spectrum (in red) of glass V1X0

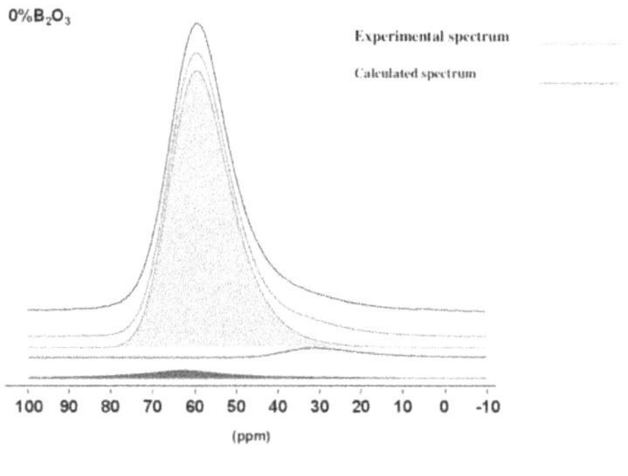

^{27}Al NMR will allow us to determine the different coordinations of Al^{3+} ions within our series of glasses, with Al(IV) units expected around +60 ppm, Al(V) between +25 and +40 ppm, and Al(VI) between -15 and +20 ppm (Walkley et al., 2018, Mabrouk et al., 2020). The ^{27}Al NMR spectra of glasses shown in Figure 9. All exhibit a signal centered around 60 ppm, which varies significantly with B_2O_3 content. Thus, regardless of the B_2O_3 concentration in the glass, aluminum in coordination 4 is predominantly present as $[AlO_4]^-$ entities, with a proportion of aluminum in coordination 5. $[AlO_4]^-$ entities require charge compensation, and there is an evolution in the spectra when adding B_2O_3, where charge compensation is provided by different modifier ions (Na^+ and La_3^+). We observe a peak around 15 ppm in the spectra of glasses with 7.5% B_2O_3 and 12.5% B_2O_3, identified as crystalline alumina. An asymmetric broadening on the low chemical shift side is also observed, which is associated with the appearance of a small proportion of Al^{3+} ions in coordination 5 within the glasses.

^{23}Na NMR

Figure 11. ^{23}Na-MAS NMR spectra of our glasses at 750 MHz

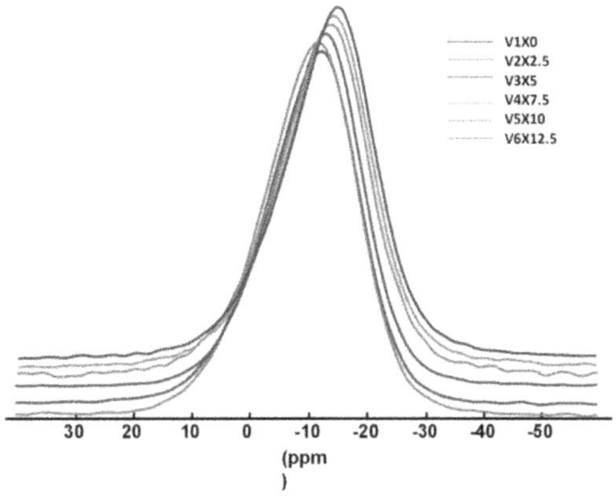

Let's now focus on the evolution of the distribution of Na^+ ions within the network as a function of boron content. There is a significant shift of the ^{23}Na MAS NMR peak position towards lower chemical shifts as the boron content increases. ^{23}Na NMR is sensitive to the local environment of Na^+ cations, particularly the average Na-O distance. We observe a linear shift with the addition of B_2O_3, reaching a value close to -20 ppm for glass V6X12.5 (Figure 11). This shift indicates an increase in the Na-O distance with the addition of B_2O_3 (Neuville et al., 2004). An increase in the coordination of sodium atoms can also be interpreted as an increase in the Na-O distance. Sodium coordination is approximately 6 in sodosilicates and increases to 8 in aluminosodosilicates (George et al., 1996; Lee et al., 2003b; Losq et al., 2014). Furthermore, the Na^+-NBO distance is smaller than the Na^+-BO distance (Lee et al., 2003b), so the observed shift could correspond to Na^+ ions transitioning from a role as network modifiers to charge compensators in fully polymerized glasses.

^{29}Si NMR

Figure 12. ^{29}Si MAS-NMR Spectra at 300 MHz of our glasses

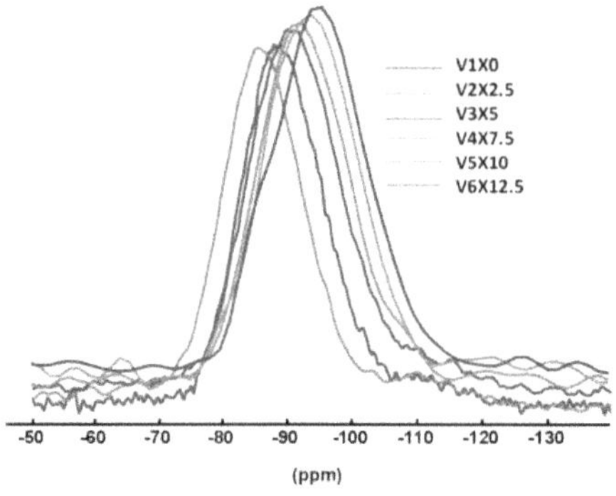

Figure 13. ^{29}Si MAS-NMR spectrum of glass V1X0 deconvoluted into various SiO_4 Q_n contributions

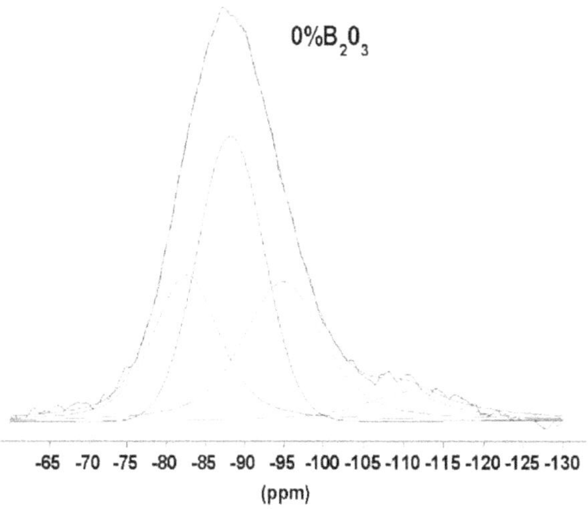

The ^{29}Si NMR spectra of silicate glasses, particularly the chemical shift δ of the signals, provide information about the degree of polymerization of the glass network, i.e., the nature of the SiO_4 Q_n units present in the glasses. Additionally, it reflects the number of neighboring Al atoms around Si, as the presence of aluminum or boron atoms in the silicon environment significantly alters the chemical shift values (substituting a SiO_4 tetrahedron with an AlO_4 tetrahedron increases δ by about 5 ppm) (Deschamps et al., 2008). The presence of non-bridging oxygen atoms within the silicon coordination polyhedron also modifies these values (Deschamps et al., 2008). The diversity of possible second neighbors (Al, B, Si) in the glasses studied in this work makes the extraction of proportions of different Q_n units extremely complex. However, in this series of glasses, where the variable parameter is the aluminum/boron substitution (with concentrations of all other oxides held constant), changes in the appearance of the ^{29}Si NMR signal can be attributed to the environment of the second coordination sphere of the silicon tetrahedra within the glass network. Substituting alumina with boron oxide in our compositions results in a shift of the signal centroid towards lower chemical shifts, as well as an increase in signal asymmetry with the appearance of a shoulder towards lower chemical shifts. This decrease in chemical shift with the addition of B_2O_3 has also been observed in glasses of $CaO-Al_2O_3-B_2O_3-SiO_2$ composition (Hung et al., 2009). The ^{29}Si NMR spectra could reflect changes in the environment around the silicon atom. Nevertheless, the chemical shift range where they appear would imply that silicon is bound to three or even four aluminum atoms, which seems consistent with the deconvolution of the V1X0 spectrum (Table 5). However, is this still the case when substituting aluminum with boron? The study of the glass microstructure revealed that peraluminous matrices exhibit phase separation as the amount of boron increases. NMR studies of the network structure have shown a network characterized by BO_3 and AlO_4 units, along with a small proportion of BO_4 and AlO_5 entities. We observed that the formation of AlO_4 entities is favored within the network with the addition of boron oxide, with these units primarily being compensated by Na^+ ions, followed by La^{3+} ions.

The Influence of Aluminum/Boron Substitution on the Network

The structural description of glasses is often complex, and this complexity increases with the addition of multiple oxides. While alkali silicates are relatively well understood, aluminoborosilicates, especially those with rare earth elements, remain less well characterized, with many questions still unresolved. One primary question is whether boron affects the environment of silica, as observed in the evolution of our spectra. To address this, a realistic analysis of the results will be discussed. Initially, a decomposition, attribution, and quantification of Si species in the spectra of aluminoborosilicate glasses was performed. The deconvolution of the signal allows for the evaluation of the relative proportions of Q_n entities in the different glasses (Table 5). According to the deconvolution of the spectra and the integration of the areas of the different contributions, four types of SiO_4 tetrahedra, Q_n(mAl), have been detected within the glasses, in the proportions indicated in Table 5, assuming that boron does not exist in the silica environment.

Table 5. Deconvolution of the silica spectra without considering the presence of boron in its environment

%B_2O_3	Q_3(3Al)	δ(ppm)	Q_4(4Al)	δ(ppm)	Q_4(3Al)	δ(ppm)	Q_4(0Al)	δ(ppm)
0	24.51	-82.59	40.92	-88.23	27.98	-94.92	6.59	-110.14
2.5	20.79	-80.9	46.63	-86.29	24.83	-92.99	7.75	-109.95
5	23.6	-86.06	45.72	-91.76	24.62	-97.77	6.02	-111.52
7.5	21.83	-84.92	48.31	-90.74	22.44	-97.55	7.41	-106.34
10	21.95	-86.29	51.81	-92.99	20.5	-100.03	5.74	-110.11
12.5	22.88	-87.39	50.15	-95.23	21.34	-102.63	5.62	-110.58

The deconvolution results indicate that the silicon spectrum can be decomposed into four broad peaks. Typically, the first peak is considered characteristic of silicon in a Q_4 environment with 4 bridging Si atoms, followed by Q_4(3Al), Q_4(4Al), and Q_3(3Al). The substitution of aluminum with boron leads to a shift of the peaks toward lower fields, with this shift increasing linearly with the concentration of B_2O_3 up to 12.5%. Figure 14a,b. Illustrates the changes in the population and positions of the four entities within the network. We observe an increase in the Q_4(4Al) species and a decrease in the Q_4(3Al) species. The populations of the Q_3(3Al) and Q_4(0Al) entities remain almost unchanged with aluminum-boron substitution. The population of Q_4(4Al) increases with substitution up to 12.5% Al_2O_3 (25% AlO_4^-).

Figure 14. a) Evolution of the population of Q_n(mAl) entities, b) evolution of the chemical shift of Q_n(mAl) entities

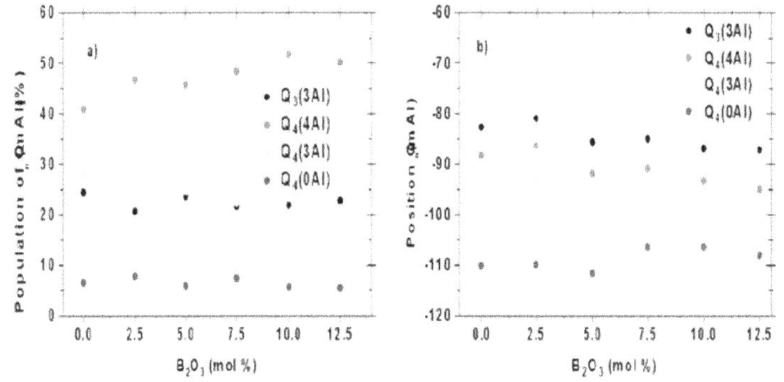

Figure 15. (a) Evolution of the population of Al(IV) and Al(V) species as a function of the boron concentration. (b) Evolution of the chemical shift of Al(IV) and Al(V) species.

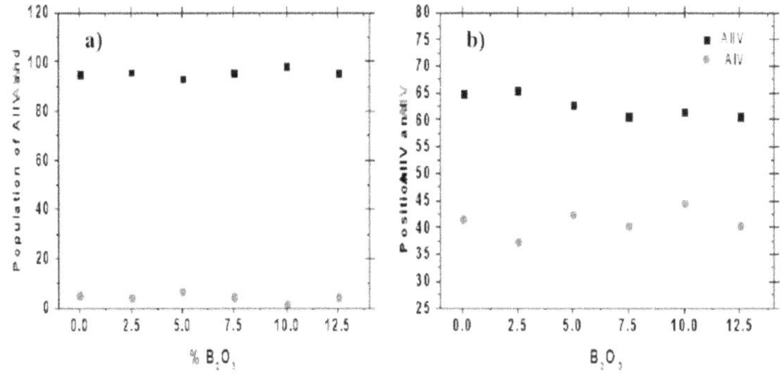

In all the glasses, Al^{3+} ions are predominantly present in a coordination of 4, as $[AlO_4]^-$ tetrahedra, with a significant proportion of aluminum atoms in a coordination of 5 (*Figure 16*) Charge modifiers such as Na^+ and La^{3+}, when present in sufficiently high concentrations, can adequately compensate for the charge of all these aluminate entities (Ollier et al., 2004)

Figure 16. Evolution of the populations of Al in coordination 3 or 4 and chemical shifts for the entire series of glasses as a function of boron concentration

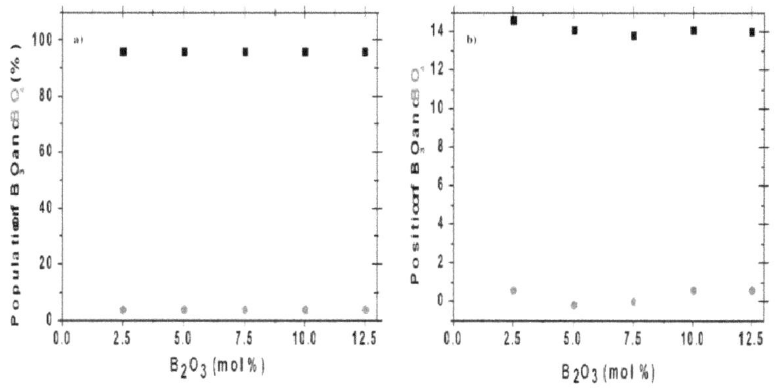

It is accepted that the structure of B_2O_3 mainly consists of boroxol rings composed of three BO_3 units (denoted R-BO_3, with "R" standing for "ring"), connected by BO_3 chains of varying lengths (denoted NR-BO_3, with "NR" standing for "non-ring")(Figure 16). (Hung et al. 2009). Therefore, it is challenging to determine the presence of one or multiple BO_3 contributions within the studied glass network. When no non-bridging oxygen atoms are expected on boron (at low modifier ion concentrations), BO_3 "ring" entities, organized into boroxol rings, are distinguishable from BO_3 "non-ring" entities integrated into the silicate network via at least one chemical bond with a SiO_4 tetrahedron (Donald et al., 2011b). At higher modifier ion concentrations, non-bridging oxygen atoms form on BO_3 entities, leading to the formation of symmetric BO_3 (with three non-bridging oxygen atoms) and asymmetric BO_3 (with one or two non-bridging oxygen atoms) (Soleilhavoup et al., 2010). The NMR provides valuable structural information about the glass network and its evolution with aluminum/boron substitution. The boron results indicate that it is predominantly present in BO_3 structural units, with the percentage of boron in BO_4 tetrahedra being very low (less than 4%). The ^{27}Al results showed that Al^{3+} ions are exclusively present in coordination 4, with a small percentage in AlO_5^{2-}. The results for the silicate network are quite surprising: as the boron content increases, a shift of the ^{29}Si NMR signals towards more negative values (Figure 12) is observed. After signal deconvolution, it becomes evident that the proportion of Q_4(4Al) units increases while Q_4(3Al) units decrease (Figure 14). Q_4(0Al) and Q_3(3Al) units remain constant. Further explanation and confirmation of these silicate network assignments and understanding the impact of boron on its environment are required. It is through the infrared study of these glasses, presented in the following section of this thesis, that we have provided the necessary explanations and confirmations.

3.2 Structural Study of Aluminoborosilicate Glasses by Infrared Spectroscopy

Infrared radiation is part of the electromagnetic spectrum, situated in a range of wavenumbers that is invisible to the human eye, between visible red light and microwaves ($5cm^{-1} < \underline{\omega} < 12500cm^{-1}$). The interaction between its electric component $\vec{E}$ and matter induces an electric displacement $\vec{D}$, the intensity of which depends on the dielectric function ε^\hat{\varepsilon}ε^ of the material and is defined by:

$$\vec{D}(\omega) = \varepsilon_0 \hat{\varepsilon}(\omega)\vec{E}(\omega) = \varepsilon_0 [1 + \hat{\chi}(\omega)]\vec{E}(\omega) = \varepsilon_0 \vec{E}(\omega) + \vec{P}(\omega)$$

In the above expression, ε_0 represents the permittivity of free space, $\hat{\chi}$ denotes the electric susceptibility, and $\vec{P}$ is the polarization of the material.

The dielectric function is a fundamental physical quantity on which infrared spectroscopy is based because it is directly accessible from experimental data and provides quantitative information about vibrational dynamics and the microstructure of the medium (De Sousa Meneses et al., 2006b). It also allows for the determination of other optical functions, such as the complex refractive index $\hat{n} = \sqrt{\hat{\varepsilon}}$ (Kischkat et al., 2012). Analyzing the evolution of the dielectric function as a function of temperature and time provides additional information, particularly about the short-range structural reorganization occurring during the annealing of a glass (a relaxation phenomenon) (Naji et al., 2013). In this chapter, we aim to measure the reflectivity of different glasses at room temperature and then fit these spectra using a dielectric function model to understand the effect of composition (boron content) on their spatial organization at a local scale. The fitting of the experimental spectra was performed using Fresnel's relation that links the reflectivity R at the air/glass interface to the dielectric function $\hat{\varepsilon}$ characterizing the glass:

$$R(\omega) = \left| \frac{\sqrt{\hat{\varepsilon}(\omega)} - 1}{\sqrt{\hat{\varepsilon}(\omega)} + 1} \right|^2$$

Reflection measurements were conducted on a series of six samples cut into parallel-faced slides and polished, aiming to characterize their infrared response and observe the effect of increasing substitution of aluminum by boron. Figure 17. Illustrates this evolution from the boron-free reference aluminosilicate glass to a composition with 12.5% boron oxide concentration, corresponding to the substitution of half of the aluminum.

Figure 17. Evolution of reflectivity spectra of different groups as a function of boron content

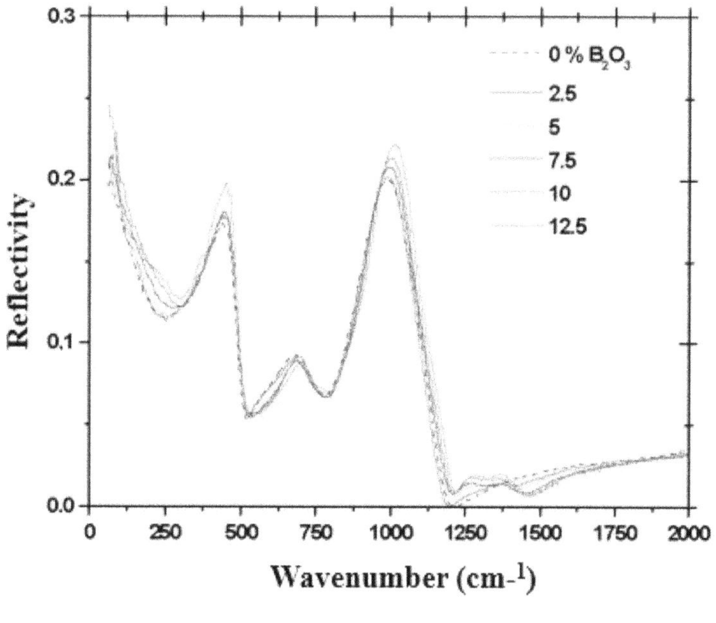

The spectrum of the boron-free glass composition exhibits multiple reflection bands distributed across wavenumbers below 1200 cm^{-1}. The response at very low frequencies (< 300 cm^{-1}) is dominated by the vibrational modes of cations within their oxygen cage, involving sodium or lanthanum in these glasses. In the range between 300 and 500 cm^{-1}, absorption bands are induced by bending motions within [AlO_4] or [SiO_4] tetrahedra. The most intense bands, located in the region between 800 and 1200 cm^{-1}, correspond to the stretching vibrations of Si-O bonds in [SiO_4] tetrahedra. Contributions between 500 and 800 cm^{-1} include several features generated by the aluminosilicate network.

As shown in Figure 17. The introduction of boron oxide and its increasing concentration progressively shifts some reflection bands towards higher frequencies, with new absorption bands appearing at higher frequencies between 1200 and 1400 cm^{-1}. These contributions, also observed in various borates (El-Egili et al., 2003), are induced by vibration modes involving boron. To aid in understanding the vibrational dynamics of glasses in this system and provide a more quantitative description of the impact of boron on their microstructure, it was necessary to reproduce their infrared response using a dielectric function model. The results from fitting the spectra, the imaginary part of the dielectric function, and its decomposition into Gaussian contributions are depicted in the reference (Mabrouk et al., 2019) For the boron-free glass (V1X0), ten components were necessary to accurately reproduce the high-frequency part of the spectrum. For boron-containing glasses, two additional components were required between 1200 and 1400 cm^{-1}. The precise origin of these components remains to be identified (Mabrouk et al., 2019). To aid in identifying the components in the studied glasses, We investigated boron's vibrational response using two types of glasses: a borate ($80B_2O_3$-$20Na_2O$) and a commercial borosilicate, Borofloat. Spectral analysis (Wereszczak et al., 2014) revealed a distinct band at around 1420 cm^{-1} in the borate glass, indicating BO_3 vibrations in boroxol rings, which is absent in Borofloat and the other glasses studied. This absence suggests that boroxol rings are either missing or present in very low concentrations in these samples. Additional bands related to BO_3^- groups and B-O bonds in BO_3^- groups were observed just below the BO_3 band. In the entire series (boron-free and boron-containing glasses), the region between 800 and 1200 cm^{-1} corresponds to antisymmetric stretching vibrations of SiO_4 tetrahedra. The presence of multiple components in this range is related to the presence of different species in the silica environment. Spectral simulations for silica and the K_2O-SiO_2 system (De Sousa Meneses et al.,2013) allowed us to assign five Gaussian components to simulate reflectance spectra. The region between 550 and 800 cm^{-1} corresponds to bands due to vibrational modes of aluminosilicate groups, which is challenging to interpret. The same methodology used for borate band assignment will be applied here. The spectrum decomposition of the binary glass without silica, with a composition of $50Al_2O_3$-50CaO (De Sousa Meneses et al.,2013) was chosen to compare the bands with those in our system. This fitting reveals the presence of an Al-O-Al band around 850 cm^{-1} and an Al-O band around 800 cm^{-1}. In gehlenite, these two bands are observed around 850 and 800 cm^{-1}, and an additional band around 750 cm^{-1} is attributed to the Si-O-Al linkage vibration.

Figure 18. Depicts the simulation results for the entire series of glasses, overlaying the various contributions to the imaginary part of the dielectric function. Particularly notable in the response of glass V1X0 is the absence, at higher frequencies, of the two bands associated with boron. After studying the infrared response of glasses with simplified compositions, we were able to attribute the majority of the bands.

Figure 18. Imaginary parts of the dielectric function of the glass series and their decompositions

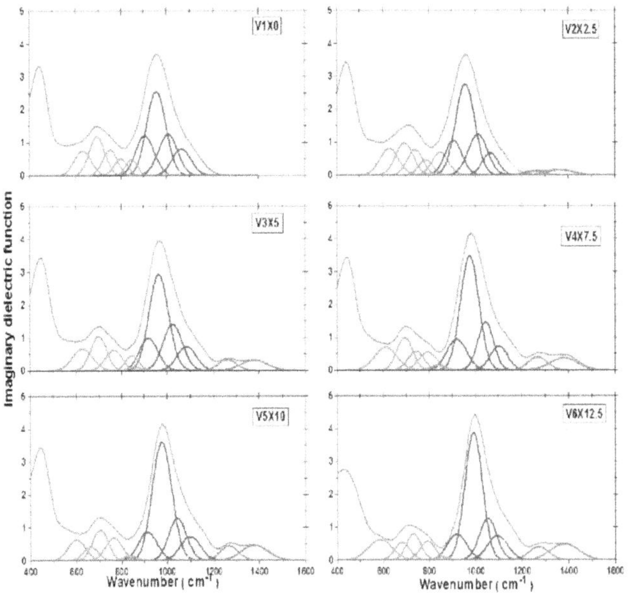

The results from NMR and Infrared spectroscopy have shown the predominant presence of SiO_4 (Q_4, Q_3), BO_3, and AlO_4 units within the peraluminous glass network derived from the composition $55SiO_2$-(25-x)Al_2O_3-xB_2O_3-$15Na_2O$-$5La_2O_3$. The comprehensive NMR results provided insights into the organization of the glass network and the nature and evolution of its constituents (Si, B, Al, Na). Within these glasses, comprising SiO_4 tetrahedra (Q_4, Q_3), BO_3 triangles, a small amount of BO_4^- tetrahedra, and $[AlO_4]^-$ entities, it was observed that varying the boron content led to notable structural modifications within the silicate network. These changes indicate that higher boron content in the glass alters the environment of these species. Infrared spectroscopy studies highlighted that aluminum tetrahedra are replaced by BO_3 triangles as more B_2O_3 is added. All glasses in the series exhibited shifts in the bands between 800-1200 cm^{-1} towards higher frequencies, potentially indicating the formation of Si-O-B bonds in the silicate network, replacing Si-O-Al bonds. These results underscore the specific role of boron across all glasses. Two techniques were instrumental in understanding the substitution of boron for aluminum, revealing an evolution in the matrix. We have already shown that this substitution promotes the occurrence of phase separation in this system. To further our understanding, we will present the results obtained from Brillouin scattering spectroscopy.

4. CHARACTERIZATION OF MEDIUM-RANGE ORDER BY BRILLOUIN SPECTROSCOPY

4.1 Brillouin Light Scattering

Brillouin spectroscopy probes vibrational energy levels through the inelastic scattering of light. This process results from the interaction of a photon from a monochromatic source with the vibrations of atoms or molecules in the sample. Approximately 1 photon in 10^4 is elastically scattered, with no change in energy, known as Rayleigh scattering, which is associated with static fluctuations in polarizability. Conversely, about 1 photon in 10^8 is inelastically scattered with either a gain or loss of energy. The observed energy difference is due to the activation of one of the material's vibrational modes: Raman scattering (optical vibrational modes) or Brillouin scattering (acoustic vibrational modes), which are linked to dynamic fluctuations in polarizability. These values (10^4 and 10^8) are rough estimates and are highly dependent on the nature and texture of the sample being analysed (Comez et al.,2012; Dragic et al., 2013).

4.2 Brillouin Light Scattering Study of Aluminoborosilicate Glasses

In this context, the spectra presented in the reference (Mabrouk et al., 2021) are in backscattering geometry. In this case, the Brillouin spectrum of an isotropic medium consists of three lines: a central Rayleigh scattering line and two Brillouin lines symmetrically positioned around the Rayleigh line, known as Stokes and anti-Stokes lines. These two Brillouin lines correspond to the longitudinal acoustic wave propagating through the material. Macroscopically, the physical characteristic that governs the propagation of this wave is the bulk modulus of elasticity C_{11}.

$$\nu_{Bl} = \frac{\nu_0}{c} n \sqrt{2V_l} = \frac{\nu_0}{c} n \sqrt{\frac{2C_{11}}{\rho}}$$

c: speed of light in vacuum, ν_0: frequency of the incident light wave, n: refractive index of the medium (our glass), ρ: density of the medium, V_l: velocity of the longitudinal acoustic wave, C_{11} : bulk modulus of elasticity of the medium (our glass). The Rayleigh line is due to static fluctuations in refractive index (controlled by density and chemical composition fluctuations), and it has a much higher intensity than the Brillouin lines. Brillouin light scattering provides access to three types of information about phonons. Before discussing the results and delving deeper into the description, it is essential to consider the definition of a phonon. The concept of a phonon is rigorously defined in crystalline systems. It is the particle, or rather the quasi-particle, associated with a specific mode of vibration of the lattice. Its energy is distributed throughout the crystal (Grimsditch et al., 1974). In glasses, the absence of periodicity does not rigorously allow for the use of the phonon concept; however, it remains a practical and heuristic tool, which we will use for convenience. (Mabrouk et al., 2021) Present the spectra recorded as a function of temperature for the as-quenched V2X2.5 glass. For clarity, only a few spectra are depicted.

The evolution of the Brillouin frequency (ν_B) and the full width at half maximum as a function of temperature for the entire series of glasses (annealed glass (red points) and as-quenched glass (black points)) is presented in ((Mabrouk et al., 2021)). *The evolution of Brillouin frequency ν_B and linewidth as a function of temperature:* for the six compositions. We observe that these frequencies exhibit a sharp change in behavior for both non-annealed glasses (black points) and annealed glasses (red points) at a temperature close to the glass transition temperature. *Below Tg: A relaxation phenomenon appears several hundred degrees below the glass transition temperature.* All non-annealed glasses show a linear decrease in frequency until a temperature we designate here as the relaxation temperature, T_R. This behavior has been observed in binary alkali silicate glasses (Na_2O, Rb_2O, Cs_2O) (Vaills et al., 2001). In the interval [T_R-T_g], a non-linear increase in Brillouin frequency is noticed. This behavior is also observed in phosphate glasses (Malfait et al., 2013), alkali binary glasses (Vaills et al., 2001), and is clearly due to a structural relaxation process triggered at the temperature T_R below the glass transition temperature. The increase in Brillouin frequency for annealed glasses results from the elimination of residual stresses (Hwa et al., 2000). Once annealed, the behavior of Brillouin frequency with temperature becomes reversible, indicating that in this temperature range, glasses behave like elastic solids. The linewidth, which reflects wave attenuation, varies almost insignificantly below Tg.

Above Tg: At the glass transition, there is a sudden change in material characteristics, entering the viscoelastic domain. Brillouin peak frequencies then rapidly decrease with increasing temperature, while linewidths begin to increase vigorously, marking the decrease in viscosity and acoustic wave attenuation.

Effect of boron-aluminum substitution: Analysis of the results reported in (Mabrouk et al., 2021). Shows that even at low boron levels, glass properties change dramatically: Brillouin peak frequency increases significantly, indicating an increase in the compression elastic constant C_{11}. Linewidth decreases with 2.5% boron oxide, and for all compositions where boron substitutes aluminum, the temperature dependence of linewidth, reflecting sound wave absorption, is nearly absent. Moreover, measurements of linewidth show almost no dispersion, indicating great dynamic stability of these glasses below the glass transition. It is notable that for the two glasses where no phase separation was observed (at 2.5% and 5% boron oxide), the Brillouin linewidth remains exactly the same before and after annealing. This demonstrates that the structural relaxation induced by annealing has no effect on the medium characteristics controlling sound wave attenuation. Additionally, these two 'homogeneous' glasses show the smallest variation in elastic constant due to structural relaxation during annealing.

In conclusion, it is clear that the mechanical properties of boron-containing glasses are superior to those without boron, with the glass containing 2.5% boron oxide exhibiting the best properties from this perspective. Furthermore, glasses containing 7.5%, 10%, and 12.5% boron oxide also show remarkable mechanical properties, coinciding with the confinement of lanthanum within small spherical inclusions of the phase separation. Such lanthanum configuration favors long-term stability of its confinement.

In this study using Brillouin spectroscopy across temperature ranges, we have highlighted the sensitivity of this spectroscopy to annealing effects below the glass transition and the aluminum/boron substitution effect. Brillouin frequencies and linewidths demonstrate that the propagation of acoustic phonons is influenced by annealing. Annealing glasses above T_g induces structural relaxation, which manifests as a change in glass configuration over a long-distance scale. In addition to structural relaxation effects.

5. CONCLUSION

The objective of this chapter was the synthesis and characterization of vitreous matrices composed of five oxides in the composition '$55SiO_2$-(25-x)Al_2O_3-xB_2O_3-15Na_2O-5La_2O_3'. Throughout this primarily experimental work, our focus was on multi-scale characterization combining nuclear magnetic resonance (NMR), infrared spectroscopy, and Brillouin spectroscopy. Given the limited studies describing the structure of lanthanum-aluminoborosilicate oxide glasses, detailed microstructural and structural studies were conducted on this series to provide insights into the organization of the glass network across various scales. The first part of the study involved microstructural and structural analysis of simplified compositions of peraluminous glasses. The influence of increasing B_2O_3 content from 0 to 12.5 mol% was evaluated concerning the appearance of phase separation at the nanoscale (TEM). It was observed that the homogeneity domain of peraluminous matrices after quenching extends over a narrow composition range from 0% to 5% B_2O_3, with the formation of a lanthanum-enriched phase above 5% B_2O_3. The combination of NMR and infrared spectroscopy in the second part of this work was successful and led to a comprehensive characterization of elementary groups comprising silicate, aluminate, and borate networks, in terms of their populations (local order) and medium-range order. From this work, it is hoped that we have demonstrated the description of aluminoborosilicate glasses as a network of $Q_n^{m(Al),p(B)}$ units rather than solely $Q_n^{m(Al)}$ units. In the glass transition domain, temperature experiments conducted using Brillouin spectroscopy highlighted the effect of boron on the mechanical properties of these glasses. Brillouin frequency and full width at half maximum demonstrated that the propagation of acoustic phonons is affected by aluminum/boron substitution. Generally, there is still much to be done to enhance the value of multispectroscopic approaches that systematically leverage the complementarities of spectroscopies (Brillouin, Infrared, NMR).

REFERENCES

Angeli, F., Charpentier, T., Molières, E., Soleilhavoup, A., Jollivet, P., & Gin, S. (2013). Influence of lanthanum on borosilicate glass structure: A multinuclear MAS and MQMAS NMR investigation. *Journal of Non-Crystalline Solids*, 376, 189–198. DOI: 10.1016/j.jnoncrysol.2013.05.042

Angeli, F., Gaillard, M., Jollivet, P., & Charpentier, T. (2006). Influence of glass composition and alteration solution on leached silicate glass structure: A solid-state NMR investigation. *Geochimica et Cosmochimica Acta*, 70(10), 2577–2590. DOI: 10.1016/j.gca.2006.02.023

Champagnon, B., Chemarin, C., Duval, E., & Parc, R. L. (2000). Glass structure and light scattering. *Journal of Non-Crystalline Solids*, 274(1–3), 81–86. DOI: 10.1016/S0022-3093(00)00207-6

Comez, L., Masciovecchio, C., Monaco, G., & Fioretto, D. (2012). Progress in Liquid and Glass Physics by Brillouin Scattering Spectroscopy. In *Solid state physics* (pp. 1–77). https://doi.org/DOI: 10.1016/B978-0-12-397028-2.00001-1

Da, N., Grassmé, O., Nielsen, K. H., Peters, G., & Wondraczek, L. (2011). Formation and structure of ionic (Na, Zn) sulfophosphate glasses. *Journal of Non-Crystalline Solids*, 357(10), 2202–2206. DOI: 10.1016/j.jnoncrysol.2011.02.037

De Sousa Meneses, D., Eckes, M., Del Campo, L., Santos, C. N., Vaills, Y., & Echegut, P. (2013). Investigation of medium range order in silicate glasses by infrared spectroscopy. *Vibrational Spectroscopy*, 65, 50–57. DOI: 10.1016/j.vibspec.2012.11.015

De Sousa Meneses, D., Malki, M., & Echegut, P. (2006). Optical and structural properties of calcium silicate glasses. *Journal of Non-Crystalline Solids*, 352(50–51), 5301–5308. DOI: 10.1016/j.jnoncrysol.2006.08.022

De Sousa Meneses, D., Malki, M., & Echegut, P. (2006a). Structure and lattice dynamics of binary lead silicate glasses investigated by infrared spectroscopy. *Journal of Non-Crystalline Solids*, 352(8), 769–776. DOI: 10.1016/j.jnoncrysol.2006.02.004

De Sousa Meneses, D., Malki, M., & Echegut, P. (2006b). Structure and lattice dynamics of binary lead silicate glasses investigated by infrared spectroscopy. *Journal of Non-Crystalline Solids*, 352(8), 769–776. DOI: 10.1016/j.jnoncrysol.2006.02.004

Deschamps, M., Fayon, F., Hiet, J., Ferru, G., Derieppe, M., Pellerin, N., & Massiot, D. (2008). Spin-counting NMR experiments for the spectral editing of structural motifs in solids. *Physical Chemistry Chemical Physics*, 10(9), 1298. DOI: 10.1039/b716319c PMID: 18292865

Deters, H., De Camargo, A. S. S., Santos, C. N., Ferrari, C. R., Hernandes, A. C., Ibanez, A., Rinke, M. T., & Eckert, H. (2009). Structural Characterization of Rare-Earth Doped Yttrium Aluminoborate Laser Glasses Using Solid State NMR. *The Journal of Physical Chemistry. C, Nanomaterials and Interfaces*, 113(36), 16216–16225. DOI: 10.1021/jp9032904

Donald, I. W., Mallinson, P. M., Metcalfe, B. L., Gerrard, L. A., & Fernie, J. A. (2011b). Recent developments in the preparation, characterization and applications of glass- and glass–ceramic-to-metal seals and coatings. *Journal of Materials Science*, 46(7), 1975–2000. DOI: 10.1007/s10853-010-5095-y

Dragic, P., Kucera, C., Furtick, J., Guerrier, J., Hawkins, T., & Ballato, J. (2013). Brillouin spectroscopy of a novel baria-doped silica glass optical fiber. *Optics Express*, 21(9), 10924. DOI: 10.1364/OE.21.010924 PMID: 23669949

El-Egili, K. (2003). Infrared studies of Na2O–B2O3–SiO2 and Al2O3–Na2O–B2O3–SiO2 glasses. *Physica B, Condensed Matter*, 325, 340–348. DOI: 10.1016/S0921-4526(02)01547-8

F Florian, P., Sadiki, N., Massiot, D., & Coutures, J. (2007). 27Al NMR Study of the Structure of Lanthanum- and Yttrium-Based Aluminosilicate Glasses and Melts. The Journal of Physical Chemistry B, 111(33), 9747–9757. https://doi.org/DOI: 10.1021/jp072061q

Gasnier, E., Bardez-Giboire, I., Montouillout, V., Pellerin, N., Allix, M., Massoni, N., Ory, S., Cabie, M., Poissonnet, S., & Massiot, D. (2014). Homogeneity of peraluminous SiO 2 –B 2 O 3 –Al 2 O 3 –Na 2 O–CaO–Nd 2 O 3 glasses: Effect of neodymium content. *Journal of Non-Crystalline Solids*, 405, 55–62. DOI: 10.1016/j.jnoncrysol.2014.08.032

George, A. M., & Stebbins, J. F. (1996). Dynamics of Na in sodium aluminosilicate glasses and liquids. *Physics and Chemistry of Minerals*, 23(8). Advance online publication. DOI: 10.1007/BF00242002

Greaves, G., Smith, W., Giulotto, E., & Pantos, E. (1997). Local structure, microstructure and glass properties. *Journal of Non-Crystalline Solids*, 222(1-2), 13–24. DOI: 10.1016/S0022-3093(97)00420-1

Grimsditch, M., & Ramdas, A. (1974). Brillouin scattering in diamond. *Physics Letters. [Part A]*, 48(1), 37–38. DOI: 10.1016/0375-9601(74)90216-3

Guo, L., Liu, S., & Shan, Z. (2023b). The influence of Mo sources, concentrations and the content of Na2O on the tendency of devitrification of modified R7T7 borosilicate glasses. *Journal of Non-Crystalline Solids*, 617, 122501. DOI: 10.1016/j.jnoncrysol.2023.122501

Heinz, M., Srabionyan, V. V., Bugaev, A. L., Pryadchenko, V. V., Ishenko, E. V., Avakyan, L. A., Zubavichus, Y. V., Ihlemann, J., Meinertz, J., Pippel, E., Dubiel, M., & Bugaev, L. A. (2016). Formation of silver nanoparticles in silicate glass using excimer laser radiation: Structural characterization by HRTEM, XRD, EXAFS and optical absorption spectra. *Journal of Alloys and Compounds*, 681, 307–315. DOI: 10.1016/j.jallcom.2016.04.214

Hung, I., Howes, A. P., Parkinson, B. G., Anupõld, T., Samoson, A., Brown, S. P., Harrison, P. F., Holland, D., & Dupree, R. (2009). Determination of the bond-angle distribution in vitreous B2O3 by 11B double rotation (DOR) NMR spectroscopy. *Journal of Solid State Chemistry*, 182(9), 2402–2408. DOI: 10.1016/j.jssc.2009.06.025

Hwa, L. G., Lu, C. L., & Liu, L. C. (2000). Elastic moduli of calcium alumino-silicate glasses studied by Brillouin scattering. *Materials Research Bulletin*, 35(8), 1285–1292. DOI: 10.1016/S0025-5408(00)00317-2

Jiang, Z. H., & Zhang, Q. Y. (2014). The structure of glass: A phase equilibrium diagram approach. *Progress in Materials Science*, 61, 144–215. DOI: 10.1016/j.pmatsci.2013.12.001

Kaneko, S., Tokuda, Y., & Masai, H. (2017). Additive Effects of Rare-Earth Ions in Sodium Aluminoborate Glasses Using ²³Na and ²⁷Al Magic Angle Spinning Nuclear Magnetic Resonance. *New Journal of Glass and Ceramics*, 07(03), 58–76. DOI: 10.4236/njgc.2017.73006

Kischkat, J., Peters, S., Gruska, B., Semtsiv, M., Chashnikova, M., Klinkmüller, M., Fedosenko, O., Machulik, S., Aleksandrova, A., Monastyrskyi, G., Flores, Y., & Masselink, W. T. (2012). Mid-infrared optical properties of thin films of aluminum oxide, titanium dioxide, silicon dioxide, aluminum nitride, and silicon nitride. *Applied Optics*, 51(28), 6789. DOI: 10.1364/AO.51.006789 PMID: 23033094

Kiyono, H., Matsuda, Y., Shimada, T., Ando, M., Oikawa, I., Maekawa, H., Nakayama, S., Ohki, S., Tansho, M., Shimizu, T., Florian, P., & Massiot, D. (2012). Oxygen-17 nuclear magnetic resonance measurements on apatite-type lanthanum silicate (La9.33(SiO4)6O2). *Solid State Ionics*, 228, 64–69. DOI: 10.1016/j.ssi.2012.09.016

Kroeker, S., & Stebbins, J. F. (2001). Three-Coordinated Boron-11 Chemical Shifts in Borates. *Inorganic Chemistry*, 40(24), 6239–6246. DOI: 10.1021/ic010305u PMID: 11703125

Lee, S. K., & Stebbins, J. F. (2003). The distribution of sodium ions in aluminosilicate glasses: A high-field Na-23 MAS and 3Q MAS NMR study. *Geochimica et Cosmochimica Acta*, 67(9), 1699–1709. DOI: 10.1016/S0016-7037(03)00026-7

Lee, S. K., & Stebbins, J. F. (2003b). The distribution of sodium ions in aluminosilicate glasses: A high-field Na-23 MAS and 3Q MAS NMR study. *Geochimica et Cosmochimica Acta*, 67(9), 1699–1709. DOI: 10.1016/S0016-7037(03)00026-7

Li, H., Li, L., Vienna, J., Qian, M., Wang, Z., Darab, J., & Peeler, D. (2000). Neodymium(III) in alumino-borosilicate glasses. *Journal of Non-Crystalline Solids*, 278(1–3), 35–57. DOI: 10.1016/S0022-3093(00)00327-6

Li, L., Strachan, D. M., Li, H., Davis, L. L., & Qian, M. (2000b). Crystallization of gadolinium- and lanthanum-containing phases from sodium alumino-borosilicate glasses. *Journal of Non-Crystalline Solids*, 272(1), 46–56. DOI: 10.1016/S0022-3093(00)00117-4

Liu, J., Luo, Z., Lin, C., Han, L., Gui, H., Song, J., Liu, T., & Lu, A. (2019). Influence of Y2O3 substitution for B2O3 on the structure and properties of alkali-free B2O3-Al2O3-SiO2 glasses containing alkaline-earth metal oxides. *Physica B, Condensed Matter*, 553, 47–52. DOI: 10.1016/j.physb.2018.10.024

Losq, . (2014). Losq, C. L., Neuville, D. R., Florian, P., Henderson, G. S., & Massiot, D. (2014). The role of Al3+ on rheology and structural changes in sodium silicate and aluminosilicate glasses and melts. *Geochimica et Cosmochimica Acta*, 126, 495–517. DOI: 10.1016/j.gca.2013.11.010

Lu, X., Weller, Z. D., Gervasio, V., & Vienna, J. D. (2024). Glass design using machine learning property models with prediction uncertainties: Nuclear waste glass formulation. *Journal of Non-Crystalline Solids*, 631, 122907. DOI: 10.1016/j.jnoncrysol.2024.122907

Lu, X., Weller, Z. D., Gervasio, V., & Vienna, J. D. (2024b). Glass design using machine learning property models with prediction uncertainties: Nuclear waste glass formulation. *Journal of Non-Crystalline Solids*, 631, 122907. DOI: 10.1016/j.jnoncrysol.2024.122907

Lu, X., Weller, Z. D., Gervasio, V., & Vienna, J. D. (2024c). Glass design using machine learning property models with prediction uncertainties: Nuclear waste glass formulation. *Journal of Non-Crystalline Solids*, 631, 122907. DOI: 10.1016/j.jnoncrysol.2024.122907

Mabrouk, A., Bachar, A., Fatani, I. F., & Vaills, Y. (2021). High temperature brillouin scattering study of lanthanum and sodium aluminoborosilicate glasses. *Materials Chemistry and Physics*, 257, 123790. DOI: 10.1016/j.matchemphys.2020.123790

Mabrouk, A., Bachar, A., Vaills, Y., Canizarès, A., & Hampshire, S. (2024). Effect of Sodium Oxide on Structure of Lanthanum Aluminosilicate Glass. *Ceramics*, 7(3), 858–872. DOI: 10.3390/ceramics7030056

Mabrouk, A., De Sousa Meneses, D., Pellerin, N., Véron, E., Genevois, C., Ory, S., & Vaills, Y. (2019). Effects of boron on structure of lanthanum and sodium aluminoborosilicate glasses studied by X-ray diffraction, transmission electron microscopy and infrared spectrometry. *Journal of Non-Crystalline Solids*, 503–504, 69–77. DOI: 10.1016/j.jnoncrysol.2018.09.030

Mabrouk, A., Vaills, Y., Pellerin, N., & Bachar, A. (2020). Structural study of lanthanum sodium aluminoborosilicate glasses by NMR spectroscopy. *Materials Chemistry and Physics*, 254, 123492. DOI: 10.1016/j.matchemphys.2020.123492

Malfait, W. J., & Halter, W. E. (2008). Structural relaxation in silicate glasses and melts: High-temperature Raman spectroscopy. *Physical Review B: Condensed Matter and Materials Physics*, 77(1), 014201. Advance online publication. DOI: 10.1103/PhysRevB.77.014201

Malfait, W. J., & Sanchez-Valle, C. (2013). Effect of water and network connectivity on glass elasticity and melt fragility. *Chemical Geology*, 346, 72–80. DOI: 10.1016/j.chemgeo.2012.04.034

Massiot, , Messinger, R. J., Cadars, S., Deschamps, M. Ë., Montouillout, V., Pellerin, N., Veron, E., Allix, M., Florian, P., & Fayon, F. (2013). Massiot, D., JMessinger, R., Cadars, S., Deschamps, M., Montouillout, V., Pellerin, N., Veron, E., Allix, M., Florian, P., & Fayon, F. (2013). Topological, Geometric, and Chemical Order in Materials: Insights from Solid-State NMR. *Accounts of Chemical Research*, 46(9), 1975–1984. DOI: 10.1021/ar3003255 PMID: 23883113

Massiot, D., Fayon, F., Capron, M., King, I., Calvé, S. L., Alonso, B., Durand, J., Bujoli, B., Gan, Z., & Hoatson, G. (2001). Modelling one- and two-dimensional solid-state NMR spectra. *Magnetic Resonance in Chemistry*, 40(1), 70–76. DOI: 10.1002/mrc.984

Mastelaro, V., & Zanotto, E. (2018). X-ray Absorption Fine Structure (XAFS) Studies of Oxide Glasses—A 45-Year Overview. *Materials (Basel)*, 11(2), 204. DOI: 10.3390/ma11020204 PMID: 29382102

Morin, E. I., & Stebbins, J. F. (2016). Separating the effects of composition and fictive temperature on Al and B coordination in Ca, La, Y aluminosilicate, aluminoborosilicate and aluminoborate glasses. *Journal of Non-Crystalline Solids*, 432, 384–392. DOI: 10.1016/j.jnoncrysol.2015.10.035

Morin, E. I., Wu, J., & Stebbins, J. F. (2014). Modifier cation (Ba, Ca, La, Y) field strength effects on aluminum and boron coordination in aluminoborosilicate glasses: The roles of fictive temperature and boron content. *Applied Physics. A, Materials Science & Processing*, 116(2), 479–490. DOI: 10.1007/s00339-014-8369-4

Naji, M., Piazza, F., Guimbretière, G., Canizarès, A., & Vaills, Y. (2013). Structural Relaxation Dynamics and Annealing Effects of Sodium Silicate Glass. *The Journal of Physical Chemistry B*, 117(18), 5757–5764. DOI: 10.1021/jp401112s PMID: 23574051

Neuville, D. R., Cormier, L., Flank, A. M., Briois, V., & Massiot, D. (2004a). Al speciation and Ca environment in calcium aluminosilicate glasses and crystals by Al and Ca K-edge X-ray absorption spectroscopy. *Chemical Geology*, 213(1–3), 153–163. DOI: 10.1016/j.chemgeo.2004.08.039

Neuville, D. R., Cormier, L., & Massiot, D. (2004). Al environment in tectosilicate and peraluminous glasses: A 27Al MQ-MAS NMR, Raman, and XANES investigation. *Geochimica et Cosmochimica Acta*, 68(24), 5071–5079. DOI: 10.1016/j.gca.2004.05.048

Neuville, D. R., & Losq, C. L. (2022). Link between Medium and Long-range Order and Macroscopic Properties of Silicate Glasses and Melts. *Reviews in Mineralogy and Geochemistry*, 87(1), 105–162. DOI: 10.2138/rmg.2022.87.03

Ollier, N., Charpentier, T., Boizot, B., Wallez, G., & Ghaleb, D. (2004). A Raman and MAS NMR study of mixed alkali Na–K and Na–Li aluminoborosilicate glasses. *Journal of Non-Crystalline Solids*, 341(1–3), 26–34. DOI: 10.1016/j.jnoncrysol.2004.05.010

Parc, R. L., Champagnon, B., Guenot, P., & Dubois, S. (2001). Thermal annealing and density fluctuations in silica glass. *Journal of Non-Crystalline Solids*, 293–295, 366–369. DOI: 10.1016/S0022-3093(01)00835-3

Qian, M., Li, L., Li, H., & Strachan, D. M. (2004). Partitioning of gadolinium and its induced phase separation in sodium-aluminoborosilicate glasses. *Journal of Non-Crystalline Solids*, 333(1), 1–15. DOI: 10.1016/j.jnoncrysol.2003.09.056

Quintas, A., Caurant, D., Majérus, O., Charpentier, T., & Dussossoy, J. L. (2009). Effect of compositional variations on charge compensation of AlO4 and BO4 entities and on crystallization tendency of a rare-earth-rich aluminoborosilicate glass. *Materials Research Bulletin*, 44(9), 1895–1898. DOI: 10.1016/j.materresbull.2009.05.009

Quintas, A., Charpentier, T., Majérus, O., Caurant, D., Dussossoy, J. L., & Vermaut, P. (2007). NMR Study of a Rare-Earth Aluminoborosilicate Glass with Varying CaO-to-Na2O Ratio. *Applied Magnetic Resonance*, 32(4), 613–634. DOI: 10.1007/s00723-007-0041-0

Reibstein, S., Wondraczek, L., De Ligny, D., Krolikowski, S., Sirotkin, S., Simon, J. P., Martinez, V., & Champagnon, B. (2011). Structural heterogeneity and pressure-relaxation in compressed borosilicate glasses by in situ small angle X-ray scattering. *The Journal of Chemical Physics*, 134(20), 204502. Advance online publication. DOI: 10.1063/1.3593399 PMID: 21639451

Saoût, G. L., Simon, P., Fayon, F., Blin, A., & Vaills, Y. (2008). Raman and infrared structural investigation of (PbO)x(ZnO)(0.6−x)(P2O5)0.4 glasses. *Journal of Raman Spectroscopy : JRS*, 40(5), 522–526. DOI: 10.1002/jrs.2158

Schaller, T., & Stebbins, J. F. (1998). The Structural Role of Lanthanum and Yttrium in Aluminosilicate Glasses: A27Al and17O MAS NMR Study. *The Journal of Physical Chemistry B*, 102(52), 10690–10697. DOI: 10.1021/jp982387m

Shamzhy, M., De, O., & Ramos, F. S. (2015). Tuning of acidic and catalytic properties of IWR zeolite by post-synthesis incorporation of three-valent elements. *Catalysis Today*, 243, 76–84. DOI: 10.1016/j.cattod.2014.06.041

Soleilhavoup, A., Delaye, J. M., Angeli, F., Caurant, D., & Charpentier, T. (2010). Contribution of first-principles calculations to multinuclear NMR analysis of borosilicate glasses. *Magnetic Resonance in Chemistry*, 48(S1), S159–S170. DOI: 10.1002/mrc.2673 PMID: 20818801

Stevensson, B., & Edén, M. (2013). Structural rationalization of the microhardness trends of rare-earth aluminosilicate glasses: Interplay between the RE3+ field-strength and the aluminum coordinations. *Journal of Non-Crystalline Solids*, 378, 163–167. DOI: 10.1016/j.jnoncrysol.2013.06.013

Vaills, Y., Luspin, Y., & Hauret, G. (2001). Annealing effects in SiO2–Na2O glasses investigated by Brillouin scattering. *Journal of Non-Crystalline Solids*, 286(3), 224–234. DOI: 10.1016/S0022-3093(01)00523-3

Walkley, B., Rees, G. J., Nicolas, R. S., Van Deventer, J. S. J., Hanna, J. V., & Provis, J. L. (2018). New Structural Model of Hydrous Sodium Aluminosilicate Gels and the Role of Charge-Balancing Extra-Framework Al. *The Journal of Physical Chemistry. C, Nanomaterials and Interfaces*, 122(10), 5673–5685. DOI: 10.1021/acs.jpcc.8b00259

Wang, S., & Stebbins, J. F. (1999). Multiple-Quantum Magic-Angle Spinning 17O NMR Studies of Borate, Borosilicate, and Boroaluminate Glasses. *Journal of the American Ceramic Society*, 82(6), 1519–1528. DOI: 10.1111/j.1151-2916.1999.tb01950.x

Wereszczak, A. A., & Anderson, C. E.Jr. (2014). Borofloat and Starphire Float Glasses: A Comparison. *International Journal of Applied Glass Science*, 5(4), 334–344. DOI: 10.1111/ijag.12095

Wheaton, B. R., & Clare, A. G. (2007). Evaluation of phase separation in glasses with the use of atomic force microscopy. *Journal of Non-Crystalline Solids*, 353(52–54), 4767–4778. DOI: 10.1016/j.jnoncrysol.2007.06.073

Wondraczek, L., Behrens, H., Yue, Y., Deubener, J., & Scherer, G. W. (2007). Relaxation and Glass Transition in an Isostatically Compressed Diopside Glass. *Journal of the American Ceramic Society*, 90(5), 1556–1561. DOI: 10.1111/j.1551-2916.2007.01566.x

Wondraczek, L., Krolikowski, S., & Behrens, H. (2010). Kinetics of pressure relaxation in a compressed alkali borosilicate glass. *Journal of Non-Crystalline Solids*, 356(35–36), 1859–1862. DOI: 10.1016/j.jnoncrysol.2010.06.009

Yiannopoulos, Y. D., Chryssikos, G. D., & Kamitsos, E. I. (2001). Structure and properties of alkaline earth borate glasses. *Physics and Chemistry of Glasses*, 42(3), 164–172. https://helios-eie.ekt.gr/EIE/handle/10442/7036

Zachariasen, W. H. (1932). THE ATOMIC ARRANGEMENT IN GLASS. *Journal of the American Chemical Society*, 54(10), 3841–3851. DOI: 10.1021/ja01349a006

Zheng, K., Yang, F., Wang, X., & Zhang, Z. (2014). Investigation of Self-Diffusion and Structure in Calcium Aluminosilicate Slags by Molecular Dynamics Simulation. *Materials Sciences and Applications*, 05(02), 73–80. DOI: 10.4236/msa.2014.52011

Zheng, Q., Potuzak, M., Mauro, J. C., Smedskjaer, M. M., Youngman, R. E., & Yue, Y. (2012). Composition–structure–property relationships in boroaluminosilicate glasses. *Journal of Non-Crystalline Solids*, 358(6–7), 993–1002. DOI: 10.1016/j.jnoncrysol.2012.01.030

Zhou, X., Niu, C., Li, K., Lin, P., & Xu, K. (2024). Vitrification of lead–bismuth alloy nuclear waste into a glass waste form. *International Journal of Applied Glass Science*, 15(2), 139–147. DOI: 10.1111/ijag.16656

Chapter 3
Slow Sand Filtration for Water Treatment

Yelriza Nurakimkyzy Yeszhan
https://orcid.org/0000-0002-4523-1211
Satbayev University, Kazakhstan

Shynggyskhan Sultakhan
https://orcid.org/0000-0003-2195-755X
Satbayev University, Kazakhstan

Erzhan Kuldeyev
Satbayev University, Kazakhstan

Bostandyk Khalkhabay
Laboratory of Engineering Profile, Satbayev University, Kazakhstan

Qin Xu
Yangzhou University, China

Seitkhan Azat
Satbayev University, Kazakhstan

ABSTRACT

Slow sand filtration (SSF) is one of the traditional methods of water treatment yet gaining attention as a promising method. The rising demand is owing to its simplicity, environmental friendliness and effectiveness, which can reach 90-95%. This chapter reviews the principles and mechanisms underlying SSF, particularly addressing the physical, biological, and chemical processes that contribute to its effectiveness. Furthermore, the chapter delves into the design and construction of SSF systems, with a particular emphasis on the critical components and factors that facilitate the attainment of optimal performance. The efficacy of the method in enhancing water quality is assessed through a variety of analytical methods, including the removal of turbidity, the reduction of pathogens, and the removal of organic matter. The chapter also encompasses the environmental impact assessment of SSF to guarantee its long-term sustainability.

I INTRODUCTION

Slow sand filtration can be considered as a technology that is easy to develop, process, control and adapt to communities that lack a safe water supply for human consumption and that face economic, educational and geographical constraints. Since this system has an efficiency of more than 95% in the removal of organic material and pathogenic microorganisms that cause gastrointestinal diseases, it can

DOI: 10.4018/979-8-3693-7505-1.ch003

be considered as an economically, technically, socially and environmentally sustainable and viable alternative to water purification.

The World Health Organization (WHO) estimates that about 80% of diseases are transmitted through contaminated water. Given that the problem of water supply is widespread and that in most cases it leads to diseases mainly of the gastrointestinal tract, a viable option is the introduction of decentralized water treatment systems that are affordable and easy to maintain. Thus, SSF becomes an alternative adapted to the environmental and social context that needs to be taken into account when working.

SSF is an established and effective method for treating drinking water, removing pathogens and particles. It has been used since the early 1800s and is still widely used globally (Haig et al., 2015). This technique has been documented to remove significant amounts of enteric bacteria from water, demonstrating its effectiveness in controlling microbiological contamination (Haig et al., 2015; Bai et al., 2024b). It is particularly suitable for use in small water systems, such as those found in rural areas and developing countries. SSF is recognized as the most suitable technology for producing safe drinking water in these regions (Ndi et al., 2008; Elhaya et al., 2024).

Research has shown that slow sand filters are an efficient way to provide safe drinking water for communities that do not have access to centralized water supply systems (Abdiyev et al., 2023). This is due to the fact that SSF is effective in removing pathogens, as demonstrated by the absence of Cryptosporidium oocysts in water treated with these filters, even when this parasite is present in raw water entering the system (Le Dantec et al., 2002). Additionally, slow sand filters effectively remove Giardia lamblia cysts and other harmful microorganisms, making them a preferred method for water treatment by federal and state authorities due to their simplicity of use and historical reliability (Haig et al., 2015; Bai et al., 2024b).

While SSF is primarily used for water treatment, it can also be adapted for wastewater disinfection (Alnahhal et al., 2024). Due to its simplicity and passive nature, this method is especially suitable for small water systems. It has been shown to be highly effective at removing bacteria and pathogens from water (Bellamy et al., 1985).

To ensure the effectiveness of slow sand filtration, it is important to use sterilized sand and regularly test the quality of the filtered water to ensure its safety (Karon et al., 2011).

The aim of this chapter is to provide a comprehensive understanding of SSF as a viable water treatment method, emphasizing its historical significance, operational principles, and modern innovations. It seeks to explore how SSF can be implemented effectively to address water contamination issues in both developed and developing regions, with particular attention to the removal of pathogens, suspended solids, and emerging pollutants such as pharmaceuticals and microplastics.

The specific objectives include:

1. To outline the principles and mechanisms underlying the effectiveness of slow sand filtration, including both physical and biological processes.
2. To examine various design and operational considerations of SSF systems, such as filter media, flow rates, and maintenance procedures.
3. To highlight the role of SSF in the removal of pathogenic microorganisms, turbidity, and other water pollutants, as well as its application in decentralized water systems.
4. To present advancements and modifications in SSF technology aimed at improving efficiency, particularly for emerging contaminants.

5. To discuss the environmental impact and sustainability of SSF as a water treatment solution, especially in resource-constrained settings.

This chapter will provide the necessary insights for the adoption and optimization of SSF in various contexts, with an emphasis on practicality and sustainability.

II PRINCIPLES OF SLOW SAND FILTRATION METHOD

The effectiveness of slow sand filters in reducing turbidity and pathogenic levels in water has been widely documented. Studies have shown that these filters can reduce turbidity by as much as 99% in well water, with varying depths of sand filter media ranging from 30 to 90 centimeters (Yusuf et al., 2019). Additionally, research has shown that slow sand filters are a simple and efficient technology for removing pathogens and particulates from drinking water (Bagundol et al., 2013).

Slow sand filtration (SSF) is a combination of physical and biological processes that destroy pathogenic microorganisms present in water unsuitable for human consumption. Due to this characteristic, it can be considered a clean technology that purifies water without creating an additional source of pollution for the environment and the consumer (Valencia, 2000).

The slow filter is a simple, clean and at the same time effective water purification system. As already mentioned, this method is a process that develops naturally, without the use of any chemicals, but requires good design, as well as proper operation and periodic maintenance, so as not to affect the microbiological layer of the filter and not reduce efficiency. microbiological and physico-chemical removal (Solsana & Méndez, 2003). In 1974, Heisman and Wood (1974) described a method of disinfection by slow filtration, for example, the circulation of raw water at low speed through a porous mantle of sand of different granulometry. During the process, impurities come into contact with the surface of the particles of the filter material and are retained for some time, further developing chemical and biological degradation processes that convert the retained substances into simpler forms that pass into solution or remain an inert material until subsequent removal or purification. Based on these preliminary studies, the hydraulic design conditions of the filters were determined depending on their filtration rate.

The raw water entering the plant remains on the filter material for 3 to 12 hours, depending on the accepted filtration rate. During this time, heavier suspended particles settle, and lighter particles tend to agglutinate, which facilitates their subsequent removal during processing. During the day, algae can grow in direct sunlight, which absorbs carbon dioxide, nitrates, phosphates, and other nutrients from the water to form cellular material and oxygen. The oxygen produced in this case is dissolved in water. It reacts chemically with organic impurities and makes them more digestible by microorganisms that have settled in a layer of organic origin.

In the upper layer of the filter surface corresponding to fine sand, a layer consisting of a material of organic origin, the so-called Schmutzdeke or "filter shell", must form through which water must pass before entering the filter material itself. This Schmutzdecke biological layer is mainly formed by algae and numerous other life forms such as plankton, diatoms, protozoa, rotifers and bacteria contained in contaminated water, which are stabilized in the pores of the granulometry of the filter material under the influence of gravity. The intense action of these microorganisms capture, digest and decompose organic substances contained in water. In this case, nitrogenous compounds decompose, nitrogen is saturated with oxygen, part of the coloring is removed, and a significant part of the inert particles in the suspension is preserved during sieving (Huisman, 1974).

Filtration mechanism

The working principle of this method based on the two main stages consisting of mechanical filtration and biological activity.

The mechanical filtration (adhesion) stage of disinfection, supported by hydraulic processes, illustrates the mechanisms by which a collision (collision) occurs between particles contained in contaminated water and grains of sand. These mechanisms are sieving, interception, precipitation, diffusion and interstitial flow.

- **Sieving:** In this mechanism, particles larger than the gaps in the filter material are trapped and retained on the surface of the filter medium.
- **Interception:** Using this mechanism, particles can collide with grains of sand.
- **Precipitation:** This mechanism allows particles to be attracted by gravity to the grains of sand, causing them to collide. This phenomenon is significantly enhanced by the action of electrostatic forces and the forces of mass attraction.
- **Diffusion:** occurs when the trajectory of a particle changes due to micro-changes in thermal energy in water and gases dissolved in it, which can lead to its collision with a grain of sand.
- **Interstitial flow:** This mechanism refers to collisions between particles due to the connection and separation of flow lines that occur due to the swirling voids of the filter medium. This constant change in flow direction creates more opportunities for collision.

Biological Activity

As mentioned above, the complete removal of particles in this process is due to the combined action of the adhesion (mechanical filtration) mechanism and the biological layer formation. In order for the filter to function as a real disinfection system, it is necessary that the biological layer be well-established and well-formed. Only when such a situation arises, the filtration system works as a cleaner from microbiological contamination and can function properly.

At the beginning of the disinfection process, microorganisms in contaminated water can use organic matter as a food source. They selectively multiply and contribute to the formation of a biological filter film - schmutzdecke (from the German word "Schmutzdecke", which translates as "mud layer"). These microorganisms oxidize organic substances to obtain energy for their metabolism and convert some of these substances into material needed for growth. This process transforms dead and organic matter into living matter, and metabolic products are transported to great depths where they are used by other organisms.

The bacteriological content of the water is limited by the amount of organic matter in the raw water, and it directly depends on these microorganisms' ability to find a suitable substrate. During this process, organic matter is released from deeper layers for use by bacteria, and the cycle continues sequentially.

Following the process described above, organic matter in the contaminated water gradually decomposes into water, carbon dioxide, and salts such as sulfates, nitrates, and phosphates. These substances are released into the wastewater that passes through the filter. Bacteriological activity is usually most pronounced in the upper layer of the filter and decreases as the depth and accessibility of the substrate increase. When cleaning the top layers of the filter, bacteria that have become stabilized are removed. A new period of filter maturation is required until the necessary bacteriological activity for disinfection processes is reached. From a depth of 0.50 meters, the bacteriological activity either decreases or stops. This leads to biochemical reactions that turn the products of microbial decomposition into ammonia and nitrites into nitrates through a process called nitrification. The performance of a slow filter largely depends on the formation of a biological layer. If this layer is not formed consistently, the efficiency of the filter will be low. In this case, the system should not be considered a purifier of organic substances, but rather an improver of water quality, primarily with regard to organoleptic characteristics.

III DESIGN AND CONSTRUCTION

The basic operation of all slow sand filters is the same and mainly consists of:

1. the supernatant water reservoir;
2. the filter bed;
3. the filter bottom and under-drainage system;
4. the filter control system.

Depending on the need, the design of the slow sand filter can be changed. This depends on the performance, degree of purification and the purpose of purification.

In designing these components, the treatment plant as a whole is considered, the methods of raw water delivery and collection of treated water are decided upon, whether and to what extent pretreatment is required, and the number and size of filters required are selected(Abdiyev et al., 2023). The likelihood that future expansions will be required should also be taken into account so that additional filters can be incorporated into the original layout. Such expansions must be carried out without interrupting the operation of the original plant. Access for construction equipment and materials to future sites must therefore be ensured, and the pipelines must have sufficient capacity to handle the future load.

Depending on the use, slow sand filters can be divided into: household (i.e., 1–10 people) slow filtration system, for schools or small communities (i.e., 10–1000 people), for small municipalities (i.e., 1000–5000 people). for industry and modified design.

Traditional slow sand filter design

Slow sand filters have been the subject of research and practical application for more than a century and a half. Their effectiveness in reducing the risk of waterborne diseases was established empirically even before the cause-and-effect relationships between pathogens and the corresponding diseases were fully understood. Modern SSF designs are largely based on the principles developed more than a century ago.

Figure 1. Scheme of traditional slow sand filtration station (Water, 1990)

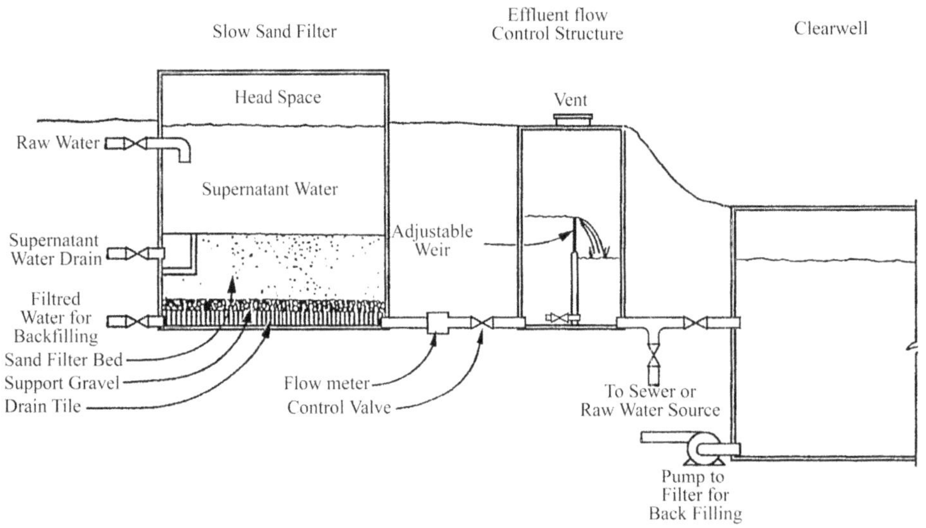

Current methodological recommendations for the design, operation and maintenance of SSFs are presented in the study by (Barrett et al., n.d.). The basic principles of SSF operation include:

A multi-layer filter structure consisting of a lower drainage layer of coarse-grained material and one or two overlying layers of finer granulometry. The design provides for the prevention of penetration of filter sand into the gravel drainage layer.

A significant depth of the filter sand layer, usually exceeding one meter. This parameter exceeds the minimum value required for virus inactivation (approximately 0.8 m) and allows for repeated cleaning of the filter by removing the top 5 cm layer before the need arises to replenish the filter material.

The optimum filtration rate through the SSF is set at 0.3 $m^3/m^2/h$ (equivalent to 300 $l/m^2/h$). This parameter is regulated by the granulometric composition of the filter material (effective particle size 0.35 mm, power factor ≤ 3), the depth of the filter layer and the hydraulic pressure determined by the difference between the maximum water level in the filter and the height of the weir. If the filter material has recently been filled, the filtration rate may decrease to 100 $l/m^2/h$. The minimum limit of the filtration rate is not regulated, but increasing the decrease may lead to an irrational increase in the dimensions of the filter. The design of the filter provides for water supply in any way that checks for damage to the upper layer of the filter material. The depth of the water layer can be one meter or more. Filtration is carried out by the next flow through the filter layer with subsequent removal of the cleaning water through the bottom drainage. The water can then be disinfected and stored for further distribution.

Table 1. Recommended parameters of slow sand filtration

Bed Dept (m)	Effective medium size (mm)	Darcy Filtration Rate (m/h)	Sand Uniformity coefficient	Suppurt Bed Dept (m)	Supernatant water Dept (m)	Recommended by
1.2	0.15-0.35	0.1-0.4	3.0	-	1-1.5	(Huisman & Wood, 1974)
0.9	0.15-0.3	0.1-0.2	-	0.3-0.5	1	(Maiyo et al., 2023)
0.8	0.4-0.45	0.08-0.24	-	0.4-0.6	0.9	(Maiyo et al., 2023)
0.6-1	0.1-0.3	0.1-0.3	3		-	(Gimbel et al., 2006)

Click or tap here to enter text. Click or tap here to enter text. The efficiency of removal of suspended particles and objects (including parasites, bacteria and viruses) decreases. The initial efficiency of destruction is about 60%, parasites - apparently close to 100%. It is assumed that the dynamics of virus removal correlates with the removal of deaths. The efficiency of bacterial purification, carried out over time (often for several weeks), reaches values close to 100%. This process is associated with biofilm techniques (Schmutzdecke) on the surface of the filter layer.

Good accumulation of organic matter on the filter surface leads to a decrease in throughput. When carrying out subsequent results, cleaning the filter by removing the upper 5-cm layer (scraping process) is required. The procedure includes stopping the filter, drainage, mechanical removal of the upper filling layer and subsequent cleaning water in the opposite direction behind the filter to prevent air inclusions in the filter material. This process can take several days even for a relatively small SSF. After cleaning, the efficiency of bacterial filtration is temporarily reduced, but recovery continues for several days. Continuity of flow through the filter is an emergency condition for the presence of SSFs. Schmutzdecke is an aerobic ecosystem dependent on the dissolved water flow in the filter. Interruption or supply of flow leads to oxygen starvation, which causes their death or transition to a state of suspended animation. This is an immediate decision to apply bacterial cleaning to the level preceding the formation of Schmutzdecke. Restoration of aerobic microflora and, accordingly, maximum filter efficiency after the resumption of flow can take two or more days.

Household Slow Sand Filtration.

A standard home slow filter consists of a concrete or plastic base that is generally cylindrical in shape. The use of concrete is generally accompanied by risks such as concrete erosion and cracking during operation. In addition, concrete is quite heavy, which causes inconvenience during transportation. Because of this, slow filters made of plastic have become widespread (Freitas et al., 2022).

The filter layer primarily consists of sand obtained from various sources such as soil, quarry, and river, while the separation and drainage layers are formed from pebbles and gravel of specific particle size distribution. Key characteristics of the filter media, including effective particle size (d10), uniformity coefficient (UC = d10/d60), percentage of fine fractions, density, and porosity, are determined by parameters developed for optimal performance of slow filters (Jenkins et al., 2011). To enhance suspended particle retention efficiency and increase the duration of filtration cycles, it is recommended to use fine particles with d10 around 0.15 mm as the main filter layer. Ideally, the filter material should have a d10 within 0.15-0.20 mm, UC in the range of 1.5 to 2.5, and fine particle content less than 4%. For drainage and separation layers, particle sizes from 1 to 12 mm are preferred to provide proper support for the filter layer. Although there is consistency in filter material characteristics, the depth of the filter layer remains a variable requiring further research. The Center for Affordable Water and Sanitation Technology (CAWST) recommends (Chan et al., 2015) a filter layer depth of 53 cm for standard household slow sand filters (HSSF), however, various values from 10 cm to 80 cm (Ghebremichael et al., 2012) are found in the literature. Several laboratory studies have examined HSSF with layer depths greater than 40 cm (Andreoli & Sabogal-Paz, 2020; Baig et al., 2011), while other works investigated systems with reduced depth (e.g., ≤25 cm) to evaluate compact systems (Adeyemo et al., 2015). Despite positive results from laboratory studies with new HSSF designs, they need to be validated in field conditions to assess long-term sustainability. Additionally, to ensure stable support and prevent sand washout into the outlet pipe, it is recommended to use a supporting layer with a thickness of 10 to 18 cm (Elliott et al., 2008).

Modified Design Slow Filtration

One of the developing directions of the slow filtration system is modified slow filters for various tasks. SSF demonstrates the ability to effectively remove pathogenic oocysts and cysts (Jobb et al., 2007). Physical stress and predation occurring in the schmutzdecke and upper layers of the slow sand filter. SSF is less effective at removing viruses and killing these organisms (Bai et al., 2024) and increases with bed level, temperature, and later stages of filter evolution (Abdiyev et al., 2023; Jobb et al., 2007). However, bacteria and viruses are relatively easily killed or inactivated by the required post-treatment disinfection processes. Slow sand and biologically active carbon filtration include the elimination or reduction of damage to certain classes of pharmaceutical compounds and microcystins. (Li et al., 2022;Gidstedt et al., 2022;Ma et al., 2018;Xu et al., 2021)

The search for practical, safe and effective solutions has led to the development and implementation of many new technologies, some of which are very complex and expensive to operate. Modification of slow sand filters is one of the simplest and most effective water treatment methods available for small systems (Kuldeyev et al., 2023; Mermillod-Blondin et al., 2005; Yang et al., 2023). Modification and adaptation of this technology allows for the effective treatment of many modern pathogens and pollutants (Zhou et al., 2024).

Slow sand filtration (SSF) is a sustainable wastewater treatment method that has demonstrated particular effectiveness in the purification of drinking water in remote areas. The mechanism of action of SSF is based on the complex interaction of biological and physicochemical processes. This technology has demonstrated high efficiency in the elimination of a wide range of pathogenic microorganisms, including protozoa (Giardia, Cryptosporidium), bacteria (Salmonella, Escherichia coli, total coliforms and faecal coliforms, faecal streptococci), viruses (bacteriophages, MS2 virus, Echovirus 12, PRD-1 virus) (Verma et al., 2017). The versatility of SSF application is noteworthy, which allows this method to be used to purify not only drinking water, but also wastewater and storm water.

IV OPERATIONAL ASPECTS AND PERFORMANCE EVALUATION

Like all filter devices, SSFs have their own operational characteristics. The main factors influencing the filter operation are the composition of the filter material, the quality of the source water and the operating conditions. The water entering the filter must be pre-cleaned from large suspended particles to prevent rapid clogging of the filter. Regular monitoring of the filter operation and the quality of the water at the outlet ensures a long service life of the system and stable quality of purified water. This is mainly due to the clogging of the filter with mechanical particles, which in turn reduces the speed and efficiency of filtration. When clogging occurs, pressure losses through the filter increase, and after a certain point, maintaining the flow rate becomes so difficult that the filter stops working (Verma et al., 2017). The accumulation of suspended particles and the formation of a biofilm on the surface of the filter medium contribute to clogging. It is known that dried residual particles in the pore space are also a major cause of clogging. Therefore, removing the top layer of sand is a common method for solving the problem. In general, methods such as removing the top layer, as well as wetting and drying cycles, can reduce the frequency of clogging of the sand filter.

Souza et. al (2021) compared two cleaning methods such as backwashing and scraping the top layer of the filter. Overall, scraping and backwashing had different effects on both slow sand filters, resulting in distinct biomass accumulation and bacterial communities. Both filters were able to improve water quality, but backwashing was easier to operate. Therefore, this method is recommended for small and community filters as an alternative to conventional scraping to obtain good quality wastewater with less labor-intensive cleaning processes.

Filter efficiency depends on whether it is operated intermittently or continuously, even with identical hydraulic residence time (HRT) and filter design. Although the smaller size of typical intermittent filters does not in itself affect their performance, the lack of continuous water dispersion in these devices, unlike continuous flow filters, does matter. If dissolved organic matter is not adsorbed directly onto the sand biofilm during the initial period of water supply, it will take longer for it to diffuse to the surface of the medium via Brownian motion. It is at the surface of the medium that the highest concentration of microorganisms is observed and the conversion of organic matter to CO_2 occurs.

Studies have shown that biosand filters demonstrate higher efficiency in continuous operation compared to intermittent mode. Thus, the removal of E. coli was 3.71 $\log_{10}$ versus 1.67 $\log_{10}$, MS_2 bacteriophages - 2.25 $\log_{10}$ versus 0.85 $\log_{10}$, and the turbidity reduction reached 96% versus 87% (Young-Rojanschi & Madramootoo, 2014).

To improve the efficiency of intermittent filters, a modification is proposed: adding a small water pump to recirculate water through the filter during the pause period. This will create an advective-dispersive flow during the downtime. This modification can also solve the problem of cleaning water with a high level of soluble carbon and nitrogen BOD, which is typical, for example, for household gray water (Maiyo et al., 2023). Water recirculation during the pause provides additional HRT, determined by the recirculation flow rate, which is significantly shorter than the total HRT determined by the pause period.

The effectiveness of SSF is evaluated based on several criteria, each of which plays an important role in determining the system's efficiency and productivity. These parameters include: removal of turbidity (suspended solids); microbiological purification; removal of organic contaminants; productivity and operational duration.

The efficiency of suspended solids removal mainly depends on the sand granularity, the HRT parameter, and the filtration rate. Removal of suspended solids primarily occurs through mechanical retention and adhesion to sand particles (Abdiyev et al., 2023). Filtration efficiency is typically measured using the nephelometric method (NTU or FTU). According to WHO recommendations, the turbidity of treated water should not exceed 1 NTU, and ideally should be 0.2 NTU.

Another indicator of SSF effectiveness is the removal of pathogenic microorganisms, which makes this purification method valuable in combating waterborne diseases. The main microbial indicators are E. coli, total coliform bacteria, and enterococci. Removal efficiency mainly depends on the formed schmutzdecke layer, whose mechanism involves biological predation, adsorption, and natural die-off. Evaluation methods include membrane filtration, the most probable number method, or other microorganism detection techniques.

Additionally, the removal of organic contaminants can be another indicator of filtration quality. SSF systems are capable of removing dissolved organic substances, improving the taste, odor, and color of water. Indicators for this include total organic carbon and dissolved organic carbon. The mechanisms may involve biodegradation and adsorption on the biofilm.

V CHARACTERIZATION TECHNIQUES FOR SSF

Grain size Distribution, Porosity and Hydraulic Retention Time

Distribution of sand grains by size Slow sand filters require well-sifted sand, which ensures proper filtration and biological activity. An important factor affecting filtration efficiency is the granulometric composition of the sand, which affects both the physical removal of suspended particles and the formation of a biologically active layer known as schmutzdecke (Ranjan & Prem, 2018). The grain size must be carefully selected to balance filtration efficiency and hydraulic conductivity, ensuring an effective filtration rate while maximizing the removal of contaminants.

The uniformity of the particles is determined by the uniformity coefficient, which for sand is calculated as the ratio of the grain size through which 60% of the sample mass passes to the grain size through which 10% of the same sample passes. Coefficient formula: K60/10 = D60/D10. If the uniformity coefficient is equal to one, this indicates the uniform size of all grains. As the uniformity of the particles increases, the filtration efficiency is enhanced. In the case of a significant difference in the size of the grains of sand, smaller grains fill the space between the large ones, which can lead to clogging of the filter (Logan et al., 2001). The effective grain size (D10), at which 10% of the sand by weight is finer. As a rule, for SFD10 it is from 0.15 to 0.35 mm. The smaller size of the D10 means finer sand, which improves filtration, but can reduce the filtration rate. The uniformity coefficient (UC) is the ratio of D60 to D10 (D60 is the grain size for which 60% of the sand is finer). A UC of less than 2 to 3 is considered ideal for slow sand filters because it indicates that the sand grains are of similar size, promoting more consistent filtration (Nassar & Hajjaj, 2013).

Porosity in SSFis an important factor affecting the efficiency of water purification. It depends on the uniformity of the size of the grains of sand, their shape and packing density, and also affects water consumption and the activity of biological processes in the filter. Porosity is a fraction of the volume of the sand layer occupied by voids or gaps between grains through which water is filtered.

Porosity was determined as follows: a certain mass of dry sand (m_{sand}), was poured into a measuring cylinder, pre-weighted (m_{empty}), and then filled with water. The cylinder was gently shaken to achieve a porosity similar to that of the filter columns. After the sand settled, the water was removed, the volume of saturated sand was measured (V_{total}) and the cylinder was weighed again (m_{full}). At an estimated water density of 1 kg/l at room temperature, the porosity p as a percentage was calculated as follows: Removal efficiency of metallic trace elements by slow sand filtration: study of the effect of sand porosity and diameter column (Barkouch et al., 2019):

$$p = \frac{V_{Pores}}{V_{total}} = \frac{m_{full} - m_{empty} - m_{sand}}{V_{total}} \times 100 \quad (1)$$

The Hydraulic Retention Time (HRT) is a function of the flow rate (Q, m^3/h), total volume of the sand (V, m^3), and sand porosity (n), and is mathematically defined by:

$$HRT = V \cdot n/Q. \quad (2)$$

where, Q is the water volume flow rate, m^3/h; V is the total sand volume m^3; and n is the sand porosity. The porosity of sand usually ranges from 0.35 to 0.50. (Maiyo et al., 2023) This means that between 35 and 50% of the active filter volume consists of water in contact with microorganisms attached to the sand grains. Reducing the particle size of the sand increases the surface area of contact between the water and the sand, as well as the material's porosity. On the other hand, a wide range of particle sizes decreases the porosity of the sand layer, which leads to a reduction in (HRT). Therefore, the sand must have a sufficiently high level of uniformity (Maiyo et al., 2023).

The specific sand surface area (As) was approximated by:

$$A_s = \frac{6000}{d_s}\left(1 - p\right) \quad (3)$$

with ds being the specific grain diameter:

$$d_s = d_{10}\left(1 + 2\, logU\right) \quad (4)$$

For any filter bed depth, the total sand surface area (A) that the water has passed through up to a sampling port can be calculated by:

$$A = A_s \frac{\pi d^2}{4} l \quad (5)$$

with d being the inner diameter of the filter column and l the corresponding bed depth.

VI ADVANCES AND INNOVATIONS

The recent developments in the field of slow sand filtration (SSF) have highlighted its potential as a promising and scalable approach to water purification, particularly in areas with limited resources. The advancements in filter design, the composition of media, and operational techniques have significantly improved the efficiency and efficacy of SSF systems, making them an attractive option for addressing diverse water quality issues.

One of the most promising areas of development in SSF (sewage treatment) technology is the integration of alternative filtration materials to enhance the efficiency of contaminant removal. Traditional SSF systems rely on layers of sand and gravel for filtration, but recent research has explored the use of innovative media such as biochar and geotextiles to improve performance.

Biochar, a high-carbon material produced through pyrolysis, has shown promise in enhancing the adsorption of both organic and inorganic contaminants in SSF systems. Kaetzl et al. (2020) found that incorporating biochar into the filtration medium resulted in improved removal of contaminants like organic matter, heavy metals, and nutrients. However, there are challenges associated with using biochar, such as the accumulation of biomass within the filter bed, which can lead to overflow and reduced operational capacity. This suggests that while biochar has the potential to enhance SSF performance, careful management of biomass is crucial to prevent clogging and ensure optimal filter efficiency.

Yusuf et al. (2019), in their research, reported that a streamlined SSF design incorporating altered filtration media demonstrated the ability to significantly enhance water quality. Their study showed that the system reduced turbidity from 6.7 NTU to 0.91 NTU, demonstrating the effectiveness of modified media in physically removing particulate matter.

Geotextiles, an alternative medium, has also been explored for its potential in enhancing SSF performance. Fitriani et al. (2020) found that geotextile materials can enhance the retention of suspended solids and other pollutants, as well as promote the development of a *schmutzdecke* layer forming on the filter surface. This layer is crucial for the overall filtration process as it supports microbial communities responsible for breaking down organic pollutants and removing pathogens.

In addition to innovations in the composition of media, the operational strategies of slow sand filters have undergone significant evolution to enhance their performance and adaptability. A notable advancement in this regard is the implementation of intermittent slow sand filtration (ISSF), a method that allows for intermittent pauses in the filtration process. This approach prolongs the contact time between wastewater and the *schmutzdecke* layer. Fitriani et al. (2020) conducted a study that demonstrated the superiority of intermittent submerged sand filters (ISSFs) over traditional continuous-flow sand filters (SSFs) in terms of treatment performance, particularly in scenarios involving high organic loadings, such as wastewater generated by food courts. This finding underscores the adaptability and versatility of ISSFs, making them a promising solution for a wide range of environmental applications. Furthermore, the implementation of pre-treatment systems has been investigated as a strategy to address one of the most prevalent issues associated with sewage treatment facilities (STFs) — clogging. The accumulation of particulate matter over time on the surface of a filter can result in decreased flow rates and reduced treatment efficiency.

Lunardi et al. (2022) proposed that pre-treatment systems, such as gravel or sediment filters, can be employed to eliminate larger particles before they enter the STF. This approach extends the operational lifespan of the STF and reduces maintenance requirements. Such pre-treatment measures have been demonstrated to alleviate the burden on STFs, enabling them to concentrate on removing finer particles. This approach facilitates the continuous development of the sludge layer without the risk of clogging.

In addition to traditional pollutants such as turbidity and pathogenic microorganisms, SSF technology has been adapted to tackle emerging contaminants of concern, including pharmaceuticals and personal care products (PPCPs). These substances, which encompass compounds like caffeine, triclosan, and over-the-counter medications, are increasingly being detected in water sources globally, raising alarms regarding their potential impact on human health and environmental sustainability. (Li et al., 2022) explored the utilization of granular activated carbon (GAC) within SSF systems, particularly GAC sandwich slow sand filters, aimed at enhancing the elimination of PPCPs. Their findings showcased the effectiveness of these modified SSF setups in eliminating a diverse range of PPCPs, surpassing the efficacy of conventional SSF methods. This breakthrough extends the applicability of SSF technologies to encompass the treatment of novel contaminants, rendering them a comprehensive solution for addressing contemporary water quality issues (Table 2).

Table 2. Typical PPCPs removal from water using sand filtration

Compound	Initial concentration	Average removal (%)	Filter type	Scale	Reference
Acetaminophen (paracetamol)	306 ± 142 ng/l 12 ng/l 2 μ/l	59-79 66 65.2	RSF SSF BSF	Lab Lab Lab	(Zearley & Summers, 2012) (Li et al., 2018) (Pompei et al., 2017)
Amoxicillin	5 μ/l	15-50	SSF	Lab	(Xu et al., 2021)
Diclofenac	252±90 ng/l 2 μ/l 100 μ/l	21-28 91 33±12	RSF BSF SSF	Lab Pilot Lab	(Zearley & Summers, 2012) (Pompei et al., 2019) (Escolà Casas et al., 2022)
DEET (N, N-diethyl-m-toluamide)	25 μ/l	25.7	SSF	Lab	(Li et al., 2018)

The two main mechanisms for PPCPs removal by sand filtration are biodegradation and adsorption. Other mechanisms, such as biosorption, may also play a role in the removal of PPCPs. Considering the fact that the removal of PPCPs during sand filtration varies significantly, some studies have been conducted to improve the adsorption process by combining it with GAC, which has a larger surface area. According to the structure of the filter, filters can be roughly divided into three types: connected, double-layer, and sandwich (Figure 2).

Figure 2. A - serially connected sand-GAC filtration system, B - dual-layer GAC-sand filtration system, C - typical GAC sandwich slow sand filter (Li et al., 2022)

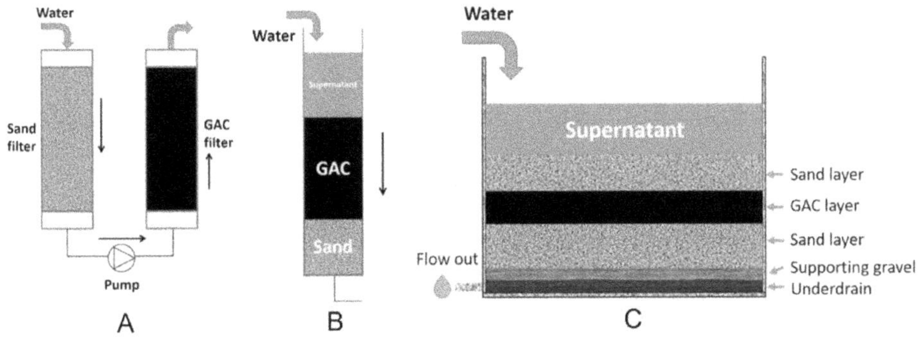

In the work of (Gidstedt et al., 2022), serially connected filters were used, and a pilot-scale study was conducted using serially connected sand-GAC filtration (media parameters not shown) to treat 12 PPCPs from tertiary-treated wastewater. The GAC reactor (surface area of 875 m^2/g) achieved a maximum removal of 62% at the start of the process, which subsequently decreased to 24% constantly after 14 h. In the work of Babaei et al., dual-layer filters were used. They reported that 86.7% of linear alkylbenzene sulfonate was removed through dual-layer sand-GAC filtration (EBCT: 50–100 min). (Ma et al., 2018) investigated removal of PPCPs (e.g., atenolol) using GAC-sand filters and anthracite-sand filters (4.88–9.76 m/h or 117.12–234.24 m/d, media parameters not specified). A higher mean removal of 49.1–94.4% was achieved using GAC-sand filters compared to a removal of 0–66.1% using anthracite-sand filters due to a combination of adsorption and biodegradation mechanisms.

A typical sandwich-layer filter consists of an upper sand layer, a middle GAC layer, and a lower sand layer. In the work of (Xu et al., 2021), good performance of GAC sandwich SSF (0.06 m/h) in removing antibiotics was observed. An average removal of 97 ± 2% was achieved for amoxicillin, clarithromycin, oxytetracycline, sulfamethoxazole, and trimethoprim, compared to a removal of just 20 ± 19% with conventional SSF.

The capacity of SSFs to eliminate a wide range of contaminants is of paramount significance considering the increasing awareness of PPCPs as persistent pollutants. The incorporation of GAC into SSFs enhances not only the adsorption of organic micropollutants, but also the overall performance of the filter. This makes it a viable solution for areas where conventional water treatment facilities may not be feasible or practical.

VII APPLICATION PROSPECTIVES OF SSF SYSTEMS

Slow Sand Filtration (SSF) is a requirement with a wide application potential, characterized by its simplicity of implementation, cost effectiveness and low yield. The uniqueness of this method can be found in regions where access to centralized water treatment systems is limited. This situation is typical for storage and stay in an unfamiliar country. The efficiency of SSF combined with its cost effectiveness makes this proposal especially attractive for countries with limited possibilities. The key advantages of the method are:

- The possibility of manufacturing filters from locally available materials.
- No need for expensive chemical reagents.
- Easy operation and maintenance.

The following priority areas of research can be identified in the slow sand filtration (SSF) field:

- Synergistic combination of SSF with disinfection methods aimed at increasing the efficiency of elimination of microbiological contaminants. Particular emphasis should be placed on the development of methods using locally available materials.
- Investigation of alternative filter substrates for use in regions with limited access to traditional materials. Inert or semi-inert materials such as ash from energy processes and biomaterials from agro-industrial waste seem promising.
- Optimization of SSF for the efficient removal of emergent pollutants, including protozoa, cyanobacteria, surfactants and microplastics. It is necessary to determine the optimal design parameters (particle size distribution, filter bed thickness, retention time, temperature regime) for different types of pollutants.
- Develop recommendations on the operational characteristics of SSF, including the life cycle, operating modes and maintenance procedures (regeneration of the filter medium, backwash). Particular attention should be paid to adapting the systems to the conditions of developing countries, taking into account seasonal fluctuations in water availability.
- Analyze the experience of implementing SSF in developing countries to identify factors limiting the widespread use of the technology. The study should cover aspects of material selection, cleaning efficiency, and filter regeneration frequency. It is advisable to assess the potential contribution

of SSF to achieving the UN Sustainable Development Goals, especially in the context of ensuring access to safe water resources.

These research areas are aimed at improving the efficiency, availability and adaptability of SSF to local conditions, which is of paramount importance for developing and least developed countries.

Industrial Application of SSF

SSF is characterized by its simplicity, low operational costs, and effectiveness in removing a wide range of contaminants, making it particularly suitable for industries seeking sustainable water treatment solutions. One of the significant industrial application of SSF is in the treatment of greywater. Oteng-Peprah et al. (2018) highlighted the importance of treating greywater, which contains a variety of contaminants, including organic matter and pathogens, before reuse. SSF can effectively reduce these contaminants, thus facilitating the safe reuse of greywater in irrigation and other applications. The method's ability to remove pathogens, such as E. coli and other harmful microorganisms, has been well-documented, making it a suitable choice for industries that require high-quality water for agricultural purposes (Leyton, 2024). Moreover, recent studies have explored the integration of SSF with alternative filtration media, such as biochar, to enhance its performance. Kaetzl et al. (2020) demonstrated that using biochar as a filtration medium in SSF could improve the removal efficiency of organic matter and nutrients from raw wastewater. This adaptation not only enhances the filtration process but also contributes to the sustainability of wastewater treatment by utilizing waste materials as filtration media.

Another promising application of SSF is in the removal of antibiotic resistance genes (ARGs) from secondary effluents. Sun et al. (2021) found that SSF, when combined with nanofiltration, exhibited significant removal rates of ARGs, which are increasingly recognized as a public health concern. This capability is particularly relevant for industries involved in pharmaceuticals and healthcare, where the presence of ARGs in wastewater can lead to environmental and health risks.

Additionally, SSF has been shown to be effective in treating industrial wastewater, particularly in polishing effluents to meet regulatory standards. Bhutiani et al. (2021) reported that SSF could significantly reduce total dissolved solids and turbidity in industrial wastewater, making it suitable for discharge or reuse. This application is critical for industries aiming to minimize their environmental impact and adhere to stringent wastewater discharge regulations.

The versatility of SSF extends to its potential use in rural and less-developed areas, where access to advanced water treatment technologies may be limited. Abdiyev et al. (2023) emphasized that SSF can provide safe drinking water in such regions, aligning with global efforts to achieve the United Nations Sustainable Development Goals. The method's low maintenance requirements and operational simplicity make it an attractive option for communities lacking centralized water supply systems.

VIII ENVIRONMENTAL IMPACT OF SSF APPLICATION

SSF is a critical technology in the field of water treatment that relies on natural filtration mechanisms and biological processes. The fundamental principle behind this method is the passage of water through a layer of sand, which undergoes both mechanical and biological purification. This technique effectively removes various contaminants from water sources, particularly in areas where access to advanced treatment systems is limited. However, it is crucial to acknowledge that this technology can have both positive and negative environmental impacts, necessitating a comprehensive evaluation to ensure its appropriate implementation (Figure 3).

Figure 3. Environmental impact of SSF application

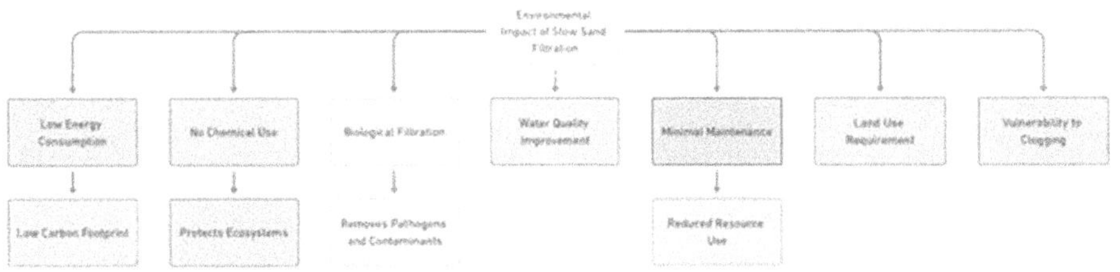

SSF is a highly effective process for removing pathogens and microbial pollutants from water, making it an invaluable tool for improving water quality. Research has shown SSF systems can remove up to 99% of fecal coliform bacteria, significantly reducing the risk of waterborne diseases such as cholera, dysentery, and other infections (Haig et al., 2015; Karon et al., 2011). The effectiveness of filtration is attributed to the formation of a biofilm on the surface of the sand bed, known as the schmutzdecke. This biofilm, composed of a diverse array of microorganisms, plays a crucial role in degrading organic matter and destroying pathogenic microorganisms like Giardia and Cryptosporidium which makes the method essential for ensuring the safety of drinking water in areas with poor sanitation conditions (Joel et al., 2018).

Schmutzdecke, a naturally occurring layer, serves as a crucial biological medium that exhibits a remarkable capacity for reliable filtration. This remarkable property significantly reduces the necessity for the use of additional chemical agents for disinfection processes, thereby contributing to the creation of a more environmentally friendly approach. The absence of need for chemical during the process reduces risks for chemical pollution that may occur in consequences of improper disposal. In the context of the global movement towards minimizing the reliance on chemical substances in water treatment procedures, SSF emerges as an increasingly appealing technology, particularly in regions characterized by limited access to advanced and costly water treatment systems, such as in developing countries (Abdiyev et al., 2023). Another important moment that should be noted that the SSF technology does not require intensive use of energy. As abovementioned, the working principle of SSF based on the natural filtration process, unlike complex systems such as membrane filtration or reverse osmosis, reduces the environmental burden and makes it a more sustainable method of water treatment (Aslan & Cakici, 2007). SSF systems can operate with virtually no energy use, as water moves through the sand under the influence of gravity, which significantly reduces the carbon footprint and energy costs of operation.

Despite the fact that the SSF offers many advantages there are some drawbacks that also should be considered during the implementation. One of the main limitations is the need for the initial period of adaptation for biofilm "schmutzdecke" formation. During this period, which can last from several weeks to several months, filtration efficiency may be below optimal and may lead to insufficient purification of water (Ndi et al., 2008). As in this period, SSF systems may not provide complete removal of microbial pollutants, which requires additional measures to monitor water quality and, possibly, the temporary use of other disinfection methods.

Another important moment that should be taken into account that the dependence of the efficiency of filtration on the size of the grains of sand and the depth of the filter. The size of the grains of sand is directly related to the filtration rate and the degree of removal of pollutants. Fine grains of sand provide better mechanical filtration, as they create a denser filter layer, which contributes to better removal of suspended particles and microorganisms. However, too fine particles can lead to rapid clogging of the filter and a decrease in the filtration rate ((Ferreira et al., 2012).

The depth of the filter also plays an important role in the efficiency of the system. The deeper the sand layer, the more time the water spends in contact with the filter material, which contributes to more efficient removal of pollutants. However, increasing the filter depth may lead to an increase in hydraulic resistance and a decrease in filtration rate, which may require optimizing the system design to achieve a balance between efficiency and productivity.

CONCLUSION

Considered to be a highly effective and sustainable water treatment method for small-scale applications in low-resource settings, SSF is increasingly gaining attention among the public. The method can remove 95% of organic materials and pathogens, making it more appealing to areas without centralized systems of purifying their drinking water. The effectiveness of SSF is based on its distinctive combination of physical and biological processes, most notably the role played by the schmutzdecke layer in removing pollutants. Moreover, some recent developments such as using alternative filter media and intermittent Slow sand filtration (ISSF), have improved its performance and flexibility. SSF has also shown promise in addressing new issues, including the removal of pharmaceuticals and personal care products. Thus, this feature emphasizes its importance in contemporary water treatment issues.

Nonetheless, the successful functioning SSF system depend on having proper design, mode of operation as well as maintenance. Performance is influenced by factors like filter material made up composition, basics source water quality and mode of operating. In order to achieve a perfect long-term periodical working, regular monitoring and maintenance is needed which entails clean up methods.

Future researches, researchers should explore ways of optimizing SSF for emerging pollutants; look into synergies with disinfection methods; and come up with guidelines for use in different geographical areas and social economic backgrounds. Moreover, alternative locally available filter substrates which can be employed by inhabitants from less developed countries should also be studied so that the application of this process will be easier for those regions.

Finally, SSF holds promise for helping to ensure global water security and public health in general; particularly with regard to the United Nations (UN) Sustainable Development Goal on providing access to safe drinking water. In terms of sustainable water treatment solutions, it is a cheap option particularly for small range populations or less industrialized nations due to its low complexity, confirmatory nature and inherent versatility.

ACKNOWLEDGMENT

This work was supported by the Science Committee of the Ministry of Science and Higher Education of the Republic of Kazakhstan (Grant No. AP23489574, 2024-2026).

REFERENCES

Abdiyev, K., Azat, S., Kuldeyev, E., Ybyraiymkul, D., Kabdrakhmanova, S., Berndtsson, R., Khalkhabai, B., Kabdrakhmanova, A., & Sultakhan, S. (2023). Review of Slow Sand Filtration for Raw Water Treatment with Potential Application in Less-Developed Countries. *Water (Basel)*, 15(11), 2007. DOI: 10.3390/w15112007

Adeyemo, F., Kamika, I., & Momba, M. (2014). Comparing the effectiveness of five low-cost home water treatment devices for Cryptosporidium, Giardiaand somatic coliphages removal from water sources. *Desalination and Water Treatment*, 56(9), 2351–2367. DOI: 10.1080/19443994.2014.960457

Alnahhal, S. Y., Elfari, A. A., Afifi, S. A., & Aljubb, A. E. R. (2024). Using slow sand filter for organic matter and suspended solids removal as post-treatment unit for wastewater effluent. *Environmental Science. Water Research & Technology*, 10(2), 490–497. DOI: 10.1039/D3EW00467H

Andreoli, F., & Sabogal-Paz, L. (2020). Household slow sand filter to treat groundwater with microbiological risks in rural communities. *Water Research*, 186, 116352. DOI: 10.1016/j.watres.2020.116352 PMID: 32916617

Aslan, S., & Cakici, H. (2007). Biological denitrification of drinking water in a slow sand filter. *Journal of Hazardous Materials*, 148(1–2), 253–258. DOI: 10.1016/j.jhazmat.2007.02.012 PMID: 17363163

Bagundol, T. B., Awa, A. L., & Enguito, M. R. C. (2013). Efficiency of slow sand filter in purifying well water. *Journal of Multidisciplinary Studies*, 2(1). Advance online publication. DOI: 10.7828/jmds.v2i1.402

Bai, X., Samari-Kermani, M., Schijven, J., Raoof, A., Dinkla, I. J. T., & Muyzer, G. (2024). Enhancing slow sand filtration for safe drinking water production: Interdisciplinary insights into Schmutzdecke characteristics and filtration performance in mini-scale filters. *Water Research*, 262, 122059. DOI: 10.1016/j.watres.2024.122059 PMID: 39059201

Baig, S. A., Mahmood, Q., Nawab, B., Shafqat, M. N., & Pervez, A. (2011). Improvement of drinking water quality by using plant biomass through household biosand filter – A decentralized approach. *Ecological Engineering*, 37(11), 1842–1848. DOI: 10.1016/j.ecoleng.2011.06.011

Barkouch, Y., El Fadeli, S., Flata, K., Ait Melloul, A., Khadiri, M. E., & Pineau, A. (2019). Removal efficiency of metallic trace elements by slow sand filtration: Study of the effect of sand porosity and diameter column. *Modeling Earth Systems and Environment*, 5(2), 533–542. DOI: 10.1007/s40808-018-0542-x

Barrett, J. M., Bryck, J., Collins, M. R., Janonis, B. A., & Logsdon, G. S. (n.d.). *Manual of Design for Slow Sand Filtration. AWWA Research Foundation and American Water Works Association, USA, 1991.*

Bhutiani, R., Ahamad, N. F., & Ruhela, M. (2021). Effect of composition and depth of filter-bed on the efficiency of Sand-intermittent-filter treating the Industrial wastewater at Haridwar, India. *Journal of Applied and Natural Science*, 13(1), 88–94. DOI: 10.31018/jans.v13i1.2421

Casas, M. E., Larzabal, E., & Matamoros, V. (2022). Exploring the usage of artificial root exudates to enhance the removal of contaminants of emerging concern in slow sand filters: Synthetic vs. real wastewater conditions. *The Science of the Total Environment*, 824, 153978. DOI: 10.1016/j.scitotenv.2022.153978 PMID: 35181359

Chan, C. C. V., Neufeld, K., Cusworth, D., Gavrilovic, S., & Ngai, T. (2015). Investigation of the effect of grain size, flow rate and diffuser design on the CAWST Biosand Filter performance. *International Journal for Service Learning in Engineering. Humanitarian Engineering and Social Entrepreneurship*, 10(1), 1–23. DOI: 10.24908/ijsle.v10i1.5705

de Souza, F. H., Roecker, P. B., Silveira, D. D., Sens, M. L., & Campos, L. C. (2021). Influence of slow sand filter cleaning process type on filter media biomass: Backwashing versus scraping. *Water Research*, 189, 116581. DOI: 10.1016/j.watres.2020.116581 PMID: 33186813

Elhaya, N., Fadeli, S. E., Erraji, E., & Barkouch, Y. (2024). Removal process of cadmium from unsafe water by slow sand filtration: Study of water feed flow rate effect. *Euro-Mediterranean Journal for Environmental Integration*. Advance online publication. DOI: 10.1007/s41207-024-00576-2

Elliott, M. A., Stauber, C. E., Koksal, F., DiGiano, F. A., & Sobsey, M. D. (2008). Reductions of E. coli, echovirus type 12 and bacteriophages in an intermittently operated household-scale slow sand filter. *Water Research*, 42(10–11), 2662–2670. DOI: 10.1016/j.watres.2008.01.016 PMID: 18281076

Ferreira, M. A., Alfenas, A. C., Binoti, D. H. B., Machado, P. S., & Mounteer, A. H. (2012). Slow sand filtration eradicates eucalypt clonal nursery plant pathogens from recycled irrigation water in Brazil. *Tropical Plant Pathology*, 37(5), 319–325. DOI: 10.1590/S1982-56762012000500003

Fitriani, N., Kusuma, M. N., Wirjodirdjo, B., Hadi, W., Hermana, J., Ni'matuzahroh, , Kurniawan, S. B., Abdullah, S. R. S., & Mohamed, R. M. S. R. (2020). Performance of geotextile-based slow sand filter media in removing total coli for drinking water treatment using system dynamics modelling. *Heliyon*, 6(9), e04967. DOI: 10.1016/j.heliyon.2020.e04967 PMID: 33015386

Freitas, B. L. S., Terin, U. C., Fava, N. M. N., Maciel, P. M. F., Garcia, L. A. T., Medeiros, R. C., Oliveira, M., Fernandez-Ibañez, P., Byrne, J. A., & Sabogal-Paz, L. P. (2022). A critical overview of household slow sand filters for water treatment. *Water Research*, 208, 117870. DOI: 10.1016/j.watres.2021.117870 PMID: 34823084

Ghebremichael, K., Wasala, L. D., Kennedy, M., & Graham, N. J. D. (2012). Comparative treatment performance and hydraulic characteristics of pumice and sand biofilters for point-of-use water treatment. *Journal of Water Supply: Research & Technology - Aqua*, 61(4), 201–209. DOI: 10.2166/aqua.2012.100

Gidstedt, S., Betsholtz, A., Falås, P., Cimbritz, M., Davidsson, Å., Micolucci, F., & Svahn, O. (2021). A comparison of adsorption of organic micropollutants onto activated carbon following chemically enhanced primary treatment with microsieving, direct membrane filtration and tertiary treatment of municipal wastewater. *The Science of the Total Environment*, 811, 152225. DOI: 10.1016/j.scitotenv.2021.152225 PMID: 34921873

Gimbel, R., Graham, N., & Collins, M. R. (2015). Recent progress in slow sand and alternative biofiltration processes. *Water Intelligence Online*, 5(0), 9781780402451. DOI: 10.2166/9781780402451

Haig, S.-J., Schirmer, M., D'Amore, R., Gibbs, J., Davies, R. L., Collins, G., & Quince, C. (2015). Stable-isotope probing and metagenomics reveal predation by protozoa drives E. coli removal in slow sand filters. *The ISME Journal*, 9(4), 797–808. DOI: 10.1038/ismej.2014.175 PMID: 25279786

Huisman, L., & Wood, W. E. (1974). *Slow sand filtration*. World Health Organization.

Jenkins, M. W., Tiwari, S. K., & Darby, J. (2011). Bacterial, viral and turbidity removal by intermittent slow sand filtration for household use in developing countries: Experimental investigation and modeling. *Water Research*, 45(18), 6227–6239. DOI: 10.1016/j.watres.2011.09.022 PMID: 21974872

Jobb, D. B., Anderson, W. B., LeCraw, R. A., & Collins, M. R. (2007). Removal of emerging contaminants and pathogens using modified slow sand filtration: an overview. *Proceedings of the 2007 AWWA Annual Conference,* Toronto, Ontario. American Water Works Association, Denver, Colorado.

Joel, C., Mwamburi, L. A., & Kiprop, E. K. (2018). Use of slow sand filtration technique to improve wastewater effluent for crop irrigation. *Microbiology Research*, 9(1). Advance online publication. DOI: 10.4081/mr.2018.7269

Kaetzl, K., Lübken, M., Nettmann, E., Krimmler, S., & Wichern, M. (2020). Slow sand filtration of raw wastewater using biochar as an alternative filtration media. *Scientific Reports*, 10(1), 1229. DOI: 10.1038/s41598-020-57981-0 PMID: 31988298

Karon, A. E., Hanni, K. D., Mohle-Boetani, J. C., Beretti, R. A., Hill, V. R., Arrowood, M., Johnston, S. P., Xiao, L., & Vugia, D. J. (2010). Giardiasis outbreak at a camp after installation of a slow-sand filtration water-treatment system. *Epidemiology and Infection*, 139(5), 713–717. DOI: 10.1017/S0950268810001573 PMID: 20587126

Kuldeyev, E., Seitzhanova, M., Tanirbergenova, S., Tazhu, K., Doszhanov, E., Mansurov, Z., Azat, S., Nurlybaev, R., & Berndtsson, R. (2023). Modifying Natural Zeolites to Improve Heavy Metal Adsorption. *Water (Basel)*, 15(12), 2215. DOI: 10.3390/w15122215

Le Dantec, C., Duguet, J.-P., Montiel, A., Dumoutier, N., Dubrou, S., & Vincent, V. (2002). Occurrence of mycobacteria in water treatment lines and in water distribution systems. *Applied and Environmental Microbiology*, 68(11), 5318–5325. DOI: 10.1128/AEM.68.11.5318-5325.2002 PMID: 12406720

Leyton, J., Fernández, J., Acosta, P., Quiroga, A., & Codony, F. (2024). Reduction of Helicobacter pylori cells in rural water supply using slow sand filtration. *Environmental Monitoring and Assessment*, 196(7), 619. Advance online publication. DOI: 10.1007/s10661-024-12764-2 PMID: 38878080

Li, J., Campos, L. C., Zhang, L., & Xie, W. (2022). Sand and sand-GAC filtration technologies in removing PPCPs: A review. *The Science of the Total Environment*, 848, 157680. DOI: 10.1016/j.scitotenv.2022.157680 PMID: 35907530

Li, J., Zhou, Q., & Campos, L. C. (2018). The application of GAC sandwich slow sand filtration to remove pharmaceutical and personal care products. *The Science of the Total Environment*, 635, 1182–1190. DOI: 10.1016/j.scitotenv.2018.04.198 PMID: 29710573

Li, J., Zhou, Q., & Campos, L. C. (2018). The application of GAC sandwich slow sand filtration to remove pharmaceutical and personal care products. *The Science of the Total Environment*, 635, 1182–1190. DOI: 10.1016/j.scitotenv.2018.04.198 PMID: 29710573

Logan, A. J., Stevik, T. K., Siegrist, R. L., & Rønn, R. M. (2001). Transport and fate of Cryptosporidium parvum oocysts in intermittent sand filters. *Water Research*, 35(18), 4359–4369. DOI: 10.1016/S0043-1354(01)00181-6 PMID: 11763038

Lunardi, S., Martins, M., Pizzolatti, B. S., & Soares, M. B. D. (2022). Pre-filtration followed by slow double-layered filtration: Media clogging effects on hydraulic aspects and water quality. *Water Environment Research*, 94(4), e10709. Advance online publication. DOI: 10.1002/wer.10709 PMID: 35362183

Maiyo, J. K., Dasika, S., & Jafvert, C. T. (2023). Slow Sand Filters for the 21st Century: A Review. *International Journal of Environmental Research and Public Health 2023, Vol. 20, Page 1019, 20*(2), 1019. DOI: 10.3390/ijerph20021019

Mermillod-Blondin, F., Mauclaire, L., & Montuelle, B. (2005). Use of slow filtration columns to assess oxygen respiration, consumption of dissolved organic carbon, nitrogen transformations, and microbial parameters in hyporheic sediments. *Water Research*, 39(9), 1687–1698. DOI: 10.1016/j.watres.2005.02.003 PMID: 15899267

Nassar, A. M., & Hajjaj, K. (2013). Purification of stormwater using sand filter. *Journal of Water Resource and Protection*, 05(11), 1007–1012. DOI: 10.4236/jwarp.2013.511105

Ndi, K., Dihang, D., Aimar, P., & Kayem, G. J. (2008). Retention of bentonite in granular natural pozzolan: Implications for water filtration. *Separation Science and Technology*, 43(7), 1621–1631. DOI: 10.1080/01496390801974712

Oteng-Peprah, M., Acheampong, M. A., & deVries, N. K. (2018). Greywater Characteristics, treatment systems, reuse strategies and user perception—A review. *Water, Air, and Soil Pollution*, 229(8), 255. Advance online publication. DOI: 10.1007/s11270-018-3909-8 PMID: 30237637

Pompei, C. M. E., Ciric, L., Canales, M., Karu, K., Vieira, E. M., & Campos, L. C. (2017). Influence of PPCPs on the performance of intermittently operated slow sand filters for household water purification. *The Science of the Total Environment*, 581–582, 174–185. DOI: 10.1016/j.scitotenv.2016.12.091 PMID: 28041695

Ranjan, P., & Prem, M. (2018). Schmutzdecke- A Filtration Layer of Slow Sand Filter. *International Journal of Current Microbiology and Applied Sciences*, 7(07), 637–645. DOI: 10.20546/ijcmas.2018.707.077

Solsana, F., & Méndez, J. P. (2003). *Water Disinfection, Pan American Center for Sanitary Engineering and Environmental Sciences*. Pan American Health Organization.

Sun, L., Ma, Y., Ding, Y., Mei, X., Liu, Y., & Feng, C. (2021). Removal of antibiotic resistance genes in secondary effluent by slow filtration-NF process. *Water Science and Technology*, 85(1), 152–165. DOI: 10.2166/wst.2021.607 PMID: 35050873

Valencia, J. A. (2000). *Teoría y práctica de la purificación del agua Tercera edición*. https://z-lib.io/book/13759703

Verma, S., Daverey, A., & Sharma, A. (2017). Slow sand filtration for water and wastewater treatment–a review. *Environmental Technology Reviews*, 6(1), 47–58. DOI: 10.1080/21622515.2016.1278278

Water, U. States. E. P. Agency. O. of D. (1990). *Technologies for Upgrading Existing or Designing New Drinking Water Treatment Facilities*. US Environmental Protection Agency.

Xu, L., Campos, L. C., Li, J., Karu, K., & Ciric, L. (2021). Removal of antibiotics in sand, GAC, GAC sandwich and anthracite/sand biofiltration systems. *Chemosphere*, 275, 130004. DOI: 10.1016/j.chemosphere.2021.130004 PMID: 33640744

Yang, H., Xu, L., Li, Y., Liu, H., Wu, X., Zhou, P., Graham, N. J. D., & Yu, W. (2023). FexO/FeNC modified activated carbon packing media for biological slow filtration to enhance the removal of dissolved organic matter in reused water. *Journal of Hazardous Materials*, 457, 131736. DOI: 10.1016/j.jhazmat.2023.131736 PMID: 37295334

Young-Rojanschi, C., & Madramootoo, C. (2014). Intermittent versus continuous operation of biosand filters. *Water Research*, 49(1), 1–10. DOI: 10.1016/j.watres.2013.11.011 PMID: 24316177

Yusuf, K. O., Adio-Yusuf, S. I., & Obalowu, R. O. (2019). Development of a simplified slow sand filter for water purification. *Journal of Applied Science & Environmental Management*, 23(3), 389. DOI: 10.4314/jasem.v23i3.3

Zearley, T. L., & Summers, R. S. (2012). Removal of trace organic micropollutants by drinking water biological filters. *Environmental Science & Technology*, 46(17), 9412–9419. DOI: 10.1021/es301428e PMID: 22881485

Zhou, T., Zhai, T., Ma, J., Wu, F., & Zhang, G. (2024). Sodium chloride-activated polymeric ferric sulfate-modified natural zeolite (Z-Na-Fe): Application of bio-slow filtration for the removal of ammonia and phosphorus from micro-polluted cellar water. *Journal of Water Process Engineering*, 61, 105367. DOI: 10.1016/j.jwpe.2024.105367

Chapter 4
New Generation of Glass Materials for Biomedical Applications:
Properties, Structure, Bioactivity, and Viability

Ahmed Bachar
Laboratory of Process Engineering, Faculty of Applied Science Ait Melloul, Ibn Zohr University, Agadir, Morocco

Assia Mabrouk
https://orcid.org/0000-0001-6399-0644
Science and Technology Research Laboratory, The Higher School of Education and Training, Ibn Zohr University, Agadir, Morocco

Cyrille Mercier
CERAMATHS, University Polytechnique Hauts-de- France, France

Claudine Follet
https://orcid.org/0000-0002-3102-5935
CERAMATHS, University Polytechnique Hauts-de- France, France

Franck Bouchart
CERAMATHS, University Polytechnique Hauts-de- France, France

Arnaud Tricoteaux
CERAMATHS, University Polytechnique Hauts-de- France, France

Stuart Hampshire
https://orcid.org/0000-0002-8993-2570
The Bernal Institute, University of Limerick, Ireland

ABSTRACT

The limited strength of bioglasses has confined their application to non-load-bearing contexts, thus necessitating enhancements in their mechanical properties. A viable approach to surmount this hurdle involves integrating nitrogen into the silicate network of the glass. This outlines the effect of nitrogen addition in bioglasses of the system SiO2-Na2O-CaO. The purpose is to determine the effects of nitrogen addition on the physical and mechanical properties and the structure of oxynitride bioglasses based on the system Na2O-CaO-SiO2-Si3N4. Properties were all observed to increase linearly with nitrogen content. These increases are consistent with N in the glass structure in 3-fold coordination with silicon and extra cross-linking of the glass network. These increases are consistent with the incorporation of

DOI: 10.4018/979-8-3693-7505-1.ch004

N into the glass structure in three-fold coordination with silicon with result in extra cross-linking of the glass network.

1. INTRODUCTION

Understanding the physico-chemical properties of biomaterials is crucial in the design of bone implants. The biomaterials used can include metals, alloys (such as 316L), polymers (e.g., silicones), natural materials (such as corals), and ceramics (e.g., hydroxyapatite, calcium phosphates, and glasses). The most well known glass in this context is Hench's Bioglass®, which has a mass chemical composition of $45SiO_2$-24.5CaO-24.5Na_2O-6P_2O_5. These glasses are notable for being reactive biomaterials, also known as bioactive materials. This means that these materials can establish an intimate and stable chemical bond with living tissues. Currently, bioactive glasses are primarily applied in the field of auditory implants. However, their use as bone substitutes is pending due to their lower mechanical properties compared to bioceramics, leading to the innovative idea of enhancing bioactive glasses by introducing nitrogen. Oxy-nitride glasses were discovered in the 1980s as an intergranular phase in silicon nitride-based compounds densified in the presence of sintering additives. The liquid phase, formed with silica and the necessary sintering additives, facilitates the dissolution-diffusion-precipitation of beta-silicon nitride and appears at grain boundaries in the form of glass. Analysis of this glass revealed the presence of nitrogen substituting oxygen in the network, with a valence of three instead of two, which induces network densification. Numerous studies on these oxy-nitride glasses have shown that small additions of nitrogen lead to increases in the glass transition temperature, thermal expansion coefficient, density, viscosity, elastic modulus, hardness, and toughness of the glass. More recent research has also demonstrated a significant increase in flexural strength. There is an increasing focus on developing bioactive glasses for applications in repairing or replacing bone (Hench., 1991). These materials are designed to foster a positive tissue response, stimulate new tissue growth, and interact with cells effectively (Kokubo et al., 2003; Fujibayashi et al., 2003; Liang et al., 2006). The clinical success of Bioglass© (Hench et al., 1971) and dense hydroxyapatite ceramics (Badr et al., 2024) has spurred efforts to find new bioactive glasses and glass-ceramics with improved properties. Bioactive materials (Kokubo, 1991; Bachar et al., 2013) are known to bond with living bone by forming an apatite layer on their surface (Neo et al., 1992; Hill, 2004). Specifically, in vivo studies on phosphate-containing Bioglass© particles (100–300 µm) used to fill bone defects have demonstrated more significant new bone growth in the initial weeks post-surgery compared to hydroxyapatite particles (Kozo et al., 2000). Comparable bone in-growth has been reported for phosphate-free glass particles of the same size in the SiO_2–CaO–Na_2O system (Fujibayashi et al., 2003). Nevertheless, the limited mechanical strength and inherent brittleness of these bioactive glasses have constrained their use to non-load-bearing applications, such as ossicles in the middle ear (Mabrouk et al., 2018). One approach to addressing the low strength of these glasses is to incorporate nitrogen into the silicate network (Mabrouk et al., 2018; Hampshire, 2008). Research has indicated that substituting oxygen with nitrogen in alumino-silicate glasses leads to a linear increase in glass transition temperature, elastic modulus, and hardness with higher nitrogen content (Bachar et al., 2016; Hampshire and Pomeroy, 2008). Further studies have confirmed that nitrogen enhances the mechanical properties of potentially bioactive glasses (Bachar et al., 2016). It has been concluded that nitrogen acts as a network-forming anion, with its effects on glass properties being independent of other modifiers (Mabrouk et al., 2018). The addition of fluorine, such as in the form of CaF_2, to bioglass coatings lowers the melting temperature (Bachar et al., 2013) and significantly affects the release of certain ions from the glass matrix. Fluorine is also known to improve bone density (Vestergaard et al., 2008), and despite its narrow therapeutic range and some debate over its effectiveness in preventing fractures (Aaseth et al., 2004), fluoride-releasing implants could be beneficial for osteoporosis patients. Studies (Hanifi et al., 2012) have shown that in alumino-silicate glasses containing both F and N, fluorine's network-terminating effect leads to substantial reductions in both glass melting temperatures (T_m) and glass transition temperatures (T_g). Meanwhile, nitrogen substitution for oxygen increases the elastic modulus and microhardness, though these properties are minimally affected by the addition of fluorine. Fluorine also aids in the dissolution of nitrogen into the melt (Hanifi et al., 2011). This chapter explores the in vitro bioactivity of oxy-nitride and oxy-fluoro-nitride glasses to assess their reactivity and bioactivity. In addition to tests conducted in Simulated Body Fluid (SBF), we discuss preliminary biological approaches to evaluate the cytocompatibility of nitrogen-doped glasses with cells. Furthermore, we review microbiological tests, known as biofilm tests, which examine the formation of a polymeric matrix that resists standard decontamination methods. These tests are crucial for determining whether the surfaces of the glasses promote the development of bacterial biofilms. The significance of this research lies in the fact that biofilm-associated bacteria on materials used in healthcare settings contribute to nosocomial infections acquired during hospitalization.

1.1. Nitrogen Ceramics and Oxynitride Glasses

According to the definition provided at the '2nd Consensus Conference on Definitions in Biomaterials (Hench., 1998), a biomaterial is a material intended to be in contact with living tissues and/or physiological fluids to evaluate, treat, modify forms, or replace any tissue, organ, or body function. These materials must, on the one hand, satisfy appropriate physico-chemical characteristics for the implantation sites and the intended function, and on the other hand, be biocompatible. The concept of biocompatibility of a biomaterial is defined by the tissue acceptance of the implant by the organism (Iken., 2023). Biomaterials have been developed to maintain the integrity and quality of life for individuals suffering from severe functional deficiencies or accidents. The goal of their development is to enable the creation of body-assistive devices capable of substituting for the functions of damaged organs.Currently, biomaterials represent a significant international concern. Indeed, more than 5% of the population has an implanted biomaterial. In France, 3.2 million people have a biomaterial, encompassing all types (Hench., 1994). The preservation of bodily integrity and individual autonomy is a major issue in our society. Several research directions are being developed to design and advance new materials intended to perform tasks that the body can no longer manage. These materials find applications in cardiac surgery (heart valves, stents), ophthalmology (contact lenses), as well as in orthopedic, maxillofacial, or dental surgery (prosthetics, bone filling, substitution).

Among the osteoconductive biomaterials commonly used in orthopedic surgery are calcium phosphate, hydroxyapatite, fluoroapatite, and more recently, biologically active glasses. These biologically active glasses fall into two distinct categories: bioactive glasses and crystallized glass-ceramics. The first and most well-known bioactive glass was developed in the early 1970s by Hench and his colleagues. It is named 45S5 and contains 45% by mass SiO_2, 24.5% by mass each of CaO and Na_2O, and 6% by mass P_2O_5. This bioactive glass is marketed under the name Bioglass® (Hench et al., 1971). A glass is defined as biologically active by its ability, when in contact with cellular tissue, to exhibit in vitro and in vivo biocompatibility, an absence of inflammatory and toxic processes, and a predisposition to osteoconductivity in the presence of osteogenic precursors that can promote a biological bond at the bone/glass interface (Lin et al., 1991). The bonding of bioactive glasses to bone is due to the formation of a carbonated hydroxyapatite layer that develops on the surface of the glass both in vitro and in vivo. In vitro studies have established the kinetics of a series of surface reactions leading to the formation of apatite on material substrates. Recent clinical interest in bioactive glasses has expanded to include specific substances for bone graft applications (Greenspan et al.,1994). The significance of these glasses is related to their ability, when in contact with body fluids, to simulate the growth of carbonated hydroxyapatite on their surface. Carbonated hydroxyapatite is the main mineral component of bone, and its growth is attributed to an ionic exchange mechanism between the glass surface and body fluids (D'Alessio et al.,2001).

- **Oxynitride glasses**

Silicon nitride has been the subject of extensive research in the field of ceramic materials in recent years due to its high mechanical strength, wear resistance, high decomposition temperature, oxidation resistance, good chemical durability, low friction coefficient, and excellent thermal shock resistance. Due to these advantageous characteristics, Si_3N_4 is highly recommended for high-temperature applications. Si_3N_4 exists in two crystallographic forms: α, which is the low-temperature phase, and β, which exists at high temperatures. Si_3N_4 is difficult to densify, partly due to the covalent Si-N bonds and partly due to a sintering temperature limited to 1700-1800°C, because of its sublimation around 1850°C. The sublimation reaction is as follows:

$Si_3N_4 \rightarrow 3\,Si + 2\,N_2\ P_{N2}=0{,}1$ MPa

One solution is to add sintering additives such as Y_2O_3, Al_2O_3, CaO, MgO, Na_2O (Hanifi et al.,2012), which react with the silica layer on the surface of Si_3N_4 and create liquid phases that enhance the densification of silicon nitride (Jack.,1976). When these liquid phases cool, oxy-nitride glasses (Mulfinger.,1966) appear at the grain boundaries and triple points of the silicon nitride (Figure 1).

Figure 1. High-resolution micrograph (TEM) of silicon nitride showing a triple point, an intergranular film, and an oxy-nitride glass

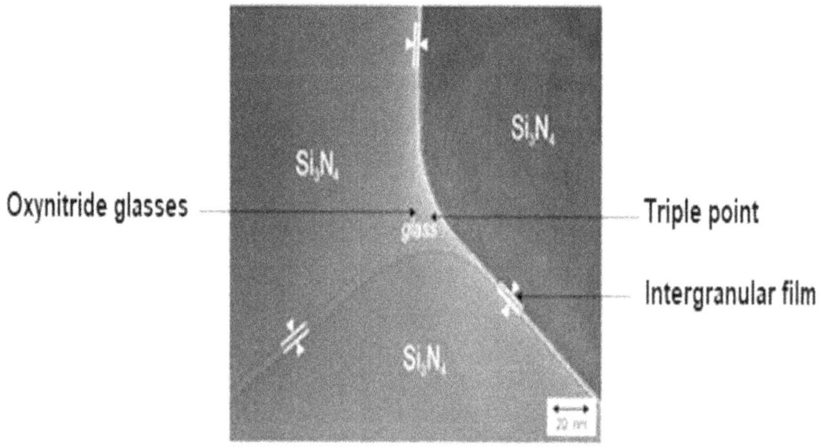

This sintering technique enabled the first demonstration of the formation of oxy-nitride glasses.

- **Synthesis Method of Oxynitride Glasses**

Literature reviews (Hanifi et al.,2012; Jack.,1976) indicate that the synthesis of oxynitride glasses is more complex than that of conventional oxide glasses for several reasons. Specifically, it is highly dependent on the synthesis conditions: **Reagents:** A limited list of nitrides is used as the nitrogen source (Mulfinger.,1966). **Atmosphere:** The need for a very low oxygen content during melting to prevent glass oxidation (Mulfinger.,1966b). **Temperature:** The requirement for high temperatures (up to 1950°C) for melting and homogenizing the reagents. **Crucibles:** Chemically inert materials are necessary. Si-O-N (Leng-Ward et al.,1989) is the basic composition of oxynitride glasses that can be formed by melting SiO_2 and Si_3N_4 at temperatures above 1800°C. This high temperature is essential to achieve a homogeneous melt of these glasses. Under these conditions, silicon nitride decomposes into Si_4^+ and N_3^-, while SiO_2 decomposes into SiO(g) and O_2(g), leading to mass loss. The incorporation of a network-modifying cation reduces the melting temperature and viscosity, facilitating the dissolution of Si_3N_4 without significant decomposition. Alkali and alkaline earth metals assist in the formation of oxynitride glasses at temperatures below 1600°C (Bachar et al.,2012). The addition of aluminum has significant effects on the formation of oxynitride glasses in the M-Si-Al-O-N system, where M is the modifying cation. Aluminum extends the vitrification range of oxynitride glasses compared to the M-Si-O-N system. It lowers the melting temperature, increases nitrogen solubility, reduces glass weight loss, and eliminates phase separation. It also enhances the glass's strength and chemical durability and decreases the tendency of the molten material to crystallize (Saka et al.,1983b). Jack (Jack.,1976). investigated oxynitride glasses in the Si_3N_4-Al_2O_3-SiO_2, Si_3N_4-MgO-SiO_2, and AlN-Y_2O_3-SiO_2 systems, containing at least 10% by mass of nitrogen. He observed a similarity between the structure of SiO_4 units in silicates and SiN_4 units in silicon nitride, where the oxygen or nitrogen atoms are tetrahedrally coordinated around the central silicon atom. He deduced that, due to the similarity in bond lengths of Si-N (0.174 nm), Si-O (0.170 nm), and Al-O (0.175 nm), nitrogen can be incorporated into the aluminosilicate structure according to the following reaction:

$$Si^{4+} + N^{3-} \dashrightarrow Al^{3+} + O^{2-}$$

This incorporation leads to the formation of new oxynitride glasses. These new glasses can also be used for high mechanical strength applications due to their high viscosity.

Nitrogen Sources: Several sources can be used to incorporate nitrogen into oxide glasses: Gaseous form: N_2 or NH_3. Solid crystalline form: SiO_2N_2 (Loehman.,1980). Nitrides: Compounds such as AlN, Mg_3N_2, Ca_3N_2, and Si_3N_4. According to Romanczuk et al. (Romanczuk et al.,2019) the nitrogen content achieved in oxide glasses is relatively low when using N_2 or NH_3 as nitrogen sources. These authors observed that the rate of nitrogen insertion is extremely slow; for example, it takes 24 hours of melting at 1550°C in a N_2 atmosphere to achieve a maximum of 2.5% by mass of nitrogen in a CaO–Al_2O_3–SiO_2 glass. In contrast, when silicon nitride (Si_3N_4) is included in the reagents (Romanczuk et al.,2019) and the mixture is melted in molybdenum crucibles at 1600°C for only 1.5 to 2.5 hours under a nitrogen atmosphere, the insertion of nitrogen occurs more rapidly and reaches significantly higher levels (Wusirika.,1984).

Preparation of Oxynitride Glasses: To highlight the importance of the operational protocol for these types of glasses, Wusirika (Wusirika.,1984) compared two operational protocols across different glass systems. In the first protocol, the oxides and silicon nitride were melted together in molybdenum crucibles under a nitrogen sweep. In the second protocol, the same compositions were melted in two stages: the oxides were initially melted in platinum crucibles at 1650°C, then ground and mixed with the appropriate amount of Si_3N_4 and remelted at 1650°C in molybdenum crucibles under a nitrogen sweep. The initial nitrogen content added was 4% by mass for all compositions.

- **Properties of Oxynitride Glasses**

One of the earliest attempts to introduce nitrogen into glass was made by Mulfinger (Mulfinger.,1966). He found that incorporating 0.33 wt% of nitrogen into silicate glasses resulted in an increase in both the melting temperature and viscosity. Numerous researchers have investigated the effect of nitrogen substitution on other properties of glasses (Brow et al.,1984). For example, the evolution of several properties in glasses from the Y-Si-Al-O-N system was studied as a function of nitrogen content ranging from 1 to 10 mol% (Hampshire et al.,2019). The authors observed that nitrogen addition led to an increase in the glass transition temperature (Tg), hardness, and toughness, but a decrease in the coefficient of thermal expansion. Loehman (Loehman.,1980) was the first to report a direct correlation between the mechanical properties and the nitrogen content in oxynitride glasses. However, these studies did not account for changes in silicon concentration during glass preparation, making it difficult to isolate the effect of nitrogen alone. Brow et al. (Brow et al.,1984) investigated the effect of nitrogen on the mechanical properties while keeping cation ratios constant, thus making the N/O ratio the sole variable. The same authors (Ali et al.,2024) proposed studying the effects of anions and cations on the properties of oxynitride glasses by separating and independently examining the effects of each group. Systematic substitution of oxygen with nitrogen leads to improvements in the physical, chemical, mechanical, and thermal properties of glasses in the M-Si-O-N and M-Si-Al-O-N systems (where M = Ca^{2+}, Mg^{2+}, Sr^{2+}, Na^{+}, Y^{3+}, Ln^{3+}, etc.) (Jankowski et al.,1980). Significant studies conducted to date have shown increases in density, compactness, chemical durability, refractive index, and dielectric constant with nitrogen substitution for oxygen. Further research has demonstrated enhancements in glass transition temperatures, Young's modulus, hardness, and toughness (Bachar et al.,2019). Hampshire et al. (Hampshire.,1992) have shown that the addition of nitrogen to glasses in the M-Si-Al-O-N systems, with M = Ca, Mg, Nd, or Y, results in variations in both physicochemical and mechanical properties.

1.2. Fluoride Ceramics and Oxyfluoride Glasses

- **Oxyfluoride glasses**

Oxyfluoride glasses are oxide glasses containing various cations (Si, Ca, Na, Ge, B, P, etc.), oxygen, and fluorine. The addition of fluorine to the glass composition lowers the synthesis temperature, significantly reduces the viscosity of the molten liquids, and consequently facilitates the molding and quenching of the glasses (Bueno et al.,2005). Fluorine is also a good anti-caries agent. Indeed, when fluorine is released, it inhibits microbial growth and exhibits antibacterial activity in acidic environments, and it can prevent demineralization by forming fluoroapatite with the formula $Ca_{10}(PO_4)_6(F)_2$ (Furtos et al.,2005). Fluoride ions released by dental materials are easily exchanged with OH^- ions in hydroxyapatite over a broad concentration range, forming a wide solid solution of fluoridated hydroxyapatite (FHA: $Ca_{10}(PO_4)_6(OH,F)_2$) up to fluoroapatite (FA: $Ca_{10}(PO_4)_2F_2$). FA, like HAC, has chemical composition, biocompatibility, and bioresorbability equivalent to the inorganic matrix of bone (Kim et al.,2004). However, it is chemically, thermodynamically, and mechanically more stable than hydroxyapatite. Given the possibility of apatite phase deposits, which is an important condition, on the surface of fluorine-containing glasses, these materials hold particular interest for use as biomaterials.

- **Atomic structure of oxyfluoride glasses**

The types of bonds formed by fluorine determine the properties of oxy-fluoride glasses. Nuclear Magnetic Resonance (NMR) is an important technique for studying the atomic structure of amorphous materials. It provides information on the coordination states of atoms, types of bonds, and nearest-neighbor atoms. Many studies have used this technique to identify the types of fluorine bonds in oxy-fluoride glasses. Kumar et al. (Kumar et al.,1961) demonstrated that fluorine exists in the form of Si-F bonds in fluoro-silicate glasses. Later, studies on alumino-silicate glasses specified that fluorine forms Si-F and Al-F bonds, respectively, in the form of SiO_3F and AlO_3F groups (Mabrouk et al.,2017). Other authors (Kohn et al.,1991) have shown that additional bonds can exist in alumino-silicate glasses in the presence of calcium, such as Al-F and Ca-F bonds. Liu et al. (Liu et al.,2002) confirmed that the main fluorine bonds in sodium alumino-silicate glasses are Al-F-Na(n), where (n) indicates the number of modifier cations coordinated by fluorine, with the presence of other minor bonds Si-F-Na(n). Furthermore, Hill et al. (Hill et al.,2006) also conducted a study on $2SiO_2$-Al_2O_3-(2-x)CaO-xCaF_2 glasses using ^{19}F NMR. These authors observed a single peak for the F-Ca(n) bond at around -100 ppm for glasses with low fluorine content. However, as the fluorine content increases, a second peak appears at -150 ppm, corresponding to the Al-F-Ca(n) bond (Figure 2).

Figure 2. ^{19}F MAS NMR spectra of glasses LG33 (x=0.25), LG34 (x=0.5), and LG35 (x=1)

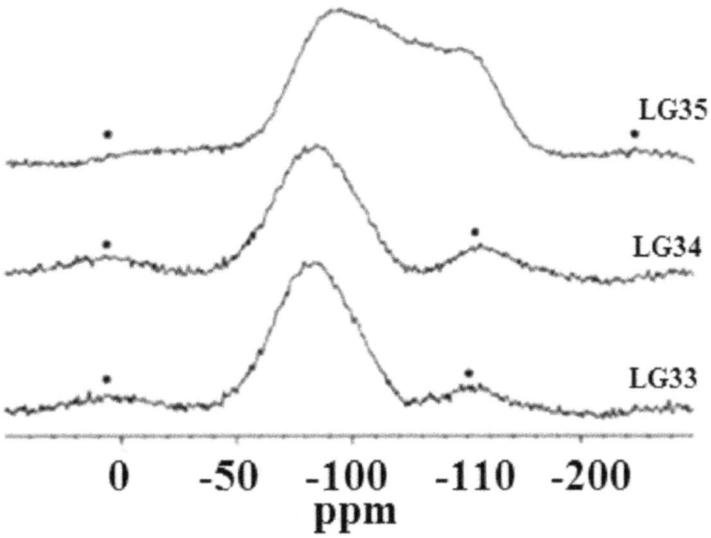

De Pablos-Martín et al. (De Pablos-Martín et al.,1980) also observed, using ^{19}F MAS NMR, F-Ca(n) and Al-F-Ca(n) bonds in glasses of the $2SiO_2$-Al_2O_3-0.5CaO-$0.5CaF_2$ system. In the presence of fluorine and silicon, various Si-F bonds are formed. However, among these bonds, the formation of a volatile gas, silicon tetrafluoride (SiF_4), is undesirable. In its volatile state, it can be hydrolyzed, leading to the formation of silica and hydrofluoric acid (equation below), which is hazardous to both the environment and human health due to its toxicity (Freer et al.,1986).

$SiF_4 + 2H_2O \rightarrow SiO_2 + 4\ HF$

Therefore, the fabrication of oxy-fluoride glasses must be conducted carefully to prevent the loss of fluorine as SiF_4. Sukenaga (Sukenaga et al.,2020) also noted that incorporating a network modifier forms non-bridging oxygen with silicon. This can suppress the formation of Si-F bonds, including SiF_4. Indeed, network modifiers break the bonds between the polyhedra in the glass network, causing depolymerization. They convert bridging oxygens into non-bridging oxygens. Free fluoride ions can then bond with other ions, reducing fluorine volatilization.

- **Properties of oxyfluoride glasses**

Fluorine is the most electronegative element in the periodic table and is always found bonded to other elements. It is known as a powerful network disruptor in vitreous materials. Substitution of non-bridging oxygen (NBO) with fluorine results in a decrease in viscosity and glass transition temperature (Tg). Fluorine also facilitates crystallization by lowering viscosity, which enhances atomic mobility and rearrangement to form crystals. Conductivity and the coefficient of thermal expansion increase with the amount of fluorine present in oxyfluoride glasses (Shelbyet al.,1990). Figure 3 illustrates the effect of replacing oxygen with fluorine through the difference in global charge density in silicon tetrahedra. Solid lines represent an increase in charge density, while dashed lines denote a decrease.

Figure 3. Difference in charge density between: a) SiO_4 tetrahedra and b) the same group with oxygen replaced by fluorine (Painter et al., 2002)

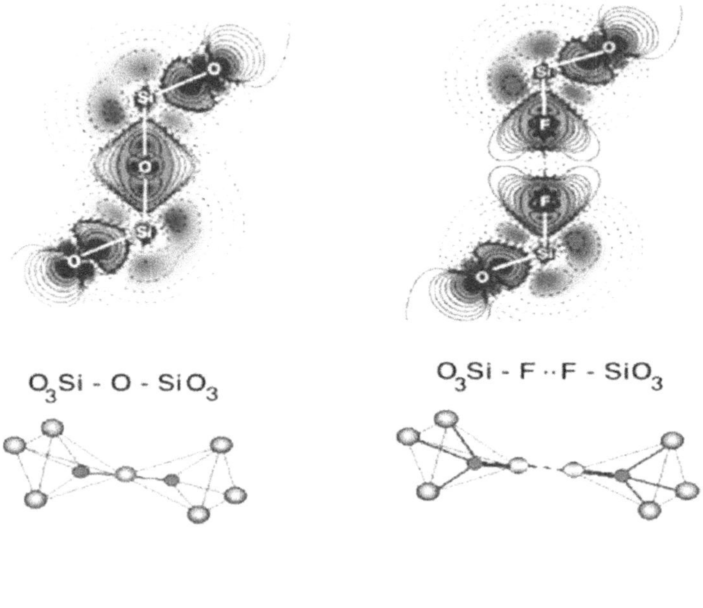

The difference in electronegativity between Si and F is greater than that between Si and O. Consequently, Si-F bonds (E -F = 610 kJ/mol) are more ionic and stronger than Si-O bonds (E -O = 398 kJ/mol). The attraction force between silicon and fluorine increases with the substitution of oxygen by fluorine. However, a strong Coulombic repulsive force between SiO_3F groups leads to the separation of these units, while SiO_4 groups remain strongly bonded. This phenomenon results in a decrease in network connectivity within the system (Kumar et al.,1961). Greene et al. (Greene et al.,2003) were the first to systematically study the effect of fluorine on the physical, mechanical, and thermal properties of glasses in the K_2O-BaO-MgO-SiO_2-Al_2O_3-B_2O_3-MgF_2 system. They found that, with a constant cation composition and substitution of oxygen with fluorine, density and molar volume remained constant but observed a slight decrease in microhardness. The authors concluded that the decrease in microhardness could be attributed to phase separation caused by the increased amount of fluorine.

- **Oxynitride Fluoride Glasses**

Hanifi et al. (Hanifi et al.,2007) investigated the formation domains of glasses within the Ca-Si-Al-O-N-F system containing 20 wt% nitrogen (N) and 5 wt% fluorine (F), melted at 1600°C. The different compositions of their glasses are illustrated in the ternary diagram (Figure 4).

Figure 4. Formation domains of glasses in the Ca-Si-Al-O-N-F system containing 20 wt% nitrogen and 5 wt% fluorine, melted at various temperatures (Hanifi et al., 2007)

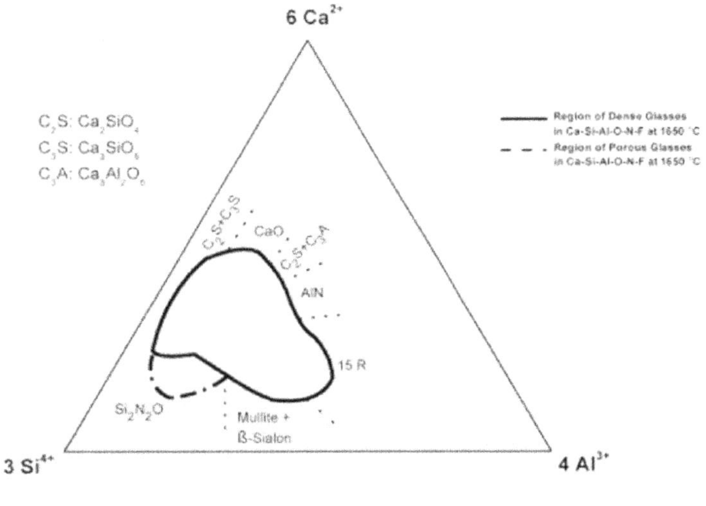

This diagram illustrates two formation zones: one dense zone with homogeneous glasses containing few surface bubbles, and another zone with porous glasses. All of these glasses are opaque and appear gray or black. The authors compared the formation zone of these glasses with those melted at 1700°C. They demonstrated that the presence of fluorine in these glasses expands the vitrification domain, even with a decrease in melting temperature. According to these authors, the bubbles present in the glass may be due to the release of SiF_4 gas. Wang et al. (Wang et al.,1982) investigated the effect of fluorine on the formation zone of glasses containing high nitrogen levels. They found that glasses with the composition $Ca_{28}Si_{57}Al_{15}O_{65}N_{35}$ are fully crystalline, whereas the composition $Ca_{28}Si_{56}Al_{16}O_{74}N_{26}$ remains entirely amorphous. The authors suggested that fluorine, by lowering the melting temperature, would facilitate the dissolution of nitrogen in the glasses. Additionally, the presence of fluorine would aid in the bonding of nitrogen atoms to silicon, thereby reducing the likelihood of SiF_4 formation.

- **Properties of Oxynitride Fluoride Glasses**

Oxynitride fluoride systems and their properties are sparsely referenced in the literature. Bachar et al. (Bachar et al.,2016) investigated the effect of fluorine and nitrogen on the thermal properties of glasses in the Ca-Si-Al-O-N-F system. They found that adding 5 wt% fluorine decreased the glass transition temperature (Tg) by 94°C. This decrease is attributed to the disruptive role of fluorine, which replaces oxygen and facilitates the movement of structural units within the glass. In contrast, the addition of nitrogen cross-links the network, thus increasing the thermal energy required for partial mobility and raising Tg (for 20 wt% nitrogen, Tg increases by approximately 38°C). Vaughn et al. (Vaughn et al.,1984) incorporated nitrogen into glasses from the Zr-Ba-Al-Y-O-F system to enhance thermal stability and mechanical properties. They observed an increase in both glass transition and crystallization temperatures, as well as an improvement in hardness. Rawlings et al. (Rawling et al.,2007) used nitrogen to enhance the stability and chemical durability of certain fluoride-phosphate glasses in the M-Al-P-O-F-N system (M = Ba, Na). Hanifi et al. (Hanifi et al.,2007) studied the effects of nitrogen and fluorine on the thermal, physical, and mechanical properties of glasses in the $Ca_{28}Si_{57}Al_{15}$ system. They found that the addition of nitrogen increases the glass transition temperature (Tg) due to the enhanced network cross-linking density. In contrast, the addition of fluorine decreases Tg, as well as the crystallization and melting temperatures. These authors clearly demonstrate that fluorine substitution for oxygen has a negligible effect on density, compactness, and Young's modulus, although there appears to be a slight decrease in Vickers hardness. However, in terms of thermal properties, fluorine significantly lowers Tg and Tc. Rouxel (Rouxel.,2007) recently showed that when the glass transition temperature decreases, Young's modulus also decreases for all silicate, aluminosilicate, and oxynitride glasses. However, oxynitride fluoride glasses exhibit different trends, as Young's modulus values are unaffected by fluorine substitution even though Tg is substantially reduced. Hanifi et al. (Hanifi et al.,2007) investigated the effect of cations on the properties of oxynitride fluoride glasses in the Ca-Si-Al-O-N-F system. Contrary to expectations regarding calcium as a network modifier, these authors found that increasing the calcium content led to an increase in Tg, while Young's modulus and microhardness did not follow this trend. To understand the actual role of calcium in the structure and properties of these glasses, NMR-MAS studies enabled these authors to explain this increase. They observed that Young's modulus, microhardness, and Tg increased with the rise in aluminum content due to enhanced network connectivity.

2. EXPERIMENTAL METHODS

2.1. Preparation of the Glasses and Glass-Ceramics

Two systematic glass compositional series, labeled Series (I) and Series (II), were prepared and analyzed. These series are denoted GNx and GFNx, respectively, and their compositions in mole percent are: Series (I): (55−3x)SiO_2−13.5CaO−31.5Na_2O−xSi_3N_4, Series (II): (55−3x)SiO_2−8.5CaO−31.5Na_2O−5CaF_2−xSi_3N_4, where x ranges from 0 to 4, and 4x represents the amount of nitrogen introduced as Si_3N_4. Atomic percentages of Si, Ca, Na, O, F, and N for these two series are detailed in Table 1. Series (I) features a "bioglass" composition (excluding P_2O_5) within the quaternary system SiO_2−Na_2O−CaO−Si_3N_4. In contrast, Series (II) simply substitutes CaF_2 for CaO in Series (I), while maintaining a constant Na ratio. This design ensures that cation ratios remain unchanged regardless of nitrogen content, making the N ratio the sole compositional variable. Consequently, any potential effects of cation ratio changes on the glass structure and properties are effectively controlled. The preparation of the glasses involved three main steps. First, glasses based on Si–Na–Ca–O and Si–Na–Ca–O–F were synthesized by reacting and melting Na_2CO_3, $CaCO_3$, and SiO_2 in platinum crucibles at 1450°C in air for 4 hours for the Si–Na–Ca–O compositions. For Si–Na–Ca–O–F compositions, CaF_2 was added, and the mixture was melted at 1350°C for 30 minutes to prevent fluorine loss, as fluorine lowers the melting temperature of the oxide glass.

In the second step, the oxide and oxyfluoride glasses were ground and mixed with varying amounts of silicon nitride (Si_3N_4) powder. The oxygen content (2 wt%) present on the surface of the silicon nitride particles was considered when calculating the required amount of silicon nitride. Samples weighing 10 g were mixed in a glass dish with a magnetic stirrer using 50 ml of isopropanol, which was then evaporated using a hot plate. In the final step, the powder mixtures were pressed under 300 MPa pressure to form pellets with a height of 10 mm and a diameter of 26 mm. These samples were then melted in a boron nitride-lined graphite crucible in a vertical tube furnace under a flow of high-purity nitrogen at 1400°C for 15 minutes. The crucible was rapidly withdrawn from the hot zone. The resulting Si–Na–Ca–O–N and Si–Na–Ca–O–F–N glasses were annealed at a temperature slightly below their glass transition temperature for 6 hours before being slowly cooled to ambient temperature to relieve any stresses induced by rapid cooling.

Table 1. Atomic percentages of elements in synthesized Na-Ca-Si-O-N (GNx) and Na-Ca-Si-O-F-N (GFNx) bioactives glasses (T- Theoretical; E-Experimental)

Element (at%)	**Si**		**Ca**		**Na**		**O**		**F**		**N**	
	T	**E**	**T**	**E**	**T**	**E**	**T**	**E**	**T**	**E**	**T**	**E**
GN0	**19.20**	**19.21**	**4.71**	**4.71**	**21.99**	**21.91**	**54.10**	**54.17**	**-**	**-**	**0**	**0**
GN1	19.34	19.22	4.74	4.76	22.14	21.71	52.38	53.11	-	-	1.40	1.20
GN2	19.47	19.82	4.77	4.97	22.30	21.46	50.62	51.65	-	-	2.83	2.10
GN3	19.61	20.44	4.81	5.18	22.46	21.37	48.84	50.31	-	-	4.28	2.70
GN4	19.75	20.71	4.85	5.14	22.62	21.6	47.04	49.25	-	-	5.74	3.30
GFN0	18.87	21.87	4.63	3.87	21.61	22.87	51.46	48.97	3.43	2.42	0	0
GFN1	19.00	22.49	4.66	3.29	21.76	22.19	49.77	48.00	3.46	2.73	1.38	1.30

continued on following page

Table 1. Continued

Element (at%)	Si		Ca		Na		O		F		N	
	T	E	T	E	T	E	T	E	T	E	T	E
GN0	**19.20**	**19.21**	**4.71**	**4.71**	**21.99**	**21.91**	**54.10**	**54.17**	**-**	**-**	**0**	**0**
GFN2	19.13	22.77	4.70	3.24	21.92	21.9	48.29	46.75	3.48	3.04	2.48	2.30
GFN3	19.26	22.89	4.73	3.57	22.07	21.87	46.24	45.08	3.50	2.79	4.20	3.80
GFN4	19.40	23.00	4.76	3.58	22.32	21.58	44.45	44.56	3.51	2.78	5.65	4.50

- **Microscopical examination**

Cross-sectional analysis of the glasses was performed using Scanning Electron Microscopy to assess their homogeneity and identify the presence of bubbles. Figure 5 illustrates the condition of nitrogen-free glasses (GN0 and GFN0) and glasses with maximum nitrogen content (GN4 and GFN4). The glasses that were free of nitrogen but contained fluorine (F) showed no bubbles. In contrast, the glasses without fluorine, which experienced CO_2 loss during melting, displayed some fine residual bubbles. Fluorine lowers both melting temperatures and viscosities, facilitating the release of gases during the melting process. On the other hand, glasses with nitrogen always exhibited evidence of bubbles, resulting from both CO_2 loss and some nitrogen loss, which led to macroporosity in the glasses. These observations have implications for the subsequent characterization of their mechanical properties.

Figure 5. Low magnification SEM micrographs of cross sections of glasses: (a) GN0, (b) GFN0), (c) GN4, (d) GFN4, showing presence of bubbles

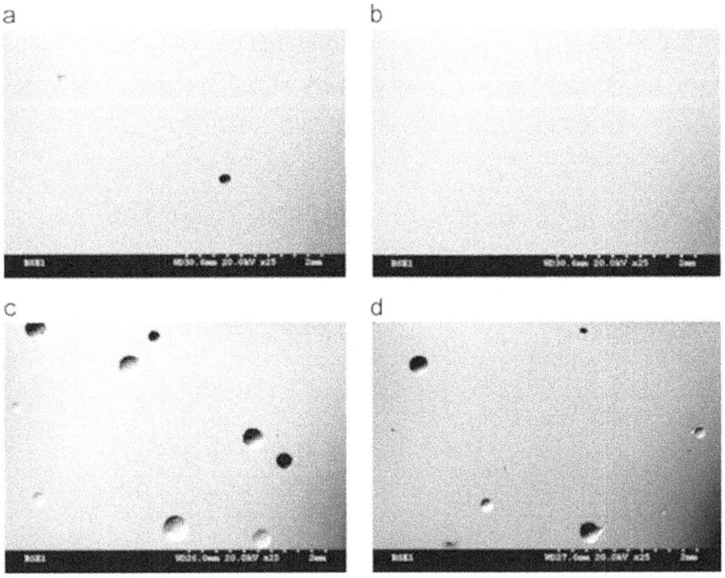

2.2. X-Ray Diffraction Analysis (XRD)

To identify the presence of crystalline phases within the samples, X-ray diffraction (XRD) analysis was conducted. The analysis was performed using a PanAlytical X-ray diffractometer, manufactured in the Netherlands. The instrument employed monochromated CuKα radiation with a wavelength of λ = 1.54056 Å. The diffraction patterns were recorded over a 2θ range of 20° to 80°, with a scan speed of 2.41° per minute. The resulting data were subsequently processed and analyzed utilizing X'pert Quantify software to elucidate the crystalline structure and phase composition of the samples.

2.3. Density, Compactness and Molar Volume

The density of the samples was determined employing a Helium pycnometer. Prior to the density measurement, the glasses were subjected to crushing to eliminate any potential internal porosity, such as bubbles, which could otherwise influence the accuracy of the results. This preparation ensured that the measurement reflected the true density of the material, free from the effects of internal voids or imperfections.

2.4. Elastic Moduli

The elastic modulus of the material was evaluated using two distinct methodologies: an ultrasonic technique and the Knoop indentation method. The ultrasonic method follows the procedures outlined by Hanifi et al. (2012). In conjunction with this, the Knoop indentation method was employed, as described by Marshall et al. (1980). This method involves calculating the ratio of hardness (HK) to elastic modulus (E) based on measurements of the diagonals of the Knoop indentations.

Marshall et al. (1980) detail a procedure for deriving the HK/E ratio from the minor and major diagonals of the Knoop indent. Specifically, the reduction in the length of the indent's minor diagonal, attributable to the elastic recovery of the material, is related to the hardness-to-modulus ratio through the following equation:

$$\frac{W'}{L'} = \frac{W}{L} - \frac{HK}{E}$$

Here, w' and L' represent the minor and major diagonals of the actual Knoop indent, respectively, while w and L denote the minor and major diagonals of the ideal Knoop indent. For the ideal Knoop indent, the ratio L/w is set at 7.11, and α is a constant with a value of 0.45, as specified by Marshall et al. (1980). The Knoop hardness (HK) is computed using the following relationship:

$$HK = 14.229\frac{P}{L^2}$$

where P is the applied load .

Ultrasonic Measurement: The speed of ultrasonic waves in solids, similar to density and compressibility, primarily depends on intermolecular and intramolecular interaction potentials. In the case of glasses, this potential largely depends on the grain consolidation, which refers to the number of bubbles formed during the melting process of these glasses. The objective is to measure the propagation speed of both transverse and longitudinal ultrasonic waves to determine mechanical properties such as Young's modulus (E). The system used for this measurement includes a pulse generator that provides the initial electrical signal. This generator is connected to an ultrasonic probe and a digital oscilloscope. The ultrasonic probe serves two main functions: converting the electrical signal into an ultrasonic wave and capturing the reflected ultrasonic waves to transform them back into electrical signals. The available probes include longitudinal wave transducers with various frequencies and transverse wave transducers with different frequencies. The digital oscilloscope is used to visualize both the initial electrical signal and the electrical signals corresponding to the echoes, producing what is known as an echogram. A vernier caliper is used for the precise measurement of the thicknesses of the analyzed glasses.

2.5. Microhardness

The microhardness of the samples under investigation was determined using a Vickers hardness indenter. The preparation of the specimens involved cutting with a low-speed diamond saw, followed by dry grinding using 1200-grit silicon carbide (SiC) paper. The specimens were then polished meticulously with 6, 3, and 1 μm diamond pastes to achieve smooth, flat, and parallel surfaces, ensuring optimal conditions for accurate indentation measurements. To investigate the effect of test load on hardness known as the Indentation Size Effect (ISE) a range of loads from 25 g to 500 g was employed, with a fixed loading duration of 15 seconds for each measurement. For each load level, at least five indentations were measured per sample to ensure reliability. However, at the highest load of 500 g, ten indentations were performed to facilitate a robust calculation of the mean true Vickers hardness number (HV_0) and its standard deviation. HV_0 was determined based on the consistency of measured values with varying loads. All measurements were conducted under standard atmospheric conditions. The Vickers hardness values, expressed in gigapascals (GPa), were calculated using the following equation:

$$HV = 1.8544\frac{P}{d^2}$$

Where P is the applied load and d is the average length of the two diagonals of the indentation. The measurements of these diagonals were carried out using an eyepiece on the microscope of the Vickers hardness testing equipment, which provided an estimated accuracy of 0.5 μm for the diagonal measurements.

2.6. Thermal Analysis

Differential Thermal Analysis (DTA) was performed using a Setaram Setsys 16/18 simultaneous TG/DTA analyzer to determine the glass transition temperature (Tg) of the samples. Each sample, weighing approximately 50 mg, was subjected to a controlled heating rate of 10°C/min, extending to a maximum temperature of 1200°C. The analysis was conducted in alumina crucibles within a flowing nitrogen atmosphere to prevent oxidation and ensure accurate results. Alumina (Al_2O_3) served as the reference material for calibration. The glass transition temperature, Tg, was identified from the DTA curves by locating the inflection point where the endothermic drift occurs. This point represents the temperature at which the sample transitions from a hard and brittle state to a more pliable and rubber-like state, providing crucial information on the thermal properties of the glass.

3. STRUCTURE AND PROPERTIES OF BIOGLASSES

3.1. Structural Characterizations by ^{29}Si NMR Spectroscopy

- **^{29}Si MAS-NMR**

Figure 6 presents the ^{29}Si Magic Angle Spinning Nuclear Magnetic Resonance (MAS-NMR) spectra for GN0, GFN0, GN4, and GFN4 glasses, with Table 2 summarizing the results of the deconvolution analysis. This analysis provides insight into the relative contributions of different Q^n structural units within the spectra of the two-glass series.

In the spectrum of the GN0 glass, which does not contain nitrogen (Figure 6a), the primary peaks are observed around −78.1 ppm and −85.3 ppm. These peaks are attributed to Q^2 (37%) and Q^3 (63%) units, respectively, as described by Mackenzie and Smith (2002). The introduction of nitrogen into the anionic network results in the emergence of additional silicon environments, evidenced by two new bands around −70 ppm and −59 ppm. These bands are assigned to [SiO_3N] and [SiO_2N_2] units, respectively (Li et al., 2024). The chemical shifts and corresponding percentages for each structural unit are detailed in Table 2.

Figure 6. Deconvolution of ^{29}Si NMR MAS spectra of (a) GN0 and GN4 and (b) GFN0 and GFN4 glasses

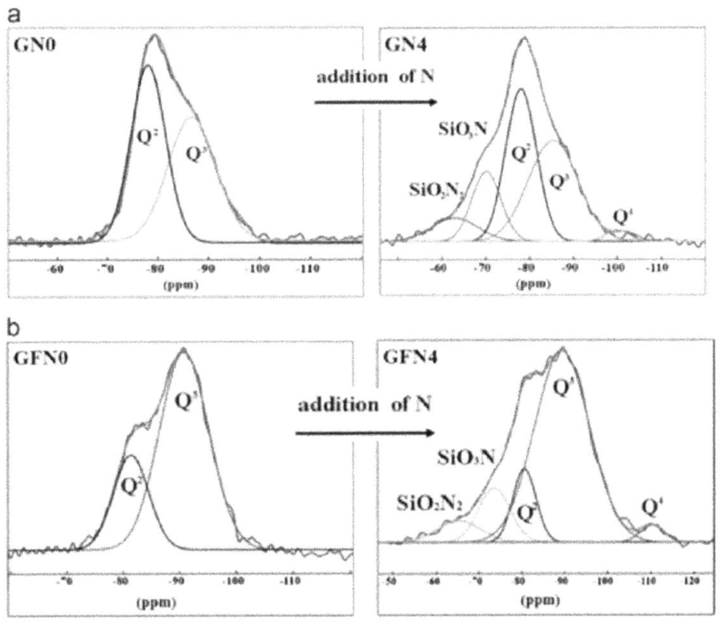

Table 2. Chemical shifts (δ) and percentages of Q^2,Q^3,Q^4, SiO_3N and SiO_2N_2 structural units in GNx and GFNx bioactive glasses

Glasses	Structural	δ_{iso}iso (^{29}Si) ppm	Percentage (%)	Glasses	Structural	δ_{iso}iso (^{29}Si) Ppm	Percentage (%)	Population (±5%)
GN0	Q^2	-78	37.1	**GFN0**	Q^2	-81	31.3	31.1
	Q^3	-85	62.9		Q^3	-88	68.9	68.9
GN1	SiO_3N	-71	5.9	**GFN1**	SiO_3N	-71	5.1	5.1
	Q^2	-78	36.7		Q^2	-81	28.7	28.7
	Q^3	-85	57.4		Q^3	-89	61.4	61.4
GN2	SiO_2N_2	-59	1.9		Q^4	-100	4.8	4.8
	SiO_3N	-70	16.4	**GFN2**	SiO_3N	-71	16.4	16.4
	Q^2	-78	34.8		Q^2	-80	24.8	24.8
	Q^3	-85	43.8		Q^3	-89	53.7	53.7
	Q^4	-95	3.1		Q^4	-104	5.1	5.1
GN3	SiO_2N_2	-59	3.6	**GFN3**	SiO_2N_2	-63	3.6	3.6
	SiO_3N	-70	22.2		SiO_3N	-72	20.2	20.2
	Q^2	-78	33.2		Q^2	-80	23.2	23.2
	Q^3	-85	35.7		Q^3	-90	47.7	47.7
	Q^4	-95	5.3		Q^4	-106	5.3	5.3
GN4	SiO_2N_2	-60	9.4	**GFN4**	SiO_2N_2	-60	9.4	9.4
	SiO_3N	-70	18.6		SiO_3N	-70	18.6	18.6
	Q^2	-78	32.7		Q^2	-78	22.7	22.7
	Q^3	-85	33.5		Q^3	-85	43.5	43.5
	Q^4	-95	5.8		Q^4	-95	5.8	5.8

Figure 7 illustrates the changes in the populations of Q^2, Q^3, Q^4, SiO_3N, and SiO_2N_2 units as a function of nitrogen content for the GNx and GFNx glass series. Initially, the incorporation of nitrogen into the GNx series (Figure 7a) leads to the formation of SiO_3N tetrahedra and a corresponding decrease in Q^3 units. With further nitrogen addition, there is an increase in SiO_3N tetrahedra, a reduction in Q^3 units, and the appearance of small amounts of Q^4 units and SiO_2N_2 tetrahedra. As nitrogen content increases further, there is a continued, albeit slight, increase in Q4 units and SiO_2N_2 tetrahedra, accompanied by a decrease in SiO_3N tetrahedra and Q^3 units. The quantity of Q_2 units remains relatively stable despite the increasing nitrogen content.

Figure 7. Effect of nitrogen content on percentages of Q^2, Q^3, Q^4, SiO_3N and SiO_2N_2 for (a) GNx and (b) GFNx glasses

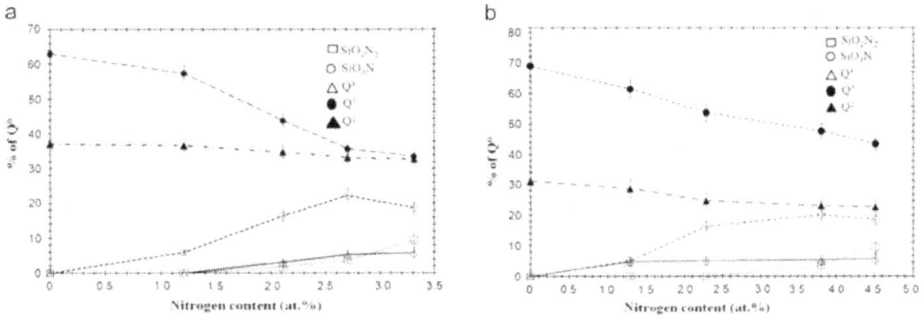

Figure 6b. displays the ^{29}Si MAS-NMR spectra for GFN0 and GFN4 glasses. In the spectrum of GFN0, the main contributions are found around −81 ppm and −88 ppm, with these peaks attributed to Q^2(31%) and Q^3(69%) units, respectively. Although these values are slightly shifted compared to those observed for GN0 glass, the proportions remain similar (Mackenzie and Smith, 2002). The introduction of nitrogen into the anionic network of these oxyfluoride glasses results in the appearance of three new bands around −100 ppm, −71 ppm, and −63 ppm, which are attributed to Q^4, [SiO_3N], and [SiO_2N_2] units, respectively (Mercier et al., 2011). As shown in Figure 7b, the initial addition of nitrogen to these oxyfluoride glasses leads to the formation of SiO_3N tetrahedra and some Q^4 units, with a reduction in Q^3 units. Further nitrogen addition results in an increase in SiO_3N tetrahedra, a decrease in Q^3 units, similar amounts of Q^4 units, and the appearance of SiO_2N_2 tetrahedra. Continued nitrogen incorporation causes additional small increases in Q^4 units and SiO_2N_2 tetrahedra, while the quantity of Q^2 units decreases only marginally.

• ^{9}F MAS NMR

Figure 8. displays the ^{19}F Magic Angle Spinning Nuclear Magnetic Resonance (MAS-NMR) spectra for the GFNx glasses, as well as for their oxide precursor glasses that were prepared with fluorine prior to the addition of Si_3N_4. The spectra of these glasses are consistent before and after the introduction of nitrogen, indicating that nitrogen content does not significantly alter the chemical environment of fluorine within the glasses.

Figure 8. ^{19}F MAS NMR spectra of GFNx glasses (before addition of nitrogen) and of GFNx glasses (after addition of nitrogen)

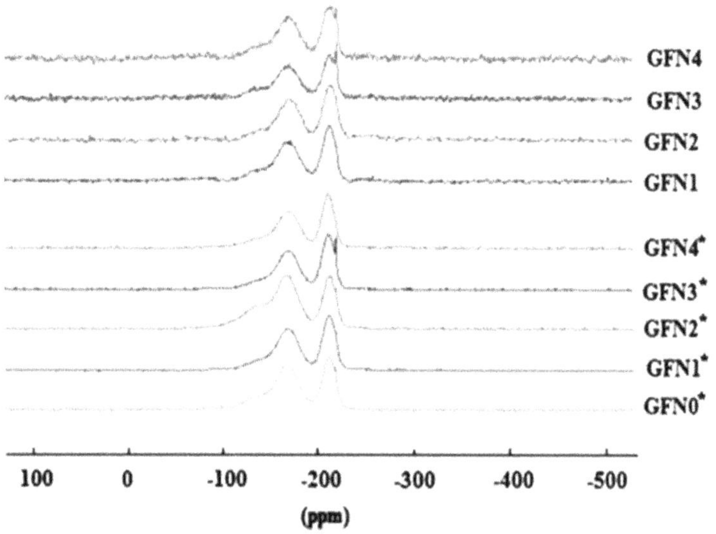

The spectra reveal several distinct chemical environments around the fluorine atoms. Specifically, three primary peaks are observed at approximately −140 ppm, −174 ppm, and −216 ppm. Additionally, a smaller peak is observed at −224 ppm, although its intensity varies with composition and is virtually absent in the GFN4 sample. The peak at −140 ppm is assigned to the F-Ca(3)Na(1) and F-Ca(2)Na(2) species, while the peak at −174 ppm is attributed to F-Ca(1)Na(4). The peak at −216 ppm corresponds to the F-Na(6) species (Brauer et al. 2009). The narrow peak at −224 ppm is tentatively attributed to nano-crystalline NaF dispersed within the glass network.

3.2. Physicochemical Properties of Si-Ca-Na-O-N and Si-Ca-Na-O-F-N Glasses

- **Density and Molar Volume**

Figure 9. demonstrates the effect of substituting oxygen with nitrogen and fluorine on the density of GNx and GFNx glasses. Substituting nitrogen significantly increases the density: in the GNx series, density rises from 2.585 g/cm^3 at 0 at% N to 2.612 g/cm^3 at 3.3 at% N (a 0.9% increase), and further to 2.634 g/cm^3 at 4.5 at% N in the GFNx series (a 1.7% increase). Conversely, substituting oxygen with fluorine leads to negligible changes in density. This indicates that nitrogen incorporation has a more pronounced effect on increasing glass density compared to fluorine substitution.

Figure 9. Density of Na–Ca–Si–O–N (GNx) and Na-Ca-Si-O-F-N (GFNx) bioactive glasses as a function of nitrogen content

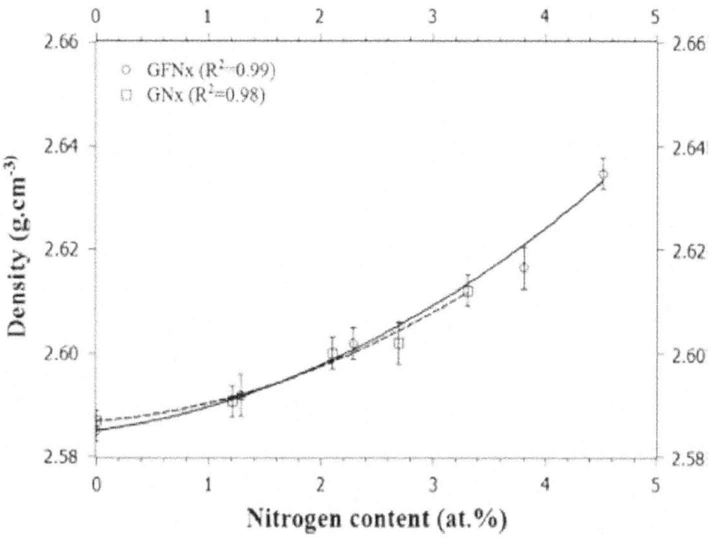

The observed density increase is associated with a decrease in the glass molar volume, which signifies a rise in the glass's compactness or atomic packing density. These findings corroborate previous studies (Hampshire et al., 1985; Hanifi et al., 2012) that demonstrate nitrogen's role in enhancing the cross-linking of oxynitride glasses. Nitrogen predominantly forms tri-coordinated bonds, whereas oxygen bonds with only two tetrahedral units. Consequently, for each percentage of oxygen replaced by nitrogen, an equivalent number of additional cross-links are introduced into the glass network. This increment in cross-linking should lead to a linear correlation with the nitrogen content, as discussed in subsequent sections. The introduction of fluorine, while expected to decrease network connectivity due to its heavier atomic weight compared to oxygen and nitrogen, results in only minimal changes in density. This is due to the limited impact of fluorine on the overall density relative to the effects of nitrogen.

- **Glass Transition Temperature (Tg)**

The glass transition temperatures (Tg) were determined from Differential Thermal Analysis (DTA) experiments. As depicted in Figure 10, the Tg exhibits a linear increase with rising nitrogen content, whereas it decreases with increasing fluorine content in the glass matrix. Specifically, in the GNx series, Tg increases from 508°C at 0 at% nitrogen to 546°C at 3.3 at% nitrogen. Similarly, in the oxyfluoronitride GFNx series, Tg rises from 475°C at 0 at% nitrogen to 538°C at 4.5 at% nitrogen. The data for both glass series were analyzed to derive empirical expressions relating Tg to nitrogen content. A best-fit line through the experimental data allows the formulation of these expressions, reflecting the dependence of Tg on the concentration of nitrogen in the glass composition.

For GNx series: Tg (°C)= 506+11.13[N]

For GFNx Tg (°C) = 407+14.04[N]

Where [N] represents the nitrogen content in atomic percent (at%).

Figure 10. Glass transition temperature of Na–Ca–Si–O–N (GNx) and Na–Ca–Si–O–F–N (GFNx) bioactive glasses as a function of nitrogen content

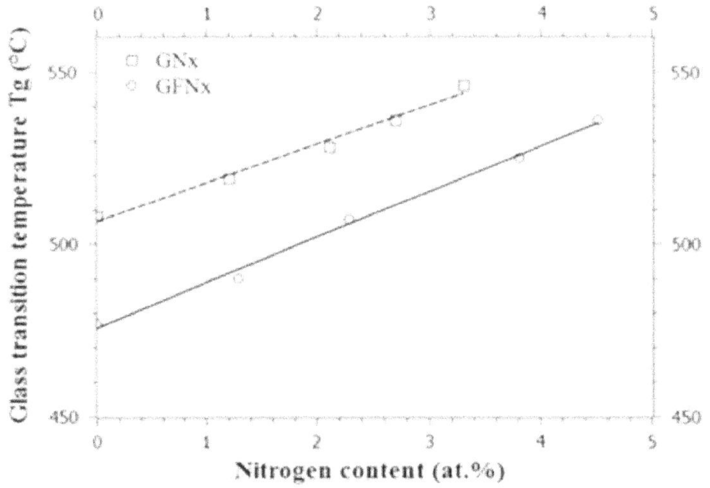

Similar behavior has been previously observed for oxyfluoride glasses (Hill et al., 2005) and oxyfluoronitride glasses (Hanifi et al., 2009, 2011). In these systems, non-bridging fluorine ions tend to disrupt the glass network structure. The presence of these weaker bonds results in a reduction in Tg. On the other hand, the observed increase in Tg with nitrogen content is in line with previous studies of oxynitride glasses (Hampshire et al., 1985; Hampshire and Pomeroy, 2008). These studies indicate that nitrogen incorporation leads to the formation of additional cross-linkages within the glass network, thereby enhancing its rigidity and increasing Tg. The decrease in Tg with increasing fluorine content suggests that the addition of fluorine facilitates the formation of lower viscosity melts. This characteristic makes the glass easier to process at lower temperatures, thus simplifying the glass production process.

3.3. Mechanical Properties Si-Ca-Na-O-N and Si-Ca-Na-O-F-N

Figure 11. illustrates the Vickers microhardness (HV) as a function of nitrogen content for GNx and GFNx glasses. The data demonstrate a linear increase in HV with increasing nitrogen content, consistent with previous observations for oxynitride glasses (Hampshire, 2008; Hampshire et al., 1985). Furthermore, the addition of fluorine does not significantly influence microhardness, corroborating findings from studies on oxyfluoronitride glasses (Hanifi et al., 2009, 2012).

Figure 11. Vickers hardness of Na–Ca–Si–O–N (GNx) and Na–Ca Si–O–F–N (GFNx) bioactive glasses as a function of nitrogen content

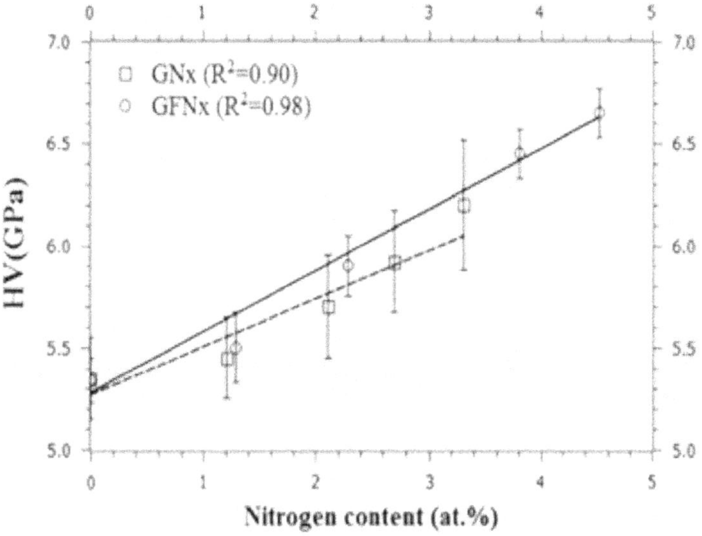

Specifically, for the GNx glass series, the microhardness increases from 5.37 ± 0.2 GPa at 0 at% nitrogen (with and without fluorine) to 6.27 ± 0.3 GPa at 3.3 at% nitrogen, reflecting a 15% increase. Similarly, for the GFNx glass series, HV rises from 5.37 ± 0.2 GPa at 0 at% nitrogen to 6.77 ± 0.1 GPa at 4.5 at% nitrogen, indicating a 24% increase. The standard deviations of the microhardness measurements also increase with nitrogen content. This variation is attributed to the presence of bubbles trapped within the glass, which tend to become more prevalent with higher nitrogen content. Despite efforts to perform indentation tests away from visible surface bubbles, some interaction between trapped bubbles and the indentation deformation zone may still influence the results. Microhardness can be modeled using the following empirical relationships:

For GNx series: HV(GPa) = 5.3+0.25[N]

For GFNx series: HV(GPa) = 5.3+0.30[N]

Where [N] is N content in at%.

Table 3 presents the elastic modulus values measured using both the Knoop indentation and ultrasonic methods. Figure 12. illustrates these values as a function of nitrogen content for GNx and GFNx glasses. The data reveal a linear increase in elastic modulus (E) with nitrogen content for both measurement techniques, consistent with findings from previous studies on oxynitride glasses (Bachar et al., 2011; Hampshire, 2008; Hampshire et al., 1985).

Table 3. Values of elastic modulus (E) measured by Knoop indentation and an ultrasonic method for GNx and GFNx bioactive glasses

GNx glasses	**GN0**	**GN1**	**GN2**	**GN3**	**GN4**
E knoop (GPa)	64±3	69±4	76±8	81±11	85±5
E Ultrasonic (GPa)	63±2	66±1	69±3	71±3	-
GFNx glasses	**GFN0**	**GFN1**	**GFN2**	**GFN3**	**GFN4**
E knoop (GPa)	64±2	72±3	81±3	90±4	93±6
E Ultrasonic (GPa)	63±1	70±2	79±1	87±2	91±2

For the GN4 glass, obtaining a value using the ultrasonic method proved challenging, which may be attributed to the higher quantity of bubbles present in these samples compared to the oxyfluoronitride glasses. Despite this difficulty, the elastic modulus values obtained by the ultrasonic method are generally within the standard deviation range of those measured by the indentation method, indicating that both methods provide equivalent results. However, the data for GNx glasses measured by the ultrasonic method show greater deviation, and the experimental errors are notably higher. This increased variability is likely due to the higher volume of bubbles present in these glasses. The standard deviations for the elastic modulus measurements obtained using the indentation method are substantial, reflecting the impact of bubble content on measurement accuracy.

Figure 12. Elastic modulus of Na–Ca–Si–O–N (GNx) and Na–Ca Si–O–F–N (GFNx) bioactive glasses as a function of nitrogen content

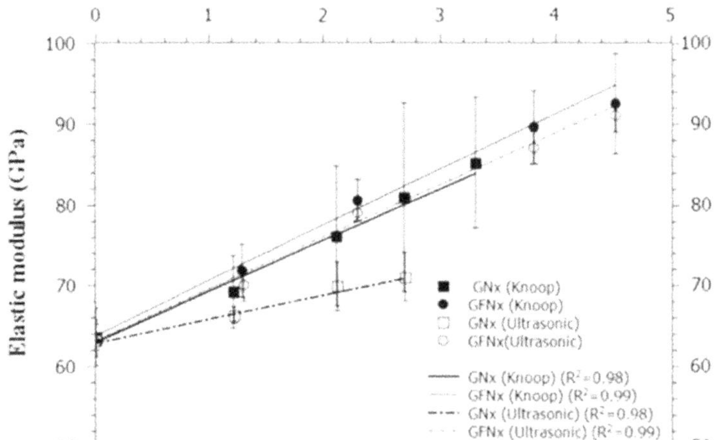

The increased variability in elastic modulus measurements, particularly with the indentation method, is attributed to the challenges associated with accurately measuring the smaller diagonal of some Knoop indents, even with the aid of image analysis. Specifically, the elastic modulus values measured using the indentation method show a clear trend of increasing with nitrogen content. For the GNx series, the elastic modulus rises from 64.73 GPa at 0 at% nitrogen (both with and without fluorine) to 85.78 GPa at 3.0 at% nitrogen, reflecting a 33% increase. In the GFNx series, the modulus increases from 64.73 GPa at 0 at% nitrogen to 93.76 GPa at 4.5 at% nitrogen, representing a 45% increase. This trend appears to be linear and can be modeled by the following empirical relationships:

For GNx series: E(GPa)= 64+6.55[N]

For GFNx series: E(GPa)= 64+4.30[N]

Where [N] represents the nitrogen content in atomic percent.

Similar to microhardness, the addition of fluorine does not have a significant impact on the elastic modulus (E). The observed increases in the maximum values of both properties for the oxyfluoronitride glasses (GFNx series) are attributed to the higher nitrogen content present in the fluorine-containing glasses.

4. BIOACTIVITY AND VIABILITY OF GLASSES

4.1. Bioactivity of Si-Ca-Na-O-N and Si-Ca-Na-O-F-N Glasses

To facilitate bonding between a biomaterial and living bone, it is crucial that an apatite layer forms on the biomaterial's surface within the body fluid environment. Figure 13. presents scanning electron micrographs (SEM) illustrating the surfaces of two bioactive glass series, specifically GN2 and GFN2, before and after 15 days of immersion in simulated body fluid (SBF) solution. Images (a) and (b) depict the surfaces of GN2 and GFN2, respectively, showing the development of a crystalline layer following immersion. Energy-dispersive spectroscopy (EDS) analysis reveals the presence of calcium (Ca) and phosphorus (P), with Ca ratios of 1.64 for GN2 and 1.72 for GFN2.

Figure 13. EDS spectra and micrographs of (a) GN2 and (b) GFN2 bioactive glasses before and after 15 days soaking in SBF solution

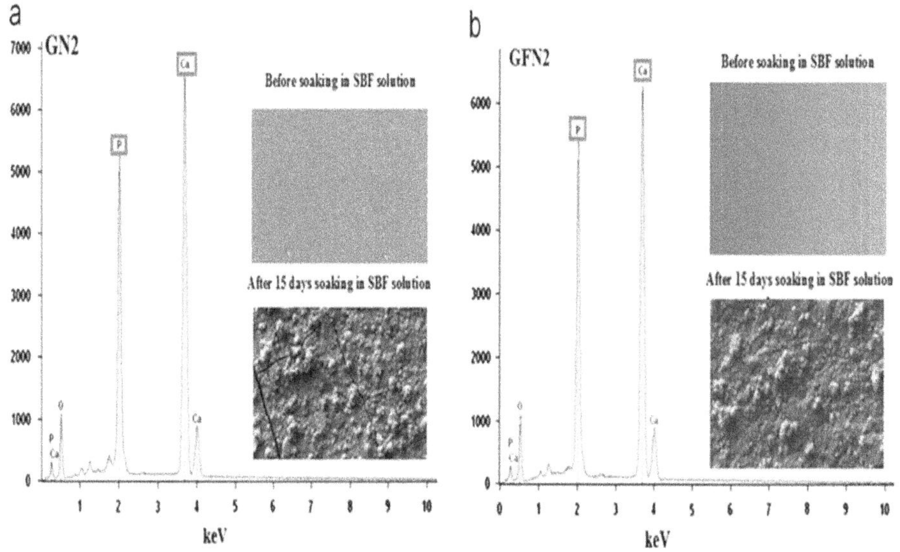

These ratios align closely with the theoretical calcium ratio of 1.67 expected for hydroxyapatite. The calcium phosphate crystals that formed on both the base oxide and oxynitride glasses are roughly 200-300 nm in diameter, showing a consistent and homogeneous morphology across the surface of the glasses. X-ray diffraction (XRD) patterns of the glasses prior to SBF immersion display an amorphous halo, indicating that the glasses were initially amorphous. Post-immersion XRD patterns for all GNx and GFNx samples are illustrated in Figure 14. and 15. A reference pattern for natural hydroxyapatite (JCPDS pattern no. 90432) is used for comparison. The diffraction patterns confirm the formation of a hydroxyapatite layer on all glass samples, as evidenced by the three most prominent diffraction peaks at 26.301°, 32.271°, and 53.151° 2θ, corresponding to the (002), (112), and (004) planes, respectively. Notably, the intensities of the major peaks (002) and (112) diminish with increasing nitrogen content, suggesting a reduction in crystallinity with higher nitrogen levels and indicating that nitrogen may adversely affect bioactivity.

Figure 14. X-ray diffraction patterns of GNx bioactive glasses after soaking in SBF solution for 15 days

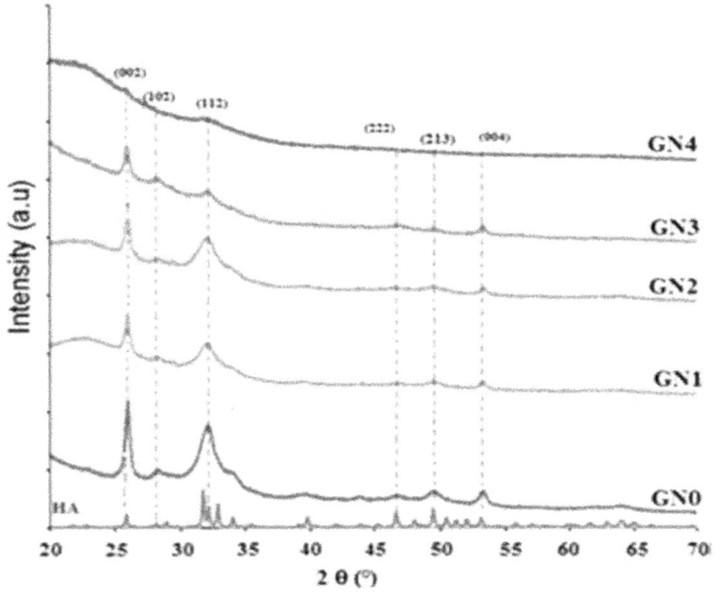

Figure 15. X-ray diffraction patterns of GFNx bioactive glasses after soaking in SBF solution for 15 days

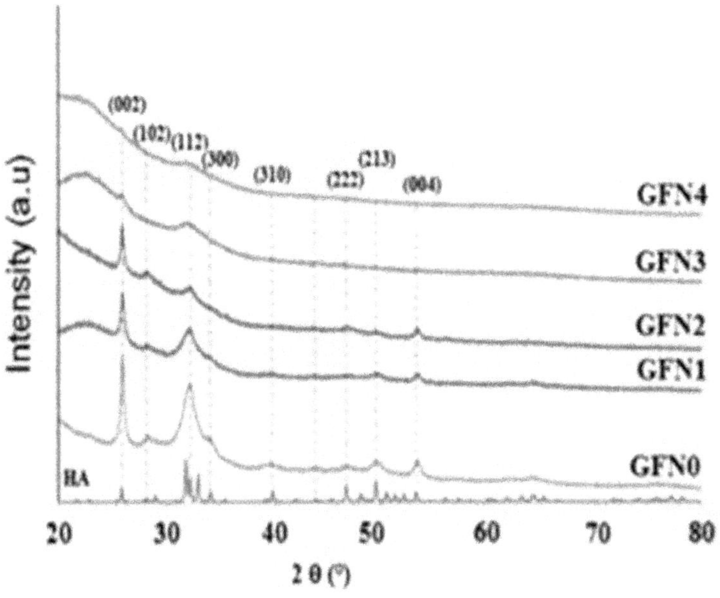

4.2. Cytotoxicity of Glasses

By definition, the organs and tissues of the human body maintain a specific three-dimensional architecture. Effective replacement or repair of such tissues therefore requires signals that enable cells to assemble and organize into a functional structure. Functional biomaterials, such as glasses, can be designed to provide these signals, which involve appropriate morphogenesis. In practice, this necessitates interaction between cells and the implanted substrate [156]. It has been established that for an implantable material to chemically bond to living bone, the formation of a biologically active hydroxyapatite (HA) layer on its surface within the living environment is necessary. These so-called active materials, such as glasses and glass-ceramics, can then influence cell attachment, proliferation, differentiation, and the integration of the material into the host tissue. Indeed, bioactive materials are capable of releasing ions that can affect cellular responses (Loty et al.,2001). The field addressed in this chapter pertains to a biological approach to bio-glasses. The vitreous materials produced are tested using cell cultures. The analyses performed include cytotoxicity assays. These are fundamental biological tests that provide an initial evaluation of a material's compatibility with living tissue. Specifically, it is necessary to ensure that bio-glasses support the maintenance of human bone cell viability. In vitro tests for evaluating the cytotoxicity of implantable materials involve culturing cells on these materials and assessing their toxicity. The direct contact test, the simplest method, involves placing a sample of the material on top of a cell layer to observe potential toxic effects on cell growth and integrity. Cytocompatibility tests have been conducted on 45S5 glass (Bouami et al.,2024), a well-studied bio-glass, which demonstrated that it induces cellular alkalinization due to the dissolution of the glass. This alkalinization results from an enrichment in Ca^{2+} and Na^{+} cations and a decrease in H^{+} ions but does not affect osteoblast viability or proliferation. In vitro cytotoxicity tests are governed by international standards (ISO 10993-5/EN 30993-511) and assess cellular viability to determine the compatibility of biomaterials.

- **Study of Biofilm Formation on Oxynitride and Oxyfluoronitride Glasses**

Biofilms are complex communities of microorganisms attached to a surface and linked by an organic polysaccharide matrix. Their formation depends on the surface and the surrounding physicochemical environment. Biofilms can develop in various settings, including riverbeds, inside water pipes, intestinal mucosa, venous catheters, and heart valves. They can also be referred to as consortia when they include a variety of species working together and capable of diverse metabolic transformations, such as biodegradation. The study of biofilm formation is crucial in microbiology because bacteria prefer to reproduce on surfaces rather than in liquid phases. Biofilms provide a protective structure against environmental stresses. Although Van Leeuwenhoek was the first to examine bacterial biofilms on his own teeth in the 17th century, the first scientific study dedicated to biofilms dates back to 1943 (Zobell.,1943). Since then, biofilms have been observed in numerous environments. However, it was not until 1978 that the prevalence of the biofilm lifestyle was established. It has been shown that over 99% of bacteria develop in biofilms on a wide range of surfaces, including metals, plastics, living tissues (human tissues, plant leaves and roots), and mineral surfaces (stones, concrete) (Costerton et al.,2003). A biofilm can consist of a single bacterial species or multiple bacterial species, as well as eukaryotic cells such as microscopic fungi, algae, and protozoa (Buzalewicz et al.,2024).

- **Biofilm Formation Tests**

The biofilm formation tests utilize visible spectroscopy to measure the amount of biofilm adhered to a surface, based on the staining of Gram-negative bacteria with crystal violet (Bakke et al.,2001). This dye binds to the cytoplasmic components of the bacteria, and its retention or loss after alcohol treatment helps distinguish Gram-positive bacteria (which retain the dye) from Gram-negative bacteria (which lose the dye). The testing protocol involves incubating a material in a culture medium containing Gram-negative bacteria, followed by washing and staining with 1% crystal violet. The amount of biofilm is quantified using a spectrophotometer to measure absorbance at 570 nm. Additionally, control samples are used to account for crystal violet binding to the material itself. This allows for accurate assessment of biofilm quantity relative to the material's specific surface area.

- **Application to Oxynitrided Bioactive Glasses of Si-Ca-Na-N and Si-Ca-Na-N-F Systems**

Biofilm formation tests were conducted on oxynitrided and oxyfluoronitrided glasses with varying nitrogen content, as well as on Hench glass, a reference material (Bachar et al.,2024). The bacterial strains used were *Escherichia coli* and *Serratia marcescens*, both Gram-negative. Bacterial cultures were grown in LB medium at 37°C until reaching the stationary growth phase. Glass samples were sterilized with 0.05% bleach and rinsed. Biofilms were formed by exposing the samples to a bacterial suspension. Bacterial concentration was measured by optical density (OD) at 620 nm, with dilutions used to promote biofilm formation. The tests included control samples (without bacteria) and test samples (with bacterial suspension) for statistical analysis. The results of tests performed on the bioactive glasses, using both *E. coli* and *Serratia marcescens*, are analyzed and discussed to understand the behavior of the glasses in contact with the bacteria. The results are detailed in Tables 4 (*E. coli*) and 5 (*Serratia marcescens*).

Table 4. Results of the tests conducted on resin-coated glasses with Escherichia coli

Glasses	Average OD (Controls)	Average OD (Tests)	Average OD (Biofim)	Average Surface (cm²)	Biofilm/cm²
Hench	2,200	2,228	0,028	1,792	0,0156
G1N0	0,294	0,861	0,568	1,792	0,317
G1N2	0,387	1,449	1,064	2,777	0,383
G1N4	0,293	2,181	1,888	2,654	0,711
G2FN0	2,157	1,994	0	1,792	0
G2FN2	0,576	0,82	0,243	2,987	0,0814
G2FN4	2,149	2,022	0	2,857	0

Table 5. Results of the tests conducted on resin-coated glasses with Serratia marcescens

Glasses	Average OD (Controls)	Average OD (Tests)	Average OD (Biofim)	Average Surface (cm²)	Biofilm/cm²
Hench	2,200	2,202	0,002	1,792	0,00130
G1N0	0,294	0,463	0,169	1,792	0,0943
G1N2	0,386	0,999	0,614	2,777	0,233

continued on following page

Table 5. Continued

Glasses	Average OD (Controls)	Average OD (Tests)	Average OD (Biofim)	Average Surface (cm^2)	Biofilm/cm^2
G1N4	0,293	0,336	0,0433	2,654	0,0186
G2N0	2,157	1,707	0	1,792	0
G2FN2	0,576	1,919	1,343	2,987	0,420
G2FN4	2,149	2,205	0,0555	2,857	0,0434

Figure 16. Shows the progression of biofilm formation by the two tested bacteria on the glasses G1N0, G1N2, G1N4, G2N0, G2N2, G2N4, and the Hench glass

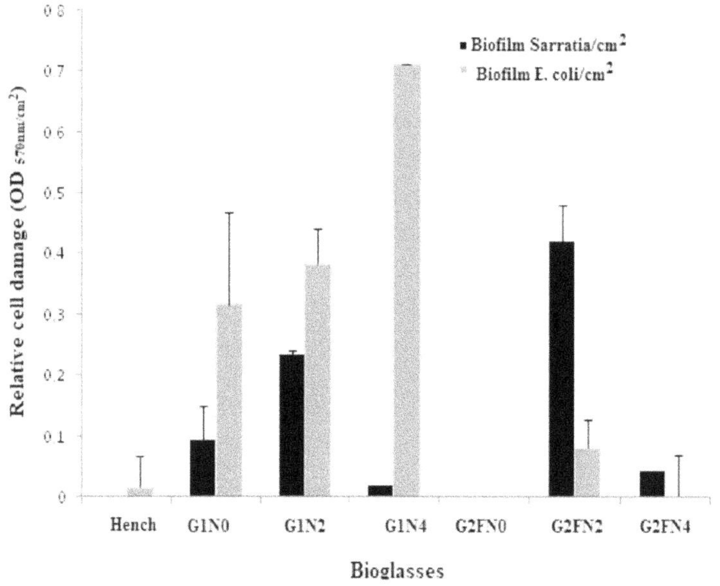

The results indicate that both bacteria produce almost no biofilm on Hench glass, consistent with previous studies on this material. This confirms the validity of the control glass tests. The G1Nx glasses show increased biofilm formation with Escherichia coli as the nitrogen content rises. However, Serratia marcescens forms very little biofilm on G1N4 glasses. Conversely, for G2FNx glasses, the behavior differs. G2FN2 glass generates the most biofilm when exposed to S. marcescens, while E. coli has a lesser effect. G2FN0 shows no biofilm development with either bacterium, or G2FN4 exhibits minimal adhesion with S. marcescens. Additionally, Serratia marcescens produces a red pigment, prodigiosin, under stress conditions, such as unfavorable environments.

To summarize, two series of oxynitride bioglasses were examined, maintaining constant Si:Ca ratios while varying the O ratio and including or excluding fluorine substitution for oxygen. The compositions were as follows (mole %): (I) $(55-3x)SiO_2-13.5CaO-31.5Na_2O-xSi_3N_4$ and (II) $(55-3x)SiO_2-8.5CaO-31.5Na_2O-5CaF_2-xSi_3N_4$ (with x = 0, 1, 2, 3, 4). The main findings are: Density: The density of the glasses increases with nitrogen substitution for oxygen due to the enhanced compactness of the glass network, leading to a reduced molar volume. Fluorine substitution does not significantly affect the density. Mechanical Properties: The glass transition temperature, hardness, elastic modulus, and indentation fracture resistance all increase linearly with nitrogen content. This trend indicates that nitrogen, which forms tri-coordinated bonds with silicon compared to the two-fold coordination of oxygen, introduces additional cross-linking and strengthens the glass network. Effect of Fluorine: Substituting fluorine for oxygen in oxynitride glasses lowers the glass transition temperature, allowing for the synthesis of lower viscosity melts at reduced temperatures, which facilitates the glass production process. Microhardness: The microhardness increases by 15% with nitrogen content in the oxynitride glass series and by 24% in the oxyfluoronitride glass series. Elastic Modulus: The elastic modulus increases by 33% with nitrogen content in the oxynitride glass series and by 45% in the oxyfluoronitride glass series. NMR Analysis: Characterization using ^{29}Si MAS-NMR indicates that the increased rigidity of the glass network is due to the formation of SiO_3N and SiO_2N_2 tetrahedra, along with Q^4 units with additional bridging anions replacing Q^3 units. The presence of nitrogen does not alter the chemical environment of fluorine, but fluorine is found in various environments: F-Ca(3)Na(1), F-Ca(2)Na(2), F-Ca(1)Na(4), and F-Na(6), as well as in nano-crystalline NaF dispersed within the glass network. Bioactivity: After immersing the glasses in simulated body fluid (SBF) for 15 days, a uniform layer of hydroxyapatite crystals with diameters of 200–300 nm formed on the surfaces of both series of glasses. The crystallinity of this layer decreases with increased nitrogen content, suggesting that nitrogen might inhibit bioactivity.

5. CONCLUSIONS

In this chapter, we comprehensively investigated the incorporation of nitrogen into bioactive glasses, specifically those with the molar compositions $(55\text{-}3x)SiO_2\text{-}13.5CaO\text{-}31.5Na_2O\text{-}xSi_3N_4$ and $(55\text{-}3x)SiO_2\text{-}8.5CaO\text{-}31.5Na_2O\text{-}5CaF_2\text{-}xSi_3N_4$. The study adopted a multifaceted approach, encompassing the synthesis and various characterizations—physicochemical, structural, mechanical, and biological—of the synthesized oxy-nitride and oxy-fluoro-nitride glasses. The results highlight significant advancements in understanding how nitrogen incorporation can alter both the properties and bioactivity of these materials, which is crucial for their application in healthcare settings.

Synthesis and Characterization

The synthesis of both oxynitride and oxyfluoronitride glasses was meticulously developed, resulting in materials with an amorphous and homogeneous structure. Wavelength Dispersive Spectroscopy (WDS) analyses confirmed successful nitrogen incorporation, showcasing how the addition of nitrogen influenced the glass structure. Notably, the inclusion of fluorine in oxynitride glasses facilitated the retention of higher nitrogen levels, which has important implications for the development of glass materials with tailored properties. Structural studies, particularly through ^{29}Si Magic-Angle Spinning Nuclear Magnetic Resonance (MAS-NMR), provided profound insights into the silicate network. The results revealed that oxide glasses without nitrogen were primarily composed of Q^2 and Q^3 silicate entities, whereas nitrogen-doped glasses exhibited additional silicate entities such as SiO3N and SiO_2N_2. These findings indicate that nitrogen not only alters the glass composition but also significantly impacts the bonding structure, with all incorporated nitrogen forming Si-N linkages. Furthermore, ^{19}F MAS-NMR analyses indicated that nitrogen insertion does not affect the chemical environment of fluorine, highlighting its role in maintaining the stability of the silicate network.

Impact of Nitrogen on Properties

The investigation into the impact of nitrogen on the physicochemical and mechanical properties of the glasses yielded compelling results. The incorporation of nitrogen resulted in substantial improvements across several key parameters, including an increase in glass transition temperature, density, hardness, and Young's modulus. The enhancement of these properties is attributed to the increased cross-linking within the glass network, as nitrogen, with its trivalent coordination, replaces divalent oxygen. This substitution leads to a more interconnected and rigid vitreous network, which not only bolsters the structural integrity of the glass but also enhances its mechanical strength and thermal stability. Quantitatively, the relationship between nitrogen content and glass transition temperature (Tg) was clearly defined. For glasses without fluorine, T_g was found to be correlated with nitrogen content through the equation T_g (G1Nx)=505.1+18.27×(%N), while for glasses with fluorine, the relationship was described by T_g (G2FNx)=473+21.14×(%N). The introduction of fluorine appeared to lower the glass transition temperature, indicating that while fluorine facilitates higher nitrogen incorporation, it may also introduce some trade-offs in thermal properties.

Bioactivity Evaluation

A significant aspect of this research was the thorough evaluation of the bioactivity of the synthesized glasses using in vitro tests in simulated physiological environments over various immersion durations and nitrogen content levels. The tests confirmed that the glasses maintained their bioactivity, as demonstrated by the formation of a calcium phosphate layer on their surfaces—an essential indicator of bioactivity. Interestingly, a slight reduction in bioactivity was observed with increasing nitrogen content. This finding is crucial as it suggests a balance must be struck between enhancing mechanical properties and maintaining bioactivity, particularly in applications where biocompatibility is vital. Complementary in vitro cellular studies further validated the bioactivity results, revealing that both oxynitride and oxyfluoronitride glasses exhibited no cytotoxic effects, regardless of nitrogen content. This is a pivotal outcome, as it confirms that the enhanced mechanical properties achieved through nitrogen doping do not compromise the safety and effectiveness of the glasses in biological settings.

Significance in the Broader Field

The findings presented in this chapter hold significant implications for the broader field of bioactive materials. The dual benefit of nitrogen doping—enhancing mechanical properties while preserving bioactivity—positions these glasses as promising candidates for a variety of applications in healthcare, particularly in the development of materials intended to mitigate nosocomial infections associated with biofilms. Biofilms, which form resilient polymeric matrices, are a significant challenge in healthcare, as they contribute to infections that are difficult to treat. The incorporation of nitrogen into bioactive glasses offers a potential strategy for developing materials that not only resist biofilm formation but also possess superior mechanical properties for surgical and implant applications. Moreover, this research contributes to a growing body of knowledge on how compositional modifications can lead to materials that meet the stringent requirements of modern healthcare applications. As the demand for more effective biocompatible materials increases, the insights gained from this study will be invaluable for guiding future research and development efforts.

In conclusion, the investigation into nitrogen-doped bioactive glasses underscores the potential for innovation in the field of biomaterials. By carefully balancing mechanical enhancements and bioactivity, we can develop advanced materials that better serve the needs of patients and healthcare providers alike, paving the way for improved patient outcomes and more effective treatment strategies.

REFERENCES

Aaseth, J., Shimshi, M., Gabrilove, J. L., & Birketvedt, G. S. (2004). Fluoride: A toxic or therapeutic agent in the treatment of osteoporosis? *The Journal of Trace Elements in Experimental Medicine*, 17(2), 83–92. DOI: 10.1002/jtra.10051

Ali, S., Wójcik, N. A., Hakeem, A. S., Gueguen, Y., & Karlsson, S. (2024). Effect of composition on the thermal properties and structure of M-Al-Si-O-N glasses, M = Na, Mg, Ca. *Progress in Solid State Chemistry*, 74, 100461. DOI: 10.1016/j.progsolidstchem.2024.100461

Bachar, A., Catteaux, R., Duée, C., Désanglois, F., Lebecq, I., Mercier, C., & Follet-Houttemane, C. (2019). Synthesis and Characterization of Doped Bioactive Glasses. In Elsevier eBooks (pp. 69–123). DOI: 10.1016/B978-0-08-102196-5.00003-3

Bachar, A., Mabrouk, A., Amrousse, R., Azat, S., Follet, C., Mercier, C., & Bouchart, F. (2024). Properties, Bioactivity and Viability of the New Generation of Oxyfluoronitride Bioglasses. *Eurasian Chemico-Technological Journal*, 26(1), 43–52. DOI: 10.18321/ectj1565

Bachar, A., Mercier, C., Follet, C., Bost, N., Bentiss, F., & Hampshire, S. (2016). An introduction of the fluorine and Nitrogen on properties of Ca-Si-Al-O glasses. https://lilloa.univ-lille.fr/handle/20.500.12210/8718

Bachar, A., Mercier, C., Tricoteaux, A., Hampshire, S., Leriche, A., & Follet, C. (2013). Effect of nitrogen and fluorine on mechanical properties and bioactivity in two series of bioactive glasses. Journal of the Mechanical Behavior of Biomedical Materials. *Journal of the Mechanical Behavior of Biomedical Materials*, 23, 133–148. DOI: 10.1016/j.jmbbm.2013.03.010 PMID: 23676624

Bachar, A., Mercier, C., Tricoteaux, A., Leriche, A., Follet, C., & Hampshire, S. (2016). Bioactive oxynitride glasses: Synthesis, structure and properties. *Journal of the European Ceramic Society*, 36(12), 2869–2881. DOI: 10.1016/j.jeurceramsoc.2015.12.017

Bachar, A., Mercier, C., Tricoteaux, A., Leriche, A., Follet, C., Saadi, M., & Hampshire, S. (2012). Effects of addition of nitrogen on bioglass properties and structure. *Journal of Non-Crystalline Solids*, 358(3), 693–701. DOI: 10.1016/j.jnoncrysol.2011.11.036

Bachar, A., Mercier, C., Tricoteaux, A., Leriche, A., Follet-Houttemane, C., Saadi, M., & Hampshire, S. (2013). Effects of nitrogen on properties of oxyfluoronitride bioglasses. *Process Biochemistry (Barking, London, England)*, 48(1), 89–95. DOI: 10.1016/j.procbio.2012.05.024

Badr, H. A., Reda, A. E., Zawrah, M. F., Khattab, R. M., & Sadek, H. E. H. (2024). Effect of hydroxyapatite on sinterability, mechanical properties, and bioactivity of chemically synthesized Alumina–Zirconia composites. *Journal of Materials Engineering and Performance*. Advance online publication. DOI: 10.1007/s11665-024-09531-2

Bouami, H. E., Mabrouk, A., Mercier, C., Mihoubi, W., Meurice, E., Follet, C., Faska, N., & Bachar, A. (2024). The effect of CuO dopant on the bioactivity, the biocompatibility, and the antibacterial properties of bioactive glasses synthesized by the sol-gel method. *Journal of Sol-Gel Science and Technology*, 111(2), 347–361. Advance online publication. DOI: 10.1007/s10971-024-06445-2

Brauer, D. S., Karpukhina, N., Law, R. V., & Hill, R. G. (2009). Structure of fluoride-containing bioactive glasses. *Journal of Materials Chemistry*, 19(31), 5629. DOI: 10.1039/b900956f

Brow, R., Pantano, C., & Boyd, D. (1984). Nitrogen Coordination in Oxynitride Glasses. *Journal of the American Ceramic Society*, 67(4). Advance online publication. DOI: 10.1111/j.1151-2916.1984.tb18834.x

Bueno, L., Messaddeq, Y., Filho, F. D., & Ribeiro, S. (2005). Study of fluorine losses in oxyfluoride glasses. *Journal of Non-Crystalline Solids*, 351(52–54), 3804–3808. DOI: 10.1016/j.jnoncrysol.2005.10.007

Buzalewicz, I., Kaczorowska, A., Fijałkowski, W., Pietrowska, A., Matczuk, A. K., Podbielska, H., Wieliczko, A., Witkiewicz, W., & Jędruchniewicz, N. (2024). Quantifying the dynamics of bacterial biofilm formation on the surface of soft contact lens materials using digital holographic tomography to advance biofilm research. *International Journal of Molecular Sciences*, 25(5), 2653. DOI: 10.3390/ijms25052653 PMID: 38473902

Costerton, W., Veeh, R., Shirtliff, M., Pasmore, M., Post, C., & Ehrlich, G. (2003). The application of biofilm science to the study and control of chronic bacterial infections. *The Journal of Clinical Investigation*, 112(10), 1466–1477. DOI: 10.1172/JCI200320365 PMID: 14617746

D'Alessio, L., Ferro, D., Marotta, V., Santagata, A., Teghil, R., & Zaccagnino, M. (2001). Laser ablation and deposition of Bioglass® 45S5 thin films. *Applied Surface Science*, 183(1–2), 10–17. DOI: 10.1016/S0169-4332(01)00466-4

De Pablos-Martín, A., Muñoz, F., Mather, G. C., Patzig, C., Bhattacharyya, S., Jinschek, J. R., Höche, T., Durán, A., & Pascual, M. J. (2013). KLaF4 nanocrystallisation in oxyfluoride glass-ceramics. *CrystEngComm*, 15(47), 10323. DOI: 10.1039/c3ce41345d

Freer, R., & Dennis, P. F. (1986). Kinetics and mass transport in silicate and oxide systems : proceedings of a meeting, held in London in September, 1984. In Trans Tech Publications eBooks. http://ci.nii.ac.jp/ncid/BA0087969X

Fujibayashi, S. (2003). A comparative study between in vivo bone ingrowth and in vitro apatite formation on Na2O–CaO–SiO2 glasses. *Biomaterials*, 24(8), 1349–1356. DOI: 10.1016/S0142-9612(02)00511-2 PMID: 12527276

Furtos, G., Cosma, V., Prejmerean, C., Moldovan, M., Brie, M., Colceriu, A., Vezsenyi, L., Silaghi-Dumitrescu, L., & Sirbu, C. (2005). Fluoride release from dental resin composites. *Materials Science and Engineering C*, 25(2), 231–236. DOI: 10.1016/j.msec.2005.01.016

Greene, K., Pomeroy, M., Hampshire, S., & Hill, R. (2003). Effect of composition on the properties of glasses in the K2O–BaO–MgO–SiO2–Al2O3–B2O3–MgF2 system. *Journal of Non-Crystalline Solids*, 325(1–3), 193–205. DOI: 10.1016/S0022-3093(03)00337-5

Greenspan, D., Zhong, J., & LaTorre, G. (1994). Effect of Surface Area to Volume Ratio on In Vitro Surface Reactions of Bioactive Glass Particulates. In Elsevier eBooks (pp. 55–60). DOI: 10.1016/B978-0-08-042144-5.50012-1

Hampshire, S. (1992). Oxynitride Glasses and Glass Ceramics. MRS Proceedings, 287. DOI: 10.1557/PROC-287-93

Hampshire, S. (2008). Oxynitride glasses. *Journal of the European Ceramic Society*, 28(7), 1475–1483. DOI: 10.1016/j.jeurceramsoc.2007.12.021

Hampshire, S., Bachar, A., Albert-Mercier, C., Tricoteaux, A., Leriche, A., & Follet-Houttemane, C. (2019). Oxynitride Glasses for Potential Biomedical Usage. In Ceramic engineering and science proceedings (pp. 33–46). DOI: 10.1002/9781119543381.ch4

Hampshire, S., & Pomeroy, M. J. (2008). Oxynitride Glasses. *International Journal of Applied Ceramic Technology*, 5(2), 155–163. DOI: 10.1111/j.1744-7402.2008.02205.x

Hanifi, A. R., Genson, A., Pomeroy, M. J., & Hampshire, S. (2007). An Introduction to the Glass Formation and Properties of Ca-Si-Al-O-N-F Glasses. *Materials Science Forum*, 554, 17–23. . DOI: 10.4028/www.scientific.net/MSF.554.17

Hanifi, A. R., Genson, A., Pomeroy, M. J., & Hampshire, S. (2011). Independent but Additive Effects of Fluorine and Nitrogen Substitution on Properties of a Calcium Aluminosilicate Glass. *Journal of the American Ceramic Society*, 95(2), 600–606. DOI: 10.1111/j.1551-2916.2011.05001.x

Hanifi, A. R., Genson, A., Redington, W., Pomeroy, M. J., & Hampshire, S. (2012). Effects of nitrogen and fluorine on crystallisation of Ca–Si–Al–O–N–F glasses. *Journal of the European Ceramic Society*, 32(4), 849–857. DOI: 10.1016/j.jeurceramsoc.2011.10.026

Hanifi, A. R., Pomeroy, M. J., & Hampshire, S. (2010). Novel Glass Formation in the Ca–Si–Al–O–N–F System. *Journal of the American Ceramic Society*, 94(2), 455–461. DOI: 10.1111/j.1551-2916.2010.04147.x

Hench, L. (1994). Bioactive Ceramics: Theory and Clinical Applications. In Elsevier eBooks (pp. 3–14). https://doi.org/DOI: 10.1016/B978-0-08-042144-5.50005-4

Hench, L. L. (1991). Bioceramics: From Concept to Clinic. *Journal of the American Ceramic Society*, 74(7), 1487–1510. DOI: 10.1111/j.1151-2916.1991.tb07132.x

Hench, L. L. (1998). Bioceramics. *Journal of the American Ceramic Society*, 81(7), 1705–1728. DOI: 10.1111/j.1151-2916.1998.tb02540.x

Hench, L. L., Splinter, R. J., Allen, W. C., & Greenlee, T. K. (1971). Bonding mechanisms at the interface of ceramic prosthetic materials. *Journal of Biomedical Materials Research*, 5(6), 117–141. DOI: 10.1002/jbm.820050611

Hill, R., Stamboulis, A., Law, R., Clifford, A., Towler, M. R., & Crowley, C. (2004). The influence of strontium substitution in fluorapatite glasses and glass-ceramics. *Journal of Non-Crystalline Solids*, 336(3), 223–229. DOI: 10.1016/j.jnoncrysol.2004.02.005

Hill, R. G., Stamboulis, A., & Law, R. V. (2006). Characterisation of fluorine containing glasses by 19F, 27Al, 29Si and 31P MAS-NMR spectroscopy. *Journal of Dentistry*, 34(8), 525–532. DOI: 10.1016/j.jdent.2005.08.005 PMID: 16522349

Iken, A. R., Poolman, R. W., Nelissen, R. G. H. H., & Gademan, M. G. J. (2023). Challenges in developing national orthopedic health research agendas in the Netherlands: Process overview and recommendations. *Acta Orthopaedica*, 94, 230–235. DOI: 10.2340/17453674.2023.12402 PMID: 37194475

Jack, K. H. (1976). Sialons and related nitrogen ceramics. *Journal of Materials Science*, 11(6), 1135–1158. DOI: 10.1007/BF02396649

Jankowski, P. E., & Risbud, S. H. (1980). Synthesis and Characterization of an Si-Na-B-O-N Glass. *Journal of the American Ceramic Society*, 63(5–6), 350–352. DOI: 10.1111/j.1151-2916.1980.tb10742.x

Kim, H. W., Kong, Y. M., Bae, C. J., Noh, Y. J., & Kim, H. E. (2004). Sol–gel derived fluor-hydroxyapatite biocoatings on zirconia substrate. *Biomaterials*, 25(15), 2919–2926. DOI: 10.1016/j.biomaterials.2003.09.074 PMID: 14967523

Kohn, S. C., Dupree, R., Mortuza, M. G., & Henderson, C. M. B. (1991). NMR evidence for five- and six-coordinated aluminum fluoride complexes in F-bearing aluminosilicate glasses. 76, 309–312

Kokubo, T., Kim, H. M., & Kawashita, M. (2003). Novel bioactive materials with different mechanical properties. *Biomaterials*, 24(13), 2161–2175. DOI: 10.1016/S0142-9612(03)00044-9 PMID: 12699652

Kokubo, T., Kushitani, H., Sakka, S., Kitsugi, T., & Yamamuro, T. (1990). Solutions able to reproduce in vivo surface-structure changes in bioactive glass-ceramic A-W3. *Journal of Biomedical Materials Research*, 24(6), 721–734. DOI: 10.1002/jbm.820240607 PMID: 2361964

Kozo, O., Yamamuro, T., Nakamura, T., & Kokubo, T. (1990). Quantitative study on osteoconduction of apatite-wollastonite containing glass ceramic granules, hydroxyapatite granules and alumina granules. *Biomaterials*, 11(4), 265–271. DOI: 10.1016/0142-9612(90)90008-E PMID: 2383622

Kumar, D., Ward, R. G., & Williams, D. J. (1961). Effect of fluorides on silicates and phosphates. *Discussions of the Faraday Society*, 32, 147. DOI: 10.1039/df9613200147

Leng-Ward, G., & Lewis, M. H. (1989). Oxynitride glasses and their glass-ceramic derivatives. In Springer eBooks (pp. 106–155). DOI: 10.1007/978-94-009-0817-8_4

Li, X., Wang, Y., Yang, P., Yin, X., Han, T., Han, B., & He, K. (2024). Topological models of yttrium aluminosilicate glass based on molecular dynamics and structure characterization analysis. *Journal of the American Ceramic Society*, jace.20118. Advance online publication. DOI: 10.1111/jace.20118

Liang, W., Rüssel, C., Day, D. E., & Völksch, G. (2006). Bioactive comparison of a borate, phosphate and silicate glass. Journal of Materials Research/Pratt's Guide to Venture Capital Sources, 21(1), 125–131. DOI: 10.1557/jmr.2006.0025

Lin, F. H., Huang, Y. Y., Hon, M. H., & Wu, S. C. (1991). Fabrication and biocompatibility of a porous bioglass ceramic in a Na2O CaO SiO2 P2O5 system. *Journal of Biomedical Engineering*, 13(4), 328–334. DOI: 10.1016/0141-5425(91)90115-N PMID: 1890828

Liu, Y., & Nekvasil, H. (2002). Si-F bonding in aluminosilicate glasses: Inferences from ab initio NMR calculations. *The American Mineralogist*, 87(2–3), 339–346. DOI: 10.2138/am-2002-2-317

Loehman, R. E. (1980). Oxynitride glasses. *Journal of Non-Crystalline Solids*, 42(1–3), 433–445. DOI: 10.1016/0022-3093(80)90042-3

Loty, C., Sautier, J. M., Tan, M. T., Oboeuf, M., Jallot, E., Boulekbache, H., Greenspan, D., & Forest, N. (2001). Bioactive Glass Stimulates In Vitro Osteoblast Differentiation and Creates a Favorable Template for Bone Tissue Formation. *Journal of Bone and Mineral Research : the Official Journal of the American Society for Bone and Mineral Research*, 16(2), 231–239. DOI: 10.1359/jbmr.2001.16.2.231 PMID: 11204423

Mabrouk, A., Bachar, A., Atbir, A., Follet-Houttemane, C., Albert-Mercier, C., Tricoteaux, A., Leriche, A., & Hampshire, S. (2018). Mechanical properties, structure, bioactivity and cytotoxicity of bioactive Na-Ca-Si-P-O-(N) glasses. Journal of the Mechanical Behavior of Biomedical Materials. *Journal of the Mechanical Behavior of Biomedical Materials*, 86, 284–293. DOI: 10.1016/j.jmbbm.2018.06.023 PMID: 30006277

Mabrouk, A., Bachar, A., Follet, C., Mercier, C., Atbir, A., Boukbir, L., Marrouche, A., Bellajrou, R., Billah, S. M., & Hadek, M. E. (2017). Bioactivity and Cytotoxicity of Oxynitride Glasses. *Research & Reviews Journal of Material Sciences*, 05(02). Advance online publication. DOI: 10.4172/2321-6212.1000173

MacKenzie, K. J. D., & Smith, M. E. (2002). Multinuclear Solid-State NMR of Inorganic Materials. In Pergamon materials series. DOI: 10.1016/S1470-1804(02)X8001-8

Marshall, D. B., Noma, T., & Evans, A. G. (1982). A Simple Method for Determining Elastic-Modulus–to-Hardness Ratios using Knoop Indentation Measurements. *Journal of the American Ceramic Society*, 65(10). Advance online publication. DOI: 10.1111/j.1151-2916.1982.tb10357.x

Mercier, C., Follet-Houttemane, C., Pardini, A., & Revel, B. (2011). Influence of P2O5 content on the structure of SiO2-Na2O-CaO-P2O5 bioglasses by 29Si and 31P MAS-NMR. *Journal of Non-Crystalline Solids*, 357(24), 3901–3909. DOI: 10.1016/j.jnoncrysol.2011.07.042

Mulfinger, H. O. (1966). Physical and Chemical Solubility of Nitrogen in Glass Melts. *Journal of the American Ceramic Society*, 49(9), 462–467. DOI: 10.1111/j.1151-2916.1966.tb13300.x

Neo, M., Kotani, S., Nakamura, T., Yamamuro, T., Ohtsuki, C., Kokubo, T., & Bando, Y. (1992). A comparative study of ultrastructures of the interfaces between four kinds of surface-active ceramic and bone. *Journal of Biomedical Materials Research*, 26(11), 1419–1432. DOI: 10.1002/jbm.820261103 PMID: 1447227

Painter, G. S., Becher, P. F., Kleebe, H. J., & Pezzotti, G. (2002). First-principles study of the effects of halogen dopants on the properties of intergranular films in silicon nitride ceramics. *Physical Review B: Condensed Matter*, 65(6), 064113. Advance online publication. DOI: 10.1103/PhysRevB.65.064113

Rawlings, R. D., Wu, J. P., & Boccaccini, A. R. (2006). Glass-ceramics: Their production from wastes—A Review. *Journal of Materials Science*, 41(3), 733–761. DOI: 10.1007/s10853-006-6554-3

Romanczuk, E., Perkowski, K., & Oksiuta, Z. (2019). Microstructure, Mechanical, and Corrosion Properties of Ni-Free Austenitic Stainless Steel Prepared by Mechanical Alloying and HIPping. *Materials (Basel)*, 12(20), 3416. DOI: 10.3390/ma12203416 PMID: 31635345

Rouxel, T. (2007). Elastic Properties and Short-to Medium-Range Order in Glasses. *Journal of the American Ceramic Society*, 90(10), 3019–3039. DOI: 10.1111/j.1551-2916.2007.01945.x

Sakka, S., Kamiya, K., & Yoko, T. (1983b). Preparation and properties of Ca Al Si O N oxynitride glasses. *Journal of Non-Crystalline Solids*, 56(1–3), 147–152. DOI: 10.1016/0022-3093(83)90460-X

Shelby, J. E., & Lord, C. E. (1990). Formation and Properties of Calcia-Calcium Fluoride-Alumina Glasses. *Journal of the American Ceramic Society*, 73(3), 750–752. DOI: 10.1111/j.1151-2916.1990.tb06586.x

Sukenaga, S., Ogawa, M., Yanaba, Y., Ando, M., & Shibata, H. (2020). Viscosity of Na–Si–O–N–F Melts: Mixing Effect of Oxygen, Nitrogen, and Fluorine. *ISIJ International*, 60(12), 2794–2806. DOI: 10.2355/isijinternational.ISIJINT-2020-326

Vaughn, W. L., & Risbud, S. H. (1984). New fluoronitride glasses in zirconium-metal-F-N systems. *Journal of Materials Science Letters*, 3(2), 162–164. DOI: 10.1007/BF00723104

Vestergaard, P., Jorgensen, N. R., Schwarz, P., & Mosekilde, L. (2007b). Effects of treatment with fluoride on bone mineral density and fracture risk - a meta-analysis. *Osteoporosis International*, 19(3), 257–268. DOI: 10.1007/s00198-007-0437-6 PMID: 17701094

Wang, C., Tao, Y., & Wang, S. (1982). Effect of nitrogen ion-implantation on silicate glasses. *Journal of Non-Crystalline Solids*, 52(1–3), 589–603. DOI: 10.1016/0022-3093(82)90336-2

Wusirika, R. (1984). Problems Associated with the Melting of Oxynitride Glasses. *Journal of the American Ceramic Society*, 67(11). Advance online publication. DOI: 10.1111/j.1151-2916.1984.tb19492.x

Zobell, C. E. (1943). The Effect of Solid Surfaces upon Bacterial Activity. *Journal of Bacteriology*, 46(1), 39–56. DOI: 10.1128/jb.46.1.39-56.1943 PMID: 16560677

Chapter 5
Next-Gen Energy Storage Devices:
Carbon-Based Materials for Flexible Supercapacitors

Krishna Kumar
https://orcid.org/0009-0001-9707-3155
Guru Gobind Singh Indraprastha University, India

Uplabdhi Tyagi
Guru Gobind Singh Indraprastha University, India

Saurav Kumar Maity
https://orcid.org/0000-0001-8953-7100
Guru Gobind Singh Indraprastha University, India

Gulshan Kumar
Guru Gobind Singh Indraprastha University, India

ABSTRACT

The rapid growth of portable electronics and an increased need for sustainable energy sources have grown interest in developing next-generation energy storage technologies. This chapter deals with the development and optimization of supercapacitors synthesized with carbon materials such as graphene, carbon nanotubes, activated carbon, and carbon aerogels. In this regard, the recent progress in the synthesis of carbon-based materials is reviewed relevance to capacitance, conductivity, and mechanical stability. Moreover, this chapter gives overall coverage of issues relating to the production and fabrication of carbon-based flexible supercapacitors. Additionally, the applicability of these carbon-based flexible electrodes in wearable electronics, flexible displays, and next-generation smart devices is explained in detail. The chapter also discussed the general overview of the present status of carbon-based flexible supercapacitors in next-generation energy storage devices, in addition to the future perspectives.

DOI: 10.4018/979-8-3693-7505-1.ch005

1. INTRODUCTION

The rapid growth in the development and manufacturing of portable electronics, wearable, and flexible displays has given rise to the requirement for advanced energy storage solutions having the properties of both high efficiency and long life. For example, an energy storage device must have high power density, quick charge/discharge, and mechanical flexibility, features that traditional batteries cannot provide (Jiang et al., 2024). Consequently, there is an increased interest in supercapacitors, especially those that can be flexible and lightweight which do not degrade their performance with mechanical deformation. Supercapacitors lie in between conventional capacitors and batteries, offering an excellent combination of both high charge density and long life cycle. In contrast to batteries, in which energy flows due to chemical reactions, supercapacitors store energy physically from adsorbed ions on the electrode surface, hence allowing very fast charge/discharge cycles (J. Ma et al., 2024). However, the high energy density in supercapacitors remains quite a challenge to attain without the loss of other vital performance metrics. To overcome this, advanced electrode materials can be utilized for the design of superior energy storage devices. In that respect, based on high conductivity, large area of contact, and adjustable surface properties, carbon-based electrodes came out as a suitable candidate for flexible supercapacitors (Ghafoor et al., 2024). For ultra-high-performance supercapacitor devices, many forms of carbon have been well explored, from graphene to carbon nanotubes and nanofibers, activated carbon, carbon aerogels, and carbon obtained from biomass. Such materials emerge with exclusive advantages in the high surface area, good mechanical strength, and the potential to be fabricated into flexible and lightweight structures (Cheng et al., 2024). They store energy through charge accumulation on the electrode and electrolyte interface. The various carbon-based materials that have been majorly researched and studied for use in flexible supercapacitors include activated carbon, CNTs, graphene and CNFs. These materials are greatly valued due to high porosity and large surface area that enable a large accumulation of charges. Such as graphene is often applied in supercapacitor designs because of its single-layer structure of carbon atoms that give it exceptional conductivity (Aruchamy et al., 2023). Another valuable approach can be given by CNTs due to their high mechanical strength that besides flexibility contribute to their good conductivity in the electrodes of the supercapacitor. As a result, these nanotubes can be integrated into fabrics or coated on flexible substrates, which renders the synthesis very promising for the realization of wearable technologies (Xiao et al., 2024). Also, the combination of CNTs with polymers or metal oxides creates new materials capable of storing energy more efficiently. Carbon nanofibers are another option for flexible supercapacitors, synthesized by using various techniques such as electrospinning. These fibers may randomly orient or align, forming mats or films, flexible with a high surface area. Such carbon nanofibers, doped with heteroatoms or combined with other conductive materials, can offer an improved capacitive behavior and become competitive toward high-performance flexible supercapacitors (Y. Wang et al., 2023). The design of flexible supercapacitors has also focused on incorporation of carbon materials into architectures that can maintain their mechanical integrity even after bending, twisting, or stretching. This usually requires the development of new fabrication techniques, like printing or coating, onto flexible substrates to prepare electrodes so that their performance can be sustained under mechanical deformation. Besides that, the flexible and efficient operation of the device significantly depends on the proper electrolyte and separator materials (Su et al., 2024).

Therefore, this chapter covers the developments of carbon-based materials for the next generation of flexible supercapacitors. It introduces the working principles of a supercapacitor in the introductory part, followed by an overview of various carbon allotropes that are currently being investigated for this application. The chapter concludes with a critical overview of the current state of material synthesis and electrode fabrication techniques, elaborating on how flexibility can balance electrochemical performance. Further, the chapter describes several major challenges in the scale-up process of carbon-based flexible supercapacitors associated with their cost and the need for sustainable material sources. The chapter further underlines their potential in such emerging technologies as wearable electronics and smart textiles, underscoring the transformative effect that carbon-based flexible supercapacitors can have on the future of energy storage. The chapter systematically reviews the current status of carbon-based materials in flexible supercapacitors, together with the associated opportunities and challenges in the development of next-gen energy storage devices that would be both powerful and adaptable to the evolving demands of technology in its modern era.

2. OVERVIEW OF FLEXIBLE SUPERCAPACITORS

Flexible supercapacitors are a class of supercapacitor systems that have the ability to bend, fold, or curve into different forms without losing their performance. Flexible supercapacitors are capable of storing and releasing power electrochemically. As compared to traditional capacitors, they have higher energy density, while as compared to batteries, they have faster charge/discharge rates with high cycle stability. The development of flexible supercapacitors is thus quite important because it meets the need of energy storage solutions that can be applied easily on unconventional and dynamic surfaces without loss of efficiency.

2.1 Comparison with Other Energy Storage Devices

Compared to batteries and traditional capacitors, flexible supercapacitors have several unique advantages and limitations:

- **Batteries:** The lithium-ion and lead-acid batteries boast high energy density and are capable of storing energy for a long time however their charge-discharge rates are relatively small and life cycles are limited. In comparison to batteries, the charge-discharge cycle rate of flexible supercapacitors is extremely high, can undergo millions of cycles with negligible degradation, and thus finds applications where rapid delivery of energy is required (Jha et al., 2023).
- **Traditional Capacitors:** Traditional capacitors have quick charge and discharge cycles but low energy density in comparison to flexible supercapacitors. Therefore, Flexible supercapacitors combine the properties of both batteries and traditional capacitors. They have high power density like batteries and quick charge and discharge cycles like traditional capacitors (Sirimanne et al., 2023).

2.2 Key Requirements in Developing Flexible Supercapacitors

The following are some of the key requirements in the development of effective flexible supercapacitors:

2.2.1 Material Selection

Material selection is crucial in the development of flexible supercapacitors. Material selection and the adoption of advanced preparation methods are crucial to develop a structure that captures favorable properties from various materials while diminishing their drawbacks (H. A. Khan et al., 2024). Frequently used materials include conductive polymers, graphene, carbon nanotubes, and activated carbon, all of which are found to exhibit excellent electrical conductivity, flexibility, and chemical stability.

2.2.2 Mechanical Flexibility and Durability

Among the major problems in developing flexible supercapacitors is to balance mechanical flexibility with electrochemical performance in such a way that these will not lead to a loss of potentiality in their function to store and deliver energy effectively after bending, stretching, or twisting. This, in turn, requires materials able to sustain structure and performance under various modes of mechanical stress. Materials used should have high tensile strength and be flexible in a way that it will easily spring back to its original form once deformed (Liu et al., 2023). At the same time, such materials should have great electrochemical performance exhibiting both high energy and power density and should be capable of fast charging/discharging at a long cycle life. This kind of performance normally arises from large surface area materials with good electrical conductivity, coupled with electrochemical stability.

2.2.3 Electrolyte Compatibility

One of the key elements in a flexible supercapacitor is electrolyte. Electrolytes used for flexible supercapacitors have to be highly conductive and operable within a wide range of temperature. Electrolytes allow for high ionic conductivity and thereby enable fast energy delivery at increased power density (**Figure 1**). It should also be very stable against oxidation, reduction, and other chemical reactions to ensure that the device operates reliably for a long time and in a safe manner. The electrolyte should operate within a wide temperature range-from very cold to very hot. It enhances the capability of the supercapacitor to work in conditions that may be quite demanding (Selvam & Yim, 2023). Many devices use solid-state electrolytes because of their safety and flexibility compared to their liquid counterparts. Moreover, in flexible supercapacitors, the use of solid-state electrolytes reduces the chances of short circuits, hence safer, and can be fabricated into thin and flexible films for any shape or surface. One of the challenges associated with solid-state electrolytes is ionic conductivity. Most of the solid-state electrolytes conducted ions less than their liquid counterparts, which may seriously affect the supercapacitor performance. Because of this fact, most of the recent research concentrates on the development of new materials and techniques which could enhance the ionic conductivity of solid-state electrolytes without compromising their strength and stability.

Figure 1. (a) Electrolyte movement during charging and discharging (b) CV curves of the supercapacitors in various electrolytes (c) Cycle stability after 10 000 cycles (Pal et al., 2019)

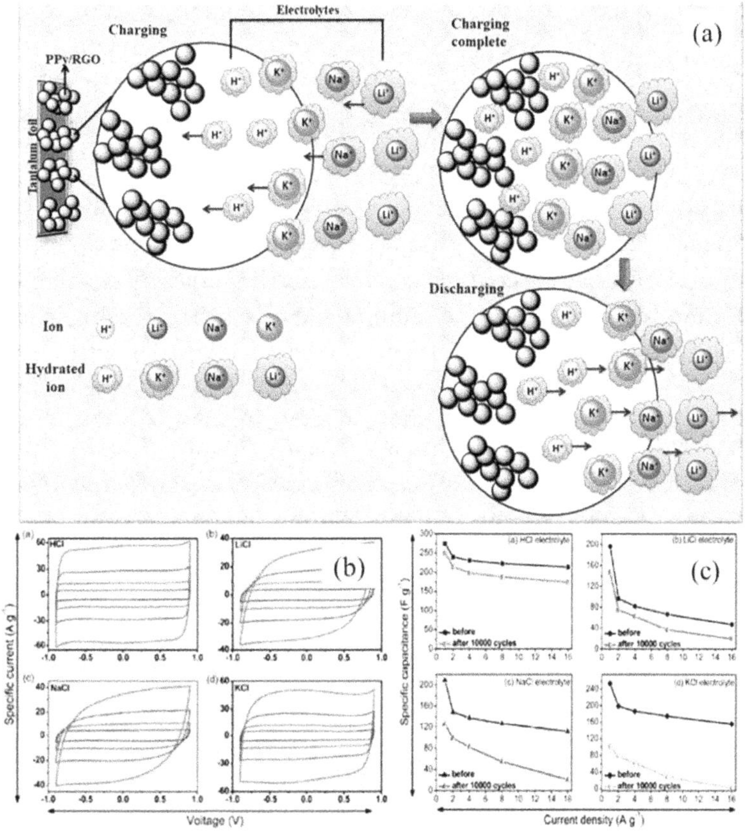

2.2.4 Fabrication Techniques

The development of flexible supercapacitors is connected with the search for new ways of their preparation with an aim for their use on different surfaces. Traditional methods of such device elaboration are not likely to be very effective because of a lack of easy allowance for flexibility, especially taking into account large-scale device fabrication and cost reduction. That is why nowadays many techniques are under consideration to overcome these challenges. Probably one of the best fabrication techniques involves inkjet printing (**Figure 2**), wherein ink with active materials is deposited to a substrate through a precision printing process. Intranet printing impressively has versatility in creating complex designs and patterns on flexible media. It does have the added benefit of being low cost and scalable, thus can be used for the production of large-area devices. The improvements in ink mixtures and printing settings enable supercapacitors to be so fine and consistent for good performance (K.-J. Wu et al., 2023). Other methods used to make flexible electronics include roll-to-roll processing, which allows a high production rate. The term "roll-to-roll processing" simply means that a flexible material is fed through rollers where materials are added and then coated and patterned in many different ways. This is a unique and low-cost method for the fabrication of flexible film or textile supercapacitors, as it is a large-scale production technique, known as roll-to-roll processing, that can be adapted for many material types and device structures. Another supercapacitor preparation involves the spray coating of liquid solutions containing supercapacitor active material onto a surface. This method is convenient for coating large areas in the shortest time and for applying a material uniformly onto flexible surfaces. The primary advantages of spray coating are that it is easy and fast, and that it generates thin uniform layers. Each of these fabrication techniques has its relative advantages and limitations. For instance, inkjet printing provides excellent pattern but can be limited by the range of materials that may be used. Although roll-to-roll processing is very easily scalable, it does require sophisticated equipment and process control. On the other hand, spray coating offers versatility and is fast, but some optimal tuning of this technique may be necessary for the desired layer properties. One can fabricate high-quality flexible supercapacitors with both efficiency and affordability to meet the requirements for different applications just by choosing and optimizing the right method.

Figure 2. (a) Inkjet printing of MXene and graphene composite, (b) Fabricated flexile electrode (c) Current-voltage curves (d) SEM images of synthesized composite (e) reaction pathway for synthesis of Mxene/graphene based composite (Kumar et al., 2023)

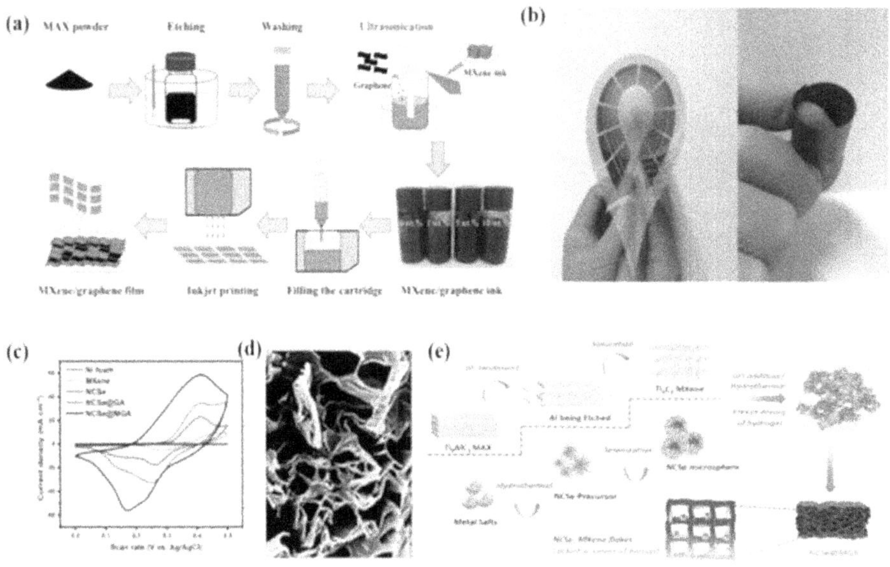

2.2.5 Performance Optimization

It is a challenging task in itself to pursue the quest of improving energy density, power density, and cycle stability of flexible supercapacitors without having their structural stability prejudiced. All these properties are important for the supercapacitor performance viewpoint; however, optimization for them would always require meticulous attention and advanced strategies. The energy density is the quantity of energy that is storable in a supercapacitor per unit volume or mass. High energy density remains a hot pursuit for flexible supercapacitors due to the ambition of competing with batteries in storage capacity. Advanced nanomaterials and composite materials are being sought to further enhance energy density via a greater charge storage capacity. For instance, the increase of the surface area in materials like graphene or in porous carbon structures directly contributes to charge storage enhancement on a unit-area basis. On the other hand, power density tests the ability to provide energy at a very high rate of time. In applications where high power densities are demanded for frequent charging and discharging, it solves the need for quick service. Improved power density is more often than not achieved through improvement in electrical conductivity and reduction of electrical resistance in the supercapacitor. Advanced nanomaterials enhance power densities, owing to their enhanced electrical conductivity properties for example, carbon nanotubes and graphene. Besides that, power delivery performance can also be enhanced by designing electrodes with high surface area and effective charge transport pathways. The other very important parameter of the supercapacitors is cycle stabilities, which represent the performance for many charge/discharge cycles. High cycle stability is a very important parameter that assures long-term reliability and lifetime of the supercapacitor. The focus of researchers nowadays is on upgrading the cycle stability by developing robust materials that resist degradation and keep structural integrity under high mechanical stress. This is via materials selection that will bear repeated bending and stretching with minimal loss in performance, together with a fabrication technique that assures good bonding of the material and stability of the interfaces. To this extent, vast research is put into the investigation on advanced carbon-based materials such as graphene, carbon nanotubes, activated carbon and their hybrids with transition metal oxides, with respect to their potential in enhancing supercapacitor performance. These have top properties that would suffice specific performance needs such as having a high surface area for energy storage and good electrical conductivity for delivering power (Gadipelli et al., 2023). This makes enormous research in the field of developing flexible supercapacitors dynamic and the development process characterized by speed owing to the urgency of advanced energy storage applications that accommodate future new and emerging technologies. This will be critical to realizing the full potential of next-generation energy storage systems.

3. CARBON-BASED MATERIALS FOR SUPERCAPACITOR ELECTRODES

Carbon-based materials are crucial in various industries due to their high surface area, conductivity, and porosity. They significantly enhance energy storage, purification, catalysis, composites, and electronics. As depicted in **Table 1**, key industrial sectors for application of carbon based materials include energy storage (~100,000 tons annually), water/air purification (~500,000 tons), catalysis (~10,000 tons), composites (~200,000 tons), and electronics (~5,000 tons). These materials drive technological advancements and support sustainable industrial growth especially in the field of next gen energy storage devices such as supercapacitor electrode synthsis. The carbon-based materials alleviate the bare simplicity of supercapacitor electrodes in view of their property diversity and, in many cases, tunability. The choice of carbon material depends on the specific application, such as the energy density requirement, the power density requirement, and mechanical strength and flexibility requirements. Thus, the advances in the material synthesis techniques and that of composite formation enhance the performance of carbon-based supercapacitors to make them fit for a multitude of applications in portable gadgets, electric vehicles, and grid energy storage.

Table 1. Importance of carbon-based materials in industry

Industry	Application	Importance of Carbon-Based Materials	Approximate Annual Consumption	Ref.
Energy Storage and Conversion	Supercapacitors, Batteries, Fuel Cells	High surface area (~1500–3000 m^2g^{-1}) and conductivity (~10^4 Sm^{-1}) enhance energy storage capacity and efficiency.	~100,000 tons (including batteries and capacitors)	(Ajmal et al., 2024)
Water and Air Purification	Filtration Systems, Adsorption	High porosity (~0.5–1.2 cm^3g^{-1}) and large surface area (~500–2000 m^2g^{-1}) allow efficient removal of contaminants.	~500,000 tons (including activated carbon)	(Nawaz et al., 2024)
Catalysis	Industrial Catalysis, Hydrogen Production	Chemical stability and electron transfer properties are crucial for catalytic efficiency.	~10,000 tons (including carbon-supported catalysts)	(Dhiman et al., 2024)
Composite Materials	Aerospace, Automotive, Construction	Lightweight (~1.3–2.0 gcm^{-3}) and strong, carbon materials improve the performance and durability of composites.	~200,000 tons (including carbon fibers and composites)	(F. Khan et al., 2024)
Electronics	Transistors, Sensors, Flexible Electronics	Superior electrical properties (conductivity ~10^4–10^5 Sm^{-1}) and mechanical flexibility enable advanced electronics manufacturing.	~5,000 tons (including graphene and nanotubes)	(Shi et al., 2024)

3.1 Types of Carbonaceous Materials

Carbonaceous materials encompass a very large variety of very different substances, but mostly consisting of carbon atoms with different structures and special properties (**Table 2**) that find applicability, particularly in the area of energy storage. Out of these, graphene is prominent because of having a single layer of carbon atoms, arranged on a 2D lattice: therefore, it is very well known for its exceptionally high electrical conductivity and high surface area. CNTs are another important type of nanotubes, known cylindrical nanostructures, where graphene sheets are rolled. They harbor the potential to demonstrate extraordinary electrical and thermal transport properties. High porosity is an attribute of activated carbon, hence having an enormous surface area proper for adsorption processes, which makes it found in supercapacitors. Carbon aerogels have the special features of light, porous materials with a sponge-like structure that unites low density with high conductivity, thus being proper for use in advanced energy storage applications (Ayati et al., 2023). Each of these carbonaceous materials boasts its specific advantages, which leads to further development and optimization in the pursuit of next-generation energy storage technologies.

Table 2. Carbon based materials: Types, synthesis method, physical and electrochemical properties

Material Type	Synthesis Method	Surface Area (m^2/g)	Pore Size (nm)	Electrical Conductivity (S/cm)	Capacitance (F/g)	Energy Density (Wh/kg)	Power Density (W/kg)	Ref.
Activated Carbon	Physical-Chemical Activation	1000 - 3000	1.0 - 2.5	1-10	150 - 300	5-15	500 - 2000	(Alhebshi et al., 2021)
Graphene	Chemical Vapor Deposition (CVD)	1500 - 2630	2.5 - 3.5	10^2 - 10^4	100 - 400	10-20	5000 - 10,000	(Banciu et al., 2022)
Porous Carbon	Soft/Hard Template Method	500 - 1500	2.0 - 5.0	10 - 100	200 - 300	10-30	3000 - 8000	(Zhang et al., 2021)

Material Type	Synthesis Method	Surface Area (m^2/g)	Pore Size (nm)	Electrical Conductivity (S/cm)	Capacitance (F/g)	Energy Density (Wh/kg)	Power Density (W/kg)	Ref.
Carbon Aerogels	Sol-Gel Process	600 - 1200	5.0 - 50.0	1-10	100 - 250	5-20	1000 - 5000	(Verma et al., 2021)
Graphene Oxide (GO)	Hummers' Method	500 - 1000	3.0 - 6.0	0.01 - 1.0	150 - 250	5-10	1000 - 4000	(Karbak et al., 2022)

Carbon materials offers higher specific surface area, enhanced electrical conductivity, and better electrochemical stability, leading to improved charge-discharge cycles and greater energy and power density. Furthermore, their sustainable and cost-effective nature, coupled with the tunability of their porosity and surface functionalities, makes them ideal candidates for next-generation flexible supercapacitors (as depicted in **Table 3**). Overall, carbon materials demonstrate significant potential in surpassing the limitations of conventional materials in advanced energy storage technologies.

Table 3. Performance of carbon materials over conventional materials for energy storage applications

Material	Specific Capacitance (F/g)	Energy Density (Wh/kg)	Power Density (W/kg)	Cycle Stability (%)	Ref.
Biomass-Derived Carbon	250-350	20-30	10,000-15,000	90-95	(Gao et al., 2024)
Graphene	150-250	10-20	5,000-10,000	85-90	(Tale et al., 2024)
Activated Carbon	100-150	5-10	1,000-5,000	80-85	(Bosco Franklin et al., 2024)
Carbon Nanotubes (CNTs)	200-300	15-25	7,000-12,000	85-92	(Rajamanickam et al., 2024)
Conducting Polymers (PEDOT, PSS)	200-300	10-20	3,000-6,000	80-85	(Dubey et al., 2024)
Vanadium Redox Flow Batteries	20-30	20-30	200-500	90-95	(Basavaraj Chavati et al., 2024)
Lithium-Ion Batteries	100-200	150-250	500-1,000	80-90	(Saikia et al., 2024)
Nickel-Metal Hydride (NiMH) Batteries	60-100	60-100	200-500	70-80	(B. Ma et al., 2024)

3.1.1 Graphene

Graphene comprises an atomic layer of carbon atoms arranged on a two-dimensional honeycomb lattice. It has very high electrical conductivity, which means high efficiency in electron transport and is thus very useful for many electronic applications. Besides, the mechanical strength of this material is exceptionally high and equals about 200 times that of steel, while it is only a few atoms thick. It has mechanical robustness with flexibility, which gives it the integrity to withstand the vagaries of many conditions. The second most important property of graphene is the existence of the large surface area. This enormous surface area comes handy in applications that involve extensive surface interactions, as in energy storage devices. These make graphene an ideal material for different uses, especially in energy storage. For supercapacitors, graphene can boast extremely large surface areas with excellent conductivity, greatly increasing the energy density and power density (Urade et al., 2023). Its large surface area will provide the space for more electrochemical reactions to occur, which will raise the amount of energy that is stored in it. Meanwhile, superior conductivity will make sure that this energy can be quickly charged or discharged to generate high power density. Accordingly, graphene-based supercapacitors can provide rapid transfer of energy; therefore, they are quite viable for use in applications involving fast charge/discharge cycles, as in the case of electric vehicles, portable electronics, and renewable energy systems. In this regard, Bai et al. (https://doi.org/10.1016/j.jpowsour.2024.234545), synthesized a high-performance fibrous electrode with a triple gradient structure, featuring Ni_3S_2 nanoflakes on Ni-coated graphene fibers. This design provides excellent electrical conductivity, mechanical strength, and flexibility. The electrode delivers a high areal specific capacitance of 1387.6 mF.cm^{-2} and an energy density of 15.6 μW.h.cm^{-2}, with impressive cycle stability, retaining 86.7% of capacitance after 10,000 cycles. It represents a significant advancement for flexible, energy-dense supercapacitors in wearable technology.

3.1.2 Carbon Nanotubes (CNTs)

CNTs are cylindrical nanostructures formed by rolling up graphene layers into seamless tubes and may include either one or multiple graphene layers, correspondingly. This unique structure gives exceptional properties, such as high electrical conductivity, a result of their pristine carbon lattice and few defect sites that facilitate rapid electron transport. Equally marvelous is their mechanical strength, with tensile strengths running into orders of magnitude higher than steel, hence highly resilient and durable. Besides, CNTs possess very good thermal stability that is, they can take high temperatures without degradation. Therefore, this will be very important where intense conditions are involved in the operation. In supercapacitors, CNTs will play a vital role in enhancing their performance. Their high electrical conductivity improves the overall conductivity of the electrode materials, hence reducing resistive losses and enabling faster cycles of charge discharge; this ultimately results in higher power densities. The need to mention this is obvious, as it is a very critical factor in determining the performance of supercapacitors (Malozyomov et al., 2023). In addition, structurally, the CNTs hold other active materials together, such as activated carbon or metal oxides, avoiding the collapse of their porous structures and thus maintaining the integrity of the electrodes while cycling. This synergy not only improves electrochemical performance but also enhances the cycle life and stability of the supercapacitor. Therefore, the incorporation of CNTs into supercapacitor electrodes shows a very big leap forward in the development of high-performance energy storage devices. In this context, Wu et al. (https://doi.org/10.1016/j.cej.2024.151589), fabricated a V_2CTx/carbon nanotube electrode, the device achieves high capacity and stability, demonstrating an area-specific capacity of 117.2 mF.cm^{-2} and an optimal capacitance of 246.88 mF.cm^{-2}. Integrated with V2CTx-based pressure sensors, it remains stable over 50,000 cycles, showcasing its potential for efficient, self-powered wearable technologies.

3.1.3 Activated Carbon

Among all, activated carbon stands out mainly because it demonstrates an extremely large surface with ultra-high porosity. This amazing fact has emerged from its making process in which carbonaceous materials like wood, coal, or agricultural by-products are put under activation to develop a highly porous structure. During the activation process, the carbon source is treated with activating agents such as steam, carbon dioxide, or chemical reagents that produce micro- and mesopores. Such a porous network appreciably enlarges the surface area of the material, therefore guaranteeing large space for adsorption and charge storing. In the application of supercapacitor electrodes, activated carbon is considered very important due to the low cost and excellent performance. Accordingly, the large surface area of the activated carbon enables a large electrochemical interface that can support high capacitance and efficient energy storage (Mandal et al., 2023). The high porosity allows quick ion movement, thereby improving charge-discharge rates and, therefore, supercapacitor performance. Thus, active carbon is employed in many supercapacitor electrode developments to balance the price with performance and provide effective solutions for energy storage applications. In this regard, Ding et al. (https://doi.org/10.1021/acsaem.4c00954), found that MAC-2 activated carbon, with its optimal pore structure, offers superior performance in EDLCs at ultralow temperatures, achieving a specific capacitance of 30.94 $F.g^{-1}$ at −60 °C and 91.3% capacitance retention at −35 °C after 10,000 cycles. This highlights its potential for improved low-temperature EDLC applications.

3.1.4 Carbon Aerogels

Carbon aerogels are a lightweight, open-porous material class with an interconnected network of carbon nanospheres or nanoparticles. This structure imparts several remarkable properties to the carbon aerogel. These include extremely large surface areas and very low densities. The high surface area provides ample space for charge storage, which is crucial in energy storage devices. Further, low density makes them lightweight, which is an added advantage in weight-critical applications. In this context, Yang et al. (https://doi.org/10.1039/D3RA07014J), synthesized 3D hierarchical porous carbon aerogel with a specific surface area of 616.97 m^2g^{-1} and a specific capacitance of 138 Fg^{-1} was developed. This carbon aerogel-acetic acid demonstrated excellent cycle stability, retaining 102% capacitance after 5000 cycles at 1 Ag^{-1}, and achieved an energy density of 10.06 Wh/kg and a power density of 181.06 W/kg, making it a promising material for supercapacitors. Other conductive properties include good electrical conductivity, which may be desirable for quick charge transfer in supercapacitors. With this set of special characteristics, carbon aerogels become of special value in high energy and power density applications (T. Wang et al., 2023). In particular, they become potential candidates to improve the performance of supercapacitors by offering great possibilities to increase both energy storage capacity and power delivery in accordance with their ability to combine high surface area with low density and good conductivity.

3.1.5 Carbon Nanofibers (CNFs)

Carbon nanofibers represent unique carbon-based materials, and a definition could be given: a type of carbon-based material of fibrous structures on the nanoscale in diameter. The fibers are of very high aspect ratio, which generates special mechanical and electrical properties. The fibers' large aspect ratio contributes to a large surface area and, coupled with efficient charge transport, makes CNF highly conductive. It also shows very high mechanical strength and flexibility, which enhance its durability and performance in most of the applications. In supercapacitor electrodes, CNFs are used to improve the mechanical strength drastically along with electrical conductivity (Acharya et al., 2023). The corresponding composite would show improved performance parameters like energy and power densities, and stability during the charge-discharge cycles by using CNFs in electrode materials. This, in turn, enables more efficient energy storage devices with increased lifetime and reliability. In this context, Radhakrishnan et al. (https://doi.org/10.1016/j.electacta.2024.144683), using a simple electrodeposition method synthesized a NiCoS/CNF supercapacitor with 70 Fg^{-1} capacitance, 86% stability after 10,000 cycles, and 56.33 $Wh.kg^{-1}$ energy density. This approach shows promise for versatile energy storage applications.

3.1.6 Biomass-Derived Carbon

Biomass-derived carbon provides a sustainable and inexpensive process for the generation of high-performance carbon material for supercapacitors. They are mainly obtained from agricultural waste, such as banana peels and potato peels, and from other biomass resources like coconut shells and wood. The main advantages of using biomass-derived carbon were the large surface areas harnessed in order to maximize electrochemical performance in supercapacitor electrodes. Moreover, these carbon materials are not only economically feasible but also environmentally friendly because they consume waste products and reduce the burden of using expensive and less sustainable alternatives (**Table 4**). In this regard, Patil et al. (https://doi.org/10.1002/pat.6362), synthesized a doped polyaniline and Pea derived activated carbon composite for supercapacitors. The composite showed high specific capacitances of 446 Fg^{-1} and 517 Fg^{-1} on different substrates, with excellent cyclic stability and specific energies of up to 101 $Wh.kg^{-1}$.

On the issue of sustainability, the renewable nature of biomass-derived carbon makes this material very promising for sustainable supercapacitor applications with both ecological and economic benefits in store (Yuan et al., 2024). Such materials will help researchers and manufacturers develop supercapacitors that are not only highly efficient but also fit for the broader goals of sustainability and resource conservation.

Table 4. Comparison of biomass-derived carbon materials with conventional carbon materials

Parameter	Biomass-Derived Carbon	Conventional Carbon	Advantage
Specific Surface Area (m^2/g)	800-1200 (e.g., from banana peel, potato peel)	1000-2000 (e.g., activated carbon, graphene)	Biomass-derived carbons offer comparable surface areas, often at lower costs.
Capacitance (F/g)	150-250 (e.g., from bamboo, agricultural waste)	200-350 (e.g., activated carbon, graphene)	Biomass-derived carbons provide competitive capacitance values.
Pore Size (nm)	2-4 (micropores)	1-3 (micropores in activated carbon, larger in graphene)	Biomass-derived carbons can offer well-structured microporosity.
Electrical Conductivity (S/cm)	0.1-1.0 (varies with treatment and type)	1000-2000 (e.g., graphene)	Conventional carbons typically have higher conductivity, but advances in treatment can enhance biomass-derived carbons.
Energy Density (Wh/kg)	10-20 (varies based on material and treatment)	20-30 (e.g., activated carbon, graphene)	Biomass-derived carbons show promising energy densities, with potential for improvement.
Power Density (W/kg)	500-1000	1000-2000	Biomass-derived carbons can achieve significant power densities, though conventional materials currently lead.
Cycle Stability (Cycles)	1000-2000	2000-5000	Biomass-derived carbons have good cycle stability but may need further optimization.

continued on following page

Table 4. Continued

Parameter	Biomass-Derived Carbon	Conventional Carbon	Advantage
Sustainability	High (renewable resources, lower environmental impact)	Variable (depends on source, often energy-intensive)	Biomass-derived carbons are more sustainable and environmentally friendly.
Cost	Low (cost-effective, often from waste)	High (expensive synthesis and processing)	Biomass-derived carbons are generally more cost-effective and derived from waste.

3.2 Synthesis Methods and their Impact on Electrode Performance

The synthesis method applied to electrode materials makes a difference in the obtained physical and chemical properties, which will then directly impact their performance for energy storage and conversion devices. Each type of synthesis methodology has its own influence on the physical and chemical characteristics of electrode materials; optimality is dependent upon the trade-off among factors such as performance requirements, cost, complexity, and scalability (**Figure 3**).

Figure 3. Synthesis of various forms of carbon with their practical applications (a) CNF/CNT/GO carbon aerogels (b) C-NGD carbon aerogels (c) MXene/rGO composite (d) Carbon framework along with PEDOT-PF-6 (Xiao et al., 2024)

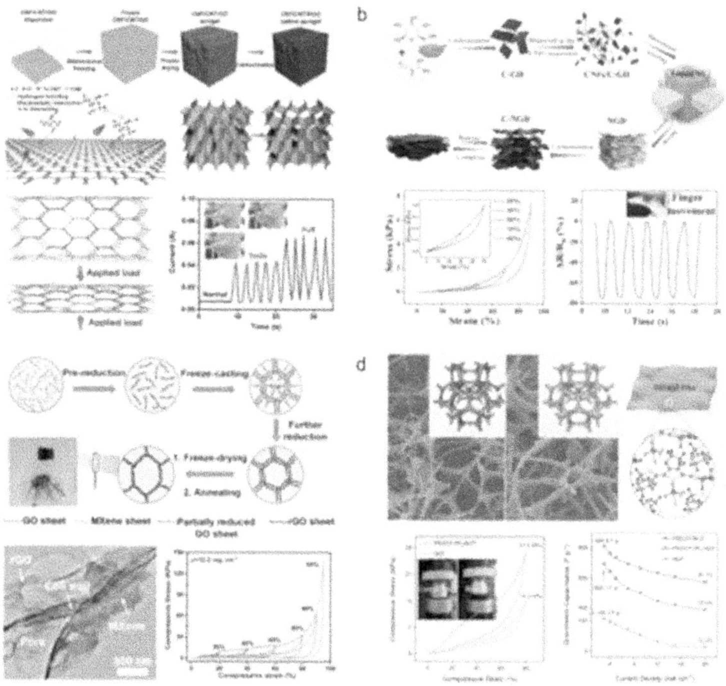

3.2.1 Chemical Vapor Deposition (CVD)

CVD is one of the more complex techniques used in the development of thin films, coatings, and nanostructures by means of chemical reduction of precursors in the gaseous state on a substrate. The CVD process provides for introducing volatile reactants into the reaction chamber, followed by decomposition or reaction to form a solid material joined to the substrate. It allows control over all sorts of material properties, from thickness and purity to structural alignment. CVD shows maximum utility in depositing high-quality carbon-based materials, like graphene or carbon nanotubes, onto supercapacitor electrodes with uniform and tailored properties (**Figure 4**). This is important to optimize the electrochemical performance of supercapacitors. The thickness control optimizes the quantity of active material in the electrodes, and high purity ensures minimal impurities that could hamper performance (Sabzi et al., 2023). Additionally, alignment and quality of structure in the deposited material may play a huge role in the conductivity, thus the general efficiency of the electrodes, in improving energy and power density. As such, CVD is a very important tool in boosting functionality to supercapacitor electrodes by providing materials with exact specifications needed for advanced energy storage applications.

Figure 4. Schematic representation of various CVD growth of graphene (a) simultaneous in situ CVD growth (b) sequential in situ CVD growth (c) lithography-assisted growth (d) conversion growth (Baig et al., 2021)

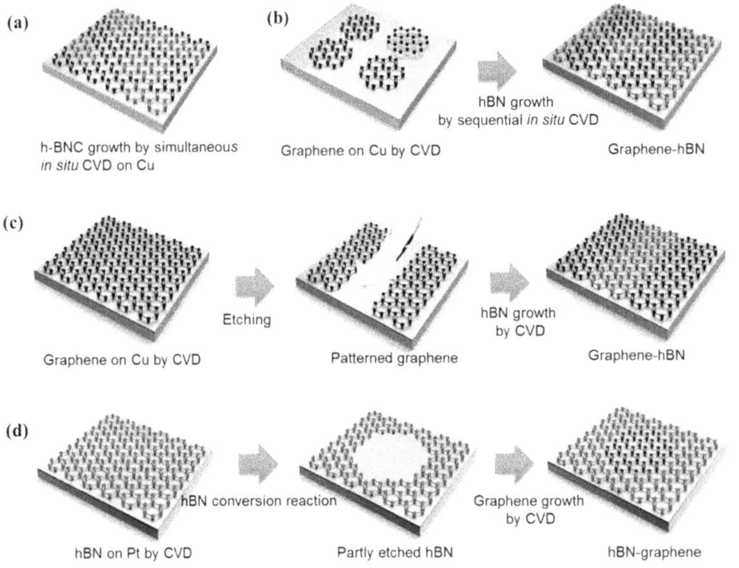

3.2.2 Sol-gel Method

The sol-gel process is a very flexible wet chemical method for creating advanced materials from a liquid "sol" that transitions into a solid "gel" phase. First of all, prepare a solution of metal alkoxides or other precursors; then, it undergoes hydrolysis and condensation reactions to obtain a gel network. This gel undergoes further heat treatment to finally convert it into a solid, porous material. The main advantages of the sol-gel technique include the possibility of very accurate incorporation for all kinds of dopants and modifiers directly in the course of gelation. This feature allows synthesizing nanostructured materials that have high surface area and controlled porosity. The features that will be mentioned herein will increase the electrochemical performance of supercapacitor electrodes by providing increased electrochemical surface area and improved ion transport pathways (Nikam et al., 2024). The resulting materials often have an increased charge storage capacity and cycling stability, and thus the sol-gel method represents a very important route to high-performance supercapacitor electrodes.

3.2.3 Hydrothermal and Solvothermal Synthesis

Hydrothermal and solvothermal syntheses are advanced methods of crystallization of materials from high-temperature aqueous solutions under elevated vapor pressure. In such methods, the reactants are dissolved in a given solvent, and the reaction takes place within a sealed container called an autoclave. These elevated temperature and pressure conditions inside the autoclave raise the dissolution of precursors and promote the formation of nanostructured materials with uniform sizes and shapes. This controlled environment is important in tailoring the structural and morphological properties of the ensuing materials. Together with solvothermal synthesis, hydrothermal synthesis has a great influence on the performance of supercapacitor electrodes. These methods can generate materials with large active surface areas and optimized pore structures by controlling the size, shape, and crystallinity of nanomaterials (Aruchamy et al., 2023). With large surface areas, an increased number of electrochemical active sites are made available for charge storage, and with improved morphology, the pathways of ion transport become easier. These, therefore, sum up to give improved electrochemical performance with higher capacitance, enhanced energy and power densities, and cycle stability in supercapacitor applications.

3.2.4 Electrodeposition

Basically, electrodeposition is a process in which electrochemical reactions deposit material onto an electrode to form a thin film or coating. Simply put, an applied voltage across a substrate immersed in an electrolyte solution makes the targeted material from the solution reduce and deposit onto the electrode surface. It provides very good control of the deposition process and material composition this is why it finds extensive use in the creation of homogeneous thin films and coatings with specially designed properties for a lot of applications, including electrodes for supercapacitors. The technique allows good control of film thickness, composition, and morphology, which can be critical in adjusting performance in different electronic devices. Some of the disadvantages are that the quality of the deposited film can easily be influenced by current density, composition of electrolyte, and temperature (Mohapatra et al., 2024). Changes in these parameters could give rise to some inconsistencies in the uniformity of the films and their material properties, something that could affect the general performance and reliability of the electrode. Even with these difficulties, electrodeposition remains one of the effective techniques for the synthesis of advanced materials with controlled features, which are further required for the creation of electrochemical devices with improved performance.

3.2.5 Activation Processes (Physical and Chemical)

For this reason, the development of high-performance carbon-based materials for supercapacitor electrodes is related to activation processes that include physical and chemical activation. Activation involves treatment with agents like steam, carbon dioxide, or potassium hydroxide. These create small holes in the carbon material and increase the surface area and porosity of the carbon material many times over. This is of great importance in supercapacitors, as it increases the area available to store charge and aids ion flow in the electrode material. Generally, during physical activation, the carbon material is treated with steam or CO_2 at high temperatures. This kind of process leads to a network of pores and channels, causing an increase in surface area. During chemical activation, it involves soaking of carbon material with an activating agent such as KOH or H_3PO_4, then heating (Serafin & Dziejarski, 2023). This provides not only a large surface area but also creates other functional groups that might increase the electrochemical performance of the material. It is possible to tailor the pore structure and the surface chemistry of the carbon material to achieve higher capacitance and an increased energy storage capacity, thus improving overall efficiency and performance of supercapacitors by optimizing these activation processes.

3.3 Structural and Morphological Considerations for Enhancing Electrochemical Properties

Electrochemical properties can be enhanced by structural and morphological considerations to portray the optimization of the features of materials to enhance their performance within electrochemical applications. This can be realized by the creation of surface area, porosity, and particle size, among other things, leading to a material with enhanced conductivity and reactivity. For instance, increased surface area and optimized pore size will enhance ion and electron transport, further enhancing its charge storage and transfer capabilities (Ravina et al., 2023). The former can even end up increasing the active surface site while the diffusion path gets reduced, hence an appended essentiality in optimizing the electrochemical performance for use in batteries, supercapacitors, or fuel cells. Additionally, tailor-made structural properties of a material for a given application can culminate in more efficient and stable electrochemical systems.

3.3.1 Surface Area and Porosity

The surface area and porosity are the two most important factors governing supercapacitor electrode performance. A high surface area ensures that the electrochemical surface available for charge storage is large, which linearly improves the capacitance of the supercapacitor. In addition, well-developed porosity is quite important in creating efficient ion transport channels that allow quick movement of ions during charge and discharge cycles. Such an efficient way of ion transport is highly desirable in supercapacitors to provide high power density and fast response times. These features can be further optimized by the methods shown next. The activation processes, either physical or chemical, are easy ways to build a porous structure, and it increases surface area through the formation of micropores and mesopores. In particular, template methods with sacrificial templates or molds will help to realize better performance because one can exactly control the size and distribution of pores (Li et al., 2024). More importantly, a hierarchical pore structure, having different sizes of pores within the same material, further optimizes the surface area and connectivity of the pores, thus improving ion diffusion and storage capabilities. It will thus be possible, by associating these techniques, to reach this ideal balance between a high surface area and optimized porosity of supercapacitor electrodes for increased efficiency and effectiveness.

3.3.2 Electrical Conductivity

One of the basic properties required to be possessed by supercapacitor electrodes for them to exhibit satisfactory performance is electrical conductivity, as it is responsible for ensuring efficient charge transfer and minimizing resistive losses. High electrical conductivity will allow for fast movement of charge carriers within the electrode material; therefore, this is a precondition for the realization of high power density and fast charge-discharge cycles in supercapacitors. Several methods can be applied to improve the conductivity of carbon-based electrodes. One of the effective ways is to incorporate high-conductivity materials, such as graphene and carbon nanotubes, into the electrode matrix. Graphene has a two-dimensional structure, while CNTs have a one-dimensional structure, thus providing both exceptional electrical pathways that could increase the conductivity of electrodes enormously. Further doping by heteroatoms like N or S can improve conductivity even more. Heteroatoms introduce some additional charge carriers and change the electronic structure of the carbon material, improving electrical performance (Yang et al., 2023).

3.3.3 Mechanical Stability

Mechanical stability is very critical to keep supercapacitor electrodes robust and functional over a long period, specifically under continuous charge-discharge cyclic operations. Conventionally, supercapacitors have been employed with high frequency cycling under normal operations that may change the physical structure of the electrodes resulting in damage or even separation. This underscores the increasing need for enhancing mechanical stability to maintain the structural integrity of electrodes so that their functionality is kept intact. One of the very popular ways of achieving such mechanically stable electrodes involves the use of composite materials. Coupling carbon-based materials with either polymers or metal oxides significantly enhances their mechanical properties (Pathaare et al., 2023). For example, the addition of polymers allows for flexibility to handle mechanical stress, while the addition of metal oxides increases strength and durability of the structure. These composite materials allow the maintenance of size, function, and performance of electrodes over a long period, a very essential issue in view of lifetime and reliability expectations of supercapacitors.

3.3.4 Electrochemical Stability

One important factor that helps supercapacitor electrodes work well for a long time is their electrochemical stability. When electrodes go through many charge and discharge cycles, they face different types of stress, which can lower their performance over time. Ensuring electrochemical stability therefore means that electrodes could bear through these stresses without resultant significant loss in capacity or efficiency. This stability is very key to be maintained in order for it not to affect high performance, which will have generally factored into the overall lifespan and reliability of the device over the period of use. Some strategies could be utilized in enhancing electrochemical stability. Functional groups and modification of the electrode surface functionalization can lead to better interaction with the electrolyte and abate degradation processes. This will improve the anti-corrosion characteristics and inhibit undesirable side reactions at the electrode. This can also be achieved through doping, which is the introduction of different elements into the carbon matrix (Sahoo et al., 2023). Doping with nitrogen or sulfur can improve electrochemical properties and conductivity and improve resistance against chemical attack. Besides, optimization at the electrode-electrolyte interface is very essential to enhance stability (**Table 5**). This can be realized by readjusting the structure and adjusting the surface property of the electrode material, enhancing its compatibility with the electrolyte to reduce problems from poor adhesion or instability of materials during cycling. Therefore, these strategies will help to increase durability and longevity for the supercapacitor electrodes to work continuously in cycles.

Table 5. Methods for enhancing the electrochemical performance of carbon-based supercapacitors

Method	Carbon Material	Electrochemical Performance	Challenges/Limitations
Doping (e.g., N-doping, B-doping)	Graphene, Carbon Nanotubes (CNTs), Activated Carbon	Increased capacitance (20-50%), enhanced electrical conductivity, improved charge transfer	Precise control of doping level and uniformity, potential introduction of defects
Hybridization	Graphene/CNT, Activated Carbon/Metal Oxides	Enhanced energy density (30-70%), synergistic effects between materials, improved electrochemical stability	Complex synthesis processes, potential agglomeration of materials, cost
Surface Functionalization	Activated Carbon, CNTs, Graphene	Improved wettability, increased specific capacitance (10-30%), better interaction with electrolytes	Stability of functional groups under operational conditions, potential loss of conductivity
Nanostructuring	Graphene, CNTs, Carbon Nanofibers	Increased surface area, enhanced ion diffusion rates, improved energy density	High-cost fabrication techniques, challenges in maintaining structural integrity
Composite Formation	Carbon Aerogels/Metal Oxides, Graphene/Polymers	Improved mechanical strength, enhanced capacitance and energy density, better flexibility for wearable applications	Complexity in composite fabrication, potential mismatch in thermal expansion coefficients
Pore Size Tuning	Activated Carbon, Carbon Aerogels	Optimization of pore size distribution for better ion accessibility, increased capacitance	Trade-off between surface area and pore size, challenging control over uniform pore size
Heteroatom Incorporation	N, S, P-doped Graphene	Enhanced pseudocapacitance, improved charge storage capability, better electrochemical performance	Control over doping concentration, potential degradation of material properties
Electrochemical Activation	Activated Carbon, Graphene	Increased specific surface area, enhanced capacitance, improved electrolyte access	Possible reduction in mechanical stability, complexity in activation process

4. HYBRIDIZATION STRATEGIES: INTEGRATING CARBON-BASED MATERIALS WITH OTHER ACTIVE MATERIALS

Hybridization strategies involve the combination of carbon-based materials with other active materials to enhance the overall performance and functionality of the formed composites. This will enable the exploitation of the intrinsic features of the carbon material, such as high surface area, excellent electrical conductivity, and chemical stability, along with the characteristic features of other active materials to yield synergistic effects that would result in superior performance in a variety of applications.Bottom of Form

4.1 Carbon-Metal Oxide Hybrids

One of the most investigated strategies is the incorporation of carbon materials with metal oxides, which exhibit complementary properties. That is because metal oxides, such as TiO_2, MnO_2, Fe_3O_4, and ZnO, exhibit unique electrochemical features that enable application in energy storage, catalysis, and environmental remediation. Generally, the hybrids fabricated by incorporating carbon materials like graphene, carbon nanotubes, or activated carbon exhibit increased electrical conductivity, larger surface area, and stronger electrochemical stability. In supercapacitors, mixing carbon with metal oxides can increase capacitance and cycling stability enormously. For example, hybrids of MnO_2 with graphene or CNTs were reported to yield high-performance supercapacitors with better energy density. This synergy originates from the fact that metal oxides provide high capacitance through faradaic reactions, while the carbon component contributes high electrical conductivity and structural stability to result in superior overall performance. Carbon-based hybrids of metal oxides such as Fe_3O_4 or TiO_2 enhance the capacity, rate capability, and cycle life of anode materials in lithium-ion batteries (Gaikwad et al., 2024). The metal oxides provide high theoretical capacity, while the carbon materials reduce problems associated with volume changes and conductivity. Further integration enhances the performance of batteries, which makes them more effective and reliable for many applications, as well as durability. The integration of carbon materials with metal oxides thus offers enormous opportunities for the advancement of energy storage, catalysis, and environmental remediation technologies by playing off the strengths of the two material classes for superior functionality.

4.2 Carbon-Polymer Hybrids

Composites with tailored mechanical, thermal, and electrical properties can be achieved by combining polymers with carbon materials. Among them, conducting polymers, such as PANI, polypyrrole, and polythiophene, have attracted special interest because of their intrinsic conductivity and flexibility. These unique features of the carbon-polymer hybrids make them highly versatile for many applications. Carbon-polymer hybrids are ideal in flexible and wearable electronics. Such synergistic combinations of conductive polymers with flexible carbon materials, like graphene or CNTs, have been utilized in the development of flexible conductors, sensors, and transistors. In this way, electronic devices that bend, stretch, and adopt various shapes will be realized, and therefore the pathway for developing wearable technology and flexible displays will be opened. These hybrids can strongly enhance the sensitivity and selectivity of electrochemical sensors. For instance, PANI-graphene composites find use in the detection of several biomolecules and pollutants. The interactions between the high surface area/conductivity of graphene and the electroactive properties of PANI produce advanced performance sensors (Ruan et al., 2023). This makes them quite good at environmental monitoring, medical diagnosis, and biochemical analysis.

4.3 Carbon-Metal Nanoparticle Hybrids

Carbon-metal nanoparticle hybrids are a new class of materials that couple synergistically the unique properties of metal nanoparticles with the advantageous features of carbon materials. The unique optical, catalytic, and electronic properties characterize metal nanoparticles, for example, gold, silver, platinum, palladium, iron (Fe), cobalt (Co), and nickel (Ni) as shown in **Figure 5**. Carbon materials combined with these nanoparticles form composites that give enhanced performance in catalysis, sensing, and energy conversion applications. Carbon-metal nanoparticle hybrids are realizing central applications in catalysis. Such hybrids act as very efficient catalysts in rather quite a few types of reactions. For example, platinum nanoparticles supported on carbon materials have been used extensively as catalysts in fuel cells, as they exhibit very good catalytic activity and stability (Cai et al., 2023). Also, these hybrids find applications in hydrogen production and environmental remediation, wherein their catalytic features drive various chemical reactions to result in improved efficiency and a reduction in energy consumption. Carbon-metal nanoparticle hybrids have shown enormous potential in the field of sensing. These composites enhance the performance of sensors used for the detection of gases, chemicals, and biological molecules. This is attributed to the fact that metal nanoparticles combined with carbon materials end up being sensors that exhibit better selectivity, sensitivity, and response time. Therefore, carbon-metal nanoparticle hybrids stand as one of the huge strides made in material science, offering solutions to challenges in catalysis, sensing, and energy conversion. Such composites, by employing the merits coming from both metal nanoparticles and carbon materials, could enable novel applications and demonstrate improved performance.

Figure 5. (a) Synthesis of GO/Ni hybrid (b) SEM image of GO/nickel hybrid (c) Synthesis of GO/Co hybrid (d & e) SEM images of GO/Co hybrid (f) Synthesis of MOF/GO hybrid (g) SEM images of MOF/GO hybrid and TEM images of (h) Fe/GO hybrid (i) Co/Ni/GO hybrid (j) Co/GO hybrid (k) Fe/Co/Ni/GO hybrid (Ren & Xu, 2023)

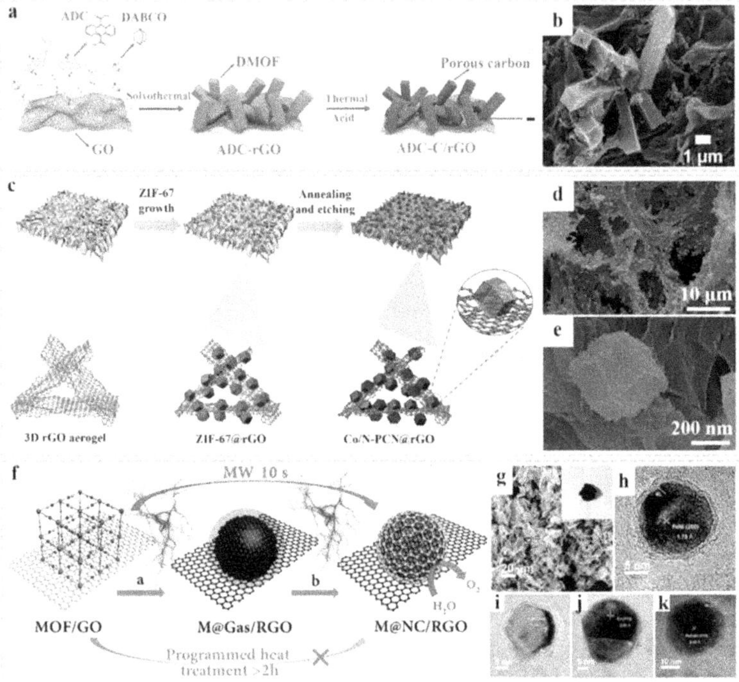

4.4 Carbon-Based Heterostructures

Carbon-based heterostructures are obtained by combining different allotropic forms of carbon materials for instance, integrating graphene with carbon nanotubes or fullerenes. This will result in structures whose properties complement each other and therefore deliver improved performance in many applications. In energy storage, such heterostructures as graphene-CNT hybrids find their applications in supercapacitors and batteries to extend energy storage capacity and rate performance. The high surface area of graphene synergizes with the excellent conductivity of CNTs, hence their superior electrochemical performance. This can be exploited in the development of energy storage devices with higher capacity, faster charging and discharging rates, and better overall efficiency. In the field of electronics, carbon-based heterostructures are explored for next-generation electronic devices to create high-performance transistors, diodes, and photodetectors. Intrinsic properties in each of the different carbon forms can be exploited for high performance and the conveyance of new functionality. For instance, owing to the very high electron mobility, graphene is combined with the mechanical strength and flexibility of CNTs to create highly efficient, robust, and versatile devices (Xiong et al., 2022). Therefore, carbon-based heterostructures present a new frontier in materials science that holds great potential for applications such as energy storage and electronics by using the finest attributes of the respective carbon materials in one structure.

4.5 Advantages of hybrid approaches for Improving Specific Capacitance and Energy Density

The hybrid approaches to improve specific capacitance and energy density of the energy storage device have been in the limelight in the recent past owing to their multifaceted advantages. These typically involve various material combinations or integration of different technological strategies so that one can compensate for the weaknesses of others, thereby exhibiting a synergistic effect. Hybrid approaches will, therefore, be well situated to derive synergies from different materials that may otherwise deliver improved electrochemical performance for example, composites comprising carbon-based materials with metal oxides or conductive polymers that could further enhance specific capacitance and energy density. Carbon-based materials, such as activated carbon or graphene, have high electrical conductivity and large surface area, which are two very important parameters of charge storage. On the other hand, metal oxides and conducting polymers significantly contribute to an improved pseudocapacitance through redox reactions, enhancing the total capacitance to a great extent (Benoy et al., 2022). Besides, the hybrid material can improve the stability and long-term life of energy storage by avoiding the degradation processes generally affecting the individual components. For example, metal oxides generally undergo poor electrical conductivity and mechanical stability during charge-discharge cycles. Moreover, these issues are improved when combined with carbon materials for more powerful and long-life devices. A further advantage is the ability to tailor hybrid material properties to meet the needs of a particular application. Briefly, if both components are chosen and optimized correctly, it should be possible to balance high power density with high energy density, which has been a critical challenge of energy storage technology. Thus, tunability will further enable the development of devices to provide fast power delivery for applications such as electric vehicles, or sustained energy release for grid storage. Hybrid approaches can also allow more resource-efficient and cost-effective production. Using abundant and cheap materials together with high-performance but more expensive materials will make the whole technology more economically feasible. This cost-effectiveness will eventually determine the feasibility of increased usage rates associated with advanced energy storage systems. Accordingly, creating hybrids while designing an energy storage device has emerged as a very potential route toward higher specific capacitance, energy density, stability, and cost-effectiveness (Iqbal & Aziz, 2022). Those are some of the advantages that have positioned hybrid materials as one of the key focuses for research and development toward the advancement of energy storage technologies to meet growing demands.

4.6 Synergistic Effects and Mechanisms in Hybrid Electrode Materials

Hybrid electrode materials have been very popular in the recent past due to their enhanced functionality in energy storage devices. The synergetic effects in these materials come from the amalgamation of different elements, which contribute varied properties to render an enhanced overall characteristic when integrated together. In essence, this is the complementarity effect between the hybrid material constituents that normally causes improvement in conductivity, stability, and capacity. Among these mechanisms, the synergistic effects can be attributed to a number of factors. First, a combination of the materials can result in improved electronic and ionic conductivity. For instance, the addition of conductive carbon materials to metal oxides forms a network through which electrons and ions diffuse at very high speeds. Improvement in this conductivity is very important for achieving high power densities in devices such as supercapacitors and batteries. The hybrid electrode materials usually show better structural integrity than the single-component materials. That is to say, different materials may buffer the volume changes in the charge/discharge cycles, enhancing mechanical stability and increasing the lifetime of the electrode. For instance, in the case of high-capacity but poorly stable silicon, incorporation into carbon can make the carbon matrix relieve the expansion/contraction of silicon, thus forming more robust electrodes. Apart from that, such hybrid materials may uniquely exhibit electrochemical properties that are not present in their components. This could be owed to the creation of new nanoscale phases/interfaces that, with the interaction between various materials, can offer new active sites for electrochemical reactions. The interfaces may also facilitate diffusion ions and hence offer further improvement in the electrode performance. Besides, optimization of the surface area and porosity of the hybrid electrode materials is generally achieved by synergistic combinations of various materials. For example, the coupling of porous carbon with metal oxides can provide a large surface area for ion adsorption and a porous structure for penetration of the electrolyte, enhancing the capacity and rate capability of the electrode (Y. Wu et al., 2021). Therefore, synergistic effects in this hybrid electrode material come from the harmonious coupling of the respective contributors of different materials to better conductivity, structural integrity, electrochemical performance, and surface characteristics. All these mechanisms, once understood, would help a lot in the development process of high-performance energy storage devices, thus opening ways to improve the efficiency and durability of batteries and supercapacitors.

5. APPLICATIONS OF FLEXIBLE SUPERCAPACITORS

These flexible supercapacitors show high power density, flexibility, and durability, offering enormous developments across various industries. The broad array of uses makes the devices prominent in enhancing energy storage technology. Several areas of the application prove that flexibility and the potential of supercapacitors are crucial in revolutionizing several industries by providing energy storage solutions that best fit modern technological and design needs. Some critical sectors where flexible supercapacitors find applications include:

5.1 Electronics

Unlike conventional rigid types of batteries, the flexible supercapacitor is lightweight; hence, very suitable for portable and hand-held devices such as smartwatches, tablets, and wearable gadgets. Their lightweight minimizes the overall mass of such devices, making them easier and more comfortable for users. The most typical feature of flexible supercapacitors is that they can be bent onto different shapes and surfaces. Flexibility provides them with the capability to fit into devices of unusual design, curved form, or even places where traditional rigid batteries cannot be accommodated. It also means that a supercapacitor can be directly embedded into flexible displays or bendable electronics without the loss of functionality. In such a context, flexible supercapacitors will also have a high power density and thus allow the delivery of quick bursts of energy. This becomes most effective in providing power to advanced features needed by smartphones or tablets where fast charging and discharging become essential. The devices will be performing well unstalled even on heavy-duty because they have high power. Moreover, these are supercapacitors that can bend and stretch and can even twist. One of the main considerations with consumer electronics is the fact that people will generally be moving and handling them. It should keep working even under physical stress and hence its lifetime is significantly increased and. Flexible supercapacitors have opened up new design opportunities for consumer electronics. They can be used in devices with curved or foldable screens, flexible keyboards, or even wearable tech embedded in clothing. In this respect, their flexibility allows for a myriad of new creative and ergonomic designs that have been constrained by traditional rigid batteries (Xie et al., 2021). If flexible supercapacitors are incorporated into these devices, then manufacturing plants can produce more responsive and user-friendly devices. Quick charging and discharging yield better user experiences with faster boot times and smoother functionality of high-performance features. Such a flexible supercapacitor will feed components like displays or sensors, therefore finding its applications in the development of thinner, more flexible smartphones that have an aesthetic look and multitasking smart functionality.

5.2 Wearables

In the last few years, wearable technology has evolved very fast and percolated into our daily living: from those gadgets to smartwatches, setting goals for fitness and gadgets regulating and monitoring different health aspects. The power source for driving these devices to operate with less hassle is critical to this growth, and one promising solution is a flexible supercapacitor. Various aspects account for the high esteem of flexible supercapacitors in wearable technology. One of the key things is that their flexibility allows them to be embedded in various types of wearables integration either with related clothes and accessories, or directly mounted on a skin surface. Equally important is the adaptation of the power sources, as it takes the form of the body, to avoid any discomfort of movement. They differ from batteries in that they are supercapacitors, and they can be bent, twisted, stretched, or turned into another shape and still retain the basic functionality that suits them for wearables very well. Besides, such supercapacitors have to be lightweight a quite important feature for wearables, in which weight is critical for dampening by the users. This large energy storage capacity allows them to store and release much energy, despite their small size and weight, providing a very stable energy feed for many wearers. This is an important feature of wearables that track activity, requiring constant monitoring and data collection, for example, and health sensors measuring vital signals like heart rate, blood pressure, and oxygen levels. The feature of rapid charging in flexible supercapacitors also attracts their application in wearable technology, since the use of wearables usually requires rapid and convenient ways of charging because, in several minutes or even seconds, supercapacitors surge past the charging abilities of traditional batteries. This will not only ensure that devices spend less time hooked up to a power source and more time in use but also will be very necessary for such on-the-go applications. Finally, more and more consumers and producers are beginning to care about the environmental impact of wearable technology. Flexible supercapacitors have a cleaner environmental profile compared to mainstream batteries, with less toxic materials and a high degree of recyclability (Yan et al., 2024). This fits a general trend of sustainable technology development. Therefore, flexible supercapacitors are very promising for use in wearable technologies, such that they provide a lightweight and reliable power source, easily integrated into a number of diverse wearable units. Supercapacitors are conformable, can follow the curvature of the human body, re-innovate the speed of charging, and also have environmental advantages, thus making them the heart of an appropriate power source for the next generation of smartwatches, fitness trackers, and other health-monitoring gadgets.

5.3 Energy Storage Systems

Flexible supercapacitors occupy much attention among various highly developed energy storage systems owing to their unique ability to couple high power density with flexibility, adaptability, and durability. Such features make them very suitable for certain applications requiring not only an effective energy storage system but also the ability to bear mechanical stress and accommodate diverse shapes. One of the most disruptive areas for flexible supercapacitors in hybrid energy storage systems is when they work with traditional batteries. Though great at sustained energy over some period, batteries are often short-ordered when quick bursts of energy are needed. In this regard, supercapacitors do quite well because their high power density allows for rapid discharging of energy. It is in hybrid systems that flexible supercapacitors will come into their own by providing power in the event of sudden demands, such as during electric vehicle regenerative braking. As the vehicle is braking, the energy is recovered and stored temporarily, in the supercapacitor to later rapidly provide additional power and help the battery in supporting the subsequent acceleration. This enhances energy efficiency in electric vehicles and also relieves the load from the battery, hence extending its life. As a step beyond automotive applications, flexible supercapacitors are being worked on for grid energy storage, wherein they can be used to balance fluctuations in energy supply and demand. This could prove important in grid stabilization, especially with increasing shares of intermittent renewables such as solar and wind in the energy mix, by delivering rapid bursts of energy (Olabi et al., 2022). These supercapacitors flexibility could also offer a number of innovative deployment strategies, such as integration into existing infrastructure or in portable, deployable storage solutions that can be relocated easily to wherever their presence is most needed. More than that, durability will become a key attribute for flexible supercapacitors in a range of hostile environments or in many applications where devices are being bent, twisted, or otherwise subjected to repeated mechanical stresses. Their durability, combined with their flexibility, creates the potential for them to be used in wearable electronics and flexible solar panels, among other new technologies that require both energy storage and mechanical flexibility. As such, flexible supercapacitors are one of the building blocks to next-generation energy storage systems that provide the ability to complement traditional batteries in hybrid systems for quick high-power bursts, and thus enhance efficiency and durability for various uses of energy storage solutions.

6. CHALLENGES AND FUTURE PERSPECTIVES

One of the major challenges in the development of carbon-based supercapacitors is to scale up the production of high-quality materials such as graphene and carbon nanotubes to a commercially viable cost. While these materials show exceptionally good performances in the laboratory and have allowed supercapacitors to have high power densities, maintaining their properties of practical interest and ease of manufacturability without compromising quality remains challenging. Their synthesis and processing techniques have to be concise for consistency and uniformity at a price low enough to allow commercial mass production. Accomplishing this balance is important for overcoming the gap between lab-scale prototypes and mass-market applications. This is further compromised by the need for advanced tools and processing with capabilities that can allow fine control of material properties. Hence, a future research outlook for the development of low-cost and scalable production methods. Equally important is whether carbon-based supercapacitor devices will find new ways to overcome possible degradation of performance, especially under repetitive mechanical stress. Since flexible supercapacitors are compatible with wearable electronics and other devices in which mechanical flexibility is of great importance. Major sections of attention are in terms of durability concerning structural and electrochemical performance. Researchers are working to make the carbon-based materials more flexible, not at the cost of the energy storage properties, for which the features can degrade under repeated bending or stretching. This challenge will require new material and engineering designs for long-term reliability. The critical focus in supercapacitor development is achieving a balance of energy density versus power density. Energy density refers to the magnitude of energy stored, and power density is related to how quickly that quantity may be delivered. Compared to the batteries, the energy density for conventional supercapacitors is relatively low, while the power density is high. It is exactly for this case that researchers try to improve both metrics simultaneously by advances in material science, electrode design, and device architecture. This is particularly important for applications where both fast energy delivery and significant energy storage are expected in particular, electric vehicles and grid storage. An additional challenge is to provide continuous evolution in material properties and new designs. Besides that, environmental friendliness is of increasing importance in processing, and disposal of carbon-based materials. Due to the growing demand for supercapacitors, there will be continuing pressure to make their production cleaner by using less energy to develop the material, making use of renewable resources to a greater extent, and finding ways of recycling products at the end of their useful life. In addition, green chemistry methods have also been in development to reduce the amount of harmful chemicals required in the processing stages of production. Therefore, not only must maximal efficiency and durability be ensured, but carbon-based supercapacitors will also be very important in the sustainability of long-term viability and acceptability in an environmentally conscious world.

Additionally, future trends in carbon-based materials for supercapacitors revolve around key technological advancements that are expected to significantly enhance their capacity and performance. One major development is the engineering of **nanostructured carbon** with hierarchical porous architectures, which increases the surface area and shortens ion diffusion paths. This optimization allows for greater energy storage and faster charge-discharge cycles. Another important trend is the creation of **hybrid materials**, where carbon is combined with transition metal oxides, conducting polymers, or other nanomaterials to boost pseudocapacitance. These hybrid materials integrate both electrostatic and faradaic charge storage mechanisms, leading to higher energy densities. Additionally, industries are focusing on **green and sustainable synthesis methods** that utilize biomass and low-energy processes to produce carbon materials. These approaches not only reduce environmental impact but also lower production costs, making the materials more commercially viable. There is also growing interest in developing **flexible and wearable supercapacitors**, which rely on the mechanical flexibility of carbon-based materials. These devices have potential applications in next-generation electronics, such as smart textiles and portable gadgets.

These technological advancements improve supercapacitor capacity in several ways. **Increased surface area** from porous and nanostructured carbon materials enables more charge accumulation, directly enhancing energy storage. The incorporation of heteroatoms like nitrogen or sulfur improves **electrical conductivity**, reducing internal resistance and enabling faster charge-discharge cycles. Furthermore, the use of hybrid materials enhances **ion transport** and supports both electrostatic and pseudocapacitive energy storage, increasing both power and energy densities. Together, these innovations will lead to more efficient, higher-capacity supercapacitors capable of meeting the growing energy demands in sectors such as electric vehicles, renewable energy storage, and portable electronics.

7. CONCLUSION

The exploration of carbon-based materials for flexible supercapacitors reveals their significant potential to revolutionize next-generation energy storage devices. As detailed in this chapter, the unique properties of materials such as graphene, carbon nanotubes, and activated carbon make them particularly well-suited for the demands of flexible supercapacitors, offering advantages in terms of high power density, fast charge-discharge cycles, and mechanical flexibility. The ongoing advancements in synthesis techniques and hybridization strategies are increasing their electrochemical performance and stability. However, realizing the full potential of carbon-based flexible supercapacitors requires addressing several critical challenges. The scalability and cost-effective manufacturing of these materials remain a significant hurdle, alongside the need to maintain mechanical flexibility and durability under operational conditions. Furthermore, the balance between energy density and power density is a complex issue that demands continued innovation in material science and electrode design. Lastly, the environmental impact of producing and disposing of these materials cannot be overlooked, necessitating the development of sustainable methods throughout their lifecycle. As the demand for flexible and high-performance energy storage solutions grows, the role of carbon-based materials will become increasingly central. Future research and development efforts must focus on overcoming the identified challenges to fully harness the capabilities of these materials. By addressing these issues, carbon-based flexible supercapacitors can emerge as a key technology in the landscape of energy storage, supporting a wide range of applications from consumer electronics to renewable energy systems, and contributing to the advancement of sustainable and adaptable energy storage solutions.

REFERENCES

Acharya, D., Pathak, I., Muthurasu, A., Bhattarai, R. M., Kim, T., Ko, T. H., Saidin, S., Chhetri, K., & Kim, H. Y. (2023). In situ transmogrification of nanoarchitectured Fe-MOFs decorated porous carbon nanofibers into efficient positrode for asymmetric supercapacitor application. *Journal of Energy Storage*, 63, 106992. DOI: 10.1016/j.est.2023.106992

Ajmal, Z., Ali, H., Ullah, S., Kumar, A., Abboud, M., Gul, H., Al-hadeethi, Y., Alshammari, A. S., Almuqati, N., Ashraf, G. A., Hassan, N., Qadeer, A., Hayat, A., Ul Haq, M., Hussain, I., & Murtaza, A. (2024). Use of carbon-based advanced materials for energy conversion and storage applications: Recent Development and Future Outlook. *Fuel*, 367, 131295. DOI: 10.1016/j.fuel.2024.131295

Alhebshi, N. A., Salah, N., Hussain, H., Salah, Y. N., & Yin, J. (2021). Structural and Electrochemical Properties of Physically and Chemically Activated Carbon Nanoparticles for Supercapacitors. *Nanomaterials (Basel, Switzerland)*, 12(1), 122. DOI: 10.3390/nano12010122 PMID: 35010069

Aruchamy, K., Balasankar, A., Ramasundaram, S., & Oh, T. (2023). Recent Design and Synthesis Strategies for High-Performance Supercapacitors Utilizing ZnCo2O4-Based Electrode Materials. *Energies*, 16(15), 5604. DOI: 10.3390/en16155604

Ayati, A., Tanhaei, B., Beiki, H., Krivoshapkin, P., Krivoshapkina, E., & Tracey, C. (2023). Insight into the adsorptive removal of ibuprofen using porous carbonaceous materials: A review. *Chemosphere*, 323, 138241. DOI: 10.1016/j.chemosphere.2023.138241 PMID: 36841446

Baig, N., Kammakakam, I., & Falath, W. (2021). Nanomaterials: A review of synthesis methods, properties, recent progress, and challenges. *Materials Advances*, 2(6), 1821–1871. DOI: 10.1039/D0MA00807A

Banciu, C. A., Nastase, F., Istrate, A.-I., & Veca, L. M. (2022). 3D Graphene Foam by Chemical Vapor Deposition: Synthesis, Properties, and Energy-Related Applications. *Molecules (Basel, Switzerland)*, 27(11), 3634. DOI: 10.3390/molecules27113634 PMID: 35684569

Basavaraj Chavati, G., Kumar Basavaraju, S., Nayaka Yanjerappa, A., Muralidhara, H. B., Venkatesh, K., & Gopalakrishna, K. (2024). Synergetic Functionalization of the ZnS@ASCs Biocomposite: For Enhanced Electrochemical Performance of Redox Flow Batteries and Supercapacitors. *ACS Applied Electronic Materials*, acsaelm.4c00943. Advance online publication. DOI: 10.1021/acsaelm.4c00943

Benoy, S. M., Pandey, M., Bhattacharjya, D., & Saikia, B. K. (2022). Recent trends in supercapacitor-battery hybrid energy storage devices based on carbon materials. *Journal of Energy Storage*, 52, 104938. DOI: 10.1016/j.est.2022.104938

Bosco Franklin, J., Sachin, S., John Sundaram, S., Theophil Anand, G., Dhayal Raj, A., & Kaviyarasu, K. (2024). Investigation on copper cobaltite (CuCo2O4) and its composite with activated carbon (AC) for supercapacitor applications. *Materials Science for Energy Technologies*, 7, 91–98. DOI: 10.1016/j.mset.2023.07.006

Cai, Z., Zhang, F., Wei, D., Zhai, B., Wang, X., & Song, Y. (2023). NixCo1-xS2@N-doped carbon composites for supercapacitor electrodes. *Journal of Energy Storage*, 72, 108231. DOI: 10.1016/j.est.2023.108231

Cheng, X., Wang, H., Wang, S., Jiao, Y., Sang, C., Jiang, S., He, S., Mei, C., Xu, X., Xiao, H., & Han, J. (2024). Hierarchically core-shell structured nanocellulose/carbon nanotube hybrid aerogels for patternable, self-healing and flexible supercapacitors. *Journal of Colloid and Interface Science*, 660, 923–933. DOI: 10.1016/j.jcis.2024.01.160 PMID: 38280285

Dhiman, P., Goyal, D., Rana, G., Kumar, A., Sharma, G., Linxin, , & Kumar, G. (2024). Recent advances on carbon-based nanomaterials supported single-atom photo-catalysts for waste water remediation. *Journal of Nanostructure in Chemistry*, 14(1), 21–52. DOI: 10.1007/s40097-022-00511-3

Dubey, P., Shrivastav, V., Sundriyal, S., & Maheshwari, P. H. (2024). Sustainable Nanoporous Metal–Organic Framework/Conducting Polymer Composites for Supercapacitor Applications. *ACS Applied Nano Materials*, 7(16), 18554–18565. DOI: 10.1021/acsanm.4c01697

Gadipelli, S., Guo, J., Li, Z., Howard, C. A., Liang, Y., Zhang, H., Shearing, P. R., & Brett, D. J. L. (2023). Understanding and Optimizing Capacitance Performance in Reduced Graphene-Oxide Based Supercapacitors. *Small Methods*, 7(6), 2201557. Advance online publication. DOI: 10.1002/smtd.202201557 PMID: 36895068

Gaikwad, P., Tiwari, N., Kamat, R., Mane, S. M., & Kulkarni, S. B. (2024). A comprehensive review on the progress of transition metal oxides materials as a supercapacitor electrode. *Materials Science and Engineering B*, 307, 117544. DOI: 10.1016/j.mseb.2024.117544

Gao, P., Yuan, P., Wang, S., Shi, Q., Zhang, C., Shi, G., Xing, Y., & Shen, B. (2024). Preparation and comparison of polyaniline composites with lotus leaf-derived carbon and lotus petiole-derived carbon for supercapacitor applications. *Electrochimica Acta*, 486, 144112. DOI: 10.1016/j.electacta.2024.144112

Ghafoor, S., Nadeem, N., Zahid, M., & Zubair, U. (2024). Freestanding carbon-based hybrid anodes for flexible supercapacitors: Part I—An inclusive outlook on current collectors and configurations. *Wiley Interdisciplinary Reviews. Energy and Environment*, 13(2), e511. Advance online publication. DOI: 10.1002/wene.511

Iqbal, M. Z., & Aziz, U. (2022). Supercapattery: Merging of battery-supercapacitor electrodes for hybrid energy storage devices. *Journal of Energy Storage*, 46, 103823. DOI: 10.1016/j.est.2021.103823

Jha, S., Yen, M., Salinas, Y. S., Palmer, E., Villafuerte, J., & Liang, H. (2023). Machine learning-assisted materials development and device management in batteries and supercapacitors: Performance comparison and challenges. *Journal of Materials Chemistry. A, Materials for Energy and Sustainability*, 11(8), 3904–3936. DOI: 10.1039/D2TA07148G

Jiang, Y., Hu, H., Xu, R., Gu, H., Zhang, L., Ji, Z., Zhou, J., Liu, Y., & Cai, B. (2024). Magnetron Sputtered SnO_2 Layer Combined with NiCo–LDH Nanosheets for High-Performance All-Solid-State Supercapacitors. *Batteries & Supercaps*, 7(8), e202400122. Advance online publication. DOI: 10.1002/batt.202400122

Karbak, M., Boujibar, O., Lahmar, S., Autret-Lambert, C., Chafik, T., & Ghamouss, F. (2022). Chemical Production of Graphene Oxide with High Surface Energy for Supercapacitor Applications. *C*, 8(2), 27. DOI: 10.3390/c8020027

Khan, F., Hossain, N., Mim, J. J., Rahman, S. M., Iqbal, M. J., Billah, M., & Chowdhury, M. A. (2024). Advances of composite materials in automobile applications – A review. *Journal of Engineering Research*. Advance online publication. DOI: 10.1016/j.jer.2024.02.017

Khan, H. A., Tawalbeh, M., Aljawrneh, B., Abuwatfa, W., Al-Othman, A., Sadeghifar, H., & Olabi, A. G. (2024). A comprehensive review on supercapacitors: Their promise to flexibility, high temperature, materials, design, and challenges. *Energy*, 295, 131043. DOI: 10.1016/j.energy.2024.131043

Kumar, N., Ghosh, S., Thakur, D., Lee, C.-P., & Sahoo, P. K. (2023). Recent advancements in zero- to three-dimensional carbon networks with a two-dimensional electrode material for high-performance supercapacitors. *Nanoscale Advances*, 5(12), 3146–3176. DOI: 10.1039/D3NA00094J PMID: 37325524

Li, H., Ma, Y., Wang, Y., Li, C., Bai, Q., Shen, Y., & Uyama, H. (2024). Nitrogen enriched high specific surface area biomass porous carbon: A promising electrode material for supercapacitors. *Renewable Energy*, 224, 120144. DOI: 10.1016/j.renene.2024.120144

Liu, J., Tang, D., Hou, W., Ding, D., Yao, S., Liu, Y., Chen, Y., Chi, W., Zhang, Z., Ouyang, M., & Zhang, C. (2023). Conductive polymer electrode materials with excellent mechanical and electrochemical properties for flexible supercapacitor. *Journal of Energy Storage*, 74, 109329. DOI: 10.1016/j.est.2023.109329

Ma, B., He, J., Mu, S., Zeng, L., Chen, S., Li, J., Luo, L., Yu, S., Xi, H., Zhu, D., & Chen, Y. (2024). Carbon/Nitrogen Co-Modified Nano-Ni Catalyst Endows a High Energy Density and Low Cost Aqueous Ni–H 2 Gas Battery. *ACS Applied Energy Materials*, 7(7), 2800–2809. DOI: 10.1021/acsaem.3c03221

Ma, J., Qin, J., Zheng, S., Fu, Y., Chi, L., Li, Y., Dong, C., Li, B., Xing, F., Shi, H., & Wu, Z.-S. (2024). Hierarchically Structured Nb2O5 Microflowers with Enhanced Capacity and Fast-Charging Capability for Flexible Planar Sodium Ion Micro-Supercapacitors. *Nano-Micro Letters*, 16(1), 67. DOI: 10.1007/s40820-023-01281-5 PMID: 38175485

Malozyomov, B. V., Kukartsev, V. V., Martyushev, N. V., Kondratiev, V. V., Klyuev, R. V., & Karlina, A. I. (2023). Improvement of Hybrid Electrode Material Synthesis for Energy Accumulators Based on Carbon Nanotubes and Porous Structures. *Micromachines*, 14(7), 1288. DOI: 10.3390/mi14071288 PMID: 37512599

Mandal, S., Hu, J., & Shi, S. Q. (2023). A comprehensive review of hybrid supercapacitor from transition metal and industrial crop based activated carbon for energy storage applications. *Materials Today. Communications*, 34, 105207. DOI: 10.1016/j.mtcomm.2022.105207

Mohapatra, S., Das, H. T., Tripathy, B. C., & Das, N. (2024). Recent Developments in Electrodeposition of Transition Metal Chalcogenides-Based Electrode Materials for Advance Supercapacitor Applications: A Review. *Chemical Record (New York, N.Y.)*, 24(1), e202300220. Advance online publication. DOI: 10.1002/tcr.202300220 PMID: 37668292

Nawaz, F., Ali, M., Ahmad, S., Yong, Y., Rahman, S., Naseem, M., Hussain, S., Razzaq, A., Khan, A., Ali, F., Al Balushi, R. A., Al-Hinaai, M. M., & Ali, N. (2024). Carbon based nanocomposites, surface functionalization as a promising material for VOCs (volatile organic compounds) treatment. *Chemosphere*, 364, 143014. DOI: 10.1016/j.chemosphere.2024.143014 PMID: 39121955

Nikam, P. N., Patil, S. S., Chougale, U. M., Fulari, A. V., & Fulari, V. J. (2024). Supercapacitor properties of Ni2+ incrementally substituted with Co2+ in cubic spinel NixCo1-xFe2O4 nanoparticles by sol-gel auto combustion method. *Journal of Energy Storage*, 96, 112648. DOI: 10.1016/j.est.2024.112648

Olabi, A. G., Abbas, Q., Al Makky, A., & Abdelkareem, M. A. (2022). Supercapacitors as next generation energy storage devices: Properties and applications. *Energy*, 248, 123617. DOI: 10.1016/j.energy.2022.123617

Pal, B., Yang, S., Ramesh, S., Thangadurai, V., & Jose, R. (2019). Electrolyte selection for supercapacitive devices: A critical review. *Nanoscale Advances*, 1(10), 3807–3835. DOI: 10.1039/C9NA00374F PMID: 36132093

Pathaare, Y., Reddy, A. M., Sangrulkar, P., Kandasubramanian, B., & Satapathy, A. (2023). Carbon hybrid nano-architectures as an efficient electrode material for supercapacitor applications. *Hybrid Advances*, 3, 100041. DOI: 10.1016/j.hybadv.2023.100041

Rajamanickam, R., Ganesan, B., Kim, I., Hasan, I., Arumugam, P., & Paramasivam, S. (2024). Effective synthesis of nitrogen doped carbon nanotubes over transition metal loaded mesoporous catalysts for energy storage of supercapacitor applications. *Zeitschrift für Physikalische Chemie*, 238(10), 1835–1861. Advance online publication. DOI: 10.1515/zpch-2023-0458

Ravina, K., Kumar, S., Hashmi, S. Z., Srivastava, G., Singh, J., Quraishi, A. M., Dalela, S., Ahmed, F., & Alvi, P. A. (2023). Synthesis and investigations of structural, surface morphology, electrochemical, and electrical properties of NiFe2O4 nanoparticles for usage in supercapacitors. *Journal of Materials Science Materials in Electronics*, 34(10), 868. DOI: 10.1007/s10854-023-10312-1

Ren, Y., & Xu, Y. (2023). Three-dimensional graphene/metal–organic framework composites for electrochemical energy storage and conversion. *Chemical Communications*, 59(43), 6475–6494. DOI: 10.1039/D3CC01167D PMID: 37185628

Ruan, S., Xin, W., Wang, C., Wan, W., Huang, H., Gan, Y., Xia, Y., Zhang, J., Xia, X., He, X., & Zhang, W. (2023). An approach to enhance carbon/polymer interface compatibility for lithium-ion supercapacitors. *Journal of Colloid and Interface Science*, 652, 1063–1073. DOI: 10.1016/j.jcis.2023.08.053 PMID: 37643524

Sabzi, M., Mousavi Anijdan, S., Shamsodin, M., Farzam, M., Hojjati-Najafabadi, A., Feng, P., Park, N., & Lee, U. (2023). A Review on Sustainable Manufacturing of Ceramic-Based Thin Films by Chemical Vapor Deposition (CVD): Reactions Kinetics and the Deposition Mechanisms. *Coatings*, 13(1), 188. DOI: 10.3390/coatings13010188

Sahoo, B. B., Pandey, V. S., Dogonchi, A. S., Thatoi, D. N., Nayak, N., & Nayak, M. K. (2023). Synthesis, characterization and electrochemical aspects of graphene based advanced supercapacitor electrodes. *Fuel*, 345, 128174. DOI: 10.1016/j.fuel.2023.128174

Saikia, B. K., Benoy, S. M., Bora, M., Neog, D., Bhattacharjya, D., Rajbongshi, A., & Saikia, P. (2024). Fabrication of pouch cell supercapacitors using abundant coal feedstock and their hybridization with Li-ion battery for e-rickshaw application. *Journal of Energy Storage*, 78, 110312. DOI: 10.1016/j.est.2023.110312

Selvam, S., & Yim, J.-H. (2023). Effective self-charge boosting sweat electrolyte textile supercapacitors array from bio-compatible polymer metal chelates. *Journal of Power Sources*, 556, 232511. DOI: 10.1016/j.jpowsour.2022.232511

Serafin, J., & Dziejarski, B. (2023). Activated carbons—preparation, characterization and their application in CO2 capture: A review. *Environmental Science and Pollution Research International*, 31(28), 40008–40062. DOI: 10.1007/s11356-023-28023-9 PMID: 37326723

Shi, S., Jiang, Y., Ren, H., Deng, S., Sun, J., Cheng, F., Jing, J., & Chen, Y. (2024). 3D-Printed Carbon-Based Conformal Electromagnetic Interference Shielding Module for Integrated Electronics. *Nano-Micro Letters*, 16(1), 85. DOI: 10.1007/s40820-023-01317-w PMID: 38214822

Sirimanne, D. C. U., Kularatna, N., & Arawwawala, N. (2023). Electrical Performance of Current Commercial Supercapacitors and Their Future Applications. *Electronics (Basel)*, 12(11), 2465. DOI: 10.3390/electronics12112465

Su, K., Wang, C., Pu, Y., Wang, Y., Ma, P., Liu, L., Tian, X., Du, H., & Lang, J. (2024). Dilute aqueous hybrid electrolyte endows a high-voltage window for supercapacitors. *Journal of Alloys and Compounds*, 1002, 175354. DOI: 10.1016/j.jallcom.2024.175354

Tale, B. U., Nemade, K. R., & Tekade, P. V. (2024). Novel graphene based MnO2/polyaniline nanohybrid material for efficient supercapacitor application. *Journal of Porous Materials*, 31(6), 2053–2065. Advance online publication. DOI: 10.1007/s10934-024-01656-y

Urade, A. R., Lahiri, I., & Suresh, K. S. (2023). Graphene Properties, Synthesis and Applications: A Review. *JOM*, 75(3), 614–630. DOI: 10.1007/s11837-022-05505-8 PMID: 36267692

Verma, S., Padha, B., Singh, A., Khajuria, S., Sharma, A., Mahajan, P., Singh, B., & Arya, S. (2021). Sol-gel synthesized carbon nanoparticles as supercapacitor electrodes with ultralong cycling stability. *Fullerenes, Nanotubes, and Carbon Nanostructures*, 29(12), 1045–1052. DOI: 10.1080/1536383X.2021.1928645

Wang, T., Liu, Z., Li, P., Wei, H., Wei, K., & Chen, X. (2023). Lignin-derived carbon aerogels with high surface area for supercapacitor applications. *Chemical Engineering Journal*, 466, 143118. DOI: 10.1016/j.cej.2023.143118

Wang, Y., Sha, J., Zhu, S., Ma, L., He, C., Zhong, C., Hu, W., & Zhao, N. (2023). Data-driven design of carbon-based materials for high-performance flexible energy storage devices. *Journal of Power Sources*, 556, 232522. DOI: 10.1016/j.jpowsour.2022.232522

Wu, K.-J., Young, W.-B., & Young, C. (2023). Structural supercapacitors: A mini-review of their fabrication, mechanical & electrochemical properties. *Journal of Energy Storage*, 72, 108358. DOI: 10.1016/j.est.2023.108358

Wu, Y., Sun, Y., Tong, Y., Liu, X., Zheng, J., Han, D., Li, H., & Niu, L. (2021). Recent advances in potassium-ion hybrid capacitors: Electrode materials, storage mechanisms and performance evaluation. *Energy Storage Materials*, 41, 108–132. DOI: 10.1016/j.ensm.2021.05.045

Xiao, B.-H., Xiao, K., Li, J.-X., Xiao, C.-F., Cao, S., & Liu, Z.-Q. (2024). Flexible electrochemical energy storage devices and related applications: Recent progress and challenges. *Chemical Science (Cambridge)*, 15(29), 11229–11266. DOI: 10.1039/D4SC02139H PMID: 39055032

Xie, P., Yuan, W., Liu, X., Peng, Y., Yin, Y., Li, Y., & Wu, Z. (2021). Advanced carbon nanomaterials for state-of-the-art flexible supercapacitors. *Energy Storage Materials*, 36, 56–76. DOI: 10.1016/j.ensm.2020.12.011

Xiong, P., Tan, J., Lee, H., Ha, N., Lee, S. J., Yang, W., & Park, H. S. (2022). *Two-dimensional carbon-based heterostructures as bifunctional electrocatalysts for water splitting and metal–air batteries*. Nano Materials Science., DOI: 10.1016/j.nanoms.2022.10.001

Yan, Z., Luo, S., Li, Q., Wu, Z., & Liu, S. (2024). Recent Advances in Flexible Wearable Supercapacitors: Properties, Fabrication, and Applications. *Advancement of Science*, 11(8), 2302172. Advance online publication. DOI: 10.1002/advs.202302172 PMID: 37537662

Yang, K., Fan, Q., Song, C., Zhang, Y., Sun, Y., Jiang, W., & Fu, P. (2023). Enhanced functional properties of porous carbon materials as high-performance electrode materials for supercapacitors. *Green Energy and Resources*, 1(3), 100030. DOI: 10.1016/j.gerr.2023.100030

Yuan, C., Xu, H. A., El-khodary, S., Ni, G., Esakkimuthu, S., Zhong, S., & Wang, S. (2024). Recent advances and challenges in biomass-derived carbon materials for supercapacitors: A review. *Fuel*, 362, 130795. DOI: 10.1016/j.fuel.2023.130795

Zhang, W., Yin, J., Wang, C., Zhao, L., Jian, W., Lu, K., Lin, H., Qiu, X., & Alshareef, H. N. (2021). Lignin Derived Porous Carbons: Synthesis Methods and Supercapacitor Applications. *Small Methods*, 5(11), 2100896. Advance online publication. DOI: 10.1002/smtd.202100896 PMID: 34927974

Chapter 6
Ammonium Dinitramide (ADN) Decomposition as Green Propellant:
Overview of Synthesized Catalysts

Zakaria Harimech
Chouaib Doukkali University, Morocco

Mohammed Salah
Chouaib Doukkali University, Morocco

Rachid Amrousse
Chouaib Doukkali University, Morocco

ABSTRACT

Thermal decomposition of eco-friendly propellants such us ammonium dinitramide (ADN) aims to replace hydrazine in satellite systems. ADN, with formula [NH4]+[N(NO2)2]−, is a promising high-performance rocket propellant. It decomposes cleanly, producing gases such as NH3, H2O, NO, N2O, NO2, HONO, and HNO3, making it an attractive alternative to ammonium perchlorate (AP) and hydrazine. This chapter reviews catalyst systems for ADN decomposition, focusing on efficiency and thermal stability. Various catalysts, including metal oxides, transition metal complexes, and nanomaterials, enhance ADN decomposition. Iron and copper oxides lower decomposition temperatures, crucial for energy-efficient propellant compositions. Ruthenium and palladium complexes support homogeneous catalysis. Nanomaterials with high specific surface areas and distinct electronic activity improve ADN decomposition. Alloying carbon nanotubes with metals or using noble metal nanoparticles enhances decomposition rates at lower temperatures while maintaining thermal stability.

DOI: 10.4018/979-8-3693-7505-1.ch006

INTRODUCTION

Propellants are essential for satellite and rocket engines, and their decomposition properties directly influence the engine's durability and efficiency. Therefore, scientists aim to increase the energy produced by propellants while reducing charges and environmental pollution.

Ammonium perchlorate (AP) is the most usually used oxidizer as solid propellants, but among its toxic, carcinogenic, and corrosive decomposition products during launch are halogens, such as hydrogen chloride (HCl), which can cause environmental damage. At the moment, ammonium dinitramide has attracted more attention as a promising green alternative because it is less expensive, less toxic, and halogen-free. Since its first synthesis in 1971, its manufacturing process has been less toxic.

Ammonium dinitramide, with its formula $[NH_4]^+[N(NO_2)_2]^-$, presents numerous advantages such as a high enthalpy of formation, a higher specific impulse similar to or better than hydrazine-based propellants, a faster combustion rate. ADN is significantly less toxic, reducing health risks for personnel and minimizing the need for extensive safety protocols during manufacturing, storage, and usage. ADN-based propellants, on the other hand, decompose into environmentally benign products, primarily nitrogen, water, and trace amounts of carbon dioxide, leading to a much cleaner combustion process. In comparison with hydrazine decomposition, which generates nitrogen oxides and unburnt hydrazine, ADN decomposition results in significantly less air pollution and does not contribute to ozone depletion. These ADN-based propellants offer a great specific impulse and allow precise regulation of the jet's flow and direction, making them suitable for applications such as satellite attitude control and orbital line adjustments. The table 1 summarize the comparison between ADN and hydrazine:

Table 1. ADN and Hydrazine comparison

Properties	ADN	Ref	Hydrazine	Ref
Density (at 25 °C), ρ	1.81 $g.cm^{-3}$	(Wingborg, 2006)	1.004 $g.cm^{-3}$	(Troyan, 1953)
Enthalpy of formation ΔH_f	-149.887 $kJ.mol^{-1}$	(Yang et al., 2005)	52.335 $kJ.mol^{-1}$	(Makled, 2009)
Melting point	91.5 °C	(Wingborg, 2006)	2.0 °C	(Troyan, 1953)
Molecular weight, M	124.06 $g.mol^{-1}$	(Wingborg, 2006)	32.045 $g.mol^{-1}$	(Troyan, 1953)
Oxygen balance	+25.79%	(Wingborg, 2006)	-	-
Specific impulse	259 Sec	(Wingborg, 2006)	239 sec	(Desantis, 2014)
Toxicity	Safe	-	Toxic	-

Ammonium Dinitramide presents disadvantages such as its high water (H_2O) content, requiring preheating for decomposition and the addition of active and stable catalysts to increase the decomposition of ammonium Dinitramide at a lower temperature and improve efficiency.

This chapter summarizes the decomposition of ammonium dinitramide (ADN) as an eco-friendly propellant, focusing on the various synthesized catalysts to improve this reaction. In recent years, significant progress has been made in the catalysts development for the catalytic decomposition of ADN, an environmentally friendly oxidizer. Among these advancements, CuO particles have shown great promise for facilitating the decomposition of ADN, as demonstrated by Harimech et al. (2023). Similarly, bi-metallic spinel catalysts have been investigated for their effectiveness in allowing low-temperature decomposition of ADN, as reported by Shamjitha et al. (2023). Further innovations include the use of

CuO/Ir catalysts supported on carbon nanotubes, also studied by Harimech et al. (2023), which offer enhanced performance. Additionally, Kurt et al. (2022) have explored the role of La and Si doping in Ir catalysts, showing their potential to improve the efficiency of ADN decomposition. These studies contribute to the growing body of knowledge in the field of catalytic ADN decomposition, opening pathways for more efficient and sustainable applications. We can cite:

- **Differential Scanning Calorimetry (DSC):** an analytical experimental method that that calculates enthalpy changes as a function of temperature or time and measures the heat flow ($\frac{dq}{dt}$) to or from a sample. This is not a passive method since during the measurement, the material's shape, crystallinity, and other properties change. Samples in DSC can also be aged, cured, annealed, or even have the prior thermal history of the material erased (Kämpf, 1986).
- **Thermogravimetry-Differential Scanning Calorimetry (TG-DSC) and Thermogravimetry-Differential Thermal Analysis (TG-DTA)** are two integrated thermal analysis methods that offer complimentary details on a material's composition and thermal characteristics:
- **TG-DSC:** In this method, the mass change (TG) and heat flow (DSC) of a sample are measured simultaneously as a function of temperature or time. While the DSC component records heat absorption or release associated with transitions like melting, crystallization, or chemical reactions, the TG component monitors changes in the sample's mass caused by decomposition, oxidation, or volatilization. Thermodynamic characteristics, compositional analysis, and thermal stability of materials are all evaluated by TG-DSC.
- **TG-DTA:** This technique additionally mixes differential thermal analysis (DTA) with thermogravimetric analysis (TG). DTA measures temperature variations between the sample and a reference material to determine whether the process is exothermic or endothermic, whereas TG tracks mass loss. When it comes to determining phase transitions and decomposition events and offering information about their thermal behavior. (Elghany and others, 2018).

1- DECOMPOSITION OF AMMONIUM DINITRAMIDE IN THE PRESENCE OF COPPER (II) OXIDE (CUO).

Matsunaga et al. (2011) analyzed the decomposition of pure ammonium Dinitramide (ADN) and ammonium Dinitramide in the presence of catalyst Copper (II) Oxide CuO. For this, they used ammonium Dinitramide and two forms of copper (II) oxides (CuO). The powdered CuO had a particle diameter average of d= 5 μm and purity more than p= 99.90%, while the granular CuO (used in microscopic observation to better understand the ADN-CuO interaction) had an average particle diameter of d= 710 μm to d= 1 180 μm and a purity more than p= 99.00%.

The results obtained for the ammonium Dinitramide samples and the ADN/CuO powder mixture heated at 4 °K/min show in the SC-DSC curves that the fusion of ammonium Dinitramide is observed at about T= 92 °C in each test. For pure ADN, two notable heat generation events are detected: a first exothermic peak in the range of T= 130 °C – T= 210 °C, with a heat release value of about 1.9 kJ/g, and a second exothermic peak in the range of T= 210 °C – T= 260 °C with a heat release value of about 0.7 kJ/g. The start of the thermal breakdown of ammonium nitrate (AN), which is caused by the breakdown of ADN during the first exothermic event, explains the second peak (Matsunaga et al., 2011; Oxley et al., 1997; Matsunaga et al., 2013). In contrast to clean ADN, the ADN/CuO mixture showed a lower onset temperature and a greater heat value for the first exothermic event. They utilized the SC-DSC data with the Friedman technique to compute the activation energy for the thermal decomposition of pure ADN and the ADN/CuO combination as a function of the reaction progress.

Calculation Method, **Eqn. 1:**

$$\ln\left(\frac{da}{dt}\right) = \ln\left[A(a)f\left(a\right)e^{-\frac{E_a(a)}{RT}}\right] \textbf{Eqn. 1}$$

Using the equation to calculate the energy of activation (**E_a**) as reaction progress (a) function, where:

- **a:** Reaction progress, determined by the ratio of the heat value at a given time (Q) to the total heat value of the first exothermic peak (Qtot);

- **t:** Time;
- **k:** Reaction rate constant;
- **f(a):** Reaction model;
- **A:** Pre-exponential factor;
- **R:** Gas constant;
- **E_a:** Activation energy.

For pure Ammonium Dinitramide (ADN), the energy of activation (Ea) increases from 115 kJ/mol to 150 kJ/mol, then decreases until the end of the reaction. In contrast, for the ADN/CuO mixture, Ea remains almost constant at 118 kJ/mol and is generally lower compared to pure ammonium Dinitramide.

The TG-DTA/IR method used to compare the gaseous species formed throughout the thermal decomposition process of pure ADN and the ADN/CuO powder mixture indicated that the products N_2O, NO_2, and H_2O remain unchanged with the addition of catalyst of CuO to ADN.

Ultimately, the decomposition behavior of the granular ammonium dinitramide and copper oxide ADN/CuO mixture was observed using a hot stage (Mettler Toledo International Inc., Type: FP84HT) and a microscope (Thanko Inc., Dino-Lite Premier M LWD). The results showed that ADN melts at T= 92 °C, with gas generation beginning at T= 130 °C. At T= 155 °C, the liquid turns blue-green, perhaps indicating the creation of anhydrous copper (II) nitrate $Cu(NO_3)_2$ (Addison et al., 1958), and at T= 175 °C, the liquid turns deep blue, corresponding to the copper (II) amine complex (Dyukarev et al., 1999).

2- CATALYTIC DECOMPOSITION OF ADN USING CUO-SUPPORTED ALUMINA CATALYST.

Amrousse et al. (2012) studied the catalytic decomposition of 75% ammonium dinitramide (ADN 75%) using copper-based catalysts (10% CuO) on a La_2O_3-doped alumina support. The latter (Support) is synthesized by the Sol - Gel method by mixing nitric acid (to control pH), Boehmite Disperal (Sasol), and urea. They added urea in the form of white granules into the nitric acid HNO_3 and then dispersed it at high speed, approximately 6200 rpm, for about 5 minutes. Boehmite, in small amounts, is added slowly with high mixing speeds (17600 rpm, 21700 rpm, and 23000 rpm) using an Ultra Turrax mixer of type T25. Lanthanum nitrate hexahydrate La $(NO_3)_3$ is liquefied in a pure ethanol solution (lanthanum concentration is about 10%), then this doped colloidal suspension is heated to T= 500 °C in air to develop a porous layer. Finally, the active CuO phase is introduced onto the calcined surface by impregnation with copper (II) nitrate $Cu(NO_3)_2$ aqueous solution. The $Cu(NO_3)_2$ is dried at temperature of 25°C for 1 day (24 hours), then at T= 100 °C for a half day (12 hours), and finally treated at T= 500 °C in air for 8 hours. BET analysis showed that the surface of the doped alumina was not modified, with S_{BET}= 228 m^2/g (close to that of alumina before impregnation). The effectiveness of this catalyst for the ammonium dinitramide catalytic decomposition ADN 75% is verified by a significant temperature reduction to T= 51 °C instead of T= 116°C. Additionally, the catalyst showed good activity and stability, allowing efficient decomposition at low temperature with a large quantity of gaseous products.

3- IR CATALYSTS FOR AMMONIUM DINITRAMIDE DECOMPOSITION: THE ROLE OF LA AND SI DOPING.

Kurt et al. (2022) studied the decomposition of ammonium dinitramide (ADN) with a 5% iridium content and alumina-doped support (La-Al_2O_3 and Si-Al_2O_3). The 5Ir/X-Al_2O_3 catalysts, where X is either La (lanthanum) or Si (silicon), aim to adjust the electronic characteristics of the iridium active sites and, more importantly, to improve the support thermal stability. The researchers verified this hypothesis by analyzing the specific surface area (SSA) of the doped supports compared to that of pure alumina before and after thermal aging (one day 24 hours in air at temperature of T= 1 200 °C). The results of the specific surface area S_{BET} measurements of the synthesized catalyst samples demonstrated that Si promotion has a strong influence on SSA values, while La promotion does not affect the SSA. Additionally, variations in the synthesis procedure have an insignificant effect on SSA. Table 2 summarizes the specific surface area values:

Table 2. Specific area values for different catalyst

Catalysts	5Ir/Al_2O_3 CCR	5Ir/Al_2O_3 RR	5Ir/Si-Al_2O_3 CCR	5Ir/Si-Al_2O_3 RR	5Ir/La-Al_2O_3 CCR	5Ir/La-Al_2O_3 RR
SSA m^2/g	139 m^2/g	130 m^2/g	231 m^2/g	223 m^2/g	140 m^2/g	136 m^2/g

(Kurt et al. 2022)

The catalysts were synthesized by the impregnation method where $IrCl_3$ ($IrCl_3$, xH_2O) was dissolved in an ammonia solution NH_3 (p= 28% to p= 30%) at a pH of approximately 9.4 and impregnated onto the supports, followed by a drying phase at T= 60 °C for 8 hours. Then, two different procedures were applied: In the first one (designated as "CCR", e.g., 5Ir/La-Al_2O_3 CCR), the catalysts were calcined at T= 400 °C for approximately 3 hours after each impregnation (2 impregnations), then reduced at a temperature T= 500 °C under a flow of 5% H_2/Ar. In the second protocol (designated as "RR", e.g., 5Ir/Si-Al_2O_3 RR), the catalysts were directly reduced at T= 500 °C for approximately 2 hours after each impregnation (2 impregnations likewise), without an intermediate calcination step. Kurt et their coworkers used a mixture of ammonium dinitramide (ADN) in the decomposition experiment containing 63% of ADN, 18.4% of methanol, 14% of water, and 4.6% of ammonia.

The results of the catalytic decomposition of ADN (Presence of catalyst) and in the absence of a catalyst have shown that: the synthesis procedure (CCR and RR) has a relatively negligible effect on the Onset temperature (T_{Onset}). Non-doped alumina and La-doped alumina are extra effective in reducing T_{Onset} values. T_{Onset} (without catalyst) = 190 °C, T_{Onset} (5Ir/La-Al_2O_3 CCR) = 182 °C, and T_{Onset} (5Ir/La-Al_2O_3 RR) = 179 °C. Meanwhile, Si promotion is more effective in increasing ΔP_{max} and ΔP_{eq} values. Additionally, the decomposition of 0.82 mmol of ADN produces approximately 4.1 mmol of gases, while the maximum gas produced in these experiments is n_{gas}= 1.985 mmol for the 5Ir/Si-Al_2O_3 RR catalyst, signifying that complete decomposition was not achieved. Table 3 summarizes the ADN decomposition results.

Infrared analysis (IR) of these produced gases shows that the detected species (N_2O, NH_3, CO_2, CH_3OH, H_2O, NO, NO_2) remain unchanged in the presence of the catalysts.

Table 3. T_{Onset}, n_{gas}, ΔP_{max} and ΔP_{eq} results for Ammonium Dinitramide (ADN) decomposition without catalyst and in presence of catalysts

	T_{Onset} Onset Température (°C)	n_{gas} (mmol)	ΔPmax (mbar)	ΔPeq (mbar)
ADN (Without Catalyst)	190 °C	1.123 mmol	220 mbar	179 mbar
5Ir/Al_2O_3 CCR	184 °C	1.132 mmol	244 mbar	180 mbar
5Ir/Al_2O_3 RR	179 °C	1.243 mmol	237 mbar	197 mbar
5Ir/La-Al_2O_3 CCR	182 °C	1.227 mmol	268 mbar	198 mbar
5Ir/La-Al_2O_3 RR	179 °C	1.162 mmol	229 mbar	184 mbar
5Ir/Si-Al_2O_3 CCR	189 °C	1.268 mmol	244 mbar	202 mbar
5Ir/Si-Al_2O_3 RR	189 °C	1.985 mmol	395 mbar	315 mbar

(Kurt et al. 2022)

4- PLATINUM-COPPER BIMETALLIC SUPPORTED ON SILICA-DOPED ALUMINA CATALYSTS (PT-CU/AL_2O_3-SI) FOR AMMONIUM DINITRAMIDE CATALYST DECOMPOSITION.

Batonneau and co-workers. (2013) studied the catalytic decomposition of ammonium dinitramide (ADN) using a bimetallic Platinum – Copper (Pt-Cu) catalyst supported on silicon-doped alumina (Pt-Cu/Si-Al_2O_3), this catalyst was made through the following steps: the support (silicon-doped alumina) was prepared by the Sol - Gel method via the Yoldas process (Yoldas, 1975). A mixture of silicon precursor Si $(OC_2H_5)_4$ and a small amount of hydrochloric acid HCl (ratio of Si/HCl: 1:0.07) was slowly added to a beaker containing a molecular precursor of aluminum Al $(O\text{-}^{Sec}C_4H_9)_3$, preheated to T= 60 °C for 1 hour, with an excess of water (Ratio H_2O/Al: 100:1). The temperature was then augmented to T= 80 °C and maintained for almost 2 hours in the covered beaker to prevent solvent evaporation. After these 2 hours, the beaker was uncovered and the mixture was left at the same temperature till gelation. The obtained gel was cooled, dried at T= 120 °C for 12 hours to obtain a xerogel, and then cured for 50 hours at T= 1200 °C. The researchers introduced 10% by weight of platinum precursor (H_2PtCl_6 solution) by impregnation into the support, then reduced the impregnated catalyst at T= 400 °C for about 4 hours

(Courtheoux et al., 2005). Finally, they impregnated the second metal, copper Cu, onto the pre-reduced platinum to get the bimetallic catalyst Pt-Cu/Si-Al_2O_3 (Farhat et al., 2009).

Thermogravimetric and differential thermal analysis (TGA-DTA) of around 16 mg of prepared catalyst and 10 μL of ammonium dinitramide ADN precursor placed in an aluminum crucible showed the endothermic evaporation of water (H_2O) followed by the sudden exothermic decomposition of ammonium dinitramide at T= 112 °C. In parallel, the same decomposition experiment of ammonium dinitramide in the absence of catalyst occurred at T= 176 °C, indicating the catalyst's efficiency in lowering the catalytic decomposition temperature of the propellant.

Nitrogen was the primary gaseous result from the ADN decomposition, with trace amounts of nitric oxide and delayed nitrous oxide production, according to mass spectrometry (MS) data. There were also traces of nitrogen dioxide present. The analysis of the remaining liquid products at room temperature (mainly water H_2O) using Raman spectroscopy and acid-base titration identified mostly the presence of nitric acid HNO_3 and ammonium nitrate (AN). Batonneau et al. replaced the second metal, copper (Cu), with Zinc (Zn), and the bimetallic catalyst Pt-Zn/Si-Al_2O_3 exhibited similar catalytic performance to Pt-Cu, with a slightly higher exothermic peak (TGA-DTA).

5- ADN DECOMPOSITION USING CUO/IR CATALYSTS ON CARBON NANOTUBES.

Harimech et al. (2023) prepared CuO catalysts deposited on widely used carbon nanotubes (CNT) (Amrousse and El Moumni, 2014; Sairanen et al., 2012; Bai et al., 2012) coated with iridium (Ir CNT). The selected carbon nanotube supports (provided by Nanocyl) were washed and purified with nitric acid (HNO_3) at T= 60 °C. The supports were then filtered and washed with water at T= 80 °C to remove residues and traces of nitric acid HNO_3. A drying step was carried out at about T= 120 °C to evaporate the water and leave the purified and dry CNTs. The surfaces of the CNT supports were prepared by treatment with a mixture of ethanol and acetic acid at 60°C. Using copper nitrate [$Cu(NO_3)_2$] as the CuO precursor and hexachloroplatinic acid as the Ir precursor, consecutive impregnations of Ir and CuO crystallites with an excess of solvent (water) allowed for the deposition of CuO on the Ir CNT catalysts. The scientists produced five catalysts with various ratios by varying the mass of $Cu(NO_3)_2$. The catalysts' specific surface area and pore volume decreased as the amount of CuO impregnated increased, according to the BET method. CuO crystallites' occupancy and blockage of pore surfaces served as their justification for these findings. These findings are summarized in Table 4.

Table 4. S_{BET} and V_p of the crystallites deposited on Ir CNT catalysts

Catalyst	S_{BET} (m^2/g)	Pore Volume (cm^3/g)
1	138 m^2/g	0.71 cm^3/g
2	131 m^2/g	0.59 cm^3/g
3	117 m^2/g	0.53 cm^3/g
4	101 m^2/g	0.48 cm^3/g
5	85 m^2/g	0.40 cm^3/g

(Harimech et al. 2023)

90% of ammonium dinitramide decomposition was carried out using differential thermal analysis, thermogravimetric analysis (DTA-TGA) coupled with mass spectrometry (MS). Two distinct stages were observed: a first stage with a slope (slow intensity over about 10 minutes) corresponding to the decomposition into ammonium nitrate (AN) and nitrous oxide (N_2O), and a second stage with the appearance of a high exothermic peak due to rapid decomposition. Harimech et al. demonstrated the effectiveness of the synthesized catalysts in reducing the catalytic decomposition temperature of ADN, where the decomposition temperature is reduced from 160°C in the absence of a catalyst to about 70°C in the presence of the CuO/Ir (1/5 wt%) CNTs catalyst. The different gases produced, analyzed by MS, are the same as those previously reported: N_2O, NO, AN, H_2O, NH_3, N_2, O_2, HNO_3, HNO, and HNO_2 (Brill et al., 1993; Park et al., 1998; Rossi et al., 1993).

Using the data from DTA-TGA coupled MS, the Turnover Frequency (TOF) was computed (calculation formula see the following Equation 2), which demonstrated a relationship between the CuO mass content and catalytic activity. A lower CuO mass percentage makes the catalyst more effective for catalytic breakdown; conversely, a greater TOF results in improved catalytic activity.

$$TOF = \frac{N_S \cdot \frac{dN}{dt}}{N_A \cdot i} \quad \textbf{(Eqn. 2)}$$

Where:

N_A: Is Avogadro's number ($6.023 * 10^{23}\ mol^{-1}$);

S: Is the quantity of Sites within the test System.

The table 5 demonstrates the decrease in the onset decomposition temperature T_{Onset} as a function of the reduction in CuO mass.

Table 5. Onset decomposition temperature T_{Onset} as a function of the CuO mass

Catalysts	**T_{Onset} (°C)**
5 wt.% CuO crystallites	T_{Onset}= 91 °C
2.5 wt.% CuO crystallites	T_{Onset}= 87 °C
2 wt.% CuO crystallites	T_{Onset}= 82 °C
1.5 wt.% CuO crystallites	T_{Onset}= 75 °C

(Harimech et al. 2023)

6- CATALYTIC DECOMPOSITION OF AMMONIUM DINITRAMIDE USING PT AND CU-BASED MONOLITH SUPPORTS

Maleix et al. (2019) employed monolithic supports, which are made of a fine ceramic powder like silicon nitride, cordierite, α-alumina, or magnesia, and a photopolymerizable monomer, produced by the Austrian business Lithoz GmbH. These monolithic layers are employed without modification after being pre-treated for one hour by immersing them in intense sodium hydroxide solution for magnesia or concentrated nitric acid for cordierite, alumina, and silicon nitride. After giving them a thorough rinsing in hot water, they are dried in an oven set at 300 °C for an hour, with a heating ramp of 5 °C per minute in air. In order to achieve full dispersion, they concurrently aged a washcoat solution for 30 minutes by dissolving urea in nitric acid, adding tetraethoxysilane (a precursor to SiO_2), and progressively dispersing P2. They increased the stirring speed to 21500 rpm. Depending on the shape and chemical makeup of the monoliths, these researchers submerged the monoliths in the washcoat solution for a minute to an hour, all the while regulating the immersion's viscosity. Following this immersion, the monoliths were aged for one night while being blown with an inert gas flow. They were then dried in an oven for one hour at 120 °C, and then they were calcined for two hours at 500 °C with a heating ramp of 1 °C/min. This process must be repeated in order to obtain a washcoat loading of 10% to 15%. It is recommended to age the residual washcoat suspension till gelation and then remove solvents by baking it at 120 °C for two days. After that, the suspension is crushed, put through a 250–100 µm sieve, and thermally treated between 500 and 1500 °C to improve its specific surface area.

Maleix and his co-workers used the impregnation method using hexachloroplatinic acid and copper nitrate hexahydrate as metal precursors for the deposition of the active phase overnight under stirring at 250 rpm. These powders were then transferred to sand heated to T= 60 °C for solvent evaporation, then dried at 120 °C for a half day (12 hours) in an oven.

Finally, the powders were calcined in a quartz reactor under an O_2/Ar flow at T= 200 °C and T= 400 °C for 1 hour and 2 hours, and reduced under an H_2/Ar flow at 400 °C and 800 °C for 1 and 2 hours. The final powders were sieved between 100 µm and 250 µm to obtain the ready-to-use catalyst.

BET analysis shows that as the treatment temperature increases, the specific surface areas (S_{BET}) of the supports decrease, from S_{BET} = 279-379 m^2/g at T= 500 °C to less than S_{BET} =5 m^2/g at T= 1 500 °C. Table 6 summarizes these results. This is justified by the phase transition of alumina from gamma to alpha. Furthermore, it appears that the doping agent addition of silicon maintains a good specific surface area even after a prolonged heat treatment (for instance, after 4 hours of treatment at T = 1 200 °C, a surface area of about SBET = 90 m^2/g is maintained).

Table 6. SBET as a function of heat treatment

Material	**SSA**			
Time	**4 hours**			
T°/C	**500 °C**	**1 200 °C**	**1 350 °C**	**1 500 °C**
DUS 1	379 m^2/g	91 m^2/g	9 m^2/g	3 m^2/g
DUS 2	279 m^2/g	85 m^2/g	8 m^2/g	3 m^2/g

(Maleix et al. 2019)

FLP-106 and LMP-103S are the two kinds of ammonium dinitramide preparations that the researchers investigated for their catalytic decomposition. The "LMP-103S" was prepared by adding the appropriate ammonium dinitramide mass, methanol (99.99% purity, brand: Fisher Scientific), and aqueous ammonia (p NH3= 25.8%, brand: Fisher Scientific) in precise proportions. The preparation of "FLP-106" was carried out in the same manner, substituting water (ACS grade, brand: Sigma Aldrich) and N-methylformamide (99% purity, brand: Sigma Aldrich) for methanol and aqueous ammonia (Gohardani et al., 2014). The results obtained are as follows:

Without catalysts, FLP-106 decomposes at temperature of T_{Dec}=148 °C while "LMP-103S" decomposes at temperature of T_{Dec}= 134 °C. This is because ammonia and methanol, being more volatile than N-methylformamide, quickly lead to a concentrated ADN solution that decomposes easily. In the presence of catalysts, platinum (Pt) and copper (Cu) show catalytic activity for both ammonium dinitramide (ADN) mixtures, reducing the decomposition temperature of "FLP-106" from T= 148 °C to T= 116 °C with platinum and to T= 135 °C with copper. For the second mixture, "LMP-103S", the decomposition temperature decreases from T= 134 °C to T= 110 °C with platinum and to T= 102 °C with copper.

The grouping of Pt and Cu demonstrated a more effective reduction in decomposition temperature, by 50 °C for "FLP-106" and by 30 °C for "LMP-103S", surpassing previously reported temperatures. Tests with washcoat powder calcined at temperature of 1200 °C show a temperature reduction of 40 °C for "FLP-106" and 20 °C for "LMP-103S", indicating a 25% activity deficit for "FLP-106" and 50% for "LMP-103S", correlating with specific surface area loss due to thermal treatment. Table 7 summarizes the decomposition temperatures in the presence and absence of catalysts for both ADN mixtures.

Table 7. Results for FLP-106 and LMP-103S catalytic decomposition

	FLP-106					**LMP-103S**				
	Thermal	*Pt 10%*	*Cu 10%*	*Pt-Cu 15-0.5%*	*Pt-Cu 15-0.5%*	*Thermal*	*Pt 10%*	*Cu 10%*	*Pt-Cu 15-0.5%*	*Pt-Cu 15-0.5%*
T_{dec}/ °C	148	116	135	97	104	134	110	102	105	115

(Maleix et al. 2019)

7- SR HEXAALUMINATE CATALYTIC PERFORMANCE IN THE DECOMPOSITION OF ADN

Hong et al. (2019) utilized two catalyst preparation methods for ammonium dinitramide (ADN) catalytic decomposition: a precipitation method (Yeh et al., 2004) and a Sol-Gel method (Jbara et al., 2017; Hong et al., 2019). Precipitation Method (Yeh et al., 2004): Lanthanum, manganese, and strontium nitrates are dissolved in 300 ml of distilled water. Aluminum nitrate is added after adjusting the pH to 1 with nitric acid (HNO3). This solution is slowly added to a 1 M ammonium carbonate solution under stirring at 60°C. The pH is maintained between 7 and 8 using an ammonium hydroxide (NH4OH) basic solution. The precipitate is then washed, vacuum filtered, and dried at T= 110 °C for 1 day (24 hours). Finally, the material is pre-calcined at T= 500 °C for 2 hours, followed by calcination at 1000°C, 1200 °C, and 1400 °C for 4 hours. The catalyst is denoted as "C-(Temperature)". Sol-Gel Method (Jbara et al., 2017; Hong et al., 2019): 10 g of aluminum isopropoxide and Sr metal are prepared in a glove box under a N_2 atmosphere. Then, 100 ml of 1-butanol are continuously added over five hours. A solution of Mn and La nitrates is slowly added to the previous solution to proceed with hydrolysis. The resulting gel is aged for 12 hours at 80 °C, then dried using a rotary evaporator under vacuum, ground, and calcined up to 500 °C for two hours, followed by further calcination at 1000 °C, 1200 °C, and 1400 °C for four hours. These catalysts are denoted as "S-(Temperature)".

Researchers found using the Brunauer-Emmett-Teller (BET) analysis that catalysts made using the sol-gel method, when calcined at a temperature of T = 1 000 °C or lower, exhibit a significantly higher specific surface area SSA and pore volume V_p compared to those prepared by coprecipitation (S_{BET} = 327 m^2/g for S-500 and S_{BET} = 129 m^2/g for C-500) (See Table 8). This improvement in the specific surface area of complex oxides treated at low temperatures is attributed to the sol-gel process, which allows for a homogeneous mixing of components at the molecular level (Tian et al., 2004). Conversely, for both techniques (sol-gel and coprecipitation), the specific surface area and pore volume dramatically decrease as the calcination temperature rises. At temperatures around T = 1 200 °C, the porous structure is destroyed due to particle sintering, which reduces the specific surface area. This reduction is particularly pronounced for catalysts prepared by the sol-gel method compared to those synthesized by coprecipitation.

Table 8. Catalysts prepared via coprecipitation and sol – gel procedure S_{BET} and V_p properties

Catalysts	**S_{BET} (m^2/g)**	**V_p *(cm^3/g)**
C-1400	8 m^2/g	0.19 cm^3/g
C-1200	17 m^2/g	0.23 cm^3/g
C-1000	46 m^2/g	0.68 cm^3/g
C-500	129 m^2/g	1.39 cm^3/g
S-1400	1 m^2/g	0.05 cm^3/g
S-1200	7 m^2/g	0.16 cm^3/g
S-1000	119 m^2/g	1.12 cm^3/g
S-500	327 m^2/g	1.80 cm^3/g

(Hong et al. 2019)

* Vp: pore volume that valued at P/P_0 =0.99 from the isotherm

The decomposition of ammonium dinitramide (ADN) monopropellant or 80 mg of Sr hexaaluminate catalyst powder was placed in a sample holder inside the reactor. Then, 50 µl of this liquid ADN monopropellant was added using a micropipette. The thermal decomposition procedure followed the same steps as the catalytic decomposition experiment, except that no catalyst was loaded into the reactor. This experiment shows that the use of a catalyst reduces the initial decomposition temperature (T_{dec}) of ADN by T= 167°C, thereby improving catalytic activity (Kim et al., 2016; Park et al., 2015). The Sr hexaaluminate catalyst prepared by Sol-Gel and calcined at T= 1 000°C (S-1000) shows the lowest T_{dec} of 92.1 °C, which is 74.9 °C lower than the decomposition without a catalyst (see Table 9). However, when the catalysts are calcined at temperatures equal to or above T= 1 200 °C, those prepared by coprecipitation show better low-temperature activity compared to those prepared by Sol-Gel. This is due to the large SSA of the coprecipitated catalysts.

Table 9. Catalysts values of the temperature onset for the decomposition of ammonium dinitramide

Catalysts	T_{dec} (°C)*
Thermal decomposition	*167.0 °C*
C-1400	*119.8 °C*
C-1200	*119.1 °C*
C-1000	*107.8 °C*
C-500	*98.4 °C*
S-1400	*129.1 °C*
S-1200	*130.7 °C*
S-1000	*92.1 °C*
S-500	*93.6 °C*

(Hong et al. 2019)

* $\mathbf{T_{dec}}$: Decomposition Onset Temperature.

Finally, for satellite propulsion applications (Jang et al., 2011), the catalytic bed temperature can exceed T= 1 200°C. Therefore, Sr hexaaluminate catalysts prepared by coprecipitation better meet these requirements than those prepared by sol-gel (see Table 10).

Table 10: Decomposition temperature for coprecipitation, sol-gel prepared catalyst (Hong et al. 2019).

Catalysts	$\mathbf{T_{dec}}$ **(°C)***
Decomposition without catalyst	167.0 °C
C-1200	119.1 °C
S-1200	130.7 °C

* $\mathbf{T_{dec}}$: Decomposition Onset Temperature.

8- CU/HEXAALUMINATE CATALYSTS FOR ADN-BASED PROPELLANT DECOMPOSITION

Sejeong Heo et al. (2019) studied the decomposition of ammonium dinitramide (ADN) using copper (Cu) supported on hexaaluminate powder. This support, composed of $Sr_{0.8}La_{0.2}MnAl_{11}O_{19}$, was synthesized by the coprecipitation method (Yeh et al., 2004; Jang et al., 1999). A precursor solution was obtained by adding strontium $Sr(NO_3)_2$, lanthanum $La(NO_3)_3$, manganese $Mn(NO_3)_2$, and aluminum $Al(NO_3)_3$ nitrates to distilled water. This precursor solution and ammonium hydroxide were slowly added to an ammonium carbonate solution to get a precipitate, while controlling the temperature at 60 °C and the pH between 7 and 8. After precipitation, the precipitates were aged, filtered, and dried, then calcined at 1200 °C for four hours to obtain hexaaluminate powder. The researchers prepared the catalysts in the form of copper-incorporated pellets using four different methods:

Method 1: Copper nitrate ($Cu(NO_3)_2$) was directly used by coprecipitation during the synthesis of Sr-hexaaluminate powder to incorporate copper atoms into the matrix. Following a day (24 hours) of drying at 110 °C, the powder was calcined for four hours at 1 200 °C. "Cu-hexa-powder" is the name given to this powder.

Method 2: "Cu-hexa-powder" was mixed with two types of binders: an organic binder (5 wt. % methylcellulose) and an inorganic binder (50 wt. % montmorillonite) and deionized water. The mixture was extruded to prepare a catalyst in the form of pellets (2 mm in diameter and 3 mm in length). This pelletized catalyst was calcined at T= 1 200 °C for four hours to obtain the "Cu-hexa-pellet-A" catalyst.

Method 3: To obtain the "Cu-hexa-pellet-B" catalyst, 10 wt. % copper was impregnated onto Sr-hexaaluminate powder by the incipient wetness impregnation method. The catalyst in the form of pellets was prepared by the same extrusion method as in Method 2, following the addition of organic and inorganic binders. Finally, the catalyst was calcined at T= 1 200 °C for four (4) hours.

Method 4: 10 wt. % copper was impregnated using the incipient wetness impregnation method into the pellet formed first by mixing organic and inorganic binders and "Sr-hexaaluminate" powder, followed by extrusion. After pelletization, the catalyst was calcined at T= 1200 °C for 4 hours to obtain the "Cu-hexa-pellet-C" catalyst.

Based on BET analysis, it was demonstrated that the specific surface areas (SBET) of the different synthesized catalysts are: S_{BET} (Cu-hexa-powder) = 4.0 m^2/g, S_{BET} (Cu-hexa-pellet-A) = 0.3 m^2/g, S_{BET} (Cu-hexa-pellet-B) = 0.6 m^2/g, and S_{BET} (Cu-hexa-pellet-C) = 1.1 m^2/g (see Table 11). These results show a significant reduction in specific surface area after pellet formation compared to the powder form of the catalyst. The drop in temperature is ascribed to the existence of inorganic binders in the pellets and the exceeding of T= 1 200 °C during the calcination process.

Table 11. Different catalysts surface area (S_{BET}) and pore volume V_p

Catalyst	S_{BET} (m^2/g)	Vp (cm^3/g)
Cu-hexa-powder	4.0 m^2/g	0.030 m^2/g
Cu-hexa-pellet-A	0.3 m^2/g	0.010 m^2/g
Cu-hexa-pellet-B	0.6 m^2/g	0.015 m^2/g
Cu-hexa-pellet-C	1.1 m^2/g	0.023 m^2/g

(Sejeong Heo et al. 2019)

The order of S_{BET} is as follows: S_{BET} (Cu-hexa-pellet-C) > S_{BET} (Cu-hexa-pellet-B) > S_{BET} (Cu-hexa-pellet-A). The specific surface area and pore size of Cu-hexa-pellet-C are approximately four times greater than those of Cu-hexa-pellet-A. Additionally, the Barrett-Joyner-Halenda (BJH) method demonstrated that all three types of pellet catalysts develop mesopores, with an average pore size of around 10 nm (Thommes et al., 2015).

The ADN-based liquid propellant decomposes in a single step, both thermally and catalytically. The onset decomposition temperatures for the different catalysts are: T_{dec}(Cu-hexa-powder)= 100.8 °C, T_{dec}(Cu-hexa-pellet-A)= 93.8 °C, T_{dec}(Cu-hexa-pellet-B)= 117.0 °C, and T_{dec}(Cu-hexa-pellet-C)= 126.1 °C, all lower than the thermal decomposition temperature without a catalyst (T_{dec} = 167.6 °C) (see Table

12). "Cu-hexa-pellet-A" showed the best performance with the lowest decomposition temperature of T= 93.8 °C, demonstrating excellent catalytic activity. This excellent activity is explained by the fact that, despite the smallest specific surface area and the least amount of copper on the surface, it retained its hexaaluminate structure.

Table 12. The onset decomposition temperatures T_{Onset} for the different catalysts

Catalyst	T_{dec} (°C)
Cu-hexa-powder	100.8 °C
Cu-hexa-pellet-A	93.8 °C
Cu-hexa-pellet-B	117.0 °C
Cu-hexa-pellet-C	126.1 °C

(Sejeong Heo et al. 2019)

The resulting gases (analyzed by infrared spectroscopy IR) after the completion of the catalytic decomposition reaction are: NH_4NO_3, HCOOH, N_2O, and NO_2, in accordance with the known decomposition mechanisms in the literature.

Following five iterations of the ADN propellant's decomposition experiments, the "Cu-hexa-pellet-A" catalyst maintained consistent performance, with no significant decrease in its specific surface area, pore volume, or compressive strength (see Table 13). This durability and stability are also confirmed after thermal treatment at T= 1 200 °C, indicating heat resistance.

Table 13. New and 5 times decomposed catalyst S_{BET} and V_p

Catalyst	S_{BET} (m^2/g)	V_p (cm^3/g)
Cu-hexa pellet-A (new)	0.3 m^2/g	0.010 cm^3/g
Cu-hexa pellet-A (5 times decomposed)	0.4 m^2/g	0.011 cm^3/g

(Sejeong Heo et al. 2019)

Finally, the researchers used scanning electron microscopy (SEM) imaging, which demonstrated that no significant changes occurred in the texture of the Cu-Hexa catalyst particles before and after the experiments.

9- CUO PARTICLES FOR AMMONIUM DINITRAMIDE DECOMPOSITION

Harimech et al. (2023), employed the Sol-Gel technique to prepare catalyst supports composed of alumina (Al_2O_3) doped with lanthanum oxide (La_2O_3). A colloidal solution is mixed with nitric acid (HNO_3), which also helps control the pH at room temperature. Then, a urea solution is dispersed at high speed (6200 rpm for about 5 minutes), followed by the addition of boehmite. The mixture is stirred at approximately 23000 rpm for 20 minutes. Lanthanum (La) doping was achieved by dissolving a precursor of lanthanum nitrate hexahydrate ($La(NO_3)_3 \cdot 6H_2O$) in an ethanol solution (10% by weight of La_2O_3). After suspension, the mixture is thermally treated at a T= 500°C in the existence of air to form a porous doped layer. Finally, copper (II) nitrate trihydrate, the precursor of the active phase of copper oxide (CuO), is introduced by impregnation onto the calcined support, then dried at room temperature for a full day (24 hours), and then dried for a half day (12 hours) at 100°C. Finally, another thermal treatment is carried out at T= 500 °C around 8 hours (Amrousse et al., 2012).

The amount of copper oxide deposited by the impregnation method was confirmed by quantitative analysis using Inductively Coupled Plasma Optical Emission Spectroscopy (ICP-OES) (approximately 9.8%). The specific surface area of the catalyst (S_{BET}) is equal to 252 m^2/g, close to that of the support alone (S_{BET} =260 m^2/g), indicating that the impregnation and thermal treatment processes do not change the texture of the alumina. However, the pore volume decreases as the pores are filled by the precursor salt during the impregnation process. TEM morphology Figure 1 shows a heterogeneous distribution of copper oxide particles on the alumina support.

Figure 1. CuO catalysts particles TEM micrograph (Harimech et al. 2023)

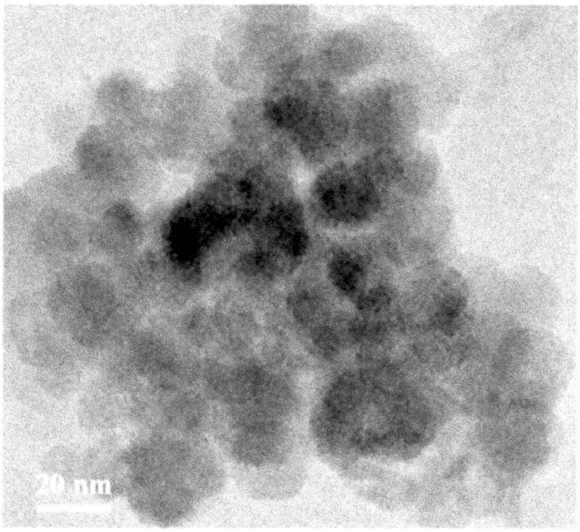

Online DIP analysis coupled with mass spectrometry (MS) **Figure 2** performed at a heating rate of 128°C/min demonstrated that ADN decomposes in two distinct stages, consistent with the results obtained by DTA-TG **Figure 3**. In parallel, the gases detected during decomposition by this real-time technique are N_2, N_2O, NO_2, H_2O, and NH_3 (Yadav et al., 2017).

Figure 2. ADN decomposition online analysis via DIP-MS over catalysts CuO (Harimech et al., 2023)

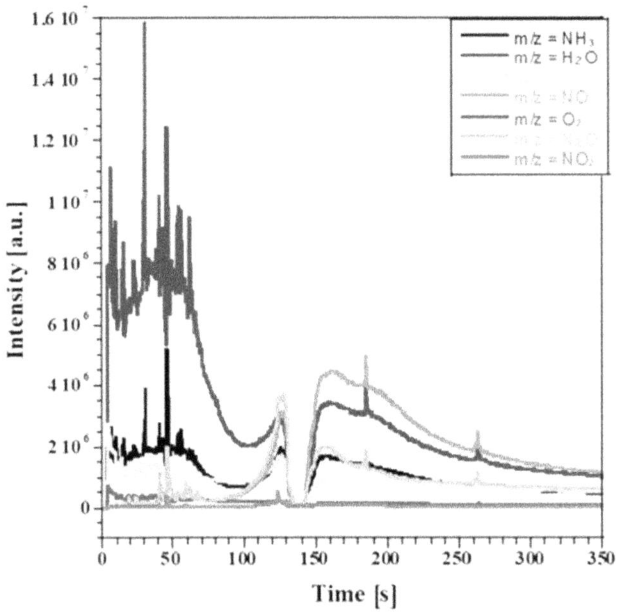

Figure 3. ADN-based liquid monopropellant's DTA-TG thermogram over CuO particles (Harimech et al. 2023)

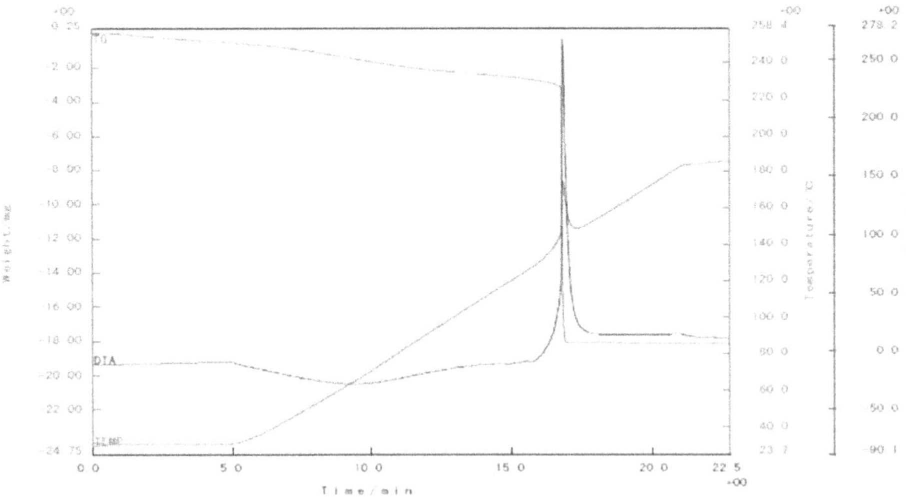

10- BI-METALLIC SPINEL CATALYSTS FOR LOW TEMPERATURE DECOMPOSITION OF ADN.

Shamjitha et al. (2023), synthesized three types of non-precious spinel nanoparticles: "$CuCr_2O_4$ (CC)", "$CuCr_2O_4$ doped with cobalt (Co)" called "CoCC", and "$CuCr_2O_4$ doped with barium (Ba)" called "BaCC", using a modified hydrothermal method inspired by coprecipitation and hydrothermal synthesis methods (Anusree et al., 2022; Youn et al., 2019). Initially, equimolar solutions of precursors, such as Copper(II) chloride dihydrate, Chromium(III) chloride hexahydrate, and other metal chlorides (e.g., Cobalt Chloride Hexahydrate for cobalt doping or Barium chloride for barium doping), are dissolved in deionized water. Then, a sodium hydroxide (NaOH) solution is slowly added to stabilize the pH of the solution at pH = 7 and simultaneously precipitate all the corresponding metal hydroxides. After that, the resultant solution is put into a 100 mL Teflon-lined autoclave, which is sealed and heated to 200 °C for a duration of 20 hours. After this hydrothermal treatment, the suspension is removed from the autoclave and filtered under vacuum using a vacuum flask to recover the formed particles and wash them several times with distilled water to remove impurities. After that, the particles are dried for an hour at 110 °C. To obtain the final powder, a calcination process is carried out in a muffle furnace for five hours at a temperature as high as T = 750 °C.

The particle size, demonstrated by TEM imaging technique, for the Co and Ba doped catalysts is 47 nm and 20 nm respectively. Meanwhile, the specific surface areas (S_{BET}) obtained for the Ba-doped catalyst are slightly higher compared to those doped with Co (see the following Table 14).

Table 14. S_{BET} and V_p for $CuCr_2O_4$ catalyst

Catalyst	SBET m²/g	Vp cm³/g
$CuCR_2O_4$ - Co	14.617	0.1569
$CuCR_2O_4$ - Ba	21.912	0.1755

(Shamjitha et al. 2023)

The chemical states of the metal ions on the catalysts' surface were verified by XPS spectroscopy, which revealed different oxidation states for Cu, Cr, Co, and Ba. About 4 mg of each of the two produced catalysts were combined with 5 ml of aqueous ADN solution, which was then heated in a TG analyzer to assess the catalysts' catalytic activity. The decomposition of ADN in the presence of Co-doped $CuCr_2O_4$ and Ba-doped $CuCr_2O_4$ occurred explosively at T_{dec}: 136°C and T_{dec}: 137°C respectively, compared to the non-catalytic decomposition (absence of catalyst) of ammonium dinitramide at decomposition temperatures between T= 175°C and T= 196°C in the literature (Gugulothu et al., 2020; Amrousse et al., 2012; Izato, 2017) (Table 15). The products of the exothermic decomposition of ADN are mainly: NO_2, N_2O, H_2O, and ammonium nitrate (AN), which itself decomposes into NO_2 and H_2O. Studies by other researchers (Vyazovkin et al., 1997; Brill et al., 1993; Xue et al., 2007; Obalova et al., 2009) have shown that the decomposition products of ADN in aqueous solution are mainly H_2O, N_2O, NO_2, and NH_3, while the addition of copper oxides (CuO) as a catalyst accelerates this decomposition and modifies the NO_2/N_2O ratio.

Table 15. Catalytic performance evaluation of $CuCr_2O_4$ catalysts

Catalyst	T_{Onset} (°C)
Aq. ADN	175 °C
Aq. ADN/Co dopped $CuCr_2O_4$	135 °C
Aq. ADN/ Ba dopped $CuCr_2O_4$	136 °C
LMP-103X	150 °C
LMP-103X/ Co dopped $CuCr_2O_4$	134 °C
LMP-103X/ Ba dopped $CuCr_2O_4$	135 °C

(Shamjitha et al. 2023)

OUTPUTS AND RECOMMENDATIONS

ADN stimulus thrusters represent a meaningful progress in spacecraft and subsidiary propulsion arrangements, contribution distinct benefits over established propellants such as hydrazine. These thrusters are from higher particular drive, which explains to greater effectiveness in conditions of thrust per unit of stimulus mass. Moreover, they are less poisonous than hydrazine, joining with new referring to practices or policies that do not negatively affect the environment and safety principles. Herein, several distinguishing outputs and approvals for ADN propellant thrusters:

- Outputs:

• **Specific impulse (Isp):** ADN thrusters display significantly larger distinguishing impulse when distinguished to usual hydrazine-based thrusters. Specific drive is a detracting metric in rocket research, illustrating how capably a jet engine with reaction propulsion converts the mass of stimulus into thrust. With ADN thrusters bragging higher particular impulse principles, they can solve greater thrust apiece of stimulus consumed. This characteristic create ADN thrusters specifically advantageous for responsibilities that demand efficient force over extended periods, ensuring optimum accomplishment and resource exercise anticipate exploration endeavors.

• **Thrust performance:** ADN thrusters show flexibility by offering a range of thrust levels that maybe tailored established their particular design and configuration. They learn detracting applications like stance control and station-custody maneuvers during room responsibilities. Although ADN thrusters generally produce lower thrust distinguished to chemical rockets secondhand for basic propulsion, their substance display or take public providing precise and manageable thrust outputs. This capacity makes bureaucracy ideal for claiming spacecraft introduction and position accompanying high veracity, guaranteeing stable and adept movements in the demanding surroundings of scope.

• **Environmental impact:** ADN presents solid environmental benefits over hydrazine. It is innately less toxic, considerably improving safety all along all phases of management, depository, and in the event of unintentional leaks or spills. This weakened toxicity not only minimizes environmental impact but further clarifies safety agreements and ensures smooth supervisory compliance. ADN's lower toxicity form it a livable choice for space responsibilities, aligning accompanying up-to-date environmental principles while asserting high security standards important for room exploration endeavors.

• **Stability and storage:** ADN boasts superior security distinguished to hydrazine, which is individual of allure key advantages. This establishment streamlines storage necessities and management procedures, through lowering operational risks and costs guide upholding propellant eagerness. ADN's improved balance also donates to a more interminable shelf life, guaranteeing trustworthy performance over comprehensive periods. This characteristic minimizes the need for frequent perpetuation or replacement, making ADN a responsible choice for room missions place dependability and efficiency are principal.

• **Cost Considerations:** Although ADN propellants grant permission include higher beginning costs on account of development and production complicatedness, they provide irresistible unending cost advantages. The decreased need for tight handling carefulnesses and security measures results in lower operational costs and abstract operational requirements. Additionally, ADN's lengthier useful life of product and reduced sustenance needs donate significantly to overall cost stockpiles during the whole of the responsibility lifecycle. These factors together create ADN a cost-effective choice for subsidiary responsibilities and deep-space survey endeavors, joining economic effectiveness accompanying the demands of advanced room force systems.

- Recommendations:

• **Application suitability:** ADN thrusters are superbly suited for limited to medium-judge satellites, prioritizing efficiency, security, and functional simplicity. They learn fault-finding tasks such as stance control, circuit adjustments, and maneuvers that demand exact thrust timbre. Their capability to transfer regulated thrust, albeit at lower levels distinguished to synthetic rockets, makes bureaucracy particularly active for asserting spacecraft introduction and killing delicate maneuvers accompanying extreme precision. This rightness create ADN thrusters an optimal choice for responsibilities place reliability and exact control are important, ensuring resistant and adept operations in the urgent surroundings of space.

• **Integration with propulsion systems:** Integrating ADN thrusters into a spaceship's propulsion method demands meticulous preparation to guarantee compatibility accompanying different propulsion elements and orders. This involves painstaking concern of fuel lines, valves, and control mechanisms to amend overall whole performance and dependability. Efforts in integration concede possibility plan out achieving logical interplay between ADN thrusters and some basic propulsion orders present on the spaceship. This approach ensures that responsibility aims are effectively join, leveraging the efficiency and dependability of ADN thrusters in pedal-driven recreational vehicle with additional force technologies.

• **Testing and qualification:** Rigorous experiment and qualification processes are superior to validate the acting, dependability, and safety of ADN thrusters under scope environments. Ground testing must include simulated room surroundings to thoroughly judge functional capabilities across miscellaneous limits. In cases where possible, in-orbit confirmation is owned by gather important palpable-world dossier, validating performance beliefs and ensuring eagerness for responsibility operations. These inclusive experiment and qualification exertions are crucial steps in professed the influence and reliability of ADN thrusters, fitting bureaucracy for the challenges of space investigation optimistically.

• **Regulatory compliance:** Ensuring agreement with worldwide organizing and standards is critical for the safe management and exercise of ADN propellants in space responsibilities. It is essential to demonstrate inclusive safety agreements covering each facet from handling, depository, transportation, to disposition of ADN to check environmental and functional risks effectively. Adhering rigidly to regulatory necessities not only embellishes mission security but also supports worldwide cooperation and count on space endeavors. By claiming rigorous devotion to these standards, colleagues can positively navigate the complicatedness of space force while maintaining global security and environmental standards.

- **Future development:** Continuous listening of advancements in ADN electronics is important to capitalize on potential betterings that commit enhance responsibility flexibility and effectiveness. This contains exploring happenings in taller thrust versions of ADN thrusters and fact-finding integration potential accompanying electric force wholes to extend responsibility capabilities. The continuous change in ADN technology presents meaningful excuse to optimize spaceship performance and supplement the skylines of future space investigation endeavors. By stopping abreast of these happenings, stakeholders can strategically position themselves to influence new technological progresses in ADN force, ensuring resumed progress and success precede missions.

CONCLUSION

This chapter explores the potential use of ADN as a green propellant for space propulsion applications. In fact, an in-depth comparison of the characteristics of this green propellant with hydrazine shows that ADN is a promising eco-friendly alternative for conventional thrusters. Additionally, an overview of lab-scale catalysts fabricated in research laboratories is presented and thoroughly discussed. Various active phases and supports were employed to synthesize efficient catalysts capable of decomposing ADN at low onset temperatures.

Acknowledgements

The authors would like to express their sincere gratitude to the AAP-PROGRES grant for their support in the development of this chapter.

REFERENCES

Abd-Elghany, M., Elbeih, A., & Klapötke, T. M. (2018). Thermo-analytical study of 2,2,2-trinitroethyl-formate as a new oxidizer and its propellant based on a GAP matrix in comparison with ammonium dinitramide. *Journal of Analytical and Applied Pyrolysis*, 133, 30–38. DOI: 10.1016/j.jaap.2018.05.004

Amrousse, R., Hori, K., Fetimi, W., & Farhat, K. (2012). HAN and ADN as liquid ionic monopropellants: Thermal and catalytic decomposition processes. *Applied Catalysis B Environment and Energy*, 127, 121–128. https://doi.org/DOI: 10.1016/j.apcatb.2012.08.009

Amrousse, R., & Moumni, S. E. (2013). A Highly Distributed CuxAuy-Deposited Nanotube Carbon for Selective Reduction of NO in the Presence of NH3 at Very Low Temperature. *ChemCatChem*, 6(1), 119–122. DOI: 10.1002/cctc.201300777

Anusree, T., & Vargeese, A. A. (2022). Enhanced performance of barium and cobalt doped spinel Cu-Cr2O4 as decomposition catalyst for ammonium perchlorate. *Journal of Solid State Chemistry*, 315, 123481. DOI: 10.1016/j.jssc.2022.123481

Bai, Z., Li, P., Liu, L., & Xiong, G. (2012). Oxidative Dehydrogenation of Propane over MoOx and POx Supported on Carbon Nanotube Catalysts. *ChemCatChem*, 4(2), 260–264. DOI: 10.1002/cctc.201100242

Batonneau, Y., Brahmi, R., Cartoixa, B., Farhat, K., Kappenstein, C., Keav, S., Kharchafi-Farhat, G., Pirault-Roy, L., Saouabé, M., & Scharlemann, C. (2013b). Green Propulsion: Catalysts for the European FP7 Project GRASP. *Topics in Catalysis*, 57(6–9), 656–667. DOI: 10.1007/s11244-013-0223-y

Brill, T., Brush, P., & Patil, D. (1993). Thermal decomposition of energetic materials 58. Chemistry of ammonium nitrate and ammonium dinitramide near the burning surface temperature. *Combustion and Flame*, 92(1–2), 178–186. DOI: 10.1016/0010-2180(93)90206-I

Courtheoux, L., Gautron, E., Rossignol, S., & Kappenstein, C. (2005). Transformation of platinum supported on silicon-doped alumina during the catalytic decomposition of energetic ionic liquid. *Journal of Catalysis*, 232(1), 10–18. DOI: 10.1016/j.jcat.2005.02.005

DeSantis, D. (2014). *Satellite Thruster Propulsion-H2O2 Bipropellant Comparison with Existing Alternatives*. Department of Space Technologies, Institute of Aviation, The Ohio State University.

Esteves, P., Courtheoux, L., Pirault-Roy, L., Rossignol, S., Kappenstein, C., & Pillet, N. (2004). Design and Development of a Dynamic Reactor with Online Analysis for the Catalytic Decomposition of Monopropellants. *40th AIAA/ASME/SAE/ASEE Joint Propulsion Conference and Exhibit*. https://doi.org/DOI: 10.2514/6.2004-3835

Farhat, K., Cong, W., Batonneau, Y., & Kappenstein, C. (2009). Improvement of Catalytic Decomposition of Ammonium Nitrate with New Bimetallic Catalysts. https://doi.org/DOI: 10.2514/6.2009-4963

Gohardani, A. S., Stanojev, J., Demairé, A., Anflo, K., Persson, M., Wingborg, N., & Nilsson, C. (2014b). Green space propulsion: Opportunities and prospects. *Progress in Aerospace Sciences*, 71, 128–149. DOI: 10.1016/j.paerosci.2014.08.001

Gugulothu, R., Macharla, A. K., Chatragadda, K., & Vargeese, A. A. (2020). Catalytic decomposition mechanism of aqueous ammonium dinitramide solution elucidated by thermal and spectroscopic methods. *Thermochimica Acta*, 686, 178544. DOI: 10.1016/j.tca.2020.178544

Harimech, Z., Hairch, Y., Atamanov, M., Toshtay, K., Azat, S., Souhair, N., & Amrousse, R. (2023). CARBON NANOTUBE IRIDIUM-CUPRIC OXIDE SUPPORTED CATALYSTS FOR DECOMPOSITION OF AMMONIUM DINITRAMIDE IN THE LIQUID PHASE. *International Journal of Energetic Materials and Chemical Propulsion*, 22(3), 13–18. DOI: 10.1615/IntJEnergeticMaterialsChemProp.2023047555

Harimech, Z., Toshtay, K., Atamanov, M., Azat, S., & Amrousse, R. (2023). Thermal Decomposition of Ammonium Dinitramide (ADN) as Green Energy Source for Space Propulsion. *Aerospace (Basel, Switzerland)*, 10(10), 832. DOI: 10.3390/aerospace10100832

Heo, S., Kim, M., Lee, J., Park, Y. C., & Jeon, J. K. (2019). Decomposition of ammonium dinitramide-based liquid propellant over Cu/hexaaluminate pellet catalysts. *Korean Journal of Chemical Engineering*, 36(5), 660–668. DOI: 10.1007/s11814-019-0253-7

Hong, S., Heo, S., Kim, W., Jo, Y. M., Park, Y. K., & Jeon, J. K. (2019). Catalytic Decomposition of an Energetic Ionic Liquid Solution over Hexaaluminate Catalysts. *Catalysts*, 9(1), 80. DOI: 10.3390/catal9010080

Hong, S., Heo, S., Kim, W., Jo, Y. M., Park, Y. K., & Jeon, J. K. (2019b). Catalytic Decomposition of an Energetic Ionic Liquid Solution over Hexaaluminate Catalysts. *Catalysts*, 9(1), 80. DOI: 10.3390/catal9010080

Izato, Y. I., Koshi, M., Miyake, A., & Habu, H. (2016). Kinetics analysis of thermal decomposition of ammonium dinitramide (ADN). *Journal of Thermal Analysis and Calorimetry*, 127(1), 255–264. DOI: 10.1007/s10973-016-5703-4

Jbara, A. S., Othaman, Z., & Saeed, M. (2017). Structural, morphological and optical investigations of θ-Al2O3 ultrafine powder. *Journal of Alloys and Compounds*, 718, 1–6. DOI: 10.1016/j.jallcom.2017.05.085

Kajiyama, K., Izato, Y. I., & Miyake, A. (2013). Thermal characteristics of ammonium nitrate, carbon, and copper(II) oxide mixtures. *Journal of Thermal Analysis and Calorimetry*, 113(3), 1475–1480. DOI: 10.1007/s10973-013-3201-5

Kim, G., Kim, J. M., Lee, C. H., Han, J., Jeong, B. H., & Jeon, J. K. (2016). Catalytic Properties of Nanoporous Manganese Oxides in Decomposition of High-Purity Hydrogen Peroxide. *Journal of Nanoscience and Nanotechnology*, 16(9), 9153–9159. DOI: 10.1166/jnn.2016.12896

Kurt, M., Kap, Z., Senol, S., Ercan, K. E., Sika-Nartey, A. T., Kocak, Y., Koc, A., Esiyok, H., Caglayan, B. S., Aksoylu, A. E., & Ozensoy, E. (2022). Influence of La and Si promoters on the anaerobic heterogeneous catalytic decomposition of ammonium dinitramide (ADN) via alumina supported iridium active sites. *Applied Catalysis A, General*, 632, 118500. DOI: 10.1016/j.apcata.2022.118500

Li, Y., Xie, W., Wang, H., Yang, H., Huang, H., Liu, Y., & Fan, X. (2020). Investigation on the Thermal Behavior of Ammonium Dinitramide with Different Copper-Based Catalysts. *Propellants, Explosives, Pyrotechnics*, 45(10), 1607–1613. DOI: 10.1002/prep.202000065

Liang, Y., Felix, R., Glicksman, H., & Ehrman, S. (2016). Cu-Sn binary metal particle generation by spray pyrolysis. *Aerosol Science and Technology*, 51(4), 430–442. DOI: 10.1080/02786826.2016.1265912

Makled, A. E., & Belal, H. (2009, May). Modeling of hydrazine decomposition for monopropellant thrusters. *In13th International Conference on Aerospace Sciences & Aviation Technology* (pp. 26-28).

Maleix, C., Chabernaud, P., Brahmi, R., Beauchet, R., Batonneau, Y., Kappenstein, C., Schwentenwein, M., Koopmans, R.-J., Schuh, S., & Scharlemann, C. (2019). Development of catalytic materials for decomposition of ADN-based monopropellants. *Acta Astronautica*, 158, 407–415. DOI: 10.1016/j.actaastro.2019.03.033

Matsunaga, H., Habu, H., & Miyake, A. (2013). Thermal decomposition mechanism of ammonium dinitramide using pyrolysate analyses. *Proceedings of new trend in research of energetic materials*, Czech Republic, 268-276.

Matsunaga, H., Yoshino, S., Kumasaki, M., Habu, H., & Miyake, A. (2011). Aging characteristics of the energetic oxidizer ammonium dinitramide. *Kōgyō Kayaku Kyōkaishi/Kayaku Gakkaishi/Science and Technology of Energetic Materials/Kōgyō Kayaku*, 72(5), 131–135. Retrieved from https://ci.nii.ac.jp/naid/10029979927

Obalová, L., Karásková, K., Jirátová, K., & Kovanda, F. (2009). Effect of potassium in calcined Co–Mn–Al layered double hydroxide on the catalytic decomposition of N2O. *Applied Catalysis B Environment and Energy*, 90(1–2), 132–140. https://doi.org/DOI: 10.1016/j.apcatb.2009.03.002

Oxley, J. C., Smith, J. L., Zheng, W., Rogers, E., & Coburn, M. D. (1997). Thermal Decomposition Studies on Ammonium Dinitramide (ADN) and 15N and 2H Isotopomers. *The Journal of Physical Chemistry A*, 101(31), 5646–5652. DOI: 10.1021/jp9625063

Park, G. O., Shon, J. K., Kim, Y. H., & Kim, J. M. (2015). Synthesis of Ordered Mesoporous Manganese Oxides with Various Oxidation States. *Journal of Nanoscience and Nanotechnology*, 15(3), 2441–2445. DOI: 10.1166/jnn.2015.10263 PMID: 26413684

Park, J., Chakraborty, D., & Lin, M. (1998). Thermal decomposition of gaseous ammonium dinitramide at low pressure: Kinetic modeling of product formation with ab initio MO/cVRRKM calculations. *Symposium (International) on Combustion*, 27(2), 2351–2357. https://doi.org/DOI: 10.1016/S0082-0784(98)80086-6

Rossi, M. J., Bottaro, J. C., & McMillen, D. F. (1993). The thermal decomposition of the new energetic material ammoniumdinitramide (NH4N(NO2)2) in relation to Nitramide (NH2NO2) and NH4NO3. *International Journal of Chemical Kinetics*, 25(7), 549–570. DOI: 10.1002/kin.550250705

Sairanen, E., Karinen, R., Borghei, M., Kauppinen, E. I., & Lehtonen, J. (2012). Preparation Methods for Multi-Walled Carbon Nanotube Supported Palladium Catalysts. *ChemCatChem*, 4(12), 2055–2061. DOI: 10.1002/cctc.201200344

Shamjitha, C., & Vargeese, A. A. (2023). Development of bimetallic spinel catalysts for low-temperature decomposition of ammonium dinitramide monopropellants. *Defence Technology*, 30, 47–54. DOI: 10.1016/j.dt.2023.05.007

Thommes, M., Kaneko, K., Neimark, A. V., Olivier, J. P., Rodriguez-Reinoso, F., Rouquerol, J., & Sing, K. S. (2015). Physisorption of gases, with special reference to the evaluation of surface area and pore size distribution (IUPAC Technical Report). *Pure and Applied Chemistry*, 87(9–10), 1051–1069. DOI: 10.1515/pac-2014-1117

Tian, M., Wang, X. D., & Zhang, T. (2016). Hexaaluminates: A review of the structure, synthesis and catalytic performance. *Catalysis Science & Technology*, 6(7), 1984–2004. DOI: 10.1039/C5CY02077H

Troyan, J. E. (1953). Properties, Production, and Uses of Hydrazine. *Industrial & Engineering Chemistry*, 45(12), 2608–2612. DOI: 10.1021/ie50528a020

Vyazovkin, S., & Wight, C. A. (1997). Ammonium Dinitramide: Kinetics and Mechanism of Thermal Decomposition. *The Journal of Physical Chemistry A*, 101(31), 5653–5658. DOI: 10.1021/jp962547z

Wingborg, N. (2006). Ammonium Dinitramide–Water: Interaction and Properties. *Journal of Chemical & Engineering Data*, 51(5), 1582–1586. DOI: 10.1021/je0600698

Xue, L., Zhang, C., He, H., & Teraoka, Y. (2007). Catalytic decomposition of N2O over CeO2 promoted Co3O4 spinel catalyst. *Applied Catalysis B Environment and Energy*, 75(3–4), 167–174. https://doi.org/ DOI: 10.1016/j.apcatb.2007.04.013

Yadav, A. K., Chowdhury, A., & Srivastava, A. (2017). Interferometric investigation of methanol droplet combustion in varying oxygen environments under normal gravity. *International Journal of Heat and Mass Transfer*, 111, 871–883. DOI: 10.1016/j.ijheatmasstransfer.2017.03.125

Yang, R., Thakre, P., & Yang, V. (2005). Thermal Decomposition and Combustion of Ammonium Dinitramide [Review]. *Combustion, Explosion, and Shock Waves*, 41(6), 657–679. DOI: 10.1007/s10573-005-0079-y

Yeh, T. F., Lee, H. G., Chu, K. S., & Wang, C. B. (2004). Characterization and catalytic combustion of methane over hexaaluminates. *Materials Science and Engineering A*, 384(1–2), 324–330. DOI: 10.1016/S0921-5093(04)00835-4

Yoldas, B. E. (1975). Alumina gels that form porous transparent Al2O3. *Journal of Materials Science*, 10(11), 1856–1860. DOI: 10.1007/BF00754473

Youn, Y., Miller, J., Nwe, K., Hwang, K. J., Choi, C., Kim, Y., & Jin, S. (2019). Effects of Metal Dopings on CuCr2O4 Pigment for Use in Concentrated Solar Power Solar Selective Coatings. *ACS Applied Energy Materials*, 2(1), 882–888. DOI: 10.1021/acsaem.8b01976

Zheng, J., Ren, X., Song, Y., & Ge, X. (2009). Catalytic combustion of methane over iron- and manganese-substituted lanthanum hexaaluminates. *Reaction Kinetics and Catalysis Letters*, 97(1), 109–114. DOI: 10.1007/s11144-009-0013-5

Chapter 7
Molecular Interaction of Lactams With Mild Steel in Hydrochloric Acid Environment:
Corrosion Inhibition Efficiency and Surface Adsorption Mechanisms

Nadia Faska
https://orcid.org/0000-0001-6203-6904
Faculty of Applied Sciences Ait Melloul, Ibn Zohr University, Morocco

Soukayna Maitouf
Ibn Zohr University, Morocco

Brahim Orayech
Powerco Battery, Spain

ABSTRACT

The inhibition effect of some lactams (Pyrrolidin-2-one, δ-valerolactam, and ε-caprolactam) on the corrosion behaviour of mild steel in 1M Hydrochloric acid solution was studied by weight loss and electrochemical techniques. The results demonstrated that both δ-valerolactam, and ε-caprolactam significantly inhibit corrosion. Specifically, δ-valerolactam achieved an inhibition efficiency of 85.2%, while ε-caprolactam exhibited a higher inhibition efficiency of 91.5%. The thermodynamic parameters governing the adsorption process such as adsorption heat, adsorption entropy, and adsorption free energy were determined and discussed. The adsorption of lactam compounds on the mild steel surface in 1M HCl follows the Langmuir adsorption isotherm model.

1. INTRODUCTION

The extensive utilization of mild steel in the industry is attributed to several factors, including its affordability, making it an economically viable choice (Fayomi et al., 2021), excellent mechanical properties, easy availability, cold working ability (Kaya et al., 2023), and the capacity to be hot worked

DOI: 10.4018/979-8-3693-7505-1.ch007

without compromising mechanical properties (Sedik et al., 2020; Oyekunle et al., 2019). mild steel is a common option for a container that holds aggressive solutions (Yadav et al., 2016), acids, bases, salts, reactions reagents, and tanks for handling corrosive liquids (Hussain et al., 2023; Matad et al., 2014). Acidic solutions play a vital role in processes, such as descaling, cleaning, and various petrochemical processes (Al-Moubaraki et al., 2021; Kuren et al., 2023). Acids including sulfuric, acetic, nitric, and hydrochloric acids are frequently used in pickling (Ding et al., 2016). Moreover, these acids are widely employed for drilling fracturing and acid simulations during various stages in oil exploration (Abd-El-Nabey et al., 2024), production, and descaling operations, as well as in numerous other industrial applications (Askari et al., 2021).

Building on this basis, Park et al. studied the long-term corrosion performance of three tested steels in an acidic environment. Their research underscored the remarkable features of mild steel, including its cost-effectiveness, widespread availability, and high tensile strength, which makes it a preferred choice in industries dealing with corrosive substances. The authors reiterated the critical role of acids in various applications including drilling and oil exploration, further supporting the notion that mild steel remains a vital material in aggressive environments (Park et al., 2022).

Most recently, (Alharbi et al., 2024) focused on low-carbon steel (AISI 1010), and its corrosion behavior was influenced by grain size. This study provides insights into the microstructural aspects that affect corrosion resistance, and demonstrates that variations in grain size can significantly affect the mechanical properties and performance of low-carbon steel. The findings indicated that a finer grain structure, achieved through heat treatment, enhances corrosion resistance, which is essential for materials used in industrial processing equipment. This study adds a nuanced understanding of how microstructural characteristics can be optimized to improve the longevity and reliability of mild steel in corrosive environments. Collectively, these studies illustrate the critical relationship between the material properties, corrosion behavior, and industrial applications of mild steel. They highlighted the ongoing need for research to enhance the performance of mild steel in aggressive environments, thereby ensuring its continued relevance in various industrial sectors.

Corrosion is a natural and electrochemical phenomenon that results in the degradation of materials through their interactions with the surrounding environment (Fernandes et al., 2014; Rinky et al., 2023). In acidic solutions, H^+ ions and dissolved oxygen serve as inherent catalyst for corrosion. It is a continuous process that cannot be entirely halted, but can be decelerated. Corrosion inhibitors represent the most efficacious method for slowing down the reaction and protect metals in various industries and fields (Amin et al., 2006). Corrosion inhibitors are among the most recognized and effective methods for preventing the degradation or destruction of metal surfaces in industrial applications. Due to its low cost and practicality, this method has become popular (Pokhmurs'kyi et al., 2004; Sivakumar et al., 2021).

The classification of inhibitors based on their chemical functionality can be categorized into two primary types. Inorganic and organic inhibitors (Oyekunle et al., 2019; Sanaei et al., 2017). Inorganic inhibitors are characterized by their capacity to facilitate the oxidation of metals, thereby forming a passive layer on the metal surface that provides protection against corrosion. The efficacy of these inhibitors is primarily attributed to the negative anions present in the crystalline salts, such as calcium metasilicate, sodium chromate, phosphate, molybdate, and zinc oxide, which play a significant role in mitigating metal corrosion (Sanaei et al., 2017).

Conversely, organic inhibitors are typically composed of complex structures that include π-bonds and possess active centers or functional groups (Aslam et al., 2021; Goyal et al., 2018). These inhibitors can be subdivided into two categories: (i) organic anionic inhibitors, which include mercaptobenzotriazole, sodium sulfonates, and phosphonates, and are commonly employed in applications such as cooling water systems, where they function as antifreeze agents. (ii) Organic cationic inhibitors are characterized by the presence of large aliphatic or aromatic groups with positively charged amine functionalities. These inhibitors can exist in either liquid or solid form, such as wax. Notable examples of organic cationic inhibitors are sodium benzoate and cinnamate (Kahkesh et al., 2023; Nataraja et al., 2011).

2. ORGANIC COMPOUNDS AS CORROSION INHIBITORS

Inhibitors are substances that retard or decrease the rate of electrochemical reactions. A corrosion inhibitor specifically reduces the rate at which metals are attacked by their environment (Askari et al., 2021; Li et al., 2022). To avoid severe corrosion, small amounts of corrosion inhibitors are typically added to acids, cooling water, steam, and other environments (Fiori-Bimbi et al., 2014; Avdeev et al., 2022). Numerous corrosion inhibitors have been investigated under acidic conditions to examine the corrosion inhibition of mild steel. Most widely recognized corrosion inhibitors are those organic compounds with aromatic and heterocyclic rings and functional groups of oxygen, nitrogen, phosphorus, and sulphate (Solmaz et al., 2008; Nataraja et al., 2011; Murmu et al., 2020). The effectiveness of these inhibitors is largely due to lone pairs and the loosely bonded electrons in these functional groups, which facilitate interaction with the metal surface through adsorption (Sengupta et al., 2021; Chen et al., 2022).

Poly(aniline-formaldehyde) was tested as an organic corrosion inhibitor of mild steel in acidic media. These findings indicate that poly(aniline-formaldehyde) functions as a mixed inhibitor, effectively reducing mild steel corrosion via an adsorption mechanism, achieving an inhibition efficiency greater than 90% at a concentration of 10 ppm. poly(aniline-formaldehyde) clearly revealed that the surface roughness of the inhibited mild steel sample was significantly lower than that of the uninhibited mild steel (Quraishi et al., 2008).

Ahmed et al. investigated a study on the substituted benzotriazoles, evaluating its efficacy as novel organic inhibitor for mild steel corrosion in hydrochloric acid solution. Their results indicated that the inhibition efficiencies increased with higher concentrations of the inhibitors, but decreased with increasing temperature. The highest observed inhibition efficiency (67%) was recorded at the maximum concentration tested. Furthermore, the adsorption behavior of the synthesized organic inhibitor was shown to be influenced by various factors, including its electronic properties, characteristics of metal surface, steric effect, temperature, and diverse surface-site activity. Adsorption process aligns with the Langmuir isotherm model (Ahmed et al., 2018).

Kahkesh and Zargar investigated the anticorrosive potency of 3-nitrobenzene-1,2-dicarboxylic acid for mild steel corrosion in hydrochloric acid environment. Their results confirmed the anticorrosive properties of 3-nitrobenzene-1,2-dicarboxylic acid, as evidenced by the enhanced surface smoothness and the creation of hydrophobic layer resulting from the adsorption of inhibitor on mild steel surface, which adhered to the Langmuir adsorption model (Kahkesh et al., 2023).

Heterocyclic compound derivatives have been reviewed as organic corrosion inhibitors, especially thiourea, urea, amino alcohols, and sulphonamide, for various metals and alloys, as well as for various factors that affect their performance in various aggressive environments (Samal et al., 2024; Elshakre et al., 2017; Caihong et al., 2022; Ahmadi et al., 2023). There are still some inhibitors that act in the vapor phase (volatile corrosion inhibitors). Examples include diisopropylammonium nitrite, benzoate, ethanolamine benzoate, and carbonate (Aslam et al., 2021).

3. INHIBITION MECHANISM

Various mechanisms for inhibitors have been identified, and many of these mechanisms can occur simultaneously. According to Mohamed et al., these mechanisms can be classified into two categories: interface and interphase inhibition (Mohamed et al., 2023). In interface inhibition, the inhibitor is directly adsorbed onto the metal surface, forming a thin two-dimensional film. This category can be further subdivided into nonselective and selective physisorption and chemisorption (Chaitra et al., 2015; Bouklah et al., 2005). Nonselective physisorption, also known as geometrical blocking or adsorption screening, is a rapid and reversible process in wherein an inhibitor interacts with the surface via Van der Waals or electrostatic forces, resulting in significant coverage. Selective physisorption, or deactivating coverage, refers to the targeted blockage of active sites on metal surface by a non-specific inhibitor at relatively low coverage (Ouadi et al., 2020). Chemisorption, or reactive coverage, is characterized by specific and slow adsorption processes that are not fully reversible and include charge sharing or transfer (Chen et al., 2022; Verma et al., 2021). Interphase Inhibition occurs when a three-dimensional layer forms on the surface of a metal as a consequence of a chemical reaction between the inhibitor and corrosion products, leading to the formation of organometallic complexes (Qian et al., 2023).

The adsorption of organic molecules onto metal surfaces is commonly proposed to take place according to one of the following mechanisms: (i) electrostatic attraction between a charged metal and a charged inhibitor; (ii) interaction between lone pair electrons of heteroatoms such as nitrogen, oxygen, or sulfur and unoccupied d-orbitals of iron atoms in metals (Alaneme et al., 2015; Solomon et al., 2021); (iii) the interaction between π electrons in aromatic rings and the unoccupied d-orbitals of iron atoms when feasible; and (iv) a combination of some of these mechanisms (Yurt et al., 2004 ; Li et al., 2024).

The application of inhibitors for metal corrosion protection is frequently associated to chemical adsorption, causing a variation in the charge of the adsorbed substance, and a transfer a variation in the charge of adsorbed substance and charge transfer between phases. Consequently, the molecular structures of these inhibitors are of particular significance. The electron density present at the atoms of the functional groups that constitute the reaction center significantly influences the strength of the adsorption bond. Furthermore, bond force is influenced by the characteristics of the metal. In addition, the polarizability of the functional groups warrants consideration. Most organic inhibitors are characterized by the presence of at least one polar group, typically containing atoms such as nitrogen, sulfur, oxygen, selenium or phosphorus. Extensive research conducted by Hackerman and colleagues established the foundation for the adsorption theory of organic inhibitors. According to this theory, the inhibiting properties of numerous compounds are determined by the electron density at the atom serving as the main reaction center. An increase in the electron density at this reaction center correlates with a strengthening of the chemisorption bonds between the inhibitor and metal (McCafferty 2010).

Recently, Shwetha et al. examined the mechanisms by which corrosion inhibitors inhibit metal deterioration (KM et al., 2024). In their study, the authors categorized several distinct inhibition mechanisms. The first category, electronic adsorption, describes the process by which charged inhibitor molecules are electrostatically attracted to a metal surface. The second category includes the attraction of uncharged electrons from inhibitors to the charged surface of the metal. Singh et al. further noted that this attraction can occur through ion acceptance and donation processes (Singh et al., 2023). The third mechanism pertains to the adsorption of π-bond orbitals, which is characterized by the interaction among conjugated organic inhibitor and metal surface. It is also plausible that inhibition mechanisms arise from a combination of two or more of these interactions.

Moreover, Dariva et al. identified several criteria for identifying effective inhibitors. One such criterion, chemisorption, has been frequently cited by various researchers. According to Sharma et al. and Prasad et al., chemisorption creates a strong interaction between metal and inhibitors, leading to the development of a protective layer on metal surface (Sharma et al., 2023; Prasad et al., 2021). However, significant uncertainty remains surrounding the notion of chemisorption as a definitive criterion, particularly because some corrosion inhibitors function solely through physisorption onto the metal surface. This was demonstrated by experiments conducted by Haris et al., who explored the corrosion inhibition potential of empty fruit bunches of oil palms. Although these inhibitors rely on physisorption, which is weaker than chemisorption, they still achieve remarkable inhibition efficiencies, exceeding, which results in a significant reduction in the degree of corrosion. Consequently, the presence of a protective film has been proposed as a secondary criterion for effective corrosion inhibition (Haris et al., 2019).

Acids play a significant role in various industrial applications, including petrochemical processes, oil-well acidizing, industrial cleaning, acid descaling, and the exposure of metals to corrosive agents (Heydari et al., 2018; Prasad et al., 2021). Numerous studies have been published on the issues of inhibitory protection of metals and alloys in acidic environments, focusing on critical issues within this field. Modern ideas regarding the corrosion of metals in acidic media, as well as the mechanism underlying the protective effects of corrosion inhibitors, have been explored in recent literature (Xiong et al.,2024; Avdeev et al., 2022). Comprehensive data on corrosion inhibitors derived from nitrogen-containing heterocyclic molecules (Vikneshvaran et al., 2019; Zhang et al., 2024), unsaturated organic compounds (Tebbji et al., 2007; Verma et al., 2021), ionic liquids (Li et al., 2022), polymeric organic compounds (Basik et al., 2022), pharmaceuticals (Rinky et al., 2023), and other corrosion inhibitors have been reported. Additionally, the industrial implications of using metal corrosion inhibitors in acidic environments have been discussed (Chen et al., 2022). Hydrochloric acid, a commonly used industrial acid, corrodes metals through chemical or electrochemical processes (Noor et al., 2008; Khatib et al., 2020).

Research on metallic corrosion has been conducted in various environments, including neutral, acidic, and alkaline conditions, using several electrochemical techniques. Farelas and his co-workers analyzed the effectiveness of bis-imidazoline and Imidazoline compounds using linear polarization resistance and electrochemical impedance spectroscopy techniques. Their findings indicated that imidazoline derivatives produced more cohesive inhibitor film, which improved corrosion protection (Farelas et al., 2010).

The aqueous extract of Lavendula was investigated in 1M hydrochloric acid media using various method, including weight loss, electrochemical impedance spectroscopy, and potentiodynamic polarization measurements, and it showed efficiency at by weight aqueous extract concentration. Polarization studies indicate that phytochemicals present in Lavendula extract act as mixed types of inhibitors (Bouammali et al., 2013).

Harek and Larabi studied the corrosion inhibition effects of divinylbenzene-4-vinylpyridine copolymer (poly(4-vinylpyridine)) and potassium iodide (KI) on mild steel in 1M hydrochloric acid using weight loss, potentiodynamic, and polarization resistance measurements. Their results demonstrated that the inhibitory effect of poly(4-vinylpyridine) increased as the concentrations of the inhibitor increased, and that the addition of KI significantly enhanced this effect, achieving approximately 92% inhibition at a concentration of 0.5% in the presence of 0.1% KI at 328 K. The study also noted that the corrosion rate of mild steel increased with temperature within the range 298-328 K, both when exposed and non-exposed to the inhibitor. Furthermore, the presence of the inhibitor increases the activation energy of the corrosion process. The inhibition mechanism was attributed to the adsorption of inhibitor molecules onto the steel surface, consistent with the Langmuir and Temkin adsorption isotherms, with the N-atom serving as the primary adsorption site. The polarization curves indicate that poly(4-vinylpyridine) functioned as a mixed-type inhibitor. The corresponding results suggest that the presence of iodide ions in the solution stabilizes the adsorption of poly(4-vinylpyridine) molecules on the metal surface and, therefore, improves the inhibition efficiency of poly(4-vinylpyridine) (Larabi et al., 2004).

To better understand the absorption behavior of inhibitors, it is useful to use thermodynamic adsorption parameters and kinetic corrosion parameters. Previous research has revealed that by comparing the activation energies of inhibited and uninhibited corrosion reactions, the heat of inhibitor adsorption may be obtained (Sanyal et al., 1981; Bentiss et al., 2005). It was determined that the positive heat of adsorption, ΔH_{ads}^{0} (endothermic process), is unequivocally linked to chemisorption (Bentiss et al., 2005; Bouammali et al., 2013). Conversely, a negative heat of adsorption, ΔH_{ads}^{0} (exothermic process), may indicate physisorption (Oguzie et al., 2004), chemisorption (Ali et al., 2004), or a combination of both processes (comprehensive adsorption) (Ouici et al., 2017). It is generally understood that the influence of temperature on the inhibition of acid–metal interactions is highly complex because various changes occur on the metal surface, such as fast etching and desorption of the inhibitor, which may decompose and/or rearrange. It was found that a few inhibitors of acid–metal systems have specific reactions that are effective at high temperatures (or greater) and low temperatures (Espinoza-Vázquez et al., 2014; Kumar et al., 2023).

The aim of the present study was to investigate the inhibitory potential of pyrrolidin-2-one, δ-valerolactam, and ε-caprolactam on mild steel corrosion in 1M hydrochloric acid, and to evaluate the kinetic and thermodynamic properties of inhibitor adsorption onto metallic surfaces in acidic media. The corrosion analyses employed in this research included potentiodynamic polarization and electrochemical impedance spectroscopy to identify the inhibitor types. Moreover, the effect of temperature on the electrochemical parameters of the system was examined to elucidate the adsorption mechanisms of the tested inhibitors.

4. EXPERIMENTAL DETAILS

4.1. Inhibitors

The study used a series of lactams: Pyrrolidin-2-one, δ-valerolactam, and ε-caprolactam, as organic inhibitors (Figure 1), with concentrations between 10^{-6} M and 10^{-3} M. The solubility of the inhibitors in a corrosive medium was examined, and based on this, the concentration range cited above was selected. A blank solution was prepared to compare purposes. All lactam compounds used in this work were Merck products (99% purity).

Figure 1. Structures of the lactams considered and their most frequently used name

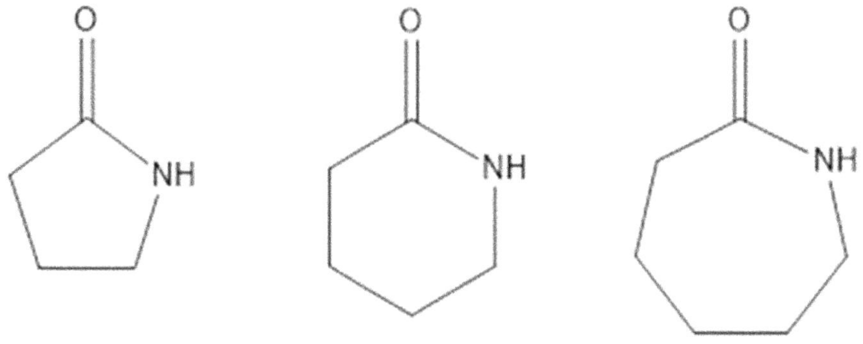

Pyrrolidin-2-one δ-valérolactam ε-caprolactam

4.2. Hydrochloric acid (HCl)

The hydrochloric acid solution used for all chemical and electrochemical analyses was obtained from hydrochloric acid (37%) of analytical grade and diluted with distilled water to form a standard solution. (Sahraoui et al., 2022; Ouici et al., 2017). It is recognized that this solution damages the steel (Vikneshvaran et al., 2019; Bouammali et al., 2013).

4.3. Metals Preparation

Corrosion tests were performed on mild steel specimens with chemical contents (in mass %) of 0.657% Mn, 0.012% P, 0.184% Si, 0.193% C, 0.005% S, 0.028% Cr, and 98.921% iron (Fe). For the experiments, mild steel was cut into several discs of cylindrical shape with an exposed surface area of for pursuing all corrosion investigations.

Initially, the steel substrate was submitted to a sequential polishing process using sandpaper with varying grit sizes, ranging from 180 to 2000. Then, the coupons were rinsed with double distilled water to remove any residues. A degreasing treatment with acetone was then applied to ensure the removal of any grease or impurities (Okey et al., 2020; Mahdi et al., 2022; Nawaz et al., 2023). Finally, cleaned coupons were dried in desiccators to remove any residual moisture (Mobin et al., 2022; Gupta et al., 2022). Following thorough preparation, the mild steel specimens were weighed before and after immersion in various solution concentrations (Petrunin et al., 2022; Likhanova et al., 2019).

4.4. Scanning Electron Microscopy Analysis of Metals

Surface morphology can be analyzed using scanning electron microscopy (SEM), a method that has been employed by several researchers (Reza et al., 2021; Bashir et al., 2019). According to the authors, scanning electron microscopy can capture high-resolution images of metal surfaces at a macroscopic level. By comparing and analyzing the images of the metal surface before and after acid immersion, the impact of acid solution on the characteristics of the metal surface can be elucidated (Karki et al., 2021). Surface morphology investigations were conducted by scanning electron microscopy (SEM) to confirm the electrochemical findings. In this study, polished mild steel specimens were immersed in a 1 M hydrochloric acid solution for 6 h when exposed and non-exposed to 10^{-3} M of the inhibitor. The specimens were rinsed with distilled water, dried in acetone, and used for further investigation.

4.5. Weight Loss Measurements

The weight loss method was realized with various concentrations of lactams, which were investigated at temperatures ranging from 303 to 343 K. In the control test, the weighed metal specimens were immersed individually in 200 ml open beakers containing 150 ml of 1M hydrochloric acid. The weighed metal coupons were immersed separately in 200 ml open beakers containing 150 ml of 1M hydrochloric acid with different concentrations of lactam compounds. After the corrosion measurement, mild steel coupons were cleaned with double-distilled water, dried, and measured. The variation in weight loss was measured at regular intervals at different temperatures, both when exposed and non-exposed to lactams at different concentrations. The coupons were removed after 1, 2, 3, 4, 5 and 6 hours and immersed in acetone. The samples were then brushed under flowing water with a steel wire brush, dried and weighed again. The weight loss (ΔW) was measured in grams as the difference between the initial mass and the mass after the removal of the corrosion product (Eq.1 1).

$$\Delta W = W_i - W_f \tag{1}$$

Where W_i and W_f represent the initial and final mass of the mild steel coupon after exposure to acidic media containing varying concentrations of inhibitor solutions, respectively.

Determination of Corrosion Rates

The corrosion rates (C_R) for mild steel specimens immersed in 1M hydrochloric acid in the presence of various concentrations of inhibitors at four distinct temperatures was determined using the following Eq.2 (Oyekunle et al., 2019).

$$C_R = \frac{\Delta W}{A.t} \tag{2}$$

Where ΔW is the weight loss. A represent the surface area of different specimen and t indicate immersion time in minute.

Determination of Inhibition Efficiency (IE %)

Inhibition efficiencies of mild steel samples immersed in 1M hydrochloric acid (HCl) solutions, containing varying concentrations of inhibitor solutions at different temperatures, have been determined using the corrosion rate values (C_R) calculated previously, according to the following Eq.3 (Fiori-Bimbi et al., 2014).

$$IE(\%) = \frac{C_{R\,0} - C_{R\,inh}}{C_{R\,0}} \times 100 \qquad (3)$$

Where $C_{R\,0}$ and $C_{R\,inh}$ are the corrosion rates when exposed and non-exposed to various inhibitors at different concentrations, respectively.

4.6. Electrochemical Corrosion Testing

Literature highlights the advantages of employing electrochemical analysis in corrosion studies. This method requires minimal testing time and can detect low corrosion rates (Abd-El-Nabey et al., 2024; Alaneme et al., 2015; Bashir et al., 2018). Potentiodynamic polarization and electrochemical impedance spectroscopy demonstrated this approach. According to Amin et al., electrochemical measurements typically use three-cell systems (Amin et al., 2006). These systems comprise three electrodes: working electrode (WE), counter electrode (CE), and reference electrode (RE). The metal under examination serves as the WE, whereas a platinum electrode and a saturated calomel electrode (SCE) commonly function as the counter electrode and reference electrode, respectively. As noted by Alaneme et al., the working electrode was initially submerged in the test solution for 1 h to achieve a steady state.

Electrochemical analyses were performed using a conventional three-electrode cell and a potentiostat-galvanostat. The experiments were conducted at a steady temperature of 303 K in ambient air, without agitation of the test solution (El-Aziz et al., 2019; Tanwer and Shukla Kumar, 2022).

4.6.1. Electrochemical Impedance Spectroscopy

Electrochemical impedance spectroscopy (EIS) measurements were conducted to examine the characteristics and kinetics of the electrochemical processes occurring at the metal/solution interface, as well as the modifications induced by pyrrolidin-2-one, δ-valerolactam, and ε-caprolactam. The corrosion behavior of mild steel in an acidic solution containing the aforementioned lactams was evaluated using electrochemical impedance spectroscopy at 303 K after a 24h immersion period. The electrochemical impedance spectroscopy measurements were conducted using specialized impedance equipment, and data analysis was performed using TACUSSEL corrosion analysis software. Impedance spectra were recorded over a frequency range of 100 kHz to 10 mHz, achieving a resolution of 10 points per decade, at the corrosion potential after 30 min of immersion in non-deaerated solutions. A sine wave with an amplitude of 10 mV was used to perturb the electrochemical system. The equivalent circuit model employed in this investigation is consistent with that reported in previous studies (Tebbji et al., 2007; Mahdi et al., 2022). The double-layer capacitance (C_{dl}) and charge-transfer resistance (R_t) were obtained from the impedance measurements, as described in prior research (Tebbji et al., 2007). In the context of electrochemical impedance spectroscopy, the inhibition efficiency (IE%) was determined based on the charge-transfer resistance, following the methodology described in the literature (Farelas et al., 2010).

4.6.2. Potentiodynamic Polarization (PDP)

Experiments were conducted using a standard three-electrode cylindrical Pyrex glass cell. The working electrode was comprised of mild steel, the auxiliary electrode consisted of platinum gauze, and the reference electrode was a saturated calomel electrode (SCE) within the three-electrode configuration.

To eliminate the oxide film from the working electrode, it was polarized at −800 mV for 10 min. The resulting oxidation and reduction curves, which elucidate the reaction mechanism from the cathode to the anode, were extrapolated to the potential axis. The intersection points of the cathodic and anodic curves yield the current density associated with corrosion and the corrosion potential, both of which are essential for understanding the corrosion mechanism (Farelas et al., 2010). The temperature was maintained at 303±1 K by using a thermostatic control system. Prior to experimentation, the working electrode was abraded with silicon carbide paper, degreased with analytical-grade ethanol and acetone, and rinsed with double-distilled water. All potential measurements were referenced to a saturated calomel electrode.

Potentiodynamic polarization studies were conducted using a potentiostat–galvanostat. Potentiodynamic current–potential curves were recorded by systematically varying the electrode potential from −800 to 0 mV at a scanning rate of 0.5 mV/s. Prior to the measurements, the working electrode was immersed in the test solution at its natural potential (open circuit potential) for approximately 0.5 hours to achieve a steady state.

5. RESULTS AND DISCUSSION

5.1. Weight Loss Tests

The corrosion behavior of mild steel in a 1M hydrochloric acid solution was investigated both when exposed and non-exposed to various concentrations of lactam compounds. This study was conducted through gravimetric analysis after a 6 h immersion period. The results pertaining to the weight loss of mild steel in 1M hydrochloric acid, with and without the incorporation of lactam compounds at varying concentrations, are presented in Table 1. The corrosion rate decreased as the concentration of the tested compounds increased. Furthermore, the efficacy of the inhibitors improved at higher concentrations, with ε-caprolactam exhibiting the highest inhibition efficiency of 91.5%.

Table 1. The corrosion rate (W) of mild steel and the inhibition efficiency (E_W) for several concentrations of pyrrolidin-2-one, δ-valerolactam, and ε-caprolactam in 1 M HCl were obtained from gravimetric analysis

Inhibitor	Concentration (M)	W(mg.cm^2.h^{-1})	E_W(%)
Blank	1	2.42	-
pyrrolidin-2-one	10^{-6}	2,10	13,29
	10^{-5}	1,26	48,03
	10^{-4}	0,92	62,14
	10^{-3}	0,73	70,03
δ-valerolactam	10^{-6}	1,74	28,14
	10^{-5}	0,90	62,69
	10^{-4}	0,62	74,52
	10^{-3}	0,36	85,2

continued on following page

Table 1. Continued

Inhibitor	Concentration (M)	W(mg.cm².h⁻¹)	E_w(%)
ε-caprolactam	10^{-6}	1,77	26,86
	10^{-5}	0,71	70,58
	10^{-4}	0,30	87,72
	10^{-3}	0,21	91,5

The results shown in Table 1 for varying concentrations of pyrrolidin-2-one, δ-valerolactam, and ε-caprolactam indicate that these compounds inhibit the corrosion of mild steel. It was observed that corrosion rates decreased with an increase in the inhibitor concentration, and the inhibition efficiency E_W(%) increases with increasing inhibitor concentration. At a concentration of 10^{-3} M for each inhibitor examined, the inhibition efficiencies were 70.03, 85.2, and 91.5 % for pyrrolidin-2-one, δ-valerolactam, and ε-caprolactam, respectively. These findings suggest that the effectiveness of inhibition is correlated with the number of methylene groups present in the lactam ring, that is with molecular size. From weight loss measurements, we can conclude that the efficiency of the tested lactams follows the order: pyrrolidin-2-one < δ-valerolactam < ε-caprolactam. Previous studies indicate that the molecular area and mass of organic molecules can influence the corrosion inhibition of metals in acidic environments (Khaled et al., 2005; Verma et al., 2021; Moloney et al., 2022). Furthermore, one can suggest that the more $C-(\underline{C}=O)-N$ and $(C=O)-\underline{N}-C$ angles tend to open that is to become larger, the more the inhibiting effectiveness increases of these lactams. The two heteroatoms (Nitrogen and oxygen) of the lactams are probably the active adsorption centers on the surface of mild steel (Muthukrishnan et al., 2013).

5.2. Polarization Results

In the present study, potentiodynamic polarization experiments were performed to obtain information about the kinetics of anodic and cathodic reactions. Fig.2 presents the polarization curves for mild steel immersed in 1M hydrochloric acid, both when exposed and non-exposed to varying concentrations of ε-caprolactam.

Figure 2. Tafel polarization curves in 1M HCl of mild steel with and without ε-caprolactam at 308 K for various concentrations

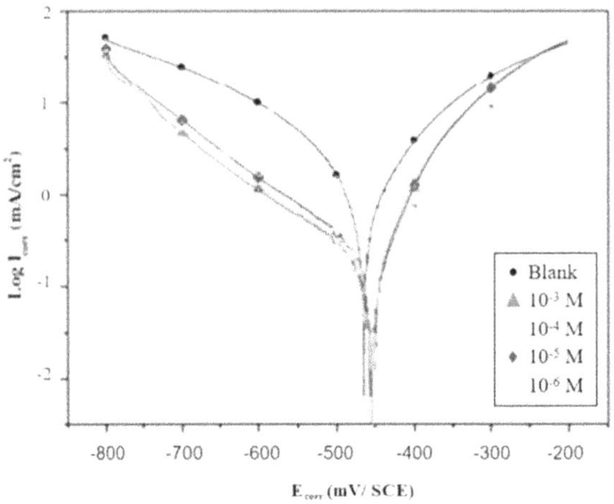

Table 2 presents the numerical values corresponding to variations in corrosion current density (I_{corr}), corrosion potential (E_{corr}), cathodic Tafel slope (β_c) and inhibition efficiency (E_i(%) with the concentrations of inhibitors at all studied temperatures. The E_i(%)is calculated as (Mu et al., 2014):

$$E_i\% = \frac{I_{corr} - I_{corr}^{inh}}{I_{corr}} \times 100 \quad (4)$$

I_{corr} and I_{corr}^{inh} are the corrosion current densities with and without inhibitor, respectively.

Table 2. Parameters of potentiodynamic polarization in1M for mild steel when exposed and non-exposed to pyrrolidin-2-one, δ-valerolactam, and ε-caprolactam

Inhibitor	**Concentration (M)**	**E_{corr}(mV/SCE)**	**β_c(μA/cm²)**	**I_{corr}(μA/cm²)**	**E_I%**
Blank	1	-470	-217	682	-
Pyrrolidin-2-one	10^{-4}	-503	-174	382	44
	10^{-3}	-498	-186	334	51
	10^{-2}	-481	-192	257	62
δ-valerolactam	10^{-4}	-461	-177	300	56
	10^{-3}	-459	-178	198	71
	10^{-2}	-457	-172	136	80
ε-caprolactam	10^{-4}	-452	-173	225	67
	10^{-3}	-454	-174	116	83
	10^{-2}	-453	-184	68	90

The polarization plots clearly demonstrate that the incorporation of inhibitors into the corrosive environment has a significant influence on anodic and cathodic reactions. This indicates that ε-caprolactam effectively reduced the anodic dissolution of mild steel and simultaneously inhibited the cathodic hydrogen evolution reaction (Goyal et al., 2018). Furthermore, the presence of these inhibitors limited the corrosive effects of the acid on the mild steel electrode. At the same concentration (10^{-3}M), the value of i_{corr} in the case of ε-caprolactam is smaller than those of pyrrolidin-2-one and δ-valerolactam. ε-caprolactam thus exhibits higher inhibition efficiency than pyrrolidin-2-one and δ-valerolactam in 1M hydrochloric acid; this enhanced efficiency is due to the molecular size of ε-caprolactam. Indeed, the molecule ε-caprolactam has a higher volume compared to pyrrolidin-2-one and δ-valerolactam molecules, consequently, it presents a better effectiveness in inhibiting the corrosion of mild steel in a 1M hydrochloric acid environment.

5.3. Electrochemical Impedance Spectroscopy

The corrosion characteristics of mild steel in 1M hydrochloric acid solution with and without ε-caprolactam were examined using electrochemical impedance spectroscopy. The Nyquist plot for various concentrations of ε-caprolactam as organic inhibitor are presented in Figure 3.

Figure 3. Nyquist plots of mild steel in 1 M HCl with and without ε-caprolactam

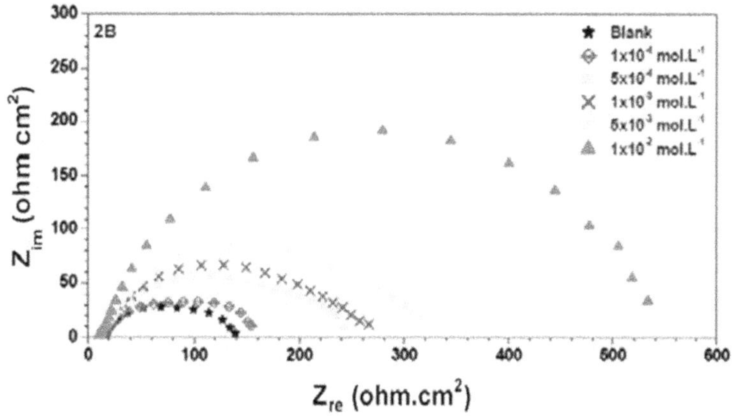

The electrochemical parameters obtained from the Nyquist plots are presented in Table 3. The Nyquist diagram show a single semicircle shifted along the real impedance axis (Z_{re}), suggesting that the corrosion of mild steel in 1 M hydrochloric acid is governed by a charge-transfer mechanism. The charge-transfer resistance values (R_t) were determined from the Z_{re}. The intersection at higher-frequency corresponds to the solution resistance (R_s), while the lower-frequency intersection corresponds to $R_s + R_{ct}$. Thus, R_t values were calculated as the difference between the high and low frequency intersection values (Liu et al., 2008). The values of double-layer capacitance (C_{dl}) were calculated using the following Eq.5:

$$C_{dl} = (1/\omega . R_t) \quad (5)$$

where $\omega = 2\pi f_{max}$ and f_{max} is the frequency at which the imaginary component of impedance is maximum. The efficiency of inhibition derived from the charge-transfer resistance is determined using the following Eq.6:

$$E_{Rt}(\%) = \frac{R^0_{ct} - R_{ct}}{R^0_{ct}} \times 100 \qquad (6)$$

Where R^0_{ct} and R_{ct} are the charge transfer resistance values without and with inhibitor, respectively. R_{ct} is the diameter of the loop.

Table 3. Impedance parameters for mild steel in 1M HCl for various concentrations of δ-valerolactam, and ε-caprolactam at 308 K

Inhibitor	Concentration (M)	$R_t(\Omega cm^{-2})$	$C_{dl}(\mu F.cm^{-2})$	$E_{Rt}(\%)$
Blank	1	90	44.23	-
δ-valerolactam	10^{-3}	788	8.08	89
	10^{-4}	475	21.21	81
	10^{-5}	133	29.93	32
ε-caprolactam	10^{-3}	1030	9.78	91
	10^{-4}	500	20.15	82
	10^{-5}	136	46.8	34

The data presented in these plots indicate that the impedance response of mild steel in uninhibited hydrochloric acid solution has significantly changed after the addition of ε-caprolactam in the corrosive solution. This observation suggests that the impedance of the inhibited substrate increases with increasing inhibitor concentration and consequently the inhibition efficiency increases. The locus of the Nyquist plots was considered as one part of a semicircle. However, deviations from a perfectly circular shape suggest a variation in the frequency dispersion of the interfacial impedance (Bonora et al., 1996). The addition of ε-caprolactam to hydrochloric acid is found to enhance R_t values and bring down C_{dl} values. These observations indicate clearly the fact that the corrosion of mild steel in 1M hydrochloric acid is governed by a charge transfer mechanism, and the inhibition of corrosion occurs through the adsorption of the inhibitor ε-caprolactam on the mild steel surface.

5.4. Effect of Temperature on Inhibition

The temperature of the environment in which a material is exposed to corrosion significantly influences its behavior (Borsari et al., 2002; Prasanna et al., 2018; Nandiyanto et al., 2017). We examined gravimetric measurements in the temperature range 303–333K both when exposed and non-exposed to ε-caprolactam at 10^{-3} M in 1 M hydrochloric acid solution to verify how temperature affects the corrosion inhibition process. The various data obtained at 1 hour immersion period from 308 K to 333 K are summarized in Table 4.

Table 4. Influence of temperature on mild steel corrosion in 1M HCl at 10^{-3} M of ε-caprolactam

T(K)	1M HCl W^0(mg.cm².h⁻¹)	ε-caprolactam W(mg.cm².h⁻¹)	E_W(%)
303	3,53	0,30	91,5
313	5,44	0.49	91
323	8,78	0.86	90,2
333	12,77	1.66	87

The corrosion rate demonstrated a significant increasing with elevated temperature in the blank solution. In the presence of ε-caprolactam, the corrosion rate was significantly reduced at the tested temperatures. It can be observed that the efficiency does not depend on the temperature.

E_W(%) remained almost constant between 308 and 333 K. This may be attributed to the stability of the adsorption process at elevated temperatures, suggesting a chemisorption mechanism. Moreover, the logarithm of the corrosion rate for mild steel W_{corr} can be expressed as a linear function of 1000/T, in accordance with the Arrhenius equation, as depicted in Figure 4, where T is the temperature in Kelvin:

$$W_{corr} = K\exp(-E_a/RT) \tag{7}$$

and

$$W^0_{corr} = K'\exp(-E'_a/RT) \tag{8}$$

The activation energy value obtained is 66.53 and 52.87 kJ/mol for ε-caprolactam and free acid, respectively. It's observed that E_a increases slightly in the presence of inhibitor that indicates the good performance of ε-caprolactam at higher temperatures. This increase in activation energy is commonly interpreted as chemisorption process of inhibitor on the mild steel surface (Szauer et al., 1983).

5.5. Thermodynamics and Adsorption Isotherms

Fundamental insights into the interaction between the studied inhibitor and mild steel surface can be derived from adsorption isotherms. Physical and chemical adsorption are the two main types of organic inhibitor adsorption on a metal surface. Chemisorption is considered the most significant type of interaction between metal surfaces and inhibitor molecule (Ali et al., 2004). It is postulated that a covalent bond formation involving electron transfer from the inhibitor to the metal, occurs during this process. Generally, organic inhibitors prosses reactive functional groups, which serve as sites for the chemisorption process. The adsorption bond strength is dependent on the electron density of the donor atom in the functional group and the group's polarizability. Also, several authors showed that adsorption depends as well on the chemical structure (Fawzy et al, 2018; Noor et al., 2008). The coverage of the metal surface (θ) by the compound is a crucial parameter in analyzing adsorption characteristics.

It is imperative to empirically determine the adsorption isotherm that best fits to the surface coverage data to utilize the corrosion rate measurements for calculating the thermodynamic parameters associated with inhibitor adsorption. Consequently, several adsorption isotherms were evaluated according to the following equations (Bentiss et al., 2002):

Temkin isotherm:

$$\exp(f.\theta) = k_{ads} C \tag{9}$$

Langmuir isotherm:

$$\theta/(1-\theta) = k_{ads} C \tag{10}$$

Frumkin isotherm:

$$\theta/(1-\theta).\exp(-2f.\theta) = k_{ads} C \tag{11}$$

where k_{ads} is the equilibrium constant of the adsorption process, C the inhibitor concentration and f the factor of energetic in homogeneity. The optimal linear regression is obtained for the plot of C_{inh}/θ versus C_{inh} with slopes around unity. The correlation coefficient (R^2) was utilized to determine the isotherm that best represented the experimental data (Ousslim et al., 2013).

Table 5. Thermodynamic parameters for the adsorption of ε-caprolactam on mild steel in 1M HCl at different temperatures

Temperature (K)	k_{ads}	ΔG^0_{ads}(kJ/mol)	ΔH^0_{ads}(kJ/mol)	ΔS^0_{ads}(kJ/mol.K)	Slope	R
303	1672	-37,61	-8.04	-115,41	0.98	0.998
313	1017.52	-35,87		-115,19	1.01	0.999
323	479.77	-32,39		-116,23	1.02	0.998
333	266.87	-31,61		-115,07	1.01	0.998

This observation suggests that the adsorption of ε-caprolactam on the metal surface adhered to the Langmuir adsorption isotherm (Figure 4).

Figure 4. Langmuir isotherm for the adsorption of ε-caprolactam on the surface of a mild steel in 1 M HCl

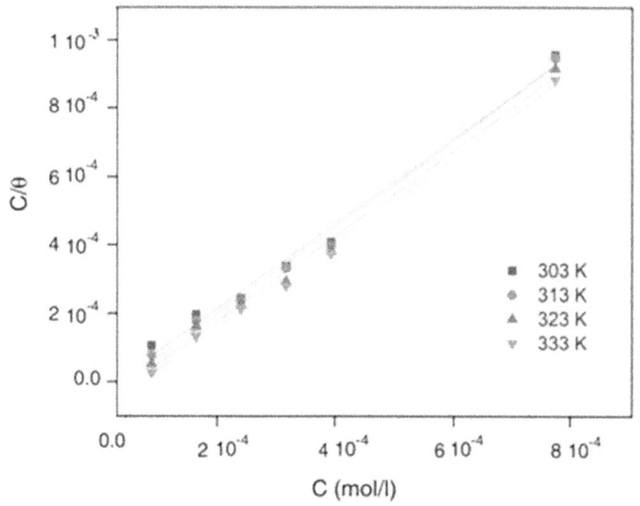

The K values were derived from the intercepts of the straight lines C_i/θ – axis, and are listed in Table 5. The adsorption constant, k_{ads}, is associated with the standard free energy of adsorption (ΔG^0_{ads}) using the following Eq.12:

$$k_{ads} = \frac{1}{55.5}\exp\left(-\frac{\Delta G^0_{ads}}{RT}\right) \tag{12}$$

The free energy of adsorption (ΔG^0_{ads}) was determined from the slope of the Langmuir isotherm. The negative values of ΔG^0_{ads} indicate that the adsorption mechanism of ε-caprolactam is spontaneous and the adsorbed film on the mild steel surface is stable. It has been established that ΔG^0_{ads} values up to -20 kJ mol^{-1} suggest physisorption mechanism, which is consistent with electrostatic interactions between charged molecules and the charged metal. In contrast, value of approximately -40 kJ mol^{-1} or higher are indicated chemisorption mechanism as a result of sharing or a transfer of electrons from organic inhibitors to the metal surface, resulting in the formation of coordinate bond (Chaitra et al., 2015; Bouklah et al., 2005). It is observed that the calculated ΔG^0_{ads} values is range from about -37.61 to -31.61 kJ.mol^{-1}, indicating, that the adsorption mechanism of the ε-caprolactam on mild steel in 1 M hydrochloric acid solution was characteristic of chemisorption (Table 5). The potential mechanisms for chemisorption can be attributed to the donation of π-electrons and the nonbinding electron pair of the oxygen and nitrogen amid group atoms as reactive centers.

Considering the values of enthalpy and entropy of the inhibition process have no distinct changes in the temperature range studied, the thermodynamic parameters ΔH^0_{ads} and ΔS^0_{ads} for the adsorption of ε-caprolactam on mild steel can be calculated from the following Eq.13:

$$\Delta G^0_{ads} = \Delta H^0_{ads} - T\Delta S^0_{ads} \tag{13}$$

where ΔH^0_{ads} and ΔS^0_{ads} are the variation of enthalpy and entropy of the adsorption process, respectively. Fig. 5 clearly displays the good dependency of ΔG^0_{ads} on temperature T, indicating a significant relationship among thermodynamic parameters. The calculated values are presented in Table 5. The negative sign of ΔH^0_{ads} indicates that the adsorption of the inhibitor is an exothermic process. Additionally, the negative values of ΔS^0_{ads} show that the adsorption process is associated with a reduction in entropy. These negative values of entropy ΔS^0_{ads} can be explained as follows: before the adsorption of inhibitors onto the mild steel surface, inhibitor molecules might freely move in the bulk solution. However, as adsorption progresses, these molecules become more orderly as they adhere to the mild steel surface, resulting in a decrease in entropy. Furthermore, ΔH^0_{ads} can also be derived from the integrated form of the Van't Hoff Eq.14, which is expressed by (Wang et al., 2003):

$$\ln K = -\frac{\Delta H^0_{ads}}{RT} + \text{constant} \tag{14}$$

Figure 5 shows the plot of $\ln k_{ads}$ versus 1/T which gives straight lines with slopes of (ΔH^0_{ads} /R) and intercepts of (ΔS^0_{ads}/R + ln(1/55,5)). The calculated ΔH^0_{ads} using the Vant'Hoff Eq.14 is −8,4 kJ/mol, confirming the exothermic behavior of the adsorption of ε-caprolactam on the steel surface. Values of ΔH^0_{ads} obtained by the both methods are in good agreement.

Figure 5. Relationship between ln k_{ads} and 1/T for ε-caprolactam

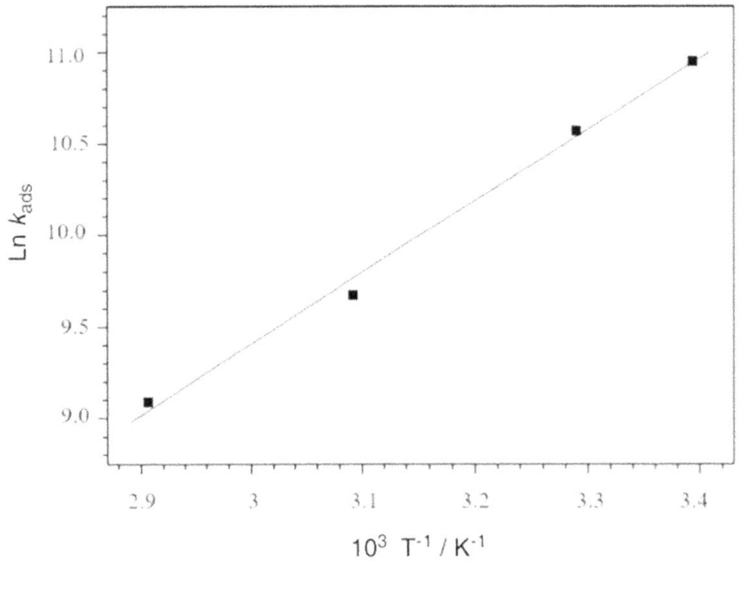

5.6. Surface Analysis by SEM

Scanning electron microscopy investigation at 308K of mild steel with and without ε-caprolactam as an organic inhibitor is presented in Fig 5. Examination of corroded metals in the acidic media utilized demonstrated notable variations in metal surface morphology with and without ε-caprolactam.

The morphology of the mild steel surface in Figure 6b indicates that in the absence of inhibitors, the surface exhibited severe damage, with regions of characteristic uniform corrosion on the mild steel surface. However, in the presence of ε-caprolactam as an organic inhibitor (Figure 6, the corrosion rate decreased, attributed to the formation of an adsorbed layer of the inhibitor on the mild steel surface. The protective nature of the layer is reflected in the inhibition efficiency measurements obtained by chemical and electrochemical methods. Figure 6 illustrates the protected mild steel surface after the addition of the inhibitor, where surface damage has decreased compared to the blank material. This observation is attributed to the formation of a protective film on the mild steel surface. This finding demonstrates the effective protective potential of ε-caprolactam which behaves as a good inhibitor for the tested surface in acidic environment.

Figure 6. SEM images of (a) non-exposed mild steel; (b) exposed mild steel in blank solution, and (c) exposed mild steel in 1 M HCl with 10^{-3} M of ε-caprolactam after 3 h immersion at 308 K

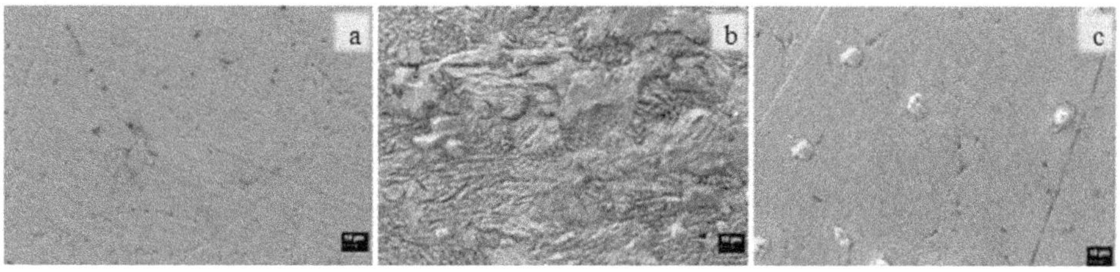

6. INDUSTRIAL IMPACT OF LACTAMS IN CORROSION INHIBITION

Corrosion inhibition using lactams in hydrochloric acid environments has significant industrial applications, particularly in sectors where mild steel is commonly used, such as construction, automotive, and manufacturing. The findings indicate that lactams, especially ε-caprolactam, can effectively reduce corrosion rates, thereby extending the lifespan of mild steel structures and components. This is particularly relevant in industries that use acidic cleaning processes or where mild steel is exposed to corrosive environments. By implementing lactam-based inhibitors, companies can enhance the durability of their products, reduce maintenance costs, and improve overall safety. Furthermore, the use of organic inhibitors, such as lactams, aligns with growing environmental concerns, as they may present an eco-friendly alternative to traditional synthetic inhibitors, contributing to sustainable industrial practices.

Lactams, as organic inhibitors for mild steel corrosion in acidic solutions, present several advantages over conventional methods. These include higher inhibition efficiencies, improved adsorption characteristics, and increased protective film formation on metal surfaces. Additionally, lactams exhibit mixed inhibition behavior, effectively reducing both anodic and cathodic reactions, which is crucial for comprehensive corrosion protection. Furthermore, their adsorption follows the Langmuir isotherm, indicating strong interactions with mild steel surfaces. Table.6 highlights the potential benefits of adopting lactam inhibitors in industrial applications, showcasing their advantages over Conventional Inhibitors such as Chromates, aluminum compounds and metallic salts (Oki et al., 2015; Obot et al., 2019; Km et al., 2024).

Table 6. Comparison of lactam-based organic inhibitors with conventional methods for mild steel corrosion protection in acidic solutions

Feature/Criteria	Conventional Inhibitors	Lactam Inhibitors
Environmental Impact	Often toxic	More eco-friendly, biodegradable
Efficiency at Low Concentration	Variable efficiency	Higher inhibition efficiency
Compatibility with mild steel	May cause surface alterations	Good compatibility
Synergistic Effects	Limited synergistic options	Can be combined with other inhibitors for enhanced performance
Application Range	Limited to specific conditions	Versatile for various acidic solution
Surface Protection	May not form protective films	Effective formation of protective layers
Temperature Stability	Performance may degrade at high temperatures	Maintains effectiveness across a range of temperatures

7. CONCLUSION

In this study, we examined the corrosion inhibition potentiality of pyrrolidin-2-one, δ-valerolactam, and ε-caprolactam in a 1 M hydrochloric acid solution on mild steel. The following conclusions can be deduced:

- Inhibition efficiency of the studied lactams increases with increasing length of the carbon chain in the lactam ring, as well as with molecular size.
- The polarization curves in acidic solution indicate that lactams function as mixed-type inhibitor, blocking anodic metal dissolution and cathodic hydrogen evolution reactions.
- Due to greater dissolution of mild steel at elevated temperatures, the inhibition efficiency decreased with increasing temperature.
- The addition of ε-caprolactam induce an increase in the activation energy of corrosion process. The adsorption equilibrium constant (k_{ads}) diminished as the temperature increased.
- The adsorption of ε-caprolactam on the mild steel surface in 1M hydrochloric acid was observed to conform to Langmuir adsorption isotherm behavior. The adsorption process is spontaneous and exothermic. The values of standard free energies of adsorption indicate that ε-caprolactam adsorb on mild steel via a chemisorption-based mechanism.

REFERENCES

Abd-El-Nabey, B., Mohamed, M., Helmy, A., Elnagar, H., & Abdel-Gaber, A. (2024). Eco-friendly corrosion inhibition of steel in acid pickling using Prunus domestica Seeds and Okra stems extracts. *International Journal of Electrochemical Science*, 19(8), 100695. DOI: 10.1016/j.ijoes.2024.100695

Ahmadi, S., & Khormali, A. (2023). Optimization of the corrosion inhibition performance of 2-mercaptobenzothiazole for carbon steel in HCl media using response surface methodology. *Fuel*, 357, 129783. DOI: 10.1016/j.fuel.2023.129783

Al-Moubaraki, A. H., & Obot, I. B. (2021). Corrosion challenges in petroleum refinery operations: Sources, mechanisms, mitigation, and future outlook. *Journal of Saudi Chemical Society*, 25(12), 101370. DOI: 10.1016/j.jscs.2021.101370

Alaneme, K., Daramola, Y., Olusegun, S., & Afolabi, A. (2015). Corrosion inhibition and adsorption characteristics of rice husk extracts on mild steel immersed in 1M H2SO4 and HCl solutions. *International Journal of Electrochemical Science*, 10(4), 3553–3567. DOI: 10.1016/S1452-3981(23)06561-6

Alharbi, S. O., Ahmad, S., Gul, T., Ali, I., & Bariq, A. (2024). The corrosion behavior of low carbon steel (AISI 1010) influenced by grain size through microstructural mechanical. *Scientific Reports*, 14(1), 5098. Advance online publication. DOI: 10.1038/s41598-023-47744-y PMID: 38429315

Ali, S., El-Shareef, A., Al-Ghamdi, R., & Saeed, M. (2004). The isoxazolidines: The effects of steric factor and hydrophobic chain length on the corrosion inhibition of mild steel in acidic medium. *Corrosion Science*, 47(11), 2659–2678. DOI: 10.1016/j.corsci.2004.11.007

Amin, M. A., El-Rehim, S. S. A., El-Sherbini, E., & Bayoumi, R. S. (2006). The inhibition of low carbon steel corrosion in HCl solutions by succinic acid. *Electrochimica Acta*, 52(11), 3588–3600. DOI: 10.1016/j.electacta.2006.10.019

Askari, M., Aliofkhazraei, M., Jafari, R., Hamghalam, P., & Hajizadeh, A. (2021). Downhole corrosion inhibitors for oil and gas production – a review. *Applied Surface Science Advances*, 6, 100128. DOI: 10.1016/j.apsadv.2021.100128

Aslam, R., Serdaroglu, G., Zehra, S., Verma, D. K., Aslam, J., Guo, L., Verma, C., Ebenso, E. E., & Quraishi, M. (2021). Corrosion inhibition of steel using different families of organic compounds: Past and present progress. *Journal of Molecular Liquids*, 348, 118373. DOI: 10.1016/j.molliq.2021.118373

Avdeev, Y., Kuznetsov, Y., & Frumkin, A. (2022). Acid corrosion of metals and its inhibition. A critical review of the current problem state. *International Journal of Corrosion and Scale Inhibition*, 11(1). Advance online publication. DOI: 10.17675/2305-6894-2022-11-1-6

Bashir, S., Singh, G., & Kumar, A. (2018). Shatavari (Asparagus racemosus) as green corrosion inhibitor of aluminium in acidic medium. *Portugaliae Electrochimica Acta*, 37(2), 83–1. DOI: 10.4152/pea.201902083

Basik, M., & Mobin, M. (2022). Metal oxide and organic polymers mixed composites as corrosion inhibitors. In Elsevier eBooks (pp. 345–355). DOI: 10.1016/B978-0-323-90410-0.00018-0

Bentiss, F., Lebrini, M., & Lagrenée, M. (2005). Thermodynamic characterization of metal dissolution and inhibitor adsorption processes in mild steel/2,5-bis(n-thienyl)-1,3,4-thiadiazoles/HCl system. *Corrosion Science*, 47(12), 2915–2931. DOI: 10.1016/j.corsci.2005.05.034

Bonora, P., Deflorian, F., & Fedrizzi, L. (1996). Electrochemical impedance spectroscopy as a tool for investigating underpaint corrosion. *Electrochimica Acta*, 41(7–8), 1073–1082. DOI: 10.1016/0013-4686(95)00440-8

Borsari, M., Ferrari, E., Grandi, R., & Saladini, M. (2002). Curcuminoids as potential new ironchelating agents: Spectroscopic, polarographic and potentiometric study on their Fe (III) complexing ability. *Inorganica Chimica Acta*, 328(1), 61–68. DOI: 10.1016/S0020-1693(01)00687-9

Bouammali, H., Ousslim, A., Bekkouch, K., Bouammali, B., Aouniti, A., Al-Deyab, S., Jama, C., Bentiss, F., & Hammouti, B. (2013). The Anti-Corrosion Behavior of Lavandula dentata Aqueous Extract on Mild Steel in 1M HCl. *International Journal of Electrochemical Science*, 8(4), 6005–6013. DOI: 10.1016/S1452-3981(23)14735-3

Bouklah, M., Hammouti, B., Lagrenée, M., & Bentiss, F. (2005). Thermodynamic properties of 2,5-bis(4-methoxyphenyl)-1,3,4-oxadiazole as a corrosion inhibitor for mild steel in normal sulfuric acid medium. *Corrosion Science*, 48(9), 2831–2842. DOI: 10.1016/j.corsci.2005.08.019

Caihong, Y., Singh, A., Ansari, K., Ali, I. H., & Kumar, R. (2022). Novel nitrogen based heterocyclic compound as Q235 steel corrosion inhibitor in 15% HCl under dynamic condition: A detailed experimental and surface analysis. *Journal of Molecular Liquids*, 362, 119720. DOI: 10.1016/j.molliq.2022.119720

Chaitra, T. K., Mohana, K. N. S., & Tandon, H. C. (2015). Thermodynamic, electrochemical and quantum chemical evaluation of some triazole Schiff bases as mild steel corrosion inhibitors in acid media. *Journal of Molecular Liquids*, 211, 1026–1038. DOI: 10.1016/j.molliq.2015.08.031

Chen, L., Lu, D., & Zhang, Y. (2022b). Organic compounds as corrosion inhibitors for carbon steel in HCl solution: A comprehensive review. *Materials (Basel)*, 15(6), 2023. DOI: 10.3390/ma15062023 PMID: 35329474

Chen, M., Chen, Y., Lim, Z. J., & Wong, M. W. (2022). Adsorption of imidazolium-based ionic liquids on the Fe(1 0 0) surface for corrosion inhibition: Physisorption or chemisorption? *Journal of Molecular Liquids*, 367, 120489. DOI: 10.1016/j.molliq.2022.120489

Ding, J., Tang, B., Li, M., Feng, X., Fu, F., Bin, L., Huang, S., Su, W., Li, D., & Zheng, L. (2016). Difference in the characteristics of the rust layers on carbon steel and their corrosion behavior in an acidic medium: Limiting factors for cleaner pickling. *Journal of Cleaner Production*, 142, 2166–2176. DOI: 10.1016/j.jclepro.2016.11.066

El-Aziz, E.-S., Ibrahem, E.-D., & Awad, S. (2019). Esomeprazole Magnesium Trihydrate drug as a potential non-toxic corrosion inhibitor for mild steel in acidic media. *Zast. Mater.*, 60(4), 245–258. DOI: 10.5937/zasmat1903245E

Elshakre, M. E., Alalawy, H. H., Awad, M. I., & El-Anadouli, B. E. (2017). On the role of the electronic states of corrosion inhibitors: Quantum chemical-electrochemical correlation study on urea derivatives. *Corrosion Science*, 124, 121–130. DOI: 10.1016/j.corsci.2017.05.015

Espinoza-Vázquez, A., Negrón-Silva, G., Angeles-Beltrán, D., Herrera-Hernández, H., Romero-Romo, M., & Palomar-Pardavé, M. (2014). electrochemical impedance spectroscopy Evaluation of Pantoprazole as corrosion inhibitor for mild steel immersed in HCl 1 M. effect of [Pantoprazole], hydrodynamic conditions, temperature and immersion times. *International Journal of Electrochemical Science*, 9(2), 493–509. DOI: 10.1016/S1452-3981(23)07734-9

Fallavena, T., Antonow, M., & Gonçalves, R. S. (2006). Caffeine as non-toxic corrosion inhibitor for copper in aqueous solutions of potassium nitrate. *Applied Surface Science*, 253(2), 566–571. DOI: 10.1016/j.apsusc.2005.12.114

Farelas, F., & Ramirez, A. (2010). Carbon dioxide corrosion inhibition of carbon steels through bis-imidazoline and imidazoline compounds studied by electrochemical impedance spectroscopy. *International Journal of Electrochemical Science*, 5(6), 797–814. DOI: 10.1016/S1452-3981(23)15324-7

Fawzy, A., Zaafarany, I., Ali, H., & Abdallah, M. (2018). New synthesized amino acids-based surfactants as efficient inhibitors for corrosion of mild steel in HCl medium: Kinetics and Thermodynamic Approach. *International Journal of Electrochemical Science*, 13(5), 4575–4600. DOI: 10.20964/2018.05.01

Fayomi, O. S. I., Olusanyan, D., Ademuyiwa, F. T., & Olarewaju, G. (2021). Progresses on mild steel protection toward surface service performance in structural industrial: An Overview. *IOP Conference Series. Materials Science and Engineering*, 1036(1), 012079. DOI: 10.1088/1757-899X/1036/1/012079

Fernandes, J. S., & Montemor, F. (2014). Corrosion. In Springer eBooks (pp. 679–716). DOI: 10.1007/978-3-319-08236-3_15

Fiori-Bimbi, M. V., Alvarez, P. E., Vaca, H., & Gervasi, C. A. (2014). Corrosion inhibition of mild steel in HCl solution by pectin. *Corrosion Science*, 92, 192–199. DOI: 10.1016/j.corsci.2014.12.002

Goyal, M., Kumar, S., Bahadur, I., Verma, C., Ebenso, E.E. (2018). Organic corrosion inhibitors for industrial cleaning of ferrous and non-ferrous metals in acidic solutions: a review. J. Mol. Liq. 256, 565–573. https://doi.org/. molliq.2018.02.045.DOI: 10.1016/j

Gupta, S. K., Mehta, R. K., & Yadav, M. (2022). Schiff bases as corrosion inhibitorson mild steel in acidic medium: Gravimetric, electrochemical, surface morphological and computational studies. *Journal of Molecular Liquids*, 368, 120747. DOI: 10.1016/j.molliq.2022.120747

Haris, N. I. N., Sobri, S., & Kassim, N. (2019). Oil palm empty fruit bunch extract as green corrosion inhibitor for mild steel in HCl solution: Central composite design optimization. *Materials and Corrosion*, 70(6), 1111–1119. DOI: 10.1002/maco.201810653

Heydari, H., Talebian, M., Salarvand, Z., Raeissi, K., Bagheri, M., & Golozar, M. A. (2018). Comparison of two Schiff bases containing O-methyl and nitro substitutes for corrosion inhibiting of mild steel in 1 M HCl solution. *Journal of Molecular Liquids*, 254, 177–187. DOI: 10.1016/j.molliq.2018.01.112

Hussain, C. M., Verma, C., Aslam, J., Aslam, R., & Zehra, S. (2023). Basics of corrosion and its impact. In Elsevier eBooks (pp. 3–30). DOI: 10.1016/B978-0-323-95185-2.00001-0

Kahkesh, H., & Zargar, B. (2023). Estimating the anti-corrosive potency of 3-nitrophthalic acid as a novel and natural organic inhibitor on corrosion monitoring of mild steel in 1 M HCl solution. *Inorganic Chemistry Communications*, 158, 111533. DOI: 10.1016/j.inoche.2023.111533

Karki, N., Neupane, S., Chaudhary, Y., Gupta, D., & Yadav, A. (2021). Equisetum hyemale: A new candidate for green corrosion inhibitor family. *International Journal of Corrosion and Scale Inhibition*, 10(1). Advance online publication. DOI: 10.17675/2305-6894-2021-10-1-12

Kaya, F., Solmaz, R., & Geçibesler, İ. H. (2023). Adsorption and corrosion inhibition capability of Rheum ribes root extract (Işgın) for mild steel protection in acidic medium: A comprehensive electrochemical, surface characterization, synergistic inhibition effect, and stability study. *Journal of Molecular Liquids*, 372, 121219. DOI: 10.1016/j.molliq.2023.121219

Khaled, K. F., Babić-Samardžija, K., & Hackerman, N. (2005). Theoretical study of the structural effects of polymethylene amines on corrosion inhibition of iron in acid solutions. *Electrochimica Acta*, 50(12), 2515–2520. DOI: 10.1016/j.electacta.2004.10.079

Khatib, L. W. E., Rahal, H. T., & Abdel-Gaber, A. M. (2020). Synergistic Effect between Fragaria ananassa and Cucurbita pepo L Leaf Extracts on Mild Steel Corrosion in HCl Solutions. *Protection of Metals and Physical Chemistry of Surfaces*, 56(5), 1096–1106. DOI: 10.1134/S2070205120050111

Km, S., Praveen, B., & Devendra, B. K. (2024b). A review on corrosion inhibitors: Types, mechanisms, electrochemical analysis, corrosion rate and efficiency of corrosion inhibitors on mild steel in an acidic environment. *Results in Surfaces and Interfaces*, 16, 100258. DOI: 10.1016/j.rsurfi.2024.100258

Kumar, A., & Das, C. (2023). A novel eco-friendly inhibitor of chayote fruit extract for mild steel corrosion in 1 M HCl: Electrochemical, weight loss studies, and the effect of temperature. *Sustainable Chemistry and Pharmacy*, 36, 101261. DOI: 10.1016/j.scp.2023.101261

Kuren, S. G., Volyanik, S. A., Savenkova, M. A., & Zaitseva, E. A. (2023). Properties of Salicylidene-Aniline as a corrosion inhibitor in oil and petroleum products transportation systems. *Safety of Technogenic and Natural Systems*, 3(3), 14–23. DOI: 10.23947/2541-9129-2023-7-3-14-23

Larabi, L., Harek, Y., Traisnel, M., & Mansri, A. (2004). Synergistic influence of Poly(4-Vinylpyridine) and potassium iodide on inhibition of corrosion of mild steel in 1M HCl. *Journal of Applied Electrochemistry*, 34(8), 833–839. DOI: 10.1023/B:JACH.0000035609.09564.e6

Li, E., Li, Y., Liu, S., & Yao, P. (2022). Choline amino acid ionic liquids as green corrosion inhibitors of mild steel in acidic medium. *Colloids and Surfaces. A, Physicochemical and Engineering Aspects*, 657, 130541. DOI: 10.1016/j.colsurfa.2022.130541

Li, X., He, J., Xie, B., He, Y., Lai, C., Wang, W., Zeng, J., Yao, B., Zhao, W., & Long, T. (2024). 1,4-Phenylenediamine-based Schiff bases as eco-friendly and efficient corrosion inhibitors for mild steel in HCl medium: Experimental and theoretical approaches. *Journal of Electroanalytical Chemistry (Lausanne, Switzerland)*, 955, 118052. DOI: 10.1016/j.jelechem.2024.118052

Likhanova, N. V., Arellanes-Lozada, P., Olivares-Xometl, O., Hernández-Cocoletzi, H., Lijanova, I. V., Arriola-Morales, J., & Castellanos-Aguila, J. (2019). Effect of organic anions on ionic liquids as corrosion inhibitors of steel in sulfuric acid solution. *Journal of Molecular Liquids*, 279, 267–278. DOI: 10.1016/j.molliq.2019.01.126

Liu, X., Xiong, J., Lv, Y., & Zuo, Y. (2008). Study on corrosion electrochemical behavior of several different coating systems by electrochemical impedance spectroscopy. *Progress in Organic Coatings*, 64(4), 497–503. DOI: 10.1016/j.porgcoat.2008.08.012

Mahdi, B., Aljibori, S., Abbass, H. S. S., Al-Azzawi, M. K., Kadhum, A. H., Hanoon, M., Isahak, M., & Al-Amiery, W. N. R. W. (2022). Gravimetric analysis and quantum chemical assessment of 4-aminoantipyrine derivatives as corrosion inhibitors. *Int. J. Corros. Scale Inhib.*, 11(3), 1191–1213. DOI: 10.17675/2305-6894-2022-11-3-17

Matad, P. B., Mokshanatha, P. B., Hebbar, N., Venkatesha, V. T., & Tandon, H. C. (2014). Ketosulfone drug as a green corrosion inhibitor for mild steel in acidic medium. *Industrial & Engineering Chemistry Research*, 53(20), 8436–8444. DOI: 10.1021/ie500232g

McCafferty, E. (2010). Corrosion Inhibitors. In *Introduction to Corrosion Science*. Springer., DOI: 10.1007/978-1-4419-0455-3_12

Mobin, M., Aslam, R., Salim, R., & Kaya, S. (2022). An investigation on the synthesis, characterization and anti-corrosion properties of choline based ionic liquids as novel and environmentally friendly inhibitors for mild steel corrosion in 5% HCl. *Journal of Colloid and Interface Science*, 620, 293–312. DOI: 10.1016/j.jcis.2022.04.036 PMID: 35429708

Mohamed, A., Martin, U., Visco, D. P.Jr, Townsend, T., & Bastidas, D. M. (2023). Interphase corrosion inhibition mechanism of sodium borate on carbon steel rebars in simulated concrete pore solution. *Construction & Building Materials*, 408, 133763. DOI: 10.1016/j.conbuildmat.2023.133763

Moloney, J., Kumar, D., Muralidhar, V., & Pojtanabuntoeng, T. (2022). Corrosion inhibition. Flow Assurance (Volume 1 in Oil and Gas Chemistry Management Series Pages 609-707). DOI: 10.1016/B978-0-12-822010-8.00006-4

Mu, X., Wei, J., Dong, J., & Ke, W. (2014). In Situ Corrosion Monitoring of Mild Steel in a Simulated Tidal Zone without Marine Fouling Attachment by Electrochemical Impedance Spectroscopy. *Journal of Materials Science and Technology*, 30(10), 1043–1050. DOI: 10.1016/j.jmst.2014.03.013

Murmu, M., Saha, S. K., Bhaumick, P., Murmu, N. C., Hirani, H., & Banerjee, P. (2020). Corrosion inhibition property of azomethine functionalized triazole derivatives in 1 mol L−1 HCl medium for mild steel: Experimental and theoretical exploration. *Journal of Molecular Liquids*, 313, 113508. DOI: 10.1016/j.molliq.2020.113508

Muthukrishnan, P., Jeyaprabha, B., & Prakash, P. (2013). Adsorption and corrosion inhibiting behavior of Lannea coromandelica leaf extract on mild steel corrosion. *Arabian Journal of Chemistry*, 10, S2343–S2354. DOI: 10.1016/j.arabjc.2013.08.011

Nandiyanto, A. B., Wiryani, D., Rusli, A. S., Purnamasari, A., Abdullah, A., Widiaty, A. G. I., & Churriyati, R. (2017). Extraction of curcumin pigment from Indonesian local turmeric with its infrared spectra and thermal decomposition properties. IOP Conf. Ser. Mater. Sci. Eng. (1), 012136 , 180.DOI: 10.1088/1757-899X/180/1/012136

Nataraja, S., Venkatesha, T., Manjunatha, K., Poojary, B., Pavithra, M., & Tandon, H. (2011). Inhibition of the corrosion of steel in HCl solution by some organic molecules containing the methylthiophenyl moiety. *Corrosion Science*, 53(8), 2651–2659. DOI: 10.1016/j.corsci.2011.05.004

Nawaz, S. S., Manjunatha, K., Ranganatha, S., Supriya, S., Ranjan, P., Chakraborty, T., & Ramakrishna, D. (2023). Nickel curcumin complexes: Physico chemical studies and nonlinear optical activity. *Optical Materials*, 136, 113450. DOI: 10.1016/j.optmat.2023.113450

Negm, N. A., Kandile, N. G., Badr, E. A., & Mohammed, M. A. (2012). Gravimetric and electrochemical evaluation of environmentally friendly nonionic corrosion inhibitors for carbon steel in 1 M HCl. *Corrosion Science*, 65, 94–103. DOI: 10.1016/j.corsci.2012.08.002

Noor, E. A., & Al-Moubaraki, A. H. (2008). Thermodynamic study of metal corrosion and inhibitor adsorption processes in mild steel/1-methyl-4 [4′(-X)-styryl pyridinium iodides/HCl systems. *Materials Chemistry and Physics*, 110(1), 145–154. DOI: 10.1016/j.matchemphys.2008.01.028

Obot, I., Onyeachu, I. B., Umoren, S. A., Quraishi, M. A., Sorour, A. A., Chen, T., Aljeaban, N., & Wang, Q. (2019). High temperature sweet corrosion and inhibition in the oil and gas industry: Progress, challenges and future perspectives. *Journal of Petroleum Science Engineering*, 185, 106469. DOI: 10.1016/j.petrol.2019.106469

Oguzie, E., Okolue, B., Ebenso, E., Onuoha, G., & Onuchukwu, A. (2004). Evaluation of the inhibitory effect of methylene blue dye on the corrosion of aluminium in HCl. *Materials Chemistry and Physics*, 87(2–3), 394–401. DOI: 10.1016/j.matchemphys.2004.06.003

Okey, N. C., Obasi, N. L., Ejikeme, P. M., Ndinteh, D. T., Ramasami, P., Sherif, E. M., Akpan, E. D., & Ebenso, E. E. (2020). Evaluation of some amino benzoic acid and 4-aminoantipyrine derived Schiff bases as corrosion inhibitors for mild steel in acidic medium: Synthesis, experimental and computational studies. *Journal of Molecular Liquids*, 315, 113773. DOI: 10.1016/j.molliq.2020.113773

Oki, M., & Anawe, P. (2015). A review of Corrosion in Agricultural Industries. *Physical Science International Journal*, 5(4), 216–222. DOI: 10.9734/PSIJ/2015/14847

Ouadi, Y. E., Fal, M. E., Hafez, B., Manssouri, M., Ansari, A., Elmsellem, H., Ramli, Y., & Bendaif, H. (2020). Physisorption and corrosion inhibition of mild steel in 1 M HCl using a new pyrazolic compound: Experimental data & quantum chemical calculations. *Materials Today: Proceedings*, 27, 3010–3016. DOI: 10.1016/j.matpr.2020.03.340

Ouici, H., Tourabi, M., Benali, O., Selles, C., Jama, C., Zarrouk, A., & Bentiss, F. (2017). Adsorption and corrosion inhibition properties of 5-amino 1,3,4-thiadiazole-2-thiol on the mild steel in HCl medium: Thermodynamic, surface and electrochemical studies. *Journal of Electroanalytical Chemistry (Lausanne, Switzerland)*, 803, 125–134. DOI: 10.1016/j.jelechem.2017.09.018

Ousslim, A., Chetouani, A., Hammouti, B., Bekkouch, K., Al-Deyab, S., Aouniti, A., & Elidrissi, A. (2013). Thermodynamics, quantum and electrochemical studies of corrosion of iron by piperazine compounds in sulphuric acid. *International Journal of Electrochemical Science*, 8(4), 5980–6004. DOI: 10.1016/S1452-3981(23)14734-1

Oyekunle, D., Agboola, O., & Ayeni, A. (2019). Corrosion inhibitors as building evidence for Mild steel: A review. *Journal of Physics: Conference Series*, 1378(3), 032046. DOI: 10.1088/1742-6596/1378/3/032046

Park, J. S., Kim, S. O., Jeong, Y. J., Lee, S. G., Choi, J. K., & Kim, S. J. (2022). Long-Term corrosion behavior of strong and ductile high MN-Low CR steel in acidic aqueous environments. *Materials (Basel)*, 15(5), 1746. DOI: 10.3390/ma15051746 PMID: 35268977

Petrunin, M., Rybkina, A., Yurasova, T., & Maksaeva, L. (2022). Effect of Organosilicon Self-Assembled Polymeric Nanolayers Formed during Surface Modification by Compositions Based on Organosilanes on the Atmospheric Corrosion of Metals. *Polymers*, 14(20), 4428. DOI: 10.3390/polym14204428 PMID: 36298006

Pokhmurs'kyi, V. I., Zin, I. M., & Lyon, S. B. (2004). Inhibition of corrosion by a mixture of nonchromate pigments in organic coatings on galvanized steel. *Materials Science*, 40(3), 383–390. DOI: 10.1007/PL00022002

Prasad, A. R., Arshad, M., & Joseph, A. (2021c). A sustainable method of mitigating acid corrosion of mild steel using jackfruit pectin (JP) as green inhibitor: Theoretical and electrochemical studies. *Journal of the Indian Chemical Society*, 99(1), 100271. DOI: 10.1016/j.jics.2021.100271

Prasanna, B. M., Praveen, B. M., Hebbar, N., Pavithra, M. K., Manjunatha, T. S., & Malladi, R. S. (2018). Theoretical and experimental approach of inhibition effect by sulfamethoxazole on mild steel corrosion in 1-M HCl. *Surface and Interface Analysis*, 50(8), 1–11. DOI: 10.1002/sia.6457

Qian, J., Wang, J., Zhang, W., Mao, J., Qin, H., Ling, X., Zeng, H., Hou, J., Chen, Y., & Wan, G. (2023). Corrosion-tailoring, osteogenic, anti-inflammatory, and antibacterial aspirin-loaded organometallic hydrogel composite coating on biodegradable Zn for orthopedic applications. *Biomaterials Advances*, 153, 213536. DOI: 10.1016/j.bioadv.2023.213536 PMID: 37418934

Quraishi, M., & Shukla, S. K. (2008). Poly(aniline-formaldehyde): A new and effective corrosion inhibitor for mild steel in HCl. *Materials Chemistry and Physics*, 113(2–3), 685–689. DOI: 10.1016/j.matchemphys.2008.08.028

Reza, N. A., Akhmal, N. H., Fadil, N. A., & Taib, M. F. M. (2021). A review on plants and biomass wastes as organic green corrosion inhibitors for mild steel in acidic environment. *Metals*, 11(7), 1062. DOI: 10.3390/met11071062

Riggs, O. L.Jr, & Hurd, R. M. (1967). Temperature coefficient of corrosion inhibition. *Corrosion*, 23(8), 252–260. DOI: 10.5006/0010-9312-23.8.252

Rinky, N. J., Islam, M. M., Hossen, J., & Islam, M. A. (2023). Comparative study of anti-ulcer drugs with benzimidazole ring as green corrosion inhibitors in acidic solution: Quantum chemical studies. *Current Research in Green and Sustainable Chemistry*, 7, 100385. DOI: 10.1016/j.crgsc.2023.100385

Sahraoui, M., Boulkroune, M., Chibani, A., Larbah, Y., & Abdessemed, A. (2022). Aqueous extract of Punica granatum fruit peel as an Eco-Friendly corrosion inhibitor for aluminium alloy in acidic medium. *Journal of Bio- and Tribo-Corrosion*, 8(2), 54. Advance online publication. DOI: 10.1007/s40735-022-00658-0

Samal, P. P., Singh, C. P., Tiwari, S., Shah, V., & Krishnamurty, S. (2024). Indazole-5-amine (AIA) as competing corrosion coating to Benzotriazole (BTAH) at the interface of Cu: A DFT and BOMD case study. *Computational & Theoretical Chemistry*, 1239, 114762. DOI: 10.1016/j.comptc.2024.114762

Sanaei, Z., Bahlakeh, G., & Ramezanzadeh, B. (2017). Active corrosion protection of mild steel by an epoxy ester coating reinforced with hybrid organic/inorganic green inhibitive pigment. *Journal of Alloys and Compounds*, 728, 1289–1304. DOI: 10.1016/j.jallcom.2017.09.095

Sedik, A., Lerari, D., Salci, A., Athmani, S., Bachari, K., Gecibesler, İ., & Solmaz, R. (2020). Dardagan Fruit extract as eco-friendly corrosion inhibitor for mild steel in 1 M HCl: Electrochemical and surface morphological studies. *Journal of the Taiwan Institute of Chemical Engineers*, 107, 189–200. DOI: 10.1016/j.jtice.2019.12.006

Sengupta, S., Murmu, M., Mandal, S., Hirani, H., & Banerjee, P. (2021b). Competitive corrosion inhibition performance of alkyl/acyl substituted 2-(2-hydroxybenzylideneamino)phenol protecting mild steel used in adverse acidic medium: A dual approach analysis using FMOs/molecular dynamics simulation corroborated experimental findings. *Colloids and Surfaces. A, Physicochemical and Engineering Aspects*, 617, 126314. DOI: 10.1016/j.colsurfa.2021.126314

Shamsheera, K. O.; Prasad, A. R.; Arshad, M. (2022). A sustainable method of mitigating acid corrosion of mild steel using jackfruit pectin (JP) as green inhibitor: Theoretical and electrochemical studies. Journal of the Indian Chemical Society, 99 (1), No. 100271, DOI: .DOI: 10.1016/j.jics.2021.100271

Sharma, M., Yadav, S. S., Sharma, P., Yadav, L., Abedeen, M. Z., Kushwaha, H. S., & Gupta, R. (2023). An experimental and theoretical investigation of corrosion inhibitive behaviour of 4-amino antipyrine and its schiff's base (BHAP) on mild steel in 1 M HCl solution. *Inorganic Chemistry Communications*, 157, 111330. DOI: 10.1016/j.inoche.2023.111330

Singh, A., Ansari, K., Sharma, N. R., Singh, S., Singh, R., Bansal, A., Ali, I. H., Younas, M., Alanazi, A. K., & Lin, Y. (2023). Corrosion and bacterial growth mitigation in the desalination plant by imidazolium based ionic liquid: Experimental, surface and molecular docking analysis. *Journal of Environmental Chemical Engineering*, 11(2), 109313. DOI: 10.1016/j.jece.2023.109313

Sivakumar, P. R. (2021). Corrosion protection behaviour of some unsymmetrical oxadiazoles on mild steel surface in 1 m H2SO4 acid medium. *Journal of Bio- and Tribo-Corrosion*, 8(1), 17. Advance online publication. DOI: 10.1007/s40735-021-00614-4

Solmaz R., Kardas G., Çulha M., Yazıcı B., Erbil M. (2008). Investigation of adsorption and inhibitive effect of 2-mercaptothiazoline on corrosion of mild steel in HCl media, Electrochim. Acta 53, 5941–5952, https://doi.org/. electacta.2008.03.055.DOI: 10.1016/j

Solomon, M. M., Essien, K. E., Loto, R. T., & Ademosun, O. T. (2021). Synergistic corrosion inhibition of low carbon steel in HCl and H2SO4 media by 5-methyl-3-phenylisoxazole-4-carboxylic acid and iodide ions. *Journal of Adhesion Science and Technology*, 36(11), 1200–1226. DOI: 10.1080/01694243.2021.1962091

Szauer, T., & Brandt, A. (1983). Equilibria in solutions of amines and fatty acids with relevance to the corrosion inhibition of iron. *Corrosion Science*, 23(12), 1247–1257. DOI: 10.1016/0010-938X(83)90075-6

Tanwer, S., & Shukla, S. K. (2021b). *Corrosion inhibition activity of Cefixime on mild steel surface in aqueous sulphuric acid*. Progress in Color, Colorants and Coatings., DOI: 10.30509/pccc.2021.166889.1133

Tebbji, K., Faska, N., Tounsi, A., Oudda, H., Benkaddour, M., & Hammouti, B. (2007). The effect of some lactones as inhibitors for the corrosion of mild steel in 1M HCl. *Materials Chemistry and Physics*, 106(2–3), 260–267. DOI: 10.1016/j.matchemphys.2007.05.046

Verma, C., Quraishi, M. A., Ebenso, E. E., & Hussain, C. M. (2021). Amines as corrosion inhibitors: A review. In Organic Corrosion Inhibitors: Synthesis, Characterization, Mechanism, and Applications (pp. 77-94). wiley. DOI: 10.1002/9781119794516.ch5

Vikneshvaran, S., & Velmathi, S. (2019). Schiff Bases of 2,5-Thiophenedicarboxaldehyde as Corrosion Inhibitor for Stainless Steel under Acidic Medium: Experimental, Quantum Chemical and Surface Studies. *ChemistrySelect*, 4(1), 387–392. DOI: 10.1002/slct.201803235

Wang, H., Fan, H., & Zheng, J. (2003). Corrosion inhibition of mild steel in HCl solution by a mercapto-triazole compound. *Materials Chemistry and Physics*, 77(3), 655–661. DOI: 10.1016/S0254-0584(02)00123-2

Xiong, Y., & Cao, M. (2024). Application of surfactants in corrosion inhibition of metals. *Current Opinion in Colloid & Interface Science*, 73, 101830. DOI: 10.1016/j.cocis.2024.101830

Yadav, M., Behera, D., & Sharma, U. (2012). Nontoxic corrosion inhibitors for N80 steel in HCl. *Arabian Journal of Chemistry*, 9, S1487–S1495. DOI: 10.1016/j.arabjc.2012.03.011

Yurt, A., Balaban, A., Kandemir, S., Bereket, G., & Erk, B. (2004). Investigation on some Schiff bases as HCl corrosion inhibitors for carbon steel. *Materials Chemistry and Physics*, 85(2–3), 420–426. DOI: 10.1016/j.matchemphys.2004.01.033

Zhang, X., Zhang, Y., Su, Y., & Guan, S. (2024). Enhancing the corrosion inhibition performance of Mannich base on mild steel in lactic acid solution through synergistic effect of allicin: Experimental and theoretical study. *Journal of Molecular Structure*, 1304, 137658. DOI: 10.1016/j.molstruc.2024.137658

Chapter 8
Sol-Gel Synthesis of a New Composition of Bioactive Glass

Halima El Bouami
CERAMATHS, Université Polytechnique Hauts-de-France, France & Laboratory of Process Engineering, Faculty of Applied Science Ait Melloul, Ibn Zohr University, Agadir, Morocco

Assia Mabrouk
https://orcid.org/0000-0001-6399-0644
Science and Technology Research Laboratory, The Higher School of Education and Training, Ibn Zohr University, Agadir, Morocco

Cyrille Mercier
CERAMATHS, Université Polytechnique Hauts-de-France, France

Claudine Follet
https://orcid.org/0000-0002-3102-5935
CERAMATHS, Université Polytechnique Hauts-de-France, France

Ahmed Bachar
Laboratory of Process Engineering, Faculty of Applied Science Ait Melloul, Ibn Zohr University, Agadir, Morocco

ABSTRACT

Bioactive glasses hold immense promise for tissue engineering and bone regeneration due to their ability to bond with living tissues and stimulate bone growth. This chapter explores the sol-gel method, a versatile technique offering advantages like lower processing temperatures and superior compositional control, for synthesizing a new bioactive glass composition. We discuss the fundamental principles of sol-gel chemistry and precursor selection for achieving the desired elements in the final glass. The chapter details the design rationale behind the new compositions, targeting the effect of therapeutic ions promotion specific antibacterial and angiogenesis properties and enhancement in bioactivity and osteoblast production. Additionally, the chapter offers a brief overview of in vitro bioactivity assessment methods for evaluating the glass's interaction with physiological fluids. Finally, we discuss the potential applications of the newly developed bioactive glass and propose future research directions for further optimization and exploration of its functionalities.

DOI: 10.4018/979-8-3693-7505-1.ch008

I. INTRODUCTION

Bioactive glasses have emerged as a revolutionary class of biomaterials, offering unprecedented potential in the fields of tissue engineering and regenerative medicine (Owens et al., 2016) These remarkable materials possess the unique ability to form strong bonds with living tissues, particularly bone, making them invaluable in various medical applications (Deshmukh et al., 2020). The development of bioactive glasses represents a significant advancement in biomaterials science, bridging the gap between inert implants and the body's natural healing processes (Lepry and Nazhat, 2021a) Bioglass® 45S5, a melt-quenched bioactive glass with the composition 46.1% SiO_2-26.9% CaO-24.4% Na_2O-2.6% P_2O_5 (in mol %), stands out as the most thoroughly researched bioactive glass(Hench, 1991). It has been utilized in clinical settings for more than 20 years, owing to its remarkable ability to integrate with both hard and soft tissues(Kokubo and Takadama, 2006). The key to its effectiveness lies in its capacity to release ions when exposed to physiological environments, which subsequently leads to the formation of a hydroxy-carbonated apatite (HCA) layer on its surface(Jones et al., 2001; Kokubo and Takadama, 2006; Hench and Jones, 2015). This unique property enables Bioglass® 45S5 to create strong bonds with bone and surrounding tissues(Hench and Thompson, 2010).

At the forefront of bioactive glass synthesis is the sol-gel method, a versatile and powerful technique that offers several advantages over traditional melt-quench processes(Deshmukh et al., 2020). This method allows for the production of glasses with higher purity, larger surface areas, and inherent porosity, all of which contribute to enhanced bioactivity and degradability.(Owens et al., 2016) The sol-gel approach opens up new possibilities for tailoring the composition and properties of bioactive glasses, enabling researchers to optimize these materials for specific medical applications.(Drago et al., 2018; Kunwong et al., 2021; Ebrahimi et al., 2023; Kaou et al., 2023; El Bouami et al., 2024)

This document explores the intricacies of the sol-gel method for bioactive glass synthesis, delving into the fundamental chemistry, advantages, and various modifications of this technique. It also examines the design considerations for new bioactive glass compositions, including the incorporation of therapeutic ions to enhance functionality. Furthermore, it discusses the mechanisms of bioactivity, methods for assessing it, and the potential applications of these innovative materials in tissue engineering and bone regeneration.

As we continue to push the boundaries of biomaterials science, bioactive glasses synthesized through sol-gel methods stand at the forefront of medical innovation, promising to revolutionize treatments for bone defects, tissue regeneration, and beyond.

II. SOL-GEL METHOD FOR BIOACTIVE GLASS SYNTHESIS

2.1. Fundamentals of Sol-Gel Chemistry

The conventional sol-gel method, at its core, involve two primary stages: solution and gelation, where a "sol" refers to a colloidal suspension of solid particles, while a "gel" is an interconnected network of these solid particles that spans a secondary phase, typically a liquid(Owens et al., 2016). This sol-gel process chemistry fundamentally require the transformation of alkoxide precursors, like tetraethyl orthosilicate (TEOS) and the triethylphosphate (TEP), through hydrolysis and condensation reactions. (Bokov et al., 2021) The hydrolysis reaction is the substitution of alkoxide groups (-OR) with hydroxyl groups (-OH), resulting in the formation of silanol species.(Owens et al., 2016; Bokov et al., 2021) As the processes of hydrolysis and polycondensation progress, the siloxane structure expands, result in the formation of siloxane bonds, which establish a network of interconnected silicon-oxygen polyhedra that make up the gel structure.(Deshmukh et al., 2020; Bokov et al., 2021) Once the gel forms, it's usually left to aging for a specific time. During this aging phase, additional condensation processes take place within the gel's structure, enhancing its interconnectedness and durability(Owens et al., 2016). The aged gel holds a large quantity of liquid in its tiny spaces. When this liquid is removed through drying, the gel shrinks considerably, resulting in a compact, dry material called a xerogel(Zhong and Greenspan, 2000; Owens et al., 2016; Bokov et al., 2021). The dehydrated gel undergoes calcination. This heating step eliminates any leftover organic matter and can make the material more compact, improving its physical strength(Deshmukh et al., 2020).

Figure 1. Sol-gel synthesis process (Owens et al., 2016)

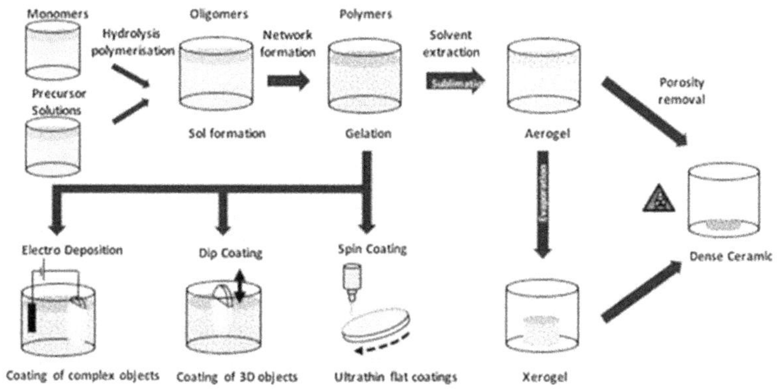

The sol-gel process stages depicted in Figure 1 can be modified or omitted, except for solvation and gelation, but hydrolysis and condensation reactions are essential for producing sol-gel-derived bioactive materials.(Owens et al., 2016)

The conventional sol-gel method offers flexibility in its approach, allowing for optimization through various modifications. These include selecting different alkoxides, adjusting the water-to-alkoxide ratio, introducing catalysts, and altering factors such as solution pH, temperature, and reactant concentration. (Owens et al., 2016; Deshmukh et al., 2020; Fiume et al., 2020) This adaptability has led to the development of multiple routes for synthesizing bioactive glass via sol-gel techniques.

One such variation is the ultrasound-assisted sol-gel process, which builds upon the conventional method by applying ultrasound waves to the solution. This is typically achieved using an ultrasonic probe or bath, which generates cavitation bubbles. These bubbles rapidly implode, creating localized hotspots with extreme temperatures and pressures, potentially enhancing the reaction kinetics (De Oliveira et al., 2014; Wu et al., 2021; Borisade et al., 2024).

Another innovative approach is the Colloidal Sol-Gel Method, which distinguishes itself by utilizing pre-synthesized colloidal particles as the foundation for the gel network. This method employs controlled hydrolysis to produce tiny, stable silica particles dispersed throughout the solvent, forming the "sol." To prevent premature particle aggregation, this technique may incorporate stabilizing agents or surface modifications.(Spirandeli et al., 2020)

2.2. Advantages of Sol-Gel for Bioactive Glass

In the production of bioactive glasses, sol-gel synthesis offers several advantages over traditional melt-quench techniques.(Owens et al., 2016) The method's lower processing temperatures result in glasses with higher purity, larger surface areas, and inherent porosity.(Owens et al., 2016)

Table 1 summarizes the key differences between the sol-gel and melt-quenched processes for producing bioactive glass. It highlights the advantages of the sol-gel process in terms of processing conditions, structural control, bioactivity, and versatility, while also noting the challenges associated with the melt-quenched process, such as higher processing temperatures and increased risk of impurities.

Table 1. Differences between the sol-gel and melt-quenched processes for producing bioactive glass

	Sol-Gel Process	Melt-Quenched Process
Synthesis Method	Hydrolysis and condensation of silicate precursors (e.g., tetraethyl orthosilicate) (El Bouami et al., 2024)	Melting and rapid cooling of glass components (typically silica, calcium oxide, and sodium oxide) (Jones, 2013)
Processing Temperature	Lower (600-700°C) (Li et al., 1991)	Higher (1250-1400°C) (Li et al., 1991)
Pore Structure	Highly porous, interconnected network	Less porous, less interconnected network
Specific Surface Area	Higher	Lower
Bioactivity	Generally higher (Li et al., 1991)	Excellent, but typically lower than sol-gel (Li et al., 1991)
Compositional Control	Precise control, allows for easy introduction of therapeutic ions (Owens et al., 2016)	Less precise control (Owens et al., 2016)
Structural Variation	Can be produced without compositional changes (Saravanapavan et al., 2003)	Primarily composition-dependent (Jones, 2013)
Purity Control	Higher control over purity	More difficult to maintain high purity (Li et al., 1991)
SiO_2 Content Range for Bioactivity	Up to 85 mol% (Saravanapavan et al., 2003)	Up to 60-65 mol% (Saravanapavan et al., 2003)
Dependence on Na_2O	Not critical (Saravanapavan et al., 2003)	Critical (Saravanapavan et al., 2003)
Dependence on P_2O_5	Not critical, can be bioactive without P_2O_5 (Saravanapavan and Hench, 2001)	required for bioactivity (Saravanapavan et al., 2003)
Production of Powders	Easier (Saravanapavan et al., 2003)	More challenging, requires additional processing steps (Li et al., 1991)
Bioactivity Control	Better control through composition or microstructure changes (Sudipta et al., 2022)	Limited control (Saravanapavan et al., 2003)
Dissolution Rate	Higher (Lepry and Nazhat, 2021a)	Lower (Hench, 1991)
Incorporation of Therapeutic Molecules	Easier due to porous structure (Mehrabi et al., 2020)	More challenging (Li et al., 1991)
Impurity Sensitivity	Less sensitive due to lower processing temperatures	More sensitive, can easily pick up impurities during melting
Compositional Limitations	Fewer limitations, can achieve higher SiO_2 content (Saravanapavan et al., 2003)	Limited by high liquidus temperature of SiO_2 and high viscosity of high-SiO_2 melts (Li et al., 1991)

continued on following page

Table 1. Continued

	Sol-Gel Process	Melt-Quenched Process
Post-processing Requirements	Minimal (Baino et al., 2016)	May require grinding, polishing, fritting, sieving, etc. (Li et al., 1991)
Risk of Contamination	Lower(Li et al., 1991)	Higher due to additional processing steps (Li et al., 1991)
Versatility in Bioactivity Range	Broader (Saravanapavan et al., 2003)	More limited (Saravanapavan et al., 2003)

While these characteristics typically enhance the degradability and bioactivity of sol-gel glasses, they often come at the cost of lower mechanical strength.(Deshmukh et al., 2020)

Despite this trade-off, sol-gel derived glasses have found numerous biomedical applications, leveraging their biocompatibility and excellent bioactivity(Zhong and Greenspan, 2000). These properties make them ideal for encapsulating proteins, enzymes, and biomolecules for controlled drug delivery and bone tissue regeneration(Huang et al., 2024). The enhanced bioactivity of sol-gel glasses stems from their molecular structure and improved textural properties, including pore size (which relates to high surface area), negative surface charge, and higher dissolution rate.(Huang et al., 2024)

These attributes are particularly beneficial in bone grafting applications, where ideal glasses should degrade at an appropriate rate, possess suitable mechanical properties, promote the formation of a hydroxyapatite (HA) layer, and stimulate beneficial biological responses.(Bejarano et al., 2015) The HA layer is crucial as it creates a strong bond between living tissue and implants(Deshmukh et al., 2020).

Scientists are exploring various ways to enhance sol-gel methods by combining them with other innovative approaches. For example, the ultrasound-assisted sol-gel method accelerates gelation, promotes finer texture and homogeneity, and can result in the formation of spherical particles with controlled size distribution(De Oliveira et al., 2014). Similarly, integrating electrospinning with sol-gel methods enables the creation of nanofibers with precisely controlled compositions and functions.(Lu et al., 2023) Another approach involves templating during the sol-gel process, which allows for the production of materials with well-defined pore structures and morphologies(Sarmast Sh et al., 2022). Additionally, when fine powders with specific particle sizes are required, spray drying of sol-gel-derived materials is utilized(Kortesuo et al., 2000). Each of these combined techniques offers unique advantages in tailoring the properties and structures of the final materials.

III. DESIGNING THE NEW BIOACTIVE GLASS COMPOSITIONS

3.1. The Different Classes of Bioactivity

Bioactivity is the ability of an implant material to create a connection with living tissues.(Hench, 1988) For successful bonding to bone, a material must form a bone-like apatite layer on its surface when in the body(Kokubo and Takadama, 2006). When materials lack bioactivity, they instead cause a layer of loose connective tissue to form around them, which doesn't stick to the implant(Hench, 1988). Larry

Hench created a system to classify materials based on their interaction with surrounding tissues after implantation.(Fiume et al., 2018)

Class A bioactive materials, like certain bioactive glasses, interact with the body at a cellular level to boost bone growth, a process called "osteoproduction"(Fiume et al., 2018). This is different from Class B bioactive materials, such as synthetic hydroxyapatite, which only allow bone to grow along their surface ("osteoconduction") when in contact with bone.(Fiume et al., 2018)

The differences between Class A and B materials have been shown through various cell studies using bone-forming cells. These cells grow better and show increased activity when in contact with bioactive glass surfaces.(Fiume et al., 2018) Studies in living organisms demonstrate that bone growth is enhanced around bioactive glass particles, matching or exceeding the growth rate seen with the body's own bone used to fill the same defect.(Fiume et al., 2018)

A key feature of Class A materials is their ability to enhance both the growth and specialization of early-stage cells. This improved bone-forming behavior seems to involve increased production of natural growth factors. These factors further stimulate cell growth and the formation of new, organized tissues at rates equal to or faster than the body's own bone.(Fiume et al., 2018)

3.2. Precursor Selection for Desired Elements

Silicate-based compositions are the most extensively studied sol-gel derived bioactive glasses. These glasses primarily use silicon alkoxides as network-forming agents in their preparation(Owens et al., 2016; Deshmukh et al., 2020). Since the development of sol-gel methods, tetraethyl orthosilicate (TEOS), chemically known as $Si(OC_2H_5)_4$ or $Si(OR)_4$, has been the predominant choice for this role, though other precursors are also available(Owens et al., 2016; Deshmukh et al., 2020).

TEOS is favored for several reasons. It forms strong, stable networks and offers moderate reactivity, allowing for significant control over the final product.(Owens et al., 2016; Deshmukh et al., 2020) When TEOS is mixed with water, it undergoes hydrolysis, creating a colloidal suspension of solid particles in liquid, known as a sol(Owens et al., 2016). Subsequently, polycondensation reactions transform this sol into a gel - a rigid, interconnected network of merged particles that forms a solid structure with submicrometer pores containing liquid. The ability to control this process makes silicate-based sol-gel glasses, particularly those using TEOS, a versatile and attractive option for creating bioactive materials with tailored properties(Owens et al., 2016).

Figure 2. Initial hydrolysis and condensation stages of TEOS in the production of silica oligomers (Owens et al., 2016)

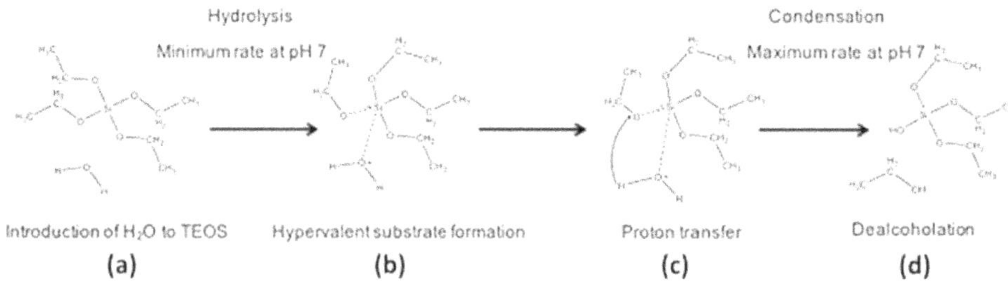

Despite their widespread use, silicate-based glasses have a significant limitation: they dissolve very slowly in the body.(Foroutan et al., 2020) This slow dissolution rate means they're primarily suitable for long-term implants.(Foroutan et al., 2020) While the stability of silicate-based glasses makes them durable, it also limits their ability to be quickly replaced by natural tissue, potentially leading to complications over time like trigging inflammatory responses in the body or risk of failing.(Foroutan et al., 2020)

Phosphate-based glasses are emerging as a promising new class of biomaterials, offering an alternative to traditional silicate-based glasses. The key advantage of those glasses is their bioresorbable nature(Kyffin et al., 2019). At the same time, they gradually dissolve completely(Foroutan et al., 2020). This dissolution process allows phosphate-based glasses to be entirely replaced by newly regenerated tissue, whether it's hard tissue like bone or soft tissue.(Foroutan et al., 2020) Nevertheless, one of the primary challenges encountered was the identification of an appropriate phosphate precursor particularly for biomedical applications.(Lepry and Nazhat, 2021b)

Phosphoric acid (H_3PO_4) is not suitable for sol-gel synthesis of phosphate materials due to its high reactivity with water, leading to precipitation rather than gelation(Livage et al., 1992; Lepry and Nazhat, 2021b). Phosphate esters ($PO(OR)_3$) were suggested as alternatives, but their slow hydrolysis rate is a drawback(Willinger et al., 2009). Phosphoryl chloride ($POCl_3$) has been suggested but led to crystallization in certain compositions(Livage et al., 1992; Clayden et al., 2001; Lepry and Nazhat, 2021b). Phosphosilicate gels containing 5-90 mol% P_2O_5 have been investigated using various precursors, including H_3PO_4, triethyl phosphate (TEP), and trimethyl phosphite(Szu et al., 1992; Lepry and Nazhat, 2021b). When heated to 200°C, phosphoric acid gels partially crystallized, while trimethyl phosphite gels fully crystallized.(Szu et al., 1992; Lepry and Nazhat, 2021b) TEP gels showed similar gelation limits but lost nearly 90% of P_2O_5 during calcination(Szu et al., 1992; Lepry and Nazhat, 2021b). A partially modified phosphorus alkoxide ($OP(OH)_3$-x$(OR)_x$) is proposed as a more suitable precursor, obtained by dissolving P_2O_5 in alcohols(Livage et al., 1992). $OP(OH)_3$-x$(OR)_x$ precursors result in lower phosphorous loss compared to $POCl_3$ or alkyl phosphite precursors(Livage et al., 1992; Lepry and Nazhat, 2021b). For biomedical sol-gel phosphate glasses $PO(OH)_3$-x(OR)x or TEP are the most commonly used precursors. (Lepry and Nazhat, 2021b)

Figure 3. Overview of common a) sol–gel phosphate precursors, b) aqueous phosphate species, and c) phosphate glass structural units (Lepry and Nazhat, 2021a)

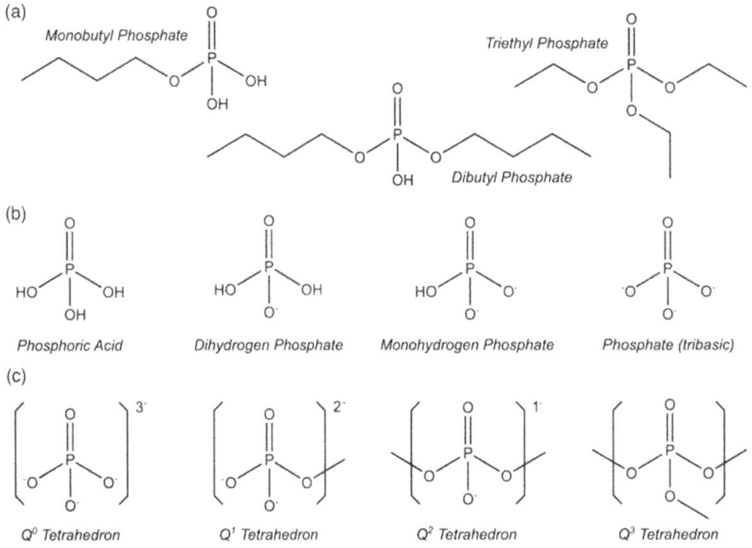

Implant usage always carries the risk of bacteria sticking to and growing on the implant's surface. This can result in implant failure and bacterial infections. The outcomes of such infections are severe, often necessitating additional surgeries and extending hospital stays. A wide range of disease-causing microorganisms can be found around implant sites, including Pseudomonas aeruginosa, Escherichia coli, Staphylococcus aureus, and Staphylococcus epidermidis.(Rabiee et al., 2015)

3.3. Effects of Therapeutic Ions on sol-gel glasses composition

Ongoing and rigorous research efforts have been undertaken to avoid implant failure by bacterial infections and enhance the bioactivity, as well as the chemical, physical, and biological properties, of bioactive glasses. This has been achieved through the incorporation of various metals and ions into their composition, thereby improving their clinical applicability(El Bouami et al., 2024). The literature has deeply investigated different compositions of doped bioactive glasses and the effects of introducing additional elements.

Copper is recognized for its role in promoting angiogenesis(Kargozar et al., 2021; El Bouami et al., 2024). The success and functionality of a bioactive glass are contingent upon its capacity to rapidly support blood vessel formation.(Kargozar et al., 2021; El Bouami et al., 2024) Copper-containing glasses and glass-ceramics have proven effective in inhibiting bacteria while maintaining the bioactivity of the glass material(Palza et al., 2013; Koohkan et al., 2018; El Bouami et al., 2024). Historically, researchers have utilized copper for its exceptional antibacterial properties(El Bouami et al., 2024). Moreover, copper-substituted bioactive glasses not only promote angiogenesis and antibacterial activity but also show potential anti-cancer properties(Kargozar et al., 2021).

Silver ions have been known for their antimicrobial properties since ancient times in Mediterranean and Asiatic cultures, and recent advances have focused on taking advantage of this antibacterial activity(Rabiee et al., 2015). A meager benefit of incorporating silver ions into a gel-glass system is that the porous glass matrix allows for controlled, sustained release of the antibacterial agent.(Rabiee et al., 2015) This controlled release helps preserve the antibacterial effectiveness of the bioactive glasses in body fluids containing chloride or proteins, which could otherwise inhibit silver's antibacterial action(Rabiee et al., 2015). While silver release from bioactive glass implants can effectively treat bone infections, it is important to note that high concentrations of silver ions may be cytotoxic(Azizabadi et al., 2021). Meanwhile, the bioactivity of the glass is not compromised by the incorporation of silver into its structure(Sharifianjazi et al., 2017). Studies have also verified that silver bioactive glass has anti-cancer effects against HepG2 cells(Azizabadi et al., 2021)

Magnesium is a promising calcium substitute in bioactive glass formulations due to its enhanced biological and mechanical properties. It is an essential mineral in the bone matrix, with concentrations of 0.72%, 1.23%, and 0.44% found in bone, dentin, and enamel, respectively.(Rabiee et al., 2015) Research indicates that incorporating Mg into bioactive glass can boost the levels of ALP, osteocalcin, collagen I, and other osteogenic markers(Ebrahimi et al., 2023). Magnesium acts as a co-factor for numerous enzymes and helps stabilize the structures of DNA and RNA(Rabiee et al., 2015). Given its functional roles and presence in bone tissue, magnesium might promote bone development and maintenance.(Rabiee et al., 2015) Additionally, bioactive glass with Mg dopants have shown improved mechanical properties and greater biocompatibility.(Ebrahimi et al., 2023)

Strontium ions are highly attractive for enhancing osteogenesis due to their biological benefits and prevalence in human tissues(Ebrahimi et al., 2023). In fact, strontium comprises 0.335% of the calcium content in the human skeleton(Rabiee et al., 2015). This alkaline-earth element plays a crucial role in reducing fracture risk and improving bone density by inhibiting osteoclast-mediated resorption while stimulating new bone formation(Ebrahimi et al., 2023). Research has demonstrated that incorporating strontium ions into bioactive glass structures can enhance osteoblast cell differentiation and increase alkaline phosphatase (ALP) activity(Rabiee et al., 2015). Consequently, strontium-doped bioactive glass shows great promise for various bone applications.(Ebrahimi et al., 2023) However, it's important to note that the optimal amount of strontium oxide (SrO) substitution in bioactive glass is approximately 5%(Sharifianjazi et al., 2020). Studies have shown that exceeding this threshold can actually decrease the bioactivity of 58S bioactive glasses.(Sharifianjazi et al., 2020)

Zinc is essential for bone formation, boosting bone cell growth in both living organisms and laboratory conditions.(Rabiee et al., 2015; Atkinson et al., 2016) It serves as a crucial element in many enzymes, facilitates protein interactions, and is necessary for DNA replication(Rabiee et al., 2015; Atkinson et al., 2016). A lack of zinc can slow down the growth of skeletal and bone tissues(Rabiee et al., 2015). Studies on zinc-doped silicate bioactive glass created through the sol-gel method revealed that adding 5 mol% zinc to bioactive glass diminished its bioactive. properties(Atkinson et al., 2016). In the glass structure, zinc oxide can function as both a network modifier and an intermediate oxide, with its role shifting as the zinc oxide content increases(Rabiee et al., 2015; Atkinson et al., 2016). Additionally, zinc exhibits antibacterial properties and enhances the mechanical durability of bioactive glass.(Atkinson et al., 2016)

IV. ASSESSING BIOACTIVITY

4.1. The Reaction Stage of Bioactivity

Bioactivity in materials is characterized by the formation of a hydroxyl carbonated apatite (HCA) layer on their surface, which closely resembles natural bone mineral(Saravanapavan et al., 2003). The development of this HCA layer is the result of a rapid series of chemical reactions that occur when the implant material comes into contact with body fluids.(Jones et al., 2001) Eleven key stage's reaction sequence are required bonding of bioactive glasses to bone(Hench and Thompson, 2010; Rabiee et al., 2015):

Stage 1: Dealkalization by Ion Exchange

This initial stage involves a rapid exchange of Na^+ and Ca^{2+} from the glass surface with H^+ or H_3O^+ from the surrounding solution. This process leads to the hydrolysis of silica groups on the glass surface, resulting in the formation of silanol (Si-OH) groups. The reaction can be represented as:

$$Si–O–Na^+ + H^+ + OH^- \rightarrow Si–OH + Na+(_{aq}) + OH^-$$

This ion exchange process is diffusion-controlled and exhibits a $t^{-1/2}$ dependence. As the exchange progresses, the pH of the solution increases due to the replacement of H^+ in the solution with cations (Na^+ or Ca^{2+}) from the glass.(Saravanapavan et al., 2003; Hench and Thompson, 2010; Sohrabi et al., 2021)

Stage 2: Silica Network Breakdown and Silanol Formation

The increased hydroxyl concentration resulting from Stage 1 leads to an attack on the silica glass network. This results in the loss of soluble silica to the solution in the form of $Si(OH)_4$. This process occurs due to the breaking of Si-O-Si bonds and the continued formation of silanols (Si-OH) at the interface between the glass and the solution. The reaction can be represented as: $Si\text{-}O\text{-}Si + H_2O \rightarrow Si\text{-}OH + OH\text{-}Si$

This stage is controlled by interfacial reactions and exhibits a time1.0 dependence. (Saravanapavan et al., 2003; Rabiee et al., 2015)

Stage 3: Silica-Rich Layer Formation

The $Si(OH)_4$ created during Stage 2 begin to reconnect, forming a new layer on the glass surface. This process, called repolymerization, can be represented as:

$HO\text{-}Si(OH)_2\text{-}OH + HO\text{-}Si(OH)_{2\text{-}}OH \rightarrow HO\text{-}Si(OH)_2\text{-}O\text{-}Si(OH)_2\text{-}OH + H_2O$

This reaction occurs because Stage 2 creates a high concentration of $Si(OH)_4$ near the glass surface. While silicic acid is stable when dilute, it quickly polymerizes at higher concentrations. These larger molecules then autocondense, resulting in a layer rich in silica on the glass surface.(Saravanapavan et al., 2003; Hench and Thompson, 2010; Rabiee et al., 2015)

Stage 4: Formation and Growth of Calcium Phosphate Layer

Ca^{2+} and PO_4^{3-} move through the silica-rich layer created in Stage 3, migrating towards the surface of the material. These ions combine to form a film rich in calcium and phosphorus ($CaO\text{-}P_2O_5$) on top of the silica-rich layer. Once this initial calcium phosphate film forms, it continues to grow. It does this by incorporating additional calcium and phosphate ions from the surrounding solution. This growing layer is amorphous.(Saravanapavan et al., 2003; Hench and Thompson, 2010; Rabiee et al., 2015)

Stage 5: Formation of Biologically Active Apatite Layer

The amorphous calcium phosphate ($CaO-P_2O_5$) film formed in Stage 4 undergoes crystallization. This process occurs as OH^- or CO_3^{2-} from the surrounding solution are incorporated into the film. The result is the formation of a mixed hydroxyl, carbonate apatite layer, often referred to as HCA .(Hench and Thompson, 2010; Rabiee et al., 2015)

Figure 4. The surface stages (1–5) reactions on bioactive glass, forming double SiO_2 rich and Ca, P-rich layers (Farag, 2023)

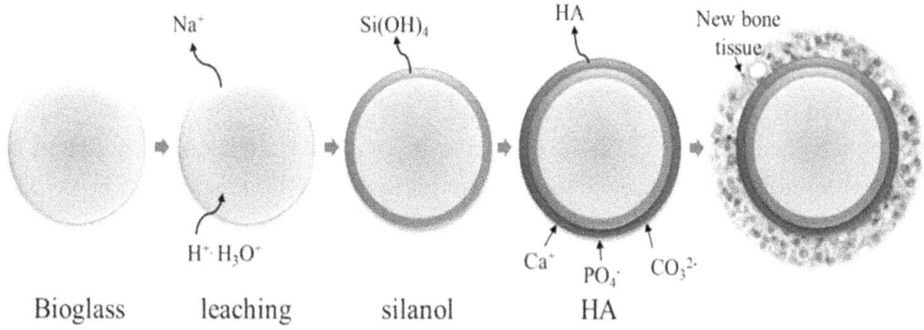

Stage 6: Biological Factor Interaction and Stem Cell Activation

In this ongoing stage, the newly formed HCA layer interacts with biological molecules. Specifically, various growth factors are absorbed into and released from the HCA layer in a continuous process. These growth factors play a crucial role in stimulating the differentiation of stem cells.(Jones et al., 2001; Hench and Thompson, 2010)

Stage 7: Site Cleaning and Cell Colonization

Specialized immune cells called macrophages become active at the implant site. Their primary role is to clear away any debris or dead cells that may have accumulated during the earlier stages of the bioactive process. This cleaning action by macrophages is crucial as it prepares the area for new cell growth. (Jones et al., 2001; Hench and Thompson, 2010)

Stage 8: Stem Cell Attachment

The bioactive surface, which has been prepared through the previous stages, now provides an ideal environment for stem cells to adhere. (Jones et al., 2001; Hench and Thompson, 2010)

Stage 9: Stem Cell Differentiation

The stem cells that have attached to the bioactive surface begin to transform. Specifically, they differentiate into bone-forming cells called osteoblasts.(Jones et al., 2001; Hench and Thompson, 2010)

Stage 10: Bone Matrix Production

The osteoblasts that developed in the previous stage begin their primary function: creating new bone tissue. These cells produce and secrete various proteins and other molecules that form the extra cellular matrix of bone.(Jones et al., 2001; Hench and Thompson, 2010)

Stage 11: Bone Matrix Crystallization and Cell Encapsulation

In this final stage, the organic extracellular matrix produced by osteoblasts undergoes crystallization. Inorganic calcium phosphate crystals form within and around the matrix, creating the hard, mineralized structure of bone. As this process occurs, bone self-cells become enclosed within the mineralized matrix.

This results in a living composite structure - the mature bone tissue - where living cells are surrounded by and integrated with the crystallized matrix.(Jones et al., 2001; Hench and Thompson, 2010)

This stage finalizes the transformation of the initial bioactive material into fully developed, living bone tissue. The outcome is a robust and dynamic structure capable of ongoing remodeling and adaptation over time.

4.2. In Vitro Bioactivity

Evaluating the bioactive properties of materials is a crucial step in the development of novel biomaterials for bone tissue engineering and regenerative medicine applications. One widely used in vitro technique for assessing the bioactivity of materials is the simulated body fluid test, which aims to mimic the ionic composition of human blood plasma(Kokubo and Takadama, 2006; Duarte et al., 2009). The formation of a bone-like apatite layer on the surface of a material immersed in simulated body fluid is considered a reliable indicator of its potential to elicit a favorable bone response in vivo.(Jones et al., 2001; Kokubo and Takadama, 2006)

The preparation procedure for simulated body fluid (SBF) begins with the sequential dissolution of reagents in a 1000 ml plastic beaker containing 700 ml of distilled water under continuous stirring(Kokubo and Takadama, 2006). The reagents - NaCl, $NaHCO_3$, KCl, $K_2HPO_4{\cdot}3H_2O$, $MgCl_2{\cdot}6H_2O$, $CaCl_2$, and Na_2SO_4 with the amount display in Table 1- are added one by one, ensuring each is fully dissolved before introducing the next to prevent precipitation. After dissolving calcium chloride, distilled water is added if necessary to reach a total volume of 900 ml.(Kokubo and Takadama, 2006) The next step involves gradually adding Tris while monitoring the pH, with the aim of raising it to 7.30 ± 0.05 at 36.5 ± 0.5°C(Kokubo and Takadama, 2006). Tris addition continues until the pH is just under 7.45, with 1M HCl used to lower it if it rises too high.(Kokubo and Takadama, 2006) Once all reagents are fully dissolved, the final step is to adjust the total volume to 1000 ml with additional distilled water, resulting in a complete SBF solution.(Kokubo and Takadama, 2006)

Table 2. Order, amounts, and formula of reagents for preparing 1000ml of SBF

Order		Reagent		Amount
1		NaCl		8.035g
2		$NaHCO_3$		0.355g
3		KCl		0.225g
4		K_2HPO_4 $3H_2O$		0.231g
5		$MgCl_2$ $6H_2O$		0.311g
6		1.0M-HCl		39ml
7		$CaCl_2$		0.292g
8		Na_2SO_4		0.072g
9		Tris		6.118g
10		1.0M-HCl		0–5ml

(Kokubo and Takadama, 2006)

The comparison of the composition between SBF and human blood plasma is represented in Table 3.

Table 3. Nominal ion concentrations of SBF in comparison with those in human blood plasma

Ion	Ion concentrations (mM)	
	Blood plasma	SBF
Na^+	142.0	142.0
K^+	5.0	5.0
Mg^{2+}	1.5	1.5
Ca^{2+}	2.5	2.5
Cl^-	103.0	147.8
HCO_3^-	27.0	4.2
HPO_4^{2-}	1.0	1.0
SO_4^{2-}	0.5	0.5
pH	7.2 - 7.4	7.4

(Kokubo and Takadama, 2006)

Several studies have investigated the relationship between in vitro bioactivity in SBF and in vivo bone regeneration. For instance, a study by Cerruti et al. demonstrated that the formation of a bone-like apatite layer on the surface of a bioactive glass scaffold in SBF correlated well with its ability to promote new bone formation when implanted in a rabbit femoral defect(Nguyen et al., 2017). Similarly, Gao et al. reported that the osteogenic differentiation and bone-forming capacity of mesoporous bioactive glass scaffolds in vivo were closely related to their apatite-forming ability in SBF(Nguyen et al., 2017)

V. APPLICATIONS AND FUTURE DIRECTIONS

5.1. Potential Applications in Tissue Engineering and Bone Regeneration

Bioactive glasses possess unique properties that make them highly desirable for various biomedical applications. When in contact with biological fluids, the surface undergoes a series of reactions that lead to the release of ions, resulting in an increase in osmotic pressure and pH. This creates an environment that is inhospitable to microbial growth, conferring the materials with antimicrobial and antibiofilm properties.

These antimicrobial properties have been particularly beneficial in the management of chronic osteomyelitis, where the use of bioactive glass in conjunction with antibiotic therapy has shown promising results. (Drago et al., 2018)

The versatility of bioactive glasses extends beyond bone regeneration, with promising applications in the field of soft tissue engineering. The ability of bioactive glasses to bond with both bone and soft tissues, as well as their capacity to convert to hydroxyapatite in vivo, make them attractive candidates for a wide range of tissue engineering applications.

Furthermore, the inherent properties of bioactive glasses, such as their ability to incorporate drugs and various biomolecules, have opened up new avenues for targeted drug delivery and even cancer treatment. The mesoporous structure of these materials allows for the incorporation and controlled release of therapeutic agents, potentially enhancing the efficacy of treatments and reducing side effects.

5.2 The Cost, Toxicity, and Environmental Impact of Sol-Gel Bioactive Glasses

Sol-gel bioactive glasses have emerged as promising biomaterials for medical applications, especially in bone and tissue regeneration (Elahpour et al., 2023). These materials, produced through the sol-gel method, offer several advantages over traditional melt-derived glasses, including higher bioactivity, adjustable physicochemical properties, and the ability to incorporate various therapeutic agents (Baino et al., 2017). However, it's crucial to consider the cost, toxicity, and environmental impact of producing and using these materials.

The production costs of bioactive glasses vary significantly between melt-quenching and sol-gel processes, with each method presenting unique economic challenges. Melt-quenching, despite being a more traditional approach, incurs substantial expenses due to its high-temperature processing in platinum crucibles and multiple handling steps. These costs extend beyond energy consumption to include capital equipment, labor, maintenance, and quality control measures.(Li et al., 1991)

In contrast, the sol-gel method, while offering lower processing temperatures, introduces its own set of financial considerations. The intricate synthesis process, need for specialized equipment, and use of costly starting materials can make sol-gel bioactive glasses significantly more expensive to produce. (Baino et al., 2016) Some researchers have found that sol-gel glass can be approximately 100 times more costly than conventionally manufactured glass.(Owens et al., 2016)

This stark cost difference suggests that each method may be better suited for different applications. While lowering processing temperatures in melt-quenching could potentially reduce overall costs, the high expenses associated with sol-gel production currently limit its economic viability to specialized, high-value glass products. Despite this significant cost difference, the enhanced bioactivity and customizable properties of sol-gel bioactive glasses may justify the higher production costs for specific medical applications.(Baino et al., 2016)

In terms of toxicity, the incorporation of various dopants and therapeutic agents into sol-gel bioactive glasses raises concerns about potential cytotoxicity and biocompatibility. To ensure the safety and efficacy of these biomaterials, careful selection and evaluation of the incorporated materials are crucial. Additionally, thorough investigation of the degradation products released during implantation and resorption of sol-gel bioactive glasses is necessary to assess their impact on surrounding tissues and the overall physiological environment.

The environmental impact of sol-gel bioactive glasses is another important consideration, particularly regarding their production and disposal. The sol-gel technique offers a more environmentally-friendly approach to producing bioactive glasses compared to melt-quenching. The lower processing temperatures required can reduce the energy consumption and greenhouse gas emissions associated with manufacturing(Lei et al., 2020)

The cost and toxicity of sol-gel bioactive glasses are critical factors that require careful evaluation and mitigation. Addressing these concerns is necessary to ensure the widespread and responsible adoption of these promising biomaterials in medical applications.

Despite these challenges, ongoing research in the field of bioactive glasses is opening up exciting new possibilities. Scientists are exploring various development paths to address these concerns while simultaneously expanding the potential applications of these materials.

5.3 Future Research Directions for Further Optimization and Exploration

The rapidly evolving field of bioactive glasses is seeing researchers exploring various promising development paths (Farag, 2023). Recent advancements in the development of bioactive glasses have expanded their potential applications beyond the traditional realm of hard tissue regeneration. This expansion includes the optimization of glass compositions tailored for specific medical applications, with significant attention being directed toward the development of composite materials that merge bioactive glasses with polymers or other ceramics, creating hybrid systems with enhanced properties and functionalities (Baino et al., 2017). Additionally, bioactive glasses are effectively used as multifunctional therapeutic drug-delivery systems, allowing for drug loading and release with adjustable kinetics to match the specific requirements of the target site (Gupta et al., 2021). Illustrating this potential, a Bioactive Glass-Chitosan-gelatin composite integrated with gold nanoparticles achieved a significantly higher loading capacity for the anti-cancer drug doxorubicin, reaching 81.6%, compared to magnetic-core silica nanoparticles, which only achieved a 4% loading capacity (Gupta et al., 2021).

The intrinsic brittleness of bioactive glass, which often leads to cracks or fractures, is a significant limitation(Kunwong et al., 2021). However, when combined with poly lactic-co-glycolic acid (PLGA), a synthetic polymer known for its biocompatibility and biodegradability, these drawbacks can be mitigated(Kunwong et al., 2021). Yao et al. created PLGA/bioactive glass microspheres using a modified emulsification method, demonstrating that the resulting composite scaffolds can enhance osteogenic differentiation in rat marrow stromal cells and increase alkaline phosphatase activity.(Yao et al., 2005; Kunwong et al., 2021) Additionally, Boccaccini et al. and Yang et al. used a thermally induced phase separation method to fabricate PLGA/bioactive glass composites, showing that the incorporation of bioactive glass improves cell adhesion, proliferation, and osteogenesis in human osteosarcoma cells(Boccaccini et al., 2003; Kunwong et al., 2021).

As the research and development in the field of bioactive glasses continues to evolve, the future applications of these materials are expected to become even more diverse and impactful. In line with this evolution, the integration of bioactive glasses with advanced manufacturing techniques, such as additive manufacturing, holds great promise for the creation of complex, customized scaffolds and implants with precisely controlled architecture and geometry(Baino et al., 2017). Specifically, the integration of bioactive glasses with advanced 3D printing techniques is opening new avenues for the creation of bioactive glass scaffolds designed for 3D tissue engineering (Bellantone et al., 2002). These scaffolds provide an ideal microenvironment for cell growth, further enhancing their potential applications. Building on this potential, the unique properties of nanostructured bioactive glasses are being studied for their capacity to deliver superior performance. In particular, sol-gel-derived mesoporous bioactive glasses, which offer greater reactivity and a broader range of compositions compared to traditional melt-derived glasses, are emerging as a promising area for further research (Kaou et al., 2023).

Another exciting development in the field of bioactive glasses is their potential application in the treatment of osteomyelitis, a serious bone infection. The S53P4 bioactive glass composition has shown promising results in this regard, highlighting the versatility of these innovative material(Van Gestel et al., 2015) As the scientific community continues to explore the untapped potential of bioactive glasses, these innovative materials are poised to play a significant role in the future of biomedicine, revolutionizing the way we approach tissue regeneration, implant integration, and various other clinical challenges. (Prakasam et al., 2017) (Sprio et al., 2023)

VI. CONCLUSION

The field of bioactive glasses, particularly those synthesized through sol-gel methods, represents a significant advancement in biomaterials science with far-reaching implications for regenerative medicine and tissue engineering. This exploration of bioactive glass synthesis, composition design, and applications highlights several key points: The sol-gel method offers unique advantages in bioactive glass production, including lower processing temperatures, higher purity, larger surface areas, and inherent porosity. These characteristics enhance the bioactivity and degradability of the resulting materials, opening up new possibilities for tailored medical applications. The incorporation of therapeutic ions into bioactive glass compositions allows for the enhancement of specific properties. Elements such as copper, silver, magnesium, strontium, and zinc can impart additional functionalities like improved angiogenesis, antibacterial properties, and enhanced osteogenesis, expanding the potential applications of these materials. The bioactivity of these glasses, characterized by the formation of a hydroxyl carbonated apatite (HCA) layer, involves a complex series of reactions that bridge the gap between the synthetic material and living tissue. Understanding these mechanisms is crucial for optimizing bioactive glass performance in vivo. In vitro testing methods, particularly the use of simulated body fluid (SBF), provide valuable tools for assessing and predicting the bioactivity of new glass compositions without initial in vivo testing, streamlining the development process. The applications of bioactive glasses extend beyond bone regeneration to include potential uses in soft tissue engineering, drug delivery, and even cancer treatment, showcasing the versatility of these materials. Looking to the future, the field of bioactive glasses continues to evolve rapidly. Emerging research directions include: Further optimization of glass compositions for specific medical applications, development of composite materials combining bioactive glasses with polymers or other ceramics, exploration of nanostructured bioactive glasses for enhanced performance, investigation of bioactive glass scaffolds for 3D tissue engineering, and integration of bioactive glasses with advanced manufacturing techniques like 3D printing.

REFERENCES

Atkinson, I., Anghel, E. M., Predoana, L., Mocioiu, O. C., Jecu, L., Raut, I., Munteanu, C., Culita, D., & Zaharescu, M. (2016). Influence of ZnO addition on the structural, in vitro behavior and antimicrobial activity of sol–gel derived CaO–P 2 O 5 –SiO 2 bioactive glasses. *Ceramics International*, 42(2), 3033–3045. DOI: 10.1016/j.ceramint.2015.10.090

Azizabadi, N., Azar, P. A., Tehrani, M. S., & Derakhshi, P. (2021). Synthesis and characteristics of gel-derived SiO2-CaO-P2O5-SrO-Ag2O-ZnO bioactive glass: Bioactivity, biocompatibility, and antibacterial properties. *Journal of Non-Crystalline Solids*, 556, 120568. DOI: 10.1016/j.jnoncrysol.2020.120568

Baino, F., Fiorilli, S., & Vitale-Brovarone, C. (2016). Bioactive glass-based materials with hierarchical porosity for medical applications: Review of recent advances. *Acta Biomaterialia*, 42, 18–32. DOI: 10.1016/j.actbio.2016.06.033 PMID: 27370907

Baino, F., Fiorilli, S., & Vitale-Brovarone, C. (2017). Composite Biomaterials Based on Sol-Gel Mesoporous Silicate Glasses: A Review, 1. *Bioengineering (Basel, Switzerland)*, 4(1), 15. DOI: 10.3390/bioengineering4010015 PMID: 28952496

Bejarano, J., Caviedes, P., & Palza, H. (2015). Sol–gel synthesis and *in vitro* bioactivity of copper and zinc-doped silicate bioactive glasses and glass-ceramics. *Biomedical Materials (Bristol, England)*, 10(2), 025001. DOI: 10.1088/1748-6041/10/2/025001 PMID: 25760730

Bellantone, M., Williams, H. D., & Hench, L. L. (2002). Broad-Spectrum Bactericidal Activity of Ag2O-Doped Bioactive Glass. *Antimicrobial Agents and Chemotherapy*, 46(6), 1940–1945. DOI: 10.1128/AAC.46.6.1940-1945.2002 PMID: 12019112

Boccaccini, A. R., Notingher, I., Maquet, V., & Jérôme, R. (2003). Bioresorbable and bioactive composite materials based on polylactide foams filled with and coated by Bioglass® particles for tissue engineering applications. *Journal of Materials Science. Materials in Medicine*, 14(5), 443–450. DOI: 10.1023/A:1023266902662 PMID: 15348448

Bokov, D., Turki Jalil, A., Chupradit, S., Suksatan, W., Javed Ansari, M., Shewael, I. H., & Kianfar, E. (2021). Nanomaterial by sol-gel method: Synthesis and application. *Advances in Materials Science and Engineering*, 2021(1), 5102014.

Borisade, S. G., Owoeye, S. S., Ajayi, K. V., Enewo, S. I., & Abdullahi, A. (2024). Investigation of physical, mechanical and in-vitro bioactivity of bioactive glass-ceramics fabricated from waste soda-lime-silica glass doped P2O5 by microwave irradiation sintering. *Hybrid Advances*, 6, 100203. DOI: 10.1016/j.hybadv.2024.100203

Clayden, N. J., Esposito, S., Pernice, P., & Aronne, A. (2001). Solid state 29Si and 31P NMR study of gel derived phosphosilicate glasses. *Journal of Materials Chemistry*, 11(3), 936–943. DOI: 10.1039/b004107f

De Oliveira, A. A. R., De Carvalho, B. B., Sander Mansur, H., & De Magalhães Pereira, M. (2014). Synthesis and characterization of bioactive glass particles using an ultrasound-assisted sol–gel process: Engineering the morphology and size of sonogels via a poly(ethylene glycol) dispersing agent. *Materials Letters*, 133, 44–48. DOI: 10.1016/j.matlet.2014.06.092

Deshmukh, K., Kovářík, T., Křenek, T., Docheva, D., Stich, T., & Pola, J. (2020). Recent advances and future perspectives of sol–gel derived porous bioactive glasses: A review. *RSC Advances*, 10(56), 33782–33835. DOI: 10.1039/D0RA04287K PMID: 35519068

Drago, L., Toscano, M., & Bottagisio, M. (2018). Recent Evidence on Bioactive Glass Antimicrobial and Antibiofilm Activity: A Mini-Review. *Materials (Basel)*, 11(2), 326. DOI: 10.3390/ma11020326 PMID: 29495292

Duarte, A., Caridade, S., Mano, J. F., & Reis, R. L. (2009). Processing of novel bioactive polymeric matrixes for tissue engineering using supercritical fluid technology. *Materials Science and Engineering C*, 29(7), 2110–2115. DOI: 10.1016/j.msec.2009.04.012

Ebrahimi, M., Manafi, S., & Sharifianjazi, F. (2023). The effect of Ag2O and MgO dopants on the bioactivity, biocompatibility, and antibacterial properties of 58S bioactive glass synthesized by the sol-gel method. *Journal of Non-Crystalline Solids*, 606, 122189. DOI: 10.1016/j.jnoncrysol.2023.122189

El Bouami, H., Mabrouk, A., Mercier, C., Mihoubi, W., Meurice, E., Follet, C., Faska, N., & Bachar, A. (2024). The effect of CuO dopant on the bioactivity, the biocompatibility, and the antibacterial properties of bioactive glasses synthesized by the sol-gel method. *Journal of Sol-Gel Science and Technology*, 111(2), 347–361. DOI: 10.1007/s10971-024-06445-2

Elahpour, N., Niesner, I., Bossard, C., Abdellaoui, N., Montouillout, V., Fayon, F., Taviot-Guého, C., Frankenbach, T., Crispin, A., Khosravani, P., Holzapfel, B. M., Jallot, E., Mayer-Wagner, S., & Lao, J. (2023). Zinc-Doped Bioactive Glass/Polycaprolactone Hybrid Scaffolds Manufactured by Direct and Indirect 3D Printing Methods for Bone Regeneration. *Cells*, 12(13), 1759. DOI: 10.3390/cells12131759 PMID: 37443794

Farag, M. M. (2023). Recent trends on biomaterials for tissue regeneration applications [review]. *Journal of Materials Science*, 58(2), 527–558. DOI: 10.1007/s10853-022-08102-x

Fiume, E., Barberi, J., Verné, E., & Baino, F. (2018). Bioactive Glasses: From Parent 45S5 Composition to Scaffold-Assisted Tissue-Healing Therapies. *Journal of Functional Biomaterials*, 9(1), 24. DOI: 10.3390/jfb9010024 PMID: 29547544

Fiume, E., Migneco, C., Verné, E., & Baino, F. (2020). Comparison between Bioactive Sol-Gel and Melt-Derived Glasses/Glass-Ceramics Based on the Multicomponent SiO2–P2O5–CaO–MgO–Na2O–K2O System. *Materials (Basel)*, 13(3), 540. DOI: 10.3390/ma13030540 PMID: 31979302

Foroutan, F., Kyffin, B. A., Abrahams, I., Corrias, A., Gupta, P., Velliou, E., Knowles, J. C., & Carta, D. (2020). Mesoporous Phosphate-Based Glasses Prepared via Sol–Gel. *ACS Biomaterials Science & Engineering*, 6(3), 1428–1437. DOI: 10.1021/acsbiomaterials.9b01896 PMID: 33455383

Gupta, S., Majumdar, S., & Krishnamurthy, S. (2021). Bioactive glass: A multifunctional delivery system. *Journal of Controlled Release*, 335, 481–497. DOI: 10.1016/j.jconrel.2021.05.043 PMID: 34087250

Hench, L. L. (1988). Bioactive Ceramics. *Annals of the New York Academy of Sciences*, 523(1), 54–71. DOI: 10.1111/j.1749-6632.1988.tb38500.x PMID: 2837945

Hench, L. L. (1991). Bioceramics: From Concept to Clinic. *Journal of the American Ceramic Society*, 74(7), 1487–1510. DOI: 10.1111/j.1151-2916.1991.tb07132.x

Hench, L. L., & Jones, J. R. (2015). Bioactive Glasses: Frontiers and Challenges. *Frontiers in Bioengineering and Biotechnology*, 3. Advance online publication. DOI: 10.3389/fbioe.2015.00194 PMID: 26649290

Hench, L. L., & Thompson, I. (2010). Twenty-first century challenges for biomaterials. *Journal of the Royal Society, Interface*, 7(suppl_4), S379–S391.

Huang, X., Lou, Y., Duan, Y., Liu, H., Tian, J., Shen, Y., & Wei, X. (2024). Biomaterial scaffolds in maxillofacial bone tissue engineering: A review of recent advances. *Bioactive Materials*, 33, 129–156. DOI: 10.1016/j.bioactmat.2023.10.031 PMID: 38024227

Jones, J. R. (2013). Review of bioactive glass: From Hench to hybrids. *Acta Biomaterialia*, 9(1), 4457–4486. DOI: 10.1016/j.actbio.2012.08.023 PMID: 22922331

Jones, J. R., Sepulveda, P., & Hench, L. L. (2001). Dose-dependent behavior of bioactive glass dissolution. *Journal of Biomedical Materials Research*, 58(6), 720–726. DOI: 10.1002/jbm.10053 PMID: 11745526

Kaou, M. H., Furkó, M., Balázsi, K., & Balázsi, C. (2023). Advanced Bioactive Glasses: The Newest Achievements and Breakthroughs in the Area. *Nanomaterials (Basel, Switzerland)*, 13(16), 2287. DOI: 10.3390/nano13162287 PMID: 37630871

Kargozar, S., Mozafari, M., Ghodrat, S., Fiume, E., & Baino, F. (2021). Copper-containing bioactive glasses and glass-ceramics: From tissue regeneration to cancer therapeutic strategies. *Materials Science and Engineering C*, 121, 111741. DOI: 10.1016/j.msec.2020.111741 PMID: 33579436

Kokubo, T., & Takadama, H. (2006). How useful is SBF in predicting in vivo bone bioactivity? *Biomaterials*, 27(15), 2907–2915. DOI: 10.1016/j.biomaterials.2006.01.017 PMID: 16448693

Koohkan, R., Hooshmand, T., Mohebbi-Kalhori, D., Tahriri, M., & Marefati, M. T. (2018). Synthesis, characterization, and in vitro biological evaluation of copper-containing magnetic bioactive glasses for hyperthermia in bone defect treatment. *ACS Biomaterials Science & Engineering*, 4(5), 1797–1811.

Kortesuo, P., Ahola, M., Kangas, M., Kangasniemi, I., Yli-Urpo, A., & Kiesvaara, J. (2000). In vitro evaluation of sol–gel processed spray dried silica gel microspheres as carrier in controlled drug delivery. *International Journal of Pharmaceutics*, 200(2), 223–229. DOI: 10.1016/S0378-5173(00)00393-8 PMID: 10867252

Kunwong, N., Tangjit, N., Rattanapinyopituk, K., Dechkunakorn, S., Anuwongnukroh, N., Arayapisit, T., & Sritanaudomchai, H. (2021). Optimization of poly (lactic-co-glycolic acid)-bioactive glass composite scaffold for bone tissue engineering using stem cells from human exfoliated deciduous teeth. *Archives of Oral Biology*, 123, 105041. DOI: 10.1016/j.archoralbio.2021.105041 PMID: 33454420

Kyffin, B. A., Foroutan, F., Raja, F. N. S., Martin, R. A., Pickup, D. M., Taylor, S. E., & Carta, D. (2019). Antibacterial silver-doped phosphate-based glasses prepared by coacervation. *Journal of Materials Chemistry. B*, 7(48), 7744–7755. DOI: 10.1039/C9TB02195G PMID: 31750507

Lei, Q., Guo, J., Noureddine, A., Wang, A., Wuttke, S., Brinker, C. J., & Zhu, W. (2020). Sol–Gel-Based Advanced Porous Silica Materials for Biomedical Applications. *Advanced Functional Materials*, 30(41), 1909539. DOI: 10.1002/adfm.201909539

Lepry, W. C., & Nazhat, S. N. (2021a). A Review of Phosphate and Borate Sol–Gel Glasses for Biomedical Applications. *Advanced NanoBiomed Research*, 1(3), 2000055. DOI: 10.1002/anbr.202000055

Lepry, W. C., & Nazhat, S. N. (2021b). A Review of Phosphate and Borate Sol–Gel Glasses for Biomedical Applications. *Advanced NanoBiomed Research*, 1(3), 2000055. DOI: 10.1002/anbr.202000055

Li, R., Clark, A. E., & Hench, L. L. (1991). An investigation of bioactive glass powders by sol-gel processing. *Journal of Applied Biomaterials*, 2(4), 231–239. DOI: 10.1002/jab.770020403 PMID: 10171144

Livage, J., Barboux, P., Vandenborre, M. T., Schmutz, C., & Taulelle, F. (1992). Sol-gel synthesis of phosphates. *Journal of Non-Crystalline Solids*, 147–148, 18–23. DOI: 10.1016/S0022-3093(05)80586-1

Lu, H.-H., Zheng, K., Boccaccini, A. R., & Liverani, L. (2023). Electrospinning of cotton-like fibers based on cerium-doped sol–gel bioactive glass. *Materials Letters*, 334, 133712. DOI: 10.1016/j.matlet.2022.133712

Mehrabi, T., Mesgar, A. S., & Mohammadi, Z. (2020). Bioactive Glasses: A Promising Therapeutic Ion Release Strategy for Enhancing Wound Healing. *ACS Biomaterials Science & Engineering*, 6(10), 5399–5430. DOI: 10.1021/acsbiomaterials.0c00528 PMID: 33320556

Nguyen, T. T., Hoang, T., Can, V. M., Ho, A. S., Nguyen, S. H., Nguyen, T. T. T., Pham, T. N., Nguyen, T. P., Nguyen, T. L. H., & Thi, M. T. D. (2017). In vitro and in vivo tests of PLA/d-HAp nanocomposite. *Advances in Natural Sciences: Nanoscience and Nanotechnology*, 8(4), 045013. DOI: 10.1088/2043-6254/aa92b0

Owens, G. J., Singh, R. K., Foroutan, F., Alqaysi, M., Han, C.-M., Mahapatra, C., Kim, H.-W., & Knowles, J. C. (2016). Sol–gel based materials for biomedical applications. *Progress in Materials Science*, 77, 1–79. DOI: 10.1016/j.pmatsci.2015.12.001

Palza, H., Escobar, B., Bejarano, J., Bravo, D., Diaz-Dosque, M., & Perez, J. (2013). Designing antimicrobial bioactive glass materials with embedded metal ions synthesized by the sol–gel method. *Materials Science and Engineering C*, 33(7), 3795–3801. DOI: 10.1016/j.msec.2013.05.012 PMID: 23910279

Prakasam, M., Locs, J., Salma-Ancane, K., Loca, D., Largeteau, A., & Berzina-Cimdina, L. (2017). Biodegradable Materials and Metallic Implants—A Review. *Journal of Functional Biomaterials*, 8(4), 44. DOI: 10.3390/jfb8040044 PMID: 28954399

Rabiee, S. M., Nazparvar, N., Azizian, M., Vashaee, D., & Tayebi, L. (2015). Effect of ion substitution on properties of bioactive glasses: A review. *Ceramics International*, 41(6), 7241–7251. DOI: 10.1016/j.ceramint.2015.02.140

Saravanapavan, P., & Hench, L. L. (2001). Low-temperature synthesis, structure, and bioactivity of gel-derived glasses in the binary CaO-SiO2 system. *Journal of Biomedical Materials Research*, 54(4), 608–618. DOI: 10.1002/1097-4636(20010315)54:4<608::AID-JBM180>3.0.CO;2-U PMID: 11426607

Saravanapavan, P., Jones, J. R., Pryce, R. S., & Hench, L. L. (2003). Bioactivity of gel–glass powders in the CaO-SiO_2 system: A comparison with ternary (CaO-P_2P_5-SiO_2) and quaternary glasses (SiO_2-CaO-P_2O_5-Na_2O). *Journal of Biomedical Materials Research. Part A*, 66A(1), 110–119. DOI: 10.1002/jbm.a.10532 PMID: 12833437

Sarmast Sh, M., George, S., Dayang Radiah, A. B., Hoey, D., Abdullah, N., & Kamarudin, S. (2022). Synthesis of bioactive glass using cellulose nano fibre template. *Journal of the Mechanical Behavior of Biomedical Materials*, 130, 105174. DOI: 10.1016/j.jmbbm.2022.105174 PMID: 35344755

Sharifianjazi, F., Moradi, M., Abouchenari, A., Pakseresht, A. H., Esmaeilkhanian, A., Shokouhimehr, M., & Shahedi Asl, M. (2020). Effects of Sr and Mg dopants on biological and mechanical properties of SiO2–CaO–P2O5 bioactive glass. *Ceramics International*, 46(14), 22674–22682. DOI: 10.1016/j.ceramint.2020.06.030

Sharifianjazi, F., Parvin, N., & Tahriri, M. (2017). Synthesis and characteristics of sol-gel bioactive SiO2-P2O5-CaO-Ag2O glasses. *Journal of Non-Crystalline Solids*, 476, 108–113. DOI: 10.1016/j.jnoncrysol.2017.09.035

Sohrabi, M., Yekta, B. E., Rezaie, H., Naimi-Jamal, M. R., Kumar, A., Cochis, A., Miola, M., & Rimondini, L. (2021). The effect of magnesium on bioactivity, rheology and biology behaviors of injectable bioactive glass-gelatin-3-glycidyloxypropyl trimethoxysilane nanocomposite-paste for small bone defects repair. *Ceramics International*, 47(9), 12526–12536. DOI: 10.1016/j.ceramint.2021.01.110

Spirandeli, B. R., Campos, T. M. B., Ribas, R. G., Thim, G. P., & Trichês, E. D. S. (2020). Evaluation of colloidal and polymeric routes in sol-gel synthesis of a bioactive glass-ceramic derived from 45S5 bioglass. *Ceramics International*, 46(12), 20264–20271. DOI: 10.1016/j.ceramint.2020.05.108

Sprio, S., Antoniac, I., Chevalier, J., Iafisco, M., Sandri, M., & Tampieri, A. (2023). Editorial: Recent advances in bioceramics for health. *Frontiers in Bioengineering and Biotechnology*, 11, 1264799. DOI: 10.3389/fbioe.2023.1264799 PMID: 37593328

Sudipta, S. M., & Murugavel, S. (2022). Biomineralization behavior of ternary mesoporous bioactive glasses stabilized through ethanol extraction process. *Journal of Non-Crystalline Solids*, 589, 121630. DOI: 10.1016/j.jnoncrysol.2022.121630

Szu, S.-P., Klein, L. C., & Greenblatt, M. (1992). Effect of precursors on the structure of phosphosilicate gels: 29Si and 31P MAS-NMR study. *Journal of Non-Crystalline Solids*, 143, 21–30. DOI: 10.1016/S0022-3093(05)80548-4

Van Gestel, N. A. P., Geurts, J., Hulsen, D. J. W., Van Rietbergen, B., Hofmann, S., & Arts, J. J. (2015). Clinical Applications of S53P4 Bioactive Glass in Bone Healing and Osteomyelitic Treatment: A Literature Review. *BioMed Research International*, 2015, 1–12. DOI: 10.1155/2015/684826 PMID: 26504821

Willinger, M.-G., Clavel, G., Di, W., & Pinna, N. (2009). A general soft-chemistry route to metal phosphate nanocrystals. *Journal of Industrial and Engineering Chemistry*, 15(6), 883–887. DOI: 10.1016/j.jiec.2009.09.017

Wu, Y., Tang, L., Zhang, Q., Kong, F., & Bi, Y. (2021). A novel synthesis of monodispersed bioactive glass nanoparticles via ultrasonic-assisted surfactant-free microemulsion approach. *Materials Letters*, 285, 129053. DOI: 10.1016/j.matlet.2020.129053

Yao, J., Radin, S., Leboy, P. S., & Ducheyne, P. (2005). The effect of bioactive glass content on synthesis and bioactivity of composite poly (lactic-co-glycolic acid)/bioactive glass substrate for tissue engineering. *Biomaterials*, 26(14), 1935–1943. DOI: 10.1016/j.biomaterials.2004.06.027 PMID: 15576167

Zhong, J., & Greenspan, D. C. (2000). Processing and properties of sol-gel bioactive glasses. *Journal of Biomedical Materials Research*, 53(6), 694–701. DOI: 10.1002/1097-4636(2000)53:6<694::AID-JBM12>3.0.CO;2-6 PMID: 11074429

Chapter 9
Electrochemical Performance of Biomass-Derived AC and CNTs-Based Supercapacitors

Meiram Atamanov
Kazakh National Women's Teacher Training University, Kazakhstan

Tolganay Atamanova
Al-Farabi Kazakh National University, Kazakhstan

Azamat Taurbekov
Al-Farabi Kazakh National University, Kazakhstan

Zulkhair Mansurov
Al-Farabi Kazakh National University, Kazakhstan

ABSTRACT

This study investigates the electrochemical performance of activated carbon (AC) derived from two biomass sources, rice husk (RH) and walnut shell (WS), utilizing two activation methods: chemical activation with potassium hydroxide (KOH) and physical activation with carbon dioxide (CO2). The aim is to evaluate the potential of these bio-derived carbon materials in energy storage applications, particularly for supercapacitors. The electrochemical behavior of the activated carbon electrodes was assessed using cyclic voltammetry (CV) and galvanostatic charge-discharge (GCD) techniques, along with electrochemical impedance spectroscopy (EIS), to evaluate the specific capacitance, energy density, charge transfer resistance, and ion diffusion properties.

1. INTRODUCTION

Activated carbons (ACs) are vital materials with diverse applications in fields such as catalysis, water purification, and carbon dioxide capture, due to their high specific surface area, adjustable porous structure, and the presence or absence of functional groups (Zhao et al., 2023). The primary sources for producing ACs include coal, wood, and biomass, with biomass being the most economically advanta-

DOI: 10.4018/979-8-3693-7505-1.ch009

geous option due to its cost-effectiveness and absence of harmful contaminants (Lee et al., 2023). The production of high-quality, eco-friendly, and economically viable ACs for use in energy storage and other applications depends on selecting the appropriate activation method (Atamanov et al., 2018), reagent, temperature, and time (Ilyin et al., 2023).

Chemical activation using potassium hydroxide (KOH) is the most common and effective method for producing ACs from biomass (Hermawan et al., 2023). However, KOH presents significant environmental and safety risks, as it is a strong alkali that can cause serious health hazards during handling. Its high cost, along with the environmental impact of its production and disposal, further complicates the process. These drawbacks highlight the need for alternative activation methods (Alharbi et al., 2022).

Physical activation using carbon dioxide (CO_2) offers a safer, cheaper, and more economical alternative. However, this method has some disadvantages, including a lower specific surface area and a narrower range of pore sizes compared to chemical activation. This is due to the relatively low reactivity of CO_2 in comparison to KOH. Consequently, there is a need to identify cost-effective and readily available biomass with suitable chemical composition and structural properties for efficient activation with CO_2 (Goel et al., 2021).

The pore formation mechanism also differs between activation methods. In chemical activation, pores form through the interaction of chemical reagents with disorganized carbon and inorganic structures, whereas CO_2 activation enlarges pore diameters through physical expansion driven by chemical reactions between the activation gas and active carbon atoms. Understanding the influence of these activation methods on the final physical and chemical properties of ACs is crucial for optimizing their performance in various applications (Okayama et al., 2010).

2. ENERGY STORAGE AND CARBON MATERIALS

Energy storage is critical for technological progress, especially as industries and transportation continue to rely heavily on fossil fuels. As a result, research focuses on developing environmentally friendly, efficient energy storage systems. Batteries, fuel cells, and electrochemical capacitors (ECs) are the most common energy storage devices. While batteries and fuel cells store energy through redox reactions, ECs store energy via ion adsorption or fast surface reactions, offering the advantage of long lifespans (Kim et al., 2005).

ECs utilize powdered active materials to form porous structures that facilitate electrochemical reactions. To enhance conductivity and structural integrity, conductive carbon materials and polymer binders are added. However, despite these additions, EC electrodes often have low conductivity, necessitating the use of metal current collectors to ensure efficient electron transport and mechanical support. To maximize energy storage in ECs, efforts are focused on using active materials that store more energy while minimizing the amount of inactive components, such as binders and conductive additives (Iijima et al., 1991).

Activated carbons (ACs), derived from plant-based materials such as apricot stones and rice husks, are commonly used in ECs. Their porosity and surface area, which are crucial for storing electrical charge through ion adsorption, are influenced by the activation conditions. These ACs provide a high surface area that improves the electrode's performance at the electrode-electrolyte interface.

Another important material for advanced energy storage applications is carbon nanotubes (CNTs), which are known for their exceptional tensile strength, electrical conductivity, and stability. These properties make CNTs highly promising for various applications, including energy storage and environmental solutions. CNTs are synthesized through methods like catalytic chemical vapor deposition (CCVD), arc discharge, and peapod techniques, and the synthesis method directly impacts their quality, structure, and properties (Shen et al., 2011).

CNTs have several unique characteristics, including a high aspect ratio, large specific surface area, and outstanding mechanical and electrical properties. These attributes make them highly effective for energy storage, accumulation, and environmental cleaning applications. However, a significant challenge with CNTs is their insolubility in water, which limits practical applications. To address this issue, CNTs are often functionalized to improve their solubility and performance in various systems. Despite these efforts, challenges remain, particularly concerning CNT toxicity.

Figure 1. Illustration of main carbon nanotube (CNT) synthesis methods: (a) arc discharge method, (b) chemical vapor deposition and annealing process, (c) laser ablation method (Taurbekov et al., 2024)

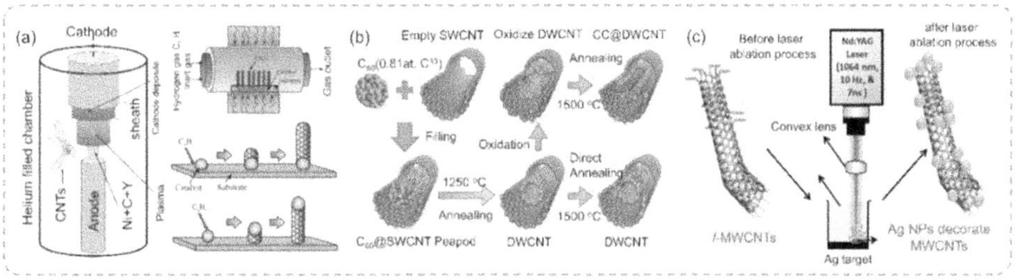

The toxicity of CNTs, though not fully understood, is believed to be influenced by factors such as surface modification, morphology, and aggregation. Strategies to reduce toxicity include functionalization, surface modification, and stabilization techniques, which aim to make CNTs safer for use. However, these methods are still being explored, and careful consideration is needed when using CNTs, especially in biological environments (Mann et al., 2022).

In addition to their potential in energy storage, CNTs have shown promise in water purification due to their high mechanical strength, large surface area, and excellent adsorption properties. Despite their effectiveness, the high cost of CNTs and the challenges of regenerating them after use limit their large-scale application. Nevertheless, their unique properties make CNTs a valuable material for a wide range of technological advancements in energy storage and environmental sustainability. Illustrations of the main CNT synthesis methods are shown in Figure 1 (Yadav et al., 2022).

3. HYBRID ELECTRODES OF AC WITH CNT

Carbon materials are vital for energy storage devices due to their roles as electrically conducting additives, substrates for active materials, electron converters, and agents for controlling thermal conductivity, porosity, and surface area. Activated carbon (AC) is the most commonly used material for supercapacitor electrodes due to its high specific capacitance and cost-effectiveness. AC is produced from carbon-rich organic materials through carbonization at temperatures between 700 and 1100°C in an inert atmosphere, followed by activation processes using CO_2, steam, or KOH to control surface area and porosity. Natural materials such as coconut shells, apricot stones, rice husks, pecan shells, or synthetic materials like polymers can serve as precursors for producing activated carbon (Ruch et al., 2010).

The final characteristics of carbon-based supercapacitors are directly linked to the physical and chemical properties of the carbon electrodes. These characteristics determine the device's overall performance, including energy density and capacitance. Hybrid electrodes, made by combining activated carbon with carbon nanotubes (CNTs), offer improved contact and allow for thicker electrodes. This configuration is more practical than thin films and increases the energy density as the metal collector content decreases. The amount of CNTs in hybrid electrodes must be optimized to balance cost and specific capacitance. Physical characteristics, such as flexibility and the full coverage of AC particles by CNTs, are crucial for the fabrication of high-performance electrodes (Jumadi et al., 2020).

The combination of activated carbon and CNTs in energy storage devices like supercapacitors is increasingly relevant for industries such as electric vehicles and renewable energy storage. As research continues, further optimization of carbon materials, activation processes, and hybrid configurations could result in more efficient, cost-effective energy storage solutions. These advancements have the potential to significantly impact the future of energy sustainability and efficiency.

3.1 Morphology and Textural Properties of Hybrid Electrodes based on AC with CNT

The N_2 adsorption-desorption isotherms for RH (rice husk) and WS (wood shavings) samples treated with KOH and CO_2 display characteristics of micromesoporous materials (type I and IV isotherms). Increased gas absorption at low pressures ($p/p0 < 0.2$) indicates a well-developed microporous structure, which is the initial indicator of porosity development. The pore size distribution further confirms the presence of developed microporosity in all samples, providing detailed insights into the pore structures formed during activation.

When comparing activation methods, KOH activation promotes the intensive development and expansion of microporosity, while CO_2 activation focuses more on expanding the existing micropores. This distinction highlights the differing impacts of the two activation methods. At higher relative pressures ($p/p0 > 0.5$), H2a hysteresis loops are observed, indicating the presence of mesoporous structures. This is particularly noticeable in the RH_KOH and RH_CO_2_2h samples, suggesting that CO_2 activation also encourages the formation of mesopores.

Figure 2. Adsorption and pore structure characterization of different activated carbons: (a) Nitrogen adsorption-desorption isotherms at 77 K for activated carbons derived from AK, WS, and RH; (b) Pore size distribution of AK, WS, and RH activated carbons determined by the DFT method; (c) Cumulative pore volume of AK, WS, and RH activated carbons as a function of pore width; (d) Nitrogen adsorption-desorption isotherms at 77 K for WS_KOH, RH_KOH, WS_CO_2_2h, and RH_CO_2_2h activated carbons; (e) Pore size distribution of WS_KOH, RH_KOH, WS_CO_2_2h, and RH_CO_2_2h activated carbons determined by the DFT method; (f) BET surface area (S_BET) as a function of activation time for WS_KOH, RH_KOH, WS_CO_2_2h, and RH_CO_2_2h activated carbons

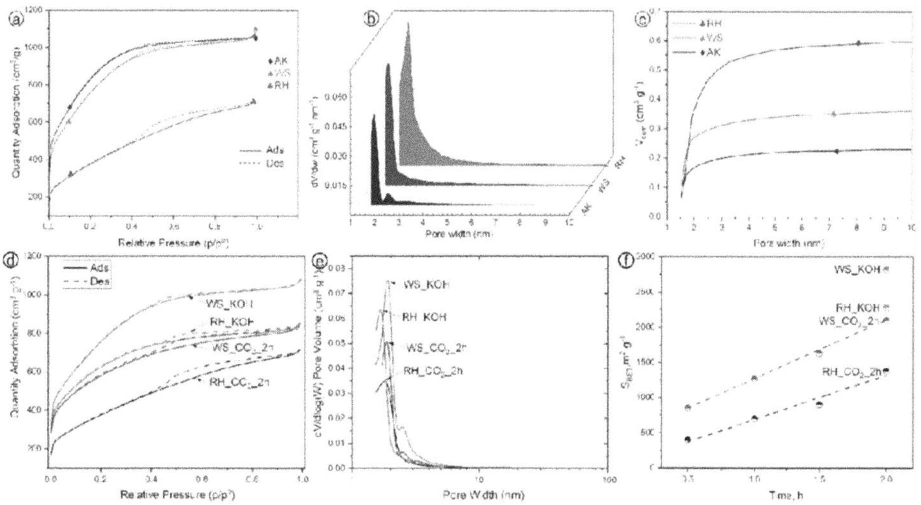

Overall, KOH activation is more effective at creating microporosity, while CO_2 activation contributes more significantly to the development of mesoporosity, particularly in RH samples. This distinction in activation effects provides valuable insights into tailoring porous structures based on activation methods.

The findings suggest that KOH activation, with its emphasis on microporosity, could lead to higher surface areas and enhanced energy storage performance in specific applications. On the other hand, CO_2 activation's ability to enhance mesoporosity may offer better ion transport and diffusion, potentially improving the overall electrochemical performance of hybrid electrodes. Therefore, selecting the appropriate activation method is crucial, depending on whether the focus is on maximizing surface area or optimizing pore size distribution for energy storage devices.

3.2 Physical and Chemical Characterization

The Raman spectroscopy analysis reveals key structural characteristics of the carbon-based materials (Figure 3). The D-band (1300-1350 cm^{-1}) indicates defects and disorganization, which are amplified by the activation process, while the G-band (1580-1600 cm^{-1}) signals the presence of graphitic carbon

(sp^2) and crystallinity (Yergaziyeva et al., 2024). The 2D (G') band provides additional insights into the material's layered structure (Lesbayev et al., 2024).

When examining the D/G ratio, KOH-treated samples show a moderate D/G ratio, indicating a balance between disorganized and graphitic carbon, meaning the content of amorphous carbon is nearly equivalent to that of graphitic carbon. In contrast, CO_2-treated samples exhibit a lower D/G ratio, suggesting a higher graphitic carbon content, with WS samples activated by CO_2 showing the most graphitized structure. The G/2D peak ratios further support this observation. CO_2-activated samples (RH = 0.45, WS = 0.49) indicate the formation of few-layer graphene structures, while KOH-activated samples (RH = 0.52, WS = 0.55) reveal a more complex, multilayer structure, indicating a higher degree of structural complexity.

The X-ray diffraction (XRD) analysis provides additional confirmation of the material's amorphous structure, with broad peaks of low intensity at 2θ values around 20° and 30°, indicative of a disordered structure with a high surface area. The presence of Fe_3O_4 (magnetite), elemental iron (Fe), and iron nitride ($FeN_{0\cdot0324}$) was detected, likely due to the carbonization process in a steel reactor. In RH samples, distinct peaks are observed at 2θ values of 30°, 35°, 57°, and 63°, which are uniquely associated with Fe_3O_4. KOH-treated samples show a peak at 45°, corresponding to elemental iron, while $FeN_{0\cdot0324}$ peaks are visible at 2θ = 43° in WS samples. A well-defined graphite peak at 26° confirms the presence of a graphitized structure in all samples, which aligns with the Raman spectroscopy findings (Pandey, 2023).

In comparing activation methods, KOH activation introduces more defects and a higher degree of disorganization, as reflected by the higher D/G ratio and the formation of multilayer graphene structures. On the other hand, CO_2 activation leads to a more graphitized structure with fewer defects and the formation of fewer-layer graphene, making it beneficial for applications requiring high conductivity and structural integrity.

In conclusion, the structural differences between KOH- and CO_2-treated samples are evident in both Raman spectroscopy and XRD analysis. KOH activation results in a higher degree of disorganization and the formation of iron-based compounds, while CO_2 activation produces a more organized, graphitized structure with fewer defects, which could be advantageous for applications involving energy storage and conductivity.

Figure 3. Structural and elemental characterization of activated carbons: (a) Raman spectra of activated carbons derived from WS, RH, and AK showing the D, G, and 2D bands with ID/IG and IG/I2D ratios; (b) XRD patterns of activated carbons from WS, RH, and AK indicating the presence of graphite, Fe_3O_4, Fe, and $FeN_{0'324}$; (c) EDX spectra and elemental composition (atomic %) of activated carbons from WS, RH, and AK; (d) Raman spectra of activated carbons from WS_CO_2_2h, RH_CO_2_2h, WS_KOH, and RH_KOH showing few-layer and multilayer structures; (e) XRD patterns of activated carbons from WS_CO_2_2h, RH_CO_2_2h, WS_KOH, and RH_KOH indicating the presence of graphite, Fe_2O_3, Fe, and $FeN_{0'324}$. (f) Particle size distribution of different samples (AS, AS10, AS30, AS30+30, AS60) based on the accumulated ratio

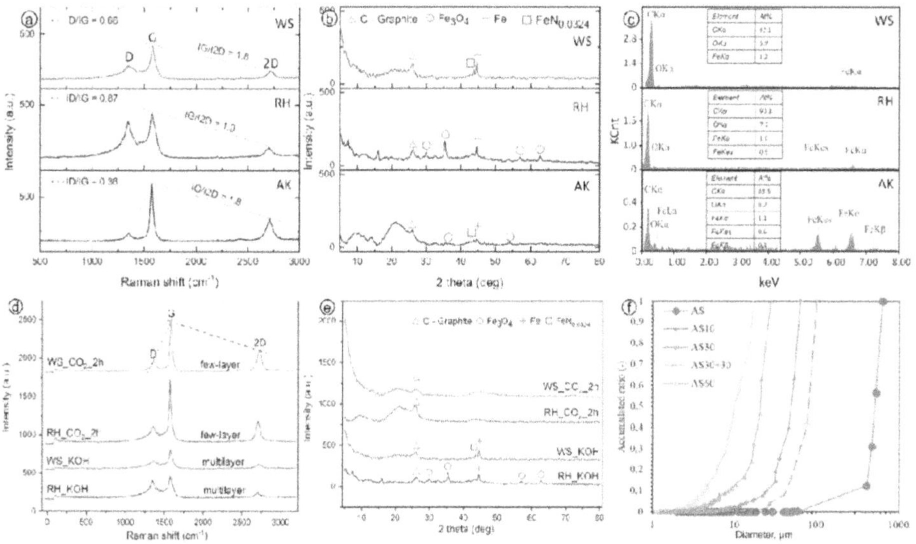

During thermolysis under nitrogen gas flow at a 5 K/min heating rate, RH (rice husk) and WS (wood shavings) undergo dehydration in the 100-150°C range, resulting in a mass loss of 5.1% for RH and 5.6% for WS due to the loss of physically bound water. Following dehydration, active pyrolysis occurs between 150-500°C, with significant mass loss attributed to the decomposition of volatile organic compounds. The decomposition stages include hemicellulose at around 300°C for WS (no distinct peak for RH), cellulose at 390-400°C for both samples, and lignin at 450-460°C for both samples.

The energy required for the decomposition of both RH and WS is approximately 23-25 J/g. During the exothermic reaction associated with the decomposition of volatile organic compounds, RH releases 16.9 J/g of energy, while WS releases 8.1 J/g, with RH releasing more energy likely due to the absence of intermediate volatile products.

This analysis demonstrates the distinct thermal behavior of RH and WS during thermolysis and pyrolysis, highlighting their differing decomposition profiles and energy releases. RH exhibits higher energy release during the decomposition of volatile organic products compared to WS, which may be attributed to the differences in their material composition and volatile gas formation during the process.

Figure 4. Thermal decomposition of WS and RH under nitrogen gas flow (100 cm3 l-1) at a 5 K min^-1^ heating rate

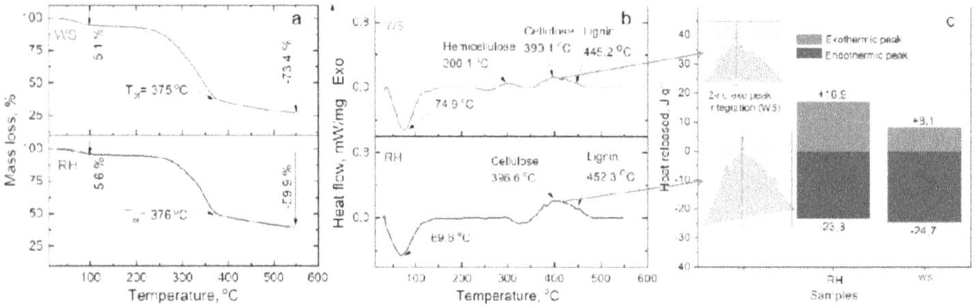

3.3 Electrochemical Characterization

The electrochemical performance of AC derived from RH and WS, using both KOH and CO_2 activation, was evaluated using cyclic voltammetry (CV) and galvanostatic charge-discharge (GCD) techniques (Figure 5). The CV tests revealed near-rectangular curves at 10 mV s^{-1} and 100 mV s^{-1}, indicating electric double-layer capacitor behavior with no Faradaic reactions. At 10 mV s^{-1}, RH_KOH exhibited a specific capacitance of 74.7 F g^{-1}, while RH_CO_2_2h showed 24.3 F g^{-1}. Similarly, for WS electrodes, WS_KOH reached 74.7 F g^{-1}, and WS_CO_2_2h achieved 59 F g^{-1}. As the scan rate increased from 5 to 500 mV s^{-1}, RH_KOH and WS_KOH maintained higher capacitance values, but they experienced substantial declines in capacitance retention. RH_KOH retained 41.68% of its initial capacitance, while RH_CO_2_2h retained about 80%, demonstrating better pore accessibility and charge transfer. WS_CO_2_2h also showed better rate capability than WS_KOH, retaining 56% of its capacitance at 1000 mV s^{-1}, compared to 31% for WS_KOH.

In the GCD tests conducted at 1 A g^{-1}, RH_KOH exhibited a peak specific capacitance of 150 F g^{-1} with an energy density of 5.2 Wh kg^{-1} at a power density of 300 W kg^{-1}, whereas RH_CO_2_2h demonstrated a much lower specific capacitance of 40 F g^{-1} and an energy density of 1.4 Wh kg^{-1} at the same power density. Similarly, WS_KOH performed well, achieving 152 F g^{-1} with an energy density of 5.3 Wh kg^{-1}, while WS_CO_2_2h achieved 109 F g^{-1} with an energy density of 3.9 Wh kg^{-1}. As the current density increased to 35 A g^{-1}, RH_KOH retained 51.5% of its capacitance, while RH_CO_2_2h retained 68%. For WS, WS_CO_2_2h retained 75.6% of its capacitance at 35 A g^{-1}, compared to 47.3% retention for WS_KOH.

The impact of the activation method on performance was clear: KOH activation resulted in higher specific capacitances and energy densities for both RH and WS electrodes due to the increased specific surface area created by chemical activation. However, CO_2 activation provided better rate capability and retention at higher scan rates and current densities, particularly for WS_CO_2_2h, which maintained a higher percentage of its capacitance as the scan rate increased. RH_CO_2_2h also exhibited improved charge transfer and pore accessibility, retaining 80% of its capacitance at higher scan rates, compared to 41.68% for RH_KOH.

Material composition played a crucial role in these differences. Rice husk (RH) contains 15-25% silicon dioxide and other inorganic compounds, which limit pore formation during CO_2 activation, resulting in lower specific surface area and poorer electrochemical performance for RH_CO_2_2h. KOH activation, however, removes these inorganic compounds, significantly increasing the specific surface area and improving the electrochemical performance of RH_KOH. In contrast, walnut shell (WS) has a low ash content (0.1-3.4%), and the performance differences between WS_KOH and WS_CO_2_2h are less pronounced. WS_CO_2_2h outperforms WS_KOH at higher scan rates and current densities due to better electrolyte penetration and charge transfer.

Figure 5. Electrochemical performance of KOH-activated and CO_2-activated AC derived from RH and WS. CV curves at 10 mV s^{-1} and 100 mV s^{-1} for (a) RH derived AC and (d) WS derived AC. GCD curves at 1 A g^{-1} and 2 A g^{-1} for (b) RH derived AC and (e) WS derived AC. Scan rate (c) and current rate (f) capabilities of KOH-activated and CO_2-activated AC derived from RH and WS

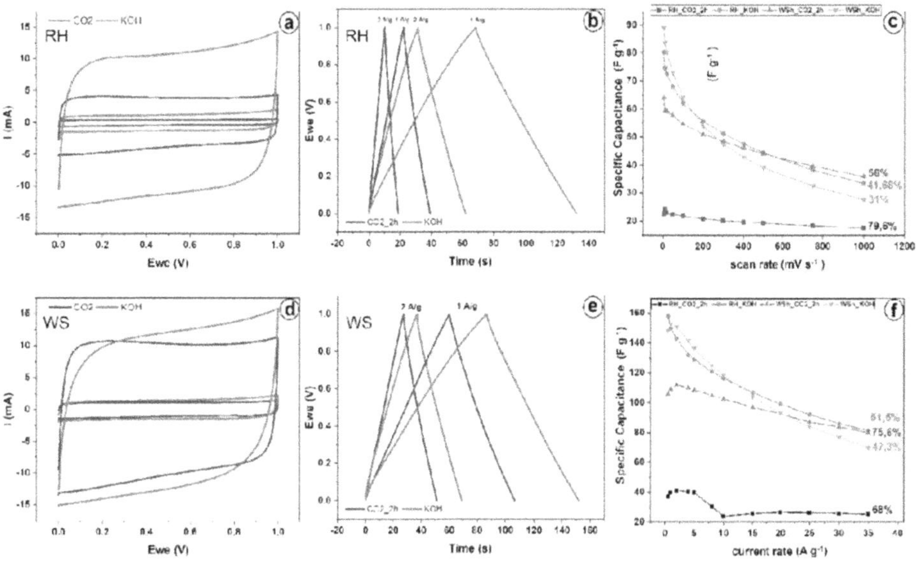

All samples, regardless of activation method, demonstrated excellent long-term stability, with near 100% Coulombic efficiency, indicating high electrical conductivity and minimal side reactions. While KOH activation led to higher initial capacitances and energy densities, CO_2 activation provided superior rate capability and capacitance retention, especially for walnut shell-based electrodes. The material composition of rice husk and walnut shell influenced the effectiveness of the activation method, with the high inorganic content in RH limiting the potential for CO_2 activation, whereas WS showed more balanced results between KOH and CO_2 activation.

Electrochemical impedance spectroscopy (EIS) tests were conducted to investigate the resistive behavior of the materials, with results presented in Nyquist and Bode plots. In Figure 6a, RH_KOH and RH_CO_2_2h exhibit similar contact resistance values (Rs ~ 0.3 Ohm), attributed to consistent electrode construction methods (active material, electrolyte, separator). However, the charge transfer resistance (Rct) differs significantly between the two samples. RH_CO_2_2h shows a much higher Rct (~0.6 Ohm) compared to RH_KOH (~0.15 Ohm), indicating slower charge transfer kinetics in the CO_2-activated material. This is further supported by the relaxation times: 266 ms for RH_CO_2_2h and 863 ms for RH_KOH. Interestingly, the spectra for RH_CO_2_2h, with a phase angle of 85 degrees (Figure 6b), show a steeper tail, reflecting improved ion diffusion and a well-structured surface area and porosity.

For walnut shell-derived electrodes, both WS_KOH and WS_CO_2_2h exhibit similar contact resistance values (Rs ~ 0.36 Ohm). However, the CO_2-activated electrode (WS_CO_2_2h) has a lower charge transfer resistance (Rct ~ 0.1 Ohm) compared to WS_KOH (Rct ~ 0.24 Ohm), suggesting faster charge transfer in the physically activated material. Relaxation times further support this, with WS_CO_2_2h showing a shorter relaxation time of 583 ms compared to 1277 ms for WS_KOH.

Electrochemical tests revealed a strong correlation between the porous structure and the specific capacitance of the electrodes. The WS_KOH sample, with its predominantly microporous structure, demonstrated a higher specific capacitance and energy density than the other samples, highlighting the importance of micropores for ion adsorption from the electrolyte, which enhances capacitance. However, at higher scan rates (starting from 200 mV s^{-1}), the specific capacitance values of WS_KOH, WS_CO_2_2h, and RH_KOH become equal. This is due to decreased ion diffusion activity in the micropores of RH_KOH and WS_KOH, as mesopores become the primary ion storage sites. WS_CO_2_2h, with its well-formed mesopores after physical activation, demonstrates superior power density at higher scan rates.

The RH_CO_2_2h sample exhibits a low specific capacity, consistent with BET analysis, which attributes this to its high ash content and the presence of inorganic elements that do not react with the activation gas, preventing the formation of new pores. This highlights the importance of selecting raw materials with an appropriate chemical composition for effective physical activation using CO_2.

These findings underscore the crucial role of micropores in providing higher specific capacitance at lower scan rates and mesopores in supporting better ion diffusion and power density at higher scan rates. Additionally, the chemical composition of the raw material significantly impacts the success of physical activation processes like CO_2 treatment.

Figure 6. Electrochemical performance of KOH-activated and CO_2-activated AC derived from rice husk and walnut shell. (a) Nyquist plot recorded in the frequency range of 100 kHz to 0.01 Hz for rice husk-based electrodes and (d) for walnut shell-based electrodes. (b, c) Bode plot for rice husk-based electrodes and (e, f) for walnut shell-based electrodes

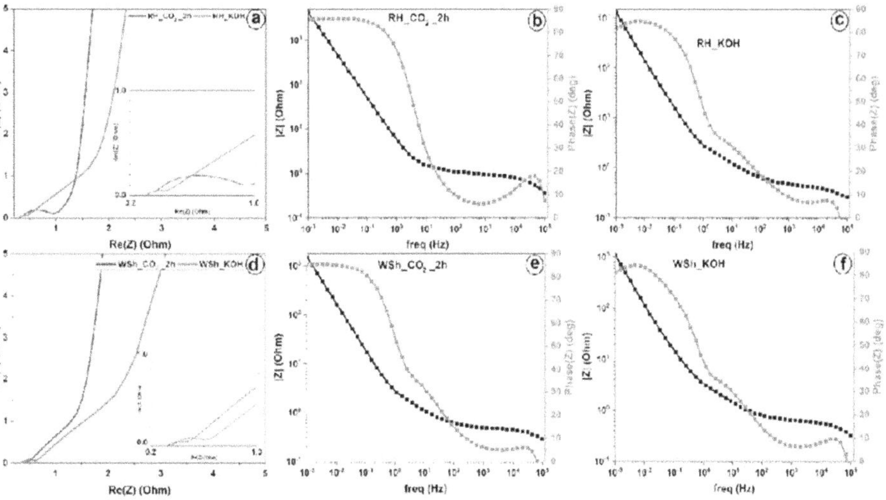

The Ragone plot, depicted in Figure 7a, presents the energy and power densities of the electrodes. The RH_KOH electrode exhibits the highest energy density of 6 Wh/kg at a power density of 59 W/kg. In contrast, the RH_CO_2_2h electrode demonstrates a much lower energy density of 1 Wh/kg at the same power density. For walnut shell-based electrodes, WS_KOH shows an energy density of 4.5 Wh/kg, while WS_CO_2_2h exhibits 2.9 Wh/kg at 59 W/kg.

Figure 7b illustrates the cycling stability tests, which involved charging and discharging the electrodes in the potential range of 0–0.9 V at a constant current density of 2 A/g. After 2000 cycles, all samples experienced only slight decreases in specific capacitance, maintaining over 97% of their initial capacitance values during an additional 2000 cycles of charging and discharging.

These results highlight the superior energy density of the RH_KOH electrode compared to the other electrodes, particularly RH_CO_2_2h, which shows the lowest energy density. Additionally, all electrodes demonstrate excellent cycling stability, retaining a significant portion of their initial capacitance even after prolonged cycling.

Figure 7. (a) Ragone plot for the rice husk and walnut shell–based electrodes. (b) Cycling GCD test at 2 A g^{-1}

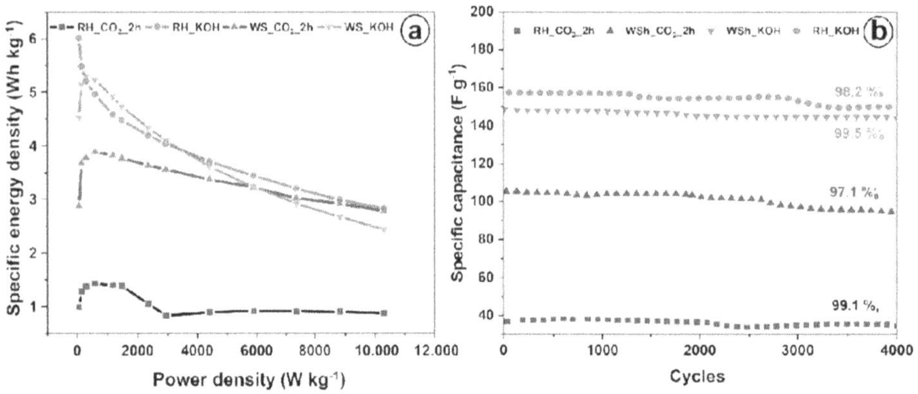

4 DEVELOPMENT OF ENERGY-INTENSIVE COMPOSITES BASED ON NANOCELLULOSE AND FUNCTIONALIZED CNTS FOR IMPROVED ENERGY STORAGE SYSTEMS

The main sources of energy storage are batteries, superconducting inductive storage devices or supercapacitors, which are widely used in various industries. Despite the progress achieved in this area, the field for new developments remains inexhaustible. New technologies for processing the materials used, and, in particular, the development of new functionalized nanomaterials, have increased the efficiency of devices. It has been established that the functionalization of CNTs with the addition of transition metal oxides to their surface allows increasing the energy capacity of electrodes. The addition of titanium, vanadium, iron, cobalt, and nickel oxides showed high values of specific energy density due to the pseudocapacitive effect. The disadvantage of CNT-based electrodes is the preparation technology, which reduces the electrochemical activity of the CNT area due to their uncontrolled agglomeration and adhesion, as well as the aggregation of doped metal oxide particles, which together greatly affects the final capacitive characteristics of the electrodes. This project proposes to develop an effective composite for the manufacture of energy storage systems based on nanocellulose and functionalized CNTs doped with transition metal oxides. Studies have shown that nanocellulose can be used as a matrix for effective dispersion of CNTs and provide adjustable ionic conductivity in the manufacture of electrodes. There is a huge amount of cellulose in the form of biowaste from production, household and wood processing industries. We propose to optimize the methods for obtaining cellulose with subsequent restructuring to crystals or fibrils of nanocellulose using chemical treatment. Commercial double-walled CNTs will be used, which showed the highest result in specific capacity. The hypothesis of this work is that the advantage of this composite will be a large number of specific sites in CNTs for both chemical interaction due to doping with transition metal oxides (NiO, CoO), and physical interaction due to uniform dispersion of CNTs in the nanocellulose matrix, which will allow getting as close as possible to theoretical values of the possible energy density.

4.1 Complex Pre-treatment (washing, grinding, drying, fractionation) of the Feedstock

In order to isolate cellulose, various cellulose precursors (office waste paper and sawdust) were selected. All selected cellulose precursors were pre-treated for further production of cellulose substrates based on them. The preparation of the raw materials included several main stages: cleaning, grinding, separation into fractions, dispersion, chemical cleaning from organic and process residues; removal of paint, glue, detergent and polymer residues from office waste paper by mechanical or chemical methods; preparation of sawdust for rough cleaning from sand, dust, dirt, earth, ethers and other contaminants for the extraction of cellulose substrates.

The work involved precursors with a high cellulose content of over 80%, low lignin content (Kappa number) and ash content: 1) elm sawdust selected during the technological processing of trees; 2) rice husks obtained as agricultural waste; 3) A4 office waste paper selected from office waste (Institute for Combustion Problems). For preliminary cleaning of rice husks and wood sawdust from various mechanical impurities in the form of dirt, dust and other solid waste, the raw materials were washed in hot water (80 °C) at least 3 times. Then, the washed samples were placed in a drying cabinet (model shs-80) in order to dehydrate the samples at a temperature of 105 °C for 12 hours. The dried samples were crushed on a Stegler LM-250 laboratory shredder for 15 minutes. After grinding, the obtained material was sieved on a vibrating sieve to select the required fraction (0.05-0.2 mm). The selected samples weighing 20 g were dispersed in 2 l of distilled water and homogenized using a PE-6500 laboratory orbital shaker at 200 vibrations per minute for 30 minutes. The resulting homogenized suspension was dehydrated using a RE-100 Pro laboratory rotary evaporator at low pressure (100 Pa) and room temperature. Pre-processing of the raw material leads to loosening of the surface of the fiber structure and to the appearance of a large number of branches consisting of smaller and thinner fibers. After washing, the office waste paper was processed using a ball mill at room temperature for defibrillation and the resulting mass for 15 minutes. Afterwards, it was dried at 105 °C for 12 hours and ground into powder on a shredder until the particle size reached no more than 200 μm.

Based on the conducted studies, it was established that the optimal conditions for washing and cleaning the raw material from various mechanical contaminants are washing at least 3 times in hot water (80 °C) with intensive mixing. When grinding samples using a laboratory shredder, the optimal fraction of the resulting samples was achieved no more than 0.2 mm. According to literary studies, it is known that when using finely dispersed raw materials with a size of less than 200 μm, the highest values for the extraction of cellulose substrate from plant materials are achieved. It was found that the homogenization process also has a beneficial effect on the splitting of cellulose fiber bundles. As a result of the work carried out, it was established that the product yield for obtaining a cellulose substrate is 91% - rice husk, 79% - office waste paper and 83% - wood sawdust.

4.2 Obtaining Cellulose from Pre-prepared Raw Materials Using the Method of Chemical-Thermal Treatment

To obtain cellulose from sawdust, delignification and desilication were carried out with the selection of the optimal ratio of reagents and their concentrations. Sawdust was subjected to alkaline cooking in a 17.5% NaOH solution at a raw material:alkaline solution ratio of 1:10 by weight, the cooking time was 2.5 hours at a temperature of 120 °C. After alkaline cooking, the resulting extract was subjected to acid hydrolysis in an autoclave using 9% H_2SO_4 at an extract:acid solution ratio of 1:20 by weight, the cooking time was 60 minutes at a temperature of 100 °C. The extraction of cellulose from office waste paper was carried out using the process of acid hydrolysis in an autoclave using 9% H_2SO_4 at an extract:acid solution ratio of 1:10 by weight, cooking time of 60 minutes at a temperature of 100 °C. The advantage of office waste paper is that the raw material is not subjected to preliminary aggressive treatment with alkali, desilicified and bleached before use, which significantly simplified the process of cellulose extraction. During the alkaline cooking process, the following components were removed: pentosans, lignin, ash and substances soluble in hot water, respectively

The resulting suspensions were washed with water to a neutral pH with centrifugal treatment and dried using a laboratory rotary evaporator RE-100 Pro at low pressure (100 Pa) and room temperature. Based on the conducted studies, optimal conditions for the chemical-thermal treatment of sawdust for the extraction of cellulose were developed. It was found that the optimal processing conditions for these precursors are alkaline cooking with a 17.5% NaOH solution at a raw material:alkaline solution ratio of 1:10 by weight for 2.5 hours at a temperature of 120 °C. It was found that when processing raw materials with a 10% alkaline solution, the efficiency of cellulose extraction is low due to the fact that various inorganic compounds, such as silicon dioxide, remain during such processing. However, with an increase in the concentration of the alkaline solution to 17.5%, the maximum value for the extraction of inorganic parts is achieved. As a result of alkaline cooking, the release of the bulk of lignin, silicon dioxide, inorganic compounds and esters is noted. In addition to delignification, the destruction of polysaccharides (hemicellulose, pectin and others) contained in the composition of the original raw material occurs. According to the IR spectra, the following main peaks are noted at frequencies of 3391, 2906, 1649, 1443, 1162, 1061, 898 and 609 cm^{-1}. According to these peaks, it can be noted that cellulose is the main component after chemical-thermal treatment of the raw materials. The process of treating the obtained extract with acid hydrolysis leads to the formation of finely dispersed cellulose fibers. Based on optical microscopy images, it was found that the obtained samples are characterized by a homogeneous structure and size with a fiber diameter in the range from 50 to 10 μm. As a result of chemical-thermal treatment of the pre-prepared raw materials, it was found that the product yield is 48 and 40%, respectively. It has been experimentally established that the optimal conditions for acid hydrolysis of waste paper in an autoclave are treatment with a 9% H_2SO_4 solution at an extract:acid solution ratio of 1:10 by weight, cooking time of 60 minutes at a temperature of 100 °C. As a result, a light-yellow powdery material was obtained. The mass yield of powder samples after hydrolysis was 45%. According to optical microscopy images, the obtained samples have an average fiber diameter in the range from 3 to 30 μm. The obtained X-ray diffraction patterns (intensity at 15, 17, 22 and 34°) indicate that as a result of acid hydrolysis, a cellulose structure is formed, the supramolecular structure of which corresponds to the structure of cellulose.

Figure 8. IR spectra of cellulose from waste paper

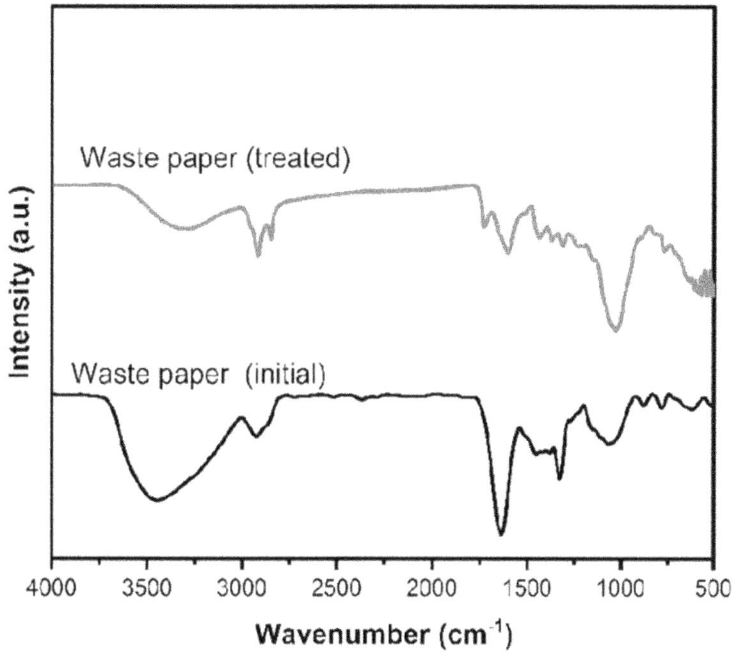

4.3 Development of the Method for Obtaining Crystalline and Fibrial Nanocellulose

To obtain crystalline cellulose, the first step was to remove unwanted components by washing the initial masses in distilled water at a temperature of 90 °C, followed by drying (60 °C, 12 hours). Then the samples are placed in a Soxhlet extractor to remove esters, paraffin and other organic compounds. Next, 300 ml of a solution of toluene (chemical purity 99.5%) and ethanol (chemical purity 99.9%) in a ratio of 2:1 are added to the resulting mass, heated to boiling and boiled for 8 hours. Dewaxing of organic compounds such as wax, chlorophyll and residual sugars occurs. Then the sample is washed with distilled water in a centrifuge at 8000 rpm and dried (110 °C). The samples are ground in a Stegler RM250 knife mill for 5 minutes to a particle size of 0.5-1 mm.

To obtain cellulose from wastepaper, a nanocellulose production technique was used with the following reagents (weight/volume): 5% NaOH, 10% H2O2, 10% Na2SiO3 at 60 °C for 90 minutes. The mixture is stirred for 30 minutes at 1000 rpm, then kept for 60 minutes. The solid is washed several times with distilled water in a centrifuge at 8000 rpm until a neutral pH is achieved. The sample is left to dry at 105 °C overnight.

Acid hydrolysis destroys the polymer structure of cellulose and leads to the formation of nanocellulose crystals. During acid hydrolysis, sulfuric acid breaks the glycosidic bonds between the monosaccharide units of cellulose, resulting in the separation of the polymer chains into shorter fragments. It was found that the conditions of acid hydrolysis for all samples can be equally the same, and the use of a similar technique does not affect the quality of the final pulp and its bleaching. Samples of cellulose after processing and bleaching in an amount of 1 g are mixed with 100 ml of H2SO4 (w/r 65%) at 90 °C with constant stirring for 60 minutes, then distilled water is added to the mixture to stop the reaction between the acid and the pulp. The suspension of the mixture is centrifuged to remove the acid. The resulting cellulose is washed with distilled water in a centrifuge at 10,000 rpm until a neutral pH is achieved. The resulting samples are ultrasonicated for 20 minutes at 120 kHz to disperse them. Scanning microscopy examination revealed that the rice husk samples had nanosized particles, i.e. crystalline cellulose was obtained.

To obtain fibril cellulose, the oxidation method with 2,2,6,6-tetramethylpiperidine-1-oxyl (TEMPO) was used. Cellulose previously obtained from rice husk, sawdust, waste paper was used as the starting material, and commercial cellulose (Cellulose C6288) was used as a comparison. The cellulose obtained earlier was dissolved in distilled water and stirred until a suspension was formed on a magnetic stirrer. Then 2,2,6,6-tetramethylpiperidine-1-oxyl (TEMPO) and sodium bromide (NaBr) were added to the suspension, and the mixture was stirred for 10 minutes at room temperature. Then, 10 mmol sodium hypochlorite (NaClO) was added for 60, 120 and 180 minutes to initiate the oxidation process (TEMPO) of the cellulose fibers. During the oxidation process, the pH was constantly monitored and maintained at 10.5 by slowly adding sodium hydroxide (NaOH) solution. After oxidation (TEMPO), the resulting fibers were washed with distilled water in a centrifuge at 10,000 rpm until neutral pH. The samples were sonicated at 120 kHz for 5 minutes to break down any remaining fiber aggregates and obtain well-dispersed nanofibers.

Figure 9. SEM images of fibrillar cellulose obtained from cellulose by TEMPO oxidation method

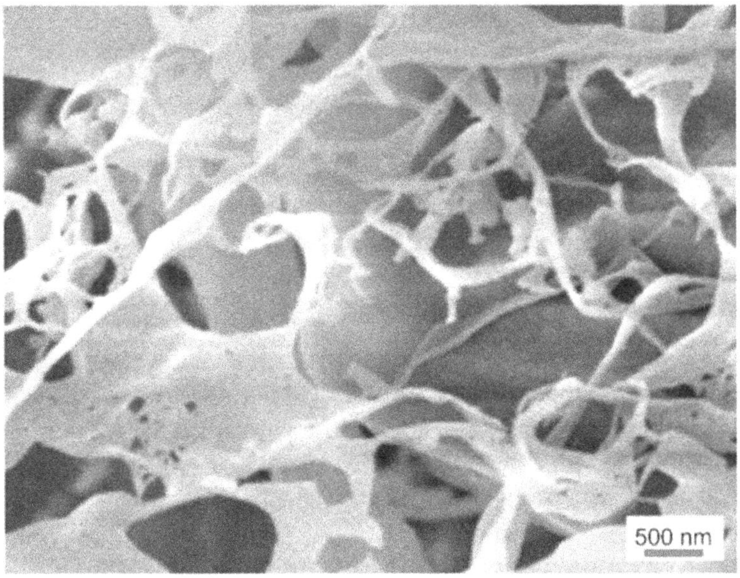

4.4 Determination of the Optimal Ratio Between Nanocellulose, Nitrogen-functionalized CNTs and Activated Carbons in the Manufacture of an Energy-intensive Composite

The most efficient ratio of the energy-intensive composite was determined. The method of manufacturing electrodes for use in flexible supercapacitors without a current collector was also determined. The composition of all electrodes was standardized and included: Activated carbon (AC) with a mass of 85 mg, Carbon black (C45) with a mass of 5 mg, Carbon nanotubes (CNT) (OCSiAl Tuball 01RW03. N1) with a mass of 5 mg, and a variable amount of nanocellulose solution (3, 5, 7 and 10 ml). The procedure for collecting NC was the same for all methods: NC dissolved in distilled water was stirred for 5 min in an ultrasonic bath (40 kHz), after which the solution was poured into a 50 ml graduated cylinder, sedimented for 5 min, and then the NC solution was collected from the 35th mark of the graduated cylinder, resulting in 3, 5, 7 and 10 ml of each sample. Three methods for producing electrodes with different component concentrations were developed. The first method involved dispersing carbon nanotubes (CNTs) in isopropyl alcohol, followed by adding activated carbon (AC), conductive carbon black C45 and stirring the resulting solution in an ultrasonic bath. Then, nanocellulose (NC) was directly added to the AC/C45/CNT mixture and subjected to vacuum filtration. The second method differed in the method of adding NC: it was introduced slowly with constant stirring of the AU/C45/CNT solution on a magnetic stirrer at 600 rpm, after which the resulting suspension was subjected to vacuum filtration. The third method differed in the sequence of adding the components: first, NC was slowly introduced into the solution of dispersed CNTs with constant stirring, after which the mixture was combined with AU and C45 and subjected to ultrasonic treatment for 20 minutes, followed by vacuum filtration. As a result of the study, it was determined that the most uniform morphology of the electrode is achieved using the third method with a content of 5 and 7 ml of NC solution, due to which the optimal electrode ratio is achieved. It was determined that the electrode with the ratio AU(85 mg)/C45(5 mg)/CNT(5 mg)/NC (3 ml) showed lower strength, and the electrode AU(85 mg)/C45(5 mg)/CNT (5 mg)/NC (10 ml) had too many irregularities and lumps caused by a high content of NC. It is worth noting that the other two methods also showed the presence of irregularities and bulges on the surface of the electrodes, which is probably due to imperfect mixing of the components in these methods. According to the project calendar plan, various methods for obtaining electrodes were tested and optimal ratios between nanocellulose and functionalized CNTs were determined in the manufacture of an energy-intensive composite. This is followed by cutting the electrodes to form supercapacitor cells and studying the electrochemical characteristics of the resulting composites.

4.5 Study of Electrochemical Characteristics of Nanocellulose-CNT Composite Material

Electrochemical impedance spectroscopy (EIS) tests were conducted to study the resistance of the materials. The figure shows the Nyquist graph obtained as a result of EIS tests for various samples of activated carbon (AC) with different additives.

Figure 10. Nyquist diagram for various activated carbon samples with additives and modifications

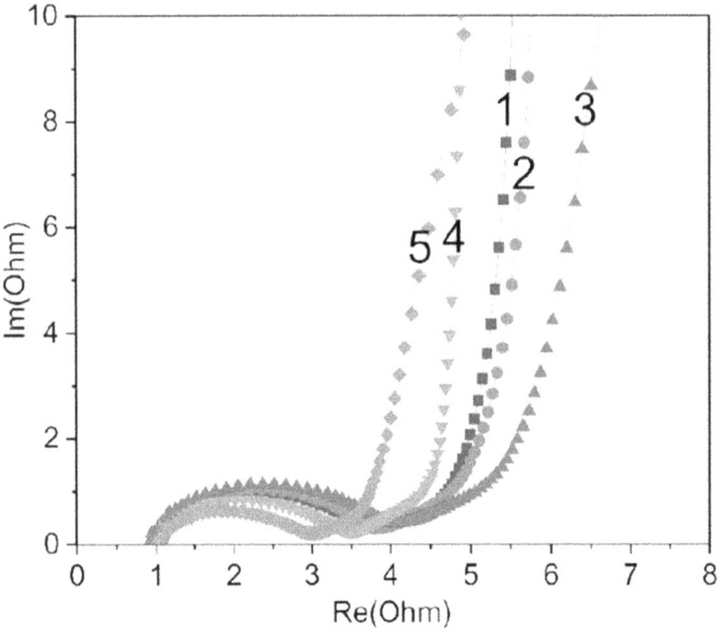

It can be seen from Figure 10 that the equivalent distributed resistance (EDR) varies for different samples. Sample (1) Au(RS)-C45-UNT shows a relatively high impedance value. The intersection of the curve with the Zr axis at high frequencies is in the range from 0.1 to 0.3 Ohm, which indicates a significant resistance of the material, probably due to limited ion diffusion.

Sample (2) Au(AC)-C45-UNT shows lower impedance values compared to Au(RS)-C45-UNT. The intersection of this curve with the Zr axis also indicates improved properties compared to the previous sample, indicating better ion diffusion.

Sample (3) Au(SG)-C45-UNT has an even lower impedance value, which indicates a decrease in material resistance. This is also confirmed by the intersection of the curve with the Zr axis at lower values. Nitrogen-doped sample (4) Au(AC)-C45-Unt-Ni demonstrates similar impedance values to Au(-CO)-C45-Unt, however, its curves indicate the presence of additional effects associated with nitrogen doping. Finally, sample (5) Au(AC)-C45-Unt-Ni, doped with nickel, shows the lowest impedance value among all the presented samples. The intersection of the curve with the Zr axis indicates a significant decrease in resistance, which can be associated with improved conductivity and better ion diffusion as a result of nickel doping.

The mechanism of the influence of nanocellulose on the electrode can be explained as follows: nanocellulose, due to its high surface area and ability to form complex structures, can significantly change the morphology and porosity of the material. In small amounts, nanocellulose can help improve the mechanical properties and stability of the electrode. However, with an increase in the content of nanocellulose, it begins to form dense networks that hinder the effective diffusion of ions and worsen the conductivity.

These results indicate that the addition of nanocellulose to the composition of materials leads to an increase in impedance. The higher the content of nanocellulose, the higher the impedance, which may be due to the deterioration of the conductivity and diffusion properties of ions in the material. These results are important for understanding the effect of additives on the electrical properties of materials and can be useful for further research and optimization of the composition of materials for various electrical and electrochemical applications.

Cyclic voltammetry (CV) tests were carried out to study the electrode properties of various materials. The figure 10 shows the CV curves for various samples of activated carbon (Au) with nickel (Ni) and nitrogen additives, as well as for different starting materials of activated carbon. The graph shows examples of CV curves obtained by scanning the potential from 0 to 1 V. Their almost rectangular shape indicates the behavior of supercapacitors (SC), but it is worth noting the differences between the curves, which indicate the influence of additives and the starting material on the electrode properties. The Au(RS)-C45-UNT sample demonstrates a relatively symmetrical and almost rectangular shape of the CV curve, which indicates good reversibility of electrode processes and low resistance. This sample consists of activated carbon from rice husk and carbon nanotubes (CNT). The Au(AK)-C45-UNT sample shows slightly lower current values compared to Au(RS)-C45-UNT.

Figure 11. CV curves for different samples of activated carbon with nickel (Ni) and nitrogen (N) additives

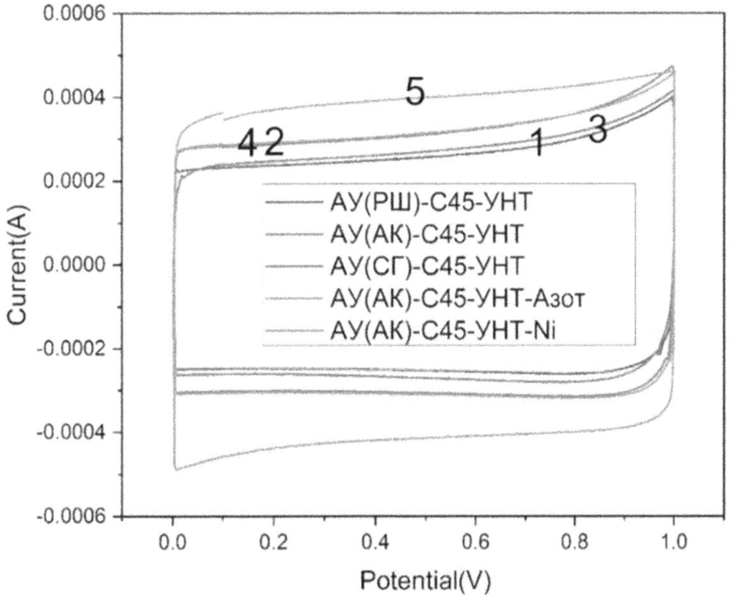

This may indicate slightly higher resistance and less efficient ion diffusion in the apricot kernel material. The Au(CO)-C45-Cnt sample exhibits similar behavior but with slightly lower current values. This may be due to the properties of walnut shell activated carbon, which affects its electrode properties. The Au(AC)-C45-Cnt-Nitrogen sample exhibits slightly improved current values compared to regular Au(AC)-C45-Cnt. This indicates the positive effect of nitrogen doping, which can improve conductivity and electrode performance by introducing additional active sites. The Au(AC)-C45-Cnt-Ni sample exhibits significantly increased current values and a more rectangular CV curve shape, indicating improved electrode properties. Ni doping results in lower resistance and improved conductivity, which contributes to more efficient ion diffusion and an increase in the overall capacitance. These results indicate that different additives and starting materials of activated carbon significantly affect their electrode properties. Doping with nitrogen and nickel leads to improved conductivity and reduced resistance, which improves the overall performance of supercapacitors. This is important to consider when developing and optimizing materials for supercapacitors and other electrochemical devices.

Electrochemical studies of the samples were carried out to study the specific capacity of the obtained electrodes at different currents. Figure 12 shows the dependence of the specific capacity (F/g) on the current (mA/g) for different samples of activated carbon (Au) with carbon nanotubes (CNT) and various additives.

Figure 12. Dependence of the specific capacity of various samples of activated carbon with additives on the current

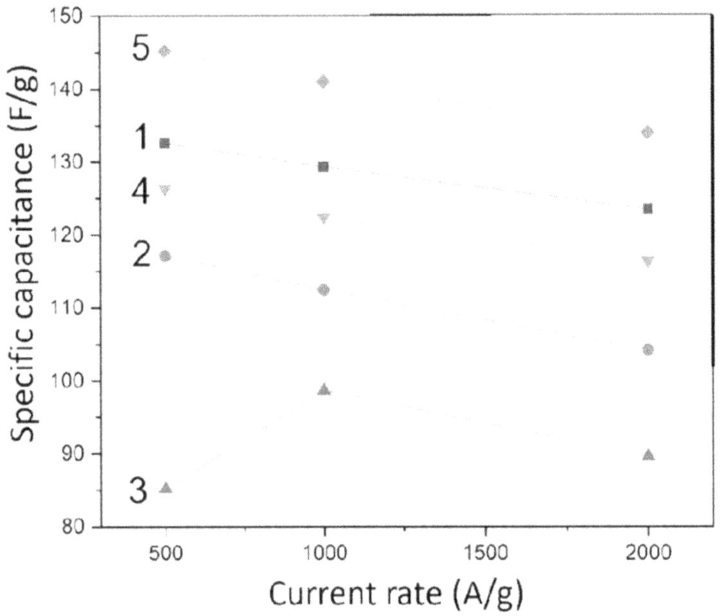

Sample (1) Au(RS)-C45-UNT shows high specific capacitance, which decreases with increasing current. This sample consists of activated carbon from rice husk. At low currents, the specific capacitance is about 130 F/g, indicating good electrode activity. However, as the current increases to 2000 mA/g, the capacitance decreases to about 115 F/g, which can be due to the limitation of ion diffusion at high currents. Sample (2) Au(AK)-C45-UNT shows similar behavior, but with slightly lower specific capacitance values. This sample consists of activated carbon from apricot kernels. At low currents, the specific capacitance is about 120 F/g, and at high currents it decreases to 110 F/g. This also indicates an increase in resistance and limitation of ion diffusion at high currents. Sample (3) Au(SG)-C45-Unt has even lower values of specific capacitance compared to the previous samples. This material consists of activated carbon from walnut shell. At low currents, the specific capacitance is about 90 F/g, and at high currents it decreases to about 85 F/g. This indicates a significant increase in resistance and deterioration of diffusion properties.

Sample (4) Au(AK)-C45-Unt-Ni demonstrates improved values of specific capacitance compared to conventional Au(AK)-C45-Unt. Doping with nitrogen leads to an increase in specific capacitance to 125 F/g at low currents, which indicates improved electrode properties and an increase in the number of active centers. At high currents, the specific capacitance decreases to 110 F/g, which is still better compared to the undoped sample.

Sample (5) Au(AK)-C45-Unt-Ni shows the best results. This sample exhibits the highest specific capacitance, which is about 145 F/g at low currents. Doping with nickel significantly improves the electrode properties, increasing conductivity and reducing resistance. At high currents, the specific capacitance of this sample decreases to 130 F/g, which is still the highest among all the presented samples.

Comparison of different samples shows that nitrogen and nickel doping significantly improves the electrode properties of activated carbon. The addition of nickel has the greatest positive effect, improving conductivity and reducing resistance. These results are important for understanding the effect of different additives on the electrical properties of materials and can be useful in the design and optimization of materials for supercapacitors and other electrochemical devices.

It has been established that additives and modifications significantly affect the electrical properties of materials used in supercapacitors. As a result of the conducted electrochemical impedance spectroscopy (EIS) and cyclic voltammetry (CV) tests, it was revealed that doping with nickel (Ni) significantly improves conductivity and reduces resistance of the materials. Samples containing activated carbon (Au) and carbon nanotubes (Cnt) doped with nickel showed the best electrode properties, demonstrating high specific capacitance and good characteristics of CV curves. It was determined that the addition of nanocellulose (NC) to the electrode material leads to an increase in resistance and deterioration of their electrode properties. With small additions of nanocellulose, some improvement in mechanical stability is observed, however, with an increase in the nanocellulose content, the specific capacitance and conductivity decrease significantly. Nanocellulose forms dense networks that impede the effective movement of ions, which leads to an increase in the overall resistance of the material. Comparison of different samples showed that activated carbon from different sources (rice husk, apricot kernels, walnut shells) also affects the electrical properties, but the greatest improvement is observed when using nickel doping and the addition of carbon nanotubes.

CONCLUSION

The electrochemical performance of activated carbon (AC) derived from rice husk (RH) and walnut shell (WS) using KOH and CO_2 activation methods demonstrates the significant influence of both activation processes and raw material composition on specific capacitance, energy density, and charge transfer behavior. KOH activation consistently produces higher specific capacitances and energy densities, attributable to the substantial increase in surface area, as evidenced by the RH_KOH electrode achieving an energy density of 6 Wh/kg at a power density of 59 W/kg, far surpassing the 1 Wh/kg of the RH_CO_2_2h sample. Similarly, WS_KOH shows better energy storage performance at low scan rates, achieving an energy density of 4.5 Wh/kg, compared to 2.9 Wh/kg for WS_CO_2_2h.

While KOH activation excels in creating high-capacitance electrodes, CO_2 activation offers distinct advantages in rate capability and capacitance retention at elevated scan rates and current densities. RH_CO_2_2h retains 80% of its initial capacitance at high scan rates, in contrast to the 41.68% retention of RH_KOH. Additionally, WS_CO_2_2h exhibits superior ion diffusion and charge transfer behavior, with a lower charge transfer resistance (Rct ~ 0.1 Ohm) compared to WS_KOH (Rct ~ 0.24 Ohm). These properties make CO_2-activated materials, particularly WS_CO_2_2h, more suitable for high-power applications, where mesopores play a critical role in ensuring efficient ion transport and storage under rapid charge-discharge conditions.

The composition of the precursor materials plays a crucial role in their electrochemical performance. Rice husk, with its 15-25% silicon dioxide content, limits the effectiveness of CO_2 activation, resulting in lower specific surface area and capacitance for RH_CO_2_2h. In contrast, walnut shell, with a low ash content (0.1-3.4%), provides a more favorable structure for both activation methods, with WS_CO_2_2h showing enhanced performance at high scan rates and current densities.

All electrodes demonstrate excellent cycling stability, maintaining over 97% of their initial capacitance after 2000 charge-discharge cycles, confirming their durability for long-term energy storage applications.

In summary, KOH activation proves highly effective for enhancing specific capacitance and energy density at low scan rates, while CO_2 activation offers superior rate capability and retention at high scan rates, making it ideal for rapid charge-discharge cycles. The performance differences observed between rice husk and walnut shell emphasize the importance of selecting appropriate raw materials for tailored activation strategies, with walnut shell emerging as a more versatile precursor across both activation techniques.

REFERENCES

Alharbi, H. A., Hameed, B. H., Alotaibi, K. D., Al-Oud, S. S., & Al-Modaihsh, A. S. (2022). Recent methods in the production of activated carbon from date palm residues for the adsorption of textile dyes: A review. *Frontiers in Environmental Science*, 10, 942059. DOI: 10.3389/fenvs.2022.996953

Atamanov, M. K., Amrousse, R., Hori, K., Kolesnikov, B. Y., & Mansurov, Z. A. (2018). Influence of activated carbon on the thermal decomposition of hydroxylammonium nitrate. *Combustion, Explosion, and Shock Waves*, 54(3), 316–324. DOI: 10.1134/S0010508218030085

Goel, C., Mohan, S., & Dinesha, P. (2021). CO2 capture by adsorption on biomass-derived activated char: A review. *The Science of the Total Environment*, 798, 149296. DOI: 10.1016/j.scitotenv.2021.149296 PMID: 34325142

Hermawan, A., Destyorini, F., Hardiansyah, A., Alviani, V. N., Mayangsari, W., Wibisono, , Septiani, N. L. W., Yudianti, R., & Yuliarto, B. (2023). High energy density asymmetric supercapacitors enabled by La-induced defective MnO2 and biomass-derived activated carbon. *Materials Letters*, 351, 135031. DOI: 10.1016/j.matlet.2023.135031

Iijima, S. (1991). Helical microtubules of graphitic carbon. *Nature*, 354(6348), 56–58. DOI: 10.1038/354056a0

Ilyin, Yu. V., Kudaibergenov, K. K., Sharipkhanov, S. D., Mansurov, Z. A., Zhaulybayev, A. A., & Atamanov, M. K. (2023). Surface modifications of CuO doped carbonaceous nanosorbents and their CO2 sorption properties. *Eurasian Chemico-Technological Journal*, 25(1), 33–38. DOI: 10.18321/ectj1493

Jumadi, J., Kamari, A., Hargreaves, J. S. J., & Yusof, N. (2020). A review of nano-based materials used as flocculants for water treatment. *International Journal of Environmental Science and Technology*, 17(7), 3571–3594. DOI: 10.1007/s13762-020-02723-y

Kim, Y. T., Tadai, K., & Mitani, T. (2005). High surface area carbons prepared from saccharides by chemical activation with alkali hydroxides for electric double-layer capacitors. *Journal of Materials Chemistry*, 15, 4914–4921. DOI: 10.1039/b511869g

Lee, J. H., Kang, Y. M., & Roh, K. C. (2023). Enhancing gravimetric and volumetric capacitance in supercapacitors with nanostructured partially graphitic activated carbon. *Electrochemistry Communications*, 154, 107560. DOI: 10.1016/j.elecom.2023.107560

Lesbayev, B., Rakhymzhan, N., Ustayeva, G., Maral, Y., Atamanov, M., Auyelkhankyzy, M., & Zhamash, A. (2024). Preparation of nanoporous carbon from rice husk with improved textural characteristics for hydrogen sorption. *Journal of Composites Science*, 8(2), 74. DOI: 10.3390/jcs8020074

Mann, F. A., Galonska, P., Herrmann, N., & Kruss, S. (2022). Quantum defects as versatile anchors for carbon nanotube functionalization. *Nature Protocols*, 17(3), 727–747. DOI: 10.1038/s41596-021-00663-6 PMID: 35110739

Okayama, R., Amano, Y., & Machida, M. (2010). Effect of nitrogen species on an activated carbon surface on the adsorption of Cu(II) ions from aqueous solution. *Carbon*, 48(10), 3000–3007. DOI: 10.1016/j.carbon.2010.03.040

Pandey, A. P., Shaz, M. A., Sekkar, V., & Tiwari, R. S. (2023). Synergistic effect of CNT bridge formation and spillover mechanism on enhanced hydrogen storage by iron doped carbon aerogel. *International Journal of Hydrogen Energy*, 48(56), 21395–21403. DOI: 10.1016/j.ijhydene.2022.02.076

Ruch, P., Cericola, D., Foelske-Schmitz, A., Koetz, R., & Wokaun, A. (2010). Aging of electrochemical double-layer capacitors with acetonitrile-based electrolyte at elevated voltages. *Electrochimica Acta*, 55(15), 4412–4420. DOI: 10.1016/j.electacta.2010.02.064

Shen, C., Brozena, A. H., & Wang, Y. (2011). Double-walled carbon nanotubes: Challenges and opportunities. *Nanoscale*, 3(2), 503–518. DOI: 10.1039/C0NR00620C PMID: 21042608

Taurbekov, A., Fierro, V., Kuspanov, Z., Abdisattar, A., Atamanova, T., Kaidar, B., Mansurov, Z., & Atamanov, M. (2024). Nanocellulose and carbon nanotube composites: A universal solution for environmental and energy challenges. *Journal of Environmental Chemical Engineering*, 12(5), 113262. DOI: 10.1016/j.jece.2024.113262

Yadav, P., Gupta, S. M., & Sharma, S. K. (2022). A review on stabilization of carbon nanotube nanofluid. *Journal of Thermal Analysis and Calorimetry*, 147(12), 6537–6561. DOI: 10.1007/s10973-021-10999-6

Yergaziyeva, G., Kuspanov, Z., Mambetova, M., Khudaibergenov, N., Makayeva, N., Daulbayev, C. (2024). Advancements in catalytic, photocatalytic, and electrocatalytic CO2 conversion processes: Current trends and future outlook. Journal of CO2 Utilization, 80, 102682. https://doi.org/DOI: 10.1016/j.jcou.2024.102682

Zhao, C., Ge, L., Zuo, M., Mai, L., Chen, S., Li, X., Li, Q., Wang, Y., & Xu, C. (2023). Study on the mechanical strength and iodine adsorption behavior of coal-based activated carbon based on orthogonal experiments. *Energy*, 282, 128450. DOI: 10.1016/j.energy.2023.128450

Chapter 10
Engineering the Invisible:
Synthesis, Characterization, and Applications of Nanoporous Materials

Krishnappa Madhu Kumar
https://orcid.org/0000-0002-7174-9110
Sir M. Visvesvaraya Institute of Technology, Bengaluru, India

Manjappa Kiran Kumar
https://orcid.org/0000-0001-5397-8254
Sir M. Visvesvaraya Institute of Technology, Bengaluru, India

Nagaraj Sasi Kumar
https://orcid.org/0000-0003-3988-4782
Sir M. Visvesvaraya Institute of Technology, Bengaluru, India

Lakshman N. Sampath Kumar
https://orcid.org/0000-0001-5004-6144
Sir M. Visvesvaraya Institute of Technology, Bengaluru, India

ABSTRACT

Nanoporous materials are a major research focus due to their unique structures and diverse applications in catalysis, drug delivery, and environmental remediation. This chapter reviews the synthesis, characterization, and practical applications of nanoporous materials. It examines various synthesis methods, detailing their principles, advantages, and limitations. Key characterization techniques such as scanning electron microscopy (SEM), X-ray diffraction (XRD), transmission electron microscopy (TEM), and atomic force microscopy (AFM) are discussed for assessing important properties like surface area and pore size. The chapter also highlights the performance of nanoporous materials in enhancing reaction rates in catalysis, improving drug delivery systems, and aiding in pollutant degradation. Ideally, this chapter contributes to providing researchers and practitioners with the knowledge necessary to advance the further development and application of nanoporous materials in various technological and industrial areas.

DOI: 10.4018/979-8-3693-7505-1.ch010

1. INTRODUCTION

Nanoporous materials have attracted significant attention in the scientific community because of their unique structural properties and vast range of applications. Characterized by their pore volume and large surface area, these materials can be engineered at the nanoscale to exhibit specific functionalities that are advantageous in several fields including catalysis, drug delivery, and environmental remediation. According to the International Union of Pure and Applied Chemistry (IUPAC), porous materials are categorized into three distinct groups: (a) microporous materials with pore sizes between 0 and 2 nm; (b) mesoporous materials, with pores between 2 and 50 nm; and (c) macroporous materials, characterized by pores larger than 50 nm (Rouquerol et al., 2011; X. S. Zhao, n.d.). Nanoporous materials fall under the broader category of porous materials, featuring bulky porosities and pore diameters ranging between 1 and 100 nm, offer a significant internal surface area, enabling chemical reactions, molecular adsorption, or controlled drug release. But occasionally, materials with pore sizes as small as 1000 nm are referred to as "nanoporous" (Polarz & Smarsly, 2002) . Nano porous materials have much finer, smaller pores compared to the other categories. This diverse spectrum of materials includes carbon-based substances like activated carbon, carbon nanotubes, polymers, zeolites, zeotypes, metal-organic frameworks, and mesoporous materials such as carbons, silicas, and metal oxides (Lu & Zhao, 2004). This chapter explores the multifaceted world of nanoporous materials, exploring their synthesis, characterization, and practical applications. Techniques such as sol-gel processing, hydrothermal synthesis, and template-assisted methods(Kołodziejczak-Radzimska & Jesionowski, 2014) will be discussed in detail, highlighting their principles, advantages, and limitations. For example, the sol-gel process is noted for its ability to produce highly homogeneous materials with controlled porosity, as demonstrated in the synthesis of titanium dioxide nanoporous films (Zha & Roggendorf, 1991; Bokov et al., 2021). Hydrothermal synthesis, which involves crystallization from high-temperature aqueous solutions under pressure, has been effectively employed to produce zeolites with precise pore structures (Byrappa & Haber, 2001; Cundy & Cox, 2005; T. Gupta et al., 2021; Kołodziejczak-Radzimska & Jesionowski, 2014). Template-assisted methods, which utilize hard and soft templates to shape the pore architecture, have shown great promise in fabricating materials with custom-made properties for specific applications (Stein, 2003).

Following the synthesis discussion, the chapter will explore key characterization techniques essential for analyzing the structural and practical properties of nanoporous materials. Methods like SEM, XRD, and gas adsorption measurements will be emphasized for their roles in determining crucial parameters like pore size, surface area and morphology. SEM and TEM provide detailed pore structure images and distribution, as illustrated in studies of mesoporous carbon. Gas adsorption techniques, like Brunauer-Emmett-Teller (BET) method, are crucial for assessing the pore volume and surface area, which are indicative of the material's potential performance in various applications (Ryoo et al., 1999; Haul, 1982; Ravikovitch & Neimark, 2001; Izhar et al., 2022).

The ability to tailor nanoporous materials for particular applications makes them indispensable in modern technology and industry, particularly their performance in catalysis, drug delivery and environmental remediation. In catalysis, for instance, the large surface area and customizable pore structure of nanoporous materials can significantly enhance reaction rates and selectivity(Yu et al., 2013; Taguchi & Schüth, 2005). According to recent research, mesoporous silica materials can be designed to increase the accessibility of active sites, hence enhancing the efficiency of catalytic reactions. (D. Zhao et al., 1998). By enabling the encapsulation and controlled release of therapeutic agents, these materials are revolutionizing drug delivery systems in the medical field. This improves efficacy and decreases side effects. Studies conducted on nanoporous silicon have indicated that it can improve the loading and release characteristics of a range of medications, including anticancer agents (M. Li & Yang, 2023; Salonen et al., 2008; (SLOWING et al., 2008). Furthermore, nanoporous materials play a very significant role in environmental remediation, where their high reactivity and adsorption capacity are utilized to remove pollutants from air and water. Studies on activated carbon and zeolites have highlighted their effectiveness in adsorbing heavy metals and organic pollutants from the contaminated water sources (Foo & Hameed, 2010; Li et al., 2012; Liu et al., 2023). These diverse applications underscore the importance of nanoporous materials and the necessity of understanding their properties and potential.

By integrating insights from synthesis, characterization, and application studies, this chapter aims to equip researchers and practitioners with a comprehensive understanding of nanoporous materials and their potential to advance various technological and industrial domains.

2. SYNTHESIS OF NANOPOROUS MATERIALS

The synthesis of nanoporous materials has emerged as a crucial area of research due to their exceptional properties and broad range of applications. Characterized by their large surface area and well-defined pore structures, these materials play an important role in various fields. In catalysis, they provide an extensive surface area for reactions, enhancing both reaction rates and selectivity. In drug delivery, their ability to encapsulate and release drugs in a regulated manner makes them ideal for targeted therapies. For environmental remediation, nanoporous materials can effectively adsorb pollutants or facilitate the breakdown of harmful substances, contributing to environmental cleanup efforts. Additionally, their unique structure enables high storage capacities and improved performance in energy storage devices such as batteries and supercapacitors. The precise manipulation of pore size, distribution, and surface properties at the nanoscale level allows for the development of materials tailored to specific applications, significantly enhancing their performance and efficiency (S. Bhattacharyya et al., 2014; Jiao Kexinand Flynn, 2016). This section is dedicated to offering an in-depth examination of the prevalent synthesis techniques used for nanoporous materials. We will delve into the core principles underlying each method, highlighting their unique advantages and inherent limitations. Additionally, we will provide examples of nanoporous materials with predefined structure that have been successfully produced using these techniques. Our goal is to provide researchers and practitioners with a thorough understanding of these methods, thereby facilitating the advancement and application of nanoporous materials across different technological and industrial sectors.

2.1. Sol-Gel Process

The sol-gel method is a versatile technique extensively used for the synthesis of various nanomaterials. This process involves the transition of a system from a liquid "sol" (mostly colloidal) into a solid "gel" phase. It provides a straightforward and cost-effective approach to produce materials with controlled porosity, large surface areas, and tailored properties for particular applications. The process generally begins with the hydrolysis and polycondensation of metal alkoxides or metal salts in a solvent like water or alcohol(Feinle et al., 2016; Liao et al., 2013; Owens et al., 2016). This transformation involves several key steps (Figure 1), each contributing to the development and characteristics of the final material:

a) **Hydrolysis:** Metal alkoxides (M(OR)x) or water mixed metal salts. This reaction breaks the metal-oxygen bonds in the precursor molecules and replaces them with metal-hydroxyl bonds:

$M(OR)_x + H_2O \rightarrow M(OH)_x + ROH$

Here, M represents a metal atom, R represents an organic group, OH represents a hydroxyl group, and ROH represents an alcohol.

b) **Polycondensation:** The metal hydroxides (M(OH)x) formed in the hydrolysis step undergo condensation reactions. During condensation, the metal hydroxides lose water molecules and form metal-oxygen-metal (M-O-M) bonds:

$M(OH)_x \rightarrow M\text{-}O\text{-}M + H_2O$

The resulting gel is a three-dimensional network structure known as a gel that entraps the solvent.

c) **Gelation**: The network of metal-oxygen bonds creates a gelatinous structure that traps the solvent used in the process (usually water or alcohol). The gel is a semi-solid material with properties between those of a liquid and a solid.
d) **Drying**: The gel undergoes a drying process to remove the trapped solvent. This step is crucial as it prevents the material from collapsing and helps to retain the porous structure formed during gelation.
e) **Calcination**: Optionally, the dried gel can be subjected to calcination, which involves heating at high temperatures (typically above 500°C). Calcination serves to further remove residual organic components, consolidate the structure, and enhance the material's properties.

Figure 1. Key steps of sol gel process (Bokov et al., 2021)

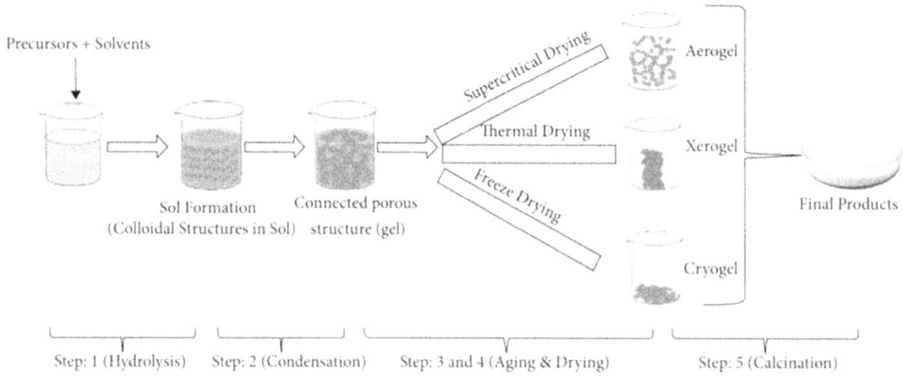

Depending on the process followed after the sol is prepared, either films or powders can be synthesized, as illustrated in Figure 2. To synthesize films using the sol-gel method, the process begins with the preparation of a colloidal sol, as shown in Figure 2(a). This involves selecting an appropriate precursor, such as a metal alkoxide like tetraethyl orthosilicate (TEOS) for silica-based films, and initiating a hydrolysis reaction by adding water. The hydrolysis is typically catalyzed by an acid or base to control the pH and the formation of colloidal particles. To ensure uniform particle size and prevent aggregation, stabilizers or surfactants may be added, resulting in a well-dispersed sol.

Once the sol is prepared, it is deposited onto a substrate, such as glass or silicon, using techniques like dip-coating, spin-coating, or spray-coating. The deposition method chosen depends on the desired film thickness and uniformity. After deposition, the solvent evaporates, leaving a thin gel-like layer on the substrate. This layer undergoes further gelation, where the sol particles condense and form a solid network structure.

Following gelation, the substrate with the gel layer is subjected to heat treatment. Initially, the film is dried to remove residual solvents and organic materials. The final step involves heating at higher temperatures to densify the film, improve its crystallinity (if required), and eliminate any remaining porosity. The end result, as depicted in Figure 2(a), is a thin, uniform film with specific properties tailored by controlling the synthesis parameters.

Figure 2. Schematic representation of Sol-Gel synthesis routes (a) Films synthesized from colloidal sol. (b) Powders synthesized from gel (Niederberger & Pinna, 2009)

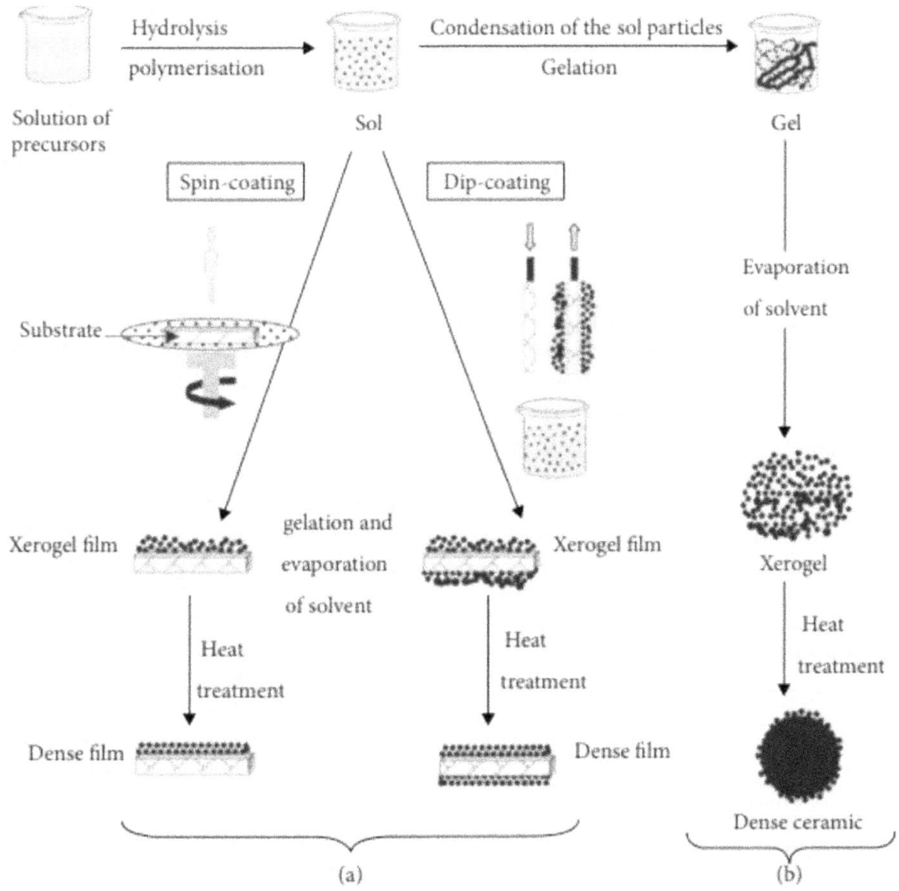

For powder synthesis using the sol-gel method, the process also starts with sol preparation, similar to film synthesis. Figure 2(b) illustrates this process, where a suitable precursor undergoes hydrolysis by adding water and adjusting the pH. As the sol undergoes further polymerization, it transforms into a gel, forming a three-dimensional network structure that encapsulates the solvent within its matrix (Biswas et al., 2001).

Once the gel is formed, it is aged to strengthen its structure. Aging is a critical step to ensure the gel can withstand drying without significant shrinkage or cracking. The gel is subsequently allowed to dry naturally or with the use of regulated techniques such as supercritical drying to extract the solvent without compromising its structural integrity. The dried gel, now known as a xerogel or aerogel, can be further processed by grinding to achieve the anticipated particle size.

The powdered gel undergoes calcination, a heat treatment process where it is heated to high temperatures to remove residual organic components and enhance crystallization. The final product, as depicted in Figure 2(b), is a fine powder with controlled phase purity, particle size, and other suitable properties for various applications, such as ceramics, catalysts, and advanced composites.

In summary, Figure 2 highlights the schematic processes for synthesizing films and powders using sol-gel method. For a variety of industrial and scientific uses, materials with definite properties can be created by meticulously regulating every stage, from sol preparation to heat treatment.

2.1.1 Types of Processes Using the Sol-Gel Method and Their Products

Table 1 and Figure 3 provide a comprehensive summary of the different processes that can be performed using the sol-gel method, detailing the types of products generated, their applications, and a brief description of each process. This overview underscores the versatility and broad applicability of the sol-gel method in creating advanced materials for diverse industrial and research purposes.

Table 1. Overview of sol-gel processes, their products, and applications

Process Type	Products	Applications	Description
Hydrolysis and Polycondensation	Metal oxides (silica, titania, zirconia)	Catalysts, adsorbents, sensors	Formation of metal-oxygen-metal network
Gelation	Gels (silica, alumina)	Insulation, drug delivery, chromatography	Transformation of sol into a semi-solid gel
Drying	Xerogels, aerogels	Insulation, structural components, absorbents	Solvent removal, yielding low-density materials
Calcination	Crystalline oxides (alumina, zirconia)	Catalysis, refractory materials, electronics	High-temperature treatment for crystallization
Dip Coating and Spin Coating	Thin films	Optical devices, corrosion protection, photovoltaics	Coating and heat treatment to form films
Electrospinning	Nanofibers	Filtration, biomedical scaffolds, sensors	Drawing sol into fine fibers using an electric field
Template Synthesis	Mesoporous materials (MCM-41, SBA-15)	Catalysts, drug delivery, adsorption	Using templates to form porous structures
Sol-Gel Synthesis with Organic-Inorganic Hybrids	Hybrid materials (ORMOSILs)	Coatings, optical materials, membranes	Incorporating organic groups into sol-gel network

Figure 3. Schematic diagram of Sol-Gel process types and their corresponding products (Niederberger & Pinna, 2009)

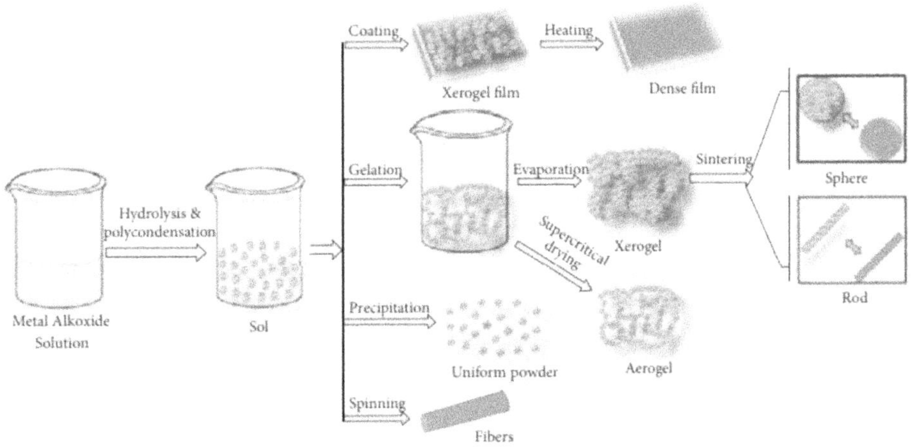

The sol-gel process offers several advantages:

- **Uniformity:** Enables precise control over chemical composition and structural properties, leading to highly uniform materials.
- **Low Temperature:** The process can occur at relatively low temperatures, which is beneficial for incorporating heat-sensitive components.
- **Versatility:** Capable of producing a large variety of materials, including oxides, hybrid organic-inorganic materials, and composites.

However, there are limitations:

- **Shrinkage and Cracking:** Significant shrinkage during drying can lead to cracking and loss of the material's porosity. To reduce these effects, the drying rate and conditions must be properly controlled.
- **Residual Solvents:** Complete removal of residual solvents and organic by-products can be difficult, potentially affecting the purity and performance of the final material. This requires careful management of the drying and calcination conditions.
- **High Temperatures:** Calcination involves high temperatures, which can sometimes cause phase changes or unwanted reactions in the material. Ensuring appropriate temperature control and selection of suitable calcination protocols is essential to maintain the desired material properties.
- **Time-Consuming:** The pace at which hydrolysis, condensation, and drying occur can cause the process to move slowly.

Numerous nanoporous materials synthesized using the sol-gel process demonstrate its broad applicability:

- **Mesoporous Silica (MCM-41 and SBA-15):** These substances are well-known for having large surface areas and ordered pore structures, which make them suitable for uses in drug delivery, adsorption, and catalysis (D. Zhao et al., 1998). Research findings indicate that mesoporous silica materials produced through the sol-gel method display pore diameters between 2 and 50 nm, along with surface areas surpassing 1000 m^2/g.
- **Titanium Dioxide (TiO2):** TiO_2 nanoporous materials synthesized via the sol-gel method are largely used in photocatalysis and solar energy conversion due to their higher surface area and light absorption properties (Zha & Roggendorf, 1991). Studies reveal that sol-gel prepared titanium dioxide nanoporous materials exhibit improved photocatalytic activity with up to 20% solar energy conversion efficiency.
- **Zirconia (ZrO2):** Zirconia nanoporous materials produced through sol-gel processes are applied in catalysis and as supports for metal catalysts, benefiting from their thermal stability and surface properties (Cundy & Cox, 2005).

The sol-gel process remains a cornerstone in the synthesis of nanoporous materials, providing a versatile and controllable route to create materials with custom made properties for specific applications. The dried gel exhibits a number of important characteristics in the SEM image (Figure 4) that highlight the efficiency of the sol-gel method. First, the image's nanoscale porosity shows that the procedure was successful in producing a material with uniformly fine pores. This level of porosity is crucial for applications, where surface interactions are significant, such as in catalysis, where a large surface area is essential for providing numerous active sites.

Figure 4. SEM image of a dried gel after the sol-gel process (Bokov et al., 2021; Niederberger & Pinna, 2009)

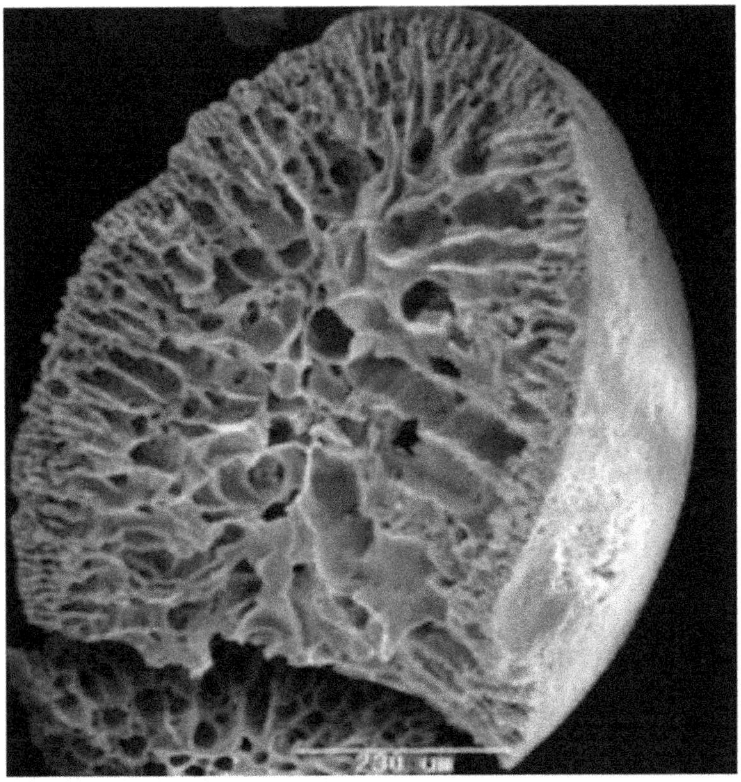

Additionally, the material exhibits a large surface area, which refers to the overall surface area available per unit of mass. This large surface area is advantageous in applications like adsorption, where the material's ability to trap and hold molecules depends heavily on its surface characteristics. By optimizing surface area, the sol-gel process improves the material's functional characteristics and makes it more capable of interacting with its surroundings. The interconnected network structure highlighted in the image is another significant feature. This structure, resulting from the gelation process, ensures that the material has consistent properties throughout. In practical applications, such as filtration or sensing, this uniformity is essential for ensuring that the material performs predictably across its entire surface, with no isolated regions that might diminish its effectiveness.

Finally, the uniform distribution of pores seen in the SEM image is crucial for the consistent performance of the material. In catalytic processes, for instance, evenly distributed pores ensure that reactant molecules can uniformly access active sites, leading to more efficient reactions. This uniformity is a hallmark of the sol-gel method's precision in creating advanced materials.

2.2. Hydrothermal and Solvothermal Methods

The hydrothermal method involves crystallizing substances from high-temperature aqueous solutions at high vapor pressures. It mimics the natural geological processes that occur deep within the Earth where minerals are formed under high temperature and pressure conditions. This technique is particularly suitable for synthesizing materials that are not stable at high temperatures or would decompose under normal conditions (Chen & Whittingham, 2006; S. Yang et al., 2001)

The hydrothermal method involves using water as the reaction medium at elevated temperatures (typically 100-1000°C) and pressures (1-1000 atm) to dissolve and recrystallize materials. The high temperature and pressure conditions inside an autoclave facilitate the nucleation and growth of crystals. This method is particularly effective for materials that require a water-rich environment to form, such as certain oxides and silicates. Hydrothermal synthesis is known for producing high-purity, well-crystallized materials (Byrappa & Haber, 2001).

Hydrothermal synthesis method procedural steps (T. Gupta et al., 2021)

- **Preparation of Precursors:** Begin by preparing the chemical precursors that will be used to form the desired material. These are typically dissolved in water or another suitable solvent. The choice of solvent depends on the nature of the precursors and the final product desired.
- **Autoclaving:** Transfer the prepared solution into a sealed autoclave, which is a strong, pressurized vessel capable of withstanding high temperatures and pressures. The autoclave ensures that the reaction environment remains sealed and controlled.
- **Heating:** Heat the autoclave to the desired temperature, typically above 100°C (the boiling point of water at standard pressure). Higher temperatures can be used depending on the stability of the precursors and the desired reaction kinetics.
- **Cooling:** After the desired reaction time, the autoclave is allowed to cool slowly to room temperature. This gradual cooling helps in the formation of crystalline materials and reduces the likelihood of thermal shock, which could affect the product's structure.
- **Product Collection:** Once cooled, the contents of the autoclave are carefully removed. The synthesized material is often collected by filtration or other separation techniques to separate it from the solvent and any unreacted precursors. The collected material may undergo further washing and drying processes to obtain the final product.

The principles of hydrothermal synthesis are extended to non-aqueous solvents through the solvothermal method. Because of its versatility, a wider variety of materials, including those that are unstable or insoluble in water, can be synthesized. The size, phase, and morphology of the synthesized materials can all be affected by the solvent selection. Common solvents include alcohols, amines, and various organic solvents. The solvothermal method is especially useful for the synthesis of metal-organic frameworks (MOFs), coordination polymers, and other complex materials (Demazeau, 2008).

Solvothermal synthesis method procedural steps

- **Solution Preparation:** Similar to the hydrothermal method, begin by preparing a solution of the chemical precursors. However, in solvothermal synthesis, the solvent used can vary widely and includes organic solvents, water, or a combination of both, depending on the solubility and compatibility of the precursors.

- **Autoclaving:** Transfer the solution containing the precursors into a sealed autoclave. This autoclave may be similar to that used in hydrothermal synthesis but is adapted to handle organic solvents and higher temperatures, as required.
- **Heating:** The principles of hydrothermal synthesis are extended to non-aqueous solvents through the solvothermal method. Because of its versatility, a wider variety of materials, including those that are unstable or insoluble in water, can be synthesized. The size, phase, and morphology of the synthesized materials can all be affected by the solvent selection.
- **Cooling:** Once the reaction time has elapsed, slowly bring the autoclave down to room temperature. The final size and morphology of the synthesized material are influenced by the nucleation and growth of crystals, which are controlled by the controlled cooling rate.
- **Product Isolation:** Once cooled, the contents of the autoclave are processed to isolate the synthesized material. This typically involves filtration or centrifugation to separate the solid product from the remaining solvent and any unreacted precursors. The isolated material may undergo washing and drying steps to remove solvent residues and obtain the final product.

Once the reaction time has elapsed, slowly bring the autoclave down to room temperature. The final size and morphology of the synthesized material are influenced by the nucleation and growth of crystals, which are controlled by the controlled cooling rate. The table 2, outlines the key differences in solvent choice and reaction conditions between hydrothermal and solvothermal methods, emphasizing their respective applications in materials synthesis. Adjustments can be made based on specific experimental requirements and desired outcomes.

Table 2. Comparison of solvent selection and reaction conditions in hydrothermal vs. solvothermal methods

Aspect	Hydrothermal	Solvothermal
Preferred Solvents	Water	Organic solvents (e.g., ethanol, dimethylformamide)
Factors to Consider	Polarity, dielectric constant, ionic strength, pH stability	Polarity, boiling point, viscosity, coordination ability
Reaction Conditions	High temperature (100-300°C), high pressure (up to 200 atm)	Moderate to high temperature (50-300°C), ambient pressure
Typical Applications	Synthesis of metal oxides, zeolites, hydroxyapatite	Nanoparticle synthesis, coordination polymers, metal-organic frameworks
Catalytic Properties	Enhances dissolution and reaction rates	Can act as ligands or catalysts depending on the solvent
Advantages	Control over crystal size and morphology	Enhanced solubility of precursors, control over particle size
Challenges	Requires robust equipment for high-pressure conditions	Solvent choice critical for reaction outcome, solvent removal
Example Reaction	Formation of TiO2 nanoparticles	Synthesis of MOFs (Metal-Organic Frameworks)

In a hydrothermal or solvothermal environment, crystal growth follows a sequence of well-defined stages, beginning with nucleation. Nucleation occurs when the solubility of the solute exceeds its limit in the solution, leading to supersaturation. At this point, the reaction becomes irreversible, and the solute precipitates into small clusters of crystals that can grow to macroscopic sizes. As these nuclei form, the conducive environment—typically characterized by high temperature and pressure—facilitates their growth through the diffusion of atoms or molecules from the surrounding solution into the crystal lattice.

As growth continues, different crystallographic planes of the crystal may develop at varying rates, resulting in distinct shapes. Ostwald ripening also plays an important role during this stage, where smaller, less stable crystals dissolve and redeposit onto larger, more stable ones, refining the crystal's overall size and morphology. The final shape and quality of the crystal are governed by a balance between kinetic factors, such as the rate at which growth units attach to different facets, and thermodynamic factors, like the minimization of surface energy. Precise control over temperature, pressure, and reaction time is essential for achieving desired crystal characteristics in these environments.

Four basic steps comprise the growth unit process: moving the units on the surface, attaching them to the crystal surface, transporting the units through the solution, and finally attaching them to designated growth sites. Figure 5 schematically represents these widely recognized mechanisms of crystal growth via hydrothermal/solvothermal methods.

Figure 5. Schematic representation of crystal growth mechanisms via hydrothermal/solvothermal methods (J. Li et al., 2016)

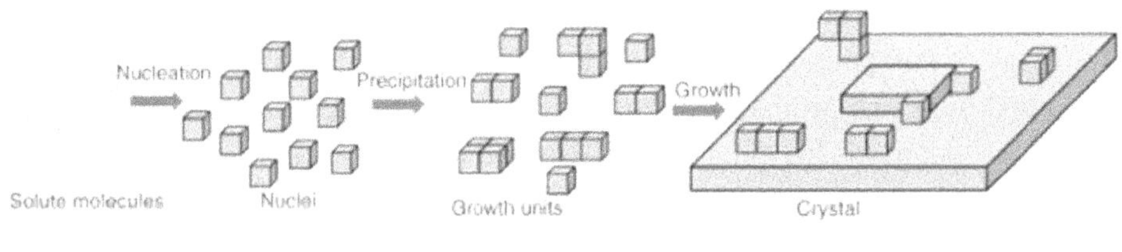

However, there is ongoing debate surrounding the exact mechanisms of crystal growth, particularly concerning the nature of growth units and the controlling steps in the process. For example, Schoeman proposed that the growth unit of a zeolite crystal could be an anionic silicate species, possibly a monomer. Others have suggested that nanoparticles might also serve as growth units for zeolite crystals under hydrothermal conditions. Each kind of growth unit may require specific sequence of governing steps to contribute effectively to crystal growth. (Demazeau, 2008; Hayashi & Hakuta, 2010; J. Li et al., 2016; Y. Xu et al., 2010)

2.2.1 Process Parameters and their Effects

- **Temperature**: Temperature is a critical parameter that affects the solubility of reactants and the kinetics of crystal growth. Greater temperatures usually cause the reactants to become more soluble, which speeds up the nucleation and the growth of larger crystals. For example, at higher temperatures, TiO2 can form well-defined nanorods and nanotubes, which are useful for photocatalytic applications due to their surface area and high aspect ratio (T. Gupta et al., 2021).
- **Pressure**: Pressure plays a significant role, especially in hydrothermal synthesis, by keeping the solvent in a liquid state at higher temperatures than the normal boiling point. In addition to improving reactant solubility, elevated pressures can also result in the formation of more homogeneous and pure crystals. In the synthesis of zeolites, increased pressure can help achieve higher crystallinity and stability (Byrappa & Haber, 2001).

- **Time**: The length of the reaction affects both the size and completeness of the crystallization. Larger, better-defined crystals can usually form with longer reaction times. Overly long reaction times, however, may end up in the formation of secondary phases or excessively large crystals, which may not be desirable in some situations. (Cundy & Cox, 2005).
- **Concentration of Precursors**: The concentration of reactants affects the nucleation rate and the resulting crystals size. Higher concentrations can lead to a higher nucleation rate and the formation of numerous small crystals, while lower concentrations may result in fewer but larger crystals. For instance, in the synthesis of MOFs, materials with different porosities and surface areas can be produced by altering the concentration of metal ions and organic linkers (Férey, 2008).
- **Solvent Choice**: In solvothermal synthesis, the solvent's properties, such as polarity, dielectric constant, and boiling point, significantly influence the reaction environment and the properties of synthesized materials. For example, using ethanol as a solvent can facilitate the formation of ZnO nanostructures with the different morphologies compared to using water (Kołodziejczak-Radzimska & Jesionowski, 2014).

2.2.2 Case Studies of Materials Synthesized using these Methods

Zeolites: These are microporous aluminosilicates and are widely used in adsorption, ion exchange, and catalysis. Their synthesis often requires precise control of crystal size, morphology, and chemical composition, making hydrothermal synthesis an ideal method. Zeolite Y, with its large surface area and strong acidity, is commonly used in fluid catalytic cracking. The synthesis involves the preparation of a gel mixture of sodium aluminate, sodium silicate, and water. This mixture is aged and then subjected to hydrothermal treatment at temperatures ranging from 80°C to 200°C for 12 to 48 hours. Adjusting the temperature and aging time allows control over the crystallinity and particle size of the zeolite. Higher synthesis temperatures increase the rate of crystallization and improve crystal size uniformity. Prolonged aging results in better-formed crystals and higher purity. Zeolite Y synthesized via this method is effective in catalytic applications due to its large surface area and pore volume (Cundy & Cox, 2005).

Metal organic frameworks (MOFs): MOFs are synthesized using the solvothermal method, which allows the coordination of metal ions with organic ligands in non-aqueous solvents. Large surface areas and particular functionalities can be formed in highly porous structures through the use of the solvothermal method. MOFs such as HKUST-1 and MOF-5 have been researched for gas storage, separation, and catalysis. HKUST-1 (Cu-BTC) is a well-known MOF with applications in gas storage and separation. The synthesis involves dissolving copper nitrate and 1,3,5-benzenetricarboxylic acid (BTC) in a mixture of ethanol and water. The solution is then heated in a Teflon-lined autoclave at 85°C for 24 hours. Using a mixture of ethanol and water enhances the solubility of BTC, leading to better crystal formation. The chosen temperature and duration result in high-quality crystals with significant porosity. HKUST-1 shows excellent performance in gas storage, particularly for methane and carbon dioxide, due to its large surface area and pore volume (Férey, 2008).

Titanium Dioxide (TiO_2) nanostructures: TiO_2 nanostructures, including nanoparticles, nanorods, and nanotubes, can be synthesized using the hydrothermal method. The reaction conditions, such as temperature, precursor concentration, and reaction time, can be adjusted to regulate the morphology and size of the TiO_2 nanostructures. TiO_2 nanotubes can be synthesized by treating titanium dioxide powder in an alkaline solution (typically NaOH) at temperatures of 120°C to 200°C. The mixture is placed in a Teflon-lined autoclave and heated for 24 hours. Post-synthesis, the product is washed with water and acid to remove residual sodium and improve purity. Higher NaOH concentrations promote the formation of well-defined nanotubes. Higher synthesis temperatures result in longer and more uniform nanotubes. TiO_2(L. Li et al., 2023; Q. Zhang & Cao, 2011).

ZnO Nanostructures: ZnO nanostructures can be produced by solvothermal and hydrothermal techniques, and these include nanorods, nanowires, and nanoparticles. The morphology and characteristics of ZnO nanostructures can be strongly influenced by the solvent selected and the synthesis conditions. ZnO nanorods can be synthesized by dissolving zinc acetate in the solvent mixture of ethanol and water, followed by the addition of a base (e.g., NaOH). The solution is transferred to a Teflon-lined autoclave and heated at 160°C for 12 hours. The resulting product is washed and dried to obtain ZnO nanorods. The creation of uniform nanorods is facilitated by the use of water and ethanol. Longer nanorod growth is encouraged by higher NaOH concentrations. Because of their superior electrical characteristics and high surface-to-volume ratio, ZnO nanorods produced by solvothermal techniques are useful for gas sensing. Sensors, photocatalysts, and optoelectronic devices are just a few of the uses for these materials (Kołodziejczak-Radzimska & Jesionowski, 2014).

The hydrothermal and solvothermal methods are versatile and powerful techniques for synthesizing nanoporous materials with controlled properties. By carefully adjusting the process parameters such as temperature, pressure, time, precursor concentration, and solvent choice, researchers can modify the morphology, size and phase of the synthesized materials to meet specific application requirements. These techniques have been successfully used to produce a wide range of materials, including zeolites, MOFs, TiO2, and ZnO nanostructures, each with unique properties and applications in catalysis, drug delivery, environmental remediation, and energy storage.

These methods' versatility and efficiency make them indispensable tools in the field of materials science, providing researchers with the means to develop cutting-edge materials with custom-made properties for various technological and industrial applications.

3. TEMPLATE ASSISTED METHODS IN NANOPOROUS MATERIAL SYNTHESIS

Template-assisted synthesis methodologies have become a focal point in nanomaterial research due to their superior ability to regulate the morphology, structure, and size of nanomaterials—factors that are often challenging to achieve through template-free methods. These synthetic approaches harness the unique properties of templates to guide the development of nanostructures, resulting in more consistent and reproducible outcomes.

The principal of the template-assisted synthesis process lies in the “template” itself, which can be any nanostructured entity characterized by specific size, morphology, and surface charge distribution (Prida et al., 2015). These characteristics are crucial as they significantly influence the structure-directing capabilities of the template.

The synthesis process typically starts with the careful preparation of the template, followed by the fabrication of the desired material. This fabrication can be executed using physical methods such as surface coating or through a range of chemical reactions, including addition, elimination, substitution, or isomerization. Once the synthesis is complete, the template may be removed using physical methods like dissolution or chemical methods such as calcination, depending on the requirements of the final material.

Templates used in these processes can be broadly categorized into hard and soft templates. Hard templates, including anodic aluminum oxide membranes, microporous zeolites, mesoporous materials and colloidal silica spheres, offer rigid and well-defined structures. These structures provide precise control over the size and morphology of the resulting nanomaterials, ensuring a high level of structural fidelity. Such rigidity makes hard templates particularly valuable in applications where uniformity in particle size and shape is essential, leading to materials with highly consistent properties (Favacho et al., 2024; D. Li et al., 2014; Luo et al., 2006).

On the other hand, soft templates, such as surfactants or block copolymers, offer more flexibility, allowing for the creation of materials with a broader range of morphologies. These soft templates, while less rigid than hard templates, still provide sufficient guidance to produce well-defined nanostructures.

The major advantage of template-assisted synthesis is its effective control over the structure, dimension, and morphology of the end product. This method has been broadly employed to synthesize various transition metal oxides (TMOs) with diverse morphologies, including one-dimensional (1D), two-dimensional (2D), and three-dimensional (3D) structures. The versatility and precision of this approach make it a preferred choice for creating nanostructured materials with specific, desirable properties. Figure 6 provides a schematic representation of the steps involved in fabricating materials using hard and soft template-assisted methods (Khun et al., 2015; Xiao et al., 2016).

Figure 6. Schematic illustration of the steps involved in fabricating materials using hard and soft template-assisted methods (Poolakkandy & Menamparambath, 2020)

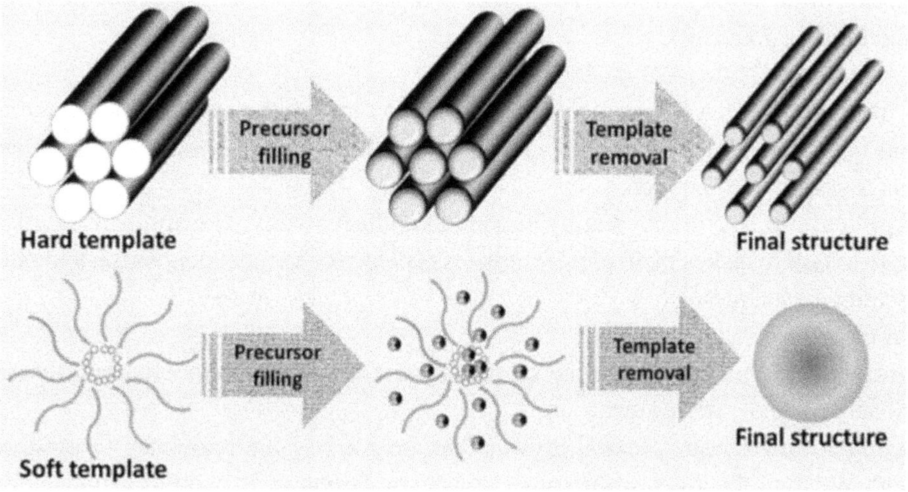

3.1 Hard Templating

Hard templating is a technique used in material science to create nanoporous structures with accurate control over pore size, shape, and distribution. By utilizing a solid template with a predetermined and well-defined architecture, this technique ensures that the resulting material inherits the structural intricacies of the template. The process involves filling the template with a precursor material, which, upon subsequent elimination of the template, results in nanoporous material that mirrors the original template's features. This method is critical in applications requiring uniform pore sizes and highly ordered structures.

3.1.1. Common hard Templates

Hard templates are the cornerstone of the hard templating technique, providing the foundational structure upon which nanoporous materials are developed. Various types of templates, such as anodic aluminum oxide (AAO) membranes, mesoporous silica, and colloidal crystals, are used to achieve different pore characteristics and structural configurations (Kawamura et al., 2014; H.-J. Liu et al., 2010; Platschek et al., 2011). Each type of template has unique properties and applications, making it suitable for specific material synthesis goals. Understanding these templates' characteristics and applications is essential for selecting the appropriate one for a given task.

1. Mesoporous Silica (e.g., SBA-15, MCM-41)
    - **Santa Barbara Amorphous-15 (SBA-15):** Mesoporous silica characterized by hexagonal arrays of uniform mesopores with a pore diameter typically between 5-30 nm. It is synthesized using a triblock copolymer as a structure directing agent in acidic conditions.
    - **Mobil Composition of Matter No. 41 (MCM-41):** Another form of mesoporous silica with a similar hexagonal pore arrangement but usually with smaller pore sizes between 2-10 nm. It is synthesized using a surfactant template under alkaline conditions.
    - **Application:** These materials can be used for synthesizing highly ordered nanoporous materials, often utilized in catalysis, adsorption, and drug delivery due to the large surface area and tunable pore sizes.
2. Anodic Aluminum Oxide (AAO) Membranes
    - **Characteristics:** AAO membranes are created through the anodization of aluminum, leading to the creation of a self-ordered array of hexagonally arranged pores with diameters ranging from 5 to 200 nm.
    - **Application:** Useful for creating nanotubes and nanowires. The uniform and highly ordered pore structure makes them ideal templates for electrochemical deposition and other nanofabrication techniques.
3. Colloidal Crystals
    - **Materials:** Typically composed of polystyrene or silica spheres that self-assemble into a closely packed, ordered array.
    - **Application:** These colloidal crystals can be used as the templates to produce materials with photonic bandgap properties, which are essential in photonics and optoelectronics. When infiltrated and subsequently removed, they create inverse opal structures with periodic porosity.

3.1.2. Synthesis Steps

The synthesis of nanoporous materials through hard templating involves a series of well-defined steps. These steps include preparing the template, infiltrating it with the precursor material, and finally removing the template to unveil the desired nanoporous structure. Each step requires careful execution to ensure that the resulting material maintains the integrity and uniformity of the template's original design. The choice of methods for infiltration and template removal significantly impacts the quality and attributes of the final material. Key steps are as follow;

1. **Template Preparation**: Choose a template based on the desired pore characteristics and final application. For example, mesoporous silica for catalytic supports or AAO membranes for nanowire arrays. Depending on the template, specific synthesis methods are employed, such as sol-gel processes for mesoporous silica or anodization for AAO membranes.
2. **Infiltration**: Introduce the precursor material into the template using sol-gel processes, chemical vapor deposition (CVD), or electrochemical deposition.
    - **Sol-Gel Processes:** A solution containing the precursor is gelled within the template pores, followed by drying and thermal treatment to form the desired material.
    - **Chemical Vapor Deposition (CVD):** Gaseous precursors are deposited onto the template surfaces, where they react and solidify.
    - **Electrochemical Deposition:** Precursor ions are reduced and deposited within the template pores through an electrochemical reaction.
3. **Template Removal**: Remove the template by chemical etching or calcination to reveal the nanoporous structure.
    - **Chemical Etching:** The template is dissolved using appropriate chemical agents that selectively remove the template without damaging the porous structure.
    - **Calcination:** The template is thermally decomposed and removed, often used for organic templates or polymer-based colloidal crystals.

The hard templating method offers numerous advantages, making it a preferred choice for creating highly ordered nanoporous materials. The primary benefits include the ability to exert exact control over pore size and uniformity, which is crucial for the applications that demand consistent and predictable material properties. Additionally, the method enables the production of highly ordered structures, enhancing the material's performance in various applications, from catalysis to drug delivery.

Despite its advantages, hard templating is not without its challenges and limitations. The process of removing the template can be particularly difficult, often leading to potential damage to the delicate nanoporous structure. Moreover, the requirement for high-quality templates can limit the scalability of the technique, making it less feasible for large-scale production. These limitations necessitate careful consideration and innovative solutions to fully leverage the benefits of hard templating in practical applications.

Case Study: Researchers used ordered mesoporous silica (SBA-15) as a template to synthesize nanoporous carbon. The process involved saturation of the silica template with a carbon precursor, followed by carbonization and removal of the template using hydrofluoric acid. The resulting nanoporous carbon exhibited large surface area and well-defined pore structure, making it suitable for applications in supercapacitors and catalysis (Ryoo et al., 1999).

3.2 Soft Templating

Soft templating is a versatile technique in material science used to create nanoporous structures by employing self-assembled soft materials as transient templates. Unlike hard templating, which uses rigid templates, soft templating relies on the ability of certain organic molecules to spontaneously organize into well-defined structures. These soft templates are then replicated by an inorganic precursor material, forming the desired nanoporous structure once the template is removed. Soft templating leverages the self-assembly properties of soft materials, such as surfactants, block copolymers, and biomolecules. These materials can form micelles, vesicles, or other complex structures that serve as temporary molds for the inorganic precursor. The formation of the nanoporous structure involves three main steps: soft template self-assembly, precursor infiltration, and template elimination. This method is flexible and can produce a variety of pore sizes and structures, based on the true nature of the soft material and the conditions under which it self-assembles (Poolakkandy & Menamparambath, 2020; Favacho et al., 2024).

3.2.1. Common Soft Templates

Common soft templates include surfactants, block copolymers, and biomolecules, each offering unique characteristics and advantages for creating nanoporous materials. The choice of template relies on the desired pore size, shape, and the application of final material. By exploring different soft templates, researchers can tailor the properties of nanoporous materials for a broad range of uses.(Poolakkandy & Menamparambath, 2020).

1. **Surfactants (e.g., cetyltrimethylammonium bromide, CTAB)**: These are amphiphilic molecules that can form micelles or liquid crystalline phases in solution. CTAB, for example, can create hexagonal, cubic, or lamellar structures relying on the concentration and conditions. These surfactant structures can be used to template mesoporous materials, commonly employed in catalysis, adsorption, and drug delivery.
2. **Block Copolymers (e.g., Pluronic P123)**: Block copolymers consist of two or more polymer blocks with various chemical properties. Pluronic P123, for instance, has hydrophilic and hydrophobic segments that can be self-assemble into various morphologies such as spheres, cylinders, or gyroids. Block copolymers are used to create mesoporous and macroporous materials with regulated pore sizes and shapes, important in applications like separations, sensors, and energy storage.
3. **Biomolecules (e.g., proteins, DNA)**: Biomolecules can form highly specific and complex structures through biological self-assembly processes. Proteins and DNA can template unique nanoporous structures due to their precise molecular recognition and folding properties. These biomolecule-templated materials are useful in biocompatible devices, biosensors, and bio-inspired materials.

3.2.2. Synthesis Steps

The synthesis of nanoporous materials via soft templating involves a series of carefully controlled steps. First, the soft template is allowed to self-assemble into a defined structure. Next, an inorganic precursor material is introduced to infiltrate the template, filling its intricate network. Finally, the soft template is removed, typically through thermal treatment or solvent extraction, leaving behind a nanoporous structure. Each step in this process is crucial for ensuring the integrity and uniformity of final material, and careful optimization is often required to attain the desired properties.

1. **Self-Assembly**: Under the specific conditions such as pH, temperature, or solvent composition, the soft template molecules self-assemble into a defined structure. This process is driven by the minimization of free energy, leading to organized arrangements such as micelles, vesicles, or liquid crystals.
2. **Infiltration**: The precursor material, often an inorganic sol or solution, is introduced into the self-assembled template. The precursor penetrates the soft template's structure, effectively replicating its morphology. Techniques like sol-gel processing or evaporation-induced self-assembly (EISA) are commonly used in this step.
3. **Template Removal**: The soft template is removed, typically through thermal treatment (calcination) or solvent extraction. Calcination involves heating the material to decompose and burn off the organic template, while solvent extraction dissolves the template away, leaving behind the porous structure.

Soft templating offers several significant advantages over other methods for creating nanoporous materials. Its versatility allows for the production of a variety of pore structures, catering to different applications and requirements. This process is often simpler and less expensive than hard templating, as it does not necessitate the use of rigid, high-quality templates. Additionally, the self-assembly of soft materials can be easily tuned and scaled, making it an attractive option for both research and industrial applications. These benefits make soft templating a valuable tool in the development of advanced materials.

Despite its many advantages, soft templating also has drawbacks that need to be addressed. One of the major challenges is the potential for less ordered pore structures compared to those produced by hard templating. The inherent flexibility and dynamic nature of soft templates can lead to variations in the final material. Moreover, achieving precise control over the nanoporous structure can be difficult, as the self-assembly process is sensitive to various factors such as temperature, pH, and solvent composition. These limitations highlight the need for careful optimization and control during the synthesis process.

Case Study: Synthesis of mesoporous silica using block copolymers: Researchers synthesized mesoporous silica using the block copolymer Pluronic P123 as a soft template. The copolymer self-assembled into micelles, which served as the template for the silica framework. After the silica is formed, the template was removed by calcination. The resulting mesoporous silica (SBA-15) had a highly ordered hexagonal pore structure, making it appropriate for applications in catalysis and drug delivery (Zhao, D., et al., 1998).

3.3 Removal of Templates

Template removal is a crucial step in the fabrication of nanoporous materials using both hard and soft templating methods. This step ensures that the template is effectively eliminated, leaving behind the desired porous structure. The choice of removal method relies on the type of template used and the attributes of the final material. Successful template removal requires careful consideration to avoid damaging the nanoporous framework.

3.3.1. Hard Template Removal

Removing hard templates involves methods that dissolve or decompose the solid template, preserving the integrity of the nanoporous structure. The two primary techniques for hard template removal are chemical etching, and calcination.

- **Chemical Etching**: Involves using strong acids or bases to dissolve the hard template. For instance, hydrofluoric acid (HF) is commonly used to dissolve silica-based templates. This method is effective for removing templates like mesoporous silica (e.g., SBA-15 or MCM-41). The acid selectively dissolves the silica without affecting the formed nanoporous structure. Handling strong acids like HF requires safety precautions due to their corrosive nature. Additionally, complete removal without damaging the porous structure can be challenging.
- **Calcination**: Calcination involves heating the material in the presence of air to decompose the organic template or burn out carbonaceous residues. This method is often used for templates like polystyrene colloidal crystals or AAO membranes. The high temperatures cause the organic template to decompose, leaving behind the inorganic framework. High temperatures can sometimes cause sintering or collapse of the porous structure, especially if the material is not thermally stable.

3.3.2 Soft Template Removal

Soft templates are typically removed through thermal decomposition (calcination) or solvent extraction. These methods are tailored to the organic nature of soft templates, such as surfactants or block copolymers.

- **Thermal decomposition (Calcination)**: Similar to hard template calcination, this method involves heating the material to high temperatures to decompose the soft template. This can be done in air or an inert atmosphere. It is commonly used for removing surfactants or block copolymers from mesoporous materials. The high temperatures break down the organic molecules, leaving behind the porous structure. Care must be taken to avoid excessive heating, to avoid the damage to material's structure. The choice of atmosphere (air or inert) can also influence the decomposition process and final material properties.
- **Solvent extraction**: This process uses appropriate solvents to dissolve and remove the soft template. This method is often used for surfactants or block copolymers that are soluble in specific solvents. Surfactants like CTAB or block copolymers like Pluronic P123 can be removed by dissolving them in solvents such as ethanol or acetone. The solvent selectively extracts the template, preserving the nanoporous structure. Ensuring complete removal of the template can be difficult, and any residual template may affect the material's properties. The choice of solvent and extraction conditions must be optimized for each specific template-material system.

Case Study: Removal of CTAB from Mesoporous Silica

In the synthesis of MCM-41, cetyltrimethylammonium bromide (CTAB) is used as a soft template to form the mesoporous structure. After the inorganic silica framework is formed around the CTAB micelles, the template must be removed to reveal the porous structure. This is achieved by calcination: The mesoporous silica material is heated to 550°C in air. High temperature causes the CTAB surfactant to decompose and burn off, leaving behind a mesoporous silica structure characterized by a large surface area and uniform pore size. This process, described by Kresge, C. T., et al. in 1992, demonstrates the effectiveness of calcination in removing soft templates and highlights the importance of template removal in achieving the desired material properties (Kresge et al., 1992).

3.4 Examples and Applications of Templated Nanoporous Materials

Templated nanoporous materials, created through hard and soft templating methods, have found numerous applications across the different fields due to their unique properties. These materials play vital roles in catalysis, adsorption, drug delivery, sensing, and energy storage. Here, we explore several examples of templated nanoporous materials and their specific applications.

3.4.1. Examples of Templated Nanoporous Materials

1. **Mesoporous Silica (e.g., SBA-15, MCM-41)**: Mesoporous silica materials like SBA-15 and MCM-41 are characterized by their highly ordered pore structures and large surface areas. These materials are synthesized using surfactant templates like CTAB or block copolymers such as Pluronic P123. These materials feature uniform pore sizes, typically in the range of 2-50 nm, high thermal stability, and chemical inertness.
2. **Mesoporous Carbons**: Mesoporous carbons are created using hard templates like silica or soft templates like block copolymers. Their structure is extremely porous and their carbon frameworks are interconnected. Large surface areas, good mechanical strength, and high electrical conductivity are among their attributes.
3. **Metal-Organic Frameworks (MOFs)**: MOFs are a class of crystalline materials composed of metal ions or clusters coordinated to organic ligands, forming porous structures. They can be templated using surfactants or block copolymers to achieve desired pore characteristics. MOFs offer tunable porosity, large surface areas, and the ability to incorporate functional groups for specific applications.
4. **Nanoporous Metals (e.g., Nanoporous Gold)**: These are typically produced by dealloying techniques where one component of an alloy is selectively dissolved, leaving behind a porous metal structure. Hard templates like AAO membranes can also be used. These metals feature a highly porous structure, catalytic activity and good electrical conductivity.
5. **Nanoporous Alumina**: Nanoporous alumina is synthesized using anodic aluminum oxide (AAO) templates. It has a uniform and highly ordered pore structure. It has strong mechanical properties, good chemical resistance, and high thermal stability.

3.4.2. Applications of Templated Nanoporous Materials

1. Mesoporous silica materials (e.g., SBA-15, MCM-41) are commonly used as supports for catalysts in chemical reactions. Their uniform pore structure and large surface area facilitate the effective dispersion of active catalytic species, thereby improving selectivity and reaction rates. These substances are employed in the fine chemical synthesis, environmental cleanup, and petroleum refining processes.
2. Metal-Organic Frameworks (MOFs) are highly effective for gas adsorption and separation because of the tunable pore sizes and functionalizable surfaces. MOFs are used in gas storage (e.g., hydrogen, methane), carbon capture, and purification processes.

Case study: Researchers used block copolymers as templates to synthesize hierarchical MOFs with both micropores and mesopores. The hierarchical structure enhanced the gas storage capacity and selectivity of the MOFs. In particular, these MOFs showed high uptake and selectivity for CO_2 over N_2, making them suitable for carbon capture applications [38].

3. Mesoporous silica nanoparticles are utilized as drug delivery vehicles. Their large surface area and pore volume enable high drug loading, while their biocompatibility ensures safe delivery. These nanoparticles can be engineered for controlled release of drugs, targeting specific tissues or cells, improving therapeutic efficacy, and reducing side effects.

Case study: Mesoporous silica nanoparticles (MSNs) synthesized using CTAB as a template were used for drug delivery applications. The large surface area and pore volume of MSNs allowed for high drug loading capacity. Targeted cancer therapy appears to have potential after in vitro studies revealed the controlled release profiles of the encapsulated drugs (SLOWING et al., 2008).

4. Nanoporous gold is used in biosensors and chemical sensors. Its large surface area and conductive properties enhance sensitivity and detection limits. These sensors can detect biomolecules, gases, and other analytes at very low concentrations, useful in medical diagnostics and environmental monitoring.
5. Mesoporous carbons are employed in supercapacitors and batteries. Their large surface area and conductive framework improve charge storage capacity and cycling stability. These materials are significant for developing high-performance energy storage devices, including lithium-ion batteries and electrochemical capacitors.

Case study: Carbon nanotubes were developed using AAO membranes as templates. The carbon precursor was deposited into the AAO pores via chemical vapor deposition (CVD). The AAO template was subsequently removed using a basic etching solution, resulting in free-standing high aspect ratio carbon nanotubes. These nanotubes demonstrated excellent performance as electrode materials in supercapacitors due to their high conductivity and large surface area (Yu et al., 2013).

6. Nanoporous alumina can be employed for removing contaminants from water and air due to its large surface area and adsorption capacity. It is applied in filtration systems, adsorption of heavy metals, and degradation of organic pollutants.

Table 3. Summery of examples and applications of templated nanoporous materials

Material	Description	Properties	Applications	Example Applications
Mesoporous Silica (e.g., SBA-15, MCM-41)	Highly ordered pore structures, large surface areas; synthesized using surfactant templates like CTAB or block copolymers like Pluronic P123	Uniform pore sizes (2-50 nm), high thermal stability, chemical inertness	Catalysis, adsorption, drug delivery, sensing, energy storage	Petroleum refining, environmental remediation, fine chemical synthesis, controlled drug release, gas sensors
Mesoporous Carbons	Created using hard templates (silica) or soft templates (block copolymers); interconnected carbon frameworks	High electrical conductivity, large surface areas, good mechanical strength	Energy storage, catalysis, adsorption	Supercapacitors, batteries, fuel cells, adsorbents for pollutants
Metal-Organic Frameworks (MOFs)	Crystalline materials with metal ions/clusters coordinated to organic ligands; templated using surfactants or block copolymers	Tunable porosity, large surface areas, functionalizable surfaces	Gas adsorption, separation, catalysis, sensing	Hydrogen storage, carbon capture, gas purification, chemical sensors
Nanoporous Metals (e.g., Nanoporous Gold)	Produced by dealloying techniques or using hard templates (AAO membranes)	Highly porous structure, good electrical conductivity, catalytic activity	Catalysis, sensing, energy storage	Biosensors, chemical sensors, fuel cells, catalysts for chemical reactions
Nanoporous Alumina	Synthesized using anodic aluminum oxide (AAO) templates; uniform and highly ordered pore structure	High thermal stability, chemical resistance, mechanical strength	Adsorption, filtration, catalysis, sensing	Water purification, air filtration, adsorbents for pollutants, catalyst supports

4 ELECTROCHEMICAL SYNTHESIS

Electrochemical synthesis is a precise, versatile, and economical method among the many available approaches for creating nanoporous materials. Electrochemical synthesis involves the use of electric currents to propel molecular processes that lead to the creation of a porous structure. This method allows for the controlled manipulation of the material's microstructure and composition, enabling the creation of nanoporous materials with specific and desired properties (Abel Santos, 2018).

Electrochemical deposition is a highly efficient technique for the preparation of metal nanoparticles, yet it remains less frequently employed compared to traditional wet-chemical methods. While this method may present certain challenges regarding the control of nanomaterial dimensions and the diversity of achievable morphologies, it offers significant advantages that make it a valuable alternative. These advantages include a notably rapid synthesis process, the absence of chemical reductants or oxidants, and the elimination of unwanted by-products, which are common in other synthesis methods (Cioffi et al., 2011).

One of the distinct benefits of electrochemical deposition is the enhanced adhesion it provides when the modifier film is directly deposited onto the electrode surface. This improved adhesion is crucial for the performance and stability of the resulting materials in numerous applications (Scavetta et al., 2014). The technique is versatile, being compatible with several electrochemical methods such as cyclic voltammetry, potential step, and double-pulse deposition, each offering unique control over the deposition process (Domínguez-Domínguez et al., 2008).

A key feature of electrochemical deposition is the ability to precisely control the size of the nanoparticles by adjusting parameters such as current density, applied potential, and the duration of electrolysis. This level of control is essential for tailoring the properties of the nanoparticles for specific applications (Rodríguez-Sánchez et al., 2000). Additionally, when used in conjunction with a template, electrochemical synthesis allows for the creation of complex three-dimensional nanostructures, such as networks formed through mesoporous silica films. An example of this is the fabrication of noble metal nanowires (D. Wang et al., 2004). This approach, therefore, not only broadens the scope of achievable nanostructures but also enhances the functionality and applicability of the materials produced through electrochemical deposition. The technique illustrates the process of forming three-dimensional (3D) continuous macroscopic networks composed of metal or semiconductor nanowires using a templated electrodeposition technique. This technique is broken down into three key stages:

Stage A represents the initial phase where a 3D cubic mesoporous template is employed. This template is critical as it provides the structural framework within which the nanowires will form. The mesoporous structure, characterized by its uniform and interconnected pores, guides the deposition of the nanomaterial, ensuring that the resulting network maintains the desired architecture and connectivity.

Stage B depicts the intermediate product—a 3D nanowire/silica nanocomposite. At this point, the nanowires have begun to form within the confines of the mesoporous template, resulting in a composite material that combines the properties of both the nanowires and the silica template. This composite stage is crucial as it determines the final morphology and connectivity of the nanowire network.

Stage C shows the final outcome, which is a 3D nanowire network. Once the silica template is removed, a continuous network of interconnected nanowires remains. This network is characterized by its macroscopic scale and continuity, making it suitable for a wide range of applications, including in electronics, sensors, and catalysis.

4.1 Mechanism of Electrochemical Deposition

Electrochemical deposition, also known as electrodeposition, is a process where ions in a solution are reduced at the surface of an electrode, leading to the formation of a solid material on the electrode's surface. This process is fundamental in the creation of thin films, coatings, and the synthesis of nanostructured materials, including nanoporous materials. The mechanism of electrochemical deposition involves several key steps, which are governed by electrochemical principles and can be influenced by various factors such as electrolyte composition, temperature, applied potential, and current density (B. Bhattacharyya, 2015; Hu et al., 2021; Ubaidah Saidin et al., n.d.).

1. **Electrolyte Preparation:** The electrolyte is a solution containing the metal ions or other species that will be deposited onto the electrode. The composition of the electrolyte is crucial as it determines the availability of the ions for deposition and can influence the quality and properties of the deposited layer. Common electrolytes include metal salts dissolved in water or organic solvents, which dissociate into metal cations (e.g., Cu^{2+}, Ni^{2+}) and anions (e.g., SO_4^{2-}, Cl^-).
2. **Electrode setup:** In a typical electrochemical deposition setup, two electrodes are used: the working electrode (cathode) where the deposition occurs and the counter electrode (anode). A reference electrode may also be used to control and measure the potential accurately. The working electrode is often made of a conductive material like copper, platinum, or gold, but it can also be an inert or semiconductor substrate, depending on the application.

3. **Ion transport:** An electric potential between the counter electrode and the working electrode causes metal cations to migrate in the direction of the working electrode (cathode). This ion transport occurs through the bulk of the electrolyte via diffusion, migration, and convection processes:
    - **Diffusion:** Movement of ions from regions of high concentration to low concentration.
    - **Migration:** Movement of ions in response to the electric field.
    - **Convection:** Movement of ions due to stirring or natural convection currents in the electrolyte.
4. **Electrode surface reactions:** At the surface of the working electrode, the metal cations undergo a reduction reaction. This reduction is a key step in the electrochemical deposition process:

$Mn^{+} + ne^{-} \rightarrow M$ (Solid)

In this reaction, Mn^{+} represents the metal ion in the electrolyte, ne^{-} represents the number of electrons transferred, and M is the deposited metal on the electrode surface.

- **Nucleation:** Initially, metal atoms are deposited onto the electrode surface, forming nuclei. These nuclei serve as the starting points for further growth. The nucleation process can be instantaneous or progressive, affecting the structure and grain size of the deposited layer.
- **Growth:** Once nucleation sites are established, additional metal atoms are reduced and deposited onto these sites, leading to the growth of a continuous layer or the formation of nanostructures. The growth rate and morphology of the deposited layer can be influenced by the applied potential, current density, and electrolyte composition.

5. **Film formation and morphology:** The morphology of the deposited film, whether it is smooth, rough, dense, or porous, depends on the interplay between the nucleation and growth processes. In nanoporous materials, the deposition conditions are often tailored to promote the formation of voids or pores within the material. This can be achieved by:

- **Template-Assisted Deposition:** Using a template (e.g., anodic aluminum oxide) that guides the deposition into a porous structure (C. Li et al., 2018a).
- **Pulse Electrodeposition:** Applying a pulsed potential or current to create layers with controlled porosity.
- **Additives:** Adding surfactants or complexing agents to the electrolyte to influence the deposition process, such as by inhibiting growth at certain sites to encourage porosity.

6. **Post-deposition treatment:** After the electrochemical deposition, the deposited material may undergo further treatments to enhance its properties or to remove any template used during the deposition. This could involve annealing (heating), chemical etching, or other post-processing steps to refine the structure, remove impurities, or enhance the material's mechanical or electrochemical properties (C. Li et al., 2018a).

4.2 Factors Influencing Electrochemical Deposition (Kozlovskiy et al., 2017; Ubaidah Saidin et al., n.d.)

Several factors can influence the mechanism and outcome of electrochemical deposition, including:

- **Electrolyte composition:** The concentration and type of ions present, as well as the presence of additives or pH regulators.
- **Applied potential/current density:** Determines the driving force for ion reduction and deposition rate.
- **Temperature:** Affects ion mobility, reaction kinetics, and the morphology of the deposited layer.
- **Substrate material and surface condition:** Influences nucleation behavior and adhesion of the deposited layer.
- **Deposition time:** Controls the thickness and overall structure of the deposited film or nanostructure.

4.3 Types of Nanoporous Materials Synthesized

Nanoporous materials can be synthesized from a wide variety of materials, including metals, oxides, semiconductors, polymers, and composites. The synthesis of nanoporous materials spans a broad range of chemical compositions and structures, each tailored to specific applications by controlling the synthesis parameters and methods. Whether metals, oxides, semiconductors, polymers, carbons, or composites, nanoporous materials continue to be at the forefront of advanced materials research, driving innovation in fields ranging from energy storage and catalysis to environmental remediation and biomedical devices. Below are the main types of nanoporous materials commonly synthesized:

4.3.1 Nanoporous Metals

Nanoporous metals are among the most common materials synthesized via electrochemical deposition, and they are valued for their large surface area, conductivity, and catalytic properties. One prominent example is nanoporous gold (np-Au), which is typically synthesized through a process called dealloying. In this method, an alloy of gold and silver is subjected to electrochemical etching, where silver is selectively dissolved, leaving behind a porous gold structure. This material is widely used in catalysis, particularly in reactions where large surface areas and active sites are crucial. Nanoporous gold also finds applications in biosensing, where its biocompatibility and electrical conductivity are leveraged, and in surface-enhanced Raman spectroscopy (SERS), where its plasmonic properties enhance the detection of molecular vibrations ;(Cioffi et al., 2011).

Hierarchical flowerlike gold (Au) microstructures have been successfully synthesized directly on indium tin-oxide (ITO) substrates, notably without the need for any template or surfactant. This synthesis method underscores the efficiency and simplicity of the approach, which leverages the natural tendency of gold nanoplates or nanoprisms to self-assemble into complex microstructures. The resulting Au microstructures, resembling intricate floral patterns, are composed of these gold nanoplates or nanoprisms as fundamental building blocks. The size and morphology of the flowerlike microstructures were found to be highly dependent on the conditions of the electrodeposition process, particularly the deposition time and the applied deposition potential. By carefully controlling these parameters, the diameter of the microstructures could be tailored, demonstrating the versatility of this synthesis technique. The electrodeposition was conducted in a 24.3 mM solution of chloroauric acid (HAuCl4) at an applied potential of 0.5 V versus a silver/silver chloride (Ag/AgCl) reference electrode, over a period of 30 minutes. This specific set of conditions facilitated the growth of well-defined Au microstructures on the ITO substrate (Guo et al., 2007).

Similarly, nanoporous platinum (np-Pt) is another key material, often synthesized through template-assisted electrodeposition. In this process, platinum is deposited into a pre-formed porous template, which is later removed to reveal the nanoporous metal. Nanoporous platinum is extensively used in the fuel cells, where its high catalytic activity is crucial for the efficient conversion of chemical energy into electrical energy. Its resistance to corrosion and large surface area makes it an ideal catalyst in various electrochemical applications, including hydrogen evolution and oxygen reduction reactions (Adane et al., 2024; C. Li et al., 2018b; J. Wang et al., 2008).

Nickel and copper are also synthesized into nanoporous forms through electrochemical deposition, often via template methods or dealloying of alloys like nickel-aluminum or copper-zinc. Nanoporous nickel (np-Ni) is employed in hydrogen evolution reactions (HER) and as an electrode material in batteries and supercapacitors, where its large surface area improves charge storage capacity and reaction kinetics. Nanoporous copper (np-Cu), on the other hand, is utilized in catalysis and as current collectors in energy storage devices. Its excellent electrical conductivity and the ability to form stable, high-surface-area structures make it a valuable material in these applications (C. Li et al., 2018b; Ubaidah Saidin et al., n.d.).

4.3.2 Nanoporous Oxides

Nanoporous oxides are distinguished by their unique structures and properties, achieved by reducing a porous framework down to nanoscale. These materials have an intricate web of interconnected voids or pores, usually with sizes between a few and several hundred nanometers. The nanoscale reduction of these structures imparts them with a significantly increased surface area and distinctive physical and chemical characteristics.

Nanoporous oxides can be categorized into two primary structural types. Oxides that have internal pores in their bulk material by nature fall into the first category. Porous materials are divided into three distinct groups based on pore size in the IUPAC classification system: microporous materials, with pore diameters of less than 2 nanometers; mesoporous materials, with pore diameters ranging from 2 to 50 nanometers; and macroporous materials, with pore diameters greater than 50 nm (Thommes et al., 2015). These pores can be arranged in either a random or highly ordered manner, leading to the formation of various oxide structures characterized by a substantial surface area.

The second category of nanoporous oxides includes structures where pores are created within the matrix's volume due to nanostructured architectures. This category is broader and includes a variety of morphological forms such as nanoparticles (Knecht & Hutchison, 2023), nanowires (Košiček et al., 2022), nanotubes (Du et al., 2007), and nanosheets (Park et al., 2019; Zhu et al., 2011). These nanostructured oxides can be further diversified into distinct configurations, which enhance the surface-to-volume ratio significantly. This increase in surface area is particularly beneficial in applications where a large surface area is crucial, such as catalysis, sensing, and energy storage.

Moreover, the nanoscale porosity and the diverse morphologies of these oxides contribute to the efficient transport of mediators, such as charge carriers and photons, by providing accelerated pathways. This characteristic makes nanoporous oxides highly advantageous for applications in fields requiring rapid charge transfer, effective light absorption, or enhanced chemical reactivity.

Similarly. nanoporous titanium dioxide (np-TiO_2) can be created by depositing titanium onto an electrode followed by anodization or oxidation, resulting in a porous oxide layer. This material is widely recognized for its photocatalytic properties, making it a key component in dye-sensitized solar cells (DSSCs) and in environmental applications such as water purification and air cleaning. The ability of

TiO_2 to generate reactive oxygen species under UV light is harnessed in these applications to degrade organic pollutants and bacteria (J. H. Park et al., 2019; Pauline et al., 2024).

Zinc oxide (ZnO) is another oxide that can be synthesized into nanoporous structures via electrochemical deposition. By controlling the deposition conditions, such as pH, temperature, and potential, ZnO can be formed with a large surface area and a porous morphology. Nanoporous ZnO is used in gas sensors and photodetectors due to its semiconducting properties, which are enhanced by the large surface area provided by the nanoporous structure. Additionally, ZnO's ability to absorb light and generate electron-hole pairs makes it an effective photocatalyst for processes like water splitting, where it facilitates the conversion of solar energy into chemical energy (Bai et al., 2013; Zhuang et al., 2023).

4.3.3 Nanoporous Semiconductors

Electrochemical deposition is a crucial technique in the synthesis of nanoporous semiconductors, materials that are essential in electronics and optoelectronics. Nanoporous silicon (np-Si) is one of the most well-known examples, typically created through electrochemical etching in a fluoride-containing solution. The resulting material has a large surface area and tunable pore sizes which are exploited in applications ranging from photonics to sensors. In lithium-ion batteries, nanoporous silicon is employed as an anode material because of its ability to accommodate large amounts of lithium ions, significantly enhancing the battery's capacity and cycle life. Nanoporous cadmium sulfide (np-CdS) can also be synthesized through electrochemical deposition, often by depositing CdS into a porous template or directly onto a substrate. The resulting material has excellent optoelectronic properties, making it suitable for use in photocatalysis, particularly in processes like water splitting and CO_2 reduction. Furthermore, np-CdS is utilized in solar cells, where its broad absorption of light increases the efficiency of solar radiation conversion to electrical energy (J. H. Park et al., 2019).

4.3.4 Nanoporous Polymers and Polymer-Metal Composites

While metals and oxides are commonly associated with electrochemical deposition, nanoporous polymers and polymer-metal composites can also be synthesized using this technique. Nanoporous polyaniline (np-PANI) is one example, where the polymerization of aniline monomers is initiated electrochemically, often in the presence of a template that guides the formation of a porous structure. Nanoporous PANI is used in sensors, where its conductive properties change in response to environmental stimuli, making it an effective material for detecting gases, pH changes, and other chemical species. Additionally, PANI's large surface area and electrical conductivity make it a suitable material for supercapacitors, where it serves as an electrode material with a large energy storage capacity.

Nanoporous polymer-metal composites, where metal nanoparticles are electrochemically deposited within a polymer matrix, represent another class of materials with unique properties. These composites combine the flexibility and processability of polymers with the metals catalytic or conductive properties. They are used in a variety of applications, including catalysis, where the metal nanoparticles provide active sites for chemical reactions, and in electronic devices, where the composite material can be engineered to have specific electrical or optical properties.

4.3.5 Nanoporous Carbon Materials

Although nanoporous carbons are synthesized through pyrolysis and activation processes, electrochemical deposition can also be utilised to create hybrid nanoporous carbon materials, particularly in composite forms. For example, metal nanoparticles can be electrochemically deposited onto a carbon substrate, leading to a nanoporous carbon-metal composite. These materials are highly valuable in energy storage devices, such as batteries and supercapacitors, where the carbon provides a conductive framework that supports the electrochemical activity of the metal nanoparticles. The large surface area of the nanoporous structure enhances the material's ability to store charge, while the metal nanoparticles improve conductivity and catalytic activity, making these composites highly efficient in their respective applications (Asimakopoulos et al., 2020).

4.4 Applications of Electrochemical Deposition (Ubaidah Saidin et al., n.d.)

Electrochemical deposition is commonly used in various industries and research fields. Some key applications include:

- **Microelectronics:** Fabrication of conductive pathways and interconnects in semiconductor devices.
- **Coatings:** Protective or decorative coatings on metals, such as gold plating or corrosion-resistant layers.
- **Nanotechnology:** Synthesis of nanowires, nanodots, and nanoporous materials for catalysis, sensors, and energy storage.
- **Energy storage:** Production of battery electrodes with tailored porosity for enhanced performance.

5 CHARACTERIZATION OF NANOPOROUS MATERIALS

These advanced materials have garnered significant attention in recent years due to their unique characteristics. These characteristics make them suitable for a variety of applications, including catalysis, gas storage, drug delivery, and separation technologies (Z. Chen et al., 2020; Zanco et al., 2017). However, to fully exploit the potential of these materials, a thorough understanding of their structural properties is essential. Structural characterization provides critical insights into the internal architecture of nanoporous materials, revealing details about their crystallinity, morphology, pore structure, and surface properties. These characteristics are directly linked to the material's performance in various applications, making structural characterization a fundamental step in the development and optimization of nanoporous materials (Thommes & Schlumberger, 2021).

This section highlights the key techniques used for the structural characterization of nanoporous materials, namely X-ray Diffraction (XRD), Transmission Electron Microscopy (TEM), Scanning Electron Microscopy (SEM), and Atomic Force Microscopy (AFM). Each of these methods offers unique and complementary information about the material's structure, from atomic-scale imaging to surface topography and crystallographic analysis. By using these techniques, researchers can gain a comprehensive understanding of the structural features that define the behavior and functionality of nanoporous materials, enabling them to tailor these materials for specific industrial and technological applications.

5.1 X-ray Diffraction (XRD)

Determining the crystalline structure of materials can be accomplished with great effectiveness and regularity using this technique. It works by directing a beam of X-rays onto a material and measuring the angles and intensities of the X-rays that are scattered by the atoms within the material. Resulting diffraction pattern is unique to the arrangement of atoms in the crystal lattice, allowing researchers to identify the phases present, determine the crystallographic structure, and estimate the size of crystalline domains. XRD is particularly valuable in the study of nanoporous materials because it can reveal information about the overall crystallinity of the sample, detect any phase impurities, and provide insights into the arrangement of pores within the material. This technique is essential for confirming the success of material synthesis and for understanding how the crystalline structure impacts the material's properties and performance (Sakamoto et al., 1998; Z. Liu et al., 2013; Suga et al., 2014; Ma et al., 2019).

X-ray powder diffraction (XRPD) and single-crystal X-ray diffraction (SC-XRD) are two non-destructive methods for examining the long-range order of crystal structures. These methods provide significant information on crystal structures, phase compositions, and microstructural details such as crystallite size, space groups, lattice parameters, and micro strain. However, many porous materials, including metal-organic frameworks (MOFs), covalent organic frameworks (COFs), and zeolites, are normally synthesized as polycrystalline powders, which limits the use of SC-XRD for structural determination. Consequently, XRPD is the primary method employed for structural investigations in these materials (Huang, Grape, et al., 2021; Huang, Willhammar, et al., 2021).

Converting diffraction data into 1D patterns is one of the challenges associated with XRPD. Structural analysis is difficult because of this and other issues such as bulky unit cell parameters, overlapping reflections, and frequently disordered porous structures. However, using high-resolution data, it is possible to solve structures and improve crystal structures. The study of nanoporous compounds can benefit greatly from synchrotron radiation's high brilliance, horizontal polarization, excellent signal-to-noise ratio, and faster acquisition times (Lo et al., 2018; Petersen & Weidenthaler, 2022).

When used in conjunction with other analytical methods, XRPD is especially helpful in determining the relationships between structural property and performance, particularly when conducting operando or in situ experiments in a variety of sample settings. For instance, combining in situ XRPD with gas adsorption–desorption measurements can offer vital information about the sorption processes inside of specific pores as well as the spatial distribution of gaseous species. In these kinds of experiments, heatable gas lines are required to prevent condensation of the gas or vapor being investigated (Lo et al., 2018; Bon et al., 2020).

An illustrative case is the investigation of "gate opening" mechanism in DUT-8 (Ni) during the adsorption of N2 and n-butane. This was monitored through a combination of Extended X-ray Absorption Fine Structure (EXAFS) spectroscopy and in situ XRPD. The results demonstrated a transition from the closed pore (cp) structure to a larger pore (lp) configuration, which was identifiable through distinct (100) and (110) reflections associated with the cp and lp phases, respectively, as depicted in Figures 7a and 7b. During gas adsorption, a significant deformation of the Ni-paddle wheel and the surrounding secondary building units (SBUs) caused a 254% expansion of the unit cell. In the lp structure, the N–Ni–Ni–N atoms and dabco molecules are arranged on a same plane (Figure 7c), while in the cp structure, they form a zigzag pattern (Figure 7d). Additionally, the interplanar angles between the carboxylates and the Ni paddle-wheel change dramatically from 3.73° in the cp structure (Figure 7f) to 47.59° in the lp structure (Figure 7e) (Bon et al., 2015, 2020).

Figure 7. XRPD analysis of DUT-8(Ni) showing the transition from closed pore (cp) structure to large pore (lp) structure during gas adsorption (Bon et al., 2015)

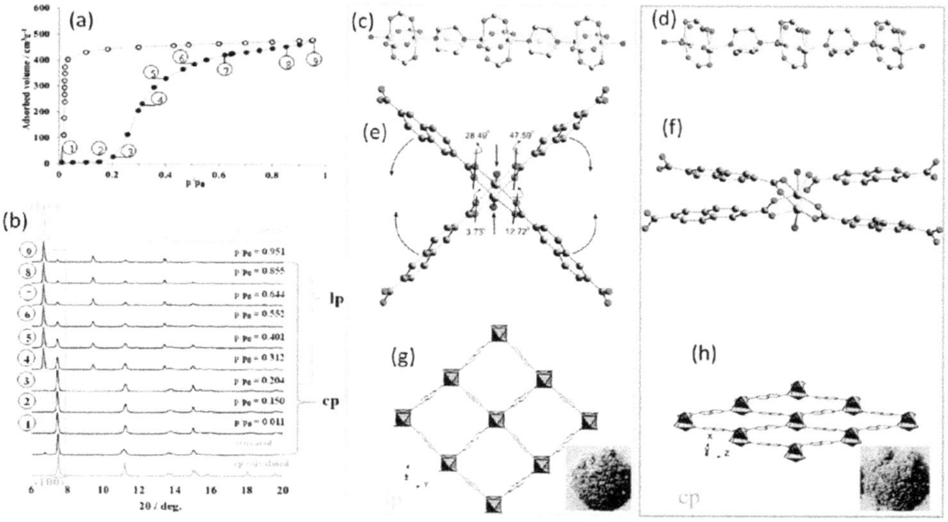

Another significant finding involved the study of DUT-49, where Krause et al. observed "negative gas adsorption" (NGA), an unexpected phenomenon where gas molecules like methane and n-butane spontaneously desorb within a defined temperature and pressure range, accompanied by structural deformation and pore contraction (Krause et al., 2016). Furthermore, operando studies combining XRPD with UV-vis spectroscopy have been utilized to investigate zeolitic catalysts featuring CHA, DDR, and LEV topologies during the methanol-to-olefins conversion process. These studies employed operando UV-vis spectroscopy to monitor hydrocarbon formation, while operando XRPD was used to observe lattice expansion as the zeolitic framework adapted to the formation of hydrocarbons (Goetze et al., 2018).

5.2 Transmission Electron Microscopy (TEM)

Transmission Electron Microscopy (TEM) is an advanced imaging technique that delivers incredibly detailed images at the atomic or molecular scale. Unlike light microscopy, which is limited by the wavelength of visible light, TEM uses a beam of electrons that pass through an ultra-thin sample. The electrons interact with the atoms in the sample, producing a highly magnified image or diffraction pattern that can be analyzed to study the materials internal structure. In the context of nanoporous materials, TEM is particularly useful for visualizing the fine details of the pore structure, such as the size, shape, and distribution of pores. It also allows researchers to observe defects, dislocations, and grain boundaries that can affect the material's properties. TEM is indispensable for providing a direct view of the nanostructure, which is highly important for tailoring materials for particular applications (Sakamoto et al., 1998; Suga et al., 2014).

Over the past decade, transmission electron microscopy (TEM) has undergone significant technological improvements, including enhanced corrections for spherical and chromatic aberrations, improved performance of electron gun, upgraded detectors, and more sophisticated imaging algorithms. These progressions have greatly improved the energy and spatial resolution capabilities of TEM, making it possible to observe finer details in materials at the atomic level. These improvements have proven especially valuable in the study of porous materials, which are often challenging to analyze due to their complex structures and small crystallite sizes (Q. Chen et al., 2020; C. Zhang et al., 2020; Wu et al., 2022).

Traditional methods, like single-crystal X-ray diffraction, often fall short when analyzing polycrystalline porous materials such as zeolites, metal-organic frameworks (MOFs), and covalent organic frameworks (COFs). These materials have small crystallite sizes and complex structures, which complicate their analysis using conventional X-ray techniques. Additionally, the disordered nature of these materials adds another layer of difficulty. Electrons, because of their strong Coulomb interactions with matter, can yield comprehensive structural data from even small sample sizes. However, these interactions also pose challenges, particularly with beam-sensitive materials. The strong electron interactions can provoke structural changes, such as atomic displacements or electronic excitations, which can alter the material under investigation (Q. Chen et al., 2020b; Huang, Grape, et al., 2021; Huang, Willhammar, et al., 2021). To address the issue of beam damage, techniques like cryo-TEM and environmental TEM (ETEM) have been developed. Cryo-TEM (Cryo-Transmission Electron Microscopy) involves imaging at low temperatures to minimize radiolytic and thermal damage to sensitive materials. This technique has proven effective in real-time monitoring of processes like zeolite crystallization and growth (Refer figure 8) (Jasim et al., 2021). For example, Cryo-TEM was successfully utilized to observe the crystallization and growth of zeolites by freezing a small portion of the synthesis solution in ethane and imaging it at regular intervals. This approach allowed to capture snapshots of the synthesis process over time (refer figure 8). Additionally, Cryo-TEM was used to study the formation of hollow zeolite crystals. The technique documented how multiple pores in ZSM-5-like compounds merge into a single cavity during the desilication process, as illustrated in Figure 8 (T. Li et al., 2019).

Figure 8. Cryo-TEM imaging of zeolite formation and hollow crystal development in ZSM-5 (T. Li et al., 2019)

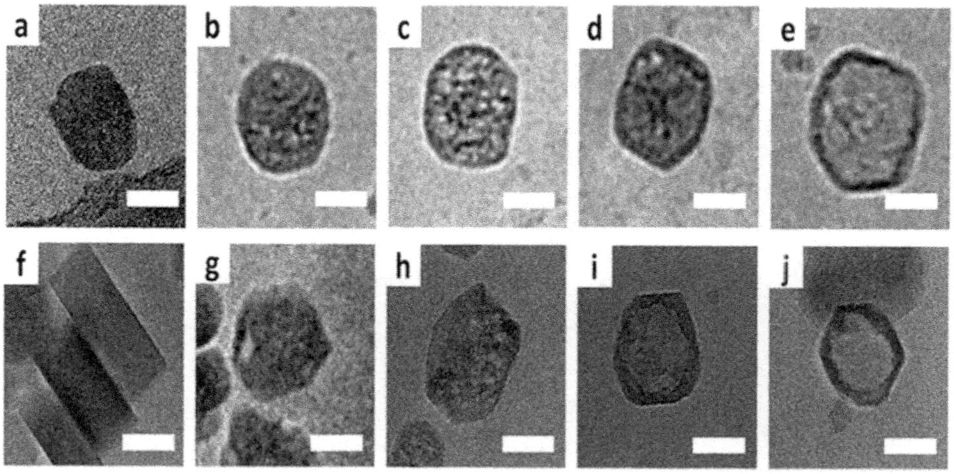

Modern TEM instruments, often equipped with advanced sample holders, facilitate both in situ and operando studies. These setups enable researchers to examine not just the overall structure of materials but also the intricate relationships between local structures and their properties under different conditions, such as varying temperatures or heating. For example, by integrating Environmental TEM (ETEM), electron diffraction (ED), and molecular dynamics (MD) simulations, researchers can directly observe the changes in a single MIL-53(Cr) nanocrystal during water adsorption and desorption (refer to Figure 9). The investigation uncovered notable structural transformations, including shifts in symmetry in response to different water vapor pressures and temperatures (Parent et al., 2017).

Figure 9. In situ ETEM analysis of structural breathing in MIL-53(Cr) nanocrystals under varying environmental conditions (Parent et al., 2017)

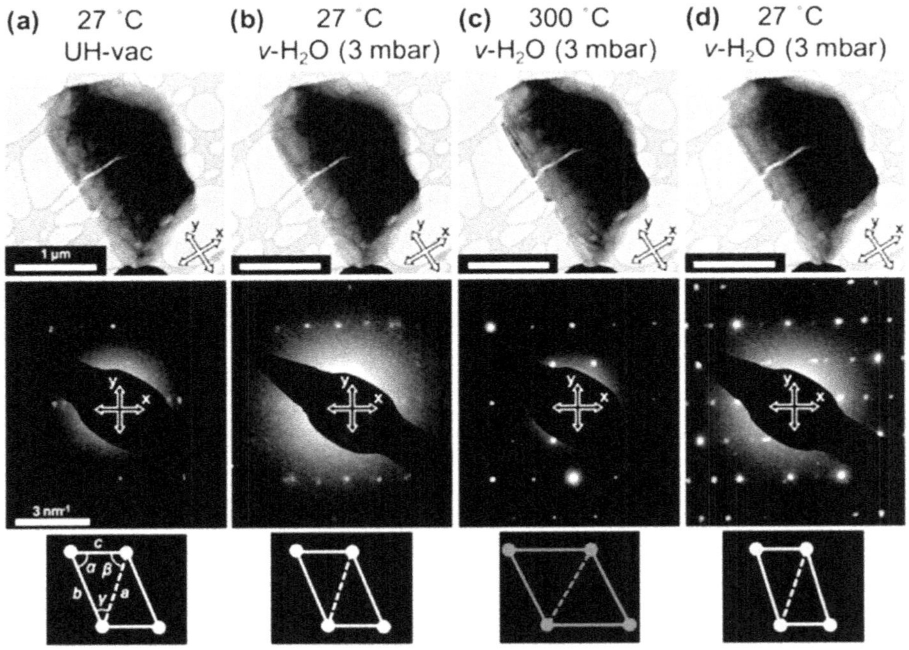

The figure 9 presents a series of in situ Environmental TEM (ETEM) images and corresponding diffraction patterns of a MIL-53(Cr) nanocrystal under various conditions, illustrating its structural changes during a "breathing" process. The top row shows the nanocrystal's physical appearance at four different states: at 27 °C in ultra-high vacuum (UH-vacuum), at 27 °C with water vapor, during annealing to 300 °C in water vapor, and after cooling back to 27 °C in water vapor. The middle row features the electron diffraction patterns obtained at each stage, revealing the changes in the crystal's internal structure. The bottom row illustrates the corresponding lattice parameters (a, b, c) and angles (α, β, γ), which quantify these structural changes. This figure demonstrates the dynamic nature of MIL-53(Cr) nanocrystals, showing how their structure adapts to different environmental conditions, a characteristic that is vital for applications in areas like gas storage and catalysis.

In situ liquid cell transmission electron microscopy (LC-TEM) is a powerful technique for studying beam-sensitive porous materials by allowing fluid flow through the TEM holder. This holder, featuring electron-transparent SixNy windows, can be equipped with heating and electrochemical measurement capabilities. For instance, Patterson et al. utilized LC-TEM to observe the real-time growth of metal-organic frameworks (MOFs), specifically ZIF-8. By imaging the particles during growth, they were able to measure the growth rate directly and determine the crystal structure of the product through electron diffraction (ED). Their study showed that the electron beam had no significant impact on crystal growth, as evidenced by comparisons with similar systems under standard conditions. This allowed them to attribute variations in particle growth and size to the influence of metal precursors and ligands (Figure 10).

Figure 10. In situ LC-TEM analysis of growth and structural characteristics of ZIF-8 particles (Patterson et al., 2015)

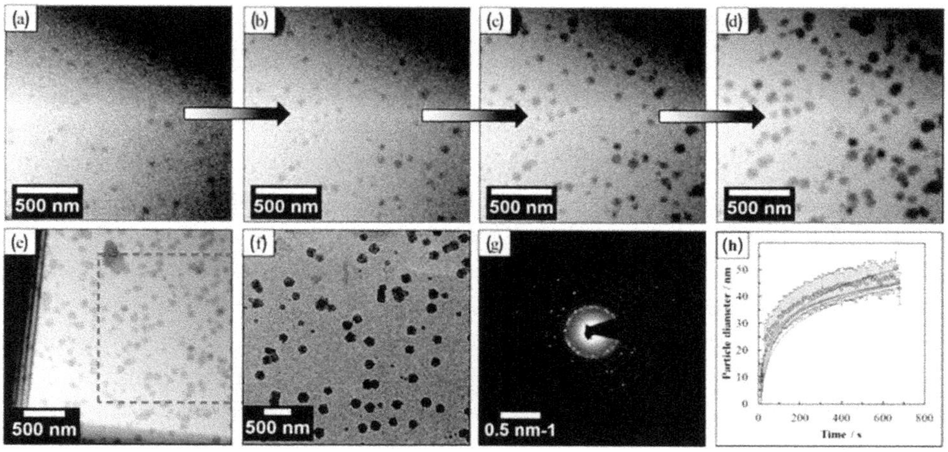

The figure 10 illustrates the in-situ Liquid-Cell Transmission Electron Microscopy (LC-TEM) analysis of ZIF-8 particles, highlighting their growth and structural characteristics. Panels (a) to (d) display snapshots capturing the progressive growth of individual ZIF-8 particles over time. Panel (e) shows an image of the particles in solution, while panel (f) presents the particles after drying, with the red box indicating the specific area observed in panels (a) to (d). Panel (g) includes electron diffraction pattern of the dried particles in the selected area, providing insight into their crystallographic structure. Finally, panel (h) presents a plot depicting the mean growth kinetics of individual ZIF-8 particles, illustrating how these particles evolve over time. This figure underscores the dynamic process of ZIF-8 particle formation and growth, as well as their structural properties both in solution and after drying.

The advancement of three-dimensional electron diffraction (3DED) techniques has created new opportunities for the study of materials with nanoporous structure. These methods allow for the determination of unit cells and space groups from much smaller crystal entities compared to X-ray diffraction. The ability to rotate crystals stepwise during the process of data collection has been particularly groundbreaking, enabling more precise and faster measurements (Gruene et al., 2021). Despite the benefits of 3DED, electron beam sensitivity remains a concern, especially for porous materials. Continuous rotation data collection, combined with low electron dose rates and fast detectors, has been shown to mitigate some of these challenges, allowing for high-quality data acquisition even from beam-sensitive samples.

The continuous advancements in TEM technology, particularly in aberration corrections, imaging algorithms, and the development of specialized techniques like cryo-TEM, ETEM, and 3DED, have expanded our ability to analyze and understand the structure of complex porous materials. These methods not only make it possible to thoroughly examine materials that were previously challenging to study, but they also offer fresh perspectives on the dynamic processes that take place at the atomic level within these materials. As TEM technology continues to evolve, it is likely to play an increasingly important role in materials science and nanotechnology research.

5.3 Scanning Electron Microscopy (SEM)

It is a versatile technique that provides high-resolution images of the surface of materials. By focusing an electron beam on a sample surface and allowing it to travel across it, SEM creates images by detecting secondary electron emissions (Suga et al., 2014). This technique is particularly useful for examining the surface morphology and topography of nanoporous materials. SEM can reveal the detailed information about the surface structure, including the size and distribution of pores, the roughness of the surface, and the presence of any surface defects or irregularities. Additionally, when equipped with an energy-dispersive X-ray spectroscopy (EDS) detector, SEM can also perform elemental analysis, providing insights into the composition of the material. Because of this, SEM has become an essential tool to comprehend the surface properties of nanoporous materials, which are frequently crucial to their performance in applications like catalysis, filtration, and sensing (Alvarez-Fernandez et al., 2020; Sakamoto et al., 1998).

Scanning electron microscopy (SEM) offers important benefits for analyzing porous materials, particularly in revealing particle morphology and surface details, such as mesopore openings and accessibility, given adequate resolution. However, when examining electrically nonconductive materials, conventional SEM faces challenges due to electron-charging, which often requires a conductive coating. Lowering the accelerating voltage can reduce sample charging but also leads to a larger electron beam at the surface, diminishing resolution. The development of low voltage high-resolution SEM (LV-HRSEM) has overcome this challenge, allowing for the detailed visualization of fine structures that were previously challenging to achieve with other methods (Che et al., 2003; Tüysüz et al., 2008).

SEM is also, an essential tool for analyzing nanoporous materials like zeolites, though these materials are particularly prone to damage from high-energy electron beams, which can hinder accurate surface observations. Traditional SEM techniques, which involve increasing the accelerating voltage to enhance resolution, often result in primary electrons penetrating the sample, leading to reduced surface detail and potential sample damage. To address this challenge, the retarding method in low-voltage ultra-high-spatial-resolution SEM has been developed. This advanced technique applies a negative retarding voltage to the sample, decelerating the electron beam just before it reaches the surface. This approach allows for high-resolution surface imaging without penetrating the sample, enabling the acquisition of detailed surface information with nanometer-scale resolution.

Figure 11. SEM images of USY zeolite with and without retarding voltage at 1 kV (Toshiyuki Yokoi, n.d.)

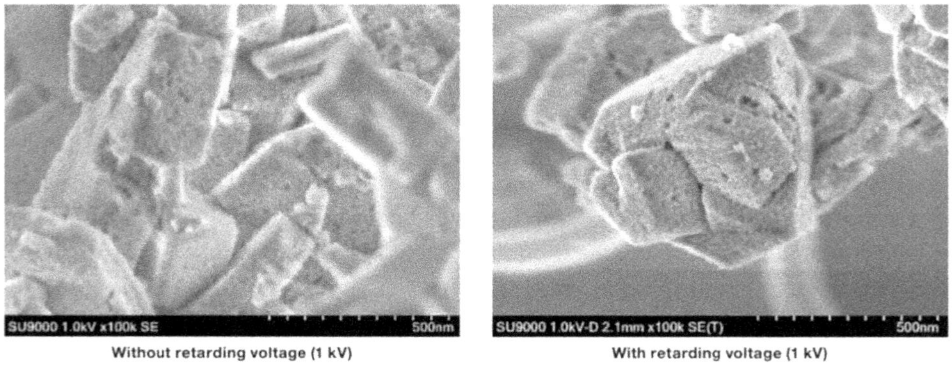

In a study focusing on FAU-type zeolites, specifically Y and ultra-stable Y (USY), the retarding method was used to obtain SEM images at a low accelerating voltage of 1 kV. Figure 11 illustrates the SEM images of USY zeolite, showing a marked improvement in spatial resolution when a retarding voltage was applied. The enhanced images reveal surface features that were not visible using non-retarding methods. Similarly, Figure 12 presents SEM images of Y-zeolite particles, where the application of the retarding voltage allowed for the observation of crystal structure steps on the surface—details that were previously inaccessible with conventional SEM techniques. Furthermore, Figure 13 contrasts the retarding-field SEM images of Y and USY zeolites, highlighting the differences caused by aluminum removal in USY, such as the formation of non-crystalline silica regions and mesopores. These figures collectively demonstrate the efficacy of the retarding method in providing comprehensive and detailed surface information for nanoporous materials like zeolites (*Toshiyuki Yokoi*, n.d.).

Figure 12. Retarding field SEM images of Y-zeolite at 1 kV (Toshiyuki Yokoi, n.d.)

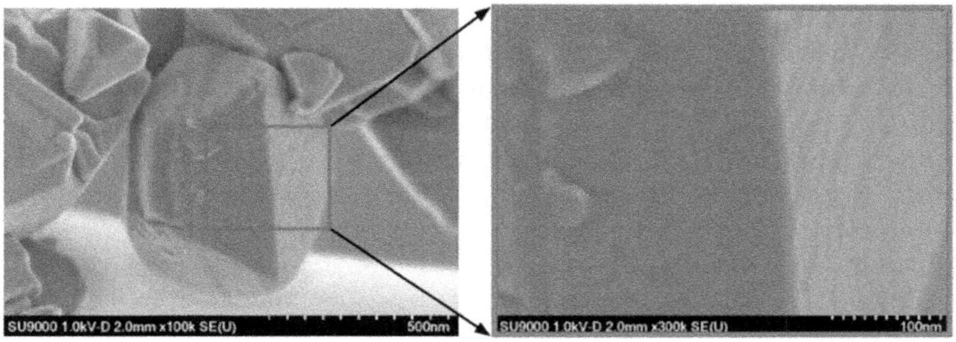

Figure 13. Comparison of retarding-field SEM images between Y and USY zeolites (Toshiyuki Yokoi, n.d.)

5.4 Atomic Force Microscopy (AFM)

This is a high-resolution imaging technique that maps the surface of a material at the nanometer scale. AFM operates by scanning a sharp probe across the sample surface. The interaction between the sample surface and the probe generates a topographical map that reveals comprehensive details about the surface structure. AFM is particularly well-suited for analyzing the surface topography of nanoporous materials, providing three-dimensional images that illustrate features such as pore size, distribution, and surface roughness. Beyond imaging, AFM can also evaluate mechanical properties like hardness, elasticity, and adhesion at the nanoscale, making it a powerful tool for characterizing the physical properties of nanoporous materials. The detailed surface information provided by AFM is crucial for optimizing the performance of these materials in various applications, especially where surface interactions play a critical role (Alvarez-Fernandez et al., 2020).

Atomic Force Microscopy (AFM) techniques offer a powerful, nondestructive method for determining the elastic properties of ultralow-k dielectric films with exceptional lateral and depth resolution. For instance, in a study force modulation AFM was used to map Young's modulus across a cross-section of a 150 nm thick UV-cured organosilicate glass film. The resulting image revealed variations in Young's modulus across the film thickness, indicating local stiffening of the material due to a standing wave effect from the monochromatic UV radiation used during curing. Another AFM technique, known as contact resonance AFM (CR-AFM), was applied to measure Young's modulus in 500 nm thick films of low-k materials. The studies reported a range of Young's modulus values, with the first study showing variations between 16 to 22 GPa and the second study reporting a broader range from 10 to 150 GPa, highlighting the sensitivity and versatility of AFM in characterizing material properties at the nanoscale (Stan et al., 2009).

The Contact Resonance Atomic Force Microscopy (CR-AFM) technique is a refined version of Atomic Force Acoustic Microscopy (AFAM), a dynamic AFM method. AFAM operates by analyzing the vibrations of an AFM cantilever, which are excited by an ultrasonic source positioned below the sample. This technique measures the effective contact stiffness between the AFM tip and the sample. A wide variety of materials can be evaluated by AFAM, and its adaptability offers insightful information about the characteristics of materials. AFAM can be used to determine the indentation modulus (M) through various approaches, including single-point measurements, statistical grid measurements, and generating stiffness and calibrated M-modulus maps. The method is especially useful for characterizing the elastic properties of films with thicknesses ranging from several hundred nanometers to a few nanometers because of its high lateral and depth resolution. This capability allows researchers to obtain detailed and accurate measurements of material stiffness and mechanical properties at the nanoscale, making AFAM a critical tool in materials science for analyzing ultrathin films and other nanoscale structures (Hurley et al., 2006; Kopycinska-Müller et al., 2009, 2011).

6 APPLICATIONS OF NANO POROUS MATERIALS

The applications of nanoporous materials are vast and continue to expand as research advances. Their unique properties enable innovative solutions across various domains, contributing to improved industrial processes, healthcare advancements, environmental sustainability, and energy efficiency. The ongoing exploration of these materials promises to unlock even more potential, addressing some of the world's most pressing challenges. Let us have the overview of the major applicational areas.

6.1 Catalysis

Nanoporous materials are critical in catalysis because of their ability to provide a large surface area for reactions and tunable pore structures, which can be optimized for specific catalytic processes. In industrial settings, zeolites are commonly used as catalysts in the production of fuels and chemicals, including the conversion of methanol to gasoline and the isomerization of hydrocarbons. On the other hand, because of their movable metal sites and functional groups, MOFs have become a highly versatile class of catalytic materials, capable of facilitating a wide variety of reactions, including hydrogenation and oxidation. The enhanced selectivity and activity of these materials make them invaluable in developing more efficient and sustainable chemical processes. Additionally, their role in environmental catalysis extends to the removal of volatile organic compounds (VOCs) and particulate matter from industrial emissions, thereby improving quality of air and reducing environmental impact (Kadja et al., 2022; Nasrollahzadeh et al., 2019; Sneddon et al., 2014; Zabukovec Logar & Kaučič, 2006).

6.2 Gas Storage and Separation

The gas storage capabilities of nanoporous materials, particularly MOFs and covalent organic frameworks (COFs), are being harnessed for critical applications in energy and environmental management. For hydrogen storage, the high gravimetric and volumetric capacities of these materials allow for the safe storage of hydrogen under relatively low pressures, which is essential for fuel cell technology. In natural gas applications, nanoporous materials facilitate the storage of methane, enabling natural gas vehicles to operate more efficiently with reduced tank sizes. Moreover, in carbon capture technologies, nanoporous materials can selectively adsorb CO_2 from the flue gases, significantly reducing the carbon footprint of power plants and other industrial sources. This technology is being integrated into existing infrastructure, showcasing its potential for real-world application in combating climate change (Cychosz & Thommes, 2018; Kostoglou et al., 2017; J. Park et al., 2021; Sharma, 2018; Z. Yang et al., 2021).

6.3 Drug Delivery

In drug delivery systems, nanoporous materials are engineered to enhance bioavailability and therapeutic efficacy. These materials can be functionalized with targeting ligands that bind to specific cells, ensuring that the drug is delivered precisely where it is needed. For instance, chemotherapeutic agents can be encapsulated in nanoparticles made of silica or polymer-based nanoporous materials, which will release the agents under controlled conditions in response to pH or temperature changes. This method minimizes the side effects and enhances treatment effectiveness. Additionally, the porous structure can be tailored to allow for multi-drug loading, providing combination therapy options that are highly important in treating diseases like cancer (França et al., 2021; Gruber et al., 2013; Perrone Donnorso et al., 2012; Samipour et al., 2024; Yan et al., 2009; Zhou et al., 2020).

6.4 Sensors and Detection

Nanoporous materials are at the forefront of sensor technology due to their high sensitivity and rapid response times. In chemical sensors, materials such as graphene oxide and MOFs can detect gases like ammonia, hydrogen sulfide, and carbon dioxide at low concentrations, making them useful for industrial safety monitoring and environmental assessments (Yong et al., 2022; Zhou et al., 2019). In the medical field, nanoporous biosensors utilize nanostructured materials to amplify signals, enabling the identification of biomarkers linked with the diseases such as diabetes, cancer, and infectious diseases. These sensors can provide real-time data and are essential in point-of-care diagnostics, improving healthcare outcomes through early detection and monitoring (Adiga et al., 2009; Esteve-Sánchez et al., 2024; Malhotra & Ali, 2018; Prasad Mishra et al., 2021; P. Xu et al., 2014).

6.5 Water Purification

The filtration capabilities of nanoporous materials are revolutionizing water purification technologies. Advanced filtration membranes made from materials like graphene and zeolites can effectively remove nanoparticles, salts, and pathogens, resulting in high-quality drinking water. Moreover, nanoporous materials are being used in forward osmosis and reverse osmosis systems, which are more energy-efficient compared to traditional desalination techniques. The ability to customize pore sizes and surface properties allows for the selective removal of specific contaminants, including heavy metals and organic pollutants, thereby addressing water scarcity and ensuring safe drinking water in regions with limited freshwater resources (Krishna Prasad et al., n.d.; Narayan, 2010; Priya et al., 2022).

6.6 Energy Storage

In the energy sector, nanoporous materials are enhancing the performance of energy storage devices. Supercapacitors made with carbon-based nanoporous materials exhibit high power densities and rapid charge-discharge capabilities, making them highly suitable for applications requiring quick bursts of energy. They are being integrated with traditional batteries to build hybrid systems that leverage the benefits of both technologies, improving overall energy efficiency. Research into the use of nanoporous materials in lithium-ion batteries focuses on increasing the surface area of electrodes, which leads to higher capacity and faster charging times. Innovations in this area could lead to breakthroughs in electric vehicle technology and renewable energy storage solutions (B.-T. Liu et al., 2018; Sang et al., 2022; W. Xu et al., 2019; J. W. Yang et al., 2024).

6.7 Electronics and Photonics

The integration of nanoporous materials into electronics and photonics is leading to advancements in device performance and functionality. Nanoporous silicon has demonstrated potential in semiconductor applications for improving photovoltaic cell light absorption, increasing their efficiency in converting sunlight to electricity (Fu et al., 2023; K. W. Guo, 2024). In photonics, materials like photonic crystals utilize the unique optical properties of nanoporous structures to manipulate light, leading to applications in optical communications, sensors, and display technologies. Research is ongoing to explore the use of nanoporous materials in quantum dots and light-emitting diodes (LEDs), where their tunable properties can enhance light emission and color purity (Ghoshal & Tewari, 2010; Patel et al., 2012; Shi & Kioupakis, 2015).

6.8 Environmental Remediation

Nanoporous materials are largely being deployed for environmental remediation efforts. Their high adsorption capability allows for the effective removal of heavy metals such as lead, arsenic, and mercury from contaminated soil and water sources. Techniques such as using activated carbon and metal oxides in nanoporous form have proven effective in adsorbing these toxic substances. Moreover, they can aid in the degradation of organic pollutants, including pesticides and industrial chemicals, through advanced oxidation processes facilitated by the catalytic properties of these materials. This application not only helps in restoring contaminated sites but also enhances ecological health and safety (Ansari et al., 2020; Gomis-Berenguer et al., 2017; Jalali Alenjareghi et al., 2021; Memetova et al., 2022; Priya et al., 2024).

7 LIFE CYCLE ASSESSMENT (LCA) OF NANO POROUS MATERIALS

The life cycle of nanoporous materials begins with the extraction of raw materials, which can have significant environmental impacts. This phase involves the mining and processing of various resources such as metals (e.g., aluminum, titanium) and non-metals (e.g., silica, carbon). The environmental consequences include habitat destruction, soil erosion, and resource depletion. Additionally, the energy consumption associated with extracting and refining these materials contributes to greenhouse gas emissions and other forms of pollution. Evaluating these impacts is crucial for understanding the overall environmental footprint of nanoporous materials.

During the manufacturing phase, various synthesis techniques are employed to produce nanoporous materials, such as sol-gel processes or chemical vapor deposition. These processes require substantial energy and resources, and they often generate emissions and waste products. The efficiency of material use and the handling of by-products are critical factors in assessing the environmental impact of manufacturing. High energy consumption and waste generation during this phase highlight the need for more sustainable production practices.

In their application phase, nanoporous materials are evaluated based on their performance, durability, and functionality. Their efficiency in applications like filtration, catalysis, or energy storage is critical. Over time, these materials may degrade or wear out, potentially leading to the release of nanoparticles or other components into the environment. Assessing how these materials perform under operational conditions and their potential for environmental release is essential for understanding their overall impact.

The end-of-life phase involves the disposal or recycling of nanoporous materials. Methods of disposal, such as landfill or incineration, can have varying environmental impacts, including leachate formation and emissions. Recycling opportunities should also be explored to reduce waste and resource consumption. Understanding the environmental consequences of different disposal methods helps in managing the end-of-life phase effectively and minimizing adverse effects.

The biodegradability of nanoporous materials depends on their behavior in different environments, such as soil, water, and air. Understanding how these materials break down and the nature of their degradation products is crucial for assessing their environmental impact. Degradation pathways can vary based on environmental conditions like pH, temperature, and microbial activity, which influence the rate and extent of breakdown. To measure biodegradability, both laboratory simulations and field studies are conducted. Laboratory tests accelerate environmental conditions to determine the rate of material breakdown, while field studies provide real-world observations of degradation. These tests help in understanding how nanoporous materials behave over time and their potential impact on ecosystems. Ecotoxicity studies evaluate the effects of degradation products on various organisms, including microorganisms, plants, and animals. This includes assessing potential toxicity to aquatic life and terrestrial plants. Investigating whether degradation products bioaccumulate in the food chain is also important, as it can reveal long-term impacts on higher organisms and ecosystems.

8 CHALLENGES AND FUTURE PROSPECTS OF NANOPOROUS MATERIALS

Nanoporous materials, distinguished by their large surface area and well-defined porous structures, are becoming increasingly significant in various applications, as mentioned in previous section, ranging from catalysis and gas storage to drug delivery and separation processes. However, the journey from laboratory-scale research to industrial-scale application is fraught with challenges. These challenges, along with the promising future prospects, form the backbone of ongoing research in this field.

One of the leading challenges in the area of nanoporous materials is attaining precise control over their synthesis (F. Yang et al., 2014). The ability to control pore size, distribution, and uniformity is crucial for the material's performance in specific applications. For instance, in catalysis or molecular sieving, even minor variations in the size of pore can significantly affect the efficiency of the process. Furthermore, scaling up the synthesis process from laboratory to industrial levels poses a significant challenge (S. K. Gupta & Mao, 2021). Many nanoporous materials are synthesized under very specific and controlled conditions that are difficult to replicate on a larger scale. This makes the large-scale production of these materials not only technically challenging but also economically unfeasible in many cases. The high cost of production, particularly for advanced nanoporous materials such as MOFs and zeolites, further limits their widespread application.

Another major challenge lies in the durability and stability of nanoporous materials (Z. Chen et al., 2021). Many of these materials face issues related to thermal and chemical stability, especially under extreme conditions. For example, in high-temperature applications or in environments with highly acidic or alkaline conditions, the structural integrity of nanoporous materials can be compromised. This degradation limits their use in certain applications where long-term stability is required. In addition, the mechanical stability of these materials is often a concern. Due to their high porosity, some nanoporous materials are mechanically fragile, making them difficult to handle, process, and integrate into practical applications without compromising their structure.

The characterization of nanoporous materials is a challenging task that requires sophisticated techniques. In addition to being costly, these methods need specific knowledge in order to properly interpret the data. Furthermore, there are no industry-wide standards for assessing and disclosing the characteristics of nanoporous materials. This lack of standardization leads to inconsistencies in data reporting, making it difficult for researchers to compare the results across different studies and hindering the progress of the field.

As nanoporous materials continue to be developed, worries about their effects on the environment and possible health hazards are growing in importance. We still don't fully understand how toxic these materials are, especially when they're used in consumer goods or released into the environment. Important concerns concerning their safe use, disposal, and long-term effects on the environment and human health are brought up by this (Martínez et al., 2020). Furthermore, the synthesis of nanoporous materials frequently entails the use of dangerous chemicals and high energy consumption, both of which are not environmentally friendly processes. As sustainability becomes an increasingly important factor in material development, the environmental footprint of nanoporous materials needs to be addressed. A comprehensive Life Cycle Assessment (LCA) would assess raw material extraction, energy consumption, emissions, and waste produced during manufacturing. These assessments can highlight the carbon footprint, energy demands for synthesis processes such as hydrothermal methods, and the environmental toxicity of solvents used in the production. Moreover, the end-of-life disposal of these materials, many of which are non-biodegradable, raises concerns about long-term environmental persistence and potential contamination of ecosystems. The health risks associated with nanoporous materials are also critical to consider, particularly during their synthesis and application. The chemicals involved in production, such as strong acids and organic solvents, can pose significant risks to workers. Prolonged exposure may result in respiratory issues, skin irritation, or even more severe toxic effects. In addition, the inhalation of nanoparticulate matter poses a serious concern, as these particles can penetrate deep into the lungs and bloodstream, potentially leading to pulmonary diseases, inflammation, or neurotoxicity. The potential for bioaccumulation of these materials in the human body, especially if they degrade and release toxic metal ions, is another important area of research. Proper ventilation, safety equipment, and training are necessary to mitigate these risks.

Despite these challenges, the future of nanoporous materials is filled with promising opportunities, driven by advancements in both technology and scientific understanding.

Nanoporous materials are poised to play significant role in several cutting-edge applications. Their large surface area can greatly improve the charge storage capacity and efficiency of energy storage devices such as batteries and supercapacitors, for which they are being researched in the energy sector. Additionally, with the growing interest in carbon capture and hydrogen storage, nanoporous materials like MOFs are being studies for their ability to selectively adsorb and store gases at high densities, potentially revolutionizing these fields. Nanoporous materials are exhibiting great potential as drug delivery systems in the biomedical field. By targeting specific body sites and providing controlled release profiles, they can improve treatment efficacy (D. Li et al., 2024).

Adopting green synthesis methods can reduce the environmental impact of nanoporous materials. Researchers are concentrating on creating greener synthesis techniques in order to address the environmental issues related to the production of nanoporous materials. By utilizing renewable feedstocks, consuming less energy, and relying less on dangerous chemicals, these techniques seek to lessen the production's negative environmental effects. These initiatives reduce the environmental impact of nanoporous materials and increase the viability of their large-scale production from an economic standpoint by improving the sustainability of the synthesis process (Ghajeri et al., 2018).

Improving the stability and durability of nanoporous materials is a key area of ongoing research. Improving the stability of nanoporous materials can reduce the risk of environmental and health impacts. Research into enhancing material durability and resistance to degradation can help in developing safer materials for long-term use. One promising approach is the development of hybrid materials, which combine nanoporous structures with other substances, such as polymers or metals, to enhance their mechanical strength and stability. These hybrid materials could be used in more demanding environments where traditional nanoporous materials would not be viable (Saadati pour et al., 2023; D. Zhao et al., 2024). Another approach is post-synthetic modification, where the surface of nanoporous materials is functionalized or ion-exchanged to improve their thermal and chemical stability. These modifications could extend the range of applications for nanoporous materials, particularly in harsh or extreme environments.

The integration of machine learning (ML) and artificial intelligence (AI) into the synthesis of nanoporous materials is a rapidly growing area of interest. By anticipating how structural alterations will impact performance, artificial intelligence (AI)-driven predictive modeling can assist researchers in creating new materials with specialized properties for particular uses. This can shorten the time and expense required for the development of new nanoporous materials by speeding up the discovery process and streamlining the synthesis procedure. Moreover, AI can be used to optimize the synthesis conditions, leading to better control over material properties and potentially enabling the large-scale production of nanoporous materials with the desired characteristics (Adamu et al., 2023; Lin et al., 2023; Puchi-Cabrera et al., 2023; Su et al., 2024).

Finally, As the use of nanoporous materials expands, there will be an increasing need for regulatory and standardization efforts to ensure their safe and effective application. Developing standardized testing and reporting protocols will be crucial for the broader adoption of these materials, particularly in industries such as pharmaceuticals and food processing, where safety and consistency are paramount. Clearer regulatory guidelines will also help address environmental and health concerns, providing a framework for the responsible development and use of nanoporous materials.

CONCLUSION

Nanoporous materials represent a significant advancement in the field of materials science, offering distinctive properties that are essential for a broad range of industrial and technological applications. The ability to control the pore size, surface area, and structural properties at the nanoscale has enabled the development of materials with enhanced performance in catalysis, drug delivery, and environmental cleanup. The synthesis techniques discussed, including sol-gel processing, hydrothermal methods, and template-assisted approaches, provide versatile and efficient routes for producing these materials with desired characteristics. The characterization techniques reviewed in this chapter are critical for understanding and optimizing the properties of nanoporous materials, ensuring their successful application in various fields. As research in this area continues to evolve, nanoporous materials are expected to play an increasingly important role in addressing global challenges in energy, healthcare, and environmental sustainability.

REFERENCES

Adamu, H., Abba, S. I., Anyin, P. B., Sani, Y., & Qamar, M. (2023). Artificial intelligence-navigated development of high-performance electrochemical energy storage systems through feature engineering of multiple descriptor families of materials. *Energy Advances*, 2(5), 615–645. https://doi.org/https://doi.org/10.1039/d3ya00104k. DOI: 10.1039/D3YA00104K

Adane, W. D., Chandravanshi, B. S., & Tessema, M. (2024). Hypersensitive electrochemical sensor based on thermally annealed gold–silver alloy nanoporous matrices for the simultaneous determination of sulfathiazole and sulfamethoxazole residues in food samples. *Food Chemistry*, 457, 140071. https://doi.org/https://doi.org/10.1016/j.foodchem.2024.140071. DOI: 10.1016/j.foodchem.2024.140071 PMID: 38905827

Adiga, S. P., Jin, C., Curtiss, L. A., Monteiro-Riviere, N. A., & Narayan, R. J. (2009). Nanoporous membranes for medical and biological applications. *Wiley Interdisciplinary Reviews. Nanomedicine and Nanobiotechnology*, 1(5), 568–581. DOI: 10.1002/wnan.50 PMID: 20049818

Alvarez-Fernandez, A., Reid, B., Fornerod, M. J., Taylor, A., Divitini, G., & Guldin, S. (2020). Structural Characterization of Mesoporous Thin Film Architectures: A Tutorial Overview. *ACS Applied Materials & Interfaces*, 12(5), 5195–5208. DOI: 10.1021/acsami.9b17899 PMID: 31961128

Ansari, M., Alam, A., Bera, R., Hassan, A., Goswami, S., & Das, N. (2020). Synthesis, characterization and adsorption studies of a novel triptycene based hydroxyl azo- nanoporous polymer for environmental remediation. *Journal of Environmental Chemical Engineering*, 8(2), 103558. https://doi.org/https://doi.org/10.1016/j.jece.2019.103558. DOI: 10.1016/j.jece.2019.103558

Asimakopoulos, G., Baikousi, M., Kostas, V., Papantoniou, M., Bourlinos, A. B., Zbořil, R., Karakassides, M. A., & Salmas, C. E. (2020). Nanoporous Activated Carbon Derived via Pyrolysis Process of Spent Coffee: Structural Characterization. Investigation of Its Use for Hexavalent Chromium Removal. *Applied Sciences (Basel, Switzerland)*, 10(24), 8812. DOI: 10.3390/app10248812

Bai, S., Sun, C., Guo, T., Luo, R., Lin, Y., Chen, A., Sun, L., & Zhang, J. (2013). Low temperature electrochemical deposition of nanoporous ZnO thin films as novel NO2 sensors. *Electrochimica Acta*, 90, 530–534. https://doi.org/https://doi.org/10.1016/j.electacta.2012.12.060. DOI: 10.1016/j.electacta.2012.12.060

Bhattacharyya, B. (2015). Design and Developments of Microtools. In Bhattacharyya, B. (Ed.), *Electrochemical Micromachining for Nanofabrication, MEMS and Nanotechnology* (pp. 101–122). William Andrew Publishing., https://doi.org/https://doi.org/10.1016/B978-0-323-32737-4.00006-2 DOI: 10.1016/B978-0-323-32737-4.00006-2

Bhattacharyya, S., Mastai, Y., Narayan Panda, R., Yeon, S.-H., & Hu, M. Z. (2014). Advanced Nanoporous Materials: Synthesis, Properties, and Applications. *Journal of Nanomaterials*, 2014(1), 275796. https://doi.org/https://doi.org/10.1155/2014/275796. DOI: 10.1155/2014/275796

Biswas, A., Friend, C. S., & Prasad, P. N. (2001). Ceramic Nanocomposites with Organic Phases, Optics of. In Buschow, K. H. J., Cahn, R. W., Flemings, M. C., Ilschner, B., Kramer, E. J., Mahajan, S., & Veyssière, P. (Eds.), *Encyclopedia of Materials: Science and Technology* (pp. 1072–1080). Elsevier., https://doi.org/https://doi.org/10.1016/B0-08-043152-6/00198-4 DOI: 10.1016/B0-08-043152-6/00198-4

Bokov, D., Turki Jalil, A., Chupradit, S., Suksatan, W., Javed Ansari, M., Shewael, I. H., Valiev, G. H., & Kianfar, E. (2021). Nanomaterial by Sol-Gel Method: Synthesis and Application. *Advances in Materials Science and Engineering, 2021*(1). DOI: 10.1155/2021/5102014

Bon, V., Brunner, E., Pöppl, A., & Kaskel, S. (2020). Unraveling Structure and Dynamics in Porous Frameworks via Advanced In Situ Characterization Techniques. *Advanced Functional Materials*, 30(41), 1907847. Advance online publication. DOI: 10.1002/adfm.201907847

Bon, V., Klein, N., Senkovska, I., Heerwig, A., Getzschmann, J., Wallacher, D., Zizak, I., Brzhezinskaya, M., Mueller, U., & Kaskel, S. (2015). Exceptional adsorption-induced cluster and network deformation in the flexible metal–organic framework DUT-8(Ni) observed by in situ X-ray diffraction and EXAFS. *Physical Chemistry Chemical Physics*, 17(26), 17471–17479. DOI: 10.1039/C5CP02180D PMID: 26079102

Byrappa, K. ., & Haber, Masahiro. (2001). *Handbook of Hydrothermal Technology*. Noyes Publications [Imprint] William Andrew, Inc. Elsevier Science & Technology Books [distributor].

Che, S., Lund, K., Tatsumi, T., Iijima, S., Joo, S. H., Ryoo, R., & Terasaki, O. (2003). Direct Observation of 3D Mesoporous Structure by Scanning Electron Microscopy (SEM): SBA-15 Silica and CMK-5 Carbon. *Angewandte Chemie International Edition*, 42(19), 2182–2185. https://doi.org/https://doi.org/10.1002/anie.200250726. DOI: 10.1002/anie.200250726 PMID: 12761755

Chen, J., & Whittingham, M.CHEN. (2006). Hydrothermal synthesis of lithium iron phosphate. *Electrochemistry Communications*, 8(5), 855–858. DOI: 10.1016/j.elecom.2006.03.021

Chen, Q., Dwyer, C., Sheng, G., Zhu, C., Li, X., Zheng, C., & Zhu, Y. (2020a). Imaging Beam-Sensitive Materials by Electron Microscopy. *Advanced Materials*, 32(16), 1907619. Advance online publication. DOI: 10.1002/adma.201907619 PMID: 32108394

Chen, Q., Dwyer, C., Sheng, G., Zhu, C., Li, X., Zheng, C., & Zhu, Y. (2020b). Imaging Beam-Sensitive Materials by Electron Microscopy. *Advanced Materials*, 32(16), 1907619. Advance online publication. DOI: 10.1002/adma.201907619 PMID: 32108394

Chen, Z., Li, P., Anderson, R., Wang, X., Zhang, X., Robison, L., Redfern, L. R., Moribe, S., Islamoglu, T., Gómez-Gualdrón, D. A., Yildirim, T., Stoddart, J. F., & Farha, O. K. (2020). Balancing volumetric and gravimetric uptake in highly porous materials for clean energy. *Science*, 368(6488), 297–303. DOI: 10.1126/science.aaz8881 PMID: 32299950

Chen, Z., Zhao, D., Ma, R., Zhang, X., Rao, J., Yin, Y., Wang, X., & Yi, F. (2021). Flexible temperature sensors based on carbon nanomaterials. *Journal of Materials Chemistry. B*, 9(8), 1941–1964. DOI: 10.1039/D0TB02451A PMID: 33532811

Cioffi, N., Colaianni, L., Ieva, E., Pilolli, R., Ditaranto, N., Angione, M. D., Cotrone, S., Buchholt, K., Spetz, A. L., Sabbatini, L., & Torsi, L. (2011). Electrosynthesis and characterization of gold nanoparticles for electronic capacitance sensing of pollutants. *Electrochimica Acta*, 56(10), 3713–3720. https://doi.org/https://doi.org/10.1016/j.electacta.2010.12.105. DOI: 10.1016/j.electacta.2010.12.105

Cundy, C. S., & Cox, P. A. (2005). The hydrothermal synthesis of zeolites: Precursors, intermediates and reaction mechanism. *Microporous and Mesoporous Materials*, 82(1–2), 1–78. DOI: 10.1016/j.micromeso.2005.02.016

Cychosz, K. A., & Thommes, M. (2018). Progress in the Physisorption Characterization of Nanoporous Gas Storage Materials. *Engineering (Beijing)*, 4(4), 559–566. https://doi.org/https://doi.org/10.1016/j.eng.2018.06.001. DOI: 10.1016/j.eng.2018.06.001

Demazeau, G. (2008). Solvothermal reactions: An original route for the synthesis of novel materials. *Journal of Materials Science*, 43(7), 2104–2114. DOI: 10.1007/s10853-007-2024-9

Domínguez-Domínguez, S., Arias-Pardilla, J., Berenguer-Murcia, Á., Morallón, E., & Cazorla-Amorós, D. (2008). Electrochemical deposition of platinum nanoparticles on different carbon supports and conducting polymers. *Journal of Applied Electrochemistry*, 38(2), 259–268. DOI: 10.1007/s10800-007-9435-9

Du, N., Zhang, H., Chen, B. D., Ma, X. Y., Liu, Z. H., Wu, J. B., & Yang, D. R. (2007). Porous Indium Oxide Nanotubes: Layer-by-Layer Assembly on Carbon-Nanotube Templates and Application for Room-Temperature NH 3 Gas Sensors. *Advanced Materials*, 19(12), 1641–1645. DOI: 10.1002/adma.200602128

Esteve-Sánchez, Y., Hernández-Montoto, A., Tormo-Mas, M. Á., Pemán, J., Calabuig, E., Gómez, M. D., Marcos, M. D., Martínez-Máñez, R., Aznar, E., & Climent, E. (2024). SARS-CoV-2 N protein IgG antibody detection employing nanoporous anodized alumina: A rapid and selective alternative for identifying naturally infected individuals in populations vaccinated with spike protein (S)-based vaccines. *Sensors and Actuators. B, Chemical*, 419, 136378. https://doi.org/https://doi.org/10.1016/j.snb.2024.136378. DOI: 10.1016/j.snb.2024.136378

Favacho, V. S. S., Melo, D. M. A., Costa, J. E. L., Silva, Y. K. R. O., Braga, R. M., & Medeiros, R. L. B. A. (2024). Perovskites synthesized by soft template-assisted hydrothermal method: A bibliometric analysis and new insights. *International Journal of Hydrogen Energy*, 78, 1391–1428. DOI: 10.1016/j.ijhydene.2024.06.326

Feinle, A., Elsaesser, M. S., & Hüsing, N. (2016). Sol–gel synthesis of monolithic materials with hierarchical porosity. *Chemical Society Reviews*, 45(12), 3377–3399. DOI: 10.1039/C5CS00710K PMID: 26563577

Férey, G. (2008). Hybrid porous solids: Past, present, future. *Chemical Society Reviews*, 37(1), 191–214. DOI: 10.1039/B618320B PMID: 18197340

Foo, K. Y., & Hameed, B. H. (2010). Insights into the modeling of adsorption isotherm systems. *Chemical Engineering Journal*, 156(1), 2–10. DOI: 10.1016/j.cej.2009.09.013

França, C. G., Plaza, T., Naveas, N., Andrade Santana, M. H., Manso-Silván, M., Recio, G., & Hernandez-Montelongo, J. (2021). Nanoporous silicon microparticles embedded into oxidized hyaluronic acid/adipic acid dihydrazide hydrogel for enhanced controlled drug delivery. *Microporous and Mesoporous Materials*, 310, 110634. https://doi.org/https://doi.org/10.1016/j.micromeso.2020.110634. DOI: 10.1016/j.micromeso.2020.110634

Fu, K., Yang, Z., Sun, H., Chen, X., Li, S., Ma, W., & Chen, R. (2023). Multiple modification on nanoporous silicon derived from photovoltaic silicon cutting waste for extraction of PbII in industrial effluents. *Materials Today. Communications*, 35, 105776. https://doi.org/https://doi.org/10.1016/j.mtcomm.2023.105776. DOI: 10.1016/j.mtcomm.2023.105776

Furukawa, H., Ko, N., Go, Y. B., Aratani, N., Choi, S. B., Choi, E., Yazaydin, A. Ö., Snurr, R. Q., O'Keeffe, M., Kim, J., & Yaghi, O. M. (2010). Ultrahigh Porosity in Metal-Organic Frameworks. *Science*, 329(5990), 424–428. DOI: 10.1126/science.1192160 PMID: 20595583

Ghajeri, F., Topalian, Z., Tasca, A., Jafri, S. H. M., Leifer, K., Norberg, P., & Sjöström, C. (2018). Case study of a green nanoporous material from synthesis to commercialisation: Quartzene®. *Current Opinion in Green and Sustainable Chemistry*, 12, 101–109. https://doi.org/https://doi.org/10.1016/j.cogsc.2018.07.003. DOI: 10.1016/j.cogsc.2018.07.003

Ghoshal, S. K., & Tewari, H. S. (2010). Photonic applications of Silicon nanostructures. *Material Science Research India*, 7(2), 381–388. DOI: 10.13005/msri/070207

Goetze, J., Yarulina, I., Gascon, J., Kapteijn, F., & Weckhuysen, B. M. (2018). Revealing Lattice Expansion of Small-Pore Zeolite Catalysts during the Methanol-to-Olefins Process Using Combined Operando X-ray Diffraction and UV–vis Spectroscopy. *ACS Catalysis*, 8(3), 2060–2070. DOI: 10.1021/acscatal.7b04129 PMID: 29527401

Gomis-Berenguer, A., Velasco, L. F., Velo-Gala, I., & Ania, C. O. (2017). Photochemistry of nanoporous carbons: Perspectives in energy conversion and environmental remediation. *Journal of Colloid and Interface Science*, 490, 879–901. https://doi.org/https://doi.org/10.1016/j.jcis.2016.11.046. DOI: 10.1016/j.jcis.2016.11.046 PMID: 27914582

Gruber, M. F., Schulte, L., & Ndoni, S. (2013). Nanoporous materials modified with biodegradable polymers as models for drug delivery applications. *Journal of Colloid and Interface Science*, 395, 58–63. https://doi.org/https://doi.org/10.1016/j.jcis.2012.12.052. DOI: 10.1016/j.jcis.2012.12.052 PMID: 23369801

Gruene, T., Holstein, J. J., Clever, G. H., & Keppler, B. (2021). Establishing electron diffraction in chemical crystallography. *Nature Reviews. Chemistry*, 5(9), 660–668. DOI: 10.1038/s41570-021-00302-4 PMID: 37118416

Guo, K. W. (2024). Nanoporous silicon materials for solar energy by electrochemical approach. In Kulkarni, N. V., & Kharissov, B. I. (Eds.), *Handbook of Emerging Materials for Sustainable Energy* (pp. 119–128). Elsevier., https://doi.org/https://doi.org/10.1016/B978-0-323-96125-7.00028-9 DOI: 10.1016/B978-0-323-96125-7.00028-9

Guo, S., Wang, L., & Wang, E. (2007). Templateless, surfactantless, simple electrochemical route to rapid synthesis of diameter-controlled 3D flowerlike gold microstructure with "clean" surface. *Chemical Communications*, 30(30), 3163. DOI: 10.1039/b705630c PMID: 17653375

Gupta, S. K., & Mao, Y. (2021). A review on molten salt synthesis of metal oxide nanomaterials: Status, opportunity, and challenge. *Progress in Materials Science*, 117, 100734. https://doi.org/https://doi.org/10.1016/j.pmatsci.2020.100734. DOI: 10.1016/j.pmatsci.2020.100734

Gupta, T., Samriti, , Cho, J., & Prakash, J. (2021). Hydrothermal synthesis of TiO2 nanorods: Formation chemistry, growth mechanism, and tailoring of surface properties for photocatalytic activities. *Materials Today. Chemistry*, 20, 100428. DOI: 10.1016/j.mtchem.2021.100428

Hayashi, H., & Hakuta, Y. (2010). Hydrothermal Synthesis of Metal Oxide Nanoparticles in Supercritical Water. *Materials (Basel)*, 3(7), 3794–3817. DOI: 10.3390/ma3073794 PMID: 28883312

Hu, J., Wang, H., Wang, S., Lei, Y., Qin, L., Li, X., Zhai, D., Li, B., & Kang, F. (2021). Electrochemical deposition mechanism of sodium and potassium. *Energy Storage Materials*, 36, 91–98. DOI: 10.1016/j.ensm.2020.12.017

Huang, Z., Grape, E. S., Li, J., Inge, A. K., & Zou, X. (2021). 3D electron diffraction as an important technique for structure elucidation of metal-organic frameworks and covalent organic frameworks. *Coordination Chemistry Reviews*, 427, 213583. https://doi.org/https://doi.org/10.1016/j.ccr.2020.213583. DOI: 10.1016/j.ccr.2020.213583

Huang, Z., Willhammar, T., & Zou, X. (2021). Three-dimensional electron diffraction for porous crystalline materials: Structural determination and beyond. *Chemical Science (Cambridge)*, 12(4), 1206–1219. DOI: 10.1039/D0SC05731B PMID: 34163882

Hurley, D. C., Kopycinska-Müller, M., Langlois, E. D., Kos, A. B., & Barbosa, N.III. (2006). Mapping substrate/film adhesion with contact-resonance-frequency atomic force microscopy. *Applied Physics Letters*, 89(2), 021911. Advance online publication. DOI: 10.1063/1.2221404

Jalali Alenjareghi, M., Rashidi, A., Kazemi, A., & Talebi, A. (2021). Highly efficient and recyclable spongy nanoporous graphene for remediation of organic pollutants. *Process Safety and Environmental Protection*, 148, 313–322. https://doi.org/https://doi.org/10.1016/j.psep.2020.09.054. DOI: 10.1016/j.psep.2020.09.054

Jasim, A. M., He, X., Xing, Y., White, T. A., & Young, M. J. (2021). Cryo-ePDF: Overcoming Electron Beam Damage to Study the Local Atomic Structure of Amorphous ALD Aluminum Oxide Thin Films within a TEM. *ACS Omega*, 6(13), 8986–9000. DOI: 10.1021/acsomega.0c06124 PMID: 33842769

Jiao Kexin and Flynn, K. T. and K. P. (2016). Synthesis, Characterization, and Applications of Nanoporous Materials for Sensing and Separation. In M. Aliofkhazraei (Ed.), *Handbook of Nanoparticles* (pp. 429–454). Springer International Publishing. DOI: 10.1007/978-3-319-15338-4_22

Kadja, G. T. M. Ilmi, Moh. M., Azhari, N. J., Khalil, M., Fajar, A. T. N., Subagjo, Makertihartha, I. G. B. N., Gunawan, M. L., Rasrendra, C. B., & Wenten, I. G. (2022). Recent advances on the nanoporous catalysts for the generation of renewable fuels. *Journal of Materials Research and Technology, 17*, 3277–3336. https://doi.org/https://doi.org/10.1016/j.jmrt.2022.02.033

Kawamura, G., Muto, H., & Matsuda, A. (2014). Hard template synthesis of metal nanowires. *Frontiers in Chemistry*, 2. Advance online publication. DOI: 10.3389/fchem.2014.00104 PMID: 25453031

Khun, K., Ibupoto, Z. H., Liu, X., Beni, V., & Willander, M. (2015). The ethylene glycol template assisted hydrothermal synthesis of Co3O4 nanowires; structural characterization and their application as glucose non-enzymatic sensor. *Materials Science and Engineering B*, 194, 94–100. DOI: 10.1016/j.mseb.2015.01.001

Knecht, T. A., & Hutchison, J. E. (2023). Precursor and Surface Reactivities Influence the Early Growth of Indium Oxide Nanocrystals in a Reagent-Driven, Continuous Addition Synthesis. *Chemistry of Materials*, 35(8), 3151–3161. DOI: 10.1021/acs.chemmater.2c03761

Kołodziejczak-Radzimska, A., & Jesionowski, T. (2014). Zinc Oxide—From Synthesis to Application: A Review. *Materials (Basel)*, 7(4), 2833–2881. DOI: 10.3390/ma7042833 PMID: 28788596

Kopycinska-Müller, M., Striegler, A., Hürrich, A., Köhler, B., Meyendorf, N., & Wolter, K.-J. (2009). Elastic Properties of Nano–Thin Films by Use of Atomic Force Acoustic Microscopy. *Proceedings of the Materials Research Society*, 1185(1), 7–12. DOI: 10.1557/PROC-1185-II09-04

Kopycinska-Müller, M., Striegler, A., Köhler, B., & Wolter, K.-J. (2011). Mechanical Characterization of Thin Films by Use of Atomic Force Acoustic Microscopy. *Advanced Engineering Materials*, 13(4), 312–318. https://doi.org/https://doi.org/10.1002/adem.201000245. DOI: 10.1002/adem.201000245

Košiček, M., Zavašnik, J., Baranov, O., Šetina Batič, B., & Cvelbar, U. (2022). Understanding the Growth of Copper Oxide Nanowires and Layers by Thermal Oxidation over a Broad Temperature Range at Atmospheric Pressure. *Crystal Growth & Design*, 22(11), 6656–6666. DOI: 10.1021/acs.cgd.2c00863

Kostoglou, N., Koczwara, C., Prehal, C., Terziyska, V., Babic, B., Matovic, B., Constantinides, G., Tampaxis, C., Charalambopoulou, G., Steriotis, T., Hinder, S., Baker, M., Polychronopoulou, K., Doumanidis, C., Paris, O., Mitterer, C., & Rebholz, C. (2017). Nanoporous activated carbon cloth as a versatile material for hydrogen adsorption, selective gas separation and electrochemical energy storage. *Nano Energy*, 40, 49–64. https://doi.org/https://doi.org/10.1016/j.nanoen.2017.07.056. DOI: 10.1016/j.nanoen.2017.07.056

Kozlovskiy, A. L., Shlimas, D. I., Shumskaya, A. E., Kaniukov, E. Y., Zdorovets, M. V., & Kadyrzhanov, K. K. (2017). Influence of electrodeposition parameters on structural and morphological features of Ni nanotubes. *The Physics of Metals and Metallography*, 118(2), 164–169. DOI: 10.1134/S0031918X17020065

Krause, S., Bon, V., Senkovska, I., Stoeck, U., Wallacher, D., Többens, D. M., Zander, S., Pillai, R. S., Maurin, G., Coudert, F.-X., & Kaskel, S. (2016). A pressure-amplifying framework material with negative gas adsorption transitions. *Nature*, 532(7599), 348–352. DOI: 10.1038/nature17430 PMID: 27049950

Kresge, C. T., Leonowicz, M. E., Roth, W. J., Vartuli, J. C., & Beck, J. S. (1992). Ordered mesoporous molecular sieves synthesized by a liquid-crystal template mechanism. *Nature*, 359(6397), 710–712. DOI: 10.1038/359710a0

Krishna Prasad, N. V., Babu, T. A., Sarma, S. R. K. N., Ramesh, S., Nirisha, K., Mathew, T., & Madhavi, N. (n.d.). ROLE OF POROUS NANOMATERIAL'S IN WATER PURIFICATION, ELECTRONICS, DRUG DELIVERY AND STORAGE: A COMPREHENSIVE REVIEW. In *Journal of Optoelectronic and Biomedical Materials* (Vol. 13, Issue 1).

Li, C., Iqbal, M., Lin, J., Luo, X., Jiang, B., Malgras, V., Wu, K. C.-W., Kim, J., & Yamauchi, Y. (2018a). Electrochemical Deposition: An Advanced Approach for Templated Synthesis of Nanoporous Metal Architectures. *Accounts of Chemical Research*, 51(8), 1764–1773. DOI: 10.1021/acs.accounts.8b00119 PMID: 29984987

Li, C., Iqbal, M., Lin, J., Luo, X., Jiang, B., Malgras, V., Wu, K. C.-W., Kim, J., & Yamauchi, Y. (2018b). Electrochemical Deposition: An Advanced Approach for Templated Synthesis of Nanoporous Metal Architectures. *Accounts of Chemical Research*, 51(8), 1764–1773. DOI: 10.1021/acs.accounts.8b00119 PMID: 29984987

Li, D., Sun, Y., Gao, P., Zhang, X., & Ge, H. (2014). Structural and magnetic properties of nickel ferrite nanoparticles synthesized via a template-assisted sol–gel method. *Ceramics International*, 40(10), 16529–16534. DOI: 10.1016/j.ceramint.2014.08.006

Li, D., Yadav, A., Zhou, H., Roy, K., Thanasekaran, P., & Lee, C. (2024). Advances and Applications of Metal-Organic Frameworks (MOFs) in Emerging Technologies: A Comprehensive Review. *Global Challenges (Hoboken, NJ)*, 8(2), 2300244. https://doi.org/https://doi.org/10.1002/gch2.202300244. DOI: 10.1002/gch2.202300244 PMID: 38356684

Li, J., Wu, Q., & Wu, J. (2016). Synthesis of Nanoparticles via Solvothermal and Hydrothermal Methods. In *Handbook of Nanoparticles* (pp. 295–328). Springer International Publishing., DOI: 10.1007/978-3-319-15338-4_17

Li, J.-R., Sculley, J., & Zhou, H.-C. (2012). Metal–Organic Frameworks for Separations. *Chemical Reviews*, 112(2), 869–932. DOI: 10.1021/cr200190s PMID: 21978134

Li, L., Han, J., Huang, X., Qiu, S., Liu, X., Liu, L., Zhao, M., Qu, J., Zou, J., & Zhang, J. (2023). Organic pollutants removal from aqueous solutions using metal-organic frameworks (MOFs) as adsorbents: A review. *Journal of Environmental Chemical Engineering*, 11(6), 111217. DOI: 10.1016/j.jece.2023.111217

Li, M., & Yang, L. (2023). Biomedical metallic materials based on nanocrystalline and nanoporous microstructures: Properties and applications. In Webster, T. J. (Ed.), *Nanomedicine* (2nd ed., pp. 555–584). Woodhead Publishing., https://doi.org/https://doi.org/10.1016/B978-0-12-818627-5.00030-0 DOI: 10.1016/B978-0-12-818627-5.00030-0

Li, T., Wu, H., Ihli, J., Ma, Z., Krumeich, F., Bomans, P. H. H., Sommerdijk, N. A. J. M., Friedrich, H., Patterson, J. P., & van Bokhoven, J. A. (2019). Cryo-TEM and electron tomography reveal leaching-induced pore formation in ZSM-5 zeolite. *Journal of Materials Chemistry. A, Materials for Energy and Sustainability*, 7(4), 1442–1446. DOI: 10.1039/C8TA10696G

Liao, Y., Xu, Y., & Chan, Y. (2013). Semiconductor nanocrystals in sol–gel derived matrices. *Physical Chemistry Chemical Physics*, 15(33), 13694. DOI: 10.1039/c3cp51351c PMID: 23842703

Lin, J., Liu, Z., Guo, Y., Wang, S., Tao, Z., Xue, X., Li, R., Feng, S., Wang, L., Liu, J., Gao, H., Wang, G., & Su, Y. (2023). Machine learning accelerates the investigation of targeted MOFs: Performance prediction, rational design and intelligent synthesis. *Nano Today*, 49, 101802. https://doi.org/https://doi.org/10.1016/j.nantod.2023.101802. DOI: 10.1016/j.nantod.2023.101802

Liu, B.-T., Zhao, M., Han, L.-P., Lang, X.-Y., Wen, Z., & Jiang, Q. (2018). Three-dimensional nanoporous N-doped graphene/iron oxides as anode materials for high-density energy storage in asymmetric supercapacitors. *Chemical Engineering Journal*, 335, 467–474. https://doi.org/https://doi.org/10.1016/j.cej.2017.11.001. DOI: 10.1016/j.cej.2017.11.001

Liu, H.-J., Wang, X.-M., Cui, W.-J., Dou, Y.-Q., Zhao, D.-Y., & Xia, Y.-Y. (2010). Highly ordered mesoporous carbon nanofiber arrays from a crab shell biological template and its application in supercapacitors and fuel cells. *Journal of Materials Chemistry*, 20(20), 4223. DOI: 10.1039/b925776d

Liu, Z., Fujita, N., Miyasaka, K., Han, L., Stevens, S. M., Suga, M., Asahina, S., Slater, B., Xiao, C., Sakamoto, Y., Anderson, M. W., Ryoo, R., & Terasaki, O. (2013). A review of fine structures of nanoporous materials as evidenced by microscopic methods. In *Journal of Electron Microscopy* (Vol. 62, Issue 1, pp. 109–146). DOI: 10.1093/jmicro/dfs098

Lo, B. T. W., Ye, L., & Tsang, S. C. E. (2018). The Contribution of Synchrotron X-Ray Powder Diffraction to Modern Zeolite Applications: A Mini-review and Prospects. *Chem*, 4(8), 1778–1808. DOI: 10.1016/j.chempr.2018.04.018

Lu, G. Q., & Zhao, X. S. (2004). *Nanoporous Materials: Science and Engineering* (Vol. 4). PUBLISHED BY IMPERIAL COLLEGE PRESS AND DISTRIBUTED BY WORLD SCIENTIFIC PUBLISHING CO., DOI: 10.1142/p181

Luo, Y., Hou, Z., Jin, D., Gao, J., & Zheng, X. (2006). Template assisted synthesis of Ga2O3–Al2O3 nanorods. *Materials Letters*, 60(3), 393–395. DOI: 10.1016/j.matlet.2005.08.059

Ma, Y., Han, L., Liu, Z., Mayoral, A., Díaz, I., Oleynikov, P., Ohsuna, T., Han, Y., Pan, M., Zhu, Y., Sakamoto, Y., Che, S., & Terasaki, O. (2019). Microscopy of Nanoporous Crystals. In Hawkes, P. W., & Spence, J. C. H. (Eds.), *Springer Handbook of Microscopy* (pp. 1391–1450). Springer International Publishing., DOI: 10.1007/978-3-030-00069-1_29

Malhotra, B. D., & Ali, Md. A. (2018). Nanomaterials in Biosensors. In *Nanomaterials for Biosensors* (pp. 1–74). Elsevier. DOI: 10.1016/B978-0-323-44923-6.00001-7

Martínez, G., Merinero, M., Pérez-Aranda, M., Pérez-Soriano, E., Ortiz, T., Villamor, E., Begines, B., & Alcudia, A. (2020). Environmental Impact of Nanoparticles' Application as an Emerging Technology: A Review. *Materials (Basel)*, 14(1), 166. DOI: 10.3390/ma14010166 PMID: 33396469

Memetova, A., Tyagi, I., Singh, L., & Karri, R. R. Suhas, Tyagi, K., Kumar, V., Memetov, N., Zelenin, A., Tkachev, A., Bogoslovskiy, V., Shigabaeva, G., Galunin, E., Mubarak, N. M., & Agarwal, S. (2022). Nanoporous carbon materials as a sustainable alternative for the remediation of toxic impurities and environmental contaminants: A review. *Science of The Total Environment, 838*, 155943. https://doi.org/https://doi.org/10.1016/j.scitotenv.2022.155943

Narayan, R. (2010). Use of nanomaterials in water purification. *Materials Today*, 13(6), 44–46. https://doi.org/https://doi.org/10.1016/S1369-7021(10)70108-5. DOI: 10.1016/S1369-7021(10)70108-5

Nasrollahzadeh, M., Issaabadi, Z., Sajjadi, M., Sajadi, S. M., & Atarod, M. (2019). Types of Nanostructures. In Nasrollahzadeh, M., Sajadi, S. M., Sajjadi, M., Issaabadi, Z., & Atarod, M. (Eds.), *An Introduction to Green Nanotechnology* (Vol. 28, pp. 29–80). Elsevier., https://doi.org/https://doi.org/10.1016/B978-0-12-813586-0.00002-X DOI: 10.1016/B978-0-12-813586-0.00002-X

Niederberger, M., & Pinna, N. (2009). *Metal Oxide Nanoparticles in Organic Solvents*. Springer London., DOI: 10.1007/978-1-84882-671-7

Owens, G. J., Singh, R. K., Foroutan, F., Alqaysi, M., Han, C.-M., Mahapatra, C., Kim, H.-W., & Knowles, J. C. (2016). Sol–gel based materials for biomedical applications. *Progress in Materials Science*, 77, 1–79. DOI: 10.1016/j.pmatsci.2015.12.001

Parent, L. R., Pham, C. H., Patterson, J. P., Denny, M. S.Jr, Cohen, S. M., Gianneschi, N. C., & Paesani, F. (2017). Pore Breathing of Metal–Organic Frameworks by Environmental Transmission Electron Microscopy. *Journal of the American Chemical Society*, 139(40), 13973–13976. DOI: 10.1021/jacs.7b06585 PMID: 28942647

Park, J., Cho, S. Y., Jung, M., Lee, K., Nah, Y.-C., Attia, N. F., & Oh, H. (2021). Efficient synthetic approach for nanoporous adsorbents capable of pre- and post-combustion CO2 capture and selective gas separation. *Journal of CO2 Utilization, 45*, 101404. https://doi.org/https://doi.org/10.1016/j.jcou.2020.101404

Park, J. H., Wang, Q., Zhu, K., Frank, A. J., & Kim, J. Y. (2019). Electrochemical Deposition of Conformal Semiconductor Layers in Nanoporous Oxides for Sensitized Photoelectrodes. *ACS Omega*, 4(22), 19772–19776. DOI: 10.1021/acsomega.9b02552 PMID: 31788609

Patel, P. N., Mishra, V., & Panchal, A. K. (2012). Theoretical and experimental study of nanoporous silicon photonic microcavity optical sensor devices. *Advances in Natural Sciences: Nanoscience and Nanotechnology*, 3(3), 35016. DOI: 10.1088/2043-6262/3/3/035016

Patterson, J. P., Abellan, P., Denny, M. S.Jr, Park, C., Browning, N. D., Cohen, S. M., Evans, J. E., & Gianneschi, N. C. (2015). Observing the Growth of Metal–Organic Frameworks by in Situ Liquid Cell Transmission Electron Microscopy. *Journal of the American Chemical Society*, 137(23), 7322–7328. DOI: 10.1021/jacs.5b00817 PMID: 26053504

Pauline, S. A., Karuppusamy, I., Gopalsamy, K., & Nallaiyan, R. (2024). Template assisted fabrication of nanoporous titanium dioxide coating on 316 L stainless steel for orthopaedic applications. *Journal of the Taiwan Institute of Chemical Engineers*, 105576, 105576. https://doi.org/https://doi.org/10.1016/j.jtice.2024.105576. DOI: 10.1016/j.jtice.2024.105576

Perrone Donnorso, M., Miele, E., De Angelis, F., La Rocca, R., Limongi, T., Cella Zanacchi, F., Marras, S., Brescia, R., & Di Fabrizio, E. (2012). Nanoporous silicon nanoparticles for drug delivery applications. *Microelectronic Engineering*, 98, 626–629. https://doi.org/https://doi.org/10.1016/j.mee.2012.07.095. DOI: 10.1016/j.mee.2012.07.095

Petersen, H., & Weidenthaler, C. (2022). A review of recent developments for the *in situ / operando* characterization of nanoporous materials. *Inorganic Chemistry Frontiers*, 9(16), 4244–4271. DOI: 10.1039/D2QI00977C

Platschek, B., Keilbach, A., & Bein, T. (2011). Mesoporous Structures Confined in Anodic Alumina Membranes. *Advanced Materials*, 23(21), 2395–2412. DOI: 10.1002/adma.201002828 PMID: 21484885

Polarz, S., & Smarsly, B. (2002). Nanoporous Materials. *Journal of Nanoscience and Nanotechnology*, 2(6), 581–612. DOI: 10.1166/jnn.2002.151 PMID: 12908422

Poolakkandy, R. R., & Menamparambath, M. M. (2020). Soft-template-assisted synthesis: A promising approach for the fabrication of transition metal oxides. *Nanoscale Advances*, 2(11), 5015–5045. DOI: 10.1039/D0NA00599A PMID: 36132034

Prasad Mishra, S., Dutta, S., Kumar Sahu, A., Mishra, K., & Kashyap, P. (2021). Potential Application of Nanoporous Materials in Biomedical Field. In *Nanopores*. IntechOpen., DOI: 10.5772/intechopen.95928

Prida, V. M., Vega, V., García, J., Iglesias, L., Hernando, B., & Minguez-Bacho, I. (2015). Electrochemical methods for template-assisted synthesis of nanostructured materials. In *Magnetic Nano- and Microwires* (pp. 3–39). Elsevier., DOI: 10.1016/B978-0-08-100164-6.00001-1

Priya, A. K., Gnanasekaran, L., Kumar, P. S., Jalil, A. A., Hoang, T. K. A., Rajendran, S., Soto-Moscoso, M., & Balakrishnan, D. (2022). Recent trends and advancements in nanoporous membranes for water purification. *Chemosphere*, 303, 135205. https://doi.org/https://doi.org/10.1016/j.chemosphere.2022.135205. DOI: 10.1016/j.chemosphere.2022.135205 PMID: 35667502

Priya, A. K., Muruganandam, M., & Suresh, S. (2024). Bio-derived carbon-based materials for sustainable environmental remediation and wastewater treatment. *Chemosphere*, 362, 142731. https://doi.org/https://doi.org/10.1016/j.chemosphere.2024.142731. DOI: 10.1016/j.chemosphere.2024.142731 PMID: 38950744

Puchi-Cabrera, E. S., Rossi, E., Sansonetti, G., Sebastiani, M., & Bemporad, E. (2023). Machine learning aided nanoindentation: A review of the current state and future perspectives. *Current Opinion in Solid State and Materials Science*, 27(4), 101091. https://doi.org/https://doi.org/10.1016/j.cossms.2023.101091. DOI: 10.1016/j.cossms.2023.101091

Rodríguez-Sánchez, L., Blanco, M. C., & López-Quintela, M. A. (2000). Electrochemical Synthesis of Silver Nanoparticles. *The Journal of Physical Chemistry B*, 104(41), 9683–9688. DOI: 10.1021/jp001761r

Rouquerol, J., Baron, G., Denoyel, R., Giesche, H., Groen, J., Klobes, P., Levitz, P., Neimark, A. V., Rigby, S., Skudas, R., Sing, K., Thommes, M., & Unger, K. (2011). Liquid intrusion and alternative methods for the characterization of macroporous materials (IUPAC Technical Report). *Pure and Applied Chemistry*, 84(1), 107–136. DOI: 10.1351/PAC-REP-10-11-19

Ryoo, R., Joo, S. H., & Jun, S. (1999). Synthesis of Highly Ordered Carbon Molecular Sieves via Template-Mediated Structural Transformation. *The Journal of Physical Chemistry B*, 103(37), 7743–7746. DOI: 10.1021/jp991673a

Saadati pour. M., Gilak, M. R., Pedram, M. Z., & Naikoo, G. (2023). Enhancement electrochemical properties of supercapacitors using hybrid CuO/Ag/rGO based nanoporous composite as electrode materials. *Journal of Energy Storage, 74*, 109330. https://doi.org/https://doi.org/10.1016/j.est.2023.109330

Sakamoto, Y., Inagaki, S., Ohsuna, T., Ohnishi, N., Fukushima, Y., Nozue, Y., & Terasaki, O. (1998). Structure analysis of mesoporous material 'FSM-16' Studies by electron microscopy and X-ray diffraction. *Microporous and Mesoporous Materials*, 21(4), 589–596. https://doi.org/https://doi.org/10.1016/S1387-1811(98)00053-5. DOI: 10.1016/S1387-1811(98)00053-5

Samipour, S., Setoodeh, P., Rahimpour, E., & Rahimpour, M. R. (2024). Functional nanoporous membranes for drug delivery. In Basile, A., Lipnizki, F., Rahimpour, M. R., & Piemonte, V. (Eds.), *Current Trends and Future Developments on (Bio-) Membranes* (pp. 255–288). Elsevier., https://doi.org/https://doi.org/10.1016/B978-0-323-90258-8.00023-7 DOI: 10.1016/B978-0-323-90258-8.00023-7

Sang, Q., Hao, S., Han, J., & Ding, Y. (2022). Dealloyed nanoporous materials for electrochemical energy conversion and storage. *EnergyChem*, 4(1), 100069. https://doi.org/https://doi.org/10.1016/j.enchem.2022.100069. DOI: 10.1016/j.enchem.2022.100069

Santos, A. (2018). *Electrochemically Engineering of Nanoporous Materials*. MDPI., DOI: 10.3390/books978-3-03897-269-3

Scavetta, E., Casagrande, A., Gualandi, I., & Tonelli, D. (2014). Analytical performances of Ni LDH films electrochemically deposited on Pt surfaces: Phenol and glucose detection. *Journal of Electroanalytical Chemistry (Lausanne, Switzerland)*, 722–723, 15–22. https://doi.org/https://doi.org/10.1016/j.jelechem.2014.03.018. DOI: 10.1016/j.jelechem.2014.03.018

Sharma, A. (2018). *Applications of Nanoporous Materials in Gas Separation and Storage*. https://digitalscholarship.unlv.edu/thesesdissertations/3326

Shi, G., & Kioupakis, E. (2015). Electronic and Optical Properties of Nanoporous Silicon for Solar-Cell Applications. *ACS Photonics*, 2(2), 208–215. DOI: 10.1021/ph5002999

Slowing, I., Viveroescoto, J., Wu, C., & Lin, V.SLOWING. (2008). Mesoporous silica nanoparticles as controlled release drug delivery and gene transfection carriers☆. *Advanced Drug Delivery Reviews*, 60(11), 1278–1288. DOI: 10.1016/j.addr.2008.03.012 PMID: 18514969

Sneddon, G., Greenaway, A., & Yiu, H. H. P. (2014). The Potential Applications of Nanoporous Materials for the Adsorption, Separation, and Catalytic Conversion of Carbon Dioxide. *Advanced Energy Materials*, 4(10), 1301873. Advance online publication. DOI: 10.1002/aenm.201301873

Stan, G., King, S. W., & Cook, R. F. (2009). Elastic modulus of low- *k* dielectric thin films measured by load-dependent contact-resonance atomic force microscopy. *Journal of Materials Research*, 24(9), 2960–2964. DOI: 10.1557/jmr.2009.0357

Stein, A. (2003). Advances in Microporous and Mesoporous Solids—Highlights of Recent Progress. *Advanced Materials*, 15(10), 763–775. DOI: 10.1002/adma.200300007

Su, Y., Wang, X., Ye, Y., Xie, Y., Xu, Y., Jiang, Y., & Wang, C. (2024). Automation and machine learning augmented by large language models in a catalysis study. *Chemical Science (Cambridge)*, 15(31), 12200–12233. https://doi.org/https://doi.org/10.1039/d3sc07012c. DOI: 10.1039/D3SC07012C PMID: 39118602

Suga, M., Asahina, S., Sakuda, Y., Kazumori, H., Nishiyama, H., Nokuo, T., Alfredsson, V., Kjellman, T., Stevens, S. M., Cho, H. S., Cho, M., Han, L., Che, S., Anderson, M. W., Schüth, F., Deng, H., Yaghi, O. M., Liu, Z., Jeong, H. Y., & Terasaki, O. (2014). Recent progress in scanning electron microscopy for the characterization of fine structural details of nano materials. *Progress in Solid State Chemistry*, 42(1), 1–21. https://doi.org/https://doi.org/10.1016/j.progsolidstchem.2014.02.001. DOI: 10.1016/j.progsolidstchem.2014.02.001

Thommes, M., Kaneko, K., Neimark, A. V., Olivier, J. P., Rodriguez-Reinoso, F., Rouquerol, J., & Sing, K. S. W. (2015). Physisorption of gases, with special reference to the evaluation of surface area and pore size distribution (IUPAC Technical Report). *Pure and Applied Chemistry*, 87(9–10), 1051–1069. DOI: 10.1515/pac-2014-1117

Thommes, M., & Schlumberger, C. (2021). Characterization of Nanoporous Materials. *Annual Review of Chemical and Biomolecular Engineering*, 12(1), 137–162. DOI: 10.1146/annurev-chembioeng-061720-081242 PMID: 33770464

Tüysüz, H., Lehmann, C. W., Bongard, H., Tesche, B., Schmidt, R., & Schüth, F. (2008). Direct Imaging of Surface Topology and Pore System of Ordered Mesoporous Silica (MCM-41, SBA-15, and KIT-6) and Nanocast Metal Oxides by High Resolution Scanning Electron Microscopy. *Journal of the American Chemical Society*, 130(34), 11510–11517. DOI: 10.1021/ja803362s PMID: 18671351

Ubaidah Saidin, N., Kuan Ying, K., & Inn Khuan, N. (n.d.). *ELECTRODEPOSITION: PRINCIPLES, APPLICATIONS AND METHODS ELEKTRO-PEMENDAPAN: PRINSIP, APLIKASI DAN KAEDAH.*

Wang, D., Luo, H., Kou, R., Gil, M. P., Xiao, S., Golub, V. O., Yang, Z., Brinker, C. J., & Lu, Y. (2004). A General Route to Macroscopic Hierarchical 3D Nanowire Networks. *Angewandte Chemie International Edition*, 43(45), 6169–6173. https://doi.org/https://doi.org/10.1002/anie.200460535. DOI: 10.1002/anie.200460535 PMID: 15549745

Wang, J., Holt-Hindle, P., MacDonald, D., Thomas, D. F., & Chen, A. (2008). Synthesis and electrochemical study of Pt-based nanoporous materials. *Electrochimica Acta*, 53(23), 6944–6952. https://doi.org/https://doi.org/10.1016/j.electacta.2008.02.028. DOI: 10.1016/j.electacta.2008.02.028

Wu, Z.-P., Zhang, H., Chen, C., Li, G., & Han, Y. (2022). *Applications of in situ electron microscopy in oxygen electrocatalysis*. Microstructures., DOI: 10.20517/microstructures.2021.12

Xiao, X., Song, H., Lin, S., Zhou, Y., Zhan, X., Hu, Z., Zhang, Q., Sun, J., Yang, B., Li, T., Jiao, L., Zhou, J., Tang, J., & Gogotsi, Y. (2016). Scalable salt-templated synthesis of two-dimensional transition metal oxides. *Nature Communications*, 7(1), 11296. DOI: 10.1038/ncomms11296 PMID: 27103200

Xu, P., Li, X., Yu, H., & Xu, T. (2014). Advanced Nanoporous Materials for Micro-Gravimetric Sensing to Trace-Level Bio/Chemical Molecules. *Sensors (Basel)*, 14(10), 19023–19056. DOI: 10.3390/s141019023 PMID: 25313499

Xu, W., Wang, T., Wang, H., Zhu, S., Liang, Y., Cui, Z., Yang, X., & Inoue, A. (2019). Free-standing amorphous nanoporous nickel cobalt phosphide prepared by electrochemically delloying process as a high performance energy storage electrode material. *Energy Storage Materials*, 17, 300–308. https://doi.org/https://doi.org/10.1016/j.ensm.2018.07.005. DOI: 10.1016/j.ensm.2018.07.005

Xu, Y., Fang, X., Xiong, J., & Zhang, Z. (2010). Hydrothermal transformation of titanate nanotubes into single-crystalline TiO2 nanomaterials with controlled phase composition and morphology. *Materials Research Bulletin*, 45(7), 799–804. DOI: 10.1016/j.materresbull.2010.03.016

Yan, W., Hsiao, V. K. S., Zheng, Y. B., Shariff, Y. M., Gao, T., & Huang, T. J. (2009). Towards nanoporous polymer thin film-based drug delivery systems. *Thin Solid Films*, 517(5), 1794–1798. https://doi.org/https://doi.org/10.1016/j.tsf.2008.09.080. DOI: 10.1016/j.tsf.2008.09.080

Yang, F., Wang, X., Zhang, D., Yang, J., Luo, D., Xu, Z., Wei, J., Wang, J.-Q., Xu, Z., Peng, F., Li, X., Li, R., Li, Y., Li, M., Bai, X., Ding, F., & Li, Y. (2014). Chirality-specific growth of single-walled carbon nanotubes on solid alloy catalysts. *Nature*, 510(7506), 522–524. DOI: 10.1038/nature13434 PMID: 24965654

Yang, J. W., Kwon, H. R., Seo, J. H., Ryu, S., & Jang, H. W. (2024). Nanoporous oxide electrodes for energy conversion and storage devices. *RSC Applied Interfaces*, 1(1), 11–42. https://doi.org/https://doi.org/10.1039/d3lf00094j. DOI: 10.1039/D3LF00094J

Yang, S., Zavalij, P. Y., & Stanley Whittingham, M. (2001). Hydrothermal synthesis of lithium iron phosphate cathodes. *Electrochemistry Communications*, 3(9), 505–508. DOI: 10.1016/S1388-2481(01)00200-4

Yang, Z., Du, X., Ye, X., Qu, X., Duan, H., Xing, Y., Shao, L.-H., & Chen, C. (2021). The free-standing nanoporous palladium for hydrogen isotope storage. *Journal of Alloys and Compounds*, 854, 157062. https://doi.org/https://doi.org/10.1016/j.jallcom.2020.157062. DOI: 10.1016/j.jallcom.2020.157062

Yong, Y., Zhang, W., Hou, Q., Gao, R., Yuan, X., Hu, S., & Kuang, Y. (2022). Highly sensitive and selective gas sensors based on nanoporous CN monolayer for reusable detection of NO, H2S and NH3: A first-principles study. *Applied Surface Science*, 606, 154806. https://doi.org/https://doi.org/10.1016/j.apsusc.2022.154806. DOI: 10.1016/j.apsusc.2022.154806

Yu, X., Wang, J., Huang, Z.-H., Shen, W., & Kang, F. (2013). Ordered mesoporous carbon nanospheres as electrode materials for high-performance supercapacitors. *Electrochemistry Communications*, 36, 66–70. DOI: 10.1016/j.elecom.2013.09.010

Zabukovec Logar, N., & Kaučič, V. (2006). She was visiting researcher in the Structural chemistry group at the University of Manchester, UK in 1995 and 1996. She was promoted to Assistant Professor in Chemistry at the University of Nova Gorica in 2004. She has worked in the Laboratory for Inorganic Chemistry and Technology at the National Institute of Chemistry in Ljubljana since. In *currently Acta Chim. Slov* (Vol. 53).

Zanco, S. E., Joss, L., Hefti, M., Gazzani, M., & Mazzotti, M. (2017). Addressing the Criticalities for the Deployment of Adsorption-based CO2 Capture Processes. *Energy Procedia*, 114, 2497–2505. https://doi.org/https://doi.org/10.1016/j.egypro.2017.03.1407. DOI: 10.1016/j.egypro.2017.03.1407

Zha, J., & Roggendorf, H. (1991). Sol–gel science, the physics and chemistry of sol–gel processing, Ed. by C. J. Brinker and G. W. Scherer, Academic Press, Boston 1990, xiv, 908 pp., bound—ISBN 0-12-134970-5. *Advanced Materials, 3*(10), 522–522. DOI: 10.1002/adma.19910031025

Zhang, C., Firestein, K. L., Fernando, J. F. S., Siriwardena, D., von Treifeldt, J. E., & Golberg, D. (2020). Recent Progress of In Situ Transmission Electron Microscopy for Energy Materials. *Advanced Materials*, 32(18), 1904094. Advance online publication. DOI: 10.1002/adma.201904094 PMID: 31566272

Zhang, Q., & Cao, G. (2011). Nanostructured photoelectrodes for dye-sensitized solar cells. *Nano Today*, 6(1), 91–109. DOI: 10.1016/j.nantod.2010.12.007

Zhao, D., Feng, J., Huo, Q., Melosh, N., Fredrickson, G. H., Chmelka, B. F., & Stucky, G. D. (1998). Triblock Copolymer Syntheses of Mesoporous Silica with Periodic 50 to 300 Angstrom Pores. *Science*, 279(5350), 548–552. DOI: 10.1126/science.279.5350.548 PMID: 9438845

Zhao, D., Xu, D., Wang, T., Yang, Z., Zhao, D., Xu, D., Wang, T., & Yang, Z. (2024). Nitrogen-Rich Nanoporous Carbon with MXene Composite for High-Performance Zn-ion Hybrid Capacitors. *Materials Today. Energy*, 101671, 101671. https://doi.org/https://doi.org/10.1016/j.mtener.2024.101671. DOI: 10.1016/j.mtener.2024.101671

Zhao, X. S. (n.d.). *NANOPOROUS MATERIALS-AN OVERVIEW*.

Zhou, Y., Niu, B., Wu, B., Luo, S., Fu, J., Zhao, Y., Quan, G., Pan, X., & Wu, C. (2020). A homogenous nanoporous pulmonary drug delivery system based on metal-organic frameworks with fine aerosolization performance and good compatibility. *Acta Pharmaceutica Sinica. B*, 10(12), 2404–2416. https://doi.org/https://doi.org/10.1016/j.apsb.2020.07.018. DOI: 10.1016/j.apsb.2020.07.018 PMID: 33354510

Zhou, Y., Zhang, Y., Tang, L., Long, B., & Zeng, G. (2019). Nanoporous Materials Based Sensors for Pollutant Detection. In Tang, L., Deng, Y., Wang, J., Wang, J., & Zeng, G. (Eds.), *Nanohybrid and Nanoporous Materials for Aquatic Pollution Control* (pp. 265–291). Elsevier., https://doi.org/https://doi.org/10.1016/B978-0-12-814154-0.00009-8 DOI: 10.1016/B978-0-12-814154-0.00009-8

Zhu, J., Yin, Z., Li, H., Tan, H., Chow, C. L., Zhang, H., Hng, H. H., Ma, J., & Yan, Q. (2011). Bottom-Up Preparation of Porous Metal-Oxide Ultrathin Sheets with Adjustable Composition/Phases and Their Applications. *Small*, 7(24), 3458–3464. DOI: 10.1002/smll.201101729 PMID: 22058077

Zhuang, Z., Chen, Y., Chen, K., Liu, Z., Guo, Z., & Huang, X. (2023). In-situ synthesis of ZnO onto nanoporous gold microelectrode for the electrochemical sensing of arsenic(III) in near-neutral conditions. *Sensors and Actuators. B, Chemical*, 378, 133184. https://doi.org/https://doi.org/10.1016/j.snb.2022.133184. DOI: 10.1016/j.snb.2022.133184

Chapter 11
Synthesis and Characterization of Activated Carbon From Biomass

Makhabbat Kunarbekova
Satbayev University, Kazakhstan

Rosa Busquets
Kingston University, UK

Inabat Sapargali
https://orcid.org/0009-0002-0258-8277
Satbayev University, Kazakhstan

Laura Seimukhanova
Satbayev University, Kazakhstan

Ulan Zhantikeyev
https://orcid.org/0000-0002-1200-2340
Satbayev University, Kazakhstan

Kenes Kudaibergenov
Satbayev University, Kazakhstan

Seitkhan Azat
Satbayev University, Kazakhstan

ABSTRACT

This chapter discusses current practices and new developments in the preparation of activated carbons for the adsorption of pollutants. There is widespread pollution such as pharmaceuticals and heavy metals in global surface water that needs to be mitigated and carbon sorbents made for it can be part of the solution. Preparing carbons for such challenge requires tailoring carbon's structure and surface chemistry to maximise interaction with low concentrations (part per billion level) of a variety of pollutants. The preparation of effective carbon sorbents constitutes a technical challenge. This chapter explores treatments and analytical approaches for the preparation and development of modified carbons to remove water pollutants. Effective carbons will help to control global contamination problems but should not be the source of more pollution: carbon dioxide emissions during the production and maintenance of carbon sorbents are a concern.

DOI: 10.4018/979-8-3693-7505-1.ch011

1. INTRODUCTION

Environmental pollution has a global impact and affects every environmental compartment (United Nations 2024). News today would be finding the sea, lakes, wetlands and rivers free from pollution. Indeed, they are generally polluted, and there are legal frameworks around the world that aim to protect them. Examples of potentially harmful pollutants reaching the environment are pharmaceuticals and other personal care products (Gros et al, 2022) or plastic degradation products (Soltani et al., 2022), which by the action of UV and other natural weathering agents, can leach some of their additives, will fragment (Jansen et al., 2024) and some will subsequently transform further in the environment. Effluents from industrial and household wastewater are an important source (or pathway) to the environment. The impact of exposing the environment to a mixture of pollutants at low concentrations (typically, they are parts per billion level) is very difficult to measure (Wilkinson et al., 2024). Today, surface water and soil, even food, can have traces of pollutants (FAO, United Nations 2021). Even especially protected environmental sites in developed countries can be contaminated (Boxall et al., 2024).

Wastewater treatment and waste management practices are sometimes insufficient for preventing the complete release of contaminants into the environment. This is because of insufficient combined treatments; limited treatment time; ineffective technologies; or having to deal with pollutants that are especially resilient (with structures that will not be degraded or removed by adsorption with the current treatments). The removal of pollutants by adsorption has the advantage of not generating degradation products, and one of their disadvantages is that the adsorbents need to be re-generated once saturated. A traditional sorbent used for removing pollutants from water is carbon, which is used as a final treatment stage (tertiary treatment) (Bhatnagar et al., 2013). Nevertheless, the absorption removal using carbons is affected by the very low concentration of pollutants in water (typically in the $<1\mu g/L$). This is because the sorption process is less effective when removing diluted concentrations of pollutants. Furthermore, carbons used in water treatment were not designed for the specific removal of certain contaminants. These remain in the treated effluents and can impact ecosystems.

Carbons are indeed useful materials because they can be adapted to effectively adsorb a wide range of pollutants that traditionally do not get removed; their selectivity can be enhanced (Liang, Z. et al., 2022, Sailaukhanuly, Y. et al., 2024). Activated carbons are porous and can be made to reach a high specific surface area, which can provide exceptional adsorption capacity. Traditionally, coal has been used in water treatment. Very effective carbons for water treatment can also be prepared synthetically (Busquets et al., 2014). In light of sustainable development and the need for rational use of resources, there is a growing interest in the production of activated carbons from renewable sources, in particular from biomass. The use of agricultural and industrial waste not only reduces the cost of raw materials, but also solves the problem of waste disposal, promoting a circular economy (Li et al., 2008). The relevance of the study of activated carbons is also due to the need to develop cost-effective and environmentally friendly purification methods. Traditional methods, such as membrane filtration, often require significant energy and material costs. At the same time, activated carbons can be used in a wide range of conditions, have a high regeneration capacity and are compatible with various technological processes (Gupta & Suhas, 2009). Current research is focused on developing new methods for synthesizing and modifying activated carbons to improve their efficiency, selectivity for specific pollutants, and scalability. Key challenges are to control the pore structure and surface chemistry of carbons. Physical and chemical activation techniques can control the pore size and distribution, as well as introduce functional groups on the surface that affect the interaction with pollutant molecules (Blankenship, 2013). Surface modification of activated carbons are important to consider for improving their sorption properties. The introduction of functional groups such as oxygen-containing, nitrogen-containing or sulfur-containing groups can enhance the adsorption of specific pollutants through chemical interaction mechanisms including ion exchange, complexation and hydrogen bonding (Bansal & Goyal, 2005). For example, surface oxidation facilitates the removal of heavy metals from aqueous solutions by forming coordination bonds between metal ions and oxygen-containing groups on the carbon surface (Dias et al., 2007). The application of metal or metal oxide nanoparticles to the surface of carbons offers new possibilities for combining adsorption, increasing surface area, and even providing catalytic properties (Ximo-Luminita et al. 2023). Such composite materials can be effective in the decomposition of organic pollutants, for example, through photocatalysis or ozonation (F. Zhang et al., 2005). Nevertheless, an advantage that activated carbons have, in comparison with composites containing metals, is that their re-activation and re-use are very well established. Unfortunately, the activation of the emission of CO_2, a greenhouse gas, during their activation is a limitation that needs to be mitigated.

2. PREPARATION OF ACTIVATED CARBONS (PRE- AND POST-ACTIVATION)

Activated carbons (ACs) are adsorbents whose structure has been enhanced for the uptake of small-size molecules (e.g. <500 u.m.a) such as common pharmaceuticals and pesticides. They can contribute to solving a wide range of issues in terms of purification and discolouration of air, gases, water and other liquids (Wang et al., 2022), (Kunarbekova et al., 2024). ACs can be prepared by thermally treating carbonaceous biomass (e.g. wood, nut shells, husks, fruit stones, and others) in the absence of oxygen. The resulting ACs can include suitable pores (mesopores (2-50 nm) and micropores (<2 nm) (IUPAC, 1971) for removing typical pollutants. Such pores also provide very large specific surface area per unit mass (Azat et al., 2013b).There are various types of activated carbons. Activated carbons differ from their starting raw materials and have new applications (Jandosov et al., 2022) and value. Figure 1 shows examples of the main areas where ACs are used.

Figure 1. Application of activated carbons

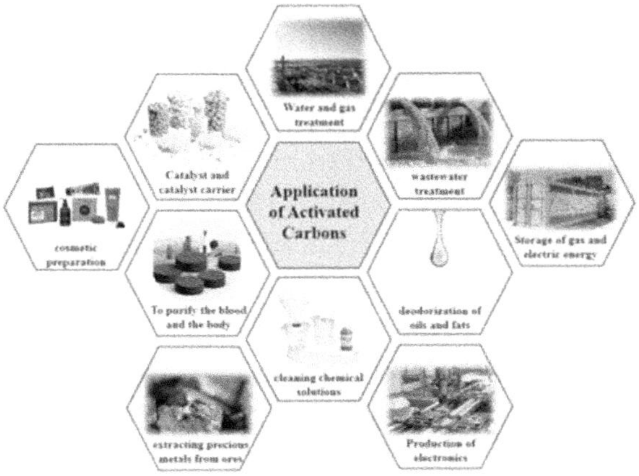

The first stages for producing activated carbons from biomass consist of washing and drying to remove impurities. Key factors that contribute to the extraction performance of activated carbons are the specific surface area, pore structure, surface functional groups, and elemental composition of the carbon (see Figure 2). The specific surface area directly influences the adsorption capacity of activated carbons, with larger surface areas providing more active sites for the adsorption of pollutants. The pore structure, including the size and distribution of micropores, mesopores, and macropores, affects the ability of the activated carbon to adsorb different sizes of molecules, as well as the rate of adsorption. Surface functional groups, such as acidic or basic groups, play a crucial role in determining the interactions between the carbon and adsorbates, influencing the selectivity and strength of adsorption. The elemental composition, particularly the carbon content and the presence of heteroatoms like oxygen, nitrogen, or sulfur, can modify the surface chemistry and reactivity, further impacting the material's performance in applications like catalysis or pollutant removal. Together, these factors work synergistically and affect the overall effectiveness of the sorbent (Abdiyev et al., 2023).

Figure 2. Key factors and primary modification methods of activated carbon and their application in adsorption of gases (Wang et al., 2022)

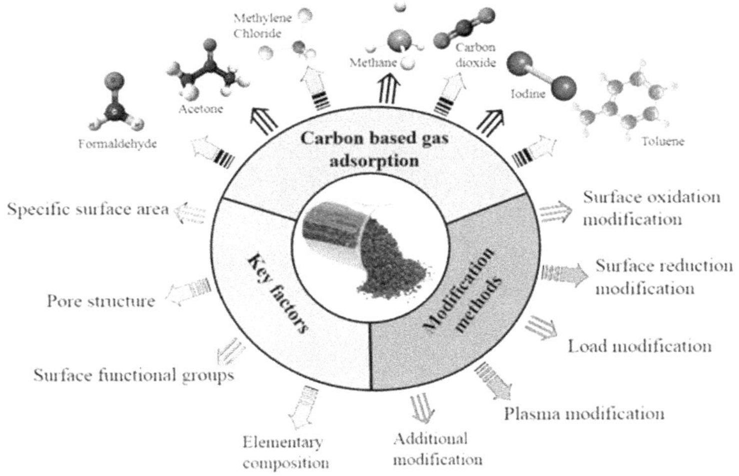

The forms of AC classified into three main groups:

Crushed activated carbon (CAC) is obtained from crushing large fractions of carbonised biomass and subsequently removing dust left in its structure. The CAC particles have an irregular shape, and their particle size is ~0.2 - 5 mm. Due to its structure and particle size distribution, this type of AC is widely used as a loading material in various filter systems (both pressured and non-pressured). The main advantage of CAC is its high efficiency in cleaning liquid media, which lies mainly on their particle size. For instance, in filters where the purification process occurs under pressure, there is a rapid flow of water through the filter bed made of CAC, while simultaneously capturing and retaining various contaminants such as organic matter, oils, chlorinated compounds and heavy metals (Chang et al., 2000). In non-pressure filters, where the liquids flow through the media by the action of gravity, CAC also demonstrates excellent adsorption properties, although they provide a slower filtration rate than CAC pressure filters.

Granular activated carbon (GAC) consists of granules with typical sizes ranging from 0.8 to 5.0 mm. GAC is characterized by a high degree of homogeneity and uniformity in size, which ensures effective interaction with the purified medium. Their main area of application is adsorption from the gas phase, where it is used to purify the air and remove harmful gases, vapours of organic compounds, toxic substances and volatile hydrocarbons. One of the key advantages of GAC is its ability to minimize the pressure drop in filter systems. Due to its cylindrical shape and stable structure of the granules, the air or gas flow easily passes through the carbon layer without creating significant resistance. This makes GAC an ideal material for gas purification in industrial ventilation systems, gas cleaning and the processing of gas emissions (Adeleke *et al.*, 2018). Another important property of granular activated carbon is its high attrition. GAC is resistant to mechanical damage, which reduces the risk of granule destruction and release of fines during operation, and this extends the service life of filters. GAC is used in packed columns (>1m) in the tertiary treatment in the production of drinking water and also in the purification of water in hospitals (Passos et al., 2023).

Powdered activated carbon (PAC) is crushed activated carbon consisting of particles usually less than 100 μm in diameter (Świątkowski, 1999). Due to its high dispersion, PAC has a specific surface area, typically about 2000 m^2/g, making it especially effective for the adsorption of various pollutants from liquid media. The main advantage of powdered activated carbon is its ability to quickly and effectively interact with pollutants due to the fast adsorption kinetics that result from their small particle size. This provides a high degree of purification even at low concentrations (ng/L) of pollutants. PAC is widely used in the final stage of the purification of liquids such as wastewater and in water purification for the production of pharmaceuticals and food. This is because such uses require high efficiency in the removal of organic and inorganic pollutants. PAC is mixed with the liquid to be purified in a pre-calculated proportion (dosage), after which the adsorption process occurs in a stirred system. Pollutants become adsorbed onto the carbon, and finally, the carbon with adsorbed substances is removed by filtration or coagulation, flocculation and sedimentation. Indeed, the separation of the fine PAC particles is a slow part of the process that becomes a disadvantage in its use. An important advantage of PAC is its versatility: it can be used in both continuous and periodic purification systems. Nevertheless, the high back pressure it can provide due to a packing bed with low particle size is a limitation. PAC's relative versatility and effectiveness make it a popular choice across industries.

3. PHYSICAL AND CHEMICAL ACTIVATION OF CARBONS

With the growing awareness of the need to protect the environment, the popularity of processing agricultural waste into activated carbon, a valuable material, is growing worldwide. A broad range of raw materials can be used to produce activated carbons. The choice of hydrocarbon raw materials significantly affects the properties of the final activated carbon (Melia et al., 2019). During carbonization and activation, which are two sequential stages that lead to the production of AC, the raw materials are transformed into a carbon matrix with a well-developed porous structure (Ao et al., 2018) . There are two main methods of producing ACs: physical and chemical activation, shown in Figure 3.

Figure 3. Processes of activated carbon production

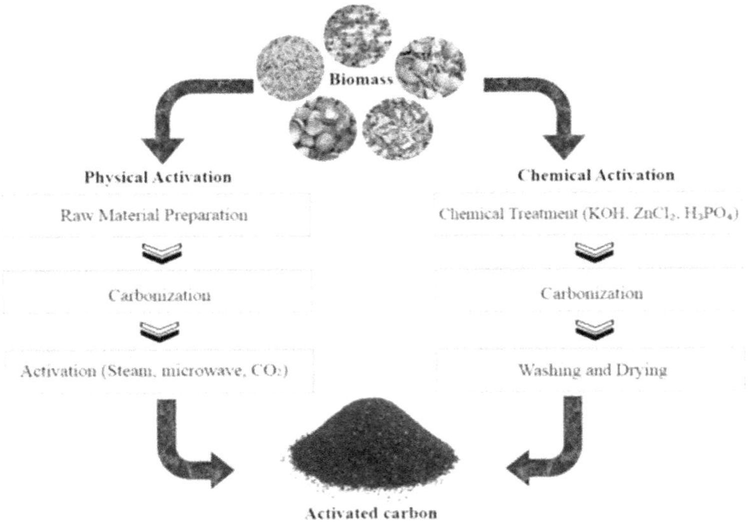

Physical activation. Physical activation usually includes two main stages: carbonization and activation. In the first stage of AC production, carbonization, the source material is subjected to heat treatment with a very low presence of oxygen. As a result of which volatiles (moisture and tars) are removed from it, the material is compacted and gains strength. During carbonization, the raw materials are exposed to high temperatures (usually from 400 to 700 °C) in an inert atmosphere (typically N_2 or Ar). This process leads to the thermal decomposition of non-carbon elements, such as hydrogen, oxygen, and nitrogen, which are present in the raw material in the form of volatile organic compounds. These elements are released as gases (e.g., water vapor, carbon dioxide, methane, and other hydrocarbons), leaving behind a carbon-rich residue. This carbonization step also results in the formation of a basic pore structure in the material, which is further developed during the activation stage (Marzuki et al., 2023).

The activation stage develops a microporous structure. The next stage, activation, increases the porosity of the carbon material. In physical activation, an oxidizing gas, such as steam or carbon dioxide, is used for this at temperatures from 700 to 1200 °C (Lan et al., 2019). The oxidizing gas reacts with the carbon surface, oxidizing carbon atoms and forming carbon oxides (CO and CO_2), which are released as gases. This process etches the carbon surface, creating a network of micropores and mesopores. Such high temperature and controlled oxidation process results in the formation of a material with a large specific surface area, which can be in the region of 1500 m^2/ g AC, which is especially important for applications requiring efficient adsorption such as the removal of pollution from water (Lua & Yang, 2004).

Microwave activation. This method relies on microwave radiation to rapidly heat the carbon feedstock. Microwave activation has several advantages, such as short processing times compared to physical activation, lower energy consumption, and more uniform heating than physical activation. This process can result in a more uniform porous structure than traditional thermal activation methods. This approach is becoming increasingly popular due to its potential to scale up activated carbon production with less environmental impact (Lua & Yang, 2004).

Chemical activation. Chemical activation is a one-step process in which the feedstock is impregnated with a chemical activating agent such as potassium hydroxide (KOH), zinc chloride ($ZnCl_2$) or phosphoric acid (H_3PO_4). The impregnated material is then heated at temperatures between 400 and 800°C. Activating agents promote the decomposition of the carbon structure, creating a highly porous material with a large specific surface area, such as activated carbon with surface areas exceeding 1500 m^2/g, which is highly desirable for applications like gas adsorption and water purification. One of the advantages of chemical activation is that it typically occurs at lower temperatures than physical activation, which can result in higher yields of activated carbon, which is economically advantageous. It also allows for the development of pores with varying sizes, ranging from micropores (less than 2 nm) to mesopores (2-50 nm), which increases the versatility of the material for different applications. However, the concentration of the activating agent and the process temperature must be carefully controlled to optimize the pore structure and prevent excessive combustion of the carbon material (Yi et al., 2021).

An example of a typical chemical activation involves KOH as an activating agent. KOH reacts with the starting carbon material, promoting the removal of non-carbon atoms, such as hydrogen, oxygen, and nitrogen, in the form of water vapor, carbon dioxide, and other gases. This process creates a highly porous structure by etching the carbon material and developing a network of micropores. After activation, the material is thoroughly washed to remove residual activating agents, resulting in a clean and highly efficient adsorbent. The assessment of different kinds of activation on key aspects of the resultant product is summarized in Table 1.

Table 1 presents a comparison of different carbon activation methods, such as physical and chemical activation. These methods vary significantly across key parameters like surface area, pore size, product yield, and costs. Specifically, chemical activation provides a much higher specific surface area than physical activation due to the use of chemical agents like KOH or H_3PO_4, though it also comes with higher costs and lower product yield (Yang et al., 2021).

Table 1. Comparison of activation methods

Parameter	Physical Activation	Chemical Activation
Surface Area	High	Very High
Pore size	Meso- and Macropores	Micro- and Mesopores
Product Yield	Moderate	Low
Costs	Low	High

4. SURFACE MODIFICATIONS

The effectivity of AC in the adsorption process is mostly affected by its surface chemistry and the pore structure of the carbons. The activation conditions and raw materials used impact activated carbons' surface chemistry and pore structure. For instance, the activation of carbon from pea peels is especially rich in amino groups (Novoseltseva et al., 2021). In addition, surface modifications can be carried out to increase particular functional groups of interest in a certain application (Bhatnagar et al., 2012). Standard surface modification methods are chemical, physical and plasma treatments.

Chemical treatments. *Oxidation* involves treating AC with oxidising chemicals (such as nitric acid, hydrogen peroxide, or ozone) to enrich the surface with oxygen-containing functional groups (such as hydroxyl, carboxyl, and carbonyl) (Jaramillo et al., 2010). These increase the hydrophilicity of the ACs and subsequently their interaction with polar molecules and heavy metal ions in liquids. Due to this, carbon that has undergone oxidation is widely used in wastewater treatment, treatment of industrial effluents and removal of heavy metals such as lead, cadmium and mercury from contaminated liquids (Thombre, 2023). Oxidation can also increase the sorption capacity of AC with respect to polar substances in the gas phase, which expands its application in gas purification and air emission control. This process of modifying the structure of carbon is an important step in the creation of materials for specialized applications where improved adsorption properties and high chemical activity are required.

Acid/base modification: modifications of AC with acids or alkalis is one of the effective methods for improving its adsorption properties. During acid treatments, such as treatment with sulfuric, phosphoric or nitric acids, acidic functional groups such as carboxyl (-COOH), phenolic (-OH) and sulfonate (-SO_3H) groups are introduced onto the carbon surface. These modifications increase the ability of carbon to adsorb polar molecules and heavy metal cations by increasing the surface acidity and increasing the number of active interaction sites. In contrast, alkaline treatments, such as those using sodium hydroxide (NaOH) or potassium hydroxide (KOH), result in the enrichment of the carbon surface with basic functional groups. Such treatments improve the adsorption of compounds of acidic nature. Alkaline functional groups increase the overall basicity of the carbon, which improves interaction with anions and acidic molecules. This is useful for the purification both gas and liquid media from acidic pollutants. This modification process is critical for creating specialized ACs used in water and gas purification, as well as in industrial filtration systems. (Liu et al., 2007).

Activation with steam or CO_2 involves high temperatures (700-1000°C) to produce new pores or enlarge the existing ones. This results in a significant increase in the surface area and overall porosity of the AC, which increases its adsorption capacity.

Annealing is a high temperature treatment during which there is a decrease on oxygen-containing functional groups from the carbon surface. This reduces the hydrophilicity of the carbon and makes it more hydrophobic, which improves its interaction with less polar substances. In addition, annealing increases the electrical conductivity of the carbon, which makes this approach especially useful for creating materials used in electrochemical devices such as supercapacitors and batteries (Chi et al., 2023). Indeed, carbons with high electrical conductivity can contribute to improved efficiency of charge storage and transfer, which is a key requirement for energy systems. Thermal treatment of AC allows for targeted modification of its properties, making it suitable for a variety of applications including pollutant adsorption, electrochemical devices, and other high-tech applications where the structural and surface characteristics of the material are important.

Physical Coating and grafting consist of applying a thin polymer layer to the surface of AC to improve its characteristics and expand its scope of application. One of the key effects achieved by polymer coating is an increase in the hydrophobicity of the carbon material (Li et al., 2020). This property is especially useful in the processes of purifying organic substances from aqueous solutions, for instance, in the case of removing hydrophobic organic compounds such as oils and hydrocarbons (Li et al., 2020b). Here, it is essential to minimize the interaction of carbon with water and enhance the adsorption of non-polar pollutants. In addition to hydrophobicity, polymer coating can increase the selectivity of AC towards certain molecules. This is achieved by modifying the surface in such a way that carbon can more effectively interact with target substances, which makes it especially useful in the processes of sorption and separation of complex mixtures. For example, a polymer coating can be especially selective for adsorbing organic pollutants, dyes or pharmaceuticals (Sun et al., 2005). Another important advantage of polymer coating is an increase in the mechanical strength of the carbon. The coating protects the carbon surface from degradation during its use, which increases its service life and resistance to mechanical stress. This is especially important for reusable systems where AC is subject to regeneration and long service life (Feng et al., 2021).

Carbon surface decorated with nanoparticles. Typical nanoparticles used to functionalised carbons are metals or metal oxides, such as silver, gold or iron oxide, to the surface of AC (Bowden et al., 2018). This modification method can improve the functional properties of AC, expanding its application in various technological processes. For instance, the addition of silver and gold nanoparticles increases the ability of carbon to participate in catalytic reactions (Molina et al., 2019). Such materials can be used in oxidation, reduction and other chemical reactions that require highly active catalysts. The catalytic activity of decorated carbon makes it indispensable in processes for cleaning water and air from organic pollutants. Silver nanoparticles also impart antibacterial properties, making the AC composite with them an effective material for disinfecting water, air or medical fluids (Xin et al., 2022). The addition of iron oxide nanoparticles imparts magnetic properties to the carbon, enabling its separation using magnetic fields. This method is particularly useful for simplifying the extraction of carbon particles from the cleaned medium after the adsorption or catalysis process is complete (Biswas & Bandyopadhyaya, 2017). Magnetic nanoparticles greatly facilitate the regeneration and reuse of AC, making the process more cost-effective and environmentally friendly (Woldeamanuel et al., 2024).

Plasma Treatment.*Cold Plasma Treatment:* Cold plasma treatment is a method of modifying the surface of carbon by exposure to a plasma field at low temperatures. This process leads to changing the composition and surface structure of AC without affecting its bulk properties. The main advantage of this method is its ability to introduce new functional groups, such as carboxyl (-COOH), hydroxyl (-OH) and other oxygen-containing groups, which significantly increases the adsorption capacity of the material for polar molecules and ions. Under the influence of cold plasma, active centres can form on the surface of carbon, which improves interaction with various pollutants. This process is especially effective for creating carbons used in water purification and gas filtration systems, where high sorption capacity and chemical activity are required. The plasma treatment can increase the carbon surface roughness or porosity. This improves the diffusion of molecules into the pores of the carbon and increases the total surface area for adsorption. At the same time, the volumetric properties of carbon, such as its strength and structure, remain unchanged, which helps maintain the durability and resistance of the material to physical impacts (L. Zhang et al., 2017).

Surface modification is usually done to saturate carbon surfaces with oxygen-containing groups. This enhances their affinity for contaminants in water. Figure 4 illustrates some examples of oxygen-containing carbon surfaces. For instance, to extract pharmaceuticals and dyes from water through hydrogen bonding (Lobato-Peralta et al., 2021). In some cases, S, N groups can be introduced to the carbon surface to increase the extraction capacity of organic contaminants and radionuclides (Xiao et al., 2020). Thiol groups interact with heavy metals through electron interaction (Zhang et al., 2012).

Figure 4. The main set of oxygen-containing functional groups in activated carbon

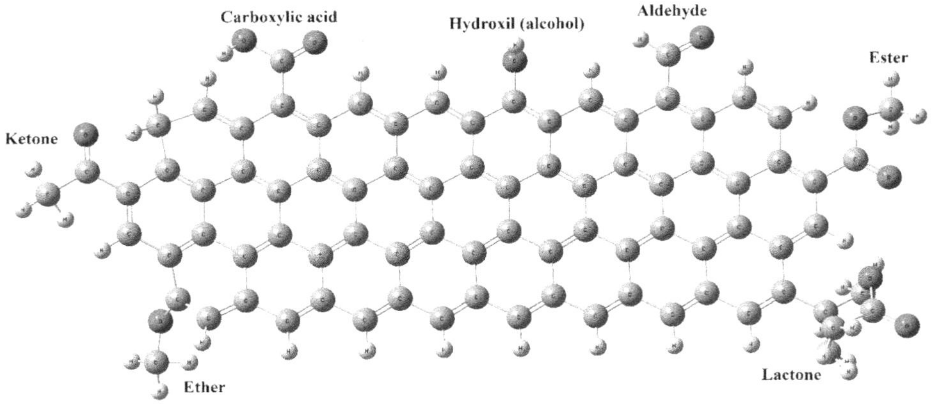

Figure 5 illustrates examples of sulfur functional groups on the surface of AC. Displayed are the following groups: sulfoxide with a methyl group (–S(=O)–CH_3), thiol (–SH), dithiocarbamate (–S(=O)$_2$N–CH_3), sulfone (–SO_2–CH_3), and sulfonic acid (–SO_3H). Each of these sulfur-containing functional groups enhances the carbon's adsorptive capacity through various chemical interactions with adsorbates. These groups are crucial for improving the adsorption of heavy metals and other pollutants due to their ability to form complexes or participate in ion exchanges (Lobato-Peralta et al., 2021b).

Figure 5. Examples of sulfur functional groups in activated carbon

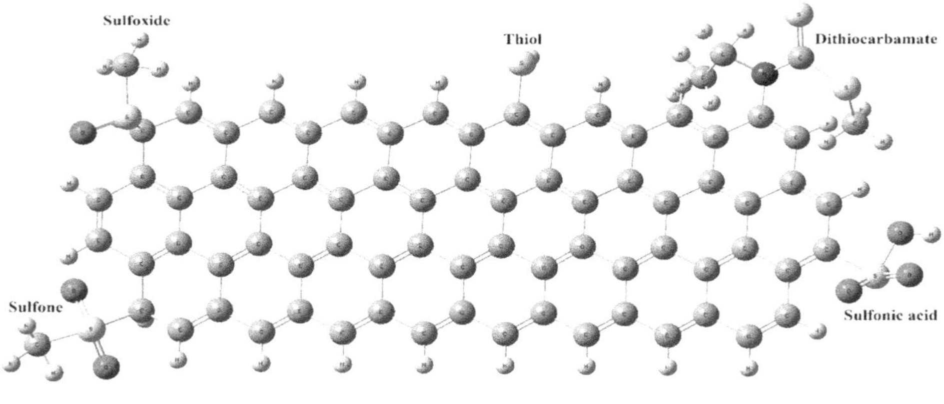

Typical nitrogen-containing functional groups on the aromatic carbon backbone of AC are shown in Figure 6. From left to right, the groups include imide, which has complexation properties due to its carbonyl groups and a methyl group; nitro, an electron-acceptor due to its double bond with oxygen; amine, capable of forming hydrogen bonds; amide, involved in hydrogen bonding and resonance stabilization; triazole, useful for metal coordination; and hydroxylamine, reactive and suitable for reduction reactions.

These groups enhance the interaction of AC with adsorbates, finding applications in environmental and chemical technologies (Lobato-Peralta et al., 2021b).

Figure 6. Examples of nitrogen functional groups in activated carbon

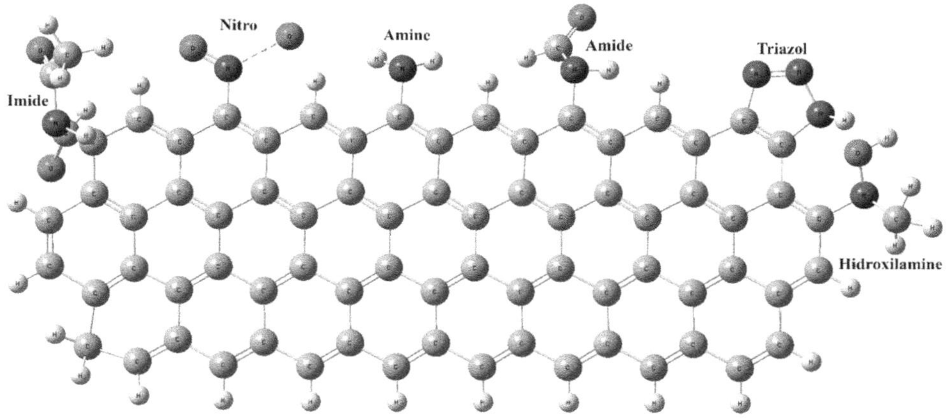

Surface modification of activated carbon is used to improve the adsorption capacity of the ACs mainly through involvement of chemical interaction between the functional groups onto the surface of the carbon with target molecules such as pollutants.

3. CHARACTERIZATION OF ACS

Complementary techniques are needed for the characterization of the carbons, this is because carbons are multifaceted materials with multiple characteristics that are linked with their performance, as shown in Figure 7. Accurate measurements of properties associated with carbons' performance will help to improve their effectivity. Such information will allow tailoring such key properties to specific applications. One can get a wealth of information related to how well the material will perform from data on its surface chemistry and structure. Thorough characterization of ACs typically includes analysis of particle size, Brunauer-Emmett-Teller (BET) surface area, porosimetry, zeta potential and well characterization of functional groups with infrared (IR) or Raman spectroscopies or even their quantification with Boehm titrations and X-ray photoelectron spectroscopy (XPS). Each of these methods offers additional information about the materials. For example, the particle size of the carbons affects overall surface area and adsorption kinetics. At the same time, BET and porosimetry provide extensive data on pore size distribution and surface area, which is critical to understand avenues for tailoring their properties and performance. Zeta potential reveals information about the surface charge, which is important for assessing the stability of ACs in suspension and their interactions with charged contaminants in solution. Boehm titration provides insight into the surface functional groups, allowing for a better understanding of how the surface chemistry of the AC affects its interactions with various contaminants. Finally, spectral techniques such as IR play a major role in identifying specific groups involved in interacting with crucial substances. Such data is critical for measuring carbons' capabilities and modifying them to achieve optimal performance in target applications. Incorporating these techniques during the development of carbon sorbents is essential to gain a comprehensive understanding of their properties and potential performance. The data obtained not only guides new development of AC materials but also expands their applications in various fields, making carbons' physicochemical characterization a fundamental aspect of ACs.

Figure 7. Schematic representation of approaches for the physical and chemical characterization of activated carbons

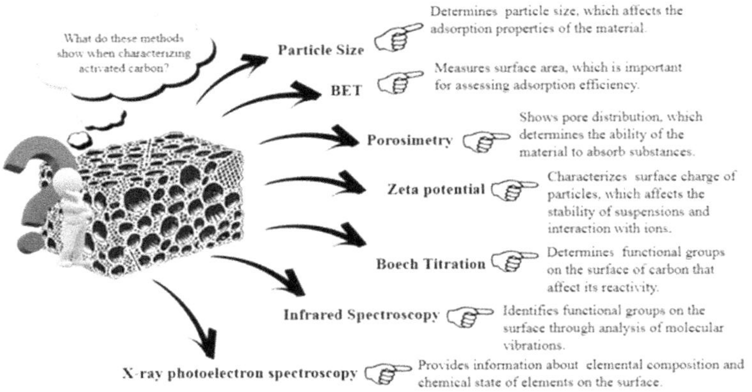

Particle Size. The particle size of AC is an important parameter that significantly affects its adsorption properties and overall efficiency in various applications. Particle size affects pore accessibility, especially in carbons that are mainly microporous and have a low abundance of transport ports. Small particle sizes, e.g. from PAC, lead to a great surface area that is exposed and available for interaction with the molecules of interest; these can be, for instance, pollutants, and this translates into greater adsorption rates. Hence, particle size is especially critical in filtration systems that require fast adsorption rates.

Therefore, particle size analysis measurements are needed. Techniques for such characterisation include laser diffraction and dynamic light scattering (DLS). ACs obtained from various raw materials such as coconut shells, wood or coal can be processed to reach particle sizes ranging from 10 to 50 microns. This kind of size provides good adsorption capacity, and the hydraulic resistance can still be overcome. However, ACs with smaller particle sizes (<10 micrometers), although have improved adsorption characteristics, lead to too high back pressure in filtration systems. A balance between particle size and porosity is important for specific applications such as water or air purification (Macías-García et al., 2019).

Specific surface area is one of the most important characteristics of ACs. The most widely used method for determining it is the BET method, which is based on the quantification of adsorbed inert gases (e.g. nitrogen) on the surface of a material at low temperatures. ACs can have a specific surface area in the range of 500 to 3000 m^2/g, depending on the activation method and feedstock. For example, AC obtained by chemical activation using KOH exhibits a specific surface area of up to 3000 m^2/g, which has high ability to adsorb small molecules and gases. BET measurements also allow estimating the volume of micropores, which plays a key role in adsorption processes from gases and liquids (Otowa et al., 1997).

Pores. The porous structure of AC is a main factor determining its efficiency in adsorption processes. Gas and mercury porosimetry are commonly used to quantify the types of micro, meso and macropores. Gas adsorption, most commonly using nitrogen at cryogenic temperatures, is used in the determination of the volume and quantity of micropores and mesopores. This method is also used to calculate the specific pore volume, which can vary depending on the raw material and the activation method. For example, coconut shell AC has typically abundant microporosity, making it particularly effective for gas adsorption. Mercury porosimetry is used to analyze the macroporous structure of ACs. This is especially important for materials used in liquid adsorption because macropores are channels where the diffusion of the target molecules towards the mesopores and micropores (adsorption sites) takes place (Muthmann et al., 2020).

Surface charge. Zeta potential (ζ-potential) informs about the surface charge of particles when suspended in a liquid. It plays an important role in determining the behavior of AC in aqueous systems. For instance, this parameter is especially important in wastewater treatment, since it informs about the interaction of carbon with pollutant ions, such as heavy metals and organic substances which are sometimes charge when in solution. The zeta potential is measured by studying the movement of AC particles in an electric field. AC with a negative zeta potential shows better adsorption capacity for cations such as lead and cadmium due to electrostatic attraction (Hong et al., 2019). For example, AC used to remove lead from aqueous solutions has a ζ-potential in the range of -30 to -50 mV, which contributes to the high efficiency of the process (Li et al., 2010b).

Surface functional groups. Boehm titrations are used to quantify the acid-base properties of the surface of ACs. Functional groups such as carboxyl, phenolic, hydroxyl, which play a key role in the adsorption of polar molecules and ions, are quantified with such method that requires incubating carbons in solutions of controlled pH and titrating the functional groups with acids and bases. ACs with abundant oxygen-containing functional groups have a better ability to adsorb basic pollutants such as ammonia or ammonium (Schönherr et al., 2018).

Infrared spectroscopy (IR) is a main method for analyzing the chemical composition of the surface of ACs. The IR spectra of ACs contain absorption bands that correspond to specific vibrations from functional groups involved in adsorption properties (Gomez-Serrano et al., 1996). IR spectra can inform about changes in the chemical composition of the carbon surface after activation or chemical modification. For example, activating carbon with KOH leads to the presence of new functional groups, such as hydroxyl and ether groups, on the surface of AC. The increase of hydroxyl groups due to the activation can be confirmed with IR (Laksaci et al., 2017).

XPS can also be used to characterize the surface of ACs. This method allows estimating the concentration of the functional groups on the surface of the carbon. XPS is widely used to analyze ACs used in adsorption and catalysis (Rossin, 1989) (Singh et al., 2019).

5. REACTIVATION OF CARBON MATERIALS

After a relatively long use, carbon needs to be reactivated so that it can return to be an effective sorbent. This is done, generally, with thermal treatment under CO_2 atmosphere. The process leads to further oxidation of the surface, pore widening and subsequent mass loss. The process of reactivating carbon takes place after several years in use. A variety of industries, for instance the water industry (Riffat, 2012) or the healthcare sector that needs to purify water for hemodialysis treatments (Passos et al., 2023), operate large scale packed carbon beds that need regular regeneration. The regeneration of carbon leads to further

CO_2 emissions, and given the global abundance of sites with such tertiary treatment, and the large scale of them, CO_2 emissions from the use and reactivation of carbon are important. Therefore, there is need to work towards minimizing CO_2 emissions from the production and regeneration of carbon adsorbents.

The advantages of activated carbon that make it attractive for these applications include its high adsorption capacity, large internal surface area, and ability to selectively remove specific contaminants. These properties enable activated carbon to effectively remove a wide range of impurities and contaminants from various substances, making it a versatile and valuable material for numerous industrial sectors, including water and gas purification, energy storage, depolarization in the food and beverage industry, cyanide leaching in the gold production, soil remediation and detoxification of humans.

6. CONCLUSION

Carbon adsorbents have been traditionally used for the purification of water, which is a global need. Preparing carbons from waste biomass is a form of revalorization that also avoids other types of pollution that result from burning biomass. Advances in material science have taught us ways to adapt carbons to effectively remove a wide range of pollutants. The form, surface composition and internal structure of the carbon affect carbon's effectivity as an adsorbent. Carbon can have different forms depending on the adsorption needs. Among them, PAC is the most effectives to control water chemical pollution despite that its use is not very practical. The structural properties of carbon can be tuned for the removal of pollutants, mainly with activation. Nevertheless, this step generates greenhouse gases, among other emissions. Surface modifications will also be important to tune the interaction with the contaminants; these can be achieved via multiple approaches. The process of making and improving the carbon adsorbent is studied with analytical complementary techniques that inform about key characteristics that affect the performance of the carbon. In particular, very important information is extracted with porosimetry, which measures carbon' surface area and the type and abundance of pores, and XPS, which gives semiquantitative information on the surface functional groups. Once the carbon has reduced part of its capacity adsorb, it is re-activated. Re-activation leads to CO_2 emissions that need to be mitigated.

Acknowledgements:

The research group of Satbayev University was supported by funding from the Ministry of Science and Higher Education of the Republic of Kazakhstan within the framework of the grant AP19577049 "Synthesis, characterization, and physicochemical study of sorbents of biomass origin for the purification of industrial waters from radionuclides".

The Marie Skłodowska-Curie Actions Staff exchanges project 101131382– CLEANWATER, supported by the European Union under Horizon-Europe Program and UK Research and Innovation, is acknowledged for supporting the research. Views and opinions expressed are, however, those of the authors only and do not necessarily reflect those of the European Union or Research Executive Agency (REA). Neither the European Union nor the REA can be held responsible for them.

REFERENCES

Abdiyev, K., Azat, S., Kuldeyev, E., Ybyraiymkul, D., Kabdrakhmanova, S., Berndtsson, R., Khalkhabai, B., Kabdrakhmanova, A., & Sultakhan, S. (2023). Review of Slow Sand Filtration for Raw Water Treatment with Potential Application in Less-Developed Countries. *Water (Basel)*, 15(11), 2007. DOI: 10.3390/w15112007

Adeleke, O. A., Latiff, A. A., Saphira, M. R., Daud, Z., Ismail, N., Ahsan, A., Aziz, N. A., Ndah, M., Kumar, V., Al-Gheethi, N. A., Rosli, M. A., & Hijab, M. (2018). Locally derived activated carbon from domestic, agricultural and industrial wastes for the treatment of palm oil mill effluent. In Elsevier eBooks (pp. 35–62). https://doi.org/DOI: 10.1016/B978-0-12-813902-8.00002-2

Ao, W., Fu, J., Mao, X., Kang, Q., Ran, C., Liu, Y., Zhang, H., Gao, Z., Li, J., Liu, G., & Dai, J. (2018). Microwave assisted preparation of activated carbon from biomass: A review. *Renewable & Sustainable Energy Reviews*, 92, 958–979. DOI: 10.1016/j.rser.2018.04.051

Azat, S., Busquets, R., Pavlenko, V., Kerimkulova, A., Whitby, R. L., & Mansurov, Z. (2013b). Applications of activated carbon sorbents based on Greek walnut. *Applied Mechanics and Materials*, 467, 49–51. . DOI: 10.4028/www.scientific.net/AMM.467.49

Bhatnagar, A., Hogland, W., Marques, M., & Sillanpää, M. (2012). An overview of the modification methods of activated carbon for its water treatment applications. *Chemical Engineering Journal*, 219, 499–511. DOI: 10.1016/j.cej.2012.12.038

Bhatnagar, A., Hogland, W., Marques, M., & Sillanpää, M. (2013). An overview of the modification methods of activated carbon for its water treatment applications. *Chemical Engineering Journal*, 219, 499–511. DOI: 10.1016/j.cej.2012.12.038

Biswas, P., & Bandyopadhyaya, R. (2017). Synergistic antibacterial activity of a combination of silver and copper nanoparticle impregnated activated carbon for water disinfection. *Environmental Science. Nano*, 4(12), 2405–2417. DOI: 10.1039/C7EN00427C

Blankenship, L. S., & Mokaya, R. (2022). Modulating the porosity of carbons for improved adsorption of hydrogen, carbon dioxide, and methane: A review. *Materials Advances*, 3(4), 1905–1930. DOI: 10.1039/D1MA00911G

Bowden, B., Davies, M., Davies, P. R., Guan, S., Morgan, D. J., Roberts, V., & Wotton, D. (2018). The deposition of metal nanoparticles on carbon surfaces: The role of specific functional groups. *Faraday Discussions*, 208, 455–470. DOI: 10.1039/C7FD00210F PMID: 29845183

Boxall, A. B. A., Collins, R., Wilkinson, J. L., Swan, C., Bouzas-Monroy, A., Jones, J., Winter, E., Leach, J., Juta, U., Deacon, A., Townsend, I., Kerr, P., Paget, R., Rogers, M., Greaves, D., Turner, D., & Pearson, C. (2024). Pharmaceutical pollution of the English National Parks. *Environmental Toxicology and Chemistry*, 43(11), 2422–2435. Advance online publication. DOI: 10.1002/etc.5973 PMID: 39138896

Busquets, R., Kozynchenko, O. P., Whitby, R. L. D., Tennison, S. R., & Cundy, A. B. (2014) Phenolic carbon tailored for the removal of polar organic contaminants from water: A solution to the metaldehyde problem? Water Research, Volume 61, 2014, Pages 46-56, ISSN 0043-1354, https://doi.org/DOI: 10.1016/j.watres.2014.04.048

Chang, C., Chang, C., & Tsai, W. (2000). Effects of Burn-off and Activation Temperature on Preparation of Activated Carbon from Corn Cob Agrowaste by CO2 and Steam. *Journal of Colloid and Interface Science*, 232(1), 45–49. DOI: 10.1006/jcis.2000.7171 PMID: 11071731

Chi, W., Wang, G., Qiu, Z., Li, Q., Xu, Z., Li, Z., Qi, B., Cao, K., Chi, C., Wei, T., & Fan, Z. (2023). Secondary High-Temperature treatment of porous carbons for High-Performance supercapacitors. *Batteries*, 10(1), 5. DOI: 10.3390/batteries10010005

Dias, J. M., Alvim-Ferraz, M. C., Almeida, M. F., Rivera-Utrilla, J., & Sánchez-Polo, M. (2007). Waste materials for activated carbon preparation and its use in aqueous-phase treatment: A review. *Journal of Environmental Management*, 85(4), 833–846. DOI: 10.1016/j.jenvman.2007.07.031 PMID: 17884280

Everett, D. H. (1972). Manual of Symbols and Terminology for Physicochemical Quantities and Units, Appendix II: Definitions, Terminology and Symbols in Colloid and Surface Chemistry. *Pure and Applied Chemistry*, 31(4), 577–638. DOI: 10.1351/pac197231040577

Feng, Z., Hu, F., Lv, L., Gao, L., & Lu, H. (2021). Preparation of ultra-high mechanical strength wear-resistant carbon fiber textiles with a PVA/PEG coating. *RSC Advances*, 11(41), 25530–25541. DOI: 10.1039/D1RA03983K PMID: 35478898

Global assessment of soil pollution: Report |Policy Support and Governance| Food and Agriculture Organization of the United Nations. (n.d.). https://www.fao.org/policy-support/tools-and-publications/resources-details/en/c/1410722/

Gomez-Serrano, V., Pastor-Villegas, J., Perez-Florindo, A., Duran-Valle, C., & Valenzuela-Calahorro, C. (1996). FT-IR study of rockrose and of char and activated carbon. *Journal of Analytical and Applied Pyrolysis*, 36(1), 71–80. DOI: 10.1016/0165-2370(95)00921-3

Gros, M., Mas-Pla, J., Sànchez-Melsió, A., Čelić, M., Castaño, M., Rodríguez-Mozaz, S., & Petrović, M. (2023). Antibiotics, antibiotic resistance and associated risk in natural springs from an agroecosystem environment. *The Science of the Total Environment*, 857, 159202.

Gupta, V., & Suhas, N. (2009). Application of low-cost adsorbents for dye removal – A review. *Journal of Environmental Management*, 90(8), 2313–2342. DOI: 10.1016/j.jenvman.2008.11.017 PMID: 19264388

Hong, M., Zhang, L., Tan, Z., & Huang, Q. (2019). Effect mechanism of biochar's zeta potential on farmland soil's cadmium immobilization. *Environmental Science and Pollution Research International*, 26(19), 19738–19748. DOI: 10.1007/s11356-019-05298-5 PMID: 31090000

Jandosov, J., Alavijeh, M., Sultakhan, S., Baimenov, A., Bernardo, M., Sakipova, Z., Azat, S., Lyubchyk, S., Zhylybayeva, N., Naurzbayeva, G., Mansurov, Z., Mikhalovsky, S., & Berillo, D. (2022). Activated Carbon/Pectin Composite Enterosorbent for Human Protection from Intoxication with Xenobiotics Pb(II) and Sodium Diclofenac. *Molecules (Basel, Switzerland)*, 27(7), 2296. DOI: 10.3390/molecules27072296 PMID: 35408695

Jansen, M. K., Andrady, A. L., Barnes, P. W., Busquets, R., Revell, L. E., Bornman, J. F., Aucamp, P. J., Bais, A. F., Banaszak, A. T., Bernhard, G. H., Bruckman, L. S., Häder, D., Hanson, M. L., Heikkilä, A. M., Hylander, S., Lucas, R. M., Mackenzie, R., Madronich, S., Neale, P. J., & Zhu, L. (2024). Environmental plastics in the context of UV radiation, climate change, and the Montreal Protocol. *Global Change Biology*, 30(4), e17279. Advance online publication. DOI: 10.1111/gcb.17279 PMID: 38619007

Jaramillo, J., Álvarez, P. M., & Gómez-Serrano, V. (2010). Oxidation of activated carbon by dry and wet methods. *Fuel Processing Technology*, 91(11), 1768–1775. DOI: 10.1016/j.fuproc.2010.07.018

Kacan, E. (2015). Optimum BET surface areas for activated carbon produced from textile sewage sludges and its application as dye removal. *Journal of Environmental Management*, 166, 116–123. DOI: 10.1016/j.jenvman.2015.09.044 PMID: 26496841

Kunarbekova, M., Busquets, R., Sailaukhanuly, Y., Mikhalovsky, S. V., Toshtay, K., Kudaibergenov, K., & Azat, S. (2024). Carbon adsorbents for the uptake of radioactive iodine from contaminated water effluents: A systematic review. *Journal of Water Process Engineering*, 67, 106174. DOI: 10.1016/j.jwpe.2024.106174

Laksaci, H., Khelifi, A., Trari, M., & Addoun, A. (2017). Synthesis and characterization of microporous activated carbon from coffee grounds using potassium hydroxides. *Journal of Cleaner Production*, 147, 254–262. DOI: 10.1016/j.jclepro.2017.01.102

Lan, X., Jiang, X., Song, Y., Jing, X., & Xing, X. (2019). The effect of activation temperature on structure and properties of blue coke-based activated carbon by CO2 activation. *Green Processing and Synthesis*, 8(1), 837–845. DOI: 10.1515/gps-2019-0054

Li, W., Yang, K., Peng, J., Zhang, L., Guo, S., & Xia, H. (2008). Effects of carbonization temperatures on characteristics of porosity in coconut shell chars and activated carbons derived from carbonized coconut shell chars. *Industrial Crops and Products*, 28(2), 190–198. DOI: 10.1016/j.indcrop.2008.02.012

Li, Y., Du, Q., Wang, X., Zhang, P., Wang, D., Wang, Z., & Xia, Y. (2010b). Removal of lead from aqueous solution by activated carbon prepared from Enteromorpha prolifera by zinc chloride activation. *Journal of Hazardous Materials*, 183(1–3), 583–589. DOI: 10.1016/j.jhazmat.2010.07.063 PMID: 20709449

Liang, Z., Guo, S., Dong, H., Li, Z., Liu, X., Li, X., Kang, H., Zhang, L., Yuan, L., & Zhao, L. (2022). Modification of activated carbon and its application in selective hydrogenation of naphthalene. *ACS Omega*, 7(43), 38550–38560. DOI: 10.1021/acsomega.2c03914 PMID: 36340089

Liu, S., Chen, X., Chen, X., Liu, Z., & Wang, H. (2007). Activated carbon with excellent chromium(VI) adsorption performance prepared by acid–base surface modification. *Journal of Hazardous Materials*, 141(1), 315–319. DOI: 10.1016/j.jhazmat.2006.07.006 PMID: 16914264

Lobato-Peralta, D. R., Duque-Brito, E., Ayala-Cortés, A., Arias, D., Longoria, A., Cuentas-Gallegos, A. K., Sebastian, P., & Okoye, P. U. (2021). Advances in activated carbon modification, surface heteroatom configuration, reactor strategies, and regeneration methods for enhanced wastewater treatment. *Journal of Environmental Chemical Engineering*, 9(4), 105626. DOI: 10.1016/j.jece.2021.105626

Lobato-Peralta, D. R., Duque-Brito, E., Ayala-Cortés, A., Arias, D., Longoria, A., Cuentas-Gallegos, A. K., Sebastian, P., & Okoye, P. U. (2021b). Advances in activated carbon modification, surface heteroatom configuration, reactor strategies, and regeneration methods for enhanced wastewater treatment. *Journal of Environmental Chemical Engineering*, 9(4), 105626. DOI: 10.1016/j.jece.2021.105626

Lua, A. C., & Yang, T. (2004). Effect of activation temperature on the textural and chemical properties of potassium hydroxide activated carbon prepared from pistachio-nut shell. *Journal of Colloid and Interface Science*, 274(2), 594–601. DOI: 10.1016/j.jcis.2003.10.001 PMID: 15144834

Macías-García, A., Torrejón-Martín, D., Díaz-Díez, M. Á., & Carrasco-Amador, J. P. (2019). Study of the influence of particle size of activate carbon for the manufacture of electrodes for supercapacitors. *Journal of Energy Storage*, 25, 100829. DOI: 10.1016/j.est.2019.100829

Melia, P. M., Busquets, R., Hooda, P. S., Cundy, A. B., & Sohi, S. P. (2019). Driving forces and barriers in the removal of phosphorus from water using crop residue, wood and sewage sludge derived biochars. *The Science of the Total Environment*, 675, 623–631. DOI: 10.1016/j.scitotenv.2019.04.232 PMID: 31035201

Molina, H. R., Muñoz, J. L. S., Leal, M. I. D., Reina, T. R., Ivanova, S., Gallego, M. Á. C., & Odriozola, J. A. (2019). Carbon supported gold nanoparticles for the catalytic reduction of 4-Nitrophenol. *Frontiers in Chemistry*, 7, 548. Advance online publication. DOI: 10.3389/fchem.2019.00548 PMID: 31475132

Muthmann, J., Bläker, C., Pasel, C., Luckas, M., Schledorn, C., & Bathen, D. (2020). Characterization of structural and chemical modifications during the steam activation of activated carbons. *Microporous and Mesoporous Materials*, 309, 110549. DOI: 10.1016/j.micromeso.2020.110549

Novoseltseva, V., Yankovych, H., Kovalenko, O., Václavíková, M., & Melnyk, I. (2021). Production of high-performance lead (II) ions adsorbents from pea peels waste as a sustainable resource. *Waste Management & Research*, 39(4), 584–593. DOI: 10.1177/0734242X20943272 PMID: 32705958

Otowa, T., Nojima, Y., & Miyazaki, T. (1997). Development of KOH activated high surface area carbon and its application to drinking water purification. *Carbon*, 35(9), 1315–1319. DOI: 10.1016/S0008-6223(97)00076-6

Passos, R. S., Davenport, A., Busquets, R., Selden, C., Silva, L. B., Baptista, J. S., Barceló, D., & Campos, L. C. (2023). Microplastics and nanoplastics in haemodialysis waters: Emerging threats to be in our radar. *Environmental Toxicology and Pharmacology*, 102, 104253. DOI: 10.1016/j.etap.2023.104253 PMID: 37604358

Riffat, R. (2012). Fundamentals of wastewater treatment and engineering. CRC Press, Taylor & Francis group, LLC, Boca Raton, Florida (US). 1st Edition. DOI: 10.1201/b12746

Rossin, J. (1989). XPS surface studies of activated carbon. *Carbon*, 27(4), 611–613. DOI: 10.1016/0008-6223(89)90012-2

Sailaukhanuly, Y., Azat, S., Kunarbekova, M., Tovassarov, A., Toshtay, K., Tauanov, Z., Carlsen, L., & Berndtsson, R. (2024). Health Risk Assessment of Nitrate in Drinking Water with Potential Source Identification: A Case Study in Almaty, Kazakhstan. *International Journal of Environmental Research and Public Health*, 21(1), 55. DOI: 10.3390/ijerph21010055 PMID: 38248520

Schönherr, J., Buchheim, J. R., Scholz, P., & Adelhelm, P. (2018). Boehm Titration Revisited (Part II): A Comparison of Boehm Titration with Other Analytical Techniques on the Quantification of Oxygen-Containing Surface Groups for a Variety of Carbon Materials. C – Journal of Carbon Research, 4(2), 22. DOI: 10.3390/c4020022

Singh, J., Bhunia, H., & Basu, S. (2019). Adsorption of CO2 on KOH activated carbon adsorbents: Effect of different mass ratios. *Journal of Environmental Management*, 250, 109457. DOI: 10.1016/j.jenvman.2019.109457 PMID: 31472376

Soltani, N., Keshavarzi, B., Moore, F., Busquets, R., Nematollahi, M. J., Javid, R., & Gobert, S. (2022). Effect of land use on microplastic pollution in a major boundary waterway: The Arvand River. *The Science of the Total Environment*, 830, 154728. DOI: 10.1016/j.scitotenv.2022.154728 PMID: 35331773

Sudaryanto, Y., Hartono, S. B., Irawaty, W., Hindarso, H., & Ismadji, S. (2005). High surface area activated carbon prepared from cassava peel by chemical activation. *Bioresource Technology*, 97(5), 734–739. DOI: 10.1016/j.biortech.2005.04.029 PMID: 15963718

Sun, Y., Chen, J., Li, A., Liu, F., & Zhang, Q. (2005). Adsorption of resorcinol and catechol from aqueous solution by aminated hypercrosslinked polymers. *Reactive & Functional Polymers*, 64(2), 63–73. DOI: 10.1016/j.reactfunctpolym.2005.03.004

Świątkowski, A. (1999). Industrial carbon adsorbents. In Studies in surface science and catalysis (pp. 69–94). https://doi.org/DOI: 10.1016/S0167-2991(99)80549-7

Thombre, N. V. (2023). Oxidation in water and used water purification. In *Springer eBooks* (pp. 1–23). https://doi.org/DOI: 10.1007/978-3-319-66382-1_174-1

United Nations Statistics Division. (n.d.). — SDG indicators. https://unstats.un.org/sdgs/report/2022/Goal06/?_gl=1*2lpenp*_ga*MTU4NjM3MzgyNy4xNzI3Njc3OTgx*_ga_TK9BQL5X7Z*MTcyNzY3Nzk4MS4xLjEuMTcyNzY3Nzk5MS4wLjAuMA. (assessed online, 30/09/2024)

Wang, X., Cheng, H., Ye, G., Fan, J., Yao, F., Wang, Y., Jiao, Y., Zhu, W., Huang, H., & Ye, D. (2022). Key factors and primary modification methods of activated carbon and their application in adsorption of carbon-based gases: A review. *Chemosphere*, 287, 131995. DOI: 10.1016/j.chemosphere.2021.131995 PMID: 34509016

Wilkinson, J. L., Thornhill, I., Oldenkamp, R., Gachanja, A., & Busquets, R. (2023). Pharmaceuticals and Personal Care Products in the Aquatic Environment: How Can Regions at Risk be Identified in the Future? *Environmental Toxicology and Chemistry*, 43(3), 575–588. DOI: 10.1002/etc.5763 PMID: 37818878

Woldeamanuel, M. M., Mohapatra, S., Senapati, S., Bastia, T. K., Panda, A. K., & Rath, P. (2024). Role of magnetic nanomaterials in environmental remediation. In *Nanostructure science and technology* (pp. 185–208). https://doi.org/DOI: 10.1007/978-3-031-44599-6_11

Xiao, K., Liu, H., Li, Y., Yang, G., Wang, Y., & Yao, H. (2020). Excellent performance of porous carbon from urea-assisted hydrochar of orange peel for toluene and iodine adsorption. *Chemical Engineering Journal*, 382, 122997. DOI: 10.1016/j.cej.2019.122997

Xin, X., Qi, C., Xu, L., Gao, Q., & Liu, X. (2022). Green synthesis of silver nanoparticles and their antibacterial effects. *Frontiers in Chemical Engineering*, 4, 941240. Advance online publication. DOI: 10.3389/fceng.2022.941240

Yang, I., Jung, M., Kim, M., Choi, D., & Jung, J. C. (2021). Physical and chemical activation mechanisms of carbon materials based on the microdomain model. *Journal of Materials Chemistry. A, Materials for Energy and Sustainability*, 9(15), 9815–9825. DOI: 10.1039/D1TA00765C

Yi, H., Nakabayashi, K., Yoon, S., & Miyawaki, J. (2021). Pressurized physical activation: A simple production method for activated carbon with a highly developed pore structure. *Carbon*, 183, 735–742. DOI: 10.1016/j.carbon.2021.07.061

Zhang, C., Sui, J., Li, J., Tang, Y., & Cai, W. (2012). Efficient removal of heavy metal ions by thiol-functionalized superparamagnetic carbon nanotubes. *Chemical Engineering Journal*, 210, 45–52. DOI: 10.1016/j.cej.2012.08.062

Zhang, F., Nriagu, J. O., & Itoh, H. (2005). Mercury removal from water using activated carbons derived from organic sewage sludge. *Water Research*, 39(2–3), 389–395. DOI: 10.1016/j.watres.2004.09.027 PMID: 15644247

Zhang, L., Sadanandam, G., Liu, X., & Scurrell, M. S. (2017). Carbon surface modifications by plasma for catalyst support and electrode materials applications. *Topics in Catalysis*, 60(12–14), 823–830. DOI: 10.1007/s11244-017-0747-7

Chapter 12
Bio-Piezoelectric Ceramic Coatings for Bone Tissue Engineering Applications

John Henao
https://orcid.org/0000-0002-8954-6039
National Council of Humanities, Science, and Technology (CONAHCYT), Mexico & CIATEQ A.C., Queretaro, Mexico

Astrid Giraldo Betancur
https://orcid.org/0000-0002-5056-7270
Center for Research and Advanced Studies of the National Polytechnic Institute, Mexico

Adriana Gallegos
https://orcid.org/0000-0001-8644-3336
National Council of Humanities, Science, and Technology (CONAHCYT), Mexico & InnovaBienestar de Mexico, San Luis Potosí, Mexico

Andrea Yamile Resendiz Mancilla
https://orcid.org/0009-0007-1002-9711
Center for Research and Advanced Studies of the National Polytechnic Institute, Mexico

Carlos Poblano Salas
CIATEQ, Queretaro, Mexico

ABSTRACT

This chapter is focused on describing the latest advances related to the development of bio-piezoelectric coatings for bone regeneration applications. It starts with the description of the main concepts about bioelectrical phenomena in the human body and its role in the regeneration of bone. It explains concepts such as dielectric and electrical responses that includes piezoelectricity, pyroelectricity, and ferroelectricity and how the human bone can present these types of phenomena. The chapter also includes the definition of bio-ceramic materials and bioactive coatings, and a summary of the main bio-ceramic coatings employed nowadays for applications in bone tissue regeneration. Also, this chapter includes a review of the latest advances in the development of bio-piezoelectric coatings for bone tissue engineering and future

DOI: 10.4018/979-8-3693-7505-1.ch012

perspectives on this topic. Overall, this chapter is focused on reviewing comprehensively the electrical response of natural tissues and the relevance of bio-piezoelectric ceramics for bone tissue regeneration.

1. INTRODUCTION

Tissue engineering is a field of biological engineering that works with biomaterials, combining scaffolds, coatings, cells, and molecules to promote the development of functional tissues. The purpose of tissue engineering is to create the conditions to restore and repair damaged tissues in the human body. This book chapter will review the main concepts related to bone tissue engineering, particularly those associated with the electrical stimulation of implanted materials that, according to normal biological processes of human bone, are able to remodel and solve local issues associated with the deterioration of hard tissues. In this section, piezoelectricity and how it appears in the human body will be introduced. In the following sections, further details about piezoelectricity and other properties of bone as well as the latest advances in biomaterials and coatings for hard tissue regeneration will be reviewed.

1.1 Piezoelectricity

Piezoelectricity was discovered in the 19th Century by Jacques and Pierre Curie (Khorsand Zak et al. 2024). They found that when a stress was applied to different materials such as sugar cane, topaz, quartz, Rochelle salt, and tourmaline, electrical charges were produced at their surface with the resultant voltage being proportional to the applied load (Thomas et al 2018). The first application of this principle occurred several decades later during World War II with the development of sonar technology employed for detecting submarines and icebergs. From that moment, a great deal of attention was given to the development of new piezoelectric materials for different applications. From a crystallography perspective, the principle of piezoelectricity obeys the asymmetry of crystal structures or molecular chains (Xu et al. 2021). When ions of different charges are asymmetrically arranged in a piezoelectric crystalline material, the formation of electric dipoles occurs. Such misalignment can be produced by the application of an external mechanical strain in compression, tension, or bending. When the dipoles are formed the polarization of the base material occurs in the stress direction (Chorsi et al. 2019; Wu, 2024) and superficial free charges are released to generate piezoelectricity. This phenomenon is known as the direct piezoelectric effect. On the other hand, when an external electrical field is applied to a piezoelectric material a deformation proportional to the electrical field is produced, this is known as the converse piezoelectric effect. The linear interaction between the electrical and mechanical states in piezoelectric materials is described by a constant of proportionality, d, which is known as the piezoelectric coefficient. There are three parameters which are relevant to describe the performance of piezoelectric materials: i) the piezoelectric coefficient which is normally expressed as d_{xy}, where x indicates the direction at which the electrical field is applied or produced and y the direction of the applied stress or resulting strain; ii) the electromechanical coupling coefficient (K) which represents the mechanical and electrical energies involved to complete the piezoelectric transformation; and iii) the mechanical quality factor (Q_m) which is the reactance to the resistance ratio in a series equivalent circuit representing the piezoelectric system (Kamel, 2022).

Ferroelectric materials are a sub-class of piezoelectric materials which are characterized for having domain structures formed by the presence of uniformly oriented electric dipole moments. When ferroelectric materials are exposed to a large electrical field, the domain orientation is aligned to a specific direction leading to an overall material polarization, which in turn enhances the piezoelectric response (Chorsi et al. 2019). The direction and magnitude of polarization can be changed by the application of an external electrical field; when such a field is removed a polarization remains in the material (known as permanent polarization) and upon a mechanical input this polarization generates the piezoelectric response.

1.2 Piezoelectric Materials.

Many materials are known to have piezoelectric properties, from a common textile like wool to sophisticated thin films employed in the fabrication of biosensors. In general, piezoelectric materials can be classified into four groups, namely crystals, polymers, composites, and ceramics. For inorganic materials, piezoelectricity shows a dependence with temperature; above a critical temperature, also known as the Curie temperature, their piezoelectric properties disappear (Tang et al. 2024). In organic materials, piezoelectricity is produced when an applied stress produces a reorientation of permanent molecular dipoles in a specific direction, creating a net polarization in the material. Crystals were the first class of materials discovered to have piezoelectric properties, with quartz and Rochelle salt being early examples of materials employed for studying the piezoelectric effect. The quest to find crystals with higher piezoelectric strength continued, and several materials such as ADP (ammonium dihydrogen phosphate) and synthetic $BaTiO_3$ were studied. Recently, ZnO crystals have been studied due to their outstanding properties such as room temperature ferroelectricity and low dielectric constant; also, it is considered a sustainable material, and it can be produced as nano rods/wires making it suitable for sensing applications. However, the piezoelectric properties of polycrystalline materials are more attractive than those find in crystals; for instance, a 4-fold increase in the piezoelectric coefficient of polycrystalline $BaTiO_3$ is observed compared with the coefficient measured for its single crystal counterpart (Willatzen, 2024).

Many ceramics having different piezoelectric performance are now available. The first discovered ceramic with piezoelectric properties was $BaTiO_3$ and is employed in different fields due to its exceptional dielectric and piezoelectric properties. Another popular ceramic is lead zirconate titanate, better known as PZT, which has a very high piezoelectric coefficient (d_{33}=500-600 pCN^{-1}) (Algueró et al. 2001; Gamboa et al. 2020) and shows ferroelectricity. This ceramic has been processed employing different routes like sol-gel (Jacob et al. 2018), chemical coprecipitation (Shirke et al. 2024), and mechanochemical synthesis (Sayagués et al. 2024). Although PZT has been employed for the fabrication of actuators, photoelectronic devices, and pyroelectric detectors, its brittle nature and lead content make it unsuitable for tissue engineering applications. As a result, lead-free ceramic substitutes like lithium sodium potassium niobate (LNKN) and potassium sodium niobate (KNN) have been developed (Martin et al. 2024; Qi et al. 2024). Because polycrystalline ceramics have crystal structures that are randomly oriented and are piezoelectrically functional only once they are oriented, a polarization procedure is required. Polarization, by the application of a high intensity electrical field, aligns the material´s dipole moments and provides the required piezoelectric properties; this procedure is known as electrical poling (Kumar et al. 2024).

Due to their biocompatibility, biodegradability and little to no toxicity, piezoelectric polymers are very attractive materials for biomedical applications (Maiti et al. 2019; Curry et al. 2018). They also show light weight, flexibility, a low dielectric constant, and are simple to fabricate (Thuau et al. 2016; Sappati et al. 2018). The fabrication and characterization of the following piezoelectric polymers have been reported in the literature: poly-L-lactic acid, poly-β-hydroxybutyrate, polyvinyl chloride, polyacrylonitrile, poly-3-hydroxybutirate-3-hydroxyvalerate, and nylon-11 (Xu et al. 2021; Rui et al. 2024). Bulk polymers normally require stretching and poling to show piezoelectricity. The arrangement of chains in the polymer microstructure, which in turn affects the alignment/misalignment of dipoles within such chains, dictates its piezoelectric response. In general, the piezoelectric performance of polymers is poor when compared with that of ceramics (100-2500 pCN^{-1}) (Kamel, 2022). Most polymers show piezoelectric coefficient values below 10 pCN^{-1}, although there are some exceptions, like the β poly(vinylidene fluoride) (PVDF) with d levels between 30-40 pCN^{-1} (Gomes et al. 2010). To improve the piezoelectric performance of polymers they have been mixed with different inorganic materials with high piezoelectric coefficients. These combinations result in composite materials able to operate at higher temperatures while keeping relatively high d values, inherited from the ceramic phase, and reasonably good plasticity. They also allow to prepare piezoelectric materials with large areas and to diversify their field of application. For instance, a $BaTiO_3$-polymethyl methacrylate composite was prepared for application as bone cement (Tang et al. 2020). The dispersed $BaTiO_3$ particles in the polymeric matrix helped to improve the osteoinductivity of the cement. A piezoelectric coefficient comparable with that of the human bone was reported by further additions of graphene, which also increased the composite conductivity and dielectric constant. Improvements in the piezoelectric performance of biomaterials have also been reported. Silva et al. 2002 studied the effect of collagen additions on the piezoelectric and dielectric properties of hydroxyapatite (HAp); the composite presented a higher piezoelectric coefficient (d_{14} = 0.040 pCN^{-1}) than the reported for the HAp (d_{14}=0.012 pCN^{-1}). Overall, different fillers have been employed in the fabrication of polymer-based piezoelectric composites; well-known examples are $BaTiO_3$, KNN, metal oxides, and salts (Surmenev et al. 2019).

1.3 Piezoelectricity in Biomolecular Materials.

Piezoelectricity is found in biological tissues like bone, hair, skin, ligaments, and tendons; it is also present in many biomolecules such as proteins, amino acids, and peptides. Due to this property, they are generally termed biomolecular piezoelectric materials (Xu et al. 2021). When they are mechanically stimulated surface charge polarization occurs, which has been associated with physiological functions like tissue regeneration, healing, and growth (Khare et al. 2020). Cells and their components are therefore affected by piezoelectricity to perform their functions like cardiac and muscular contraction, protein folding, and biomolecular interactions (Venkateshwarlu et al. 2024; Chen et al. 2024; Makhatadze GI. 2017). Such a behavior arises from the asymmetrical nature of the helical/spiral shape of biological macromolecules. For instance, Keratin is the main component of hair and nails, and its piezoelectricity comes from its characteristic alpha-helix shape formed by hydrogen-bonded carbonyl and amine groups. Proteins such as collagen, myosin, elastin, and actin also show piezoelectric activity which has been attributed to their constituent amino acids, packing of peptide chains, and the formation of electric dipoles (Kim et al. 2020). Cellulose is also another piezoelectric material which provides this property to wood. The individual cellulose chain shows a dipole; however, cellulose fibrils present in wood are geometrically organized in a non-parallel fashion, leading to an overall reduced piezoelectric behavior of wood.

Cellulose is then preferred to be employed in nanometric form, keeping its piezoelectric properties, as a reinforcing material of polymers and biological materials (Zhu et al. 2024; Janićijević et al. 2024; Choi et al. 2021). Amino acids are organic compounds that also show piezoelectric properties. Although there is a large number of amino acids in nature, the most relevant are the 22 α-amino acids incorporated into proteins. Proteins are the second largest component of human muscles and other tissues. Therefore, the piezoelectric behavior of amino acids is relevant to understand the piezoelectric properties of tissues and organs. The piezoelectric behavior of amino acids comes from their lack of crystal symmetry and because they belong to chiral symmetry groups, which implies that they crystallize in at least two optical isomer forms (Bystrov, 2024).

1.4 Relevance of Bio-piezoelectricity in Bone Tissue Formation and Regeneration.

Bone is another good example of a biological piezoelectric material. It has been shown that bone is not a static tissue as it regularly requires to be maintained, remodeled, and reshaped (Akbarov & Tillakhodjayeva, 2024). Such tasks involve three types of cells namely osteoblasts, osteocytes, and osteoclasts. Osteoblasts participate in generating new bone and repair, whereas osteocytes are inactive osteoblasts and help bone deposition and resorption. Osteoclasts help in remodeling damaged bone and in resorption. In the XIX Century, Wolf proposed a scientific law stating that the higher the mechanical stress imposed on bones over time, the denser and stronger they become, and vice versa (Kamel, 2022). This means that a healthy bone requires mechanical stress to promote bone tissue build up. Since then, important efforts have been performed to understand the role of mechanical stress on the functions and behavior of such cells. Bone is a composite material made from hydroxyapatite (70 wt.%), collagen (19 wt. %), water, cells, and other components (11 wt. %). Collagen was found to have an important role in the piezoelectric nature of bone tissue. Collagen is a piezoelectric polymer and is the most abundant protein in the human body (Ghosh et al. 2024). Collagen molecules have a three-helical unsymmetrical spiral shape which is responsible for its piezoelectric properties. Thus, collagen is thought to be the principal cause for piezoelectricity in bones. During daily physical activity, bones are subjected to different stress levels in tension and compression, such mechanical stimuli promote the development of electrical charges by the nano-scale collagen fibers present in bone tissue. The polarity of such charges is dictated by the direction of the applied mechanical stress, positive for tension and negative for compression. For human tibia, a piezoelectric coefficient ranging from 7.7 to 8.7 pCN^{-1} has been reported (Ghosh et al. 2024). The increase of bone density due to physical activity occurs because the collagen molecules are stressed and this causes the formation of electrical dipoles, which attract osteoblasts and minerals to the strained part of the bone, leading to an increase in bone density (Zhao et al. 2024; Vasquez-Sancho, 2018). In recent years it has been demonstrated that HAp also shows piezoelectricity despite crystallizing in a P63/m centrosymmetrical space group. It is thought that piezoelectricity in HAp arises because it lacks an inversion center (Rajabi et al. 2015). Moreover, it has been found that some HAp-based thin films have piezoelectric and ferroelectric properties, while some poled thick coatings have shown piezoelectric coefficient values close to that reported for collagen (Li et al. 2024).

2. BIOELECTRICITY IN THE HUMAN BODY

Bioelectric science was quantitatively described around 1790, first by Galvani and then by Volta. Galvani observed motion in frog legs when he touched the muscle using metallic wires, noting that the activity depended on the metal used to produce the stimuli (Baloh, 2024; Schofield et al. 2020). Since that discovery, bioelectricity has been considered an important parameter of biological processes (Baloh, 2024; Schofield et al. 2020).

Bioelectricity is the electric phenomenon of living organisms, defined as the electrical potentials and currents within and between living cells and tissues. This phenomenon is essential for various physiological functions, from heart rate to brain activity. Bioelectricity also influences processes such as tissue repair and immune response, highlighting its importance in the health and well-being of organisms. Additionally, bioelectricity is responsible for communication and coordination among different body parts, serving as a fundamental signal of life processes (Baloh, 2024). Furthermore, bioelectricity originates at the cell membrane and is associated with a lack of charge balance between the inter- and extracellular components, due to the exchange of ions through the channels and ionic pumps in the membrane (Bhavsar et al. 2020).

One of the most essential functions of bioelectricity is the transmission of nerve impulses; neurons rely on bioelectrical activity to communicate. This process occurs through the generation of action potentials, which are rapid changes in the electrical potential of the cell membrane. These impulses enable the transmission of information along nerve pathways, facilitating functions such as movement, sensory perception, and the regulation of involuntary processes (Park, 2023). Furthermore, bioelectricity is crucial in muscle function. Muscle cells generate electrical signals that trigger the release of calcium, which is critical to muscle fiber contraction. Bioelectricity also plays an important role in regulating heart rhythm. The sinoatrial nodule acts as a natural pacemaker, generating electrical impulses that are associated with the rate and rhythm of heartbeats. This process ensures adequate blood flow and efficient oxygenation of tissues (Park, 2023; Sinha et al. 2022).

3. SIGNIFICANCE OF BIOELECTRICITY IN BONE TISSUE ENGINEERING (BTE)

Although bioelectricity is a phenomenon that applies to the entire body, it is particularly critical in influencing the generation and maintenance of functional bone tissue. This phenomenon is related to the physiological processes that affect bone formation and remodeling, as bone cells (osteocytes, osteoblasts, and osteoclasts) respond to electrical signals. These signals generate the required stimuli to influence the proliferation, differentiation, and metabolic activity of bone cells, facilitating a balanced process between bone formation and resorption (Heng et al. 2023).

To understand the bioelectrical behavior of bone, it is necessary to establish that, in its hard and rigid structure, it is composed by an organic matrix formed mainly by collagen fibers which show dielectric, piezoelectric, pyroelectric, and ferroelectric behavior. Bone is also composed of a mineral matrix responsible for the transmission of mechanical load to the bone itself, and particularly to the collagen fibers; this mineral phase is mainly composed of HAp nanocrystals that, among others, restrict the ability of collagen to form hydrogen bonds with water molecules, thus influencing the bioelectrical response of the bone (See Figure. 1) (Heng et al. 2023).

Figure 1. Bioelectric properties associated with the bone tissue (Created in BioRender.com)

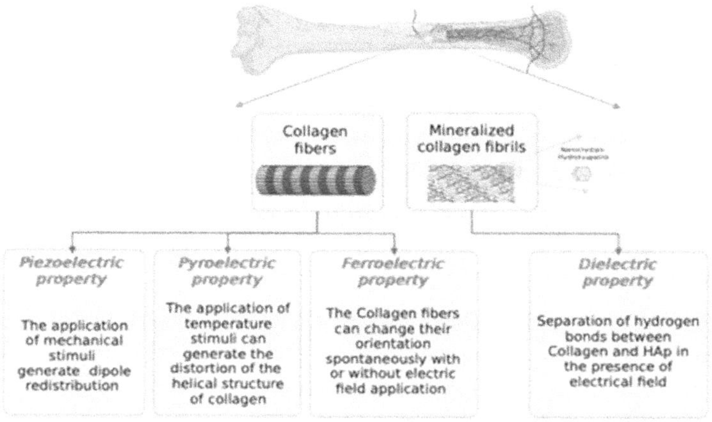

Bioelectricity in bone function and regeneration is of utmost importance. Bioelectric phenomena in natural bone, such as piezoelectricity, pyroelectricity, and ferroelectricity, are key to regulating metabolic activities such as growth, structural remodeling, and fracture healing. The non-centrosymmetric nature of the collagen molecule is the primary reason for bioelectricity in bones. The application of physiological compressive loads on bones increases negative charges due to the piezoelectric potential, which in turn promotes osteoblastic cell functionality and consequently enhances bone regeneration. Bones generate a piezoelectric potential during physical activities such as walking. This negative potential attracts osteoblastic cells, which are crucial for mineralizing the bone matrix at fracture sites, suggesting that piezoelectric stimulation plays a vital role in controlled bone regeneration (Chae et al. 2018; Khare et al. 2020).

Piezoelectricity refers to the electrical response of a material to mechanical stress; it is associated with a class of materials in which the application of mechanical stress results in electrical polarization (direct piezoelectricity) and vice versa (converse piezoelectricity), as previously mentioned. In bone, this effect is related to the sliding of collagen fibrils under mechanical stimulus. The piezoelectric effect in bone is naturally generated during physical activities that involve movements where the bone is subjected to stress (Khare et al. 2020). In fact, this property is responsible for the endogenous electrical field in bones (Da Silva et al. 2020; Luo et al. 2024; Ghosh et al. 2024; Barak 2024; Minary-Jolandan & Yu, 2010; Pais et al. 2024). Pyroelectricity is the ability of a material to polarize in response to changes in temperature, generating a temporary voltage across its structure. Natural bone exhibits pyroelectricity due to the presence of collagen. It has been proposed that the triple helix model of collagen, which consists of three parallel helical chains, is responsible for the pyroelectric properties in living bone (Li et al. 2024; Bai et al. 2024; Bai Y et al. 2024). Ferroelectric materials are those that exhibit reversible spontaneous polarization and a hysteresis loop. The collagen fibers in bone tissue can change their orientation in various directions, similar to the typical alignment of ferroelectric domains. The presence of a hysteresis loop and permanent dipoles in the bone structure confirms that ferroelectricity is a fundamental characteristic of natural bone (Joo et al. 2024; Qin et al. 2024). Finally, to cover the concepts in Figure 1, dielectric materials refer to electrical insulators that can generate dipoles due to the relative movement or separation between positive and negative charges, either in the absence or presence of an external electric field. This process is known as polarization. The polarizability of a dielectric material in response to external electric stimulation is measured in terms of a dielectric constant. In living bone, the dielectric property is related to the hydrogen bond distance separation between collagen fibrils and HAp in the presence of an electric field and moisture (Figure. 1). For this property, it has been shown that osteogenesis processes can benefit under the endogenous or external stimuli of electrical fields or impulses in a cellular environment in BTE. This result is related to the increase in osteoblast activity that promotes mineralization processes of the extracellular matrix related to this type of stimuli. In fact, according to Khare et al. (2020), changes in bone mechanical performance can be followed through the measurement of conductivity properties (Khare et al. 2020).

The regeneration process of bone is therefore closely related to the bioelectrical properties of this tissue, which are triggered by the endogenous electrical field. However, until recently, these properties, particularly the electric conductivity, have not been considered in biomaterials design. This property is now considered in the research and development of "smart" biomaterials, able to provide multiple responses that mimic the properties and behavior of bone under specific physiological conditions. This approach drives more efficient biological responses, thus favoring healing and the adaptation of the biomaterial in the tissue, offering promising prospects for the future of biomaterial design (Da Silva et al. 2020). Therefore, bioelectricity is an essential component in BTE and understanding this phenomenon opens many possibilities and allows the establishment of strategies to enhance bone regeneration processes.

4. BIOMATERIALS REPORTED BY TYPE AND APPLICATION IN BONE REGENERATION AND BIOMEDICAL APPLICATIONS

Based on the properties and functionalities of human bone, various materials intended for use in implants that emulate some of these properties have been developed over time. Below is presented a list of the most common biomaterials, along with their relevant properties:

4.1 Piezoelectric Ceramics

(Wu et al. 2014; Wu et al. 2016; Kang et al. 2018; Sun et al. 2017; Carville et al. 2015; Vaněk et al. 2016; Parravicini et al. 2011; Marchesano et al. 2015 ;Kang et al. 2018; Ladj et al. 2013; Nakayama et al. 2007; Murillo et al. 2017; Park et al. 2004; Park et al. 2010; Seo et al. 2010):

- **$MgSiO_3$ (Magnesium Silicate)**: Compounds based on $MgSiO_3$ that are used in combination with various polymers to enhance bone regeneration.
- **$LiNbO_3$ (Lithium Niobate)**: Ferroelectric material with piezoelectric and pyroelectric properties used for stimulating cell proliferation and differentiation.
- **$KNbO_3$ (Potassium Niobate)**: Piezoelectric material used as a bio-probe for disease diagnosis and prostate cancer treatment.
- **ZnO (Zinc Oxide)**: Exhibits piezoelectric and biocompatible properties, employed for stimulating osteoblastic activity and improve cell compatibility.

4.2 Synthetic Piezoelectric Polymers

(Bai et al. 2024; Dubey et al. 2014; Rajabi et al. 2015; Ribeiro et al. 2015; Ribeiro et al. 2017; Di Martino et al. 2005; Zhou et al. 2016).

- **PVDF (Polyvinylidene Fluoride):** Piezoelectric polymer used in the form of fibers, films, and microparticles to stimulate osteogenic differentiation and matrix mineralization.
- **PLLA (Poly-L-Lactic Acid):** Copolymer used in tissue engineering applications due to its piezoelectric and biodegradable properties.
- **PHB (Polyhydroxybutyrate):** Biodegradable polymer with piezoelectric properties, used in scaffolds for bone regeneration.
- **Polyamides (Nylons, Peptides, etc.):** Includes various amides with piezoelectric properties, used in biomedical applications for tissue engineering.

4.3 Natural Piezoelectric Polymers

(Luo et al. 2024; Li et al. 2024 ; Baxter et al. 2010; Ravikumar et al. 2016; Amin et al. 2019; Köse et al. 2005; Kim et al. 2006)

- **Cellulose:** Natural polysaccharide with piezoelectric properties, used in scaffolds for bone regeneration.
- **Chitosan:** Biocompatible polysaccharide with piezoelectric properties, used in combination with other materials to enhance bone regeneration.
- **Amylose (Starch)**: Natural polysaccharide with piezoelectric capability, used in biomedical applications.
- **Collagen:** Structural protein with piezoelectric properties, fundamental in the bioelectricity of natural bone.
- **Silk:** Natural protein with piezoelectric properties, used in scaffolds for tissue regeneration.
- **Keratin:** Fibrous protein with piezoelectric capability, used in biomedical applications for tissue engineering.

4.4 Hydroxyapatite, Composites and Scaffolds

(Dubey et al. 2011; Dubey et al. 2014; Wu et al. 2016; Kang et al. 2018; Sun et al. 2017; Feng et al. 2016; Ramu et al. 2018):

- **Hydroxyapatite (HAp):** Bioceramic material widely used in bone regeneration applications due to its similarity to natural bone mineral.
- **HAp-MS-CS (Hydroxyapatite-Magnesium Silicate-Chitosan) Implant:** Composite used to improve bone regeneration by combining bioceramics and polymers.
- **HAp-CS (Hydroxyapatite-Chitosan):** Bioceramic-polymer composite used in scaffolds for bone regeneration.
- **CS Scaffold (Chitosan Scaffold):** Scaffold made of chitosan, used to support cell growth and bone regeneration.

- **Poly (Butylene Succinate) or PBSu:** Polymer used in the formation of scaffolds by methods such as particle leaching.
- **Poly(ε-Caprolactone) or PCL:** Piezoelectric polymer used in scaffolds manufactured by particle leaching or rapid prototyping methods.
- **Poly (Ethylene Glycol) or PEG:** Polymer used in the formation of scaffolds by particle leaching methods.
- **Wheat Protein or WP:** Natural polymer used in scaffolds manufactured by compression molding methods.

4.5 Nanomaterials

(Lee et al. 2001; Zhi et al. 2005; Rajabi et al. 2015; Ribeiro et al. 2015; Fisher et al. 2018):

- **Boron Nitride nanotubes (BNNT):** Includes boron nitride nanotubes used to promote osteogenic differentiation and improve cell adhesion.
- **ZnO Nanorods:** oxide nanorods used to reduce macrophage inflammatory response. Employed as a coating on electroactive material surfaces to enhance cell interaction and bone regeneration.

4.6 Properties of Dielectric, Piezoelectric, Ferroelectric, Pyroelectric and Electromechanical Materials

The dielectric, piezoelectric, ferroelectric, pyroelectric, and electromechanical properties of the biomaterials are summarized in Table 1; it also contains properties reported for human bone.

Table 1. Dielectric, piezoelectric, ferroelectric, pyroelectric, and electromechanical properties of some materials

Material Biomaterials	ε_r (at 1-100 kHz)	d_{33} (pC/N)	(Qm)	P_r (μC/cm²)	k_p (%)	(p) (μC/m²K)	Ref.
$Na_{0.5}K_{0.5}NbO_3$	657	160	240	31.4	44	-	(Jaeger et al. 1962; Kakimoto et al. 2010)
$BaTiO_3$	1135	191	32	12.6	35	200	(Willatzen 2024; Sharma et al. 1998)
$Li_{0.06}(Na_{0.5}K_{0.5})_{0.94}NbO_3$	722.4	124	-	16.8	30.63	-	(Carville et al. 2015; Chen et al. 2017)
$LiNbO_3$	62	23	-	-	-	103.9	(Vaněk et al. 2016; Parravicini et al. 2011)
$MgSiO_3$	-	_ 1.74 (d_{31})	-	-	-	-	(Nakamachi et al. 2006)
$KNbO_3$	394	91.7	325	-	28	93	(Nagata et al. 2007; Birol et al. 2005)
ZnO	12.27	12.4	-	16.31	48 (k_{33})	9.4	(Lang, 2005; Chaari et al. 2011)

continued on following page

Table 1. Continued

Material Biomaterials	ε_r (at 1-100 kHz)	d_{33} (pC/N)	(Qm)	P_r (μC/cm²)	k_p (%)	(p) (μC/m²K)	Ref.
BN	2 - 4	31.2 (d_{31})	-	-	-	-	(Kim et al. 2012; Ban et al. 2018)
PVDF	6 - 12	34	17.2	13	20 (k_{33})	41	(Martins et al. 2009; Song et al. 2017)
P(VDF-TrFE)	18	38	-	9.9	29 (k_{33})	50	(Xu 2001; Omote et al. 1997)
PLLA	3 - 4	9.82 (d_{14})	30.3	-	-	-	(Ribeiro et al. 2015; Ochiai et al. 1998)
PHB	2 - 3.5	1.6-2 (d_{14})	-	-	-	-	(Jensen et al. 2007; He et al. 2014)
Polyamide	4 - 5	4 (d_{31})	8.5	5.6 - 8.6	11	5	(Ribeiro et al. 2015; Koga et al. 1986)
Collagen	2.6	0.2 - 2 (d_{14})	-	0.29	-	0.12 - 0.37 @ 40 °C	(Lima et al. 2006)
Chitosan	3.94	0.2 - 1.5 (d_{14})	-	0.178	-	7	(Hoque et al. 2018)
Human bone	**9.2**	**7.50 - 9**	**-**	**0.00068**	**-**	**0.0036 ± 0.0021**	(Bai et al. 2024; Minary-Jolandan & Yu, 2015)

4.7 Bone Healing as Result of Piezoelectric Stimulation

The process of healing damaged bone tissue can be stimulated by both piezoelectricity and direct mechanical stimulation (Khare et al. 2020):

4.7.1 Electrical Stimulation Induced by Piezoelectricity

Piezoelectric scaffolds generate electric potentials when subjected to functional loads. This electrical stimulation activates voltage-gated calcium channels (VGCC) and stretch-activated calcium channels, resulting in an increase in intracellular Ca^{2+} ions. Additionally, the local electric field modifies the configuration of membrane receptors and open receptor channels, allowing additional Ca^{2+} ions to influx from the endoplasmic reticulum. These activated Ca^{2+} ions within the cells stimulate calcium-modulated proteins, such as calcineurin, leading to gene transcription and ultimately regulating cell metabolism and extracellular matrix (ECM) synthesis.

4.7.2 Direct Mechanical Stimulation

Mechanical stimulation directly activates mechanoreceptors, such as integrins, which then translocate protein kinase C (PKC) to the cell membrane and activate MAPK (mitogen-activated protein kinase) signaling pathways. These signaling cascades propagate to the cell nucleus, where they interact with mechanosensitive transcription factors, executing gene transcription that promotes bone healing.

5. BIOCERAMICS AND COATINGS

Numerous genetic, infectious, and degenerative diseases, along with physical or chemical deterioration, can lead to cell damage in tissues or organs, resulting in alterations to their normal functions that affect people´s quality of life (Serrato Ochoa et al. 2015). Although treatments exist for these diseases, in some cases the damage is so severe that these treatments are no longer applicable. When organ or tissue damage requires surgical intervention or the use of a prosthesis, a several challenges arise due to the potential for rejection, or the development of complications associated with the response to the implant (Thomas et al. 2024).

In the case of bone tissue, depending on the lost function, surgical process and the use of prostheses are often necessary. The application of coatings on the surface of prostheses has emerged in response to issues related to fixation of implants, inflammatory responses that can become chronic, allergies due to implant wear, and the production or release of metallic ions or by-products from the surface. These challenges have created a need not only for biocompatible coatings but also for alternatives that possess characteristics as similar as possible to bone behavior, aiming to enhance the osseointegration process and promote effective healing (Axente et al. 2020). The use of bioceramic materials and coatings, which have the potential to induce favorable biological responses, may significantly improve the quality of life by restoring the functionality of damaged tissues more effectively and efficiently. In this section, the concepts of bioceramics and coating are reviewed in the context of bone tissue regeneration.

5.1 Bioceramics

The concept of bioceramics is linked to bioengineering and clinical medicine, denoted by the prefix "bio", which refers to the biological conditions under which these materials must function in human or animal models. Bioceramics comprises a family of inorganic materials employed in the treatment of both soft and hard tissues (Shekhawat et al. 2021). These materials are classified into three groups, as shown in Figure 2.

Bioinert ceramics are biocompatible materials that exhibit minimal chemical interaction with living tissues over a long period of implantation in the human body, without eliciting harmful responses (Kamboj et al. 2023). These materials are often chemically inert, biocompatible, and possess wear resistance and compressive strength. Consequently, they are commonly employed as components in moving parts of orthopedic joints and as dental implants. Examples of bioinert ceramics include silicon nitride, alumina, and zirconia. Silicon nitride-based materials often exhibit high fracture toughness, making them suitable for use in artificial joint components for hip and knee implants (Punj et al. 2021). On the other hand, zirconia-based components have high compressive strength and an elastic modulus similar to that of titanium alloys, which facilitates their mechanical adaptation to the metallic components of a prosthesis (Nikkerdar et al. 2024). Additionally, alumina is another common bioceramic material used in medicine for its high hardness, compressive strength, chemical stability, and wear resistance. This material is employed in joint components of total hip implants (Al-Moameri et al. 2020).

Figure 2. Classification of bioceramics

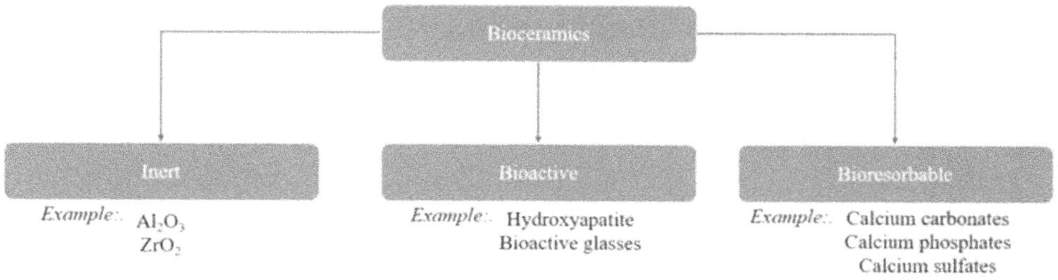

Bioactive ceramics are biocompatible materials capable of interacting with human fluids and bonding with hard tissues over a short time within the human body. Moreover, these materials do not have harmful effects on the human body; in fact, they integrate more effectively than bioinert ceramics. Bioactive ceramics are said to mimic biomineralization response of the hard tissues, and consequently, they have been employed in bone tissue engineering (Rajendran et al. 2024). The most common bioactive ceramics are hydroxyapatite, bioactive glasses, and glass-ceramics. For example, HAp ($Ca_{10}(PO_4)_6(OH)_2$) is a compound from the apatite's family and is one the constituents in human bones. It has a Ca/P atomic ratio of 1.67 but it is also found with Ca-deficient or Ca-rich ratios. HAp stability inside the human body is associated with Ca/P ratios close to 1.67. Ca-deficient HAp often results in a degree of solubility and Ca-rich HAp promotes a decrease in mechanical strength. Due to its chemical composition, HAp has a strong advantage in hard tissue engineering, especially for orthopedics and fractures fixation. In fact, when HAp is implanted in the body, it bonds rapidly with bone and promotes new bone growth (Kareem et al. 2024). On the other hand, bioactive glasses are amorphous materials obtained from a mixture of oxides such as SiO_2, Na_2O, CaO, P_2O_5 and others. The bioactive behavior of these glasses depends on the oxides contribution. The ratio of these components determines the solubility degree and bone bonding ability. These materials are also useful as bone fillers and for applications in hard tissue treatment (Baino et al. 2018). Nevertheless, the mechanical properties of bioactive glasses are not the best among bioceramics. This has led to the development of glass-ceramic composites, in which ceramic phases are added to bioactive glasses for mechanical reinforcement (Fu et al. 2020). For example, wollastonite, apatite, and bioactive glass ceramic composites are commonly found in the literature (Yılmaz et al. 2024; Katubi et al. 2024).

Bioresorbable ceramics are a family of compounds that are gradually resorbed by the human body. These materials can be resorbed either physiochemically or by cellular processes. This behavior is beneficial for tissue engineering applications, as the gradual resorption allows for the treatment of various pathologies, including the replacement and regeneration of tissues, promotion of vascularization in lost bone, and temporary treatments for hard and/or soft tissues (Sarkar, 2024; Batool et al. 2024). Most famous bioresorbable ceramics are calcium phosphates, calcium carbonates, and calcium sulfates (Safronova et al. 2013; Moreno-Gomez 2019; Civinini et al. 2017). These materials are also known for their ability to promote bone growth and regeneration. They may serve as vehicles for drug release, carry growth factors, and transport cells, enabling them to induce specific responses in the local environment within the human body. Due to their solubility, these materials lose their mechanical properties after a certain interaction time in the body. Consequently, they are employed to induce therapeutic effects rather than to fulfill a mechanical role (Bose & Tarafder, 2012; Chaudhari et al. 2024).

5.2 Bioceramics in Soft Tissue Regeneration

The use of bioceramics for soft tissue regeneration is today a well-stablished approach for treating damaged tissues in the human body (Kumar et al. 2023). The role of bioceramics in soft tissue engineering is to guide the organism through appropriate differentiation of cells for the formation of new soft tissue. Soft tissues in humans are numerous and are found in the skin, nerves, muscles, and organs. Common materials in soft tissue engineering include hydrogels, polymers, and bioceramics (Mazzoni et al. 2021). Among bioceramics, bioactive glasses and calcium phosphates can stimulate the growth of specific tissues (Negut et al. 2023; Mishchenko et al. 2023). Bioceramics have been tested in skin, nerves, organs, and epithelial damaged tissue (Figure 3) (Kumar et al. 2023). They have been explored as nano powders and mixtures with organic matrix to regenerate damaged skin (Rochdi et al. 2023). These materials are not only intended to promote the formation of new skin but also to improve its quality and firmness (Aguilera et al. 2023). Other uses include periodontal tissues, where calcium phosphates have been studied alongside growth factors and human cells to enhance the regeneration of ligaments (Sun et al. 2023.).

Bioactive glasses are also interesting materials for applications in soft tissue engineering, as they are reactive in biological fluids. One notable feature of bioactive glasses is their ability to promote gene activation and the growth of soft tissues (Shearer et al. 2023). Silicate and borosilicate bioactive glasses have been shown to stimulate soft tissue growth due to the metal ion release that occurs when these materials encounter body fluids. Ion release also has other functions such as anti-inflammatory and antibacterial activity. $Ca^{2+}$and Si^{4+} favor the reconstruction of soft tissue, while others such as Cu^{2+} and Zn^{2+} promote angiogenesis (Bai et al. 2023). Bioactive glasses have been tested in periodontal injuries, mucosal wounds, cardiac injuries, and cutaneous tissues. These are commonly studied in powder shape, nanofibers, and porous scaffolds (Zhu et al. 2024). Some bioactive glasses have also been tested in animal models and clinical trials with promising results arising from their biocompatibility, regeneration capabilities, and antibacterial properties (Abushahba et al. 2023).

Figure 3. Applications of bioceramics in soft tissue (Created with BioRender.com)

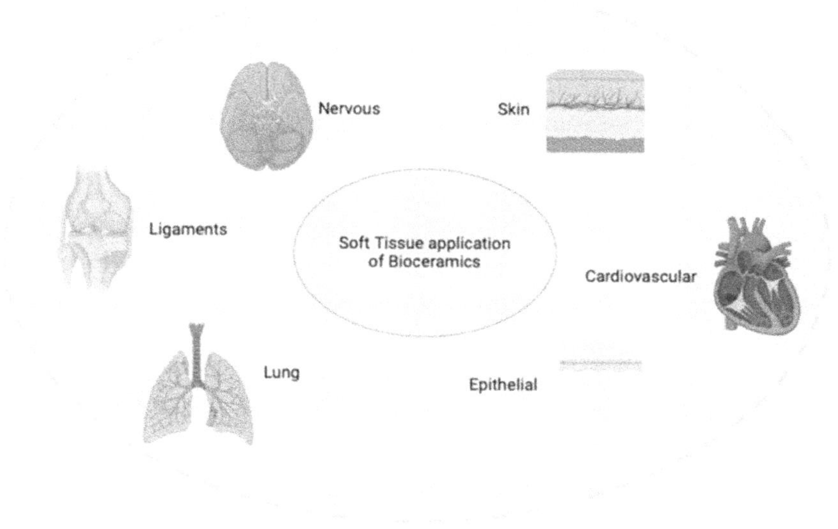

Calcium silicates have emerged as bioceramics materials for soft tissue applications (Sakthi et al. 2023). These materials comprise various compositions and exhibit interesting properties, such as promoting angiogenesis and enhancing collagen deposition in body fluids —both important features for soft tissue repair (Yuan et al. 2023). Some studies have evaluated the behavior of calcium silicates under in-vitro conditions, confirming cell proliferation and differentiation, antibacterial effects, and improvements in soft tissue wound healing (Rebolledo et al. 2023).

Composite biomaterials have also been developed to repair damaged tissues, particularly in relation to skin and organs (Farag, 2023). Since polymers are among the most common biomaterials used for scaffolds in soft tissue engineering, many studies have focused on enhancing their properties by preparing polymer/bioceramics composite scaffolds. Polymers such as polycaprolactone (PCL), polylactide acid (PLA), poly(lactic-co-glycolic acid) (PGLA), chitosan, and others have been mixed with bioceramics like hydroxyapatite, bioactive glasses, and calcium phosphates for soft tissue repair (Lu et al. 2023; Sharma et al. 2023). The use of polymers in fabricating composite materials offers the advantage of flexibility, allowing the preparation of different types of scaffolds. Additionally, the incorporation of bioceramics into the polymer matrix can enhance mechanical strength, bioactivity, cellular activity, anti-inflammatory properties, and antibacterial activity (Bhong et al. 2023).

5.3 Bioceramics in Hard Tissue Regeneration

The goal of hard tissue engineering is to regenerate and restore the function of bones, cartilage, or teeth (Kumar et al. 2018). Overall, hard tissue reconstruction is required when fractures or ablation of bones, cartilage, or teeth occur due to normal human activities, accidental traumas, and/or congenital diseases (Seppänen-Kaijansinkko, 2019). Bioceramics play an important role in repairing hard tissues, as they exhibit properties such as bioactivity or bioinertness combined with mechanical strength. For instance, materials such as ZrO_2 have been widely used in tooth repair due to their bioinertness and high mechanical strength under compressive forces (Malode et al. 2022). In cases when bone reconstruction is needed, bioceramics like HAp and bioactive glasses have been employed. Bioactive ceramics are fundamental in bone repair, as they possess conductive and inductive properties that support cell proliferation and new bone growth (Oliveira et al. 2017).

In the human body, bone reconstruction is carried out by specific cells. The cells involved in this process include mesenchymal stem cells (MSCs), osteoblasts, osteocytes, and osteoclasts. MSCs are adult stem cells capable of differentiating into various cell types and are responsive to local changes in both hard and soft tissues, allowing the organism to be prepared for remodeling and reconstruction. These cells are also known as osteoprogenitor cells, as they differentiate to give rise to osteoblasts, which are involved in the precipitation and mineralization of the bone´s extracellular matrix. Similarly, osteocytes are a type of cell that responds to local changes in bone tissue. They are formed through the differentiation of osteoblasts and are often found in mature bone tissue. Osteocytes play a role in the remodeling and repair of bones, responding to biochemical and biophysical stimuli coordinated by the actions of osteoblasts and osteoclasts (Liu et al. 2006), the latter being associated with bone resorption, Figure 4. Therefore, bioceramics used in bone tissue reconstruction must possess the ability to interact with these cells and serve as support structure for their differentiation, proliferation, and the deposition of new tissue.

Figure 4. Schematic of bone tissue remodeling and repairing

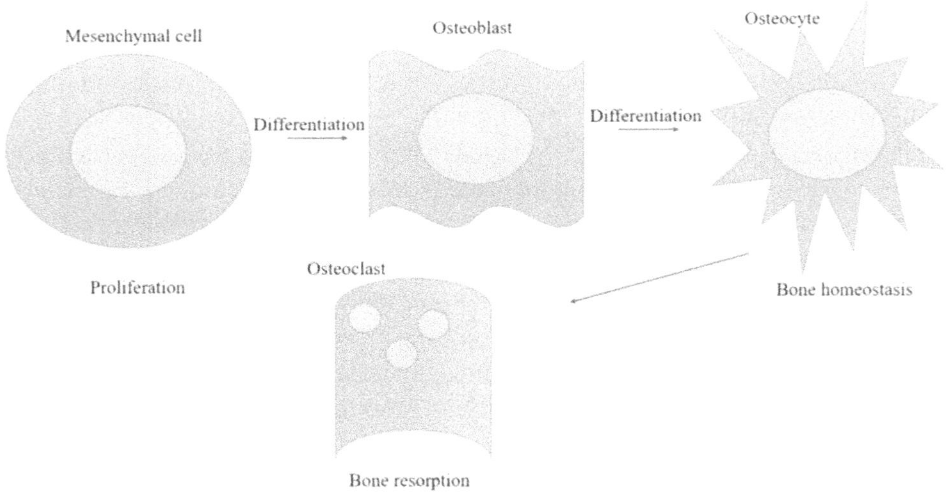

Bioceramics for bone tissue engineering have transitioned from basic research to clinical applications. The conventional clinical approach involves using bioceramics and polymer/bioceramic composites as scaffolds, leveraging their osteogenic properties (Re et al. 2023). However, alternative strategies have also been explored, such as incorporating growth factors or osteoprogenitor cells into the bioceramics to enhance bone remodeling and repair (Stamnitz et al. 2021). HAp and bioactive glasses have been extensively studied in both animal and clinical trials (Cannio et al. 2021; Bròdano et al. 2014). These studies indicate that porous HAp possesses osteoconductive and osteointegration capabilities. Additionally, some researchers have reported that adding MSCs and growth factors, such as the tripeptide Arg-Gly-Asp (RGD), enhances cell attachment, differentiation, proliferation, and bone mineralization (Chen et al. 2013). Notably, calcium phosphate-based scaffolds functionalized with human MSCs have been shown to promote new bone formation and stimulate vascularization (Chen et al. 2018). Vascularization is crucial as it provides oxygen, nutrients, neurotransmitters, and hormones necessary for bone remodeling and normal blood cell production (Filipowska et al. 2017).

5.4 Coatings

The concept of "coatings" arises from the need to modify the interaction between materials and their environment, with the goal of improving performance and extending service life (Holmberg et al. 2009). Coatings have been employed in various applications to reduce friction coefficients, increase corrosion resistance, enhance fatigue life, and provide magnetic or optical properties, among other physical and chemical characteristics (Ramezani et al. 2023). A coating is defined as the deposition of one material onto another, referred to as substrate, in such a way that it forms a layer that covers part or all the substrate's volume. This layer provides physical or chemical properties that the substrate does not possess (Martin, 2009). The use of coatings enables the development of tailored components with specific mechanical, thermal, magnetic, electrical, or chemical properties.

The selection and design of coatings are closely related to their desired properties during service. Required surface properties may include hardness, fracture toughness, compressive and shear strength, thermal insulation, corrosion resistance, ionic dissolution, electrical conduction, and others. These properties can be considered in combination, depending on the surrounding environment (Wood et al. 2024). Additionally, features related to the deposition of coatings and their interaction with the substrate include adhesion strength, roughness, and porosity, Figure 5.

Figure 5. Properties and features to consider in selection of coatings

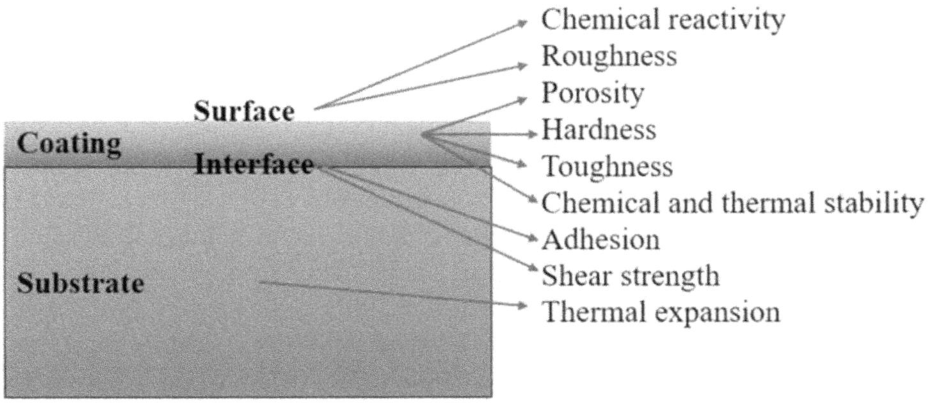

Various methods for fabricating coatings are well-document, as illustrated in the generalized classification in Figure 6 (Fotovvati et al. 2019). These methods can be categorized into two types based on thickness of the coatings: i) thin films, which have thicknesses below 10 µm and are often produced using gaseous and chemical methods, and ii) thick coatings, which range from tens of microns to millimeters in thickness and are typically prepared using techniques that promote a molten or semi-molten state of the coating material. The choice of fabrication technique can be complex and depends on the desired benefits of the coating, the geometry of the component, and associated costs. Engineering coatings are widely used in various industries, including electronics, energy generation, aeronautics, and biomedical engineering (Ou et al. 2023).

Figure 6. General classification of coating preparation methods (Created in BioRender.com)

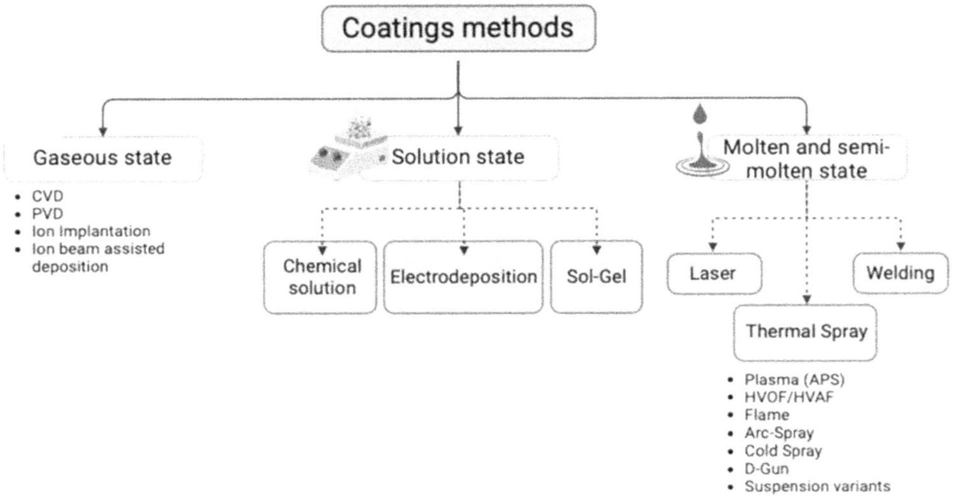

In the biomedical field, various types of coatings have been applied primarily for their biomechanical, biocompatible, and bioactive functions. Ideal biomaterials should exhibit a combination of good biocompatibility and mechanical strength, while also facilitating biological processes and integrating into the body without adverse effects (Amirtharaj Mosas et al. 2022). However, truly ideal biomaterials do not exist; thus, the development of coatings serves as a method to combine the properties of biomaterials to approach these ideal conditions. For instance, joint implants, heart valves, spinal cages, stents and similar devices require blood biocompatibility along with corrosion and wear resistance. Thin films based on nitrides, diamond-like carbon (DLC), and oxides are commonly used coatings to enhance the biocompatibility and protective properties of materials such as polymers and metals, including fluoropolymers, polyurethanes, polyketones, Ni-Ti alloys, Co-Cr alloys, stainless steels, and Ti alloys (Rajan et al. 2022).

Vacuum deposition techniques, such as physical vapor deposition (PVD) and chemical vapor deposition (CVD), have significantly advanced the production of thin coatings with specific and precise properties. The PVD process involves evaporating solid materials, which are then deposited onto a substrate to form a coating (Moore et al. 2017). PVD coatings are recognized for their high hardness and wear resistance, making them an excellent choice for enhancing the durability of medical implants, depending on the intended function. These coatings can be applied to materials like titanium and stainless steel, creating surfaces that resist abrasion and corrosion—crucial factors in biological environments (Alias et al. 2020). In contrast, the CVD process is a promising technique for uniformly depositing coatings on workpieces with complex geometries. In this method, chemical precursors in the form of gases decompose or react on the substrate surface to form a solid coating (Moore et al. 2017). CVD coatings enhance the biocompatibility of implants by modifying their surface properties, such as roughness and chemistry, thereby promoting a positive biological response (Song et al. 2021).

Vapor deposition techniques are valuable methods in tissue engineering. These techniques continue to evolve, and their application in developing medical devices and biomaterials holds great promise for creating innovative solutions in regenerative medicine and tissue engineering. Among these techniques, magnetron sputtering (MS) and pulsed laser deposition (PLD) are the most widely used for depositing bio-piezoelectric materials.

The MS technique is versatile in terms of substrate selection and produce highly dense coatings. For example, it has been reported that ZnO can be fabricated on polymeric substrates with good piezoelectric properties. In contrast, the PLD technique minimizes contamination effects due to its setup and deposition conditions, allowing to produce high-quality coatings. (Xu et al. 2021) reported the fabrication of a coating using a lead-free piezoelectric material ($(1-x)Ba(Ti_{0.8}Zr_{0.2})TiO_3-x(Ba_{0.7}Ca_{0.3})TiO_3$, x=0.45). This material exhibited good piezoelectric and biological performance, indicating its potential for use in bone tissue repair (Xu et al. 2021).

As previously mentioned, polymeric materials also exhibit piezoelectric behavior and hold significant potential for use in BTE. While physical techniques can be employed to manufacture polymeric coatings, the most common approach combines physical and chemical processes. The most widely used methods include spin coating (SC), spray coating, and dip coating. In the SC technique, a polymeric solution is deposited onto a substrate that rotates at a specific speed. The centrifugal force spreads the solution evenly across the substrate, resulting in a uniform layer. In spray coating, a solution passed through a nozzle to create a fine aerosol that forms the film. Finally, the dip coating technique involves immersing the substrate in the polymeric solution and withdrawing it at a controlled speed; as the solution dries, a uniform layer is formed (Aziz & Ismail, 2015; Nguyen et al. 2020; Ten Elshof, 2022). Challenges such as achieving smooth coatings and ensuring proper adhesion have been reported in the production of PVDF films (Nathanael & Oh, 2020). However, other studies have demonstrated successful fabrication of smooth and continuous thin films (~90nm) using the spin coating technique, which exhibited electroactive phases and ferromagnetic/superparamagnetic properties at room temperature (Nguyen et al. 2020). Furthermore, research indicate that incorporating copolymers with PVDF can enhance surface quality and adhesion (Yin et al. 2019).

Thermal spray technologies are widely used in the industry for preparing coatings. These coatings often serve as protective layers on various substrates, providing properties such as wear and corrosion resistance, electric and thermal insulation, chemical and abrasion resistance (Gaur & Kamari, 2024). Thermal spray processes involve spraying molten or semi-molten particles onto a substrate to form a coating. In tissue engineering, thermal spray techniques are employed to enhance the properties of medical devices and implants, offering benefits such as improved durability and protection. These coatings safeguard implants against mechanical and chemical deterioration, ensuring their long-term functionality and reducing the need for replacements or revision surgeries. Moreover, coatings produced through thermal spray processes can be designed to promote cell adhesion, which is crucial for integrating implants with surrounding tissues (Gaur & Kamari, 2024).

In the case of bio-piezoelectric materials, which exhibit behavior highly dependent on their microstructure, extreme thermal conditions encountered during most thermal spray technologies may affect their microstructure. However, an appropriate combination of parameters can help to minimize these microstructural changes. The most widely used technique in the literature for achieving this is atmospheric plasma spray (APS). APS has been particularly successful in depositing piezoelectric systems such as KNN- and BNT-based materials (i.e., $K_{0.44}Na_{0.52}Li_{0.04}(Nb_{0.84}Ta_{0.10}Sb_{0.06})O_3$. *10 mol% K excess;* $Bi_{0.5}Na_{0.5}TiO_3$. *10 mol% Bi excess*) prepared with excess of the most volatile elements. This approach has resulted in a high-density single-phase perovskite with a piezoelectric coefficient (d_{33}) comparable to that obtained for these materials in bulk form, produced by conventional techniques (Yao et al. 2017; Guo et al. 2017).

On the other hand, few piezoceramic coatings have been produced using the high-velocity oxygen combustion (HVOF) technique. For instance, Dent et al. (2001) reported the deposition of $BaTiO_3$ with a single phase and a thickness of approximately 150 µm, although their report did not include any piezoelectric measurement (Dent et al. 2001). More recently, Gutierrez et al. (2021) deposited $(Na_{0.5}Bi_{0.5})TiO_3$ using the Flame Spray technique. While their study primarily focused on optical properties, it demonstrated the feseability of using this technique to obtain piezoceramic coatings (Gutierrez-Pérez et al. 2021). These are among few existing works on this topic. Although the produced coatings were not specifically aimed at biomedical applications, their use is promising for creating bio-piezoelectric coatings with considerable thicknesses and the potential to respond to the stimuli required for bone recovery processes.

For bone regeneration, thick coatings based on HAp, bioactive glasses, and glass-ceramics have been extensively investigated and tested under both *in-vitro* and *in-vivo* conditions. HAp coatings and porous titanium coatings produced by plasma spray are established as the standard coatings in commercial implants worldwide (Heimann et al. 2024). Bio-piezoelectric ceramic coatings are particularly promising for biomedical applications due to their biocompatibility and ability to convert mechanical stress into electrical energy. These coatings can be applied to implantable devices, enabling them to generate electrical signals from bodily movements, which may enhance integration within the human body (Samadi et al. 2022). Their piezoelectric properties also make them suitable for developing advanced biosensors that can monitor physiological changes, providing real-time data for diagnostics and treatment. For instance, these can serve as transducers for cancer diagnosis due to their ability to interact with biomarkers in the human body (Kaur et al. 2024). Overall, bio-piezoelectric ceramic coatings hold great potential for transforming various aspects of biomedical technology.

5.5 Advances in Bioceramic Coatings for Bone Repairing Applications

Bioceramic coatings are primarily used for permanent implants, playing a crucial role in supporting the adhesion, differentiation, and proliferation of cells involved in the formation and growth of new bone (Montazerian et al. 2022). Permanent implants in contact with mature bone often require a strong mechanical integration to restore mobility in an extremity or to stabilize a fracture. Bioceramic coatings aim to achieve this integration by facilitating the formation of new bone on the implant, thereby anchoring the new tissue to the existing bone where the implant is inserted (Kaliaraj et al. 2021).

HAp coatings are commonly used bioceramic coatings for the osseointegration of permanent implants (Awasthi et al. 2021). These coatings have demonstrated the ability to fix implants within the body by supporting new bone formation. Additionally, they serve as a physical barrier, retarding the ion release from metallic substrates that could lead to toxicity. Various methods have been investigated for the preparation of HAp coatings on metallic substrates, including sol-gel, electrochemical deposition, microarc oxidation, and thermal spray techniques (Sharma 2023). The adhesion and mechanical strength of HAp coatings are crucial for permanent implants, as they are often subjected to mechanical stresses from normal human activities (Li et al. 2023). Although HAp coatings obtained by different methods exhibit good bioactive behavior, thermal sprayed HAp coatings are currently regarded as the most reliable option (Prashar et al. 2023). Research efforts focused on thermally sprayed HAp coatings have aimed to enhance their mechanical strength by optimizing process parameters, adjusting coating architecture, and introducing reinforcement phases (Levingstone et al 2017; Baheti et al. 2023). These initiatives are in response to clinical trial results that have reported medium-term mechanical failures of HAp coatings. However, the clinical survivability rate of plasma sprayed HAp coatings on metallic implants remains high, ranging from 90% to 98% (Karia et al. 2023; Studers et al. 2023).

Different authors have proposed alternative thermal spray processes to plasma spray for preparing HAp coatings, including high-velocity oxygen fuel spray (HVOF), high-velocity air fuel spray (HVAF), flame spray (FS), and cold spray (CS) (Henao et al. 2018; Henao et al. 2020; Jagadeeshanayaka et al. 2023; Henao et al. 2024). Recent investigations suggest that these techniques involve less thermal energy compared to plasma spray, which helps to retain HAp in its crystalline form, thereby reducing the presence of unwanted phases (Henao et al. 2020; Henao et al. 2024). This may enhance the chemical stability of HAp coatings and reduce the likelihood of mechanical failures in the body. Additionally, other research efforts have been focused on developing of HAp/bioactive glass composites, bioactive glass, and glass-ceramic coatings via thermal spray techniques (Henao et al. 2019; Yılmaz et al. 2024). The addition of bioactive glasses to HAp coatings aims to achieve stronger bone bonding for the implant compared to hydroxyapatite coatings alone (Poblano-Salas et al. 2024; Müller et al. 2022). This enhancement leverages the osteoconductive and osteoinductive properties of bioactive glasses. Additionally, researchers have explored the development of pure bioactive glass coatings via thermal spray techniques (Garrido et al. 2022). This approach capitalizes on the high bioactivity of bioactive glasses and allows for the formulation of specific compositions tailored to address particular surgical complications, such as infection and inflammation (Rabiee et al. 2015). However, several issues related to the devitrification of bioactive glasses when using various thermal spray techniques have been reported. Devitrification can lead to the formation of bioinert or toxic phases, which diminish the biological performance of bioactive glass coatings (Sergi et al. 2020). Conversely, some researchers have successfully produced completely amorphous plasma-sprayed coatings by employing specific bioactive glass compositions that demonstrate better glass-forming abilities than the clinically proven 45S5 bioglass® composition (Garrido et al. 2022).

To mitigate devitrification, some authors have proposed process modifications, such as using feedstocks in suspension. Techniques like suspension plasma spray (SPS) and high-velocity suspension flame spray (HVSFS) have emerged as alternative processes for preparing bioactive glass coatings via thermal spray (Cañas et al. 2017; Sergi et al. 2020). The advantage of suspension-based processes is that they deliver only the thermal energy necessary to melt and deposit the sprayed powder, thereby preventing overheating. Additionally, these methods enable the preparation of liquid suspensions containing nanometric and submicrometric powders, which cannot be processed using conventional thermal spray techniques. This is significant because chemical wet methods commonly used to produce bioactive glasses often yield nanometric and submicrometric powders. However, suspension thermal spray processes can be complex due to the necessary steps to achieve the stability and viscosity of the suspensions required for successful spraying (Mittal et al. 2022). Evidence suggests that the performance of suspension plasma spray (SPS) coatings is generally inferior to that of high-velocity suspension flame spray (HVSFS) coatings, making them less suitable for bone implant applications. This difference is linked to the reported mechanical performance of SPS coatings. Despite these challenges, numerous studies on the fabrication of bioactive glasses via plasma spray and suspension-based spraying processes consistently show positive *in-vitro* performance of these coatings. They demonstrate a bioactive response characterized by the formation of an apatite layer and support for cells involved in bone formation (Bolelli et al. 2014).

Thermally sprayed glass-ceramic coatings have emerged as a promising alternative for biomedical applications in the past decade (Yılmaz et al. 2024). For instance, apatite-wollastonite-based glass ceramics are recognized for their high bioactivity, which surpasses that of hydroxyapatite. Plasma sprayed apatite-wollastonite coatings also exhibit bioactive behavior, demonstrated by the formation of a bone like-apatite layer. Some *in-vitro* trials involving bioglass and glass-ceramic coatings have shown bioactivity in animal models, leading to osseointegration. However, the mechanical performance of these coatings is not as favorable when compared to bulk materials (Piatti et al. 2024).

In recent years, the thermal spray community has returned to HAp coatings due to several biological studies highlighting the enhanced *in-vivo* performance of HAp-based materials when subjected to electrical stimulation (Swain et al. 2020). Polarized HAp has been shown to reduce the time required for bone bonding compared to non-polarized HAp, thereby accelerating osseointegration and improving bone repair and remodeling at the implantation site (Bodhak et al. 2010; Sagawa et al. 2010). However, thermal-sprayed HAp coatings often contain cracks and pores, are non-conductive, and exhibit rough surfaces. To enable poling, these coatings typically require grinding and polishing to achieve a planar surface, which may compromise their mechanical performance due to their fragile nature. Additionally, the application of conductive paint for poling must be removed afterward, complicating the sterilization processes for implants. The non-planar surfaces of implants present a significant challenge in this regard. Interestingly, recent investigations (Prezas et al. 2023) have proposed a method called "corona triode", which allows for the electrical charging of plasma-sprayed HAp coatings without the need for paints or other substances. Electrically charged plasma-sprayed HAp coatings hold promise for enhancing the biological response of coated implants in the human body. Furthermore, other studies have focused on exploring thermal-sprayed HAp/piezoelectric ceramic composite coatings. For example, HAp/$BaTiO_3$ and HAp/$SrTiO_3$ coating systems by plasma spray (Gómez Batres et al. 2021; Senthilkumar et al. 2021; Dias et al. 2023). These coatings have shown to stimulate the growth of osteoblasts and improved mineralization process than pure HAp coatings (Senthilkumar et al. 2021). There is also evidence that the mechanical behavior of plasma sprayed HAp/$BaTiO_3$ coatings is improved with respect to that of conventional HAp coatings (Gómez Batres et al. 2021). A summary of advantages of both piezoelectric and HAp coatings for hard tissue engineering are listed in Table 2.

Table 2. Comparative advantages of hydroxyapatite and bio-piezoelectric coatings for bone tissue engineering

Hydroxyapatite coatings	Bio-piezoelectric coatings
• Biocompatibility: Hydroxyapatite closely resembles natural bone minerals, promoting a favorable environment after the introduction of the implant. • Osteointegration: These coatings enhance the bonding between the implant and existing bone due to their bioactive behavior. • Facilitates bone regeneration: Hydroxyapatite supports the biological processes necessary for bone healing due to its osteoconductive properties.	Electrical Stimulation: These coatings can enhance cellular activities that are crucial for bone formation and healing, potentially improving the fixation of implants. Mechanical Response: These coatings can respond to mechanical stress and help to stimulate natural bone growth conditions, encouraging better integration of the implant with surrounding bone.

6. SUMMARY AND FUTURE PERSPECTIVES

This book chapter thoroughly examines the role of bio-piezoelectric coatings in bone tissue engineering applications, including an overview of bone remodeling and repair in the human body, as well as the expected functions of bio-piezoelectric materials. It delves into the concepts of ferro-piezoelectric responses in materials and provides a comprehensive review of bioceramic coatings. Despite advancements in developing bio-piezoelectric materials, research on the preparation and performance of these coatings under *in-vitro* and *in-vivo* conditions is ongoing. Recent studies have highlighted the potential of pure piezoelectric coatings and bioceramic/composite coatings produced through various methods, including plasma spray, physical vapor deposition, hydrothermal methods, chemical bath deposition, pulsed laser deposition, micro-arc oxidation, and spin coating. Nevertheless, many questions remain unanswered, and further investigations are crucial to unlocking the full potential of bio-piezoelectric coatings for bone tissue engineering. This aligns with the limited literature currently available on bioceramic coatings with piezoelectric properties on metallic substrates for bone implantation. Existing studies indicate that these coatings can stimulate cellular processes related to bone resorption and repair. However, each coating technique presents its own challenges and characteristics that may not be ideal for clinical applications. For instance, some reports (Jelínek et al. 2018) regarding pulsed laser deposition for producing biomaterials suggest using bond coating layers of platinum (Pt) to deposit $BaTiO_3$ thin films on titanium alloys. While this approach may be advantageous from a fabrication perspective, it is important to note that Pt is cytotoxic to the human body. Similarly, although piezoelectric coatings can be produced using physical vapor deposition on titanium alloys, some findings indicate a need for a post-annealing process to enhance their properties. (Dahl-Hansen et al. 2024). On the other hand, hydrothermal and microarc oxidation methods have demonstrated promising biological responses; however, the thickness and microstructure of these coatings remain concerns from a mechanical perspective (Wang et al. 2015; Vandrovcova et al. 2021). Some studies have investigated polymer/piezoelectric and bioactive/piezoelectric composite coatings as alternatives to single piezoelectric coatings (Sharma et al. 2022; Senthilkumar et al. 2021).

This is motivated by the potential stability issues in the mid- to long-term, as the presence of bioinert ferroelectric phases may lead to foreign body rejection.

A key point for future studies is to identify optimal combinations of bioceramic/piezoelectric or polymer/piezoelectric materials and to develop effective strategies for producing these coatings that may be feasible for industrial implementation in biomedical products. Bioceramic/piezoelectric composite coatings, particularly those based on HAp, bioactive glass, and glass-ceramics, hold significant potential for enhancing osseointegration in permanent implants. This potential arises from the bioactive properties of the bioceramic matrix, which supports new bone formation, while the piezoelectric phase promotes cellular processes associated with mineralization and bone growth. There is still much work to be done in developing various bioceramic/piezoelectric composite coatings. Key areas of research include exploring new piezoelectric/bioceramic compositions, optimizing coating architectures, and assessing the impact of coating features on biological performance over both the short and long term. In summary, recent studies have shown promise for the use of bio-piezoelectric coatings in bone tissue engineering. Biopolymers and bioceramic/piezoelectric composites may provide an excellent platform for facilitating cellular processes involved in bone formation. Understanding the interactions between different piezoelectric and biomaterial matrix composites, along with processing methods to create mechanically and chemically stable coatings, will significantly advance the development of a new generation of implants with improved osseointegration and clinical survival rates.

ACKNOWLEDGMENT

The authors acknowledge the support from the National Council of Humanities Science and Technologies CONAHCYT, particularly through the "Cátedras" program, project 848.

REFERENCES

Abushahba, F., Areid, N., Gürsoy, M., Willberg, J., Laine, V., Yatkin, E., Hupa, L., & Närhi, T. O. (2023). Bioactive glass air-abrasion promotes healing around contaminated implant surfaces surrounded by circumferential bone defects: An experimental study in the rat. *Clinical Implant Dentistry and Related Research*, 25(2), 409–418. DOI: 10.1111/cid.13172 PMID: 36602418

Aguilera, S. B., McCarthy, A., Khalifian, S., Lorenc, Z. P., Goldie, K., & Chernoff, W. G. (2023). The Role of Calcium Hydroxylapatite (Radiesse) as a Regenerative Aesthetic Treatment: A Narrative Review. *Aesthetic Surgery Journal*, 43(10), 1063–1090. DOI: 10.1093/asj/sjad173 PMID: 37635437

Akbarov, A. N., & Tillakhodjayeva, M. M. (2024). Studying the Dynamics of Bone Tissue Remodeling in Blood Plasma by Determining the Level of Bone Alkaline Phosphatase in the Blood. *International Journal of Integrative and Modern Medicine*, 2(5), 66–69. https://medicaljournals.eu/index.php/IJIMM/article/view/291

Al-Moameri, H. H., Nahi, Z. M., Al-Sharify, N. T., & Rzaij, D. R. (2020). A review on the biomedical applications of alumina. *Journal of Engineering and Sustainable Development*, 24(5), 28–36. DOI: 10.31272/jeasd.24.5.5

Algueró, M., Cheng, B. L., Guiu, F., Reece, M. J., Poole, M., & Alford, N. (2001). Degradation of the d33 piezoelectric coffcient for PZT ceramics under static and cyclic compressive loading. *Journal of the European Ceramic Society*, 21(10–11), 1437–1440. DOI: 10.1016/S0955-2219(01)00036-X

Alias, R., Mahmoodian, R., Genasan, K., Vellasamy, K. M., Hamdi Abd Shukor, M., & Kamarul, T. (2020). Mechanical, antibacterial, and biocompatibility mechanism of PVD grown silver-tantalum-oxide-based nanostructured thin film on stainless steel 316L for surgical applications. *Materials Science and Engineering C*, 107, 110304. DOI: 10.1016/j.msec.2019.110304 PMID: 31761210

Amin, B., Elahi, M. A., Shahzad, A., Porter, E., & O'Halloran, M. (2019). A review of the dielectric properties of the bone for low frequency medical technologies. *Biomedical physics & engineering express, 5*(2), . https://doi.org/DOI: 10.1088/2057-1976/aaf210

Amirtharaj Mosas, K. K., Chandrasekar, A. R., Dasan, A., Pakseresht, A., & Galusek, D. (2022). Recent advancements in materials and coatings for biomedical implants. *Gels (Basel, Switzerland)*, 8(5), 323. DOI: 10.3390/gels8050323 PMID: 35621621

Awasthi, S., Pandey, S. K., Arunan, E., & Srivastava, C. (2021). A review on hydroxyapatite coatings for the biomedical applications: Experimental and theoretical perspectives. *Journal of Materials Chemistry. B*, 9(2), 228–249. DOI: 10.1039/D0TB02407D PMID: 33231240

Axente, E., Elena Sima, L., & Sima, F. (2020). Biomimetic coatings obtained by combinatorial laser technologies. *Coatings*, 10(5), 463. DOI: 10.3390/coatings10050463

Aziz, F., & Ismail, A. F. (2015). Spray coating methods for polymer solar cells fabrication: A review. *Materials Science in Semiconductor Processing*, 39, 416–425. DOI: 10.1016/j.mssp.2015.05.019

Baheti, W., & Lv, S., Mila, Ma, L., Amantai, D., Sun, H., & He, H. (. (2023). Graphene/hydroxyapatite coating deposit on titanium alloys for implant application. *Journal of Applied Biomaterials & Functional Materials*, 21. Advance online publication. DOI: 10.1177/22808000221148104 PMID: 36633270

Bai, L., Song, P., & Su, J. (2023). Bioactive elements manipulate bone regeneration. *Biomaterials Translational*, 4(4), 248–269. DOI: 10.12336/biomatertransl.2023.04.005 PMID: 38282709

Bai, Y., Li, X., Wu, K., Heng, B., Zhang, X., & Deng, X. (2024). Biophysical stimuli for promoting bone repair and regeneration. *Medical Review (Berlin, Germany)*. Advance online publication. DOI: 10.1515/mr-2024-0023

Bai, Y., Meng, H., Li, Z., & Wang, Z. L. (2024). Degradable piezoelectric biomaterials for medical applications. *MedMat*, 1(1), 40–49. DOI: 10.1097/mm9.0000000000000002

Baino, F., Hamzehlou, S., & Kargozar, S. (2018). Bioactive glasses: Where are we and where are we going? *Journal of Functional Biomaterials*, 9(1), 25. DOI: 10.3390/jfb9010025 PMID: 29562680

Baloh, R. W. (2024). Electricity and the Nervous System. In Brain Electricity: The Interwoven History of Electricity and Neuroscience (pp. 125-158). *Springer Nature*. https://doi.org/DOI: 10.1007/978-3-031-62994-5_5

Ban, C., Li, L., & Wei, L. (2018). Electrical properties of O-self-doped boron-nitride nanotubes and the piezoelectric effects of their freestanding network film. *RSC Advances*, 8(51), 29141–29146. DOI: 10.1039/C8RA05698F PMID: 35548006

Barak, M. M. (2024). Cortical and Trabecular Bone Modeling and Implications for Bone Functional Adaptation in the Mammalian Tibia. *Bioengineering (Basel, Switzerland)*, 11(5), 514. DOI: 10.3390/bioengineering11050514 PMID: 38790379

Batool, M., Abid, M. A., Javed, T., & Haider, M. N. (2024). Applications of biodegradable polymers and ceramics for bone regeneration: A mini-review. *International Journal of Polymeric Materials*, •••, 1–15. DOI: 10.1080/00914037.2024.2314601

Baxter, F. R., Bowen, C. R., Turner, I. G., & Dent, A. C. (2010). Electrically active bioceramics: A review of interfacial responses. *Annals of Biomedical Engineering*, 38(6), 2079–2092. DOI: 10.1007/s10439-010-9977-6 PMID: 20198510

Bhavsar, M. B., Leppik, L., Costa Oliveira, K. M., & Barker, J. H. (2020). Role of Bioelectricity During Cell Proliferation in Different Cell Types. *Frontiers in Bioengineering and Biotechnology*, 8, 603. DOI: 10.3389/fbioe.2020.00603 PMID: 32714900

Bhong, M., Khan, T. K., Devade, K., Krishna, B. V., Sura, S., Eftikhaar, H. K., & Gupta, N. (2023). Review of composite materials and applications. *Materials Today: Proceedings*, •••, 2214–7853. DOI: 10.1016/j.matpr.2023.10.026

Birol, H., Damjanovic, D., & Setter, N. (2005). Preparation and characterization of KNbO3 ceramics. *Journal of the American Ceramic Society*, 88(7), 1754–1759. DOI: 10.1111/j.1551-2916.2005.00347.x

Bodhak, S., Bose, S., & Bandyopadhyay, A. (2010). Electrically polarized HAp-coated Ti: In vitro bone cell-material interactions. *Acta Biomaterialia*, 6(2), 641–651. DOI: 10.1016/j.actbio.2009.08.008 PMID: 19671456

Bolelli, G., Bellucci, D., Cannillo, V., Lusvarghi, L., Sola, A., Stiegler, N., Müller, P., Killinger, A., Gadow, R., Altomare, L., & De Nardo, L. (2014). Suspension thermal spraying of hydroxyapatite: Microstructure and in vitro behaviour. *Materials Science and Engineering C*, 34, 287–303. DOI: 10.1016/j.msec.2013.09.017 PMID: 24268261

Bose, S., & Tarafder, S. (2012). Calcium phosphate ceramic systems in growth factor and drug delivery for bone tissue engineering: A review. *Acta Biomaterialia*, 8(4), 1401–1421. DOI: 10.1016/j.actbio.2011.11.017 PMID: 22127225

Bròdano, G. B., Giavaresi, G., Lolli, F., Salamanna, F., Parrilli, A., Martini, L., Griffoni, C., Greggi, T., Arcangeli, E., Pressato, D., Boriani, S., & Fini, M. (2014). Hydroxyapatite-Based Biomaterials Versus Autologous Bone Graft in Spinal Fusion: An In Vivo Animal Study. *Spine*, 39(11), E661–E668. DOI: 10.1097/BRS.0000000000000311 PMID: 24718060

Bystrov, V. S. (2024). Molecular self-assembled helix peptide nanotubes based on some amino acid molecules and their dipeptides: Molecular modeling studies. *Journal of Molecular Modeling*, 30(8), 257. DOI: 10.1007/s00894-024-05995-0 PMID: 38976043

Cañas, E., Vicent, M., Orts, M. J., & Sánchez, E. (2017). Bioactive glass coatings by suspension plasma spraying from glycolether-based solvent feedstock. *Surface and Coatings Technology*, 318, 190–197. DOI: 10.1016/j.surfcoat.2016.12.060

Cannio, M., Bellucci, D., Roether, J. A., Boccaccini, D. N., & Cannillo, V. (2021). Bioactive Glass Applications: A Literature Review of Human Clinical Trials. *Materials (Basel)*, 14(18), 5440. DOI: 10.3390/ma14185440 PMID: 34576662

Carville, N. C., Collins, L., Manzo, M., Gallo, K., Lukasz, B. I., McKayed, K. K., Simpson, J. C., & Rodriguez, B. J. (2015). Biocompatibility of ferroelectric lithium niobate and the influence of polarization charge on osteoblast proliferation and function. *Journal of Biomedical Materials Research. Part A*, 103(8), 2540–2548. DOI: 10.1002/jbm.a.35390 PMID: 25504748

Chaari, M., Matoussi, A., & Fakhfakh, Z. (2011). Structural and dielectric properties of sintering zinc oxide bulk ceramic. *Materials Sciences and Applications*, 2(7), 764–769. DOI: 10.4236/msa.2011.27105

Chae, I., Jeong, C. K., Ounaies, Z., & Kim, S. H. (2018). Review on Electromechanical Coupling Properties of Biomaterials. *ACS Applied Bio Materials*, 1(4), 936–953. DOI: 10.1021/acsabm.8b00309 PMID: 34996135

Chaudhari, V. S., Kushram, P., & Bose, S. (2024). Drug delivery strategies through 3D-printed calcium phosphate. *Trends in biotechnology*, S0167-7799(24)00145-8. https://doi.org/DOI: 10.1016/j.tibtech.2024.05.006

Chen, C., Ji, J., Jiao, X., Wang, A., & Ding, H. (2017). Fabrication and Properties of Lithium Sodium Potassium Niobate Lead-Free Piezoelectric Ceramics. *Journal of Advanced Microscopy Research*, 12(2), 85–88. DOI: 10.1166/jamr.2017.1323

Chen, S., Tong, X., Huo, Y., Liu, S., Yin, Y., Tan, M. L., Cai, K., & Ji, W. (2024). Piezoelectric Biomaterials Inspired by Nature for Applications in Biomedicine and Nanotechnology. *Advanced Materials*, 36(35), e2406192. DOI: 10.1002/adma.202406192 PMID: 39003609

Chen, W., Liu, X., Chen, Q., Bao, C., Zhao, L., Zhu, Z., & Xu, H. H. K. (2018). Angiogenic and osteogenic regeneration in rats via calcium phosphate scaffold and endothelial cell co-culture with human bone marrow mesenchymal stem cells (MSCs), human umbilical cord MSCs, human induced pluripotent stem cell-derived MSCs and human embryonic stem cell-derived MSCs. *Journal of Tissue Engineering and Regenerative Medicine*, 12(1), 191–203. DOI: 10.1002/term.2395 PMID: 28098961

Chen, W., Zhou, H., Weir, M. D., Tang, M., Bao, C., & Xu, H. H. (2013). Human embryonic stem cell-derived mesenchymal stem cell seeding on calcium phosphate cement-chitosan-RGD scaffold for bone repair. *Tissue Engineering. Part A*, 19(7-8), 915–927. DOI: 10.1089/ten.tea.2012.0172 PMID: 23092172

Choi, E. S., Kim, H. C., Muthoka, R. M., Panicker, P. S., Agumba, D. O., & Kim, J. (2021). Aligned cellulose nanofiber composite made with electrospinning of cellulose nanofiber—Polyvinyl alcohol and its vibration energy harvesting. *Composites Science and Technology*, 209, 108795. DOI: 10.1016/j.compscitech.2021.108795

Chorsi, M. T., Curry, E. J., Chorsi, H. T., Das, R., Baroody, J., Purohit, P. K., Ilies, H., & Nguyen, T. D. (2019). Piezoelectric Biomaterials for Sensors and Actuators. *Advanced Materials*, 31(1), e1802084. DOI: 10.1002/adma.201802084 PMID: 30294947

Civinini, R., Capone, A., Carulli, C., Matassi, F., Nistri, L., & Innocenti, M. (2017). The kinetics of remodeling of a calcium sulfate/calcium phosphate bioceramic. *Journal of Materials Science. Materials in Medicine*, 28(9), 137. DOI: 10.1007/s10856-017-5940-5 PMID: 28785889

Curry, E. J., Ke, K., Chorsi, M. T., Wrobel, K. S., Miller, A. N.III, Patel, A., Kim, I., Feng, J., Yue, L., Wu, Q., Kuo, C. L., Lo, K. W., Laurencin, C. T., Ilies, H., Purohit, P. K., & Nguyen, T. D. (2018). Biodegradable Piezoelectric Force Sensor. *Proceedings of the National Academy of Sciences of the United States of America*, 115(5), 909–914. DOI: 10.1073/pnas.1710874115 PMID: 29339509

Da Silva, L. P., Kundu, S. C., Reis, R. L., & Correlo, V. M. (2020). Electric Phenomenon: A Disregarded Tool in Tissue Engineering and Regenerative Medicine. *Trends in Biotechnology*, 38(1), 24–49. DOI: 10.1016/j.tibtech.2019.07.002 PMID: 31629549

Dahl-Hansen, R. P., Stange, M. S. S., Sunde, T. O., Ræder, J. H., & Rørvik, P. M. (2024). On the Evolution of Stress and Microstructure in Radio Frequency-Sputtered Lead-Free (Ba, Ca)(Zr, Ti)O3 Thin Films. *Actuators*, 13(3), 115. DOI: 10.3390/act13030115

Dent, A. H., Patel, A., Gutleber, J., Tormey, E., Sampath, S., & Herman, H. (2001). High velocity oxy-fuel and plasma deposition of BaTiO3 and (Ba, Sr) TiO3. *Materials Science and Engineering B*, 87(1), 23–30. DOI: 10.1016/S0921-5107(01)00653-5

Di Martino, A., Sittinger, M., & Risbud, M. V. (2005). Chitosan: A versatile biopolymer for orthopaedic tissue-engineering. *Biomaterials*, 26(30), 5983–5990. DOI: 10.1016/j.biomaterials.2005.03.016 PMID: 15894370

Dias, I. J., Pádua, A. S., Pires, E. A., Borges, J. P., Silva, J. C., & Lança, M. C. (2023). Hydroxyapatite-Barium Titanate Biocoatings Using Room Temperature Coblasting. *Crystals*, 13(4), 579. DOI: 10.3390/cryst13040579

Dubey, A. K., & Basu, B. (2014). Pulsed electrical stimulation and surface charge induced cell growth on multistage spark plasma sintered hydroxyapatite-barium titanate piezobiocomposite. *Journal of the American Ceramic Society*, 97(2), 481–489. DOI: 10.1111/jace.12647

Dubey, A. K., Basu, B., Balani, K., Guo, R., & Bhalla, A. S. (2011). Multifunctionality of perovskites BaTiO3 and CaTiO3 in a composite with hydroxyapatite as orthopedic implant materials. *Integrated Ferroelectrics*, 131(1), 119–126. DOI: 10.1080/10584587.2011.616425

Farag, M. M. (2023). Recent trends on biomaterials for tissue regeneration applications. *Journal of Materials Science*, 58(2), 527–558. DOI: 10.1007/s10853-022-08102-x

Feng, S., Li, J., Jiang, X., Li, X., Pan, Y., Zhao, L., Boccaccini, A. R., Zheng, K., Yang, L., & Wei, J. (2016). Influences of mesoporous magnesium silicate on the hydrophilicity, degradability, mineralization and primary cell response to a wheat protein based biocomposite. *Journal of Materials Chemistry. B*, 4(39), 6428–6436. DOI: 10.1039/C6TB01449F PMID: 32263451

Filipowska, J., Tomaszewski, K. A., Niedźwiedzki, Ł., Walocha, J. A., & Niedźwiedzki, T. (2017). The role of vasculature in bone development, regeneration and proper systemic functioning. *Angiogenesis*, 20(3), 291–302. DOI: 10.1007/s10456-017-9541-1 PMID: 28194536

Fisher, J. G., Thuan, U. T., Farooq, M. U., Chandrasekaran, G., Jung, Y. D., Hwang, E. C., Lee, J. J., & Lakshmanan, V. K. (2018). Prostate Cancer Cell-Specific Cytotoxicity of Sub-Micron Potassium Niobate Powder. *Journal of Nanoscience and Nanotechnology*, 18(5), 3141–3147. DOI: 10.1166/jnn.2018.14666 PMID: 29442813

Fotovvati, B., Namdari, N., & Dehghanghadikolaei, A. (2019). On coating techniques for surface protection: A review. *Journal of Manufacturing and Materials processing*, 3(1), 28. https://doi.org/DOI: 10.3390/jmmp3010028

Fu, L., Engqvist, H., & Xia, W. (2020). Glass-Ceramics in Dentistry: A Review. *Materials (Basel)*, 13(5), 1049. DOI: 10.3390/ma13051049 PMID: 32110874

Gamboa, B., Bhalla, A., & Guo, R. (2020). Assessment of PZT (Soft/Hard) Composites for Energy Harvesting. *Ferroelectrics*, 555(1), 118–123. DOI: 10.1080/00150193.2019.1691389

Garrido, B., Dosta, S., & Cano, I. G. (2022). Bioactive glass coatings obtained by thermal spray: Current status and future challenges. *Boletín de la Sociedad Española de Cerámica y Vidrio*, 61(5), 516–530. DOI: 10.1016/j.bsecv.2021.04.001

Gaur, U. P., & Kamari, E. (2024). Applications of Thermal Spray Coatings: A Review. *Journal of Thermal Spray and Engineering*, 4, 106–114. DOI: 10.52687/2582-1474/405

Ghosh, A., Ray, S. S., Orasugh, J. T., & Chattopadhyay, D. (2024). Collagen-Based Hybrid Piezoelectric Material. *Hybrid Materials for Piezoelectric Energy Harvesting and Conversion*, 283-299. https://doi.org/DOI: 10.1002/9781394150373.ch11

Gomes, J., Nunes, J. S., Sencadas, V., & Lanceros-Méndez, S. (2010). Influence of the β-phase content and degree of crystallinity on the piezo-and ferroelectric properties of poly(vinylidene fluoride). *Smart Materials and Structures*, 19(6), 1–7. DOI: 10.1088/0964-1726/19/6/065010

Gómez Batres, R., Guzmán Escobedo, Z. S., Gutiérrez, K. C., Leal Berumen, I., Hurtado Macias, A., Pérez, G. H., & Orozco Carmona, V. M. (2021). Impact evaluation of high energy ball milling homogenization process in the phase distribution of hydroxyapatite-barium titanate plasma spray biocoating. *Coatings*, 11(6), 728. DOI: 10.3390/coatings11060728

Guo, K., Chen, S., Tan, C. K. I., Sharifzadeh Mirshekarloo, M., Yao, K., & Tay, F. E. H. (2017). Bismuth sodium titanate lead-free piezoelectric coatings by thermal spray process. *Journal of the American Ceramic Society*, 100(8), 3385–3392. DOI: 10.1111/jace.14882

Gutiérrez-Pérez, A. I., Ayala-Ayala, M. T., Mora-García, A. G., Moreno-Murguía, B., Ruiz-Luna, H., & Muñoz-Saldaña, J. (2021). Visible-light photoactive thermally sprayed coatings deposited from spray-dried (Na0. 5Bi0. 5)TiO3 microspheres. *Surface and Coatings Technology*, 427, 127851. DOI: 10.1016/j.surfcoat.2021.127851

He, D., Dong, W., Tang, S., Wei, J., Liu, Z., Gu, X., Li, M., Guo, H., & Niu, Y. (2014). Tissue engineering scaffolds of mesoporous magnesium silicate and poly(ε-caprolactone)-poly(ethylene glycol)-poly(ε-caprolactone) composite. *Journal of Materials Science. Materials in Medicine*, 25(6), 1415–1424. DOI: 10.1007/s10856-014-5183-7 PMID: 24595904

Heimann, R. B. (2024). Plasma-Sprayed Osseoconductive Hydroxylapatite Coatings for Endoprosthetic Hip Implants: Phase Composition, Microstructure, Properties, and Biomedical Functions. *Coatings*, 14(7), 787. DOI: 10.3390/coatings14070787

Henao, J., Cruz-Bautista, M., Hincapie-Bedoya, J., Ortega-Bautista, B., Corona-Castuera, J., Giraldo-Betancur, A. L., Espinosa-Arbelaez, D. G., Alvarado Orozco, J. M., Clavijo-Mejía, G. A., Trápaga-Martínez, L. G., & Poblano-Salas, C. A. (2018). HVOF hydroxyapatite/titania-graded coatings: Microstructural, mechanical, and in vitro characterization. *Journal of Thermal Spray Technology*, 27(8), 1302–1321. DOI: 10.1007/s11666-018-0811-2

Henao, J., Giraldo-Betancur, A. L., Poblano-Salas, C. A., Forero-Sossa, P. A., Espinosa-Arbelaez, D. G., Gonzalez, J. V., & Corona-Castuera, J. (2024). On the deposition of cold-sprayed hydroxyapatite coatings. *Surface and Coatings Technology*, 476, 130289. DOI: 10.1016/j.surfcoat.2023.130289

Henao, J., Poblano-Salas, C., Monsalve, M., Corona-Castuera, J., & Barceinas-Sanchez, O. (2019). Bioactive glass coatings manufactured by thermal spray: A status report. *Journal of Materials Research and Technology*, 8(5), 4965–4984. DOI: 10.1016/j.jmrt.2019.07.011

Henao, J., Sotelo-Mazon, O., Giraldo-Betancur, A. L., Hincapie-Bedoya, J., Espinosa-Arbelaez, D. G., Poblano-Salas, C., Cuevas-Arteaga, C., Corona-Castuera, J., & Martinez-Gomez, L. (2020). Study of HVOF-sprayed hydroxyapatite/titania graded coatings under in-vitro conditions. *Journal of Materials Research and Technology*, 9(6), 14002–14016. DOI: 10.1016/j.jmrt.2020.10.005

Heng, B. C., Bai, Y., Li, X., Meng, Y., Lu, Y., Zhang, X., & Deng, X. (2023). The bioelectrical properties of bone tissue. *Animal Models and Experimental Medicine*, 6(2), 120–130. DOI: 10.1002/ame2.12300 PMID: 36856186

Holmberg, K., & Matthews, A. (2009). *Coatings tribology: properties, mechanisms, techniques and applications in surface engineering*. Elsevier.

Hoque, N. A., Thakur, P., Biswas, P., Saikh, M. M., Roy, S., Bagchi, B., Das, S., & Ray, P. P. (2018). Biowaste crab shell-extracted chitin nanofiber-based superior piezoelectric nanogenerator. *Journal of Materials Chemistry. A, Materials for Energy and Sustainability*, 6(28), 13848–13858. DOI: 10.1039/C8TA04074E

Jacob, J., More, N., Kalia, K., & Kapusetti, G. (2018). Piezoelectric smart biomaterials for bone and cartilage tissue engineering. *Inflammation and Regeneration*, 38(1), 2. DOI: 10.1186/s41232-018-0059-8 PMID: 29497465

Jaeger, R. E., & Egerton, L. (1962). Hot pressing of potassium-sodium niobates. *Journal of the American Ceramic Society*, 45(5), 209–213. DOI: 10.1111/j.1151-2916.1962.tb11127.x

Jagadeeshanayaka, N., Kele, S. N., & Jambagi, S. C. (2023). An Investigation into the Relative Efficacy of High-Velocity Air-Fuel-Sprayed Hydroxyapatite Implants Based on the Crystallinity Index, Residual Stress, Wear, and In-Flight Powder Particle Behavior. *Langmuir*, 39(48), 17513–17528. DOI: 10.1021/acs.langmuir.3c02840 PMID: 38050681

Janićijević, A., Pavlović, V. P., Kovačević, D., Đorđević, N., Marinković, A., Vlahović, B., Karoui, A., Pavlovic, V. P., & Filipović, S. (2024). Impact of Nanocellulose Loading on the Crystal Structure, Morphology and Properties of PVDF/Magnetite@ NC/BaTiO3 Multi-component Hybrid Ceramic/Polymer Composite Material. *Journal of Inorganic and Organometallic Polymers and Materials*, 34(5), 2129–2139. DOI: 10.1007/s10904-023-02953-w

Jelínek, M., Buixaderas, E., Drahokoupil, J., Kocourek, T., Remsa, J., Vaněk, P., Vandrovcová, M., Doubková, M., & Bačáková, L. (2018). Laser-synthesized nanocrystalline, ferroelectric, bioactive $BaTiO_3$/Pt/FS for bone implants. *Journal of Biomaterials Applications*, 32(10), 1464–1475. DOI: 10.1177/0885328218768646 PMID: 29621929

Jensen, D. B., Li, Z., Pavel, I., Dervishi, E., Biris, A. S., Biris, A. R., Lupu, D., & Jensen, P. J. (2007). Bone tissue: A relationship between micro and nano structural composition and its corresponding electrostatic properties with applications in tissue engineering. *2007 IEEE Industry Applications Annual Meeting*, 55-59. https://doi.org/DOI: 10.1109/07IAS.2007.8

Joo, S., Gwon, Y., Kim, S., Park, S., Kim, J., & Hong, S. (2024). Piezoelectrically and Topographically Engineered Scaffolds for Accelerating Bone Regeneration. *ACS Applied Materials & Interfaces*, 16(2), 1999–2011. DOI: 10.1021/acsami.3c12575 PMID: 38175621

Kakimoto, K. I., Hayakawa, Y., & Kagomiya, I. (2010). Low-temperature sintering of dense (Na, K) NbO3 piezoelectric ceramics using the citrate precursor technique. *Journal of the American Ceramic Society*, 93(9), 2423–2426. DOI: 10.1111/j.1551-2916.2010.03748.x

Kaliaraj, G. S., Siva, T., & Ramadoss, A. (2021). Surface functionalized bioceramics coated on metallic implants for biomedical and anticorrosion performance - a review. *Journal of Materials Chemistry. B*, 9(46), 9433–9460. DOI: 10.1039/D1TB01301G PMID: 34755756

Kamboj, N., Piili, H., Ganvir, A., Gopaluni, A., Nayak, C., Moritz, N., & Salminen, A. (2023). Bioinert ceramics scaffolds for bone tissue engineering by laser-based powder bed fusion: A preliminary review. *IOP Conference Series. Materials Science and Engineering*, 1296(1), 012022. DOI: 10.1088/1757-899X/1296/1/012022

Kamel, N. A. (2022). Bio-piezoelectricity: Fundamentals and applications in tissue engineering and regenerative medicine. *Biophysical Reviews*, 14(3), 717–733. DOI: 10.1007/s12551-022-00969-z PMID: 35783122

Kang, Y. G., Wei, J., Shin, J. W., Wu, Y. R., Su, J., Park, Y. S., & Shin, J. W. (2018). Enhanced biocompatibility and osteogenic potential of mesoporous magnesium silicate/polycaprolactone/wheat protein composite scaffolds. *International Journal of Nanomedicine*, 13, 1107–1117. DOI: 10.2147/IJN.S157921 PMID: 29520139

Kareem, R., Bulut, N., & Kaygili, O. (2024). Hydroxyapatite Biomaterials: A Comprehensive Review of their Properties, Structures, Medical Applications, and Fabrication Methods. *Journal of Chemical Reviews*, 6(1), 1–26. DOI: 10.48309/jcr.2024.415051.1253 PMID: 38118062

Karia, M., Logishetty, K., Johal, H., Edwards, T. C., & Cobb, J. P. (2023). 5 year follow up of a hydroxyapatite coated short stem femoral component for hip arthroplasty: A prospective multicentre study. *Scientific Reports*, 13(1), 17166. DOI: 10.1038/s41598-023-44191-7 PMID: 37821511

Katubi, K. M., Ibrahimoglu, E., Çalışkan, F., Alrowaili, Z. A., Olarinoye, I. O., & Al-Buriahi, M. S. (2024). Apatite–Wollastonite (AW) glass ceramic doped with B2O3: Synthesis, structure, SEM, hardness, XRD, and neutron/charged particle attenuation properties. *Ceramics International*, 50(15), 27139–27146. DOI: 10.1016/j.ceramint.2024.05.011

Kaur, A., Kumar, P., Gupta, A., & Sapra, G. (2024). Piezoelectric Biosensors in Healthcare. In *Enzyme-based Biosensors: Recent Advances and Applications in Healthcare* (pp. 255-271). Springer Nature. https://doi.org/DOI: 10.1007/978-981-15-6982-1_11

Khare, D., Basu, B., & Dubey, A. K. (2020). Electrical stimulation and piezoelectric biomaterials for bone tissue engineering applications. *Biomaterials*, 258, 120280. DOI: 10.1016/j.biomaterials.2020.120280 PMID: 32810650

Khorsand Zak, A., Yazdi, S. T., Abrishami, M. E., & Hashim, A. M. (2024). A review on piezoelectric ceramics and nanostructures: Fundamentals and fabrications. *Journal of the Australian Ceramic Society*, 3(3), 723–753. DOI: 10.1007/s41779-024-00990-3

Kim, D., Han, S. A., Kim, J. H., Lee, J. H., Kim, S. W., & Lee, S. W. (2020). Biomolecular Piezoelectric Materials: From Amino Acids to Living Tissues. *Advanced Materials*, 32(14), e1906989. DOI: 10.1002/adma.201906989 PMID: 32103565

Kim, H. W., Song, J. H., & Kim, H. E. (2006). Bioactive glass nanofiber-collagen nanocomposite as a novel bone regeneration matrix. *Journal of Biomedical Materials Research. Part A*, 79(3), 698–705. DOI: 10.1002/jbm.a.30848 PMID: 16850456

Kim, K. K., Hsu, A., Jia, X., Kim, S. M., Shi, Y., Dresselhaus, M., Palacios, T., & Kong, J. (2012). Synthesis and characterization of hexagonal boron nitride film as a dielectric layer for graphene devices. *ACS Nano*, 6(10), 8583–8590. DOI: 10.1021/nn301675f PMID: 22970651

Koga, K., & Ohigashi, H. (1986). Piezoelectricity and related properties of vinylidene fluoride and trifluoroethylene copolymers. *Journal of Applied Physics*, 59(6), 2142–2150. DOI: 10.1063/1.336351

Köse, G. T., Korkusuz, F., Ozkul, A., Soysal, Y., Ozdemir, T., Yildiz, C., & Hasirci, V. (2005). Tissue engineered cartilage on collagen and PHBV matrices. *Biomaterials*, 26(25), 5187–5197. DOI: 10.1016/j.biomaterials.2005.01.037 PMID: 15792546

Kumar, A., Kim, W., Sriboriboon, P., Lee, H. Y., Kim, Y., & Ryu, J. (2024). AC poling-induced giant piezoelectricity and high mechanical quality factor in [001] PMN-PZT hard single crystals. *Sensors and Actuators. A, Physical*, 372, 115342. DOI: 10.1016/j.sna.2024.115342

Kumar, P., Dehiya, B. S., & Sindhu, A. (2018). Bioceramics for hard tissue engineering applications: A review. *International Journal of Applied Engineering Research: IJAER*, 13(5), 2744–2752. http://www.ripublication.com

Kumar, R., Pattanayak, I., Dash, P. A., & Mohanty, S. (2023). Bioceramics: A review on design concepts toward tailor-made (multi)-functional materials for tissue engineering applications. *Journal of Materials Science*, 58(8), 3460–3484. DOI: 10.1007/s10853-023-08226-8

Ladj, R., Magouroux, T., Eissa, M., Dubled, M., Mugnier, Y., Le Dantec, R., Galez, C., Valour, J. P., Fessi, H., & Elaissari, A. (2013). Aminodextran-coated potassium niobate (KNbO3) nanocrystals for second harmonic bio-imaging. *Colloids and Surfaces. A, Physicochemical and Engineering Aspects*, 439, 131–137. DOI: 10.1016/j.colsurfa.2013.02.025

Lang, S. B. (2005). Pyroelectricity: From ancient curiosity to modern imaging tool. *Physics Today*, 58(8), 31–36. DOI: 10.1063/1.2062916

Lang, S. B. (2016). Review of ferroelectric hydroxyapatite and its application to biomedicine. *Phase Transitions*, 89/7-8), 678–694. https://doi.org/DOI: 10.1080/01411594.2016.1182166

Lee, C. H., Singla, A., & Lee, Y. (2001). Biomedical applications of collagen. *International Journal of Pharmaceutics*, 221(1-2), 1–22. DOI: 10.1016/S0378-5173(01)00691-3 PMID: 11397563

Levingstone, T. J., Barron, N., Ardhaoui, M., Benyounis, K., Looney, L., & Stokes, J. (2017). Application of response surface methodology in the design of functionally graded plasma sprayed hydroxyapatite coatings. *Surface and Coatings Technology*, 313, 307–318. DOI: 10.1016/j.surfcoat.2017.01.113

Li, J., Zhang, T., Liao, Z., Wei, Y., Hang, R., & Huang, D. (2023). Engineered functional doped hydroxyapatite coating on titanium implants for osseointegration. *Journal of Materials Research and Technology*, 27, 122–152. DOI: 10.1016/j.jmrt.2023.09.239

Li, Y., Chen, J., Liu, S., Wang, Z., Zhang, S., Mao, C., & Wang, J. (2024). Biodegradable piezoelectric polymer for cartilage remodeling. *Matter*, 7(4), 1631–1643. DOI: 10.1016/j.matt.2024.01.034

Lima, C., De Oliveira, R., Figueiro, S., Wehmann, C., Goes, J., & Sombra, A. (2006). DC conductivity and dielectric permittivity of collagen–chitosan films. *Materials Chemistry and Physics*, 99(2-3), 284–288. DOI: 10.1016/j.matchemphys.2005.10.027

Liu, W., Cui, L., & Cao, Y. (2006). Bone reconstruction with bone marrow stromal cells. *Methods in Enzymology*, 420, 362–380. DOI: 10.1016/S0076-6879(06)20017-X PMID: 17161706

Lu, Y., Cheng, D., Niu, B., Wang, X., Wu, X., & Wang, A. (2023). Properties of Poly (Lactic-co-Glycolic Acid) and Progress of Poly (Lactic-co-Glycolic Acid)-Based Biodegradable Materials in Biomedical Research. *Pharmaceuticals (Basel, Switzerland)*, 16(3), 454. DOI: 10.3390/ph16030454 PMID: 36986553

Luo, T., Tan, B., Liao, J., Shi, K., & Ning, L. (2024). A review on external physical stimuli with biomaterials for bone repair. *Chemical Engineering Journal*, 153749, 153749. Advance online publication. DOI: 10.1016/j.cej.2024.153749

Maiti, S., Karan, S. K., Kim, J., & Khatua, B. B. (2019). Nature driven bio-piezoelectric/triboelectric nanogenerator as next-generation green energy harvester for smart and pollution free society. *Advanced Energy Materials*, 9(9), 1803027. DOI: 10.1002/aenm.201803027

Makhatadze, G. I. (2017). Linking computation and experiments to study the role of charge-charge interactions in protein folding and stability. *Physical Biology*, 14(1), 013002. DOI: 10.1088/1478-3975/14/1/013002 PMID: 28169222

Malode, S., & Shetti, N. P. (2022). ZrO2 in biomedical applications. In K. Mondal, *Metal Oxides for Biomedical and Biosensor Applications* (pp. 471-501). Elsevier. https://doi.org/DOI: 10.1016/B978-0-12-823033-6.00016-8

Marchesano, V., Gennari, O., Mecozzi, L., Grilli, S., & Ferraro, P. (2015). Effects of Lithium Niobate Polarization on Cell Adhesion and Morphology. *ACS Applied Materials & Interfaces*, 7(32), 18113–18119. DOI: 10.1021/acsami.5b05340 PMID: 26222955

Martin, A., Khansur, N. H., Urushihara, D., Asaka, T., Kakimoto, K. i., & Webber, K. G. (2024). Effect of Li on the intrinsic and extrinsic contributions of the piezoelectric response in Lix(Na0.5K0.5)1-xNbO3 piezoelectric ceramics across the polymorphic phase boundary. *Acta Materialia*, 266, 119691. Advance online publication. DOI: 10.1016/j.actamat.2024.119691

Martin, P. M. (2009). *Handbook of deposition technologies for films and coatings: science, applications and technology*. William Andrew.

Martins, P., Nunes, J. S., Hungerford, G., Miranda, D., Ferreira, A., Sencadas, V., & Lanceros-Méndez, S. (2009). Local variation of the dielectric properties of poly (vinylidene fluoride) during the α-to β-phase transformation. *Physics Letters. [Part A]*, 373(2), 177–180. DOI: 10.1016/j.physleta.2008.11.026

Mazzoni, E., Iaquinta, M. R., Lanzillotti, C., Mazziotta, C., Maritati, M., Montesi, M., Sprio, S., Tampieri, A., Tognon, M., & Martini, F. (2021). Bioactive Materials for Soft Tissue Repair. *Frontiers in Bioengineering and Biotechnology*, 9, 613787. DOI: 10.3389/fbioe.2021.613787 PMID: 33681157

Minary-Jolandan, M., & Yu, M. F. (2010). Shear piezoelectricity in bone at the nanoscale. *Applied Physics Letters*, 97(15), 153127, 153127–3. DOI: 10.1063/1.3503965

Mishchenko, O., Yanovska, A., Kosinov, O., Maksymov, D., Moskalenko, R., Ramanavicius, A., & Pogorielov, M. (2023). Synthetic Calcium-Phosphate Materials for Bone Grafting. *Polymers*, 15(18), 3822. DOI: 10.3390/polym15183822 PMID: 37765676

Mittal, G., & Paul, S. (2022). Suspension and solution precursor plasma and HVOF spray: A review. *Journal of Thermal Spray Technology*, 31(5), 1443–1475. DOI: 10.1007/s11666-022-01360-w

Montazerian, M., Hosseinzadeh, F., Migneco, C., Fook, M. V., & Baino, F. (2022). Bioceramic coatings on metallic implants: An overview. *Ceramics International*, 48(11), 8987–9005. DOI: 10.1016/j.ceramint.2022.02.055

Moore, B., Asadi, E., & Lewis, G. (2017). Deposition methods for microstructured and nanostructured coatings on metallic bone implants: A review. *Advances in Materials Science and Engineering*, 2017(1), 5812907. DOI: 10.1155/2017/5812907

Moreno-Gomez, I.(2019). Degradation of Bioresorbable Composites: Calcium Carbonate Case Studies. *Springer Theses A Phenomenological Mathematical Modelling Framework for the Degradation of Bioresorbable Composites*, 245-266. https://doi.org/DOI: 10.1007/978-3-030-04990-4_7

Müller, V., & Djurado, E. (2022). Microstructural designed S58 bioactive glass/hydroxyapatite composites for enhancing osteointegration of Ti6Al4V-based implants. *Ceramics International*, 48(23), 35365–35375. DOI: 10.1016/j.ceramint.2022.08.138

Murillo, G., Blanquer, A., Vargas-Estevez, C., Barrios, L., Ibáñez, E., Nogués, C., & Esteve, J. (2017). Electromechanical Nanogenerator-Cell Interaction Modulates Cell Activity. *Advanced materials (Deerfield Beach, Fla.), 29*(24), . https://doi.org/DOI: 10.1002/adma.201605048

Nagata, H., Matsumoto, K., Hirosue, T., Hiruma, Y., & Takenaka, T. (2007). Fabrication and electrical properties of potassium niobate ferroelectric ceramics. *Japanese Journal of Applied Physics*, 46(10S), 7084. DOI: 10.1143/JJAP.46.7084

Nakamachi, E., Okuda, Y., Kumazawa, S., Uetsuji, Y., Tsuchiya, K., & Nakayasu, H. (2006). Development of Fabrication Technique of Bio-Compatible Piezoelectric Material MgSiO 3 by Using Helicon Wave Plasma Sputter. *Nippon Kikai Gakkai Ronbunshu A Hen(Transactions of the Japan Society of Mechanical Engineers Part A)(Japan), 18*(3), 353-355. https://doi.org/DOI: 10.1299/kikaia.72.353

Nakayama, Y., Pauzauskie, P. J., Radenovic, A., Onorato, R. M., Saykally, R. J., Liphardt, J., & Yang, P. (2007). Tunable nanowire nonlinear optical probe. *Nature*, 447(7148), 1098–1101. DOI: 10.1038/nature05921 PMID: 17597756

Nathanael, A. J., & Oh, T. H. (2020). Biopolymer Coatings for Biomedical Applications. *Polymers*, 12(12), 3061. DOI: 10.3390/polym12123061 PMID: 33371349

Negut, I., & Ristoscu, C. (2023). Bioactive glasses for soft and hard tissue healing applications—A short review. *Applied Sciences (Basel, Switzerland)*, 13(10), 6151. DOI: 10.3390/app13106151

Nguyen, A. N., Solard, J., Nong, H. T. T., Ben Osman, C., Gomez, A., Bockelée, V., Tencé-Girault, S., Schoenstein, F., Simón-Sorbed, M., Carrillo, A. E., & Mercone, S. (2020). Spin Coating and Micro-Patterning Optimization of Composite Thin Films Based on PVDF. *Materials (Basel)*, 13(6), 1342. DOI: 10.3390/ma13061342 PMID: 32187993

Nikkerdar, N., Golshah, A., Mobarakeh, M. S., Fallahnia, N., Azizi, B., Souri, R., Safaei, M., & Shoohanizad, E. (2024). Recent progress in application of zirconium oxide in dentistry. *Journal of Medicinal and Pharmaceutical Chemistry Research*, 6(8), 1042–1071. DOI: 10.48309/jmpcr.2024.432254.1069

Ochiai, T., & Fukada, E. (1998). Electromechanical properties of poly-L-lactic acid. *Japanese Journal of Applied Physics*, 37(6R), 3374. DOI: 10.1143/JJAP.37.3374

Oliveira, H. L., Da Rosa, W. L. O., Cuevas-Suárez, C. E., Carreño, N. L. V., da Silva, A. F., Guim, T. N., Dellagostin, O. A., & Piva, E. (2017). Histological Evaluation of Bone Repair with Hydroxyapatite: A Systematic Review. *Calcified Tissue International*, 101(4), 341–354. DOI: 10.1007/s00223-017-0294-z PMID: 28612084

Omote, K., Ohigashi, H., & Koga, K. (1997). Temperature dependence of elastic, dielectric, and piezoelectric properties of "single crystalline" films of vinylidene fluoride trifluoroethylene copolymer. *Journal of Applied Physics*, 81(6), 2760–2769. DOI: 10.1063/1.364300

Ou, Y. X., Wang, H. Q., Ouyang, X., Zhao, Y. Y., Zhou, Q., Luo, C. W., Hua, Q. S., Ouyang, X. P., & Zhang, S. (2023). Recent advances and strategies for high-performance coatings. *Progress in Materials Science*, 136, 101125. DOI: 10.1016/j.pmatsci.2023.101125

Pais, A., Moreira, C., & Belinha, J. (2024). The Biomechanical Analysis of Tibial Implants Using Meshless Methods: Stress and Bone Tissue Remodeling Analysis. *Designs*, 8(2), 28. DOI: 10.3390/designs8020028

Park, H. Y., Seo, I. T., Choi, J. H., Nahm, S., & Lee, H. G. (2010). Low-temperature sintering and piezoelectric properties of (Na0. 5K0. 5) NbO3 lead-free piezoelectric ceramics. *Journal of the American Ceramic Society*, 93(1), 36–39. DOI: 10.1111/j.1551-2916.2009.03359.x

Park, K. S. (2023). *Humans and Electricity: Understanding Body Electricity and Applications*. Springer Nature., DOI: 10.1007/978-3-031-20784-6

Park, S. H., Ahn, C. W., Nahm, S., & Song, J. S. (2004). Microstructure and piezoelectric properties of ZnO-added (Na0. 5K0. 5) NbO3 ceramics. *Japanese Journal of Applied Physics*, 43(8B), L1072–L1074. DOI: 10.1143/JJAP.43.L1072

Parravicini, J., Safioui, J., Degiorgio, V., Minzioni, P., & Chauvet, M. (2011). All-optical technique to measure the pyroelectric coefficient in electro-optic crystals. *Journal of Applied Physics*, 109(3), 033106. DOI: 10.1063/1.3544069

Piatti, E., Miola, M., & Verné, E. (2024). Tailoring of bioactive glass and glass-ceramics properties for *in vitro* and *in vivo* response optimization: A review. *Biomaterials Science*, 12(18), 4546–4589. DOI: 10.1039/D3BM01574B PMID: 39105508

Poblano-Salas, C. A., Henao, J., Giraldo-Betancur, A. L., Forero-Sossa, P., Espinosa-Arbelaez, D. G., González-Sánchez, J. A., Dzib-Pérez, L. R., Estrada-Moo, S. T., & Pech-Pech, I. E. (2024). HVOF-sprayed HAp/S53P4 BG composite coatings on an AZ31 alloy for potential applications in temporary implants. *Journal of Magnesium and Alloys*, 12(1), 345–360. DOI: 10.1016/j.jma.2023.12.010

Prashar, G., Vasudev, H., Thakur, L., & Bansal, A. (2023). Performance of thermally sprayed hydroxyapatite coatings for biomedical implants: A comprehensive review. *Surface Review and Letters*, 30(01), 2241001. DOI: 10.1142/S0218625X22410013

Prezas, P. R., Soares, M. J., Borges, J. P., Silva, J. C., Oliveira, F. J., & Graça, M. P. F. (2023). Bioactivity Enhancement of Plasma-Sprayed Hydroxyapatite Coatings through Non-Contact Corona Electrical Charging. *Nanomaterials (Basel, Switzerland)*, 13(6), 1058. DOI: 10.3390/nano13061058 PMID: 36985952

Punj, S., Singh, J., & Singh, K. (2021). Ceramic biomaterials: Properties, state of the art and future prospectives. *Ceramics International*, 47(20), 28059–28074. DOI: 10.1016/j.ceramint.2021.06.238

Qi, X., Ren, P., Tong, X., Wang, X., & Zhuo, F. (2024). Enhanced piezoelectric properties of KNN-based ceramics by synergistic modulation of phase constitution, grain size and domain configurations. *Journal of the European Ceramic Society*, 45(1), 116874. DOI: 10.1016/j.jeurceramsoc.2024.116874

Qin, Z., Chen, K., Sun, X., Zhang, M., Wang, L., Zheng, S., Chen, C., Tang, H., Li, H., Zou, C., & Wu, G. (2024). The Electrical Properties and In Vitro Osteogenic Properties of 3D-Printed Fe@ BT/HA Piezoelectric Ceramic Scaffold. *Ceramics International*. Advance online publication. DOI: 10.1016/j.ceramint.2024.06.371

Rabiee, S. M., Nazparvar, N., Azizian, M., Vashaee, D., & Tayebi, L. (2015). Effect of ion substitution on properties of bioactive glasses: A review. *Ceramics International*, 41(6), 7241–7251. DOI: 10.1016/j.ceramint.2015.02.140

Rajabi, A. H., Jaffe, M., & Arinzeh, T. L. (2015). Piezoelectric materials for tissue regeneration: A review. *Acta Biomaterialia*, 24, 12–23. DOI: 10.1016/j.actbio.2015.07.010 PMID: 26162587

Rajan, S. T., Subramanian, B., & Arockiarajan, A. (2022). A comprehensive review on biocompatible thin films for biomedical application. *Ceramics International*, 48(4), 4377–4400. DOI: 10.1016/j.ceramint.2021.10.243

Rajendran, A. K., Anthraper, M. S., Hwang, N. S., & Rangasamy, J. (2024). Osteogenesis and angiogenesis promoting bioactive ceramics. *Materials Science and Engineering R Reports*, 159, 100801. DOI: 10.1016/j.mser.2024.100801

Ramezani, M., Mohd Ripin, Z., Pasang, T., & Jiang, C. P. (2023). Surface engineering of metals: Techniques, characterizations and applications. *Metals*, 13(7), 1299. DOI: 10.3390/met13071299

Ramu, M., Ananthasubramanian, M., Kumaresan, T., Gandhinathan, R., & Jothi, S. (2018). Optimization of the configuration of porous bone scaffolds made of Polyamide/Hydroxyapatite composites using Selective Laser Sintering for tissue engineering applications. *Bio-Medical Materials and Engineering*, 29(6), 739–755. DOI: 10.3233/BME-181020 PMID: 30282331

Ravikumar, K., Kar, G. P., Bose, S., & Basu, B. (2016). Synergistic effect of polymorphism, substrate conductivity and electric field stimulation towards enhancing muscle cell growth in vitro. *RSC Advances*, 6(13), 10837–10845. DOI: 10.1039/C5RA26104J

Re, F., Borsani, E., Rezzani, R., Sartore, L., & Russo, D. (2023). Bone Regeneration Using Mesenchymal Stromal Cells and Biocompatible Scaffolds: A Concise Review of the Current Clinical Trials. *Gels (Basel, Switzerland)*, 9(5), 389. DOI: 10.3390/gels9050389 PMID: 37232981

Rebolledo, S., Alcántara-Dufeu, R., Luengo Machuca, L., Ferrada, L., & Sánchez-Sanhueza, G. A. (2023). Real-time evaluation of the biocompatibility of calcium silicate-based endodontic cements: An in vitro study. *Clinical and Experimental Dental Research*, 9(2), 322–331. DOI: 10.1002/cre2.714 PMID: 36866428

Ribeiro, C., Correia, D. M., Rodrigues, I., Guardão, L., Guimarães, S., Soares, R., & Lanceros-Méndez, S. (2017). In vivo demonstration of the suitability of piezoelectric stimuli for bone reparation. *Materials Letters*, 209, 118–121. DOI: 10.1016/j.matlet.2017.07.099

Ribeiro, C., Sencadas, V., Correia, D. M., & Lanceros-Méndez, S. (2015). Piezoelectric polymers as biomaterials for tissue engineering applications. *Colloids and Surfaces. B, Biointerfaces*, 136, 46–55. DOI: 10.1016/j.colsurfb.2015.08.043 PMID: 26355812

Rochdi, N., El Azhary, K., Jouti, N. T., Badou, A., Majidi, B., & Habti, N., & NAIMI, Y. (2023). In Vivo Study Of Hypersensitivity And Irritation Skin For Nanocomposite Biphasic Calcium Phosphate Synthesized Directly By Hydrothermal Process. *Journal of Namibian Studies: History Politics Culture*, 33(2), 6142–6157. DOI: 10.59670/jns.v33i.5040

Rui, G., Allahyarov, E., Zhu, Z., Huang, Y., Wongwirat, T., Zou, Q., Taylor, P., & Zhu, L. (2024). Challenges and opportunities in piezoelectric polymers: Effect of oriented amorphous fraction in ferroelectric semicrystalline polymers. *Responsive Materials*, 2(3), e20240002. DOI: 10.1002/rpm.20240002

Safronova, T. V., Putlayev, V., & Shekhirev, M. (2013). Resorbable calcium phosphates based ceramics. *Powder Metallurgy and Metal Ceramics*, 52(5-6), 357–363. DOI: 10.1007/s11106-013-9534-6

Sagawa, H., Itoh, S., Wang, W., & Yamashita, K. (2010). Enhanced bone bonding of the hydroxyapatite/beta-tricalcium phosphate composite by electrical polarization in rabbit long bone. *Artificial Organs*, 34(6), 491–497. DOI: 10.1111/j.1525-1594.2009.00912.x PMID: 20456322

Sakthi Muthulakshmi, S., Shailajha, S., & Shanmugapriya, B. (2023). Bio-physical investigation of calcium silicate biomaterials by green synthesis-osseous tissue regeneration. *Journal of Materials Research*, 38(19), 4369–4384. DOI: 10.1557/s43578-023-01149-9

Samadi, A., Salati, M. A., Safari, A., Jouyandeh, M., Barani, M., Singh Chauhan, N. P., Golab, E. G., Zarrintaj, P., Kar, S., Seidi, F., Hejna, A., & Saeb, M. R. (2022). Comparative review of piezoelectric biomaterials approach for bone tissue engineering. *Journal of Biomaterials Science. Polymer Edition*, 33(12), 1555–1594. DOI: 10.1080/09205063.2022.2065409 PMID: 35604896

Sappati, K. K., & Bhadra, S. (2018). Piezoelectric Polymer and Paper Substrates: A Review. *Sensors (Basel)*, 18(11), 3605. DOI: 10.3390/s18113605 PMID: 30355961

Sarkar, K. (2024). Research progress on biodegradable magnesium phosphate ceramics in orthopaedic applications. *Journal of Materials Chemistry. B*, 12(35), 8605–8615. DOI: 10.1039/D4TB01123F PMID: 39140212

Sayagués, M. J., Otero, A., Santiago-Andrades, L., Poyato, R., Monzón, M., Paz, R., Gotor, F. J., & Moriche, R. (2024). Fine-grained BCZT piezoelectric ceramics by combining high-energy mechanochemical synthesis and hot-press sintering. *Journal of Alloys and Compounds*, 176453, 176453. Advance online publication. DOI: 10.1016/j.jallcom.2024.176453

Schofield, Z., Meloni, G. N., Tran, P., Zerfass, C., Sena, G., Hayashi, Y., Grant, M., Contera, S. A., Minteer, S. D., Kim, M., Prindle, A., Rocha, P., Djamgoz, M. B. A., Pilizota, T., Unwin, P. R., Asally, M., & Soyer, O. S. (2020). Bioelectrical understanding and engineering of cell biology. *Journal of the Royal Society, Interface*, 17(166), 20200013. DOI: 10.1098/rsif.2020.0013 PMID: 32429828

Senthilkumar, G., Kaliaraj, G. S., Vignesh, P., Vishwak, R. S., Joy, T. N., & Hemanandh, J. (2021). Hydroxyapatite–barium/strontium titanate composite coatings for better mechanical, corrosion and biological performance. *Materials Today: Proceedings*, 44, 3618–3621. DOI: 10.1016/j.matpr.2020.09.758

Seo, H. J., Cho, Y. E., Kim, T., Shin, H. I., & Kwun, I. S. (2010). Zinc may increase bone formation through stimulating cell proliferation, alkaline phosphatase activity and collagen synthesis in osteoblastic MC3T3-E1 cells. *Nutrition Research and Practice*, 4(5), 356–361. DOI: 10.4162/nrp.2010.4.5.356 PMID: 21103080

Seppänen-Kaijansinkko, R. (2019). Hard Tissue Engineering. In *Tissue Engineering in Oral and Maxillofacial Surgery*. Springer International Publishing., DOI: 10.1007/978-3-030-24517-7_7

Sergi, R., Bellucci, D., & Cannillo, V. (2020). A comprehensive review of bioactive glass coatings: State of the art, challenges and future perspectives. *Coatings*, 10(8), 757. DOI: 10.3390/coatings10080757

Serrato Ochoa, D., Nieto Aguilar, R., & Aguilera Méndez, A. (2015). Ingeniería de tejidos. Una nueva disciplina en medicina regenerativa. *Investigación y Ciencia. Universidad Autónoma de Aguascalientes*, 64(64), 61–69. DOI: 10.33064/iycuaa2015643597

Sharma, A. (2023). *Hydroxyapatite coating techniques for Titanium Dental Implants—an overview*. Qeios., DOI: 10.32388/2E6UHN

Sharma, A., Kokil, G. R., He, Y., Lowe, B., Salam, A., Altalhi, T. A., Ye, Q., & Kumeria, T. (2023). Inorganic/organic combination: Inorganic particles/polymer composites for tissue engineering applications. *Bioactive Materials*, 24, 535–550. DOI: 10.1016/j.bioactmat.2023.01.003 PMID: 36714332

Sharma, H. B., & Mansingh, A. (1998). Sol-gel processed barium titanate ceramics and thin films. *Journal of Materials Science*, 33(17), 4455–4459. DOI: 10.1023/A:1004576315328

Sharma, V., Chowdhury, S., Bose, S., & Basu, B. (2022). Polydopamine Codoped $BaTiO_3$-Functionalized Polyvinylidene Fluoride Coating as a Piezo-Biomaterial Platform for an Enhanced Cellular Response and Bioactivity. *ACS Biomaterials Science & Engineering*, 8(1), 170–184. DOI: 10.1021/acsbiomaterials.1c00879 PMID: 34964600

Shearer, A., Montazerian, M., Sly, J. J., Hill, R. G., & Mauro, J. C. (2023). Trends and perspectives on the commercialization of bioactive glasses. *Acta Biomaterialia*, 160, 14–31. DOI: 10.1016/j.actbio.2023.02.020 PMID: 36804821

Shekhawat, D., Singh, A., Banerjee, M. K., Singh, T., & Patnaik, A. (2021). Bioceramic composites for orthopaedic applications: A comprehensive review of mechanical, biological, and microstructural properties. *Ceramics International*, 47(3), 3013–3030. DOI: 10.1016/j.ceramint.2020.09.214

Shirke, X., & Balivada, S. (2024). Lead Zirconate Titanate (PZT) Synthesis and Its Applications. *International Journal of Health Technology and Innovation*, 3(02), 15–19. https://ijht.org.in/index.php/ijhti/article/view/135

Silva, C. C., Pinheiro, A. G., Figueiró, S. D., Goes, J. C., Sasaki, J. M., Miranda, M. A. R., & Sombra, A. S. B. (2002). Piezoelectric properties of collagen-nanocrystalline hydroxyapatite composites. *Journal of Materials Science*, 37(10), 2061–2070. DOI: 10.1023/A:1015219800490

Sinha, S., Elbaz-Alon, Y., & Avinoam, O. (2022). Ca^{2+} as a coordinator of skeletal muscle differentiation, fusion and contraction. *The FEBS Journal*, 289(21), 6531–6542. DOI: 10.1111/febs.16552 PMID: 35689496

Song, J., Zhao, G., Li, B., & Wang, J. (2017). Design optimization of PVDF-based piezoelectric energy harvesters. *Heliyon*, 3(9), e00377. DOI: 10.1016/j.heliyon.2017.e00377 PMID: 28948235

Song, Q., Zhu, M., & Mao, Y. (2021). Chemical vapor deposited polyelectrolyte coatings with osteoconductive and osteoinductive activities. *Surface and Coatings Technology*, 423, 127522. DOI: 10.1016/j.surfcoat.2021.127522

Stamnitz, S., & Klimczak, A. (2021). Mesenchymal Stem Cells, Bioactive Factors, and Scaffolds in Bone Repair: From Research Perspectives to Clinical Practice. *Cells*, 10(8), 1925. DOI: 10.3390/cells10081925 PMID: 34440694

Studers, P., Jakovicka, D., Solska, J., Bladiko, U., & Radziņa, M. (2023). Mid-term clinical and radiographic outcomes after primary total hip replacement with fully hydroxyapatite-coated stem: A cross-sectional study. *Journal of Orthopaedics, Trauma and Rehabilitation*, 30(2), 241–247. DOI: 10.1177/22104917231171937

Sun, T. W., Yu, W. L., Zhu, Y. J., Yang, R. L., Shen, Y. Q., Chen, D. Y., He, Y. H., & Chen, F. (2017). Hydroxyapatite Nanowire@Magnesium Silicate Core-Shell Hierarchical Nanocomposite: Synthesis and Application in Bone Regeneration. *ACS Applied Materials & Interfaces*, 9(19), 16435–16447. DOI: 10.1021/acsami.7b03532 PMID: 28481082

Sun, Y., Zhao, Z., Qiao, Q., Li, S., Yu, W., Guan, X., Schneider, A., Weir, M. D., Xu, H. H. K., Zhang, K., & Bai, Y. (2023). Injectable periodontal ligament stem cell-metformin-calcium phosphate scaffold for bone regeneration and vascularization in rats. *Dental materials: official publication of the Academy of Dental Materials, 39*(10), 872–885. https://doi.org/DOI: 10.1016/j.dental.2023.07.008

Surmenev, R. A., Orlova, T., Chernozem, R. V., Ivanova, A. A., Bartasyte, A., Mathur, S., & Surmeneva, M. A. (2019). Hybrid lead-free polymer-based nanocomposites with improved piezoelectric response for biomedical energy-harvesting applications: A review. *Nano Energy*, 62, 475–506. DOI: 10.1016/j.nanoen.2019.04.090

Swain, S., Padhy, R. N., & Rautray, T. R. (2020). Electrically stimulated hydroxyapatite–barium titanate composites demonstrate immunocompatibility in vitro. *Journal of the Korean Ceramic Society*, 57(5), 495–502. DOI: 10.1007/s43207-020-00048-7

Tang, M., Liu, L., Gao, X., Zhang, Z., & Wang, Y. (2024). Collaborative Optimization of Curie Temperature and Piezoelectricity in Ternary $BiFeO_3$-$BiScO_3$-$PbTiO_3$. *ACS Applied Materials & Interfaces*, 16(31), 41185–41193. DOI: 10.1021/acsami.4c08300 PMID: 39069883

Tang, Y., Chen, L., Duan, Z., Zhao, K., & Wu, Z. (2020). Graphene/barium titanate/polymethyl methacrylate bio-piezoelectric composites for biomedical application. *Ceramics International*, 46(5), 6567–6574. DOI: 10.1016/j.ceramint.2019.11.142

Ten Elshof, J. E. (2022). Chemical solution deposition of oxide thin films. In *Epitaxial Growth of Complex Metal Oxides* (pp. 75–100). Woodhead Publishing., DOI: 10.1016/B978-0-08-102945-9.00012-5

Thomas, P., Arenberger, P., Bader, R., Bircher, A. J., Bruze, M., de Graaf, N., Hartmann, D., Johansen, J. D., Jowitz-Heinke, A., Krenn, V., Kurek, M., Odgaard, A., Rustemeyer, T., Summer, B., & Thyssen, J. P. (2024). A literature review and expert consensus statement on diagnostics in suspected metal implant allergy. *Journal of the European Academy of Dermatology and Venereology*, 38(8), 1471–1477. DOI: 10.1111/jdv.20026 PMID: 38606660

Thomas, S., Balakrishnan, P., & Sreekala, M. S. (2018). *Fundamental biomaterials: ceramics*. Woodhead Publishing., DOI: 10.1016/B978-0-08-102203-0.00001-9

Thuau, D., Abbas, M., Wantz, G., Hirsch, L., Dufour, I., & Ayela, C. (2016). Piezoelectric polymer gated OFET: Cutting-edge electro-mechanical transducer for organic MEMS-based sensors. *Scientific Reports*, 6(1), 38672. DOI: 10.1038/srep38672 PMID: 27924853

Vandrovcova, M., Tolde, Z., Vanek, P., Nehasil, V., Doubková, M., Trávníčková, M., Drahokoupil, J., Buixaderas, E., Borodavka, F., Novakova, J., & Bacakova, L. (2021). Beta-Titanium alloy covered by ferroelectric coating–physicochemical properties and human osteoblast-like cell response. *Coatings*, 11(2), 210. DOI: 10.3390/coatings11020210

Vaněk, P., Kolská, Z., Luxbacher, T., García, J. A. L., Lehocký, M., Vandrovcová, M., Bačáková, L., & Petzelt, J. (2016). Electrical activity of ferroelectric biomaterials and its effects on the adhesion, growth and enzymatic activity of human osteoblast-like cells. *Journal of Physics. D, Applied Physics*, 49(17), 175403. DOI: 10.1088/0022-3727/49/17/175403

Vasquez-Sancho, F., Catalan, G., & Sort Viñas, J. (2018) *Flexoelectricity in biomaterials*. Universitat Autònoma de Barcelona. Accesed: https://dialnet.unirioja.es/servlet/tesis?codigo=229380

Venkateshwarlu, A., Singh, S., & Melnik, R. (2024). Piezoelectricity and flexoelectricity in biological cells: The role of cell structure and organelles. *arXiv preprint arXiv:2403.02050*. https://doi.org//arXiv.2403.02050DOI: 10.48550

Wang, M., Zhang, G., Li, W., & Wang, X. (2015). Microstructure and Properties of BaTiO3 Ferroelectric Films Prepared by DC Micro Arc Oxidation. *Bulletin of the Korean Chemical Society*, 36(4), 1178–1182. DOI: 10.1002/bkcs.10220

Willatzen, M. (2024). *Piezoelectricity in Classical and Modern Systems*. IOP Publishing. DOI: 10.1088/978-0-7503-5557-5

Wood, R. J., & Lu, P. (2024). Coatings and surface modification of alloys for tribo-corrosion applications. *Coatings*, 14(1), 99. DOI: 10.3390/coatings14010099

Wu, J. (2024). *Piezoelectric Materials: From Fundamentals to Emerging Applications*. John Wiley & Sons. Accessed: https://www.wiley.com/

Wu, Z., Tang, T., Guo, H., Tang, S., Niu, Y., Zhang, J., Zhang, W., Ma, R., Su, J., Liu, C., & Wei, J. (2014). In vitro degradability, bioactivity and cell responses to mesoporous magnesium silicate for the induction of bone regeneration. *Colloids and Surfaces. B, Biointerfaces*, 120, 38–46. DOI: 10.1016/j.colsurfb.2014.04.010 PMID: 24905677

Wu, Z., Zheng, K., Zhang, J., Tang, T., Guo, H., Boccaccini, A. R., & Wei, J. (2016). Effects of magnesium silicate on the mechanical properties, biocompatibility, bioactivity, degradability, and osteogenesis of poly(butylene succinate)-based composite scaffolds for bone repair. *Journal of Materials Chemistry. B*, 4(48), 7974–7988. DOI: 10.1039/C6TB02429G PMID: 32263787

Xu, H. (2001). Dielectric properties and ferroelectric behavior of poly (vinylidene fluoride-trifluoroethylene) 50/50 copolymer ultrathin films. *Journal of Applied Polymer Science*, 80(12), 2259–2266. DOI: 10.1002/app.1330

Xu, Q., Gao, X., Zhao, S., Liu, Y. N., Zhang, D., Zhou, K., Khanbareh, H., Chen, W., Zhang, Y., & Bowen, C. (2021). Construction of Bio-Piezoelectric Platforms: From Structures and Synthesis to Applications. *Advanced Materials*, 33(27), e2008452. DOI: 10.1002/adma.202008452 PMID: 34033180

Xu, Q., Gao, X., Zhao, S., Liu, Y. N., Zhang, D., Zhou, K., Khanbareh, H., Chen, W., Zhang, Y., & Bowen, C. (2021). Construction of Bio-Piezoelectric Platforms: From Structures and Synthesis to Applications. *Advanced Materials*, 33(27), e2008452. DOI: 10.1002/adma.202008452 PMID: 34033180

Yao, K., Chen, S., Guo, K., Tan, C. K. I., Mirshekarloo, M. S., & Tay, F. E. H. (2017). Lead-Free Piezoelectric Ceramic Coatings Fabricated by Thermal Spray Process. *IEEE Transactions on Ultrasonics, Ferroelectrics, and Frequency Control*, 64(11), 1758–1765. DOI: 10.1109/TUFFC.2017.2748154 PMID: 28880167

Yılmaz, E., Türk, S., Semerci, A. B., Kırkbınar, M., İbrahimoğlu, E., & Çalışkan, F. (2024). Bioactive apatite-wollastonite glass ceramics coating on metallic titanium for biomedical applications: effect of boron. *Journal of biological inorganic chemistry: JBIC: a publication of the Society of Biological Inorganic Chemistry, 29*(1), 75–85. https://doi.org/DOI: 10.1007/s00775-023-02032-y

Yin, Z., Tian, B., Zhu, Q., & Duan, C. (2019). Characterization and Application of PVDF and Its Copolymer Films Prepared by Spin-Coating and Langmuir-Blodgett Method. *Polymers*, 11(12), 2033. DOI: 10.3390/polym11122033 PMID: 31817985

Yuan, X., Wu, T., Lu, T., & Ye, J. (2023). Effects of Zinc and Strontium Doping on In Vitro Osteogenesis and Angiogenesis of Calcium Silicate/Calcium Phosphate Cement. *ACS Biomaterials Science & Engineering*, 9(10), 5761–5771. DOI: 10.1021/acsbiomaterials.3c00193 PMID: 37676927

Zhao, Z., Yan, K., Guan, Q., Guo, Q., & Zhao, C. (2024). Mechanism and physical activities in bone-skeletal muscle crosstalk. *Frontiers in Endocrinology*, 14, 1287972. DOI: 10.3389/fendo.2023.1287972 PMID: 38239981

Zhi, C., Bando, Y., Tang, C., & Golberg, D. (2005). Immobilization of proteins on boron nitride nanotubes. *Journal of the American Chemical Society*, 127(49), 17144–17145. DOI: 10.1021/ja055989+ PMID: 16332036

Zhou, Z., Li, W., He, T., Qian, L., Tan, G., & Ning, C. (2016). Polarization of an electroactive functional film on titanium for inducing osteogenic differentiation. *Scientific Reports*, 6(1), 35512. DOI: 10.1038/srep35512 PMID: 27762318

Zhu, Q., Song, X., Chen, X., Li, D., Tang, X., Chen, J., & Yuan, Q. (2024). A high performance nanocellulose-PVDF based piezoelectric nanogenerator based on the highly active CNF@ ZnO via electrospinning technology. *Nano Energy*, 127, 109741. DOI: 10.1016/j.nanoen.2024.109741

Zhu, Y., Zhang, X., Chang, G., Deng, S., & Chan, H. F. (2024). Bioactive Glass in Tissue Regeneration: Unveiling Recent Advances in Regenerative Strategies and Applications. *Advanced materials (Deerfield Beach, Fla.)*, e2312964. https://doi.org/DOI: 10.1002/adma.202312964

Conclusion

The refined capacity *Innovative Materials for Industrial Applications: Synthesis, Characterization and Evaluation* provides an inclusive survey of the prominence growths in material science, specifically for industrialized use. It produces together new advancements in the combining, description, and act judgment of novel materials, contribution two together hypothetical insights and proficient requests. From nanomaterials to smart composites, procedure demonstrates how novelties in matters skill are mutating industries to a degree production.

Key themes during the whole of procedure stress the importance of tenable material design, economical combination methods, and the role of progressive description forms in guaranteeing the reliability and conduct of new fabrics under industrialized environments. Each chapter donates valuable information for investigators, engineers, and industry pros attempt to meet the always-progressing demands for enhanced material features, adeptness, and sustainability.

As corporations advance smarter, more environmental, and extreme-accomplishment solutions, the resumed progress of creative matters will be crucial. The combination of new fabrics and their all-encompassing judgment will remain alive to aggressive the edges of what is attainable in industrial uses. This book, by top a different range of materials and methods, determines a detracting means for understanding and driving forward these changes.

In summary, *Innovative Materials for Industrial Applications* serves as an up-to-date and essential citation, contribution a roadmap for future research and technical exercise, guaranteeing that the potential of next-creation materials is sufficiently fulfilled.

Compilation of References

Aaseth, J., Shimshi, M., Gabrilove, J. L., & Birketvedt, G. S. (2004). Fluoride: A toxic or therapeutic agent in the treatment of osteoporosis? *The Journal of Trace Elements in Experimental Medicine*, 17(2), 83–92. DOI: 10.1002/jtra.10051

Abd-Elghany, M., Elbeih, A., & Klapötke, T. M. (2018). Thermo-analytical study of 2,2,2-trinitroethyl-formate as a new oxidizer and its propellant based on a GAP matrix in comparison with ammonium dinitramide. *Journal of Analytical and Applied Pyrolysis*, 133, 30–38. DOI: 10.1016/j.jaap.2018.05.004

Abd-El-Nabey, B., Mohamed, M., Helmy, A., Elnagar, H., & Abdel-Gaber, A. (2024). Eco-friendly corrosion inhibition of steel in acid pickling using Prunus domestica Seeds and Okra stems extracts. *International Journal of Electrochemical Science*, 19(8), 100695. DOI: 10.1016/j.ijoes.2024.100695

Abdiyev, K., Azat, S., Kuldeyev, E., Ybyraiymkul, D., Kabdrakhmanova, S., Berndtsson, R., Khalkhabai, B., Kabdrakhmanova, A., & Sultakhan, S. (2023). Review of Slow Sand Filtration for Raw Water Treatment with Potential Application in Less-Developed Countries. *Water (Basel)*, 15(11), 2007. DOI: 10.3390/w15112007

Abushahba, F., Areid, N., Gürsoy, M., Willberg, J., Laine, V., Yatkin, E., Hupa, L., & Närhi, T. O. (2023). Bioactive glass air-abrasion promotes healing around contaminated implant surfaces surrounded by circumferential bone defects: An experimental study in the rat. *Clinical Implant Dentistry and Related Research*, 25(2), 409–418. DOI: 10.1111/cid.13172 PMID: 36602418

Acharya, D., Pathak, I., Muthurasu, A., Bhattarai, R. M., Kim, T., Ko, T. H., Saidin, S., Chhetri, K., & Kim, H. Y. (2023). In situ transmogrification of nanoarchitectured Fe-MOFs decorated porous carbon nanofibers into efficient positrode for asymmetric supercapacitor application. *Journal of Energy Storage*, 63, 106992. DOI: 10.1016/j.est.2023.106992

Adamu, H., Abba, S. I., Anyin, P. B., Sani, Y., & Qamar, M. (2023). Artificial intelligence-navigated development of high-performance electrochemical energy storage systems through feature engineering of multiple descriptor families of materials. *Energy Advances*, 2(5), 615–645. https://doi.org/https://doi .org/10.1039/d3ya00104k. DOI: 10.1039/D3YA00104K

Adane, W. D., Chandravanshi, B. S., & Tessema, M. (2024). Hypersensitive electrochemical sensor based on thermally annealed gold–silver alloy nanoporous matrices for the simultaneous determination of sulfathiazole and sulfamethoxazole residues in food samples. *Food Chemistry*, 457, 140071. https:// doi.org/https://doi.org/10.1016/j.foodchem.2024.140071. DOI: 10.1016/j.foodchem.2024.140071 PMID: 38905827

Adeleke, O. A., Latiff, A. A., Saphira, M. R., Daud, Z., Ismail, N., Ahsan, A., Aziz, N. A., Ndah, M., Kumar, V., Al-Gheethi, N. A., Rosli, M. A., & Hijab, M. (2018). Locally derived activated carbon from domestic, agricultural and industrial wastes for the treatment of palm oil mill effluent. In Elsevier eBooks (pp. 35–62). https://doi.org/DOI: 10.1016/B978-0-12-813902-8.00002-2

Adeyemo, F., Kamika, I., & Momba, M. (2014). Comparing the effectiveness of five low-cost home water treatment devices for Cryptosporidium, Giardiaand somatic coliphages removal from water sources. *Desalination and Water Treatment*, 56(9), 2351–2367. DOI: 10.1080/19443994.2014.960457

Adiga, S. P., Jin, C., Curtiss, L. A., Monteiro-Riviere, N. A., & Narayan, R. J. (2009). Nanoporous membranes for medical and biological applications. *Wiley Interdisciplinary Reviews. Nanomedicine and Nanobiotechnology*, 1(5), 568–581. DOI: 10.1002/wnan.50 PMID: 20049818

Aguilera, S. B., McCarthy, A., Khalifian, S., Lorenc, Z. P., Goldie, K., & Chernoff, W. G. (2023). The Role of Calcium Hydroxylapatite (Radiesse) as a Regenerative Aesthetic Treatment: A Narrative Review. *Aesthetic Surgery Journal*, 43(10), 1063–1090. DOI: 10.1093/asj/sjad173 PMID: 37635437

Ahmadi, S., & Khormali, A. (2023). Optimization of the corrosion inhibition performance of 2-mercaptobenzothiazole for carbon steel in HCl media using response surface methodology. *Fuel*, 357, 129783. DOI: 10.1016/j.fuel.2023.129783

Ajmal, Z., Ali, H., Ullah, S., Kumar, A., Abboud, M., Gul, H., Al-hadeethi, Y., Alshammari, A. S., Almuqati, N., Ashraf, G. A., Hassan, N., Qadeer, A., Hayat, A., Ul Haq, M., Hussain, I., & Murtaza, A. (2024). Use of carbon-based advanced materials for energy conversion and storage applications: Recent Development and Future Outlook. *Fuel*, 367, 131295. DOI: 10.1016/j.fuel.2024.131295

Akbarov, A. N., & Tillakhodjayeva, M. M. (2024). Studying the Dynamics of Bone Tissue Remodeling in Blood Plasma by Determining the Level of Bone Alkaline Phosphatase in the Blood. *International Journal of Integrative and Modern Medicine*, 2(5), 66–69. https://medicaljournals.eu/index.php/IJIMM/article/view/291

Alaneme, K., Daramola, Y., Olusegun, S., & Afolabi, A. (2015). Corrosion inhibition and adsorption characteristics of rice husk extracts on mild steel immersed in 1M H2SO4 and HCl solutions. *International Journal of Electrochemical Science*, 10(4), 3553–3567. DOI: 10.1016/S1452-3981(23)06561-6

Algueró, M., Cheng, B. L., Guiu, F., Reece, M. J., Poole, M., & Alford, N. (2001). Degradation of the d33 piezoelectric coffcient for PZT ceramics under static and cyclic compressive loading. *Journal of the European Ceramic Society*, 21(10–11), 1437–1440. DOI: 10.1016/S0955-2219(01)00036-X

Alharbi, H. A., Hameed, B. H., Alotaibi, K. D., Al-Oud, S. S., & Al-Modaihsh, A. S. (2022). Recent methods in the production of activated carbon from date palm residues for the adsorption of textile dyes: A review. *Frontiers in Environmental Science*, 10, 942059. DOI: 10.3389/fenvs.2022.996953

Alharbi, S. O., Ahmad, S., Gul, T., Ali, I., & Bariq, A. (2024). The corrosion behavior of low carbon steel (AISI 1010) influenced by grain size through microstructural mechanical. *Scientific Reports*, 14(1), 5098. Advance online publication. DOI: 10.1038/s41598-023-47744-y PMID: 38429315

Alhebshi, N. A., Salah, N., Hussain, H., Salah, Y. N., & Yin, J. (2021). Structural and Electrochemical Properties of Physically and Chemically Activated Carbon Nanoparticles for Supercapacitors. *Nanomaterials (Basel, Switzerland)*, 12(1), 122. DOI: 10.3390/nano12010122 PMID: 35010069

Alias, R., Mahmoodian, R., Genasan, K., Vellasamy, K. M., Hamdi Abd Shukor, M., & Kamarul, T. (2020). Mechanical, antibacterial, and biocompatibility mechanism of PVD grown silver-tantalum-oxide-based nanostructured thin film on stainless steel 316L for surgical applications. *Materials Science and Engineering C*, 107, 110304. DOI: 10.1016/j.msec.2019.110304 PMID: 31761210

Ali, S., El-Shareef, A., Al-Ghamdi, R., & Saeed, M. (2004). The isoxazolidines: The effects of steric factor and hydrophobic chain length on the corrosion inhibition of mild steel in acidic medium. *Corrosion Science*, 47(11), 2659–2678. DOI: 10.1016/j.corsci.2004.11.007

Ali, S., Wójcik, N. A., Hakeem, A. S., Gueguen, Y., & Karlsson, S. (2024). Effect of composition on the thermal properties and structure of M-Al-Si-O-N glasses, M = Na, Mg, Ca. *Progress in Solid State Chemistry*, 74, 100461. DOI: 10.1016/j.progsolidstchem.2024.100461

Al-Ketan, O., Soliman, A., AlQubaisi, A. M., & Al-Rub, R. K. A. (2017). Nature-Inspired Lightweight Cellular Co-Continuous Composites with Architected Periodic Gyroidal Structures. *Advanced Engineering Materials*, 20(2), 1700549. Advance online publication. DOI: 10.1002/adem.201700549

Al-Moameri, H. H., Nahi, Z. M., Al-Sharify, N. T., & Rzaij, D. R. (2020). A review on the biomedical applications of alumina. *Journal of Engineering and Sustainable Development*, 24(5), 28–36. DOI: 10.31272/jeasd.24.5.5

Al-Moubaraki, A. H., & Obot, I. B. (2021). Corrosion challenges in petroleum refinery operations: Sources, mechanisms, mitigation, and future outlook. *Journal of Saudi Chemical Society*, 25(12), 101370. DOI: 10.1016/j.jscs.2021.101370

Alnahhal, S. Y., Elfari, A. A., Afifi, S. A., & Aljubb, A. E. R. (2024). Using slow sand filter for organic matter and suspended solids removal as post-treatment unit for wastewater effluent. *Environmental Science. Water Research & Technology*, 10(2), 490–497. DOI: 10.1039/D3EW00467H

Alvarez-Fernandez, A., Reid, B., Fornerod, M. J., Taylor, A., Divitini, G., & Guldin, S. (2020). Structural Characterization of Mesoporous Thin Film Architectures: A Tutorial Overview. *ACS Applied Materials & Interfaces*, 12(5), 5195–5208. DOI: 10.1021/acsami.9b17899 PMID: 31961128

Amariei, D., Amrousse, R., Batonneau, Y., Kappenstein, C., & Cartoixa, B. (2010). Monolithic catalysts for the decomposition of energetic compounds. In *Studies in surface science and catalysis* (pp. 35–42). DOI: 10.1016/S0167-2991(10)75005-9

Amin, B., Elahi, M. A., Shahzad, A., Porter, E., & O'Halloran, M. (2019). A review of the dielectric properties of the bone for low frequency medical technologies. *Biomedical physics & engineering express, 5*(2), . https://doi.org/DOI: 10.1088/2057-1976/aaf210

Amin, M. A., El-Rehim, S. S. A., El-Sherbini, E., & Bayoumi, R. S. (2006). The inhibition of low carbon steel corrosion in HCl solutions by succinic acid. *Electrochimica Acta*, 52(11), 3588–3600. DOI: 10.1016/j.electacta.2006.10.019

Amirtharaj Mosas, K. K., Chandrasekar, A. R., Dasan, A., Pakseresht, A., & Galusek, D. (2022). Recent advancements in materials and coatings for biomedical implants. *Gels (Basel, Switzerland)*, 8(5), 323. DOI: 10.3390/gels8050323 PMID: 35621621

Amrousse, R. (2021) Thermal decomposition of HAN green propellant. In the Proceedings of the 72nd International Astronautical Congress, Dubai, UAE.

Amrousse, R. (2022) Survey on the green propulsion systems: from the lab-scale to the pilot scale-up: HAN is a good Example. In the Proceedings of the 73rd International Astronautical Congress, Paris, France.

Amrousse, R., Batonneau, Y., Kappenstein, C., Théron, M., & Bravais, P. (2010). Preparation of monolithic catalysts for space propulsion applications. In *Studies in surface science and catalysis* (pp. 755–758). DOI: 10.1016/S0167-2991(10)75153-3

Amrousse, R., Elidrissi, A. N., Bachar, A., Mabrouk, A., Toshtay, K., & Azat, S. (2024). Nanosized catalytic particles for the decomposition of green propellants as substitute for hydrazine. In Advances in chemical and materials engineering book series (pp. 195–217). DOI: 10.4018/979-8-3693-3268-9.ch009

Amrousse, R., Elidrissi, A. N., Nosseir, A. E. S., Toshtay, K., Atamanov, M. K., & Azat, S. (2024b). Thermal and Catalytic Decomposition of Hydroxylammonium Nitrate (HAN)-Based Propellant. In *Space technology library* (pp. 33–60). DOI: 10.1007/978-3-031-62574-9_2

Amrousse, R., Augustin, C., Farhat, K., Batonneau, Y., & Kappenstein, C. J. (2011). Catalytic decomposition of H_2O_2 using FeCrAl metallic foam-based catalysts. *International Journal of Energetic Materials and Chemical Propulsion*, 10(4), 337–349. DOI: 10.1615/IntJEnergeticMaterialsChemProp.2012005202

Amrousse, R., Hori, K., Fetimi, W., & Farhat, K. (2012). HAN and ADN as liquid ionic monopropellants: Thermal and catalytic decomposition processes. *Applied Catalysis B: Environmental*, 127, 121–128. DOI: 10.1016/j.apcatb.2012.08.009

Amrousse, R., Katsumi, T., Azuma, N., & Hori, K. (2017). Hydroxylammonium nitrate (HAN)-based green propellant as alternative energy resource for potential hydrazine substitution: From lab scale to pilot plant scale-up. *Combustion and Flame*, 176, 334–348. DOI: 10.1016/j.combustflame.2016.11.011

Amrousse, R., Katsumi, T., Bachar, A., Brahmi, R., Bensitel, M., & Hori, K. (2013). Chemical engineering study for hydroxylammonium nitrate monopropellant decomposition over monolith and grain metal-based catalysts. *Reaction Kinetics, Mechanisms and Catalysis*, 111(1), 71–88. DOI: 10.1007/s11144-013-0626-6

Amrousse, R., Katsumi, T., Itouyama, N., Azuma, N., Kagawa, H., Hatai, K., Ikeda, H., & Hori, K. (2015). New HAN-based mixtures for reaction control system and low toxic spacecraft propulsion subsystem: Thermal decomposition and possible thruster applications. *Combustion and Flame*, 162(6), 2686–2692. DOI: 10.1016/j.combustflame.2015.03.026

Amrousse, R., Katsumi, T., Niboshi, Y., Azuma, N., Bachar, A., & Hori, K. (2013). Performance and deactivation of Ir-based catalyst during hydroxylammonium nitrate catalytic decomposition. *Applied Catalysis A, General*, 452, 64–68. DOI: 10.1016/j.apcata.2012.11.038

Amrousse, R., Keav, S., Batonneau, Y., Kappenstein, C. J., Theron, M., & Bravais, P. (2011). Catalytic ignition of cold hydrogen/oxygen mixtures for space propulsion applications. *International Journal of Energetic Materials and Chemical Propulsion*, 10(3), 217–230. DOI: 10.1615/IntJEnergeticMaterialsChemProp.2012004877

Amrousse, R., & Moumni, S. E. (2013). A Highly Distributed CuxAuy-Deposited Nanotube Carbon for Selective Reduction of NO in the Presence of NH3 at Very Low Temperature. *ChemCatChem*, 6(1), 119–122. DOI: 10.1002/cctc.201300777

Andreoli, F., & Sabogal-Paz, L. (2020). Household slow sand filter to treat groundwater with microbiological risks in rural communities. *Water Research*, 186, 116352. DOI: 10.1016/j.watres.2020.116352 PMID: 32916617

Angeli, F., Charpentier, T., Molières, E., Soleilhavoup, A., Jollivet, P., & Gin, S. (2013). Influence of lanthanum on borosilicate glass structure: A multinuclear MAS and MQMAS NMR investigation. *Journal of Non-Crystalline Solids*, 376, 189–198. DOI: 10.1016/j.jnoncrysol.2013.05.042

Angeli, F., Gaillard, M., Jollivet, P., & Charpentier, T. (2006). Influence of glass composition and alteration solution on leached silicate glass structure: A solid-state NMR investigation. *Geochimica et Cosmochimica Acta*, 70(10), 2577–2590. DOI: 10.1016/j.gca.2006.02.023

An, S., Lee, J., Kappenstein, C., & Kwon, S. (2010). Comparison of catalyst support between monolith and pellet in hydrogen peroxide thrusters. *Journal of Propulsion and Power*, 26(3), 439–445. DOI: 10.2514/1.46075

Ansari, M., Alam, A., Bera, R., Hassan, A., Goswami, S., & Das, N. (2020). Synthesis, characterization and adsorption studies of a novel triptycene based hydroxyl azo- nanoporous polymer for environmental remediation. *Journal of Environmental Chemical Engineering*, 8(2), 103558. https://doi.org/https://doi.org/10.1016/j.jece.2019.103558. DOI: 10.1016/j.jece.2019.103558

Anusree, T., & Vargeese, A. A. (2022). Enhanced performance of barium and cobalt doped spinel CuCr2O4 as decomposition catalyst for ammonium perchlorate. *Journal of Solid State Chemistry*, 315, 123481. DOI: 10.1016/j.jssc.2022.123481

Ao, W., Fu, J., Mao, X., Kang, Q., Ran, C., Liu, Y., Zhang, H., Gao, Z., Li, J., Liu, G., & Dai, J. (2018). Microwave assisted preparation of activated carbon from biomass: A review. *Renewable & Sustainable Energy Reviews*, 92, 958–979. DOI: 10.1016/j.rser.2018.04.051

Aruchamy, K., Balasankar, A., Ramasundaram, S., & Oh, T. (2023). Recent Design and Synthesis Strategies for High-Performance Supercapacitors Utilizing ZnCo2O4-Based Electrode Materials. *Energies*, 16(15), 5604. DOI: 10.3390/en16155604

Asimakopoulos, G., Baikousi, M., Kostas, V., Papantoniou, M., Bourlinos, A. B., Zbořil, R., Karakassides, M. A., & Salmas, C. E. (2020). Nanoporous Activated Carbon Derived via Pyrolysis Process of Spent Coffee: Structural Characterization. Investigation of Its Use for Hexavalent Chromium Removal. *Applied Sciences (Basel, Switzerland)*, 10(24), 8812. DOI: 10.3390/app10248812

Askari, M., Aliofkhazraei, M., Jafari, R., Hamghalam, P., & Hajizadeh, A. (2021). Downhole corrosion inhibitors for oil and gas production – a review. *Applied Surface Science Advances*, 6, 100128. DOI: 10.1016/j.apsadv.2021.100128

Aslam, R., Serdaroglu, G., Zehra, S., Verma, D. K., Aslam, J., Guo, L., Verma, C., Ebenso, E. E., & Quraishi, M. (2021). Corrosion inhibition of steel using different families of organic compounds: Past and present progress. *Journal of Molecular Liquids*, 348, 118373. DOI: 10.1016/j.molliq.2021.118373

Aslan, S., & Cakici, H. (2007). Biological denitrification of drinking water in a slow sand filter. *Journal of Hazardous Materials*, 148(1–2), 253–258. DOI: 10.1016/j.jhazmat.2007.02.012 PMID: 17363163

Atamanov, M. K., Amrousse, R., Hori, K., Kolesnikov, B. Y., & Mansurov, Z. A. (2018). Influence of activated carbon on the thermal decomposition of hydroxylammonium nitrate. *Combustion, Explosion, and Shock Waves*, 54(3), 316–324. DOI: 10.1134/S0010508218030085

Atkinson, I., Anghel, E. M., Predoana, L., Mocioiu, O. C., Jecu, L., Raut, I., Munteanu, C., Culita, D., & Zaharescu, M. (2016). Influence of ZnO addition on the structural, in vitro behavior and antimicrobial activity of sol–gel derived CaO–P 2 O 5 –SiO 2 bioactive glasses. *Ceramics International*, 42(2), 3033–3045. DOI: 10.1016/j.ceramint.2015.10.090

Avdeev, Y., Kuznetsov, Y., & Frumkin, A. (2022). Acid corrosion of metals and its inhibition. A critical review of the current problem state. *International Journal of Corrosion and Scale Inhibition*, 11(1). Advance online publication. DOI: 10.17675/2305-6894-2022-11-1-6

Awasthi, S., Pandey, S. K., Arunan, E., & Srivastava, C. (2021). A review on hydroxyapatite coatings for the biomedical applications: Experimental and theoretical perspectives. *Journal of Materials Chemistry. B*, 9(2), 228–249. DOI: 10.1039/D0TB02407D PMID: 33231240

Axente, E., Elena Sima, L., & Sima, F. (2020). Biomimetic coatings obtained by combinatorial laser technologies. *Coatings*, 10(5), 463. DOI: 10.3390/coatings10050463

Ayati, A., Tanhaei, B., Beiki, H., Krivoshapkin, P., Krivoshapkina, E., & Tracey, C. (2023). Insight into the adsorptive removal of ibuprofen using porous carbonaceous materials: A review. *Chemosphere*, 323, 138241. DOI: 10.1016/j.chemosphere.2023.138241 PMID: 36841446

Azat, S., Busquets, R., Pavlenko, V., Kerimkulova, A., Whitby, R. L., & Mansurov, Z. (2013b). Applications of activated carbon sorbents based on Greek walnut. *Applied Mechanics and Materials*, 467, 49–51. . DOI: 10.4028/www.scientific.net/AMM.467.49

Azizabadi, N., Azar, P. A., Tehrani, M. S., & Derakhshi, P. (2021). Synthesis and characteristics of gel-derived SiO2-CaO-P2O5-SrO-Ag2O-ZnO bioactive glass: Bioactivity, biocompatibility, and antibacterial properties. *Journal of Non-Crystalline Solids*, 556, 120568. DOI: 10.1016/j.jnoncrysol.2020.120568

Aziz, F., & Ismail, A. F. (2015). Spray coating methods for polymer solar cells fabrication: A review. *Materials Science in Semiconductor Processing*, 39, 416–425. DOI: 10.1016/j.mssp.2015.05.019

Bachar, A., Catteaux, R., Duée, C., Désanglois, F., Lebecq, I., Mercier, C., & Follet-Houttemane, C. (2019). Synthesis and Characterization of Doped Bioactive Glasses. In Elsevier eBooks (pp. 69–123). DOI: 10.1016/B978-0-08-102196-5.00003-3

Bachar, A., Mercier, C., Follet, C., Bost, N., Bentiss, F., & Hampshire, S. (2016). An introduction of the fluorine and Nitrogen on properties of Ca-Si-Al-O glasses. https://lilloa.univ-lille.fr/handle/20.500.12210/8718

Bachar, A., Mabrouk, A., Amrousse, R., Azat, S., Follet, C., Mercier, C., & Bouchart, F. (2024). Properties, Bioactivity and Viability of the New Generation of Oxyfluoronitride Bioglasses. *Eurasian Chemico-Technological Journal*, 26(1), 43–52. DOI: 10.18321/ectj1565

Bachar, A., Mercier, C., Tricoteaux, A., Hampshire, S., Leriche, A., & Follet, C. (2013). Effect of nitrogen and fluorine on mechanical properties and bioactivity in two series of bioactive glasses. Journal of the Mechanical Behavior of Biomedical Materials. *Journal of the Mechanical Behavior of Biomedical Materials*, 23, 133–148. DOI: 10.1016/j.jmbbm.2013.03.010 PMID: 23676624

Bachar, A., Mercier, C., Tricoteaux, A., Leriche, A., Follet, C., & Hampshire, S. (2016). Bioactive oxynitride glasses: Synthesis, structure and properties. *Journal of the European Ceramic Society*, 36(12), 2869–2881. DOI: 10.1016/j.jeurceramsoc.2015.12.017

Bachar, A., Mercier, C., Tricoteaux, A., Leriche, A., Follet, C., Saadi, M., & Hampshire, S. (2012). Effects of addition of nitrogen on bioglass properties and structure. *Journal of Non-Crystalline Solids*, 358(3), 693–701. DOI: 10.1016/j.jnoncrysol.2011.11.036

Bachar, A., Mercier, C., Tricoteaux, A., Leriche, A., Follet-Houttemane, C., Saadi, M., & Hampshire, S. (2013). Effects of nitrogen on properties of oxyfluoronitride bioglasses. *Process Biochemistry (Barking, London, England)*, 48(1), 89–95. DOI: 10.1016/j.procbio.2012.05.024

Badr, H. A., Reda, A. E., Zawrah, M. F., Khattab, R. M., & Sadek, H. E. H. (2024). Effect of hydroxyapatite on sinterability, mechanical properties, and bioactivity of chemically synthesized Alumina–Zirconia composites. *Journal of Materials Engineering and Performance*. Advance online publication. DOI: 10.1007/s11665-024-09531-2

Bagundol, T. B., Awa, A. L., & Enguito, M. R. C. (2013). Efficiency of slow sand filter in purifying well water. *Journal of Multidisciplinary Studies*, 2(1). Advance online publication. DOI: 10.7828/jmds.v2i1.402

Baheti, W., & Lv, S., Mila, Ma, L., Amantai, D., Sun, H., & He, H. (. (2023). Graphene/hydroxyapatite coating deposit on titanium alloys for implant application. *Journal of Applied Biomaterials & Functional Materials*, 21. Advance online publication. DOI: 10.1177/22808000221148104 PMID: 36633270

Baig, N., Kammakakam, I., & Falath, W. (2021). Nanomaterials: A review of synthesis methods, properties, recent progress, and challenges. *Materials Advances*, 2(6), 1821–1871. DOI: 10.1039/D0MA00807A

Baig, S. A., Mahmood, Q., Nawab, B., Shafqat, M. N., & Pervez, A. (2011). Improvement of drinking water quality by using plant biomass through household biosand filter – A decentralized approach. *Ecological Engineering*, 37(11), 1842–1848. DOI: 10.1016/j.ecoleng.2011.06.011

Bai, L., Song, P., & Su, J. (2023). Bioactive elements manipulate bone regeneration. *Biomaterials Translational*, 4(4), 248–269. DOI: 10.12336/biomatertransl.2023.04.005 PMID: 38282709

Baino, F., Fiorilli, S., & Vitale-Brovarone, C. (2016). Bioactive glass-based materials with hierarchical porosity for medical applications: Review of recent advances. *Acta Biomaterialia*, 42, 18–32. DOI: 10.1016/j.actbio.2016.06.033 PMID: 27370907

Baino, F., Fiorilli, S., & Vitale-Brovarone, C. (2017). Composite Biomaterials Based on Sol-Gel Mesoporous Silicate Glasses: A Review, 1. *Bioengineering (Basel, Switzerland)*, 4(1), 15. DOI: 10.3390/bioengineering4010015 PMID: 28952496

Baino, F., Hamzehlou, S., & Kargozar, S. (2018). Bioactive glasses: Where are we and where are we going? *Journal of Functional Biomaterials*, 9(1), 25. DOI: 10.3390/jfb9010025 PMID: 29562680

Bai, S., Sun, C., Guo, T., Luo, R., Lin, Y., Chen, A., Sun, L., & Zhang, J. (2013). Low temperature electrochemical deposition of nanoporous ZnO thin films as novel NO2 sensors. *Electrochimica Acta*, 90, 530–534. https://doi.org/https://doi.org/10.1016/j.electacta.2012.12.060. DOI: 10.1016/j.electacta.2012.12.060

Bai, X., Samari-Kermani, M., Schijven, J., Raoof, A., Dinkla, I. J. T., & Muyzer, G. (2024). Enhancing slow sand filtration for safe drinking water production: Interdisciplinary insights into Schmutzdecke characteristics and filtration performance in mini-scale filters. *Water Research*, 262, 122059. DOI: 10.1016/j.watres.2024.122059 PMID: 39059201

Bai, Y., Li, X., Wu, K., Heng, B., Zhang, X., & Deng, X. (2024). Biophysical stimuli for promoting bone repair and regeneration. *Medical Review (Berlin, Germany)*. Advance online publication. DOI: 10.1515/mr-2024-0023

Bai, Y., Meng, H., Li, Z., & Wang, Z. L. (2024). Degradable piezoelectric biomaterials for medical applications. *MedMat*, 1(1), 40–49. DOI: 10.1097/mm9.0000000000000002

Bai, Z., Li, P., Liu, L., & Xiong, G. (2012). Oxidative Dehydrogenation of Propane over MoOx and POx Supported on Carbon Nanotube Catalysts. *ChemCatChem*, 4(2), 260–264. DOI: 10.1002/cctc.201100242

Baloh, R. W. (2024). Electricity and the Nervous System. In Brain Electricity: The Interwoven History of Electricity and Neuroscience (pp. 125-158). *Springer Nature*. https://doi.org/DOI: 10.1007/978-3-031-62994-5_5

Ban, C., Li, L., & Wei, L. (2018). Electrical properties of O-self-doped boron-nitride nanotubes and the piezoelectric effects of their freestanding network film. *RSC Advances*, 8(51), 29141–29146. DOI: 10.1039/C8RA05698F PMID: 35548006

Banciu, C. A., Nastase, F., Istrate, A.-I., & Veca, L. M. (2022). 3D Graphene Foam by Chemical Vapor Deposition: Synthesis, Properties, and Energy-Related Applications. *Molecules (Basel, Switzerland)*, 27(11), 3634. DOI: 10.3390/molecules27113634 PMID: 35684569

Barak, M. M. (2024). Cortical and Trabecular Bone Modeling and Implications for Bone Functional Adaptation in the Mammalian Tibia. *Bioengineering (Basel, Switzerland)*, 11(5), 514. DOI: 10.3390/bioengineering11050514 PMID: 38790379

Barkouch, Y., El Fadeli, S., Flata, K., Ait Melloul, A., Khadiri, M. E., & Pineau, A. (2019). Removal efficiency of metallic trace elements by slow sand filtration: Study of the effect of sand porosity and diameter column. *Modeling Earth Systems and Environment*, 5(2), 533–542. DOI: 10.1007/s40808-018-0542-x

Barrett, J. M., Bryck, J., Collins, M. R., Janonis, B. A., & Logsdon, G. S. (n.d.). *Manual of Design for Slow Sand Filtration. AWWA Research Foundation and American Water Works Association, USA, 1991.*

Basavaraj Chavati, G., Kumar Basavaraju, S., Nayaka Yanjerappa, A., Muralidhara, H. B., Venkatesh, K., & Gopalakrishna, K. (2024). Synergetic Functionalization of the ZnS@ASCs Biocomposite: For Enhanced Electrochemical Performance of Redox Flow Batteries and Supercapacitors. *ACS Applied Electronic Materials*, acsaelm.4c00943. Advance online publication. DOI: 10.1021/acsaelm.4c00943

Bashir, S., Singh, G., & Kumar, A. (2018). Shatavari (Asparagus racemosus) as green corrosion inhibitor of aluminium in acidic medium. *Portugaliae Electrochimica Acta*, 37(2), 83–1. DOI: 10.4152/pea.201902083

Basik, M., & Mobin, M. (2022). Metal oxide and organic polymers mixed composites as corrosion inhibitors. In Elsevier eBooks (pp. 345–355). DOI: 10.1016/B978-0-323-90410-0.00018-0

Batonneau, Y., Brahmi, R., Cartoixa, B., Farhat, K., Kappenstein, C., Keav, S., Kharchafi-Farhat, G., Pirault-Roy, L., Saouabé, M., & Scharlemann, C. (2013b). Green Propulsion: Catalysts for the European FP7 Project GRASP. *Topics in Catalysis*, 57(6–9), 656–667. DOI: 10.1007/s11244-013-0223-y

Batool, M., Abid, M. A., Javed, T., & Haider, M. N. (2024). Applications of biodegradable polymers and ceramics for bone regeneration: A mini-review. *International Journal of Polymeric Materials*, •••, 1–15. DOI: 10.1080/00914037.2024.2314601

Baxter, F. R., Bowen, C. R., Turner, I. G., & Dent, A. C. (2010). Electrically active bioceramics: A review of interfacial responses. *Annals of Biomedical Engineering*, 38(6), 2079–2092. DOI: 10.1007/s10439-010-9977-6 PMID: 20198510

Bejarano, J., Caviedes, P., & Palza, H. (2015). Sol–gel synthesis and *in vitro* bioactivity of copper and zinc-doped silicate bioactive glasses and glass-ceramics. *Biomedical Materials (Bristol, England)*, 10(2), 025001. DOI: 10.1088/1748-6041/10/2/025001 PMID: 25760730

Bellantone, M., Williams, H. D., & Hench, L. L. (2002). Broad-Spectrum Bactericidal Activity of Ag2O-Doped Bioactive Glass. *Antimicrobial Agents and Chemotherapy*, 46(6), 1940–1945. DOI: 10.1128/AAC.46.6.1940-1945.2002 PMID: 12019112

Benoy, S. M., Pandey, M., Bhattacharjya, D., & Saikia, B. K. (2022). Recent trends in supercapacitor-battery hybrid energy storage devices based on carbon materials. *Journal of Energy Storage*, 52, 104938. DOI: 10.1016/j.est.2022.104938

Bentiss, F., Lebrini, M., & Lagrenée, M. (2005). Thermodynamic characterization of metal dissolution and inhibitor adsorption processes in mild steel/2,5-bis(n-thienyl)-1,3,4-thiadiazoles/HCl system. *Corrosion Science*, 47(12), 2915–2931. DOI: 10.1016/j.corsci.2005.05.034

Bhatnagar, A., Hogland, W., Marques, M., & Sillanpää, M. (2012). An overview of the modification methods of activated carbon for its water treatment applications. *Chemical Engineering Journal*, 219, 499–511. DOI: 10.1016/j.cej.2012.12.038

Bhattacharyya, B. (2015). Design and Developments of Microtools. In Bhattacharyya, B. (Ed.), *Electrochemical Micromachining for Nanofabrication, MEMS and Nanotechnology* (pp. 101–122). William Andrew Publishing., https://doi.org/https://doi.org/10.1016/B978-0-323-32737-4.00006-2 DOI: 10.1016/B978-0-323-32737-4.00006-2

Bhattacharyya, S., Mastai, Y., Narayan Panda, R., Yeon, S.-H., & Hu, M. Z. (2014). Advanced Nanoporous Materials: Synthesis, Properties, and Applications. *Journal of Nanomaterials*, 2014(1), 275796. https://doi.org/https://doi.org/10.1155/2014/275796. DOI: 10.1155/2014/275796

Bhavsar, M. B., Leppik, L., Costa Oliveira, K. M., & Barker, J. H. (2020). Role of Bioelectricity During Cell Proliferation in Different Cell Types. *Frontiers in Bioengineering and Biotechnology*, 8, 603. DOI: 10.3389/fbioe.2020.00603 PMID: 32714900

Bhong, M., Khan, T. K., Devade, K., Krishna, B. V., Sura, S., Eftikhaar, H. K., & Gupta, N. (2023). Review of composite materials and applications. *Materials Today: Proceedings*, •••, 2214–7853. DOI: 10.1016/j.matpr.2023.10.026

Bhutiani, R., Ahamad, N. F., & Ruhela, M. (2021). Effect of composition and depth of filter-bed on the efficiency of Sand-intermittent-filter treating the Industrial wastewater at Haridwar, India. *Journal of Applied and Natural Science*, 13(1), 88–94. DOI: 10.31018/jans.v13i1.2421

Binner, J. (2006). Ceramic Foams In *Cellular Ceramics: Structure, Manufacturing, Properties and Applications. Ceramics Foams,* Scheffler, M., Colombo, P. (Eds.), 31–56. DOI: 10.1002/3527606696.ch2a

Birol, H., Damjanovic, D., & Setter, N. (2005). Preparation and characterization of KNbO3 ceramics. *Journal of the American Ceramic Society*, 88(7), 1754–1759. DOI: 10.1111/j.1551-2916.2005.00347.x

Biswas, A., Friend, C. S., & Prasad, P. N. (2001). Ceramic Nanocomposites with Organic Phases, Optics of. In Buschow, K. H. J., Cahn, R. W., Flemings, M. C., Ilschner, B., Kramer, E. J., Mahajan, S., & Veyssière, P. (Eds.), *Encyclopedia of Materials: Science and Technology* (pp. 1072–1080). Elsevier., https://doi.org/https://doi.org/10.1016/B0-08-043152-6/00198-4 DOI: 10.1016/B0-08-043152-6/00198-4

Biswas, P., & Bandyopadhyaya, R. (2017). Synergistic antibacterial activity of a combination of silver and copper nanoparticle impregnated activated carbon for water disinfection. *Environmental Science. Nano*, 4(12), 2405–2417. DOI: 10.1039/C7EN00427C

Blankenship, L. S., & Mokaya, R. (2022). Modulating the porosity of carbons for improved adsorption of hydrogen, carbon dioxide, and methane: A review. *Materials Advances*, 3(4), 1905–1930. DOI: 10.1039/D1MA00911G

Boccaccini, A. R., Notingher, I., Maquet, V., & Jérôme, R. (2003). Bioresorbable and bioactive composite materials based on polylactide foams filled with and coated by Bioglass® particles for tissue engineering applications. *Journal of Materials Science. Materials in Medicine*, 14(5), 443–450. DOI: 10.1023/A:1023266902662 PMID: 15348448

Bodhak, S., Bose, S., & Bandyopadhyay, A. (2010). Electrically polarized HAp-coated Ti: In vitro bone cell-material interactions. *Acta Biomaterialia*, 6(2), 641–651. DOI: 10.1016/j.actbio.2009.08.008 PMID: 19671456

Bokov, D., Turki Jalil, A., Chupradit, S., Suksatan, W., Javed Ansari, M., Shewael, I. H., Valiev, G. H., & Kianfar, E. (2021). Nanomaterial by Sol-Gel Method: Synthesis and Application. *Advances in Materials Science and Engineering, 2021*(1). DOI: 10.1155/2021/5102014

Bokov, D., Turki Jalil, A., Chupradit, S., Suksatan, W., Javed Ansari, M., Shewael, I. H., & Kianfar, E. (2021). Nanomaterial by sol-gel method: Synthesis and application. *Advances in Materials Science and Engineering*, 2021(1), 5102014.

Bolelli, G., Bellucci, D., Cannillo, V., Lusvarghi, L., Sola, A., Stiegler, N., Müller, P., Killinger, A., Gadow, R., Altomare, L., & De Nardo, L. (2014). Suspension thermal spraying of hydroxyapatite: Microstructure and in vitro behaviour. *Materials Science and Engineering C*, 34, 287–303. DOI: 10.1016/j.msec.2013.09.017 PMID: 24268261

Bonora, P., Deflorian, F., & Fedrizzi, L. (1996). Electrochemical impedance spectroscopy as a tool for investigating underpaint corrosion. *Electrochimica Acta*, 41(7–8), 1073–1082. DOI: 10.1016/0013-4686(95)00440-8

Bon, V., Brunner, E., Pöppl, A., & Kaskel, S. (2020). Unraveling Structure and Dynamics in Porous Frameworks via Advanced In Situ Characterization Techniques. *Advanced Functional Materials*, 30(41), 1907847. Advance online publication. DOI: 10.1002/adfm.201907847

Bon, V., Klein, N., Senkovska, I., Heerwig, A., Getzschmann, J., Wallacher, D., Zizak, I., Brzhezinskaya, M., Mueller, U., & Kaskel, S. (2015). Exceptional adsorption-induced cluster and network deformation in the flexible metal–organic framework DUT-8(Ni) observed by in situ X-ray diffraction and EXAFS. *Physical Chemistry Chemical Physics*, 17(26), 17471–17479. DOI: 10.1039/C5CP02180D PMID: 26079102

Borisade, S. G., Owoeye, S. S., Ajayi, K. V., Enewo, S. I., & Abdullahi, A. (2024). Investigation of physical, mechanical and in-vitro bioactivity of bioactive glass-ceramics fabricated from waste soda-lime-silica glass doped P2O5 by microwave irradiation sintering. *Hybrid Advances*, 6, 100203. DOI: 10.1016/j.hybadv.2024.100203

Borsari, M., Ferrari, E., Grandi, R., & Saladini, M. (2002). Curcuminoids as potential new ironchelating agents: Spectroscopic, polarographic and potentiometric study on their Fe (III) complexing ability. *Inorganica Chimica Acta*, 328(1), 61–68. DOI: 10.1016/S0020-1693(01)00687-9

Bosco Franklin, J., Sachin, S., John Sundaram, S., Theophil Anand, G., Dhayal Raj, A., & Kaviyarasu, K. (2024). Investigation on copper cobaltite (CuCo2O4) and its composite with activated carbon (AC) for supercapacitor applications. *Materials Science for Energy Technologies*, 7, 91–98. DOI: 10.1016/j.mset.2023.07.006

Bose, S., & Tarafder, S. (2012). Calcium phosphate ceramic systems in growth factor and drug delivery for bone tissue engineering: A review. *Acta Biomaterialia*, 8(4), 1401–1421. DOI: 10.1016/j.actbio.2011.11.017 PMID: 22127225

Bouami, H. E., Mabrouk, A., Mercier, C., Mihoubi, W., Meurice, E., Follet, C., Faska, N., & Bachar, A. (2024). The effect of CuO dopant on the bioactivity, the biocompatibility, and the antibacterial properties of bioactive glasses synthesized by the sol-gel method. *Journal of Sol-Gel Science and Technology*, 111(2), 347–361. Advance online publication. DOI: 10.1007/s10971-024-06445-2

Bouammali, H., Ousslim, A., Bekkouch, K., Bouammali, B., Aouniti, A., Al-Deyab, S., Jama, C., Bentiss, F., & Hammouti, B. (2013). The Anti-Corrosion Behavior of Lavandula dentata Aqueous Extract on Mild Steel in 1M HCl. *International Journal of Electrochemical Science*, 8(4), 6005–6013. DOI: 10.1016/S1452-3981(23)14735-3

Bouklah, M., Hammouti, B., Lagrenée, M., & Bentiss, F. (2005). Thermodynamic properties of 2,5-bis(4-methoxyphenyl)-1,3,4-oxadiazole as a corrosion inhibitor for mild steel in normal sulfuric acid medium. *Corrosion Science*, 48(9), 2831–2842. DOI: 10.1016/j.corsci.2005.08.019

Bowden, B., Davies, M., Davies, P. R., Guan, S., Morgan, D. J., Roberts, V., & Wotton, D. (2018). The deposition of metal nanoparticles on carbon surfaces: The role of specific functional groups. *Faraday Discussions*, 208, 455–470. DOI: 10.1039/C7FD00210F PMID: 29845183

Boxall, A. B. A., Collins, R., Wilkinson, J. L., Swan, C., Bouzas-Monroy, A., Jones, J., Winter, E., Leach, J., Juta, U., Deacon, A., Townsend, I., Kerr, P., Paget, R., Rogers, M., Greaves, D., Turner, D., & Pearson, C. (2024). Pharmaceutical pollution of the English National Parks. *Environmental Toxicology and Chemistry*, 43(11), 2422–2435. Advance online publication. DOI: 10.1002/etc.5973 PMID: 39138896

Brauer, D. S., Karpukhina, N., Law, R. V., & Hill, R. G. (2009). Structure of fluoride-containing bioactive glasses. *Journal of Materials Chemistry*, 19(31), 5629. DOI: 10.1039/b900956f

Brezny, R., & Green, D. J. (1989). Fracture behavior of Open-Cell ceramics. *Journal of the American Ceramic Society*, 72(7), 1145–1152. DOI: 10.1111/j.1151-2916.1989.tb09698.x

Brezny, R., Green, D. J., & Dam, C. Q. (1989). Evaluation of strut strength in Open-Cell ceramics. *Journal of the American Ceramic Society*, 72(6), 885–889. DOI: 10.1111/j.1151-2916.1989.tb06239.x

Brill, T., Brush, P., & Patil, D. (1993). Thermal decomposition of energetic materials 58. Chemistry of ammonium nitrate and ammonium dinitramide near the burning surface temperature. *Combustion and Flame*, 92(1–2), 178–186. DOI: 10.1016/0010-2180(93)90206-I

Bròdano, G. B., Giavaresi, G., Lolli, F., Salamanna, F., Parrilli, A., Martini, L., Griffoni, C., Greggi, T., Arcangeli, E., Pressato, D., Boriani, S., & Fini, M. (2014). Hydroxyapatite-Based Biomaterials Versus Autologous Bone Graft in Spinal Fusion: An In Vivo Animal Study. *Spine*, 39(11), E661–E668. DOI: 10.1097/BRS.0000000000000311 PMID: 24718060

Brow, R., Pantano, C., & Boyd, D. (1984). Nitrogen Coordination in Oxynitride Glasses. *Journal of the American Ceramic Society*, 67(4). Advance online publication. DOI: 10.1111/j.1151-2916.1984.tb18834.x

Bueno, L., Messaddeq, Y., Filho, F. D., & Ribeiro, S. (2005). Study of fluorine losses in oxyfluoride glasses. *Journal of Non-Crystalline Solids*, 351(52–54), 3804–3808. DOI: 10.1016/j.jnoncrysol.2005.10.007

Busquets, R., Kozynchenko, O. P., Whitby, R. L. D., Tennison, S. R., & Cundy, A. B. (2014) Phenolic carbon tailored for the removal of polar organic contaminants from water: A solution to the metaldehyde problem? Water Research, Volume 61, 2014, Pages 46-56, ISSN 0043-1354, https://doi.org/DOI: 10.1016/j.watres.2014.04.048

Buzalewicz, I., Kaczorowska, A., Fijałkowski, W., Pietrowska, A., Matczuk, A. K., Podbielska, H., Wieliczko, A., Witkiewicz, W., & Jędruchniewicz, N. (2024). Quantifying the dynamics of bacterial biofilm formation on the surface of soft contact lens materials using digital holographic tomography to advance biofilm research. *International Journal of Molecular Sciences*, 25(5), 2653. DOI: 10.3390/ijms25052653 PMID: 38473902

Byrappa, K. ., & Haber, Masahiro. (2001). *Handbook of Hydrothermal Technology*. Noyes Publications [Imprint] William Andrew, Inc. Elsevier Science & Technology Books [distributor].

Bystrov, V. S. (2024). Molecular self-assembled helix peptide nanotubes based on some amino acid molecules and their dipeptides: Molecular modeling studies. *Journal of Molecular Modeling*, 30(8), 257. DOI: 10.1007/s00894-024-05995-0 PMID: 38976043

Caihong, Y., Singh, A., Ansari, K., Ali, I. H., & Kumar, R. (2022). Novel nitrogen based heterocyclic compound as Q235 steel corrosion inhibitor in 15% HCl under dynamic condition: A detailed experimental and surface analysis. *Journal of Molecular Liquids*, 362, 119720. DOI: 10.1016/j.molliq.2022.119720

Cai, Z., Zhang, F., Wei, D., Zhai, B., Wang, X., & Song, Y. (2023). NixCo1-xS2@N-doped carbon composites for supercapacitor electrodes. *Journal of Energy Storage*, 72, 108231. DOI: 10.1016/j.est.2023.108231

Cañas, E., Vicent, M., Orts, M. J., & Sánchez, E. (2017). Bioactive glass coatings by suspension plasma spraying from glycolether-based solvent feedstock. *Surface and Coatings Technology*, 318, 190–197. DOI: 10.1016/j.surfcoat.2016.12.060

Cannio, M., Bellucci, D., Roether, J. A., Boccaccini, D. N., & Cannillo, V. (2021). Bioactive Glass Applications: A Literature Review of Human Clinical Trials. *Materials (Basel)*, 14(18), 5440. DOI: 10.3390/ma14185440 PMID: 34576662

Carville, N. C., Collins, L., Manzo, M., Gallo, K., Lukasz, B. I., McKayed, K. K., Simpson, J. C., & Rodriguez, B. J. (2015). Biocompatibility of ferroelectric lithium niobate and the influence of polarization charge on osteoblast proliferation and function. *Journal of Biomedical Materials Research. Part A*, 103(8), 2540–2548. DOI: 10.1002/jbm.a.35390 PMID: 25504748

Casas, M. E., Larzabal, E., & Matamoros, V. (2022). Exploring the usage of artificial root exudates to enhance the removal of contaminants of emerging concern in slow sand filters: Synthetic vs. real wastewater conditions. *The Science of the Total Environment*, 824, 153978. DOI: 10.1016/j.scitotenv.2022.153978 PMID: 35181359

Chaari, M., Matoussi, A., & Fakhfakh, Z. (2011). Structural and dielectric properties of sintering zinc oxide bulk ceramic. *Materials Sciences and Applications*, 2(7), 764–769. DOI: 10.4236/msa.2011.27105

Chae, I., Jeong, C. K., Ounaies, Z., & Kim, S. H. (2018). Review on Electromechanical Coupling Properties of Biomaterials. *ACS Applied Bio Materials*, 1(4), 936–953. DOI: 10.1021/acsabm.8b00309 PMID: 34996135

Chai, W. S., Liu, L., Sun, X., Li, X., & Lu, Y. Y. (2024). An Overview of Green Propellants and Propulsion System Applications: Merits and Demerits. In *Space technology library* (pp. 3–32). DOI: 10.1007/978-3-031-62574-9_1

Chaitra, T. K., Mohana, K. N. S., & Tandon, H. C. (2015). Thermodynamic, electrochemical and quantum chemical evaluation of some triazole Schiff bases as mild steel corrosion inhibitors in acid media. *Journal of Molecular Liquids*, 211, 1026–1038. DOI: 10.1016/j.molliq.2015.08.031

Champagnon, B., Chemarin, C., Duval, E., & Parc, R. L. (2000). Glass structure and light scattering. *Journal of Non-Crystalline Solids*, 274(1–3), 81–86. DOI: 10.1016/S0022-3093(00)00207-6

Chan, C. C. V., Neufeld, K., Cusworth, D., Gavrilovic, S., & Ngai, T. (2015). Investigation of the effect of grain size, flow rate and diffuser design on the CAWST Biosand Filter performance. *International Journal for Service Learning in Engineering. Humanitarian Engineering and Social Entrepreneurship*, 10(1), 1–23. DOI: 10.24908/ijsle.v10i1.5705

Chang, C., Chang, C., & Tsai, W. (2000). Effects of Burn-off and Activation Temperature on Preparation of Activated Carbon from Corn Cob Agrowaste by CO2 and Steam. *Journal of Colloid and Interface Science*, 232(1), 45–49. DOI: 10.1006/jcis.2000.7171 PMID: 11071731

Chaudhari, V. S., Kushram, P., & Bose, S. (2024). Drug delivery strategies through 3D-printed calcium phosphate. *Trends in biotechnology*, S0167-7799(24)00145-8. https://doi.org/DOI: 10.1016/j.tibtech.2024.05.006

Chen, C., Ji, J., Jiao, X., Wang, A., & Ding, H. (2017). Fabrication and Properties of Lithium Sodium Potassium Niobate Lead-Free Piezoelectric Ceramics. *Journal of Advanced Microscopy Research*, 12(2), 85–88. DOI: 10.1166/jamr.2017.1323

Cheng, X., Wang, H., Wang, S., Jiao, Y., Sang, C., Jiang, S., He, S., Mei, C., Xu, X., Xiao, H., & Han, J. (2024). Hierarchically core-shell structured nanocellulose/carbon nanotube hybrid aerogels for patternable, self-healing and flexible supercapacitors. *Journal of Colloid and Interface Science*, 660, 923–933. DOI: 10.1016/j.jcis.2024.01.160 PMID: 38280285

Chen, J., & Whittingham, M.CHEN. (2006). Hydrothermal synthesis of lithium iron phosphate. *Electrochemistry Communications*, 8(5), 855–858. DOI: 10.1016/j.elecom.2006.03.021

Chen, L., Lu, D., & Zhang, Y. (2022b). Organic compounds as corrosion inhibitors for carbon steel in HCl solution: A comprehensive review. *Materials (Basel)*, 15(6), 2023. DOI: 10.3390/ma15062023 PMID: 35329474

Chen, M., Chen, Y., Lim, Z. J., & Wong, M. W. (2022). Adsorption of imidazolium-based ionic liquids on the Fe(1 0 0) surface for corrosion inhibition: Physisorption or chemisorption? *Journal of Molecular Liquids*, 367, 120489. DOI: 10.1016/j.molliq.2022.120489

Chen, Q., Dwyer, C., Sheng, G., Zhu, C., Li, X., Zheng, C., & Zhu, Y. (2020a). Imaging Beam-Sensitive Materials by Electron Microscopy. *Advanced Materials*, 32(16), 1907619. Advance online publication. DOI: 10.1002/adma.201907619 PMID: 32108394

Chen, S., Tong, X., Huo, Y., Liu, S., Yin, Y., Tan, M. L., Cai, K., & Ji, W. (2024). Piezoelectric Biomaterials Inspired by Nature for Applications in Biomedicine and Nanotechnology. *Advanced Materials*, 36(35), e2406192. DOI: 10.1002/adma.202406192 PMID: 39003609

Chen, W., Liu, X., Chen, Q., Bao, C., Zhao, L., Zhu, Z., & Xu, H. H. K. (2018). Angiogenic and osteogenic regeneration in rats via calcium phosphate scaffold and endothelial cell co-culture with human bone marrow mesenchymal stem cells (MSCs), human umbilical cord MSCs, human induced pluripotent stem cell-derived MSCs and human embryonic stem cell-derived MSCs. *Journal of Tissue Engineering and Regenerative Medicine*, 12(1), 191–203. DOI: 10.1002/term.2395 PMID: 28098961

Chen, W., Zhou, H., Weir, M. D., Tang, M., Bao, C., & Xu, H. H. (2013). Human embryonic stem cell-derived mesenchymal stem cell seeding on calcium phosphate cement-chitosan-RGD scaffold for bone repair. *Tissue Engineering. Part A*, 19(7-8), 915–927. DOI: 10.1089/ten.tea.2012.0172 PMID: 23092172

Chen, Z., Li, P., Anderson, R., Wang, X., Zhang, X., Robison, L., Redfern, L. R., Moribe, S., Islamoglu, T., Gómez-Gualdrón, D. A., Yildirim, T., Stoddart, J. F., & Farha, O. K. (2020). Balancing volumetric and gravimetric uptake in highly porous materials for clean energy. *Science*, 368(6488), 297–303. DOI: 10.1126/science.aaz8881 PMID: 32299950

Chen, Z., Zhao, D., Ma, R., Zhang, X., Rao, J., Yin, Y., Wang, X., & Yi, F. (2021). Flexible temperature sensors based on carbon nanomaterials. *Journal of Materials Chemistry. B*, 9(8), 1941–1964. DOI: 10.1039/D0TB02451A PMID: 33532811

Che, S., Lund, K., Tatsumi, T., Iijima, S., Joo, S. H., Ryoo, R., & Terasaki, O. (2003). Direct Observation of 3D Mesoporous Structure by Scanning Electron Microscopy (SEM): SBA-15 Silica and CMK-5 Carbon. *Angewandte Chemie International Edition*, 42(19), 2182–2185. https://doi.org/https://doi.org/10.1002/anie.200250726. DOI: 10.1002/anie.200250726 PMID: 12761755

Chi, W., Wang, G., Qiu, Z., Li, Q., Xu, Z., Li, Z., Qi, B., Cao, K., Chi, C., Wei, T., & Fan, Z. (2023). Secondary High-Temperature treatment of porous carbons for High-Performance supercapacitors. *Batteries*, 10(1), 5. DOI: 10.3390/batteries10010005

Choi, E. S., Kim, H. C., Muthoka, R. M., Panicker, P. S., Agumba, D. O., & Kim, J. (2021). Aligned cellulose nanofiber composite made with electrospinning of cellulose nanofiber—Polyvinyl alcohol and its vibration energy harvesting. *Composites Science and Technology*, 209, 108795. DOI: 10.1016/j.compscitech.2021.108795

Chorsi, M. T., Curry, E. J., Chorsi, H. T., Das, R., Baroody, J., Purohit, P. K., Ilies, H., & Nguyen, T. D. (2019). Piezoelectric Biomaterials for Sensors and Actuators. *Advanced Materials*, 31(1), e1802084. DOI: 10.1002/adma.201802084 PMID: 30294947

Cioffi, N., Colaianni, L., Ieva, E., Pilolli, R., Ditaranto, N., Angione, M. D., Cotrone, S., Buchholt, K., Spetz, A. L., Sabbatini, L., & Torsi, L. (2011). Electrosynthesis and characterization of gold nanoparticles for electronic capacitance sensing of pollutants. *Electrochimica Acta*, 56(10), 3713–3720. https://doi.org/https://doi.org/10.1016/j.electacta.2010.12.105. DOI: 10.1016/j.electacta.2010.12.105

Civinini, R., Capone, A., Carulli, C., Matassi, F., Nistri, L., & Innocenti, M. (2017). The kinetics of remodeling of a calcium sulfate/calcium phosphate bioceramic. *Journal of Materials Science. Materials in Medicine*, 28(9), 137. DOI: 10.1007/s10856-017-5940-5 PMID: 28785889

Clayden, N. J., Esposito, S., Pernice, P., & Aronne, A. (2001). Solid state 29Si and 31P NMR study of gel derived phosphosilicate glasses. *Journal of Materials Chemistry*, 11(3), 936–943. DOI: 10.1039/b004107f

Comez, L., Masciovecchio, C., Monaco, G., & Fioretto, D. (2012). Progress in Liquid and Glass Physics by Brillouin Scattering Spectroscopy. In *Solid state physics* (pp. 1–77). https://doi.org/DOI: 10.1016/B978-0-12-397028-2.00001-1

Costerton, W., Veeh, R., Shirtliff, M., Pasmore, M., Post, C., & Ehrlich, G. (2003). The application of biofilm science to the study and control of chronic bacterial infections. *The Journal of Clinical Investigation*, 112(10), 1466–1477. DOI: 10.1172/JCI200320365 PMID: 14617746

Courtheoux, L., Gautron, E., Rossignol, S., & Kappenstein, C. (2005). Transformation of platinum supported on silicon-doped alumina during the catalytic decomposition of energetic ionic liquid. *Journal of Catalysis*, 232(1), 10–18. DOI: 10.1016/j.jcat.2005.02.005

Cundy, C. S., & Cox, P. A. (2005). The hydrothermal synthesis of zeolites: Precursors, intermediates and reaction mechanism. *Microporous and Mesoporous Materials*, 82(1–2), 1–78. DOI: 10.1016/j.micromeso.2005.02.016

Curry, E. J., Ke, K., Chorsi, M. T., Wrobel, K. S., Miller, A. N.III, Patel, A., Kim, I., Feng, J., Yue, L., Wu, Q., Kuo, C. L., Lo, K. W., Laurencin, C. T., Ilies, H., Purohit, P. K., & Nguyen, T. D. (2018). Biodegradable Piezoelectric Force Sensor. *Proceedings of the National Academy of Sciences of the United States of America*, 115(5), 909–914. DOI: 10.1073/pnas.1710874115 PMID: 29339509

Cychosz, K. A., & Thommes, M. (2018). Progress in the Physisorption Characterization of Nanoporous Gas Storage Materials. *Engineering (Beijing)*, 4(4), 559–566. https://doi.org/https://doi.org/10.1016/j.eng.2018.06.001. DOI: 10.1016/j.eng.2018.06.001

D'Alessio, L., Ferro, D., Marotta, V., Santagata, A., Teghil, R., & Zaccagnino, M. (2001). Laser ablation and deposition of Bioglass® 45S5 thin films. *Applied Surface Science*, 183(1–2), 10–17. DOI: 10.1016/S0169-4332(01)00466-4

Da Silva, L. P., Kundu, S. C., Reis, R. L., & Correlo, V. M. (2020). Electric Phenomenon: A Disregarded Tool in Tissue Engineering and Regenerative Medicine. *Trends in Biotechnology*, 38(1), 24–49. DOI: 10.1016/j.tibtech.2019.07.002 PMID: 31629549

Dahl-Hansen, R. P., Stange, M. S. S., Sunde, T. O., Ræder, J. H., & Rørvik, P. M. (2024). On the Evolution of Stress and Microstructure in Radio Frequency-Sputtered Lead-Free (Ba, Ca)(Zr, Ti)O3 Thin Films. *Actuators*, 13(3), 115. DOI: 10.3390/act13030115

Da, N., Grassmé, O., Nielsen, K. H., Peters, G., & Wondraczek, L. (2011). Formation and structure of ionic (Na, Zn) sulfophosphate glasses. *Journal of Non-Crystalline Solids*, 357(10), 2202–2206. DOI: 10.1016/j.jnoncrysol.2011.02.037

Darunte, L. A., Terada, Y., Murdock, C. R., Walton, K. S., Sholl, D. S., & Jones, C. W. (2017). Monolith-Supported Amine-Functionalized Mg_2(dobpdc) Adsorbents for CO_2 Capture. *ACS Applied Materials & Interfaces*, 9(20), 17042–17050. DOI: 10.1021/acsami.7b02035 PMID: 28440615

De Oliveira, A. A. R., De Carvalho, B. B., Sander Mansur, H., & De Magalhães Pereira, M. (2014). Synthesis and characterization of bioactive glass particles using an ultrasound-assisted sol–gel process: Engineering the morphology and size of sonogels via a poly(ethylene glycol) dispersing agent. *Materials Letters*, 133, 44–48. DOI: 10.1016/j.matlet.2014.06.092

De Pablos-Martín, A., Muñoz, F., Mather, G. C., Patzig, C., Bhattacharyya, S., Jinschek, J. R., Höche, T., Durán, A., & Pascual, M. J. (2013). KLaF4 nanocrystallisation in oxyfluoride glass-ceramics. *CrystEngComm*, 15(47), 10323. DOI: 10.1039/c3ce41345d

De Sousa Meneses, D., Eckes, M., Del Campo, L., Santos, C. N., Vaills, Y., & Echegut, P. (2013). Investigation of medium range order in silicate glasses by infrared spectroscopy. *Vibrational Spectroscopy*, 65, 50–57. DOI: 10.1016/j.vibspec.2012.11.015

De Sousa Meneses, D., Malki, M., & Echegut, P. (2006). Optical and structural properties of calcium silicate glasses. *Journal of Non-Crystalline Solids*, 352(50–51), 5301–5308. DOI: 10.1016/j.jnoncrysol.2006.08.022

De Sousa Meneses, D., Malki, M., & Echegut, P. (2006a). Structure and lattice dynamics of binary lead silicate glasses investigated by infrared spectroscopy. *Journal of Non-Crystalline Solids*, 352(8), 769–776. DOI: 10.1016/j.jnoncrysol.2006.02.004

de Souza, F. H., Roecker, P. B., Silveira, D. D., Sens, M. L., & Campos, L. C. (2021). Influence of slow sand filter cleaning process type on filter media biomass: Backwashing versus scraping. *Water Research*, 189, 116581. DOI: 10.1016/j.watres.2020.116581 PMID: 33186813

Demazeau, G. (2008). Solvothermal reactions: An original route for the synthesis of novel materials. *Journal of Materials Science*, 43(7), 2104–2114. DOI: 10.1007/s10853-007-2024-9

Deng, X., Ran, S., Han, L., Zhang, H., Ge, S., & Zhang, S. (2017). Foam-gelcasting preparation of high-strength self-reinforced porous mullite ceramics. *Journal of the European Ceramic Society*, 37(13), 4059–4066. DOI: 10.1016/j.jeurceramsoc.2017.05.009

Dent, A. H., Patel, A., Gutleber, J., Tormey, E., Sampath, S., & Herman, H. (2001). High velocity oxy-fuel and plasma deposition of BaTiO3 and (Ba, Sr) TiO3. *Materials Science and Engineering B*, 87(1), 23–30. DOI: 10.1016/S0921-5107(01)00653-5

DeSantis, D. (2014). *Satellite Thruster Propulsion-H2O2 Bipropellant Comparison with Existing Alternatives*. Department of Space Technologies, Institute of Aviation, The Ohio State University.

Deschamps, M., Fayon, F., Hiet, J., Ferru, G., Derieppe, M., Pellerin, N., & Massiot, D. (2008). Spin-counting NMR experiments for the spectral editing of structural motifs in solids. *Physical Chemistry Chemical Physics*, 10(9), 1298. DOI: 10.1039/b716319c PMID: 18292865

Deshmukh, K., Kovářík, T., Křenek, T., Docheva, D., Stich, T., & Pola, J. (2020). Recent advances and future perspectives of sol–gel derived porous bioactive glasses: A review. *RSC Advances*, 10(56), 33782–33835. DOI: 10.1039/D0RA04287K PMID: 35519068

Deters, H., De Camargo, A. S. S., Santos, C. N., Ferrari, C. R., Hernandes, A. C., Ibanez, A., Rinke, M. T., & Eckert, H. (2009). Structural Characterization of Rare-Earth Doped Yttrium Aluminoborate Laser Glasses Using Solid State NMR. *The Journal of Physical Chemistry. C, Nanomaterials and Interfaces*, 113(36), 16216–16225. DOI: 10.1021/jp9032904

Dhiman, P., Goyal, D., Rana, G., Kumar, A., Sharma, G., Linxin, , & Kumar, G. (2024). Recent advances on carbon-based nanomaterials supported single-atom photo-catalysts for waste water remediation. *Journal of Nanostructure in Chemistry*, 14(1), 21–52. DOI: 10.1007/s40097-022-00511-3

Di Martino, A., Sittinger, M., & Risbud, M. V. (2005). Chitosan: A versatile biopolymer for orthopaedic tissue-engineering. *Biomaterials*, 26(30), 5983–5990. DOI: 10.1016/j.biomaterials.2005.03.016 PMID: 15894370

Dias, I. J., Pádua, A. S., Pires, E. A., Borges, J. P., Silva, J. C., & Lança, M. C. (2023). Hydroxyapatite-Barium Titanate Biocoatings Using Room Temperature Coblasting. *Crystals*, 13(4), 579. DOI: 10.3390/cryst13040579

Dias, J. M., Alvim-Ferraz, M. C., Almeida, M. F., Rivera-Utrilla, J., & Sánchez-Polo, M. (2007). Waste materials for activated carbon preparation and its use in aqueous-phase treatment: A review. *Journal of Environmental Management*, 85(4), 833–846. DOI: 10.1016/j.jenvman.2007.07.031 PMID: 17884280

Ding, J., Tang, B., Li, M., Feng, X., Fu, F., Bin, L., Huang, S., Su, W., Li, D., & Zheng, L. (2016). Difference in the characteristics of the rust layers on carbon steel and their corrosion behavior in an acidic medium: Limiting factors for cleaner pickling. *Journal of Cleaner Production*, 142, 2166–2176. DOI: 10.1016/j.jclepro.2016.11.066

Domínguez-Domínguez, S., Arias-Pardilla, J., Berenguer-Murcia, Á., Morallón, E., & Cazorla-Amorós, D. (2008). Electrochemical deposition of platinum nanoparticles on different carbon supports and conducting polymers. *Journal of Applied Electrochemistry*, 38(2), 259–268. DOI: 10.1007/s10800-007-9435-9

Donald, I. W., Mallinson, P. M., Metcalfe, B. L., Gerrard, L. A., & Fernie, J. A. (2011b). Recent developments in the preparation, characterization and applications of glass- and glass–ceramic-to-metal seals and coatings. *Journal of Materials Science*, 46(7), 1975–2000. DOI: 10.1007/s10853-010-5095-y

Dragic, P., Kucera, C., Furtick, J., Guerrier, J., Hawkins, T., & Ballato, J. (2013). Brillouin spectroscopy of a novel baria-doped silica glass optical fiber. *Optics Express*, 21(9), 10924. DOI: 10.1364/OE.21.010924 PMID: 23669949

Drago, L., Toscano, M., & Bottagisio, M. (2018). Recent Evidence on Bioactive Glass Antimicrobial and Antibiofilm Activity: A Mini-Review. *Materials (Basel)*, 11(2), 326. DOI: 10.3390/ma11020326 PMID: 29495292

Duarte, A., Caridade, S., Mano, J. F., & Reis, R. L. (2009). Processing of novel bioactive polymeric matrixes for tissue engineering using supercritical fluid technology. *Materials Science and Engineering C*, 29(7), 2110–2115. DOI: 10.1016/j.msec.2009.04.012

Dubey, A. K., & Basu, B. (2014). Pulsed electrical stimulation and surface charge induced cell growth on multistage spark plasma sintered hydroxyapatite-barium titanate piezobiocomposite. *Journal of the American Ceramic Society*, 97(2), 481–489. DOI: 10.1111/jace.12647

Dubey, A. K., Basu, B., Balani, K., Guo, R., & Bhalla, A. S. (2011). Multifunctionality of perovskites BaTiO3 and CaTiO3 in a composite with hydroxyapatite as orthopedic implant materials. *Integrated Ferroelectrics*, 131(1), 119–126. DOI: 10.1080/10584587.2011.616425

Dubey, P., Shrivastav, V., Sundriyal, S., & Maheshwari, P. H. (2024). Sustainable Nanoporous Metal–Organic Framework/Conducting Polymer Composites for Supercapacitor Applications. *ACS Applied Nano Materials*, 7(16), 18554–18565. DOI: 10.1021/acsanm.4c01697

Du, N., Zhang, H., Chen, B. D., Ma, X. Y., Liu, Z. H., Wu, J. B., & Yang, D. R. (2007). Porous Indium Oxide Nanotubes: Layer-by-Layer Assembly on Carbon-Nanotube Templates and Application for Room-Temperature NH 3 Gas Sensors. *Advanced Materials*, 19(12), 1641–1645. DOI: 10.1002/adma.200602128

Dupère, I. D. J., Lu, T. J., & Dowling, A. P. (2005). *Acoustic Properties*. 381–399. DOI: 10.1002/3527606696.ch4e

Du, Z., Yao, D., Xia, Y., Zuo, K., Yin, J., Liang, H., & Zeng, Y. (2019). The high porosity silicon nitride foams prepared by the direct foaming method. *Ceramics International*, 45(2), 2124–2130. DOI: 10.1016/j.ceramint.2018.10.118

Ebrahimi, M., Manafi, S., & Sharifianjazi, F. (2023). The effect of Ag2O and MgO dopants on the bioactivity, biocompatibility, and antibacterial properties of 58S bioactive glass synthesized by the sol-gel method. *Journal of Non-Crystalline Solids*, 606, 122189. DOI: 10.1016/j.jnoncrysol.2023.122189

Elahpour, N., Niesner, I., Bossard, C., Abdellaoui, N., Montouillout, V., Fayon, F., Taviot-Guého, C., Frankenbach, T., Crispin, A., Khosravani, P., Holzapfel, B. M., Jallot, E., Mayer-Wagner, S., & Lao, J. (2023). Zinc-Doped Bioactive Glass/Polycaprolactone Hybrid Scaffolds Manufactured by Direct and Indirect 3D Printing Methods for Bone Regeneration. *Cells*, 12(13), 1759. DOI: 10.3390/cells12131759 PMID: 37443794

El-Aziz, E.-S., Ibrahem, E.-D., & Awad, S. (2019). Esomeprazole Magnesium Trihydrate drug as a potential non-toxic corrosion inhibitor for mild steel in acidic media. *Zast. Mater.*, 60(4), 245–258. DOI: 10.5937/zasmat1903245E

El-Egili, K. (2003). Infrared studies of Na2O–B2O3–SiO2 and Al2O3–Na2O–B2O3–SiO2 glasses. *Physica B, Condensed Matter*, 325, 340–348. DOI: 10.1016/S0921-4526(02)01547-8

Elhaya, N., Fadeli, S. E., Erraji, E., & Barkouch, Y. (2024). Removal process of cadmium from unsafe water by slow sand filtration: Study of water feed flow rate effect. *Euro-Mediterranean Journal for Environmental Integration*. Advance online publication. DOI: 10.1007/s41207-024-00576-2

Elliott, M. A., Stauber, C. E., Koksal, F., DiGiano, F. A., & Sobsey, M. D. (2008). Reductions of E. coli, echovirus type 12 and bacteriophages in an intermittently operated household-scale slow sand filter. *Water Research*, 42(10–11), 2662–2670. DOI: 10.1016/j.watres.2008.01.016 PMID: 18281076

Elshakre, M. E., Alalawy, H. H., Awad, M. I., & El-Anadouli, B. E. (2017). On the role of the electronic states of corrosion inhibitors: Quantum chemical-electrochemical correlation study on urea derivatives. *Corrosion Science*, 124, 121–130. DOI: 10.1016/j.corsci.2017.05.015

Ercan, K. E., Yurtseven, M. A., & Yilmaz, C. (2024). Development of Mono and Bipropellant Systems for Green Propulsion Applications. In *Space technology library* (pp. 249–280). DOI: 10.1007/978-3-031-62574-9_9

Espinoza-Vázquez, A., Negrón-Silva, G., Angeles-Beltrán, D., Herrera-Hernández, H., Romero-Romo, M., & Palomar-Pardavé, M. (2014). electrochemical impedance spectroscopy Evaluation of Pantoprazole as corrosion inhibitor for mild steel immersed in HCl 1 M. effect of [Pantoprazole], hydrodynamic conditions, temperature and immersion times. *International Journal of Electrochemical Science*, 9(2), 493–509. DOI: 10.1016/S1452-3981(23)07734-9

Esteve-Sánchez, Y., Hernández-Montoto, A., Tormo-Mas, M. Á., Pemán, J., Calabuig, E., Gómez, M. D., Marcos, M. D., Martínez-Máñez, R., Aznar, E., & Climent, E. (2024). SARS-CoV-2 N protein IgG antibody detection employing nanoporous anodized alumina: A rapid and selective alternative for identifying naturally infected individuals in populations vaccinated with spike protein (S)-based vaccines. *Sensors and Actuators. B, Chemical*, 419, 136378. https://doi.org/https://doi.org/10.1016/j.snb.2024.136378. DOI: 10.1016/j.snb.2024.136378

Esteves, P., Courtheoux, L., Pirault-Roy, L., Rossignol, S., Kappenstein, C., & Pillet, N. (2004). Design and Development of a Dynamic Reactor with Online Analysis for the Catalytic Decomposition of Monopropellants. *40th AIAA/ASME/SAE/ASEE Joint Propulsion Conference and Exhibit*. https://doi.org/DOI: 10.2514/6.2004-3835

Everett, D. H. (1972). Manual of Symbols and Terminology for Physicochemical Quantities and Units, Appendix II: Definitions, Terminology and Symbols in Colloid and Surface Chemistry. *Pure and Applied Chemistry*, 31(4), 577–638. DOI: 10.1351/pac197231040577

F Florian, P., Sadiki, N., Massiot, D., & Coutures, J. (2007). 27Al NMR Study of the Structure of Lanthanum- and Yttrium-Based Aluminosilicate Glasses and Melts. The Journal of Physical Chemistry B, 111(33), 9747–9757. https://doi.org/DOI: 10.1021/jp072061q

Fallavena, T., Antonow, M., & Gonçalves, R. S. (2006). Caffeine as non-toxic corrosion inhibitor for copper in aqueous solutions of potassium nitrate. *Applied Surface Science*, 253(2), 566–571. DOI: 10.1016/j.apsusc.2005.12.114

Farag, M. M. (2023). Recent trends on biomaterials for tissue regeneration applications [review]. *Journal of Materials Science*, 58(2), 527–558. DOI: 10.1007/s10853-022-08102-x

Farelas, F., & Ramirez, A. (2010). Carbon dioxide corrosion inhibition of carbon steels through bis-imidazoline and imidazoline compounds studied by electrochemical impedance spectroscopy. *International Journal of Electrochemical Science*, 5(6), 797–814. DOI: 10.1016/S1452-3981(23)15324-7

Farhat, K., Cong, W., Batonneau, Y., & Kappenstein, C. (2009). Improvement of Catalytic Decomposition of Ammonium Nitrate with New Bimetallic Catalysts. https://doi.org/DOI: 10.2514/6.2009-4963

Favacho, V. S. S., Melo, D. M. A., Costa, J. E. L., Silva, Y. K. R. O., Braga, R. M., & Medeiros, R. L. B. A. (2024). Perovskites synthesized by soft template-assisted hydrothermal method: A bibliometric analysis and new insights. *International Journal of Hydrogen Energy*, 78, 1391–1428. DOI: 10.1016/j.ijhydene.2024.06.326

Fawzy, A., Zaafarany, I., Ali, H., & Abdallah, M. (2018). New synthesized amino acids-based surfactants as efficient inhibitors for corrosion of mild steel in HCl medium: Kinetics and Thermodynamic Approach. *International Journal of Electrochemical Science*, 13(5), 4575–4600. DOI: 10.20964/2018.05.01

Fayomi, O. S. I., Olusanyan, D., Ademuyiwa, F. T., & Olarewaju, G. (2021). Progresses on mild steel protection toward surface service performance in structural industrial: An Overview. *IOP Conference Series. Materials Science and Engineering*, 1036(1), 012079. DOI: 10.1088/1757-899X/1036/1/012079

Feinle, A., Elsaesser, M. S., & Hüsing, N. (2016). Sol–gel synthesis of monolithic materials with hierarchical porosity. *Chemical Society Reviews*, 45(12), 3377–3399. DOI: 10.1039/C5CS00710K PMID: 26563577

Fend, T., Trimis, D., Pitz-Paal, R., Hoffschmidt, B., & Reutter, O. (2005). *Thermal Properties*. 342–360. DOI: 10.1002/3527606696.ch4c

Feng, C., Zhang, K., He, R., Ding, G., Xia, M., Jin, X., & Xie, C. (2020). Additive manufacturing of hydroxyapatite bioceramic scaffolds: Dispersion, digital light processing, sintering, mechanical properties, and biocompatibility. *Journal of Advanced Ceramics*, 9(3), 360–373. DOI: 10.1007/s40145-020-0375-8

Feng, S., Li, J., Jiang, X., Li, X., Pan, Y., Zhao, L., Boccaccini, A. R., Zheng, K., Yang, L., & Wei, J. (2016). Influences of mesoporous magnesium silicate on the hydrophilicity, degradability, mineralization and primary cell response to a wheat protein based biocomposite. *Journal of Materials Chemistry. B*, 4(39), 6428–6436. DOI: 10.1039/C6TB01449F PMID: 32263451

Feng, Z., Hu, F., Lv, L., Gao, L., & Lu, H. (2021). Preparation of ultra-high mechanical strength wear-resistant carbon fiber textiles with a PVA/PEG coating. *RSC Advances*, 11(41), 25530–25541. DOI: 10.1039/D1RA03983K PMID: 35478898

Férey, G. (2008). Hybrid porous solids: Past, present, future. *Chemical Society Reviews*, 37(1), 191–214. DOI: 10.1039/B618320B PMID: 18197340

Fernandes, J. S., & Montemor, F. (2014). Corrosion. In Springer eBooks (pp. 679–716). DOI: 10.1007/978-3-319-08236-3_15

Ferreira, M. A., Alfenas, A. C., Binoti, D. H. B., Machado, P. S., & Mounteer, A. H. (2012). Slow sand filtration eradicates eucalypt clonal nursery plant pathogens from recycled irrigation water in Brazil. *Tropical Plant Pathology*, 37(5), 319–325. DOI: 10.1590/S1982-56762012000500003

Filipowska, J., Tomaszewski, K. A., Niedźwiedzki, Ł., Walocha, J. A., & Niedźwiedzki, T. (2017). The role of vasculature in bone development, regeneration and proper systemic functioning. *Angiogenesis*, 20(3), 291–302. DOI: 10.1007/s10456-017-9541-1 PMID: 28194536

Fiori-Bimbi, M. V., Alvarez, P. E., Vaca, H., & Gervasi, C. A. (2014). Corrosion inhibition of mild steel in HCl solution by pectin. *Corrosion Science*, 92, 192–199. DOI: 10.1016/j.corsci.2014.12.002

Fisher, J. G., Thuan, U. T., Farooq, M. U., Chandrasekaran, G., Jung, Y. D., Hwang, E. C., Lee, J. J., & Lakshmanan, V. K. (2018). Prostate Cancer Cell-Specific Cytotoxicity of Sub-Micron Potassium Niobate Powder. *Journal of Nanoscience and Nanotechnology*, 18(5), 3141–3147. DOI: 10.1166/jnn.2018.14666 PMID: 29442813

Fitriani, N., Kusuma, M. N., Wirjodirdjo, B., Hadi, W., Hermana, J., Ni'matuzahroh, , Kurniawan, S. B., Abdullah, S. R. S., & Mohamed, R. M. S. R. (2020). Performance of geotextile-based slow sand filter media in removing total coli for drinking water treatment using system dynamics modelling. *Heliyon*, 6(9), e04967. DOI: 10.1016/j.heliyon.2020.e04967 PMID: 33015386

Fiume, E., Barberi, J., Verné, E., & Baino, F. (2018). Bioactive Glasses: From Parent 45S5 Composition to Scaffold-Assisted Tissue-Healing Therapies. *Journal of Functional Biomaterials*, 9(1), 24. DOI: 10.3390/jfb9010024 PMID: 29547544

Fiume, E., Migneco, C., Verné, E., & Baino, F. (2020). Comparison between Bioactive Sol-Gel and Melt-Derived Glasses/Glass-Ceramics Based on the Multicomponent SiO2–P2O5–CaO–MgO–Na2O–K2O System. *Materials (Basel)*, 13(3), 540. DOI: 10.3390/ma13030540 PMID: 31979302

Foo, K. Y., & Hameed, B. H. (2010). Insights into the modeling of adsorption isotherm systems. *Chemical Engineering Journal*, 156(1), 2–10. DOI: 10.1016/j.cej.2009.09.013

Foroutan, F., Kyffin, B. A., Abrahams, I., Corrias, A., Gupta, P., Velliou, E., Knowles, J. C., & Carta, D. (2020). Mesoporous Phosphate-Based Glasses Prepared via Sol–Gel. *ACS Biomaterials Science & Engineering*, 6(3), 1428–1437. DOI: 10.1021/acsbiomaterials.9b01896 PMID: 33455383

Fotovvati, B., Namdari, N., & Dehghanghadikolaei, A. (2019). On coating techniques for surface protection: A review. *Journal of Manufacturing and Materials processing*, 3(1), 28. https://doi.org/DOI: 10.3390/jmmp3010028

França, C. G., Plaza, T., Naveas, N., Andrade Santana, M. H., Manso-Silván, M., Recio, G., & Hernandez-Montelongo, J. (2021). Nanoporous silicon microparticles embedded into oxidized hyaluronic acid/ adipic acid dihydrazide hydrogel for enhanced controlled drug delivery. *Microporous and Mesoporous Materials*, 310, 110634. https://doi.org/https://doi.org/10.1016/j.micromeso.2020.110634. DOI: 10.1016/j.micromeso.2020.110634

Frank, M. B., Naleway, S. E., Haroush, T., Liu, C., Siu, S. H., Ng, J., Torres, I., Ismail, A., Karandikar, K., Porter, M. M., Graeve, O. A., & McKittrick, J. (2017). Stiff, porous scaffolds from magnetized alumina particles aligned by magnetic freeze casting. *Materials Science and Engineering C*, 77, 484–492. DOI: 10.1016/j.msec.2017.03.246 PMID: 28532056

Freer, R., & Dennis, P. F. (1986). Kinetics and mass transport in silicate and oxide systems : proceedings of a meeting, held in London in September, 1984. In Trans Tech Publications eBooks. http://ci.nii.ac.jp/ncid/BA0087969X

Freitas, B. L. S., Terin, U. C., Fava, N. M. N., Maciel, P. M. F., Garcia, L. A. T., Medeiros, R. C., Oliveira, M., Fernandez-Ibañez, P., Byrne, J. A., & Sabogal-Paz, L. P. (2022). A critical overview of household slow sand filters for water treatment. *Water Research*, 208, 117870. DOI: 10.1016/j.watres.2021.117870 PMID: 34823084

Fujibayashi, S. (2003). A comparative study between in vivo bone ingrowth and in vitro apatite formation on Na2O–CaO–SiO2 glasses. *Biomaterials*, 24(8), 1349–1356. DOI: 10.1016/S0142-9612(02)00511-2 PMID: 12527276

Fu, K., Yang, Z., Sun, H., Chen, X., Li, S., Ma, W., & Chen, R. (2023). Multiple modification on nanoporous silicon derived from photovoltaic silicon cutting waste for extraction of PbII in industrial effluents. *Materials Today. Communications*, 35, 105776. https://doi.org/https://doi.org/10.1016/j.mtcomm.2023.105776. DOI: 10.1016/j.mtcomm.2023.105776

Fu, L., Engqvist, H., & Xia, W. (2020). Glass-Ceramics in Dentistry: A Review. *Materials (Basel)*, 13(5), 1049. DOI: 10.3390/ma13051049 PMID: 32110874

Furtos, G., Cosma, V., Prejmerean, C., Moldovan, M., Brie, M., Colceriu, A., Vezsenyi, L., Silaghi-Dumitrescu, L., & Sirbu, C. (2005). Fluoride release from dental resin composites. *Materials Science and Engineering C*, 25(2), 231–236. DOI: 10.1016/j.msec.2005.01.016

Furukawa, H., Ko, N., Go, Y. B., Aratani, N., Choi, S. B., Choi, E., Yazaydin, A. Ö., Snurr, R. Q., O'Keeffe, M., Kim, J., & Yaghi, O. M. (2010). Ultrahigh Porosity in Metal-Organic Frameworks. *Science*, 329(5990), 424–428. DOI: 10.1126/science.1192160 PMID: 20595583

Gadipelli, S., Guo, J., Li, Z., Howard, C. A., Liang, Y., Zhang, H., Shearing, P. R., & Brett, D. J. L. (2023). Understanding and Optimizing Capacitance Performance in Reduced Graphene-Oxide Based Supercapacitors. *Small Methods*, 7(6), 2201557. Advance online publication. DOI: 10.1002/smtd.202201557 PMID: 36895068

Gaikwad, P., Tiwari, N., Kamat, R., Mane, S. M., & Kulkarni, S. B. (2024). A comprehensive review on the progress of transition metal oxides materials as a supercapacitor electrode. *Materials Science and Engineering B*, 307, 117544. DOI: 10.1016/j.mseb.2024.117544

Gamboa, B., Bhalla, A., & Guo, R. (2020). Assessment of PZT (Soft/Hard) Composites for Energy Harvesting. *Ferroelectrics*, 555(1), 118–123. DOI: 10.1080/00150193.2019.1691389

Gao, P., Yuan, P., Wang, S., Shi, Q., Zhang, C., Shi, G., Xing, Y., & Shen, B. (2024). Preparation and comparison of polyaniline composites with lotus leaf-derived carbon and lotus petiole-derived carbon for supercapacitor applications. *Electrochimica Acta*, 486, 144112. DOI: 10.1016/j.electacta.2024.144112

Garrido, B., Dosta, S., & Cano, I. G. (2022). Bioactive glass coatings obtained by thermal spray: Current status and future challenges. *Boletín de la Sociedad Española de Cerámica y Vidrio*, 61(5), 516–530. DOI: 10.1016/j.bsecv.2021.04.001

Gasnier, E., Bardez-Giboire, I., Montouillout, V., Pellerin, N., Allix, M., Massoni, N., Ory, S., Cabie, M., Poissonnet, S., & Massiot, D. (2014). Homogeneity of peraluminous SiO 2 –B 2 O 3 –Al 2 O 3 –Na 2 O–CaO–Nd 2 O 3 glasses: Effect of neodymium content. *Journal of Non-Crystalline Solids*, 405, 55–62. DOI: 10.1016/j.jnoncrysol.2014.08.032

Gatica, J. M., Gómez, D. M., Harti, S., & Vidal, H. (2013). Clay honeycomb monoliths for water purification: Modulating methylene blue adsorption through controlled activation via natural coal templating. *Applied Surface Science*, 277, 242–248. DOI: 10.1016/j.apsusc.2013.04.034

Gauckler, L. J., & Waeber, M. M. (1985) in Light Metals 1985, Proc. 114th Ann. Meet. Metal. Soc. AIME, 1261–1283.

Gaur, U. P., & Kamari, E. (2024). Applications of Thermal Spray Coatings: A Review. *Journal of Thermal Spray and Engineering*, 4, 106–114. DOI: 10.52687/2582-1474/405

George, A. M., & Stebbins, J. F. (1996). Dynamics of Na in sodium aluminosilicate glasses and liquids. *Physics and Chemistry of Minerals*, 23(8). Advance online publication. DOI: 10.1007/BF00242002

Ghafoor, S., Nadeem, N., Zahid, M., & Zubair, U. (2024). Freestanding carbon-based hybrid anodes for flexible supercapacitors: Part I—An inclusive outlook on current collectors and configurations. *Wiley Interdisciplinary Reviews. Energy and Environment*, 13(2), e511. Advance online publication. DOI: 10.1002/wene.511

Ghajeri, F., Topalian, Z., Tasca, A., Jafri, S. H. M., Leifer, K., Norberg, P., & Sjöström, C. (2018). Case study of a green nanoporous material from synthesis to commercialisation: Quartzene®. *Current Opinion in Green and Sustainable Chemistry*, 12, 101–109. https://doi.org/https://doi.org/10.1016/j.cogsc.2018.07.003. DOI: 10.1016/j.cogsc.2018.07.003

Ghebremichael, K., Wasala, L. D., Kennedy, M., & Graham, N. J. D. (2012). Comparative treatment performance and hydraulic characteristics of pumice and sand biofilters for point-of-use water treatment. *Journal of Water Supply: Research & Technology - Aqua*, 61(4), 201–209. DOI: 10.2166/aqua.2012.100

Ghosh, A., Ray, S. S., Orasugh, J. T., & Chattopadhyay, D. (2024). Collagen-Based Hybrid Piezoelectric Material. *Hybrid Materials for Piezoelectric Energy Harvesting and Conversion*, 283-299. https://doi.org/DOI: 10.1002/9781394150373.ch11

Ghoshal, S. K., & Tewari, H. S. (2010). Photonic applications of Silicon nanostructures. *Material Science Research India*, 7(2), 381–388. DOI: 10.13005/msri/070207

Gianella, S., Gaia, D., & Ortona, A. (2012). High temperature applications of SIðSIC Cellular ceramics. *Advanced Engineering Materials*, 14(12), 1074–1081. DOI: 10.1002/adem.201200012

Gibson, L. J., & Ashby, M. F. (1988). *Cellular Solids: Structure and Properties*. Pergamon.

Gidstedt, S., Betsholtz, A., Falås, P., Cimbritz, M., Davidsson, Å., Micolucci, F., & Svahn, O. (2021). A comparison of adsorption of organic micropollutants onto activated carbon following chemically enhanced primary treatment with microsieving, direct membrane filtration and tertiary treatment of municipal wastewater. *The Science of the Total Environment*, 811, 152225. DOI: 10.1016/j.scitotenv.2021.152225 PMID: 34921873

Gimbel, R., Graham, N., & Collins, M. R. (2015). Recent progress in slow sand and alternative biofiltration processes. *Water Intelligence Online*, 5(0), 9781780402451. DOI: 10.2166/9781780402451

Global assessment of soil pollution: Report |Policy Support and Governance| Food and Agriculture Organization of the United Nations. (n.d.). https://www.fao.org/policy-support/tools-and-publications/resources-details/en/c/1410722/

Goel, C., Mohan, S., & Dinesha, P. (2021). CO2 capture by adsorption on biomass-derived activated char: A review. *The Science of the Total Environment*, 798, 149296. DOI: 10.1016/j.scitotenv.2021.149296 PMID: 34325142

Goetze, J., Yarulina, I., Gascon, J., Kapteijn, F., & Weckhuysen, B. M. (2018). Revealing Lattice Expansion of Small-Pore Zeolite Catalysts during the Methanol-to-Olefins Process Using Combined Operando X-ray Diffraction and UV–vis Spectroscopy. *ACS Catalysis*, 8(3), 2060–2070. DOI: 10.1021/acscatal.7b04129 PMID: 29527401

Gohardani, A. S., Stanojev, J., Demairé, A., Anflo, K., Persson, M., Wingborg, N., & Nilsson, C. (2014b). Green space propulsion: Opportunities and prospects. *Progress in Aerospace Sciences*, 71, 128–149. DOI: 10.1016/j.paerosci.2014.08.001

Gomes, J., Nunes, J. S., Sencadas, V., & Lanceros-Méndez, S. (2010). Influence of the β-phase content and degree of crystallinity on the piezo-and ferroelectric properties of poly(vinylidene fluoride). *Smart Materials and Structures*, 19(6), 1–7. DOI: 10.1088/0964-1726/19/6/065010

Gómez Batres, R., Guzmán Escobedo, Z. S., Gutiérrez, K. C., Leal Berumen, I., Hurtado Macias, A., Pérez, G. H., & Orozco Carmona, V. M. (2021). Impact evaluation of high energy ball milling homogenization process in the phase distribution of hydroxyapatite-barium titanate plasma spray biocoating. *Coatings*, 11(6), 728. DOI: 10.3390/coatings11060728

Gomez-Serrano, V., Pastor-Villegas, J., Perez-Florindo, A., Duran-Valle, C., & Valenzuela-Calahorro, C. (1996). FT-IR study of rockrose and of char and activated carbon. *Journal of Analytical and Applied Pyrolysis*, 36(1), 71–80. DOI: 10.1016/0165-2370(95)00921-3

Gomis-Berenguer, A., Velasco, L. F., Velo-Gala, I., & Ania, C. O. (2017). Photochemistry of nanoporous carbons: Perspectives in energy conversion and environmental remediation. *Journal of Colloid and Interface Science*, 490, 879–901. https://doi.org/https://doi.org/10.1016/j.jcis.2016.11.046. DOI: 10.1016/j.jcis.2016.11.046 PMID: 27914582

Goyal, M., Kumar, S., Bahadur, I., Verma, C., Ebenso, E.E. (2018). Organic corrosion inhibitors for industrial cleaning of ferrous and non-ferrous metals in acidic solutions: a review. J. Mol. Liq. 256, 565–573. https://doi.org/. molliq.2018.02.045.DOI: 10.1016/j

Greaves, G., Smith, W., Giulotto, E., & Pantos, E. (1997). Local structure, microstructure and glass properties. *Journal of Non-Crystalline Solids*, 222(1-2), 13–24. DOI: 10.1016/S0022-3093(97)00420-1

Greene, K., Pomeroy, M., Hampshire, S., & Hill, R. (2003). Effect of composition on the properties of glasses in the K2O–BaO–MgO–SiO2–Al2O3–B2O3–MgF2 system. *Journal of Non-Crystalline Solids*, 325(1–3), 193–205. DOI: 10.1016/S0022-3093(03)00337-5

Greenspan, D., Zhong, J., & LaTorre, G. (1994). Effect of Surface Area to Volume Ratio on In Vitro Surface Reactions of Bioactive Glass Particulates. In Elsevier eBooks (pp. 55–60). DOI: 10.1016/B978-0-08-042144-5.50012-1

Grimsditch, M., & Ramdas, A. (1974). Brillouin scattering in diamond. *Physics Letters. [Part A]*, 48(1), 37–38. DOI: 10.1016/0375-9601(74)90216-3

Gros, M., Mas-Pla, J., Sànchez-Melsió, A., Čelić, M., Castaño, M., Rodríguez-Mozaz, S., & Petrović, M. (2023). Antibiotics, antibiotic resistance and associated risk in natural springs from an agroecosystem environment. *The Science of the Total Environment*, 857, 159202.

Gruber, M. F., Schulte, L., & Ndoni, S. (2013). Nanoporous materials modified with biodegradable polymers as models for drug delivery applications. *Journal of Colloid and Interface Science*, 395, 58–63. https://doi.org/https://doi.org/10.1016/j.jcis.2012.12.052. DOI: 10.1016/j.jcis.2012.12.052 PMID: 23369801

Gruene, T., Holstein, J. J., Clever, G. H., & Keppler, B. (2021). Establishing electron diffraction in chemical crystallography. *Nature Reviews. Chemistry*, 5(9), 660–668. DOI: 10.1038/s41570-021-00302-4 PMID: 37118416

Gugulothu, R., Macharla, A. K., Chatragadda, K., & Vargeese, A. A. (2020). Catalytic decomposition mechanism of aqueous ammonium dinitramide solution elucidated by thermal and spectroscopic methods. *Thermochimica Acta*, 686, 178544. DOI: 10.1016/j.tca.2020.178544

Guo, K. W. (2024). Nanoporous silicon materials for solar energy by electrochemical approach. In Kulkarni, N. V., & Kharissov, B. I. (Eds.), *Handbook of Emerging Materials for Sustainable Energy* (pp. 119–128). Elsevier., https://doi.org/https://doi.org/10.1016/B978-0-323-96125-7.00028-9 DOI: 10.1016/B978-0-323-96125-7.00028-9

Guo, K., Chen, S., Tan, C. K. I., Sharifzadeh Mirshekarloo, M., Yao, K., & Tay, F. E. H. (2017). Bismuth sodium titanate lead-free piezoelectric coatings by thermal spray process. *Journal of the American Ceramic Society*, 100(8), 3385–3392. DOI: 10.1111/jace.14882

Guo, L., Liu, S., & Shan, Z. (2023b). The influence of Mo sources, concentrations and the content of Na2O on the tendency of devitrification of modified R7T7 borosilicate glasses. *Journal of Non-Crystalline Solids*, 617, 122501. DOI: 10.1016/j.jnoncrysol.2023.122501

Guo, S., Wang, L., & Wang, E. (2007). Templateless, surfactantless, simple electrochemical route to rapid synthesis of diameter-controlled 3D flowerlike gold microstructure with "clean" surface. *Chemical Communications*, 30(30), 3163. DOI: 10.1039/b705630c PMID: 17653375

Gupta, S. K., & Mao, Y. (2021). A review on molten salt synthesis of metal oxide nanomaterials: Status, opportunity, and challenge. *Progress in Materials Science*, 117, 100734. https://doi.org/https://doi.org/10.1016/j.pmatsci.2020.100734. DOI: 10.1016/j.pmatsci.2020.100734

Gupta, S. K., Mehta, R. K., & Yadav, M. (2022). Schiff bases as corrosion inhibitorson mild steel in acidic medium: Gravimetric, electrochemical, surface morphological and computational studies. *Journal of Molecular Liquids*, 368, 120747. DOI: 10.1016/j.molliq.2022.120747

Gupta, S., Majumdar, S., & Krishnamurthy, S. (2021). Bioactive glass: A multifunctional delivery system. *Journal of Controlled Release*, 335, 481–497. DOI: 10.1016/j.jconrel.2021.05.043 PMID: 34087250

Gupta, T., Samriti, , Cho, J., & Prakash, J. (2021). Hydrothermal synthesis of TiO2 nanorods: Formation chemistry, growth mechanism, and tailoring of surface properties for photocatalytic activities. *Materials Today. Chemistry*, 20, 100428. DOI: 10.1016/j.mtchem.2021.100428

Gupta, V., & Suhas, N. (2009). Application of low-cost adsorbents for dye removal – A review. *Journal of Environmental Management*, 90(8), 2313–2342. DOI: 10.1016/j.jenvman.2008.11.017 PMID: 19264388

Gutiérrez-Pérez, A. I., Ayala-Ayala, M. T., Mora-García, A. G., Moreno-Murguía, B., Ruiz-Luna, H., & Muñoz-Saldaña, J. (2021). Visible-light photoactive thermally sprayed coatings deposited from spray-dried (Na0. 5Bi0. 5)TiO3 microspheres. *Surface and Coatings Technology*, 427, 127851. DOI: 10.1016/j.surfcoat.2021.127851

Hagiwara, H., & Green, D. J. (1987). Elastic behavior of Open-Cell alumina. *Journal of the American Ceramic Society*, 70(11), 811–815. DOI: 10.1111/j.1151-2916.1987.tb05632.x

Haig, S.-J., Schirmer, M., D'Amore, R., Gibbs, J., Davies, R. L., Collins, G., & Quince, C. (2015). Stable-isotope probing and metagenomics reveal predation by protozoa drives E. coli removal in slow sand filters. *The ISME Journal*, 9(4), 797–808. DOI: 10.1038/ismej.2014.175 PMID: 25279786

Hampshire, S. (1992). Oxynitride Glasses and Glass Ceramics. MRS Proceedings, 287. DOI: 10.1557/PROC-287-93

Hampshire, S., Bachar, A., Albert-Mercier, C., Tricoteaux, A., Leriche, A., & Follet-Houttemane, C. (2019). Oxynitride Glasses for Potential Biomedical Usage. In Ceramic engineering and science proceedings (pp. 33–46). DOI: 10.1002/9781119543381.ch4

Hampshire, S. (2008). Oxynitride glasses. *Journal of the European Ceramic Society*, 28(7), 1475–1483. DOI: 10.1016/j.jeurceramsoc.2007.12.021

Hampshire, S., & Pomeroy, M. J. (2008). Oxynitride Glasses. *International Journal of Applied Ceramic Technology*, 5(2), 155–163. DOI: 10.1111/j.1744-7402.2008.02205.x

Hanifi, A. R., Genson, A., Pomeroy, M. J., & Hampshire, S. (2007). An Introduction to the Glass Formation and Properties of Ca-Si-Al-O-N-F Glasses. *Materials Science Forum*, 554, 17–23. . DOI: 10.4028/www.scientific.net/MSF.554.17

Hanifi, A. R., Genson, A., Pomeroy, M. J., & Hampshire, S. (2011). Independent but Additive Effects of Fluorine and Nitrogen Substitution on Properties of a Calcium Aluminosilicate Glass. *Journal of the American Ceramic Society*, 95(2), 600–606. DOI: 10.1111/j.1551-2916.2011.05001.x

Hanifi, A. R., Genson, A., Redington, W., Pomeroy, M. J., & Hampshire, S. (2012). Effects of nitrogen and fluorine on crystallisation of Ca–Si–Al–O–N–F glasses. *Journal of the European Ceramic Society*, 32(4), 849–857. DOI: 10.1016/j.jeurceramsoc.2011.10.026

Hanifi, A. R., Pomeroy, M. J., & Hampshire, S. (2010). Novel Glass Formation in the Ca–Si–Al–O–N–F System. *Journal of the American Ceramic Society*, 94(2), 455–461. DOI: 10.1111/j.1551-2916.2010.04147.x

Harimech, Z., Hairch, Y., Atamanov, M., Toshtay, K., Azat, S., Souhair, N., & Amrousse, R. (2023). Carbon nanotube iridium-cupric oxide supported catalysts for decomposition of ammonium dinitramide in the liquid phase. *International Journal of Energetic Materials and Chemical Propulsion*, 22(3), 13–18. DOI: 10.1615/IntJEnergeticMaterialsChemProp.2023047555

Harimech, Z., Toshtay, K., Atamanov, M., Azat, S., & Amrousse, R. (2023). Thermal decomposition of Ammonium dinitramide (ADN) as green energy source for space propulsion. *Aerospace (Basel, Switzerland)*, 10(10), 832. DOI: 10.3390/aerospace10100832

Haris, N. I. N., Sobri, S., & Kassim, N. (2019). Oil palm empty fruit bunch extract as green corrosion inhibitor for mild steel in HCl solution: Central composite design optimization. *Materials and Corrosion*, 70(6), 1111–1119. DOI: 10.1002/maco.201810653

Hayashi, H., & Hakuta, Y. (2010). Hydrothermal Synthesis of Metal Oxide Nanoparticles in Supercritical Water. *Materials (Basel)*, 3(7), 3794–3817. DOI: 10.3390/ma3073794 PMID: 28883312

He, D., Dong, W., Tang, S., Wei, J., Liu, Z., Gu, X., Li, M., Guo, H., & Niu, Y. (2014). Tissue engineering scaffolds of mesoporous magnesium silicate and poly(ε-caprolactone)-poly(ethylene glycol)-poly(ε-caprolactone) composite. *Journal of Materials Science. Materials in Medicine*, 25(6), 1415–1424. DOI: 10.1007/s10856-014-5183-7 PMID: 24595904

Heimann, R. B. (2024). Plasma-Sprayed Osseoconductive Hydroxylapatite Coatings for Endoprosthetic Hip Implants: Phase Composition, Microstructure, Properties, and Biomedical Functions. *Coatings*, 14(7), 787. DOI: 10.3390/coatings14070787

Heinz, M., Srabionyan, V. V., Bugaev, A. L., Pryadchenko, V. V., Ishenko, E. V., Avakyan, L. A., Zubavichus, Y. V., Ihlemann, J., Meinertz, J., Pippel, E., Dubiel, M., & Bugaev, L. A. (2016). Formation of silver nanoparticles in silicate glass using excimer laser radiation: Structural characterization by HRTEM, XRD, EXAFS and optical absorption spectra. *Journal of Alloys and Compounds*, 681, 307–315. DOI: 10.1016/j.jallcom.2016.04.214

Henao, J., Cruz-Bautista, M., Hincapie-Bedoya, J., Ortega-Bautista, B., Corona-Castuera, J., Giraldo-Betancur, A. L., Espinosa-Arbelaez, D. G., Alvarado Orozco, J. M., Clavijo-Mejía, G. A., Trápaga-Martínez, L. G., & Poblano-Salas, C. A. (2018). HVOF hydroxyapatite/titania-graded coatings: Microstructural, mechanical, and in vitro characterization. *Journal of Thermal Spray Technology*, 27(8), 1302–1321. DOI: 10.1007/s11666-018-0811-2

Henao, J., Giraldo-Betancur, A. L., Poblano-Salas, C. A., Forero-Sossa, P. A., Espinosa-Arbelaez, D. G., Gonzalez, J. V., & Corona-Castuera, J. (2024). On the deposition of cold-sprayed hydroxyapatite coatings. *Surface and Coatings Technology*, 476, 130289. DOI: 10.1016/j.surfcoat.2023.130289

Henao, J., Poblano-Salas, C., Monsalve, M., Corona-Castuera, J., & Barceinas-Sanchez, O. (2019). Bioactive glass coatings manufactured by thermal spray: A status report. *Journal of Materials Research and Technology*, 8(5), 4965–4984. DOI: 10.1016/j.jmrt.2019.07.011

Henao, J., Sotelo-Mazon, O., Giraldo-Betancur, A. L., Hincapie-Bedoya, J., Espinosa-Arbelaez, D. G., Poblano-Salas, C., Cuevas-Arteaga, C., Corona-Castuera, J., & Martinez-Gomez, L. (2020). Study of HVOF-sprayed hydroxyapatite/titania graded coatings under in-vitro conditions. *Journal of Materials Research and Technology*, 9(6), 14002–14016. DOI: 10.1016/j.jmrt.2020.10.005

Hench, L. (1994). Bioactive Ceramics: Theory and Clinical Applications. In Elsevier eBooks (pp. 3–14). https://doi.org/DOI: 10.1016/B978-0-08-042144-5.50005-4

Hench, L. L. (1988). Bioactive Ceramics. *Annals of the New York Academy of Sciences*, 523(1), 54–71. DOI: 10.1111/j.1749-6632.1988.tb38500.x PMID: 2837945

Hench, L. L. (1991). Bioceramics: From Concept to Clinic. *Journal of the American Ceramic Society*, 74(7), 1487–1510. DOI: 10.1111/j.1151-2916.1991.tb07132.x

Hench, L. L. (1998). Bioceramics. *Journal of the American Ceramic Society*, 81(7), 1705–1728. DOI: 10.1111/j.1151-2916.1998.tb02540.x

Hench, L. L., & Jones, J. R. (2015). Bioactive Glasses: Frontiers and Challenges. *Frontiers in Bioengineering and Biotechnology*, 3. Advance online publication. DOI: 10.3389/fbioe.2015.00194 PMID: 26649290

Hench, L. L., Splinter, R. J., Allen, W. C., & Greenlee, T. K. (1971). Bonding mechanisms at the interface of ceramic prosthetic materials. *Journal of Biomedical Materials Research*, 5(6), 117–141. DOI: 10.1002/jbm.820050611

Hench, L. L., & Thompson, I. (2010). Twenty-first century challenges for biomaterials. *Journal of the Royal Society, Interface*, 7(suppl_4), S379–S391.

Heng, B. C., Bai, Y., Li, X., Meng, Y., Lu, Y., Zhang, X., & Deng, X. (2023). The bioelectrical properties of bone tissue. *Animal Models and Experimental Medicine*, 6(2), 120–130. DOI: 10.1002/ame2.12300 PMID: 36856186

Heo, S., Kim, M., Lee, J., Park, Y. C., & Jeon, J. K. (2019). Decomposition of ammonium dinitramide-based liquid propellant over Cu/hexaaluminate pellet catalysts. *Korean Journal of Chemical Engineering*, 36(5), 660–668. DOI: 10.1007/s11814-019-0253-7

Hermawan, A., Destyorini, F., Hardiansyah, A., Alviani, V. N., Mayangsari, W., Wibisono, , Septiani, N. L. W., Yudianti, R., & Yuliarto, B. (2023). High energy density asymmetric supercapacitors enabled by La-induced defective MnO2 and biomass-derived activated carbon. *Materials Letters*, 351, 135031. DOI: 10.1016/j.matlet.2023.135031

Heydari, H., Talebian, M., Salarvand, Z., Raeissi, K., Bagheri, M., & Golozar, M. A. (2018). Comparison of two Schiff bases containing O-methyl and nitro substitutes for corrosion inhibiting of mild steel in 1 M HCl solution. *Journal of Molecular Liquids*, 254, 177–187. DOI: 10.1016/j.molliq.2018.01.112

Hill, R. G., Stamboulis, A., & Law, R. V. (2006). Characterisation of fluorine containing glasses by 19F, 27Al, 29Si and 31P MAS-NMR spectroscopy. *Journal of Dentistry*, 34(8), 525–532. DOI: 10.1016/j.jdent.2005.08.005 PMID: 16522349

Hill, R., Stamboulis, A., Law, R., Clifford, A., Towler, M. R., & Crowley, C. (2004). The influence of strontium substitution in fluorapatite glasses and glass-ceramics. *Journal of Non-Crystalline Solids*, 336(3), 223–229. DOI: 10.1016/j.jnoncrysol.2004.02.005

Holmberg, K., & Matthews, A. (2009). *Coatings tribology: properties, mechanisms, techniques and applications in surface engineering*. Elsevier.

Hong, M., Zhang, L., Tan, Z., & Huang, Q. (2019). Effect mechanism of biochar's zeta potential on farmland soil's cadmium immobilization. *Environmental Science and Pollution Research International*, 26(19), 19738–19748. DOI: 10.1007/s11356-019-05298-5 PMID: 31090000

Hong, S., Heo, S., Kim, W., Jo, Y. M., Park, Y. K., & Jeon, J. K. (2019). Catalytic Decomposition of an Energetic Ionic Liquid Solution over Hexaaluminate Catalysts. *Catalysts*, 9(1), 80. DOI: 10.3390/catal9010080

Hoque, N. A., Thakur, P., Biswas, P., Saikh, M. M., Roy, S., Bagchi, B., Das, S., & Ray, P. P. (2018). Biowaste crab shell-extracted chitin nanofiber-based superior piezoelectric nanogenerator. *Journal of Materials Chemistry. A, Materials for Energy and Sustainability*, 6(28), 13848–13858. DOI: 10.1039/C8TA04074E

Hossain, M. M., Mok, Y. S., Nguyen, V. T., Sosiawati, T., Lee, B., Kim, Y. J., Lee, J. H., & Heo, I. (2022). Plasma-catalytic oxidation of volatile organic compounds with honeycomb catalyst for industrial application. *Process Safety and Environmental Protection/Transactions of the Institution of Chemical Engineers. Part B, Process Safety and Environmental Protection/Chemical Engineering Research and Design. Chemical Engineering Research & Design*, 177, 406–417. DOI: 10.1016/j.cherd.2021.11.010

http://finewayceramics.com/ceramic-foams/ Access_date: July 7, 2024 https://www.americanelements.com/silicon-carbide-honeycomb-409-21-2 Access date: July 7, 2024.

https://www.ikts.fraunhofer.de/en/departments/environmental_process_engineering/nanoporous_membranes/functional_carrier_systems_layers/honeycomb_ceramics.html access date 7 July, 2024.

https://www.rsref.com/steel-industry/foam-ceramic-filter.html?network=g&keyword=silicon%20carbide%20foam&matchtype=p&creative=593261174063&device=c&16559093407=16559093407&target=&placement=&gclid=CjwKCAjwnK60BhA9EiwAmpHZw1EUMV-FtL6kHwjPTjo3AbgAimyC9VUjm6szERKcGZoIJ647zffjnhoC9iQQAvD_BwE&gad_source=1 Access date: July 7, 2024

Huang, X., Lou, Y., Duan, Y., Liu, H., Tian, J., Shen, Y., & Wei, X. (2024). Biomaterial scaffolds in maxillofacial bone tissue engineering: A review of recent advances. *Bioactive Materials*, 33, 129–156. DOI: 10.1016/j.bioactmat.2023.10.031 PMID: 38024227

Huang, Z., Grape, E. S., Li, J., Inge, A. K., & Zou, X. (2021). 3D electron diffraction as an important technique for structure elucidation of metal-organic frameworks and covalent organic frameworks. *Coordination Chemistry Reviews*, 427, 213583. https://doi.org/https://doi.org/10.1016/j.ccr.2020.213583. DOI: 10.1016/j.ccr.2020.213583

Huang, Z., Willhammar, T., & Zou, X. (2021). Three-dimensional electron diffraction for porous crystalline materials: Structural determination and beyond. *Chemical Science (Cambridge)*, 12(4), 1206–1219. DOI: 10.1039/D0SC05731B PMID: 34163882

Huisman, L., & Wood, W. E. (1974). *Slow sand filtration*. World Health Organization.

Hu, J., Wang, H., Wang, S., Lei, Y., Qin, L., Li, X., Zhai, D., Li, B., & Kang, F. (2021). Electrochemical deposition mechanism of sodium and potassium. *Energy Storage Materials*, 36, 91–98. DOI: 10.1016/j.ensm.2020.12.017

Hung, I., Howes, A. P., Parkinson, B. G., Anupõld, T., Samoson, A., Brown, S. P., Harrison, P. F., Holland, D., & Dupree, R. (2009). Determination of the bond-angle distribution in vitreous B2O3 by 11B double rotation (DOR) NMR spectroscopy. *Journal of Solid State Chemistry*, 182(9), 2402–2408. DOI: 10.1016/j.jssc.2009.06.025

Hurley, D. C., Kopycinska-Müller, M., Langlois, E. D., Kos, A. B., & Barbosa, N.III. (2006). Mapping substrate/film adhesion with contact-resonance-frequency atomic force microscopy. *Applied Physics Letters*, 89(2), 021911. Advance online publication. DOI: 10.1063/1.2221404

Hussain, C. M., Verma, C., Aslam, J., Aslam, R., & Zehra, S. (2023). Basics of corrosion and its impact. In Elsevier eBooks (pp. 3–30). DOI: 10.1016/B978-0-323-95185-2.00001-0

Hwa, L. G., Lu, C. L., & Liu, L. C. (2000). Elastic moduli of calcium alumino-silicate glasses studied by Brillouin scattering. *Materials Research Bulletin*, 35(8), 1285–1292. DOI: 10.1016/S0025-5408(00)00317-2

Iijima, S. (1991). Helical microtubules of graphitic carbon. *Nature*, 354(6348), 56–58. DOI: 10.1038/354056a0

Iken, A. R., Poolman, R. W., Nelissen, R. G. H. H., & Gademan, M. G. J. (2023). Challenges in developing national orthopedic health research agendas in the Netherlands: Process overview and recommendations. *Acta Orthopaedica*, 94, 230–235. DOI: 10.2340/17453674.2023.12402 PMID: 37194475

Ilyin, Yu. V., Kudaibergenov, K. K., Sharipkhanov, S. D., Mansurov, Z. A., Zhaulybayev, A. A., & Atamanov, M. K. (2023). Surface modifications of CuO doped carbonaceous nanosorbents and their CO2 sorption properties. *Eurasian Chemico-Technological Journal*, 25(1), 33–38. DOI: 10.18321/ectj1493

Innocentini, M. D. M., Salvini, V. R., Pandolfelli, V. C., & Coury, J. R. (1999). Assessment of Forchheimer's Equation to Predict the Permeability of Ceramic Foams. *Journal of the American Ceramic Society*, 82(7), 1945–1948. DOI: 10.1111/j.1151-2916.1999.tb02024.x

Inocentini, M. D., Pardo, A. R., & Pandolfelli, V. C. (2002). Permeability. *Journal of the American Ceramic Society*, 85(6), 1517–1521.

Iqbal, M. Z., & Aziz, U. (2022). Supercapattery: Merging of battery-supercapacitor electrodes for hybrid energy storage devices. *Journal of Energy Storage*, 46, 103823. DOI: 10.1016/j.est.2021.103823

Izato, Y. I., Koshi, M., Miyake, A., & Habu, H. (2016). Kinetics analysis of thermal decomposition of ammonium dinitramide (ADN). *Journal of Thermal Analysis and Calorimetry*, 127(1), 255–264. DOI: 10.1007/s10973-016-5703-4

Jack, K. H. (1976). Sialons and related nitrogen ceramics. *Journal of Materials Science*, 11(6), 1135–1158. DOI: 10.1007/BF02396649

Jacob, J., More, N., Kalia, K., & Kapusetti, G. (2018). Piezoelectric smart biomaterials for bone and cartilage tissue engineering. *Inflammation and Regeneration*, 38(1), 2. DOI: 10.1186/s41232-018-0059-8 PMID: 29497465

Jaeger, R. E., & Egerton, L. (1962). Hot pressing of potassium-sodium niobates. *Journal of the American Ceramic Society*, 45(5), 209–213. DOI: 10.1111/j.1151-2916.1962.tb11127.x

Jagadeeshanayaka, N., Kele, S. N., & Jambagi, S. C. (2023). An Investigation into the Relative Efficacy of High-Velocity Air-Fuel-Sprayed Hydroxyapatite Implants Based on the Crystallinity Index, Residual Stress, Wear, and In-Flight Powder Particle Behavior. *Langmuir*, 39(48), 17513–17528. DOI: 10.1021/acs.langmuir.3c02840 PMID: 38050681

Jalali Alenjareghi, M., Rashidi, A., Kazemi, A., & Talebi, A. (2021). Highly efficient and recyclable spongy nanoporous graphene for remediation of organic pollutants. *Process Safety and Environmental Protection*, 148, 313–322. https://doi.org/https://doi.org/10.1016/j.psep.2020.09.054. DOI: 10.1016/j.psep.2020.09.054

Jandosov, J., Alavijeh, M., Sultakhan, S., Baimenov, A., Bernardo, M., Sakipova, Z., Azat, S., Lyubchyk, S., Zhylybayeva, N., Naurzbayeva, G., Mansurov, Z., Mikhalovsky, S., & Berillo, D. (2022). Activated Carbon/Pectin Composite Enterosorbent for Human Protection from Intoxication with Xenobiotics Pb(II) and Sodium Diclofenac. *Molecules (Basel, Switzerland)*, 27(7), 2296. DOI: 10.3390/molecules27072296 PMID: 35408695

Janićijević, A., Pavlović, V. P., Kovačević, D., Đorđević, N., Marinković, A., Vlahović, B., Karoui, A., Pavlovic, V. P., & Filipović, S. (2024). Impact of Nanocellulose Loading on the Crystal Structure, Morphology and Properties of PVDF/Magnetite@ NC/BaTiO3 Multi-component Hybrid Ceramic/Polymer Composite Material. *Journal of Inorganic and Organometallic Polymers and Materials*, 34(5), 2129–2139. DOI: 10.1007/s10904-023-02953-w

Jankowski, P. E., & Risbud, S. H. (1980). Synthesis and Characterization of an Si-Na-B-O-N Glass. *Journal of the American Ceramic Society*, 63(5–6), 350–352. DOI: 10.1111/j.1151-2916.1980.tb10742.x

Jansen, M. K., Andrady, A. L., Barnes, P. W., Busquets, R., Revell, L. E., Bornman, J. F., Aucamp, P. J., Bais, A. F., Banaszak, A. T., Bernhard, G. H., Bruckman, L. S., Häder, D., Hanson, M. L., Heikkilä, A. M., Hylander, S., Lucas, R. M., Mackenzie, R., Madronich, S., Neale, P. J., & Zhu, L. (2024). Environmental plastics in the context of UV radiation, climate change, and the Montreal Protocol. *Global Change Biology*, 30(4), e17279. Advance online publication. DOI: 10.1111/gcb.17279 PMID: 38619007

Jaramillo, J., Álvarez, P. M., & Gómez-Serrano, V. (2010). Oxidation of activated carbon by dry and wet methods. *Fuel Processing Technology*, 91(11), 1768–1775. DOI: 10.1016/j.fuproc.2010.07.018

Jasim, A. M., He, X., Xing, Y., White, T. A., & Young, M. J. (2021). Cryo-ePDF: Overcoming Electron Beam Damage to Study the Local Atomic Structure of Amorphous ALD Aluminum Oxide Thin Films within a TEM. *ACS Omega*, 6(13), 8986–9000. DOI: 10.1021/acsomega.0c06124 PMID: 33842769

Jbara, A. S., Othaman, Z., & Saeed, M. (2017). Structural, morphological and optical investigations of θ-Al2O3 ultrafine powder. *Journal of Alloys and Compounds*, 718, 1–6. DOI: 10.1016/j.jallcom.2017.05.085

Jelínek, M., Buixaderas, E., Drahokoupil, J., Kocourek, T., Remsa, J., Vaněk, P., Vandrovcová, M., Doubková, M., & Bačáková, L. (2018). Laser-synthesized nanocrystalline, ferroelectric, bioactive $BaTiO_3$/Pt/FS for bone implants. *Journal of Biomaterials Applications*, 32(10), 1464–1475. DOI: 10.1177/0885328218768646 PMID: 29621929

Jenkins, M. W., Tiwari, S. K., & Darby, J. (2011). Bacterial, viral and turbidity removal by intermittent slow sand filtration for household use in developing countries: Experimental investigation and modeling. *Water Research*, 45(18), 6227–6239. DOI: 10.1016/j.watres.2011.09.022 PMID: 21974872

Jensen, D. B., Li, Z., Pavel, I., Dervishi, E., Biris, A. S., Biris, A. R., Lupu, D., & Jensen, P. J. (2007). Bone tissue: A relationship between micro and nano structural composition and its corresponding electrostatic properties with applications in tissue engineering. *2007 IEEE Industry Applications Annual Meeting*, 55-59. https://doi.org/DOI: 10.1109/07IAS.2007.8

Jha, S., Yen, M., Salinas, Y. S., Palmer, E., Villafuerte, J., & Liang, H. (2023). Machine learning-assisted materials development and device management in batteries and supercapacitors: Performance comparison and challenges. *Journal of Materials Chemistry. A, Materials for Energy and Sustainability*, 11(8), 3904–3936. DOI: 10.1039/D2TA07148G

Jiang, Y., Hu, H., Xu, R., Gu, H., Zhang, L., Ji, Z., Zhou, J., Liu, Y., & Cai, B. (2024). Magnetron Sputtered SnO_2 Layer Combined with NiCo−LDH Nanosheets for High-Performance All-Solid-State Supercapacitors. *Batteries & Supercaps*, 7(8), e202400122. Advance online publication. DOI: 10.1002/batt.202400122

Jiang, Z. H., & Zhang, Q. Y. (2014). The structure of glass: A phase equilibrium diagram approach. *Progress in Materials Science*, 61, 144–215. DOI: 10.1016/j.pmatsci.2013.12.001

Jiao Kexin and Flynn, K. T. and K. P. (2016). Synthesis, Characterization, and Applications of Nanoporous Materials for Sensing and Separation. In M. Aliofkhazraei (Ed.), *Handbook of Nanoparticles* (pp. 429–454). Springer International Publishing. DOI: 10.1007/978-3-319-15338-4_22

Jobb, D. B., Anderson, W. B., LeCraw, R. A., & Collins, M. R. (2007). Removal of emerging contaminants and pathogens using modified slow sand filtration: an overview. *Proceedings of the 2007 AWWA Annual Conference,* Toronto, Ontario. American Water Works Association, Denver, Colorado.

Joel, C., Mwamburi, L. A., & Kiprop, E. K. (2018). Use of slow sand filtration technique to improve wastewater effluent for crop irrigation. *Microbiology Research*, 9(1). Advance online publication. DOI: 10.4081/mr.2018.7269

Jones, J. R. (2013). Review of bioactive glass: From Hench to hybrids. *Acta Biomaterialia*, 9(1), 4457–4486. DOI: 10.1016/j.actbio.2012.08.023 PMID: 22922331

Jones, J. R., Sepulveda, P., & Hench, L. L. (2001). Dose-dependent behavior of bioactive glass dissolution. *Journal of Biomedical Materials Research*, 58(6), 720–726. DOI: 10.1002/jbm.10053 PMID: 11745526

Joo, S., Gwon, Y., Kim, S., Park, S., Kim, J., & Hong, S. (2024). Piezoelectrically and Topographically Engineered Scaffolds for Accelerating Bone Regeneration. *ACS Applied Materials & Interfaces*, 16(2), 1999–2011. DOI: 10.1021/acsami.3c12575 PMID: 38175621

Jumadi, J., Kamari, A., Hargreaves, J. S. J., & Yusof, N. (2020). A review of nano-based materials used as flocculants for water treatment. *International Journal of Environmental Science and Technology*, 17(7), 3571–3594. DOI: 10.1007/s13762-020-02723-y

Kacan, E. (2015). Optimum BET surface areas for activated carbon produced from textile sewage sludges and its application as dye removal. *Journal of Environmental Management*, 166, 116–123. DOI: 10.1016/j.jenvman.2015.09.044 PMID: 26496841

Kadja, G. T. M. Ilmi, Moh. M., Azhari, N. J., Khalil, M., Fajar, A. T. N., Subagjo, Makertihartha, I. G. B. N., Gunawan, M. L., Rasrendra, C. B., & Wenten, I. G. (2022). Recent advances on the nanoporous catalysts for the generation of renewable fuels. *Journal of Materials Research and Technology, 17*, 3277–3336. https://doi.org/https://doi.org/10.1016/j.jmrt.2022.02.033

Kaetzl, K., Lübken, M., Nettmann, E., Krimmler, S., & Wichern, M. (2020). Slow sand filtration of raw wastewater using biochar as an alternative filtration media. *Scientific Reports*, 10(1), 1229. DOI: 10.1038/s41598-020-57981-0 PMID: 31988298

Kahkesh, H., & Zargar, B. (2023). Estimating the anti-corrosive potency of 3-nitrophthalic acid as a novel and natural organic inhibitor on corrosion monitoring of mild steel in 1 M HCl solution. *Inorganic Chemistry Communications*, 158, 111533. DOI: 10.1016/j.inoche.2023.111533

Kajiyama, K., Izato, Y. I., & Miyake, A. (2013). Thermal characteristics of ammonium nitrate, carbon, and copper(II) oxide mixtures. *Journal of Thermal Analysis and Calorimetry*, 113(3), 1475–1480. DOI: 10.1007/s10973-013-3201-5

Kakimoto, K. I., Hayakawa, Y., & Kagomiya, I. (2010). Low-temperature sintering of dense (Na, K) NbO3 piezoelectric ceramics using the citrate precursor technique. *Journal of the American Ceramic Society*, 93(9), 2423–2426. DOI: 10.1111/j.1551-2916.2010.03748.x

Kaliaraj, G. S., Siva, T., & Ramadoss, A. (2021). Surface functionalized bioceramics coated on metallic implants for biomedical and anticorrosion performance - a review. *Journal of Materials Chemistry. B*, 9(46), 9433–9460. DOI: 10.1039/D1TB01301G PMID: 34755756

Kamboj, N., Piili, H., Ganvir, A., Gopaluni, A., Nayak, C., Moritz, N., & Salminen, A. (2023). Bioinert ceramics scaffolds for bone tissue engineering by laser-based powder bed fusion: A preliminary review. *IOP Conference Series. Materials Science and Engineering*, 1296(1), 012022. DOI: 10.1088/1757-899X/1296/1/012022

Kamel, N. A. (2022). Bio-piezoelectricity: Fundamentals and applications in tissue engineering and regenerative medicine. *Biophysical Reviews*, 14(3), 717–733. DOI: 10.1007/s12551-022-00969-z PMID: 35783122

Kaneko, S., Tokuda, Y., & Masai, H. (2017). Additive Effects of Rare-Earth Ions in Sodium Aluminoborate Glasses Using ²³Na and ²⁷Al Magic Angle Spinning Nuclear Magnetic Resonance. *New Journal of Glass and Ceramics*, 07(03), 58–76. DOI: 10.4236/njgc.2017.73006

Kang, Y. G., Wei, J., Shin, J. W., Wu, Y. R., Su, J., Park, Y. S., & Shin, J. W. (2018). Enhanced biocompatibility and osteogenic potential of mesoporous magnesium silicate/polycaprolactone/wheat protein composite scaffolds. *International Journal of Nanomedicine*, 13, 1107–1117. DOI: 10.2147/IJN.S157921 PMID: 29520139

Kaou, M. H., Furkó, M., Balázsi, K., & Balázsi, C. (2023). Advanced Bioactive Glasses: The Newest Achievements and Breakthroughs in the Area. *Nanomaterials (Basel, Switzerland)*, 13(16), 2287. DOI: 10.3390/nano13162287 PMID: 37630871

Karbak, M., Boujibar, O., Lahmar, S., Autret-Lambert, C., Chafik, T., & Ghamouss, F. (2022). Chemical Production of Graphene Oxide with High Surface Energy for Supercapacitor Applications. *C*, 8(2), 27. DOI: 10.3390/c8020027

Kareem, R., Bulut, N., & Kaygili, O. (2024). Hydroxyapatite Biomaterials: A Comprehensive Review of their Properties, Structures, Medical Applications, and Fabrication Methods. *Journal of Chemical Reviews*, 6(1), 1–26. DOI: 10.48309/jcr.2024.415051.1253 PMID: 38118062

Kargozar, S., Mozafari, M., Ghodrat, S., Fiume, E., & Baino, F. (2021). Copper-containing bioactive glasses and glass-ceramics: From tissue regeneration to cancer therapeutic strategies. *Materials Science and Engineering C*, 121, 111741. DOI: 10.1016/j.msec.2020.111741 PMID: 33579436

Karia, M., Logishetty, K., Johal, H., Edwards, T. C., & Cobb, J. P. (2023). 5 year follow up of a hydroxyapatite coated short stem femoral component for hip arthroplasty: A prospective multicentre study. *Scientific Reports*, 13(1), 17166. DOI: 10.1038/s41598-023-44191-7 PMID: 37821511

Karki, N., Neupane, S., Chaudhary, Y., Gupta, D., & Yadav, A. (2021). Equisetum hyemale: A new candidate for green corrosion inhibitor family. *International Journal of Corrosion and Scale Inhibition*, 10(1). Advance online publication. DOI: 10.17675/2305-6894-2021-10-1-12

Karon, A. E., Hanni, K. D., Mohle-Boetani, J. C., Beretti, R. A., Hill, V. R., Arrowood, M., Johnston, S. P., Xiao, L., & Vugia, D. J. (2010). Giardiasis outbreak at a camp after installation of a slow-sand filtration water-treatment system. *Epidemiology and Infection*, 139(5), 713–717. DOI: 10.1017/S0950268810001573 PMID: 20587126

Katubi, K. M., Ibrahimoglu, E., Çalışkan, F., Alrowaili, Z. A., Olarinoye, I. O., & Al-Buriahi, M. S. (2024). Apatite–Wollastonite (AW) glass ceramic doped with B2O3: Synthesis, structure, SEM, hardness, XRD, and neutron/charged particle attenuation properties. *Ceramics International*, 50(15), 27139–27146. DOI: 10.1016/j.ceramint.2024.05.011

Kaur, A., Kumar, P., Gupta, A., & Sapra, G. (2024). Piezoelectric Biosensors in Healthcare. In *Enzyme-based Biosensors: Recent Advances and Applications in Healthcare* (pp. 255-271). Springer Nature. https://doi.org/DOI: 10.1007/978-981-15-6982-1_11

Kawamura, G., Muto, H., & Matsuda, A. (2014). Hard template synthesis of metal nanowires. *Frontiers in Chemistry*, 2. Advance online publication. DOI: 10.3389/fchem.2014.00104 PMID: 25453031

Kaya, F., Solmaz, R., & Geçibesler, İ. H. (2023). Adsorption and corrosion inhibition capability of Rheum ribes root extract (Işgın) for mild steel protection in acidic medium: A comprehensive electrochemical, surface characterization, synergistic inhibition effect, and stability study. *Journal of Molecular Liquids*, 372, 121219. DOI: 10.1016/j.molliq.2023.121219

Khaled, K. F., Babić-Samardžija, K., & Hackerman, N. (2005). Theoretical study of the structural effects of polymethylene amines on corrosion inhibition of iron in acid solutions. *Electrochimica Acta*, 50(12), 2515–2520. DOI: 10.1016/j.electacta.2004.10.079

Khan, F., Hossain, N., Mim, J. J., Rahman, S. M., Iqbal, M. J., Billah, M., & Chowdhury, M. A. (2024). Advances of composite materials in automobile applications – A review. *Journal of Engineering Research*. Advance online publication. DOI: 10.1016/j.jer.2024.02.017

Khan, H. A., Tawalbeh, M., Aljawrneh, B., Abuwatfa, W., Al-Othman, A., Sadeghifar, H., & Olabi, A. G. (2024). A comprehensive review on supercapacitors: Their promise to flexibility, high temperature, materials, design, and challenges. *Energy*, 295, 131043. DOI: 10.1016/j.energy.2024.131043

Khare, D., Basu, B., & Dubey, A. K. (2020). Electrical stimulation and piezoelectric biomaterials for bone tissue engineering applications. *Biomaterials*, 258, 120280. DOI: 10.1016/j.biomaterials.2020.120280 PMID: 32810650

Khatib, L. W. E., Rahal, H. T., & Abdel-Gaber, A. M. (2020). Synergistic Effect between Fragaria ananassa and Cucurbita pepo L Leaf Extracts on Mild Steel Corrosion in HCl Solutions. *Protection of Metals and Physical Chemistry of Surfaces*, 56(5), 1096–1106. DOI: 10.1134/S2070205120050111

Khorsand Zak, A., Yazdi, S. T., Abrishami, M. E., & Hashim, A. M. (2024). A review on piezoelectric ceramics and nanostructures: Fundamentals and fabrications. *Journal of the Australian Ceramic Society*, 3(3), 723–753. DOI: 10.1007/s41779-024-00990-3

Khun, K., Ibupoto, Z. H., Liu, X., Beni, V., & Willander, M. (2015). The ethylene glycol template assisted hydrothermal synthesis of Co3O4 nanowires; structural characterization and their application as glucose non-enzymatic sensor. *Materials Science and Engineering B*, 194, 94–100. DOI: 10.1016/j.mseb.2015.01.001

Kim, D., Han, S. A., Kim, J. H., Lee, J. H., Kim, S. W., & Lee, S. W. (2020). Biomolecular Piezoelectric Materials: From Amino Acids to Living Tissues. *Advanced Materials*, 32(14), e1906989. DOI: 10.1002/adma.201906989 PMID: 32103565

Kim, G., Kim, J. M., Lee, C. H., Han, J., Jeong, B. H., & Jeon, J. K. (2016). Catalytic Properties of Nanoporous Manganese Oxides in Decomposition of High-Purity Hydrogen Peroxide. *Journal of Nanoscience and Nanotechnology*, 16(9), 9153–9159. DOI: 10.1166/jnn.2016.12896

Kim, H. W., Kong, Y. M., Bae, C. J., Noh, Y. J., & Kim, H. E. (2004). Sol–gel derived fluor-hydroxyapatite biocoatings on zirconia substrate. *Biomaterials*, 25(15), 2919–2926. DOI: 10.1016/j.biomaterials.2003.09.074 PMID: 14967523

Kim, H. W., Song, J. H., & Kim, H. E. (2006). Bioactive glass nanofiber-collagen nanocomposite as a novel bone regeneration matrix. *Journal of Biomedical Materials Research. Part A*, 79(3), 698–705. DOI: 10.1002/jbm.a.30848 PMID: 16850456

Kim, K. K., Hsu, A., Jia, X., Kim, S. M., Shi, Y., Dresselhaus, M., Palacios, T., & Kong, J. (2012). Synthesis and characterization of hexagonal boron nitride film as a dielectric layer for graphene devices. *ACS Nano*, 6(10), 8583–8590. DOI: 10.1021/nn301675f PMID: 22970651

Kim, Y. T., Tadai, K., & Mitani, T. (2005). High surface area carbons prepared from saccharides by chemical activation with alkali hydroxides for electric double-layer capacitors. *Journal of Materials Chemistry*, 15, 4914–4921. DOI: 10.1039/b511869g

Kischkat, J., Peters, S., Gruska, B., Semtsiv, M., Chashnikova, M., Klinkmüller, M., Fedosenko, O., Machulik, S., Aleksandrova, A., Monastyrskyi, G., Flores, Y., & Masselink, W. T. (2012). Mid-infrared optical properties of thin films of aluminum oxide, titanium dioxide, silicon dioxide, aluminum nitride, and silicon nitride. *Applied Optics*, 51(28), 6789. DOI: 10.1364/AO.51.006789 PMID: 23033094

Kiyono, H., Matsuda, Y., Shimada, T., Ando, M., Oikawa, I., Maekawa, H., Nakayama, S., Ohki, S., Tansho, M., Shimizu, T., Florian, P., & Massiot, D. (2012). Oxygen-17 nuclear magnetic resonance measurements on apatite-type lanthanum silicate (La9.33(SiO4)6O2). *Solid State Ionics*, 228, 64–69. DOI: 10.1016/j.ssi.2012.09.016

Km, S., Praveen, B., & Devendra, B. K. (2024b). A review on corrosion inhibitors: Types, mechanisms, electrochemical analysis, corrosion rate and efficiency of corrosion inhibitors on mild steel in an acidic environment. *Results in Surfaces and Interfaces*, 16, 100258. DOI: 10.1016/j.rsurfi.2024.100258

Knecht, T. A., & Hutchison, J. E. (2023). Precursor and Surface Reactivities Influence the Early Growth of Indium Oxide Nanocrystals in a Reagent-Driven, Continuous Addition Synthesis. *Chemistry of Materials*, 35(8), 3151–3161. DOI: 10.1021/acs.chemmater.2c03761

Koga, K., & Ohigashi, H. (1986). Piezoelectricity and related properties of vinylidene fluoride and trifluoroethylene copolymers. *Journal of Applied Physics*, 59(6), 2142–2150. DOI: 10.1063/1.336351

Kohn, S. C., Dupree, R., Mortuza, M. G., & Henderson, C. M. B. (1991). NMR evidence for five- and six-coordinated aluminum fluoride complexes in F-bearing aluminosilicate glasses. 76, 309–312

Kokubo, T., Kim, H. M., & Kawashita, M. (2003). Novel bioactive materials with different mechanical properties. *Biomaterials*, 24(13), 2161–2175. DOI: 10.1016/S0142-9612(03)00044-9 PMID: 12699652

Kokubo, T., Kushitani, H., Sakka, S., Kitsugi, T., & Yamamuro, T. (1990). Solutions able to reproduce in vivo surface-structure changes in bioactive glass-ceramic A-W3. *Journal of Biomedical Materials Research*, 24(6), 721–734. DOI: 10.1002/jbm.820240607 PMID: 2361964

Kokubo, T., & Takadama, H. (2006). How useful is SBF in predicting in vivo bone bioactivity? *Biomaterials*, 27(15), 2907–2915. DOI: 10.1016/j.biomaterials.2006.01.017 PMID: 16448693

Kołodziejczak-Radzimska, A., & Jesionowski, T. (2014). Zinc Oxide—From Synthesis to Application: A Review. *Materials (Basel)*, 7(4), 2833–2881. DOI: 10.3390/ma7042833 PMID: 28788596

Koohkan, R., Hooshmand, T., Mohebbi-Kalhori, D., Tahriri, M., & Marefati, M. T. (2018). Synthesis, characterization, and in vitro biological evaluation of copper-containing magnetic bioactive glasses for hyperthermia in bone defect treatment. *ACS Biomaterials Science & Engineering*, 4(5), 1797–1811.

Kopycinska-Müller, M., Striegler, A., Hürrich, A., Köhler, B., Meyendorf, N., & Wolter, K.-J. (2009). Elastic Properties of Nano–Thin Films by Use of Atomic Force Acoustic Microscopy. *Proceedings of the Materials Research Society*, 1185(1), 7–12. DOI: 10.1557/PROC-1185-II09-04

Kopycinska-Müller, M., Striegler, A., Köhler, B., & Wolter, K.-J. (2011). Mechanical Characterization of Thin Films by Use of Atomic Force Acoustic Microscopy. *Advanced Engineering Materials*, 13(4), 312–318. https://doi.org/https://doi.org/10.1002/adem.201000245. DOI: 10.1002/adem.201000245

Kortesuo, P., Ahola, M., Kangas, M., Kangasniemi, I., Yli-Urpo, A., & Kiesvaara, J. (2000). In vitro evaluation of sol–gel processed spray dried silica gel microspheres as carrier in controlled drug delivery. *International Journal of Pharmaceutics*, 200(2), 223–229. DOI: 10.1016/S0378-5173(00)00393-8 PMID: 10867252

Köse, G. T., Korkusuz, F., Ozkul, A., Soysal, Y., Ozdemir, T., Yildiz, C., & Hasirci, V. (2005). Tissue engineered cartilage on collagen and PHBV matrices. *Biomaterials*, 26(25), 5187–5197. DOI: 10.1016/j.biomaterials.2005.01.037 PMID: 15792546

Košiček, M., Zavašnik, J., Baranov, O., Šetina Batič, B., & Cvelbar, U. (2022). Understanding the Growth of Copper Oxide Nanowires and Layers by Thermal Oxidation over a Broad Temperature Range at Atmospheric Pressure. *Crystal Growth & Design*, 22(11), 6656–6666. DOI: 10.1021/acs.cgd.2c00863

Kostoglou, N., Koczwara, C., Prehal, C., Terziyska, V., Babic, B., Matovic, B., Constantinides, G., Tampaxis, C., Charalambopoulou, G., Steriotis, T., Hinder, S., Baker, M., Polychronopoulou, K., Doumanidis, C., Paris, O., Mitterer, C., & Rebholz, C. (2017). Nanoporous activated carbon cloth as a versatile material for hydrogen adsorption, selective gas separation and electrochemical energy storage. *Nano Energy*, 40, 49–64. https://doi.org/https://doi.org/10.1016/j.nanoen.2017.07.056. DOI: 10.1016/j.nanoen.2017.07.056

Kozlovskiy, A. L., Shlimas, D. I., Shumskaya, A. E., Kaniukov, E. Y., Zdorovets, M. V., & Kadyrzhanov, K. K. (2017). Influence of electrodeposition parameters on structural and morphological features of Ni nanotubes. *The Physics of Metals and Metallography*, 118(2), 164–169. DOI: 10.1134/S0031918X17020065

Kozo, O., Yamamuro, T., Nakamura, T., & Kokubo, T. (1990). Quantitative study on osteoconduction of apatite-wollastonite containing glass ceramic granules, hydroxyapatite granules and alumina granules. *Biomaterials*, 11(4), 265–271. DOI: 10.1016/0142-9612(90)90008-E PMID: 2383622

Krause, S., Bon, V., Senkovska, I., Stoeck, U., Wallacher, D., Többens, D. M., Zander, S., Pillai, R. S., Maurin, G., Coudert, F.-X., & Kaskel, S. (2016). A pressure-amplifying framework material with negative gas adsorption transitions. *Nature*, 532(7599), 348–352. DOI: 10.1038/nature17430 PMID: 27049950

Krejci, D., Woschnak, A., Scharlemann, C., & Ponweiser, K. (2012). Structural impact of honeycomb catalysts on hydrogen peroxide decomposition for micro propulsion. *Process Safety and Environmental Protection/Transactions of the Institution of Chemical Engineers. Part B, Process Safety and Environmental Protection/Chemical Engineering Research and Design. Chemical Engineering Research & Design*, 90(12), 2302–2315. DOI: 10.1016/j.cherd.2012.05.015

Kresge, C. T., Leonowicz, M. E., Roth, W. J., Vartuli, J. C., & Beck, J. S. (1992). Ordered mesoporous molecular sieves synthesized by a liquid-crystal template mechanism. *Nature*, 359(6397), 710–712. DOI: 10.1038/359710a0

Krishna Prasad, N. V., Babu, T. A., Sarma, S. R. K. N., Ramesh, S., Nirisha, K., Mathew, T., & Madhavi, N. (n.d.). ROLE OF POROUS NANOMATERIAL'S IN WATER PURIFICATION, ELECTRONICS, DRUG DELIVERY AND STORAGE: A COMPREHENSIVE REVIEW. In *Journal of Optoelectronic and Biomedical Materials* (Vol. 13, Issue 1).

Kroeker, S., & Stebbins, J. F. (2001). Three-Coordinated Boron-11 Chemical Shifts in Borates. *Inorganic Chemistry*, 40(24), 6239–6246. DOI: 10.1021/ic010305u PMID: 11703125

Kuldeyev, E., Seitzhanova, M., Tanirbergenova, S., Tazhu, K., Doszhanov, E., Mansurov, Z., Azat, S., Nurlybaev, R., & Berndtsson, R. (2023). Modifying Natural Zeolites to Improve Heavy Metal Adsorption. *Water (Basel)*, 15(12), 2215. DOI: 10.3390/w15122215

Kumar, A., & Das, C. (2023). A novel eco-friendly inhibitor of chayote fruit extract for mild steel corrosion in 1 M HCl: Electrochemical, weight loss studies, and the effect of temperature. *Sustainable Chemistry and Pharmacy*, 36, 101261. DOI: 10.1016/j.scp.2023.101261

Kumar, A., Kim, W., Sriboriboon, P., Lee, H. Y., Kim, Y., & Ryu, J. (2024). AC poling-induced giant piezoelectricity and high mechanical quality factor in [001] PMN-PZT hard single crystals. *Sensors and Actuators. A, Physical*, 372, 115342. DOI: 10.1016/j.sna.2024.115342

Kumar, D., Ward, R. G., & Williams, D. J. (1961). Effect of fluorides on silicates and phosphates. *Discussions of the Faraday Society*, 32, 147. DOI: 10.1039/df9613200147

Kumari, S., Agnihotri, R., & Oommen, C. (2024). Cerium Oxide-Based Robust Catalyst for Hydroxylammonium Nitrate Monopropellant Thruster. In *Space technology library* (pp. 187–215). DOI: 10.1007/978-3-031-62574-9_7

Kumar, N., Ghosh, S., Thakur, D., Lee, C.-P., & Sahoo, P. K. (2023). Recent advancements in zero- to three-dimensional carbon networks with a two-dimensional electrode material for high-performance supercapacitors. *Nanoscale Advances*, 5(12), 3146–3176. DOI: 10.1039/D3NA00094J PMID: 37325524

Kumar, P., Dehiya, B. S., & Sindhu, A. (2018). Bioceramics for hard tissue engineering applications: A review. *International Journal of Applied Engineering Research: IJAER*, 13(5), 2744–2752. http://www.ripublication.com

Kumar, R., Pattanayak, I., Dash, P. A., & Mohanty, S. (2023). Bioceramics: A review on design concepts toward tailor-made (multi)-functional materials for tissue engineering applications. *Journal of Materials Science*, 58(8), 3460–3484. DOI: 10.1007/s10853-023-08226-8

Kunarbekova, M., Busquets, R., Sailaukhanuly, Y., Mikhalovsky, S. V., Toshtay, K., Kudaibergenov, K., & Azat, S. (2024). Carbon adsorbents for the uptake of radioactive iodine from contaminated water effluents: A systematic review. *Journal of Water Process Engineering*, 67, 106174. DOI: 10.1016/j.jwpe.2024.106174

Kunwong, N., Tangjit, N., Rattanapinyopituk, K., Dechkunakorn, S., Anuwongnukroh, N., Arayapisit, T., & Sritanaudomchai, H. (2021). Optimization of poly (lactic-co-glycolic acid)-bioactive glass composite scaffold for bone tissue engineering using stem cells from human exfoliated deciduous teeth. *Archives of Oral Biology*, 123, 105041. DOI: 10.1016/j.archoralbio.2021.105041 PMID: 33454420

Kuren, S. G., Volyanik, S. A., Savenkova, M. A., & Zaitseva, E. A. (2023). Properties of Salicylidene-Aniline as a corrosion inhibitor in oil and petroleum products transportation systems. *Safety of Technogenic and Natural Systems*, 3(3), 14–23. DOI: 10.23947/2541-9129-2023-7-3-14-23

Kurt, M., Kap, Z., Senol, S., Ercan, K. E., Sika-Nartey, A. T., Kocak, Y., Koc, A., Esiyok, H., Caglayan, B. S., Aksoylu, A. E., & Ozensoy, E. (2022). Influence of La and Si promoters on the anaerobic heterogeneous catalytic decomposition of ammonium dinitramide (ADN) via alumina supported iridium active sites. *Applied Catalysis A, General*, 632, 118500. DOI: 10.1016/j.apcata.2022.118500

Kyffin, B. A., Foroutan, F., Raja, F. N. S., Martin, R. A., Pickup, D. M., Taylor, S. E., & Carta, D. (2019). Antibacterial silver-doped phosphate-based glasses prepared by coacervation. *Journal of Materials Chemistry. B*, 7(48), 7744–7755. DOI: 10.1039/C9TB02195G PMID: 31750507

Ladj, R., Magouroux, T., Eissa, M., Dubled, M., Mugnier, Y., Le Dantec, R., Galez, C., Valour, J. P., Fessi, H., & Elaissari, A. (2013). Aminodextran-coated potassium niobate (KNbO3) nanocrystals for second harmonic bio-imaging. *Colloids and Surfaces. A, Physicochemical and Engineering Aspects*, 439, 131–137. DOI: 10.1016/j.colsurfa.2013.02.025

Laksaci, H., Khelifi, A., Trari, M., & Addoun, A. (2017). Synthesis and characterization of microporous activated carbon from coffee grounds using potassium hydroxides. *Journal of Cleaner Production*, 147, 254–262. DOI: 10.1016/j.jclepro.2017.01.102

Lang, S. B. (2016). Review of ferroelectric hydroxyapatite and its application to biomedicine. *Phase Transitions*, 89/7-8), 678–694. https://doi.org/DOI: 10.1080/01411594.2016.1182166

Lang, S. B. (2005). Pyroelectricity: From ancient curiosity to modern imaging tool. *Physics Today*, 58(8), 31–36. DOI: 10.1063/1.2062916

Lan, X., Jiang, X., Song, Y., Jing, X., & Xing, X. (2019). The effect of activation temperature on structure and properties of blue coke-based activated carbon by CO2 activation. *Green Processing and Synthesis*, 8(1), 837–845. DOI: 10.1515/gps-2019-0054

Larabi, L., Harek, Y., Traisnel, M., & Mansri, A. (2004). Synergistic influence of Poly(4-Vinylpyridine) and potassium iodide on inhibition of corrosion of mild steel in 1M HCl. *Journal of Applied Electrochemistry*, 34(8), 833–839. DOI: 10.1023/B:JACH.0000035609.09564.e6

Le Dantec, C., Duguet, J.-P., Montiel, A., Dumoutier, N., Dubrou, S., & Vincent, V. (2002). Occurrence of mycobacteria in water treatment lines and in water distribution systems. *Applied and Environmental Microbiology*, 68(11), 5318–5325. DOI: 10.1128/AEM.68.11.5318-5325.2002 PMID: 12406720

Lee, C. H., Singla, A., & Lee, Y. (2001). Biomedical applications of collagen. *International Journal of Pharmaceutics*, 221(1-2), 1–22. DOI: 10.1016/S0378-5173(01)00691-3 PMID: 11397563

Lee, J. H., Kang, Y. M., & Roh, K. C. (2023). Enhancing gravimetric and volumetric capacitance in supercapacitors with nanostructured partially graphitic activated carbon. *Electrochemistry Communications*, 154, 107560. DOI: 10.1016/j.elecom.2023.107560

Lee, S. K., & Stebbins, J. F. (2003). The distribution of sodium ions in aluminosilicate glasses: A high-field Na-23 MAS and 3Q MAS NMR study. *Geochimica et Cosmochimica Acta*, 67(9), 1699–1709. DOI: 10.1016/S0016-7037(03)00026-7

Lei, Q., Guo, J., Noureddine, A., Wang, A., Wuttke, S., Brinker, C. J., & Zhu, W. (2020). Sol–Gel-Based Advanced Porous Silica Materials for Biomedical Applications. *Advanced Functional Materials*, 30(41), 1909539. DOI: 10.1002/adfm.201909539

Leng-Ward, G., & Lewis, M. H. (1989). Oxynitride glasses and their glass-ceramic derivatives. In Springer eBooks (pp. 106–155). DOI: 10.1007/978-94-009-0817-8_4

Lepry, W. C., & Nazhat, S. N. (2021a). A Review of Phosphate and Borate Sol–Gel Glasses for Biomedical Applications. *Advanced NanoBiomed Research*, 1(3), 2000055. DOI: 10.1002/anbr.202000055

Lesbayev, B., Rakhymzhan, N., Ustayeva, G., Maral, Y., Atamanov, M., Auyelkhankyzy, M., & Zhamash, A. (2024). Preparation of nanoporous carbon from rice husk with improved textural characteristics for hydrogen sorption. *Journal of Composites Science*, 8(2), 74. DOI: 10.3390/jcs8020074

Levingstone, T. J., Barron, N., Ardhaoui, M., Benyounis, K., Looney, L., & Stokes, J. (2017). Application of response surface methodology in the design of functionally graded plasma sprayed hydroxyapatite coatings. *Surface and Coatings Technology*, 313, 307–318. DOI: 10.1016/j.surfcoat.2017.01.113

Leyton, J., Fernández, J., Acosta, P., Quiroga, A., & Codony, F. (2024). Reduction of Helicobacter pylori cells in rural water supply using slow sand filtration. *Environmental Monitoring and Assessment*, 196(7), 619. Advance online publication. DOI: 10.1007/s10661-024-12764-2 PMID: 38878080

Liang, W., Rüssel, C., Day, D. E., & Völksch, G. (2006). Bioactive comparison of a borate, phosphate and silicate glass. Journal of Materials Research/Pratt's Guide to Venture Capital Sources, 21(1), 125–131. DOI: 10.1557/jmr.2006.0025

Liang, Y., Felix, R., Glicksman, H., & Ehrman, S. (2016). Cu-Sn binary metal particle generation by spray pyrolysis. *Aerosol Science and Technology*, 51(4), 430–442. DOI: 10.1080/02786826.2016.1265912

Liang, Z., Guo, S., Dong, H., Li, Z., Liu, X., Li, X., Kang, H., Zhang, L., Yuan, L., & Zhao, L. (2022). Modification of activated carbon and its application in selective hydrogenation of naphthalene. *ACS Omega*, 7(43), 38550–38560. DOI: 10.1021/acsomega.2c03914 PMID: 36340089

Liao, M., Yi, X., Dai, Z., Qin, H., Guo, W., & Xiao, H. (2023). Application of metal-BDC-derived catalyst on cordierite honeycomb ceramic support in a microreactor for hydrogen production. *Ceramics International*, 49(17), 29082–29093. DOI: 10.1016/j.ceramint.2023.06.184

Liao, Y., Xu, Y., & Chan, Y. (2013). Semiconductor nanocrystals in sol–gel derived matrices. *Physical Chemistry Chemical Physics*, 15(33), 13694. DOI: 10.1039/c3cp51351c PMID: 23842703

Li, C., Iqbal, M., Lin, J., Luo, X., Jiang, B., Malgras, V., Wu, K. C.-W., Kim, J., & Yamauchi, Y. (2018a). Electrochemical Deposition: An Advanced Approach for Templated Synthesis of Nanoporous Metal Architectures. *Accounts of Chemical Research*, 51(8), 1764–1773. DOI: 10.1021/acs.accounts.8b00119 PMID: 29984987

Li, D., Sun, Y., Gao, P., Zhang, X., & Ge, H. (2014). Structural and magnetic properties of nickel ferrite nanoparticles synthesized via a template-assisted sol–gel method. *Ceramics International*, 40(10), 16529–16534. DOI: 10.1016/j.ceramint.2014.08.006

Li, D., Yadav, A., Zhou, H., Roy, K., Thanasekaran, P., & Lee, C. (2024). Advances and Applications of Metal-Organic Frameworks (MOFs) in Emerging Technologies: A Comprehensive Review. *Global Challenges (Hoboken, NJ)*, 8(2), 2300244. https://doi.org/https://doi.org/10.1002/gch2.202300244. DOI: 10.1002/gch2.202300244 PMID: 38356684

Li, E., Li, Y., Liu, S., & Yao, P. (2022). Choline amino acid ionic liquids as green corrosion inhibitors of mild steel in acidic medium. *Colloids and Surfaces. A, Physicochemical and Engineering Aspects*, 657, 130541. DOI: 10.1016/j.colsurfa.2022.130541

Li, H., Li, L., Vienna, J., Qian, M., Wang, Z., Darab, J., & Peeler, D. (2000). Neodymium(III) in alumino-borosilicate glasses. *Journal of Non-Crystalline Solids*, 278(1–3), 35–57. DOI: 10.1016/S0022-3093(00)00327-6

Li, H., Ma, Y., Wang, Y., Li, C., Bai, Q., Shen, Y., & Uyama, H. (2024). Nitrogen enriched high specific surface area biomass porous carbon: A promising electrode material for supercapacitors. *Renewable Energy*, 224, 120144. DOI: 10.1016/j.renene.2024.120144

Li, J., Campos, L. C., Zhang, L., & Xie, W. (2022). Sand and sand-GAC filtration technologies in removing PPCPs: A review. *The Science of the Total Environment*, 848, 157680. DOI: 10.1016/j.scitotenv.2022.157680 PMID: 35907530

Li, J.-R., Sculley, J., & Zhou, H.-C. (2012). Metal–Organic Frameworks for Separations. *Chemical Reviews*, 112(2), 869–932. DOI: 10.1021/cr200190s PMID: 21978134

Li, J., Wu, Q., & Wu, J. (2016). Synthesis of Nanoparticles via Solvothermal and Hydrothermal Methods. In *Handbook of Nanoparticles* (pp. 295–328). Springer International Publishing., DOI: 10.1007/978-3-319-15338-4_17

Li, J., Zhang, T., Liao, Z., Wei, Y., Hang, R., & Huang, D. (2023). Engineered functional doped hydroxyapatite coating on titanium implants for osseointegration. *Journal of Materials Research and Technology*, 27, 122–152. DOI: 10.1016/j.jmrt.2023.09.239

Li, J., Zhou, Q., & Campos, L. C. (2018). The application of GAC sandwich slow sand filtration to remove pharmaceutical and personal care products. *The Science of the Total Environment*, 635, 1182–1190. DOI: 10.1016/j.scitotenv.2018.04.198 PMID: 29710573

Likhanova, N. V., Arellanes-Lozada, P., Olivares-Xometl, O., Hernández-Cocoletzi, H., Lijanova, I. V., Arriola-Morales, J., & Castellanos-Aguila, J. (2019). Effect of organic anions on ionic liquids as corrosion inhibitors of steel in sulfuric acid solution. *Journal of Molecular Liquids*, 279, 267–278. DOI: 10.1016/j.molliq.2019.01.126

Li, L., Han, J., Huang, X., Qiu, S., Liu, X., Liu, L., Zhao, M., Qu, J., Zou, J., & Zhang, J. (2023). Organic pollutants removal from aqueous solutions using metal-organic frameworks (MOFs) as adsorbents: A review. *Journal of Environmental Chemical Engineering*, 11(6), 111217. DOI: 10.1016/j.jece.2023.111217

Li, L., Strachan, D. M., Li, H., Davis, L. L., & Qian, M. (2000b). Crystallization of gadolinium- and lanthanum-containing phases from sodium alumino-borosilicate glasses. *Journal of Non-Crystalline Solids*, 272(1), 46–56. DOI: 10.1016/S0022-3093(00)00117-4

Li, M., & Yang, L. (2023). Biomedical metallic materials based on nanocrystalline and nanoporous microstructures: Properties and applications. In Webster, T. J. (Ed.), *Nanomedicine* (2nd ed., pp. 555–584). Woodhead Publishing., https://doi.org/https://doi.org/10.1016/B978-0-12-818627-5.00030-0 DOI: 10.1016/B978-0-12-818627-5.00030-0

Lima, C., De Oliveira, R., Figueiro, S., Wehmann, C., Goes, J., & Sombra, A. (2006). DC conductivity and dielectric permittivity of collagen–chitosan films. *Materials Chemistry and Physics*, 99(2-3), 284–288. DOI: 10.1016/j.matchemphys.2005.10.027

Lin, F. H., Huang, Y. Y., Hon, M. H., & Wu, S. C. (1991). Fabrication and biocompatibility of a porous bioglass ceramic in a Na2O CaO SiO2 P2O5 system. *Journal of Biomedical Engineering*, 13(4), 328–334. DOI: 10.1016/0141-5425(91)90115-N PMID: 1890828

Lin, J., Liu, Z., Guo, Y., Wang, S., Tao, Z., Xue, X., Li, R., Feng, S., Wang, L., Liu, J., Gao, H., Wang, G., & Su, Y. (2023). Machine learning accelerates the investigation of targeted MOFs: Performance prediction, rational design and intelligent synthesis. *Nano Today*, 49, 101802. https://doi.org/https://doi.org/10.1016/j.nantod.2023.101802. DOI: 10.1016/j.nantod.2023.101802

Li, R., Clark, A. E., & Hench, L. L. (1991). An investigation of bioactive glass powders by sol-gel processing. *Journal of Applied Biomaterials*, 2(4), 231–239. DOI: 10.1002/jab.770020403 PMID: 10171144

Li, T., Wu, H., Ihli, J., Ma, Z., Krumeich, F., Bomans, P. H. H., Sommerdijk, N. A. J. M., Friedrich, H., Patterson, J. P., & van Bokhoven, J. A. (2019). Cryo-TEM and electron tomography reveal leaching-induced pore formation in ZSM-5 zeolite. *Journal of Materials Chemistry. A, Materials for Energy and Sustainability*, 7(4), 1442–1446. DOI: 10.1039/C8TA10696G

Liu, P., & Chen, G. (2014). Fabricating Porous Ceramics. In *Elsevier eBooks* (pp. 221–302). DOI: 10.1016/B978-0-12-407788-1.00005-8

Liu, Z., Fujita, N., Miyasaka, K., Han, L., Stevens, S. M., Suga, M., Asahina, S., Slater, B., Xiao, C., Sakamoto, Y., Anderson, M. W., Ryoo, R., & Terasaki, O. (2013). A review of fine structures of nanoporous materials as evidenced by microscopic methods. In *Journal of Electron Microscopy* (Vol. 62, Issue 1, pp. 109–146). DOI: 10.1093/jmicro/dfs098

Liu, B.-T., Zhao, M., Han, L.-P., Lang, X.-Y., Wen, Z., & Jiang, Q. (2018). Three-dimensional nanoporous N-doped graphene/iron oxides as anode materials for high-density energy storage in asymmetric supercapacitors. *Chemical Engineering Journal*, 335, 467–474. https://doi.org/https://doi.org/10.1016/j.cej.2017.11.001. DOI: 10.1016/j.cej.2017.11.001

Liu, H.-J., Wang, X.-M., Cui, W.-J., Dou, Y.-Q., Zhao, D.-Y., & Xia, Y.-Y. (2010). Highly ordered mesoporous carbon nanofiber arrays from a crab shell biological template and its application in supercapacitors and fuel cells. *Journal of Materials Chemistry*, 20(20), 4223. DOI: 10.1039/b925776d

Liu, J., Luo, Z., Lin, C., Han, L., Gui, H., Song, J., Liu, T., & Lu, A. (2019). Influence of Y2O3 substitution for B2O3 on the structure and properties of alkali-free B2O3-Al2O3-SiO2 glasses containing alkaline-earth metal oxides. *Physica B, Condensed Matter*, 553, 47–52. DOI: 10.1016/j.physb.2018.10.024

Liu, J., Tang, D., Hou, W., Ding, D., Yao, S., Liu, Y., Chen, Y., Chi, W., Zhang, Z., Ouyang, M., & Zhang, C. (2023). Conductive polymer electrode materials with excellent mechanical and electrochemical properties for flexible supercapacitor. *Journal of Energy Storage*, 74, 109329. DOI: 10.1016/j.est.2023.109329

Liu, S., Chen, X., Chen, X., Liu, Z., & Wang, H. (2007). Activated carbon with excellent chromium(VI) adsorption performance prepared by acid–base surface modification. *Journal of Hazardous Materials*, 141(1), 315–319. DOI: 10.1016/j.jhazmat.2006.07.006 PMID: 16914264

Liu, W., Cui, L., & Cao, Y. (2006). Bone reconstruction with bone marrow stromal cells. *Methods in Enzymology*, 420, 362–380. DOI: 10.1016/S0076-6879(06)20017-X PMID: 17161706

Liu, X., Xiong, J., Lv, Y., & Zuo, Y. (2008). Study on corrosion electrochemical behavior of several different coating systems by electrochemical impedance spectroscopy. *Progress in Organic Coatings*, 64(4), 497–503. DOI: 10.1016/j.porgcoat.2008.08.012

Liu, Y., & Nekvasil, H. (2002). Si-F bonding in aluminosilicate glasses: Inferences from ab initio NMR calculations. *The American Mineralogist*, 87(2–3), 339–346. DOI: 10.2138/am-2002-2-317

Livage, J., Barboux, P., Vandenborre, M. T., Schmutz, C., & Taulelle, F. (1992). Sol-gel synthesis of phosphates. *Journal of Non-Crystalline Solids*, 147–148, 18–23. DOI: 10.1016/S0022-3093(05)80586-1

Li, W., Yang, K., Peng, J., Zhang, L., Guo, S., & Xia, H. (2008). Effects of carbonization temperatures on characteristics of porosity in coconut shell chars and activated carbons derived from carbonized coconut shell chars. *Industrial Crops and Products*, 28(2), 190–198. DOI: 10.1016/j.indcrop.2008.02.012

Li, X., He, J., Xie, B., He, Y., Lai, C., Wang, W., Zeng, J., Yao, B., Zhao, W., & Long, T. (2024). 1,4-Phenylenediamine-based Schiff bases as eco-friendly and efficient corrosion inhibitors for mild steel in HCl medium: Experimental and theoretical approaches. *Journal of Electroanalytical Chemistry (Lausanne, Switzerland)*, 955, 118052. DOI: 10.1016/j.jelechem.2024.118052

Li, X., Wang, Y., Yang, P., Yin, X., Han, T., Han, B., & He, K. (2024). Topological models of yttrium aluminosilicate glass based on molecular dynamics and structure characterization analysis. *Journal of the American Ceramic Society*, jace.20118. Advance online publication. DOI: 10.1111/jace.20118

Li, Y., Chen, J., Liu, S., Wang, Z., Zhang, S., Mao, C., & Wang, J. (2024). Biodegradable piezoelectric polymer for cartilage remodeling. *Matter*, 7(4), 1631–1643. DOI: 10.1016/j.matt.2024.01.034

Li, Y., Du, Q., Wang, X., Zhang, P., Wang, D., Wang, Z., & Xia, Y. (2010b). Removal of lead from aqueous solution by activated carbon prepared from Enteromorpha prolifera by zinc chloride activation. *Journal of Hazardous Materials*, 183(1–3), 583–589. DOI: 10.1016/j.jhazmat.2010.07.063 PMID: 20709449

Li, Y., Xie, W., Wang, H., Yang, H., Huang, H., Liu, Y., & Fan, X. (2020). Investigation on the Thermal Behavior of Ammonium Dinitramide with Different Copper-Based Catalysts. *Propellants, Explosives, Pyrotechnics*, 45(10), 1607–1613. DOI: 10.1002/prep.202000065

Lo, B. T. W., Ye, L., & Tsang, S. C. E. (2018). The Contribution of Synchrotron X-Ray Powder Diffraction to Modern Zeolite Applications: A Mini-review and Prospects. *Chem*, 4(8), 1778–1808. DOI: 10.1016/j.chempr.2018.04.018

Lobato-Peralta, D. R., Duque-Brito, E., Ayala-Cortés, A., Arias, D., Longoria, A., Cuentas-Gallegos, A. K., Sebastian, P., & Okoye, P. U. (2021). Advances in activated carbon modification, surface heteroatom configuration, reactor strategies, and regeneration methods for enhanced wastewater treatment. *Journal of Environmental Chemical Engineering*, 9(4), 105626. DOI: 10.1016/j.jece.2021.105626

Loehman, R. E. (1980). Oxynitride glasses. *Journal of Non-Crystalline Solids*, 42(1–3), 433–445. DOI: 10.1016/0022-3093(80)90042-3

Logan, A. J., Stevik, T. K., Siegrist, R. L., & Rønn, R. M. (2001). Transport and fate of Cryptosporidium parvum oocysts in intermittent sand filters. *Water Research*, 35(18), 4359–4369. DOI: 10.1016/S0043-1354(01)00181-6 PMID: 11763038

Losq, . (2014). Losq, C. L., Neuville, D. R., Florian, P., Henderson, G. S., & Massiot, D. (2014). The role of Al3+ on rheology and structural changes in sodium silicate and aluminosilicate glasses and melts. *Geochimica et Cosmochimica Acta*, 126, 495–517. DOI: 10.1016/j.gca.2013.11.010

Loty, C., Sautier, J. M., Tan, M. T., Oboeuf, M., Jallot, E., Boulekbache, H., Greenspan, D., & Forest, N. (2001). Bioactive Glass Stimulates In Vitro Osteoblast Differentiation and Creates a Favorable Template for Bone Tissue Formation. *Journal of Bone and Mineral Research : the Official Journal of the American Society for Bone and Mineral Research*, 16(2), 231–239. DOI: 10.1359/jbmr.2001.16.2.231 PMID: 11204423

Lou, Y., Wang, F., Li, Z., Zou, Z., Fan, G., Wang, X., Lei, W., & Lu, W. (2020). Fabrication of high-performance $MgTiO_3$–$CaTiO_3$ microwave ceramics through a stereolithography-based 3D printing. *Ceramics International*, 46(10), 16979–16986. DOI: 10.1016/j.ceramint.2020.03.282

Lua, A. C., & Yang, T. (2004). Effect of activation temperature on the textural and chemical properties of potassium hydroxide activated carbon prepared from pistachio-nut shell. *Journal of Colloid and Interface Science*, 274(2), 594–601. DOI: 10.1016/j.jcis.2003.10.001 PMID: 15144834

Lu, G. Q., & Zhao, X. S. (2004). *Nanoporous Materials: Science and Engineering* (Vol. 4). PUBLISHED BY IMPERIAL COLLEGE PRESS AND DISTRIBUTED BY WORLD SCIENTIFIC PUBLISHING CO., DOI: 10.1142/p181

Lu, H.-H., Zheng, K., Boccaccini, A. R., & Liverani, L. (2023). Electrospinning of cotton-like fibers based on cerium-doped sol–gel bioactive glass. *Materials Letters*, 334, 133712. DOI: 10.1016/j.matlet.2022.133712

Lunardi, S., Martins, M., Pizzolatti, B. S., & Soares, M. B. D. (2022). Pre-filtration followed by slow double-layered filtration: Media clogging effects on hydraulic aspects and water quality. *Water Environment Research*, 94(4), e10709. Advance online publication. DOI: 10.1002/wer.10709 PMID: 35362183

Luo, T., Tan, B., Liao, J., Shi, K., & Ning, L. (2024). A review on external physical stimuli with biomaterials for bone repair. *Chemical Engineering Journal*, 153749, 153749. Advance online publication. DOI: 10.1016/j.cej.2024.153749

Luo, Y., Hou, Z., Jin, D., Gao, J., & Zheng, X. (2006). Template assisted synthesis of Ga2O3–Al2O3 nanorods. *Materials Letters*, 60(3), 393–395. DOI: 10.1016/j.matlet.2005.08.059

Lu, X., Weller, Z. D., Gervasio, V., & Vienna, J. D. (2024). Glass design using machine learning property models with prediction uncertainties: Nuclear waste glass formulation. *Journal of Non-Crystalline Solids*, 631, 122907. DOI: 10.1016/j.jnoncrysol.2024.122907

Lu, Y., Cheng, D., Niu, B., Wang, X., Wu, X., & Wang, A. (2023). Properties of Poly (Lactic-co-Glycolic Acid) and Progress of Poly (Lactic-co-Glycolic Acid)-Based Biodegradable Materials in Biomedical Research. *Pharmaceuticals (Basel, Switzerland)*, 16(3), 454. DOI: 10.3390/ph16030454 PMID: 36986553

Ma, B., He, J., Mu, S., Zeng, L., Chen, S., Li, J., Luo, L., Yu, S., Xi, H., Zhu, D., & Chen, Y. (2024). Carbon/Nitrogen Co-Modified Nano-Ni Catalyst Endows a High Energy Density and Low Cost Aqueous Ni–H 2 Gas Battery. *ACS Applied Energy Materials*, 7(7), 2800–2809. DOI: 10.1021/acsaem.3c03221

Mabrouk, A., Bachar, A., Atbir, A., Follet-Houttemane, C., Albert-Mercier, C., Tricoteaux, A., Leriche, A., & Hampshire, S. (2018). Mechanical properties, structure, bioactivity and cytotoxicity of bioactive Na-Ca-Si-P-O-(N) glasses. Journal of the Mechanical Behavior of Biomedical Materials. *Journal of the Mechanical Behavior of Biomedical Materials*, 86, 284–293. DOI: 10.1016/j.jmbbm.2018.06.023 PMID: 30006277

Mabrouk, A., Bachar, A., Fatani, I. F., & Vaills, Y. (2021). High temperature brillouin scattering study of lanthanum and sodium aluminoborosilicate glasses. *Materials Chemistry and Physics*, 257, 123790. DOI: 10.1016/j.matchemphys.2020.123790

Mabrouk, A., Bachar, A., Follet, C., Mercier, C., Atbir, A., Boukbir, L., Marrouche, A., Bellajrou, R., Billah, S. M., & Hadek, M. E. (2017). Bioactivity and Cytotoxicity of Oxynitride Glasses. *Research & Reviews Journal of Material Sciences*, 05(02). Advance online publication. DOI: 10.4172/2321-6212.1000173

Mabrouk, A., Bachar, A., Vaills, Y., Canizarès, A., & Hampshire, S. (2024). Effect of Sodium Oxide on Structure of Lanthanum Aluminosilicate Glass. *Ceramics*, 7(3), 858–872. DOI: 10.3390/ceramics7030056

Mabrouk, A., De Sousa Meneses, D., Pellerin, N., Véron, E., Genevois, C., Ory, S., & Vaills, Y. (2019). Effects of boron on structure of lanthanum and sodium aluminoborosilicate glasses studied by X-ray diffraction, transmission electron microscopy and infrared spectrometry. *Journal of Non-Crystalline Solids*, 503–504, 69–77. DOI: 10.1016/j.jnoncrysol.2018.09.030

Mabrouk, A., Vaills, Y., Pellerin, N., & Bachar, A. (2020). Structural study of lanthanum sodium aluminoborosilicate glasses by NMR spectroscopy. *Materials Chemistry and Physics*, 254, 123492. DOI: 10.1016/j.matchemphys.2020.123492

Macías-García, A., Torrejón-Martín, D., Díaz-Díez, M. Á., & Carrasco-Amador, J. P. (2019). Study of the influence of particle size of activate carbon for the manufacture of electrodes for supercapacitors. *Journal of Energy Storage*, 25, 100829. DOI: 10.1016/j.est.2019.100829

MacKenzie, K. J. D., & Smith, M. E. (2002). Multinuclear Solid-State NMR of Inorganic Materials. In Pergamon materials series. DOI: 10.1016/S1470-1804(02)X8001-8

Mahdi, B., Aljibori, S., Abbass, H. S. S., Al-Azzawi, M. K., Kadhum, A. H., Hanoon, M., Isahak, M., & Al-Amiery, W. N. R. W. (2022). Gravimetric analysis and quantum chemical assessment of 4-aminoantipyrine derivatives as corrosion inhibitors. *Int. J. Corros. Scale Inhib.*, 11(3), 1191–1213. DOI: 10.17675/2305-6894-2022-11-3-17

Maiti, S., Karan, S. K., Kim, J., & Khatua, B. B. (2019). Nature driven bio-piezoelectric/triboelectric nanogenerator as next-generation green energy harvester for smart and pollution free society. *Advanced Energy Materials*, 9(9), 1803027. DOI: 10.1002/aenm.201803027

Maiyo, J. K., Dasika, S., & Jafvert, C. T. (2023). Slow Sand Filters for the 21st Century: A Review. *International Journal of Environmental Research and Public Health 2023, Vol. 20, Page 1019, 20*(2), 1019. DOI: 10.3390/ijerph20021019

Ma, J., Qin, J., Zheng, S., Fu, Y., Chi, L., Li, Y., Dong, C., Li, B., Xing, F., Shi, H., & Wu, Z.-S. (2024). Hierarchically Structured Nb2O5 Microflowers with Enhanced Capacity and Fast-Charging Capability for Flexible Planar Sodium Ion Micro-Supercapacitors. *Nano-Micro Letters*, 16(1), 67. DOI: 10.1007/s40820-023-01281-5 PMID: 38175485

Makhatadze, G. I. (2017). Linking computation and experiments to study the role of charge-charge interactions in protein folding and stability. *Physical Biology*, 14(1), 013002. DOI: 10.1088/1478-3975/14/1/013002 PMID: 28169222

Makled, A. E., & Belal, H. (2009, May). Modeling of hydrazine decomposition for monopropellant thrusters. *In13th International Conference on Aerospace Sciences & Aviation Technology* (pp. 26-28).

Maleix, C., Chabernaud, P., Brahmi, R., Beauchet, R., Batonneau, Y., Kappenstein, C., Schwentenwein, M., Koopmans, R.-J., Schuh, S., & Scharlemann, C. (2019). Development of catalytic materials for decomposition of ADN-based monopropellants. *Acta Astronautica*, 158, 407–415. DOI: 10.1016/j.actaastro.2019.03.033

Malfait, W. J., & Halter, W. E. (2008). Structural relaxation in silicate glasses and melts: High-temperature Raman spectroscopy. *Physical Review B: Condensed Matter and Materials Physics*, 77(1), 014201. Advance online publication. DOI: 10.1103/PhysRevB.77.014201

Malfait, W. J., & Sanchez-Valle, C. (2013). Effect of water and network connectivity on glass elasticity and melt fragility. *Chemical Geology*, 346, 72–80. DOI: 10.1016/j.chemgeo.2012.04.034

Malhotra, B. D., & Ali, Md. A. (2018). Nanomaterials in Biosensors. In *Nanomaterials for Biosensors* (pp. 1–74). Elsevier. DOI: 10.1016/B978-0-323-44923-6.00001-7

Malode, S., & Shetti, N. P. (2022). ZrO2 in biomedical applications. In K. Mondal, *Metal Oxides for Biomedical and Biosensor Applications* (pp. 471-501). Elsevier. https://doi.org/DOI: 10.1016/B978-0-12-823033-6.00016-8

Malozyomov, B. V., Kukartsev, V. V., Martyushev, N. V., Kondratiev, V. V., Klyuev, R. V., & Karlina, A. I. (2023). Improvement of Hybrid Electrode Material Synthesis for Energy Accumulators Based on Carbon Nanotubes and Porous Structures. *Micromachines*, 14(7), 1288. DOI: 10.3390/mi14071288 PMID: 37512599

Mandal, S., Hu, J., & Shi, S. Q. (2023). A comprehensive review of hybrid supercapacitor from transition metal and industrial crop based activated carbon for energy storage applications. *Materials Today. Communications*, 34, 105207. DOI: 10.1016/j.mtcomm.2022.105207

Mann, F. A., Galonska, P., Herrmann, N., & Kruss, S. (2022). Quantum defects as versatile anchors for carbon nanotube functionalization. *Nature Protocols*, 17(3), 727–747. DOI: 10.1038/s41596-021-00663-6 PMID: 35110739

Mansurov, Z. A., Amrousse, R., Hori, K., & Atamanov, M. K. (2020). Combustion/Decomposition behavior of HAN under the effects of nanoporous activated carbon. In Springer eBooks (pp. 211–230). DOI: 10.1007/978-981-15-4831-4_8

Marchesano, V., Gennari, O., Mecozzi, L., Grilli, S., & Ferraro, P. (2015). Effects of Lithium Niobate Polarization on Cell Adhesion and Morphology. *ACS Applied Materials & Interfaces*, 7(32), 18113–18119. DOI: 10.1021/acsami.5b05340 PMID: 26222955

Marshall, D. B., Noma, T., & Evans, A. G. (1982). A Simple Method for Determining Elastic-Modulus–to-Hardness Ratios using Knoop Indentation Measurements. *Journal of the American Ceramic Society*, 65(10). Advance online publication. DOI: 10.1111/j.1151-2916.1982.tb10357.x

Martin, A., Khansur, N. H., Urushihara, D., Asaka, T., Kakimoto, K. i., & Webber, K. G. (2024). Effect of Li on the intrinsic and extrinsic contributions of the piezoelectric response in Lix(Na0.5K0.5)1-xNbO3 piezoelectric ceramics across the polymorphic phase boundary. *Acta Materialia*, 266, 119691. Advance online publication. DOI: 10.1016/j.actamat.2024.119691

Martínez, G., Merinero, M., Pérez-Aranda, M., Pérez-Soriano, E., Ortiz, T., Villamor, E., Begines, B., & Alcudia, A. (2020). Environmental Impact of Nanoparticles' Application as an Emerging Technology: A Review. *Materials (Basel)*, 14(1), 166. DOI: 10.3390/ma14010166 PMID: 33396469

Martin, P. M. (2009). *Handbook of deposition technologies for films and coatings: science, applications and technology*. William Andrew.

Martins, P., Nunes, J. S., Hungerford, G., Miranda, D., Ferreira, A., Sencadas, V., & Lanceros-Méndez, S. (2009). Local variation of the dielectric properties of poly (vinylidene fluoride) during the α-to β-phase transformation. *Physics Letters. [Part A]*, 373(2), 177–180. DOI: 10.1016/j.physleta.2008.11.026

Massiot, D., Fayon, F., Capron, M., King, I., Calvé, S. L., Alonso, B., Durand, J., Bujoli, B., Gan, Z., & Hoatson, G. (2001). Modelling one- and two-dimensional solid-state NMR spectra. *Magnetic Resonance in Chemistry*, 40(1), 70–76. DOI: 10.1002/mrc.984

Massiot, , Messinger, R. J., Cadars, S., Deschamps, M. Ë., Montouillout, V., Pellerin, N., Veron, E., Allix, M., Florian, P., & Fayon, F. (2013). Massiot, D., JMessinger, R., Cadars, S., Deschamps, M., Montouillout, V., Pellerin, N., Veron, E., Allix, M., Florian, P., & Fayon, F. (2013). Topological, Geometric, and Chemical Order in Materials: Insights from Solid-State NMR. *Accounts of Chemical Research*, 46(9), 1975–1984. DOI: 10.1021/ar3003255 PMID: 23883113

Mastelaro, V., & Zanotto, E. (2018). X-ray Absorption Fine Structure (XAFS) Studies of Oxide Glasses—A 45-Year Overview. *Materials (Basel)*, 11(2), 204. DOI: 10.3390/ma11020204 PMID: 29382102

Matad, P. B., Mokshanatha, P. B., Hebbar, N., Venkatesha, V. T., & Tandon, H. C. (2014). Ketosulfone drug as a green corrosion inhibitor for mild steel in acidic medium. *Industrial & Engineering Chemistry Research*, 53(20), 8436–8444. DOI: 10.1021/ie500232g

Matsunaga, H., Habu, H., & Miyake, A. (2013). Thermal decomposition mechanism of ammonium dinitramide using pyrolysate analyses. *Proceedings of new trend in research of energetic materials*, Czech Republic, 268-276.

Matsunaga, H., Yoshino, S., Kumasaki, M., Habu, H., & Miyake, A. (2011). Aging characteristics of the energetic oxidizer ammonium dinitramide. *Kōgyō Kayaku Kyōkaishi/Kayaku Gakkaishi/Science and Technology of Energetic Materials/Kōgyō Kayaku*, 72(5), 131–135. Retrieved from https://ci.nii.ac.jp/naid/10029979927

Maurath, J., & Willenbacher, N. (2017). 3D printing of open-porous cellular ceramics with high specific strength. *Journal of the European Ceramic Society*, 37(15), 4833–4842. DOI: 10.1016/j.jeurceramsoc.2017.06.001

Ma, Y., Han, L., Liu, Z., Mayoral, A., Díaz, I., Oleynikov, P., Ohsuna, T., Han, Y., Pan, M., Zhu, Y., Sakamoto, Y., Che, S., & Terasaki, O. (2019). Microscopy of Nanoporous Crystals. In Hawkes, P. W., & Spence, J. C. H. (Eds.), *Springer Handbook of Microscopy* (pp. 1391–1450). Springer International Publishing., DOI: 10.1007/978-3-030-00069-1_29

Mazzoni, E., Iaquinta, M. R., Lanzillotti, C., Mazziotta, C., Maritati, M., Montesi, M., Sprio, S., Tampieri, A., Tognon, M., & Martini, F. (2021). Bioactive Materials for Soft Tissue Repair. *Frontiers in Bioengineering and Biotechnology*, 9, 613787. DOI: 10.3389/fbioe.2021.613787 PMID: 33681157

McCafferty, E. (2010). Corrosion Inhibitors. In *Introduction to Corrosion Science*. Springer., DOI: 10.1007/978-1-4419-0455-3_12

Mehrabi, T., Mesgar, A. S., & Mohammadi, Z. (2020). Bioactive Glasses: A Promising Therapeutic Ion Release Strategy for Enhancing Wound Healing. *ACS Biomaterials Science & Engineering*, 6(10), 5399–5430. DOI: 10.1021/acsbiomaterials.0c00528 PMID: 33320556

Melia, P. M., Busquets, R., Hooda, P. S., Cundy, A. B., & Sohi, S. P. (2019). Driving forces and barriers in the removal of phosphorus from water using crop residue, wood and sewage sludge derived biochars. *The Science of the Total Environment*, 675, 623–631. DOI: 10.1016/j.scitotenv.2019.04.232 PMID: 31035201

Memetova, A., Tyagi, I., Singh, L., & Karri, R. R. Suhas, Tyagi, K., Kumar, V., Memetov, N., Zelenin, A., Tkachev, A., Bogoslovskiy, V., Shigabaeva, G., Galunin, E., Mubarak, N. M., & Agarwal, S. (2022). Nanoporous carbon materials as a sustainable alternative for the remediation of toxic impurities and environmental contaminants: A review. *Science of The Total Environment, 838*, 155943. https://doi.org/https://doi.org/10.1016/j.scitotenv.2022.155943

Mercier, C., Follet-Houttemane, C., Pardini, A., & Revel, B. (2011). Influence of P2O5 content on the structure of SiO2-Na2O-CaO-P2O5 bioglasses by 29Si and 31P MAS-NMR. *Journal of Non-Crystalline Solids*, 357(24), 3901–3909. DOI: 10.1016/j.jnoncrysol.2011.07.042

Mermillod-Blondin, F., Mauclaire, L., & Montuelle, B. (2005). Use of slow filtration columns to assess oxygen respiration, consumption of dissolved organic carbon, nitrogen transformations, and microbial parameters in hyporheic sediments. *Water Research*, 39(9), 1687–1698. DOI: 10.1016/j.watres.2005.02.003 PMID: 15899267

Micoli, L., Bagnasco, G., Turco, M., Trifuoggi, M., Sorge, A. R., Fanelli, E., Pernice, P., & Aronne, A. (2013). Vapour phase H_2O_2 decomposition on Mn based monolithic catalysts synthesized by innovative procedures. *Applied Catalysis B: Environmental*, 140–141, 516–522. DOI: 10.1016/j.apcatb.2013.04.072

Minary-Jolandan, M., & Yu, M. F. (2010). Shear piezoelectricity in bone at the nanoscale. *Applied Physics Letters*, 97(15), 153127, 153127–3. DOI: 10.1063/1.3503965

Mishchenko, O., Yanovska, A., Kosinov, O., Maksymov, D., Moskalenko, R., Ramanavicius, A., & Pogorielov, M. (2023). Synthetic Calcium-Phosphate Materials for Bone Grafting. *Polymers*, 15(18), 3822. DOI: 10.3390/polym15183822 PMID: 37765676

Mittal, G., & Paul, S. (2022). Suspension and solution precursor plasma and HVOF spray: A review. *Journal of Thermal Spray Technology*, 31(5), 1443–1475. DOI: 10.1007/s11666-022-01360-w

Mobin, M., Aslam, R., Salim, R., & Kaya, S. (2022). An investigation on the synthesis, characterization and anti-corrosion properties of choline based ionic liquids as novel and environmentally friendly inhibitors for mild steel corrosion in 5% HCl. *Journal of Colloid and Interface Science*, 620, 293–312. DOI: 10.1016/j.jcis.2022.04.036 PMID: 35429708

Mohamed, A., Martin, U., Visco, D. P.Jr, Townsend, T., & Bastidas, D. M. (2023). Interphase corrosion inhibition mechanism of sodium borate on carbon steel rebars in simulated concrete pore solution. *Construction & Building Materials*, 408, 133763. DOI: 10.1016/j.conbuildmat.2023.133763

Mohapatra, S., Das, H. T., Tripathy, B. C., & Das, N. (2024). Recent Developments in Electrodeposition of Transition Metal Chalcogenides-Based Electrode Materials for Advance Supercapacitor Applications: A Review. *Chemical Record (New York, N.Y.)*, 24(1), e202300220. Advance online publication. DOI: 10.1002/tcr.202300220 PMID: 37668292

Molina, H. R., Muñoz, J. L. S., Leal, M. I. D., Reina, T. R., Ivanova, S., Gallego, M. Á. C., & Odriozola, J. A. (2019). Carbon supported gold nanoparticles for the catalytic reduction of 4-Nitrophenol. *Frontiers in Chemistry*, 7, 548. Advance online publication. DOI: 10.3389/fchem.2019.00548 PMID: 31475132

Moloney, J., Kumar, D., Muralidhar, V., & Pojtanabuntoeng, T. (2022). Corrosion inhibition. Flow Assurance (Volume 1 in Oil and Gas Chemistry Management Series Pages 609-707). DOI: 10.1016/B978-0-12-822010-8.00006-4

Montazerian, M., Hosseinzadeh, F., Migneco, C., Fook, M. V., & Baino, F. (2022). Bioceramic coatings on metallic implants: An overview. *Ceramics International*, 48(11), 8987–9005. DOI: 10.1016/j.ceramint.2022.02.055

Moore, B., Asadi, E., & Lewis, G. (2017). Deposition methods for microstructured and nanostructured coatings on metallic bone implants: A review. *Advances in Materials Science and Engineering*, 2017(1), 5812907. DOI: 10.1155/2017/5812907

Moreira, E., Innocentini, M., & Coury, J. (2004). Permeability of ceramic foams to compressible and incompressible flow. *Journal of the European Ceramic Society*, 24(10–11), 3209–3218. DOI: 10.1016/j.jeurceramsoc.2003.11.014

Moreno-Gomez, I.(2019). Degradation of Bioresorbable Composites: Calcium Carbonate Case Studies. *Springer Theses A Phenomenological Mathematical Modelling Framework for the Degradation of Bioresorbable Composites*, 245-266. https://doi.org/DOI: 10.1007/978-3-030-04990-4_7

Morin, E. I., & Stebbins, J. F. (2016). Separating the effects of composition and fictive temperature on Al and B coordination in Ca, La, Y aluminosilicate, aluminoborosilicate and aluminoborate glasses. *Journal of Non-Crystalline Solids*, 432, 384–392. DOI: 10.1016/j.jnoncrysol.2015.10.035

Morin, E. I., Wu, J., & Stebbins, J. F. (2014). Modifier cation (Ba, Ca, La, Y) field strength effects on aluminum and boron coordination in aluminoborosilicate glasses: The roles of fictive temperature and boron content. *Applied Physics. A, Materials Science & Processing*, 116(2), 479–490. DOI: 10.1007/s00339-014-8369-4

Mulfinger, H. O. (1966). Physical and Chemical Solubility of Nitrogen in Glass Melts. *Journal of the American Ceramic Society*, 49(9), 462–467. DOI: 10.1111/j.1151-2916.1966.tb13300.x

Müller, V., & Djurado, E. (2022). Microstructural designed S58 bioactive glass/hydroxyapatite composites for enhancing osteointegration of Ti6Al4V-based implants. *Ceramics International*, 48(23), 35365–35375. DOI: 10.1016/j.ceramint.2022.08.138

Murillo, G., Blanquer, A., Vargas-Estevez, C., Barrios, L., Ibáñez, E., Nogués, C., & Esteve, J. (2017). Electromechanical Nanogenerator-Cell Interaction Modulates Cell Activity. *Advanced materials (Deerfield Beach, Fla.), 29*(24), . https://doi.org/DOI: 10.1002/adma.201605048

Murmu, M., Saha, S. K., Bhaumick, P., Murmu, N. C., Hirani, H., & Banerjee, P. (2020). Corrosion inhibition property of azomethine functionalized triazole derivatives in 1 mol L−1 HCl medium for mild steel: Experimental and theoretical exploration. *Journal of Molecular Liquids*, 313, 113508. DOI: 10.1016/j.molliq.2020.113508

Muthmann, J., Bläker, C., Pasel, C., Luckas, M., Schledorn, C., & Bathen, D. (2020). Characterization of structural and chemical modifications during the steam activation of activated carbons. *Microporous and Mesoporous Materials*, 309, 110549. DOI: 10.1016/j.micromeso.2020.110549

Muthukrishnan, P., Jeyaprabha, B., & Prakash, P. (2013). Adsorption and corrosion inhibiting behavior of Lannea coromandelica leaf extract on mild steel corrosion. *Arabian Journal of Chemistry*, 10, S2343–S2354. DOI: 10.1016/j.arabjc.2013.08.011

Mu, X., Wei, J., Dong, J., & Ke, W. (2014). In Situ Corrosion Monitoring of Mild Steel in a Simulated Tidal Zone without Marine Fouling Attachment by Electrochemical Impedance Spectroscopy. *Journal of Materials Science and Technology*, 30(10), 1043–1050. DOI: 10.1016/j.jmst.2014.03.013

Nagata, H., Matsumoto, K., Hirosue, T., Hiruma, Y., & Takenaka, T. (2007). Fabrication and electrical properties of potassium niobate ferroelectric ceramics. *Japanese Journal of Applied Physics*, 46(10S), 7084. DOI: 10.1143/JJAP.46.7084

Naji, M., Piazza, F., Guimbretière, G., Canizarès, A., & Vaills, Y. (2013). Structural Relaxation Dynamics and Annealing Effects of Sodium Silicate Glass. *The Journal of Physical Chemistry B*, 117(18), 5757–5764. DOI: 10.1021/jp401112s PMID: 23574051

Nakamachi, E., Okuda, Y., Kumazawa, S., Uetsuji, Y., Tsuchiya, K., & Nakayasu, H. (2006). Development of Fabrication Technique of Bio-Compatible Piezoelectric Material MgSiO 3 by Using Helicon Wave Plasma Sputter. *Nippon Kikai Gakkai Ronbunshu A Hen(Transactions of the Japan Society of Mechanical Engineers Part A)(Japan), 18*(3), 353-355. https://doi.org/DOI: 10.1299/kikaia.72.353

Nakayama, Y., Pauzauskie, P. J., Radenovic, A., Onorato, R. M., Saykally, R. J., Liphardt, J., & Yang, P. (2007). Tunable nanowire nonlinear optical probe. *Nature*, 447(7148), 1098–1101. DOI: 10.1038/nature05921 PMID: 17597756

Nandiyanto, A. B., Wiryani, D., Rusli, A. S., Purnamasari, A., Abdullah, A., Widiaty, A. G. I., & Churriyati, R. (2017). Extraction of curcumin pigment from Indonesian local turmeric with its infrared spectra and thermal decomposition properties. IOP Conf. Ser. Mater. Sci. Eng. (1), 012136 , 180.DOI: 10.1088/1757-899X/180/1/012136

Narayan, R. (2010). Use of nanomaterials in water purification. *Materials Today*, 13(6), 44–46. https://doi.org/https://doi.org/10.1016/S1369-7021(10)70108-5. DOI: 10.1016/S1369-7021(10)70108-5

Nasrollahzadeh, M., Issaabadi, Z., Sajjadi, M., Sajadi, S. M., & Atarod, M. (2019). Types of Nanostructures. In Nasrollahzadeh, M., Sajadi, S. M., Sajjadi, M., Issaabadi, Z., & Atarod, M. (Eds.), *An Introduction to Green Nanotechnology* (Vol. 28, pp. 29–80). Elsevier., https://doi.org/https://doi.org/10.1016/B978-0-12-813586-0.00002-X DOI: 10.1016/B978-0-12-813586-0.00002-X

Nassar, A. M., & Hajjaj, K. (2013). Purification of stormwater using sand filter. *Journal of Water Resource and Protection*, 05(11), 1007–1012. DOI: 10.4236/jwarp.2013.511105

Nataraja, S., Venkatesha, T., Manjunatha, K., Poojary, B., Pavithra, M., & Tandon, H. (2011). Inhibition of the corrosion of steel in HCl solution by some organic molecules containing the methylthiophenyl moiety. *Corrosion Science*, 53(8), 2651–2659. DOI: 10.1016/j.corsci.2011.05.004

Nathanael, A. J., & Oh, T. H. (2020). Biopolymer Coatings for Biomedical Applications. *Polymers*, 12(12), 3061. DOI: 10.3390/polym12123061 PMID: 33371349

Nawaz, F., Ali, M., Ahmad, S., Yong, Y., Rahman, S., Naseem, M., Hussain, S., Razzaq, A., Khan, A., Ali, F., Al Balushi, R. A., Al-Hinaai, M. M., & Ali, N. (2024). Carbon based nanocomposites, surface functionalization as a promising material for VOCs (volatile organic compounds) treatment. *Chemosphere*, 364, 143014. DOI: 10.1016/j.chemosphere.2024.143014 PMID: 39121955

Nawaz, S. S., Manjunatha, K., Ranganatha, S., Supriya, S., Ranjan, P., Chakraborty, T., & Ramakrishna, D. (2023). Nickel curcumin complexes: Physico chemical studies and nonlinear optical activity. *Optical Materials*, 136, 113450. DOI: 10.1016/j.optmat.2023.113450

Ndi, K., Dihang, D., Aimar, P., & Kayem, G. J. (2008). Retention of bentonite in granular natural pozzolan: Implications for water filtration. *Separation Science and Technology*, 43(7), 1621–1631. DOI: 10.1080/01496390801974712

Negm, N. A., Kandile, N. G., Badr, E. A., & Mohammed, M. A. (2012). Gravimetric and electrochemical evaluation of environmentally friendly nonionic corrosion inhibitors for carbon steel in 1 M HCl. *Corrosion Science*, 65, 94–103. DOI: 10.1016/j.corsci.2012.08.002

Negut, I., & Ristoscu, C. (2023). Bioactive glasses for soft and hard tissue healing applications—A short review. *Applied Sciences (Basel, Switzerland)*, 13(10), 6151. DOI: 10.3390/app13106151

Neo, M., Kotani, S., Nakamura, T., Yamamuro, T., Ohtsuki, C., Kokubo, T., & Bando, Y. (1992). A comparative study of ultrastructures of the interfaces between four kinds of surface-active ceramic and bone. *Journal of Biomedical Materials Research*, 26(11), 1419–1432. DOI: 10.1002/jbm.820261103 PMID: 1447227

Neuville, D. R., Cormier, L., Flank, A. M., Briois, V., & Massiot, D. (2004a). Al speciation and Ca environment in calcium aluminosilicate glasses and crystals by Al and Ca K-edge X-ray absorption spectroscopy. *Chemical Geology*, 213(1–3), 153–163. DOI: 10.1016/j.chemgeo.2004.08.039

Neuville, D. R., Cormier, L., & Massiot, D. (2004). Al environment in tectosilicate and peraluminous glasses: A 27Al MQ-MAS NMR, Raman, and XANES investigation. *Geochimica et Cosmochimica Acta*, 68(24), 5071–5079. DOI: 10.1016/j.gca.2004.05.048

Neuville, D. R., & Losq, C. L. (2022). Link between Medium and Long-range Order and Macroscopic Properties of Silicate Glasses and Melts. *Reviews in Mineralogy and Geochemistry*, 87(1), 105–162. DOI: 10.2138/rmg.2022.87.03

Nguyen, A. N., Solard, J., Nong, H. T. T., Ben Osman, C., Gomez, A., Bockelée, V., Tencé-Girault, S., Schoenstein, F., Simón-Sorbed, M., Carrillo, A. E., & Mercone, S. (2020). Spin Coating and Micro-Patterning Optimization of Composite Thin Films Based on PVDF. *Materials (Basel)*, 13(6), 1342. DOI: 10.3390/ma13061342 PMID: 32187993

Nguyen, T. T., Hoang, T., Can, V. M., Ho, A. S., Nguyen, S. H., Nguyen, T. T. T., Pham, T. N., Nguyen, T. P., Nguyen, T. L. H., & Thi, M. T. D. (2017). In vitro and in vivo tests of PLA/d-HAp nanocomposite. *Advances in Natural Sciences: Nanoscience and Nanotechnology*, 8(4), 045013. DOI: 10.1088/2043-6254/aa92b0

Nguyen, V. T., Dinh, D. K., Lan, N. M., Trinh, Q. H., Hossain, M. M., Dao, V. D., & Mok, Y. S. (2023). Critical role of reactive species in the degradation of VOC in a plasma honeycomb catalyst reactor. *Chemical Engineering Science*, 276, 118830. DOI: 10.1016/j.ces.2023.118830

Niederberger, M., & Pinna, N. (2009). *Metal Oxide Nanoparticles in Organic Solvents*. Springer London., DOI: 10.1007/978-1-84882-671-7

Nikam, P. N., Patil, S. S., Chougale, U. M., Fulari, A. V., & Fulari, V. J. (2024). Supercapacitor properties of Ni2+ incrementally substituted with Co2+ in cubic spinel NixCo1-xFe2O4 nanoparticles by sol-gel auto combustion method. *Journal of Energy Storage*, 96, 112648. DOI: 10.1016/j.est.2024.112648

Nikkerdar, N., Golshah, A., Mobarakeh, M. S., Fallahnia, N., Azizi, B., Souri, R., Safaei, M., & Shoohanizad, E. (2024). Recent progress in application of zirconium oxide in dentistry. *Journal of Medicinal and Pharmaceutical Chemistry Research*, 6(8), 1042–1071. DOI: 10.48309/jmpcr.2024.432254.1069

Noor, E. A., & Al-Moubaraki, A. H. (2008). Thermodynamic study of metal corrosion and inhibitor adsorption processes in mild steel/1-methyl-4 [4′(-X)-styryl pyridinium iodides/HCl systems. *Materials Chemistry and Physics*, 110(1), 145–154. DOI: 10.1016/j.matchemphys.2008.01.028

Nosseir, A. E. S., Cervone, A., Amrousse, R., Pasini, A., Igarashi, S., & Matsuura, Y. (2024). Green Monopropellants: State-of-the-Art. In *Space technology library* (pp. 95–134). DOI: 10.1007/978-3-031-62574-9_4

Novoseltseva, V., Yankovych, H., Kovalenko, O., Václavíková, M., & Melnyk, I. (2021). Production of high-performance lead (II) ions adsorbents from pea peels waste as a sustainable resource. *Waste Management & Research*, 39(4), 584–593. DOI: 10.1177/0734242X20943272 PMID: 32705958

Obalová, L., Karásková, K., Jirátová, K., & Kovanda, F. (2009). Effect of potassium in calcined Co–Mn–Al layered double hydroxide on the catalytic decomposition of N2O. *Applied Catalysis B Environment and Energy*, 90(1–2), 132–140. https://doi.org/DOI: 10.1016/j.apcatb.2009.03.002

Obot, I., Onyeachu, I. B., Umoren, S. A., Quraishi, M. A., Sorour, A. A., Chen, T., Aljeaban, N., & Wang, Q. (2019). High temperature sweet corrosion and inhibition in the oil and gas industry: Progress, challenges and future perspectives. *Journal of Petroleum Science Engineering*, 185, 106469. DOI: 10.1016/j.petrol.2019.106469

Ochiai, T., & Fukada, E. (1998). Electromechanical properties of poly-L-lactic acid. *Japanese Journal of Applied Physics*, 37(6R), 3374. DOI: 10.1143/JJAP.37.3374

Oguzie, E., Okolue, B., Ebenso, E., Onuoha, G., & Onuchukwu, A. (2004). Evaluation of the inhibitory effect of methylene blue dye on the corrosion of aluminium in HCl. *Materials Chemistry and Physics*, 87(2–3), 394–401. DOI: 10.1016/j.matchemphys.2004.06.003

Okayama, R., Amano, Y., & Machida, M. (2010). Effect of nitrogen species on an activated carbon surface on the adsorption of Cu(II) ions from aqueous solution. *Carbon*, 48(10), 3000–3007. DOI: 10.1016/j.carbon.2010.03.040

Okey, N. C., Obasi, N. L., Ejikeme, P. M., Ndinteh, D. T., Ramasami, P., Sherif, E. M., Akpan, E. D., & Ebenso, E. E. (2020). Evaluation of some amino benzoic acid and 4-aminoantipyrine derived Schiff bases as corrosion inhibitors for mild steel in acidic medium: Synthesis, experimental and computational studies. *Journal of Molecular Liquids*, 315, 113773. DOI: 10.1016/j.molliq.2020.113773

Oki, M., & Anawe, P. (2015). A review of Corrosion in Agricultural Industries. *Physical Science International Journal*, 5(4), 216–222. DOI: 10.9734/PSIJ/2015/14847

Olabi, A. G., Abbas, Q., Al Makky, A., & Abdelkareem, M. A. (2022). Supercapacitors as next generation energy storage devices: Properties and applications. *Energy*, 248, 123617. DOI: 10.1016/j.energy.2022.123617

Oliveira, H. L., Da Rosa, W. L. O., Cuevas-Suárez, C. E., Carreño, N. L. V., da Silva, A. F., Guim, T. N., Dellagostin, O. A., & Piva, E. (2017). Histological Evaluation of Bone Repair with Hydroxyapatite: A Systematic Review. *Calcified Tissue International*, 101(4), 341–354. DOI: 10.1007/s00223-017-0294-z PMID: 28612084

Ollier, N., Charpentier, T., Boizot, B., Wallez, G., & Ghaleb, D. (2004). A Raman and MAS NMR study of mixed alkali Na–K and Na–Li aluminoborosilicate glasses. *Journal of Non-Crystalline Solids*, 341(1–3), 26–34. DOI: 10.1016/j.jnoncrysol.2004.05.010

Omote, K., Ohigashi, H., & Koga, K. (1997). Temperature dependence of elastic, dielectric, and piezoelectric properties of "single crystalline" films of vinylidene fluoride trifluoroethylene copolymer. *Journal of Applied Physics*, 81(6), 2760–2769. DOI: 10.1063/1.364300

Oteng-Peprah, M., Acheampong, M. A., & deVries, N. K. (2018). Greywater Characteristics, treatment systems, reuse strategies and user perception—A review. *Water, Air, and Soil Pollution*, 229(8), 255. Advance online publication. DOI: 10.1007/s11270-018-3909-8 PMID: 30237637

Otowa, T., Nojima, Y., & Miyazaki, T. (1997). Development of KOH activated high surface area carbon and its application to drinking water purification. *Carbon*, 35(9), 1315–1319. DOI: 10.1016/S0008-6223(97)00076-6

Ouadi, Y. E., Fal, M. E., Hafez, B., Manssouri, M., Ansari, A., Elmsellem, H., Ramli, Y., & Bendaif, H. (2020). Physisorption and corrosion inhibition of mild steel in 1 M HCl using a new pyrazolic compound: Experimental data & quantum chemical calculations. *Materials Today: Proceedings*, 27, 3010–3016. DOI: 10.1016/j.matpr.2020.03.340

Ouici, H., Tourabi, M., Benali, O., Selles, C., Jama, C., Zarrouk, A., & Bentiss, F. (2017). Adsorption and corrosion inhibition properties of 5-amino 1,3,4-thiadiazole-2-thiol on the mild steel in HCl medium: Thermodynamic, surface and electrochemical studies. *Journal of Electroanalytical Chemistry (Lausanne, Switzerland)*, 803, 125–134. DOI: 10.1016/j.jelechem.2017.09.018

Ousslim, A., Chetouani, A., Hammouti, B., Bekkouch, K., Al-Deyab, S., Aouniti, A., & Elidrissi, A. (2013). Thermodynamics, quantum and electrochemical studies of corrosion of iron by piperazine compounds in sulphuric acid. *International Journal of Electrochemical Science*, 8(4), 5980–6004. DOI: 10.1016/S1452-3981(23)14734-1

Ou, Y. X., Wang, H. Q., Ouyang, X., Zhao, Y. Y., Zhou, Q., Luo, C. W., Hua, Q. S., Ouyang, X. P., & Zhang, S. (2023). Recent advances and strategies for high-performance coatings. *Progress in Materials Science*, 136, 101125. DOI: 10.1016/j.pmatsci.2023.101125

Owens, G. J., Singh, R. K., Foroutan, F., Alqaysi, M., Han, C.-M., Mahapatra, C., Kim, H.-W., & Knowles, J. C. (2016). Sol–gel based materials for biomedical applications. *Progress in Materials Science*, 77, 1–79. DOI: 10.1016/j.pmatsci.2015.12.001

Oxley, J. C., Smith, J. L., Zheng, W., Rogers, E., & Coburn, M. D. (1997). Thermal Decomposition Studies on Ammonium Dinitramide (ADN) and 15N and 2H Isotopomers. *The Journal of Physical Chemistry A*, 101(31), 5646–5652. DOI: 10.1021/jp9625063

Oyekunle, D., Agboola, O., & Ayeni, A. (2019). Corrosion inhibitors as building evidence for Mild steel: A review. *Journal of Physics: Conference Series*, 1378(3), 032046. DOI: 10.1088/1742-6596/1378/3/032046

Painter, G. S., Becher, P. F., Kleebe, H. J., & Pezzotti, G. (2002). First-principles study of the effects of halogen dopants on the properties of intergranular films in silicon nitride ceramics. *Physical Review B: Condensed Matter*, 65(6), 064113. Advance online publication. DOI: 10.1103/PhysRevB.65.064113

Pais, A., Moreira, C., & Belinha, J. (2024). The Biomechanical Analysis of Tibial Implants Using Meshless Methods: Stress and Bone Tissue Remodeling Analysis. *Designs*, 8(2), 28. DOI: 10.3390/designs8020028

Pal, B., Yang, S., Ramesh, S., Thangadurai, V., & Jose, R. (2019). Electrolyte selection for supercapacitive devices: A critical review. *Nanoscale Advances*, 1(10), 3807–3835. DOI: 10.1039/C9NA00374F PMID: 36132093

Palma, V., Barba, D., Vaiano, V., Colozzi, M., Palo, E., Barbato, L., Cortese, S., & Miccio, M. (2018). Honeycomb structured catalysts for H_2 production via H_2S oxidative decomposition. *Catalysts*, 8(11), 488. DOI: 10.3390/catal8110488

Palza, H., Escobar, B., Bejarano, J., Bravo, D., Diaz-Dosque, M., & Perez, J. (2013). Designing antimicrobial bioactive glass materials with embedded metal ions synthesized by the sol–gel method. *Materials Science and Engineering C*, 33(7), 3795–3801. DOI: 10.1016/j.msec.2013.05.012 PMID: 23910279

Pandey, A. P., Shaz, M. A., Sekkar, V., & Tiwari, R. S. (2023). Synergistic effect of CNT bridge formation and spillover mechanism on enhanced hydrogen storage by iron doped carbon aerogel. *International Journal of Hydrogen Energy*, 48(56), 21395–21403. DOI: 10.1016/j.ijhydene.2022.02.076

Papetti, V., Eggenschwiler, P. D., Della Torre, A., Lucci, F., Ortona, A., & Montenegro, G. (2018). Additive Manufactured open cell polyhedral structures as substrates for automotive catalysts. *International Journal of Heat and Mass Transfer. International Journal of Heat and Mass Transfer*, 126, 1035–1047. DOI: 10.1016/j.ijheatmasstransfer.2018.06.061

Parc, R. L., Champagnon, B., Guenot, P., & Dubois, S. (2001). Thermal annealing and density fluctuations in silica glass. *Journal of Non-Crystalline Solids*, 293–295, 366–369. DOI: 10.1016/S0022-3093(01)00835-3

Parent, L. R., Pham, C. H., Patterson, J. P., Denny, M. S.Jr, Cohen, S. M., Gianneschi, N. C., & Paesani, F. (2017). Pore Breathing of Metal–Organic Frameworks by Environmental Transmission Electron Microscopy. *Journal of the American Chemical Society*, 139(40), 13973–13976. DOI: 10.1021/jacs.7b06585 PMID: 28942647

Park, J., Chakraborty, D., & Lin, M. (1998). Thermal decomposition of gaseous ammonium dinitramide at low pressure: Kinetic modeling of product formation with ab initio MO/cVRRKM calculations. *Symposium (International) on Combustion*, 27(2), 2351–2357. https://doi.org/DOI: 10.1016/S0082-0784(98)80086-6

Park, J., Cho, S. Y., Jung, M., Lee, K., Nah, Y.-C., Attia, N. F., & Oh, H. (2021). Efficient synthetic approach for nanoporous adsorbents capable of pre- and post-combustion CO2 capture and selective gas separation. *Journal of CO2 Utilization, 45*, 101404. https://doi.org/https://doi.org/10.1016/j.jcou.2020.101404

Park, G. O., Shon, J. K., Kim, Y. H., & Kim, J. M. (2015). Synthesis of Ordered Mesoporous Manganese Oxides with Various Oxidation States. *Journal of Nanoscience and Nanotechnology*, 15(3), 2441–2445. DOI: 10.1166/jnn.2015.10263 PMID: 26413684

Park, H. Y., Seo, I. T., Choi, J. H., Nahm, S., & Lee, H. G. (2010). Low-temperature sintering and piezoelectric properties of (Na0. 5K0. 5) NbO3 lead-free piezoelectric ceramics. *Journal of the American Ceramic Society*, 93(1), 36–39. DOI: 10.1111/j.1551-2916.2009.03359.x

Park, J. H., Wang, Q., Zhu, K., Frank, A. J., & Kim, J. Y. (2019). Electrochemical Deposition of Conformal Semiconductor Layers in Nanoporous Oxides for Sensitized Photoelectrodes. *ACS Omega*, 4(22), 19772–19776. DOI: 10.1021/acsomega.9b02552 PMID: 31788609

Park, J. S., Kim, S. O., Jeong, Y. J., Lee, S. G., Choi, J. K., & Kim, S. J. (2022). Long-Term corrosion behavior of strong and ductile high MN-Low CR steel in acidic aqueous environments. *Materials (Basel)*, 15(5), 1746. DOI: 10.3390/ma15051746 PMID: 35268977

Park, K. S. (2023). *Humans and Electricity: Understanding Body Electricity and Applications*. Springer Nature., DOI: 10.1007/978-3-031-20784-6

Park, S. H., Ahn, C. W., Nahm, S., & Song, J. S. (2004). Microstructure and piezoelectric properties of ZnO-added (Na0. 5K0. 5) NbO3 ceramics. *Japanese Journal of Applied Physics*, 43(8B), L1072–L1074. DOI: 10.1143/JJAP.43.L1072

Parravicini, J., Safioui, J., Degiorgio, V., Minzioni, P., & Chauvet, M. (2011). All-optical technique to measure the pyroelectric coefficient in electro-optic crystals. *Journal of Applied Physics*, 109(3), 033106. DOI: 10.1063/1.3544069

Passos, R. S., Davenport, A., Busquets, R., Selden, C., Silva, L. B., Baptista, J. S., Barceló, D., & Campos, L. C. (2023). Microplastics and nanoplastics in haemodialysis waters: Emerging threats to be in our radar. *Environmental Toxicology and Pharmacology*, 102, 104253. DOI: 10.1016/j.etap.2023.104253 PMID: 37604358

Patel, P. N., Mishra, V., & Panchal, A. K. (2012). Theoretical and experimental study of nanoporous silicon photonic microcavity optical sensor devices. *Advances in Natural Sciences: Nanoscience and Nanotechnology*, 3(3), 35016. DOI: 10.1088/2043-6262/3/3/035016

Pathaare, Y., Reddy, A. M., Sangrulkar, P., Kandasubramanian, B., & Satapathy, A. (2023). Carbon hybrid nano-architectures as an efficient electrode material for supercapacitor applications. *Hybrid Advances*, 3, 100041. DOI: 10.1016/j.hybadv.2023.100041

Patterson, J. P., Abellan, P., Denny, M. S.Jr, Park, C., Browning, N. D., Cohen, S. M., Evans, J. E., & Gianneschi, N. C. (2015). Observing the Growth of Metal–Organic Frameworks by in Situ Liquid Cell Transmission Electron Microscopy. *Journal of the American Chemical Society*, 137(23), 7322–7328. DOI: 10.1021/jacs.5b00817 PMID: 26053504

Pauline, S. A., Karuppusamy, I., Gopalsamy, K., & Nallaiyan, R. (2024). Template assisted fabrication of nanoporous titanium dioxide coating on 316 L stainless steel for orthopaedic applications. *Journal of the Taiwan Institute of Chemical Engineers*, 105576, 105576. https://doi.org/https://doi.org/10.1016/j.jtice.2024.105576. DOI: 10.1016/j.jtice.2024.105576

Perrone Donnorso, M., Miele, E., De Angelis, F., La Rocca, R., Limongi, T., Cella Zanacchi, F., Marras, S., Brescia, R., & Di Fabrizio, E. (2012). Nanoporous silicon nanoparticles for drug delivery applications. *Microelectronic Engineering*, 98, 626–629. https://doi.org/https://doi.org/10.1016/j.mee.2012.07.095. DOI: 10.1016/j.mee.2012.07.095

Petersen, H., & Weidenthaler, C. (2022). A review of recent developments for the *in situ / operando* characterization of nanoporous materials. *Inorganic Chemistry Frontiers*, 9(16), 4244–4271. DOI: 10.1039/D2QI00977C

Petrunin, M., Rybkina, A., Yurasova, T., & Maksaeva, L. (2022). Effect of Organosilicon Self-Assembled Polymeric Nanolayers Formed during Surface Modification by Compositions Based on Organosilanes on the Atmospheric Corrosion of Metals. *Polymers*, 14(20), 4428. DOI: 10.3390/polym14204428 PMID: 36298006

Piatti, E., Miola, M., & Verné, E. (2024). Tailoring of bioactive glass and glass-ceramics properties for *in vitro* and *in vivo* response optimization: A review. *Biomaterials Science*, 12(18), 4546–4589. DOI: 10.1039/D3BM01574B PMID: 39105508

Platschek, B., Keilbach, A., & Bein, T. (2011). Mesoporous Structures Confined in Anodic Alumina Membranes. *Advanced Materials*, 23(21), 2395–2412. DOI: 10.1002/adma.201002828 PMID: 21484885

Poblano-Salas, C. A., Henao, J., Giraldo-Betancur, A. L., Forero-Sossa, P., Espinosa-Arbelaez, D. G., González-Sánchez, J. A., Dzib-Pérez, L. R., Estrada-Moo, S. T., & Pech-Pech, I. E. (2024). HVOF-sprayed HAp/S53P4 BG composite coatings on an AZ31 alloy for potential applications in temporary implants. *Journal of Magnesium and Alloys*, 12(1), 345–360. DOI: 10.1016/j.jma.2023.12.010

Pokhmurs'kyi, V. I., Zin, I. M., & Lyon, S. B. (2004). Inhibition of corrosion by a mixture of nonchromate pigments in organic coatings on galvanized steel. *Materials Science*, 40(3), 383–390. DOI: 10.1007/PL00022002

Polarz, S., & Smarsly, B. (2002). Nanoporous Materials. *Journal of Nanoscience and Nanotechnology*, 2(6), 581–612. DOI: 10.1166/jnn.2002.151 PMID: 12908422

Pompei, C. M. E., Ciric, L., Canales, M., Karu, K., Vieira, E. M., & Campos, L. C. (2017). Influence of PPCPs on the performance of intermittently operated slow sand filters for household water purification. *The Science of the Total Environment*, 581–582, 174–185. DOI: 10.1016/j.scitotenv.2016.12.091 PMID: 28041695

Poolakkandy, R. R., & Menamparambath, M. M. (2020). Soft-template-assisted synthesis: A promising approach for the fabrication of transition metal oxides. *Nanoscale Advances*, 2(11), 5015–5045. DOI: 10.1039/D0NA00599A PMID: 36132034

Prakasam, M., Locs, J., Salma-Ancane, K., Loca, D., Largeteau, A., & Berzina-Cimdina, L. (2017). Biodegradable Materials and Metallic Implants—A Review. *Journal of Functional Biomaterials*, 8(4), 44. DOI: 10.3390/jfb8040044 PMID: 28954399

Prasad Mishra, S., Dutta, S., Kumar Sahu, A., Mishra, K., & Kashyap, P. (2021). Potential Application of Nanoporous Materials in Biomedical Field. In *Nanopores*. IntechOpen., DOI: 10.5772/intechopen.95928

Prasad, A. R., Arshad, M., & Joseph, A. (2021c). A sustainable method of mitigating acid corrosion of mild steel using jackfruit pectin (JP) as green inhibitor: Theoretical and electrochemical studies. *Journal of the Indian Chemical Society*, 99(1), 100271. DOI: 10.1016/j.jics.2021.100271

Prasanna, B. M., Praveen, B. M., Hebbar, N., Pavithra, M. K., Manjunatha, T. S., & Malladi, R. S. (2018). Theoretical and experimental approach of inhibition effect by sulfamethoxazole on mild steel corrosion in 1-M HCl. *Surface and Interface Analysis*, 50(8), 1–11. DOI: 10.1002/sia.6457

Prashar, G., Vasudev, H., Thakur, L., & Bansal, A. (2023). Performance of thermally sprayed hydroxyapatite coatings for biomedical implants: A comprehensive review. *Surface Review and Letters*, 30(01), 2241001. DOI: 10.1142/S0218625X22410013

Prezas, P. R., Soares, M. J., Borges, J. P., Silva, J. C., Oliveira, F. J., & Graça, M. P. F. (2023). Bioactivity Enhancement of Plasma-Sprayed Hydroxyapatite Coatings through Non-Contact Corona Electrical Charging. *Nanomaterials (Basel, Switzerland)*, 13(6), 1058. DOI: 10.3390/nano13061058 PMID: 36985952

Prida, V. M., Vega, V., García, J., Iglesias, L., Hernando, B., & Minguez-Bacho, I. (2015). Electrochemical methods for template-assisted synthesis of nanostructured materials. In *Magnetic Nano- and Microwires* (pp. 3–39). Elsevier., DOI: 10.1016/B978-0-08-100164-6.00001-1

Priya, A. K., Gnanasekaran, L., Kumar, P. S., Jalil, A. A., Hoang, T. K. A., Rajendran, S., Soto-Moscoso, M., & Balakrishnan, D. (2022). Recent trends and advancements in nanoporous membranes for water purification. *Chemosphere*, 303, 135205. https://doi.org/https://doi.org/10.1016/j.chemosphere.2022.135205. DOI: 10.1016/j.chemosphere.2022.135205 PMID: 35667502

Priya, A. K., Muruganandam, M., & Suresh, S. (2024). Bio-derived carbon-based materials for sustainable environmental remediation and wastewater treatment. *Chemosphere*, 362, 142731. https://doi.org/https://doi.org/10.1016/j.chemosphere.2024.142731. DOI: 10.1016/j.chemosphere.2024.142731 PMID: 38950744

Puchi-Cabrera, E. S., Rossi, E., Sansonetti, G., Sebastiani, M., & Bemporad, E. (2023). Machine learning aided nanoindentation: A review of the current state and future perspectives. *Current Opinion in Solid State and Materials Science*, 27(4), 101091. https://doi.org/https://doi.org/10.1016/j.cossms.2023.101091. DOI: 10.1016/j.cossms.2023.101091

Punj, S., Singh, J., & Singh, K. (2021). Ceramic biomaterials: Properties, state of the art and future prospectives. *Ceramics International*, 47(20), 28059–28074. DOI: 10.1016/j.ceramint.2021.06.238

Qian, J., Wang, J., Zhang, W., Mao, J., Qin, H., Ling, X., Zeng, H., Hou, J., Chen, Y., & Wan, G. (2023). Corrosion-tailoring, osteogenic, anti-inflammatory, and antibacterial aspirin-loaded organometallic hydrogel composite coating on biodegradable Zn for orthopedic applications. *Biomaterials Advances*, 153, 213536. DOI: 10.1016/j.bioadv.2023.213536 PMID: 37418934

Qian, M., Li, L., Li, H., & Strachan, D. M. (2004). Partitioning of gadolinium and its induced phase separation in sodium-aluminoborosilicate glasses. *Journal of Non-Crystalline Solids*, 333(1), 1–15. DOI: 10.1016/j.jnoncrysol.2003.09.056

Qin, Z., Chen, K., Sun, X., Zhang, M., Wang, L., Zheng, S., Chen, C., Tang, H., Li, H., Zou, C., & Wu, G. (2024). The Electrical Properties and In Vitro Osteogenic Properties of 3D-Printed Fe@ BT/HA Piezoelectric Ceramic Scaffold. *Ceramics International*. Advance online publication. DOI: 10.1016/j.ceramint.2024.06.371

Qi, X., Ren, P., Tong, X., Wang, X., & Zhuo, F. (2024). Enhanced piezoelectric properties of KNN-based ceramics by synergistic modulation of phase constitution, grain size and domain configurations. *Journal of the European Ceramic Society*, 45(1), 116874. DOI: 10.1016/j.jeurceramsoc.2024.116874

Quintas, A., Caurant, D., Majérus, O., Charpentier, T., & Dussossoy, J. L. (2009). Effect of compositional variations on charge compensation of AlO4 and BO4 entities and on crystallization tendency of a rare-earth-rich aluminoborosilicate glass. *Materials Research Bulletin*, 44(9), 1895–1898. DOI: 10.1016/j.materresbull.2009.05.009

Quintas, A., Charpentier, T., Majérus, O., Caurant, D., Dussossoy, J. L., & Vermaut, P. (2007). NMR Study of a Rare-Earth Aluminoborosilicate Glass with Varying CaO-to-Na2O Ratio. *Applied Magnetic Resonance*, 32(4), 613–634. DOI: 10.1007/s00723-007-0041-0

Quraishi, M., & Shukla, S. K. (2008). Poly(aniline-formaldehyde): A new and effective corrosion inhibitor for mild steel in HCl. *Materials Chemistry and Physics*, 113(2–3), 685–689. DOI: 10.1016/j.matchemphys.2008.08.028

Rabiee, S. M., Nazparvar, N., Azizian, M., Vashaee, D., & Tayebi, L. (2015). Effect of ion substitution on properties of bioactive glasses: A review. *Ceramics International*, 41(6), 7241–7251. DOI: 10.1016/j.ceramint.2015.02.140

Rajabi, A. H., Jaffe, M., & Arinzeh, T. L. (2015). Piezoelectric materials for tissue regeneration: A review. *Acta Biomaterialia*, 24, 12–23. DOI: 10.1016/j.actbio.2015.07.010 PMID: 26162587

Rajamanickam, R., Ganesan, B., Kim, I., Hasan, I., Arumugam, P., & Paramasivam, S. (2024). Effective synthesis of nitrogen doped carbon nanotubes over transition metal loaded mesoporous catalysts for energy storage of supercapacitor applications. *Zeitschrift für Physikalische Chemie*, 238(10), 1835–1861. Advance online publication. DOI: 10.1515/zpch-2023-0458

Rajan, S. T., Subramanian, B., & Arockiarajan, A. (2022). A comprehensive review on biocompatible thin films for biomedical application. *Ceramics International*, 48(4), 4377–4400. DOI: 10.1016/j.ceramint.2021.10.243

Rajendran, A. K., Anthraper, M. S., Hwang, N. S., & Rangasamy, J. (2024). Osteogenesis and angiogenesis promoting bioactive ceramics. *Materials Science and Engineering R Reports*, 159, 100801. DOI: 10.1016/j.mser.2024.100801

Ramezani, M., Mohd Ripin, Z., Pasang, T., & Jiang, C. P. (2023). Surface engineering of metals: Techniques, characterizations and applications. *Metals*, 13(7), 1299. DOI: 10.3390/met13071299

Ramu, M., Ananthasubramanian, M., Kumaresan, T., Gandhinathan, R., & Jothi, S. (2018). Optimization of the configuration of porous bone scaffolds made of Polyamide/Hydroxyapatite composites using Selective Laser Sintering for tissue engineering applications. *Bio-Medical Materials and Engineering*, 29(6), 739–755. DOI: 10.3233/BME-181020 PMID: 30282331

Ranjan, P., & Prem, M. (2018). Schmutzdecke- A Filtration Layer of Slow Sand Filter. *International Journal of Current Microbiology and Applied Sciences*, 7(07), 637–645. DOI: 10.20546/ijcmas.2018.707.077

Ravikumar, K., Kar, G. P., Bose, S., & Basu, B. (2016). Synergistic effect of polymorphism, substrate conductivity and electric field stimulation towards enhancing muscle cell growth in vitro. *RSC Advances*, 6(13), 10837–10845. DOI: 10.1039/C5RA26104J

Ravina, K., Kumar, S., Hashmi, S. Z., Srivastava, G., Singh, J., Quraishi, A. M., Dalela, S., Ahmed, F., & Alvi, P. A. (2023). Synthesis and investigations of structural, surface morphology, electrochemical, and electrical properties of NiFe2O4 nanoparticles for usage in supercapacitors. *Journal of Materials Science Materials in Electronics*, 34(10), 868. DOI: 10.1007/s10854-023-10312-1

Rawlings, R. D., Wu, J. P., & Boccaccini, A. R. (2006). Glass-ceramics: Their production from wastes—A Review. *Journal of Materials Science*, 41(3), 733–761. DOI: 10.1007/s10853-006-6554-3

Rebolledo, S., Alcántara-Dufeu, R., Luengo Machuca, L., Ferrada, L., & Sánchez-Sanhueza, G. A. (2023). Real-time evaluation of the biocompatibility of calcium silicate-based endodontic cements: An in vitro study. *Clinical and Experimental Dental Research*, 9(2), 322–331. DOI: 10.1002/cre2.714 PMID: 36866428

Re, F., Borsani, E., Rezzani, R., Sartore, L., & Russo, D. (2023). Bone Regeneration Using Mesenchymal Stromal Cells and Biocompatible Scaffolds: A Concise Review of the Current Clinical Trials. *Gels (Basel, Switzerland)*, 9(5), 389. DOI: 10.3390/gels9050389 PMID: 37232981

Regufe, M. J., Ferreira, A. F., Loureiro, J. M., Rodrigues, A., & Ribeiro, A. M. (2019). Electrical conductive 3D-printed monolith adsorbent for CO_2 capture. *Microporous and Mesoporous Materials*, 278, 403–413. DOI: 10.1016/j.micromeso.2019.01.009

Reibstein, S., Wondraczek, L., De Ligny, D., Krolikowski, S., Sirotkin, S., Simon, J. P., Martinez, V., & Champagnon, B. (2011). Structural heterogeneity and pressure-relaxation in compressed borosilicate glasses by in situ small angle X-ray scattering. *The Journal of Chemical Physics*, 134(20), 204502. Advance online publication. DOI: 10.1063/1.3593399 PMID: 21639451

Remissa, I., Baragh, F., Mabrouk, A., Bachar, A., & Amrousse, R. (2024). Low-Cost Catalysts for Hydrogen Peroxide (H_2O_2) Thermal Decomposition. In *Space technology library* (pp. 61–94). DOI: 10.1007/978-3-031-62574-9_3

Remissa, I., Jabri, H., Hairch, Y., Toshtay, K., Atamanov, M., Azat, S., & Amrousse, R. (2023). Propulsion Systems, Propellants, Green Propulsion Subsystems and their Applications: A Review. *Eurasian Chemico-technological Journal*, 25(1), 3–19. DOI: 10.18321/ectj1491

Remissa, I., Souagh, A., Hairch, Y., Sahib-Eddine, A., Atamanov, M., & Amrousse, R. (2022). Thermal decomposition behaviors of hydrogen peroxide over free noble metal-synthesized solid catalysts. *International Journal of Energetic Materials and Chemical Propulsion*, 21(4), 17–29. DOI: 10.1615/IntJEnergeticMaterialsChemProp.2022043338

Ren, Y., & Xu, Y. (2023). Three-dimensional graphene/metal–organic framework composites for electrochemical energy storage and conversion. *Chemical Communications*, 59(43), 6475–6494. DOI: 10.1039/D3CC01167D PMID: 37185628

Reza, N. A., Akhmal, N. H., Fadil, N. A., & Taib, M. F. M. (2021). A review on plants and biomass wastes as organic green corrosion inhibitors for mild steel in acidic environment. *Metals*, 11(7), 1062. DOI: 10.3390/met11071062

Ribeiro, C., Correia, D. M., Rodrigues, I., Guardão, L., Guimarães, S., Soares, R., & Lanceros-Méndez, S. (2017). In vivo demonstration of the suitability of piezoelectric stimuli for bone reparation. *Materials Letters*, 209, 118–121. DOI: 10.1016/j.matlet.2017.07.099

Ribeiro, C., Sencadas, V., Correia, D. M., & Lanceros-Méndez, S. (2015). Piezoelectric polymers as biomaterials for tissue engineering applications. *Colloids and Surfaces. B, Biointerfaces*, 136, 46–55. DOI: 10.1016/j.colsurfb.2015.08.043 PMID: 26355812

Rice ,R. (2005). *Mechanical Properties*. 289–312. DOI: 10.1002/3527606696.ch4a

Rice, R. W. (1996). Evaluation and extension of physical property-porosity models based on minimum solid area. *Journal of Materials Science*, 31(1), 102–118. DOI: 10.1007/BF00355133

Riffat, R. (2012). Fundamentals of wastewater treatment and engineering. CRC Press, Taylor & Francis group, LLC, Boca Raton, Florida (US). 1st Edition. DOI: 10.1201/b12746

Riggs, O. L.Jr, & Hurd, R. M. (1967). Temperature coefficient of corrosion inhibition. *Corrosion*, 23(8), 252–260. DOI: 10.5006/0010-9312-23.8.252

Rinky, N. J., Islam, M. M., Hossen, J., & Islam, M. A. (2023). Comparative study of anti-ulcer drugs with benzimidazole ring as green corrosion inhibitors in acidic solution: Quantum chemical studies. *Current Research in Green and Sustainable Chemistry*, 7, 100385. DOI: 10.1016/j.crgsc.2023.100385

Rochdi, N., El Azhary, K., Jouti, N. T., Badou, A., Majidi, B., & Habti, N., & NAIMI, Y. (2023). In Vivo Study Of Hypersensitivity And Irritation Skin For Nanocomposite Biphasic Calcium Phosphate Synthesized Directly By Hydrothermal Process. *Journal of Namibian Studies: History Politics Culture*, 33(2), 6142–6157. DOI: 10.59670/jns.v33i.5040

Rodríguez-Sánchez, L., Blanco, M. C., & López-Quintela, M. A. (2000). Electrochemical Synthesis of Silver Nanoparticles. *The Journal of Physical Chemistry B*, 104(41), 9683–9688. DOI: 10.1021/jp001761r

Romanczuk, E., Perkowski, K., & Oksiuta, Z. (2019). Microstructure, Mechanical, and Corrosion Properties of Ni-Free Austenitic Stainless Steel Prepared by Mechanical Alloying and HIPping. *Materials (Basel)*, 12(20), 3416. DOI: 10.3390/ma12203416 PMID: 31635345

Rossi, M. J., Bottaro, J. C., & McMillen, D. F. (1993). The thermal decomposition of the new energetic material ammoniumdinitramide (NH4N(NO2)2) in relation to Nitramide (NH2NO2) and NH4NO3. *International Journal of Chemical Kinetics*, 25(7), 549–570. DOI: 10.1002/kin.550250705

Rossin, J. (1989). XPS surface studies of activated carbon. *Carbon*, 27(4), 611–613. DOI: 10.1016/0008-6223(89)90012-2

Rouquerol, J., Baron, G., Denoyel, R., Giesche, H., Groen, J., Klobes, P., Levitz, P., Neimark, A. V., Rigby, S., Skudas, R., Sing, K., Thommes, M., & Unger, K. (2011). Liquid intrusion and alternative methods for the characterization of macroporous materials (IUPAC Technical Report). *Pure and Applied Chemistry*, 84(1), 107–136. DOI: 10.1351/PAC-REP-10-11-19

Rouxel, T. (2007). Elastic Properties and Short-to Medium-Range Order in Glasses. *Journal of the American Ceramic Society*, 90(10), 3019–3039. DOI: 10.1111/j.1551-2916.2007.01945.x

Ruan, S., Xin, W., Wang, C., Wan, W., Huang, H., Gan, Y., Xia, Y., Zhang, J., Xia, X., He, X., & Zhang, W. (2023). An approach to enhance carbon/polymer interface compatibility for lithium-ion supercapacitors. *Journal of Colloid and Interface Science*, 652, 1063–1073. DOI: 10.1016/j.jcis.2023.08.053 PMID: 37643524

Ruch, P., Cericola, D., Foelske-Schmitz, A., Koetz, R., & Wokaun, A. (2010). Aging of electrochemical double-layer capacitors with acetonitrile-based electrolyte at elevated voltages. *Electrochimica Acta*, 55(15), 4412–4420. DOI: 10.1016/j.electacta.2010.02.064

Rui, G., Allahyarov, E., Zhu, Z., Huang, Y., Wongwirat, T., Zou, Q., Taylor, P., & Zhu, L. (2024). Challenges and opportunities in piezoelectric polymers: Effect of oriented amorphous fraction in ferroelectric semicrystalline polymers. *Responsive Materials*, 2(3), e20240002. DOI: 10.1002/rpm.20240002

Ryoo, R., Joo, S. H., & Jun, S. (1999). Synthesis of Highly Ordered Carbon Molecular Sieves via Template-Mediated Structural Transformation. *The Journal of Physical Chemistry B*, 103(37), 7743–7746. DOI: 10.1021/jp991673a

Saadati pour. M., Gilak, M. R., Pedram, M. Z., & Naikoo, G. (2023). Enhancement electrochemical properties of supercapacitors using hybrid CuO/Ag/rGO based nanoporous composite as electrode materials. *Journal of Energy Storage, 74*, 109330. https://doi.org/https://doi.org/10.1016/j.est.2023.109330

Sabzi, M., Mousavi Anijdan, S., Shamsodin, M., Farzam, M., Hojjati-Najafabadi, A., Feng, P., Park, N., & Lee, U. (2023). A Review on Sustainable Manufacturing of Ceramic-Based Thin Films by Chemical Vapor Deposition (CVD): Reactions Kinetics and the Deposition Mechanisms. *Coatings*, 13(1), 188. DOI: 10.3390/coatings13010188

Safronova, T. V., Putlayev, V., & Shekhirev, M. (2013). Resorbable calcium phosphates based ceramics. *Powder Metallurgy and Metal Ceramics*, 52(5-6), 357–363. DOI: 10.1007/s11106-013-9534-6

Sagawa, H., Itoh, S., Wang, W., & Yamashita, K. (2010). Enhanced bone bonding of the hydroxyapatite/beta-tricalcium phosphate composite by electrical polarization in rabbit long bone. *Artificial Organs*, 34(6), 491–497. DOI: 10.1111/j.1525-1594.2009.00912.x PMID: 20456322

Sahoo, B. B., Pandey, V. S., Dogonchi, A. S., Thatoi, D. N., Nayak, N., & Nayak, M. K. (2023). Synthesis, characterization and electrochemical aspects of graphene based advanced supercapacitor electrodes. *Fuel*, 345, 128174. DOI: 10.1016/j.fuel.2023.128174

Sahraoui, M., Boulkroune, M., Chibani, A., Larbah, Y., & Abdessemed, A. (2022). Aqueous extract of Punica granatum fruit peel as an Eco-Friendly corrosion inhibitor for aluminium alloy in acidic medium. *Journal of Bio- and Tribo-Corrosion*, 8(2), 54. Advance online publication. DOI: 10.1007/s40735-022-00658-0

Saikia, B. K., Benoy, S. M., Bora, M., Neog, D., Bhattacharjya, D., Rajbongshi, A., & Saikia, P. (2024). Fabrication of pouch cell supercapacitors using abundant coal feedstock and their hybridization with Li-ion battery for e-rickshaw application. *Journal of Energy Storage*, 78, 110312. DOI: 10.1016/j.est.2023.110312

Sailaukhanuly, Y., Azat, S., Kunarbekova, M., Tovassarov, A., Toshtay, K., Tauanov, Z., Carlsen, L., & Berndtsson, R. (2024). Health Risk Assessment of Nitrate in Drinking Water with Potential Source Identification: A Case Study in Almaty, Kazakhstan. *International Journal of Environmental Research and Public Health*, 21(1), 55. DOI: 10.3390/ijerph21010055 PMID: 38248520

Sairanen, E., Karinen, R., Borghei, M., Kauppinen, E. I., & Lehtonen, J. (2012). Preparation Methods for Multi-Walled Carbon Nanotube Supported Palladium Catalysts. *ChemCatChem*, 4(12), 2055–2061. DOI: 10.1002/cctc.201200344

Sakamoto, Y., Inagaki, S., Ohsuna, T., Ohnishi, N., Fukushima, Y., Nozue, Y., & Terasaki, O. (1998). Structure analysis of mesoporous material 'FSM-16' Studies by electron microscopy and X-ray diffraction. *Microporous and Mesoporous Materials*, 21(4), 589–596. https://doi.org/https://doi.org/10.1016/S1387-1811(98)00053-5. DOI: 10.1016/S1387-1811(98)00053-5

Sakka, S., Kamiya, K., & Yoko, T. (1983b). Preparation and properties of Ca Al Si O N oxynitride glasses. *Journal of Non-Crystalline Solids*, 56(1–3), 147–152. DOI: 10.1016/0022-3093(83)90460-X

Sakthi Muthulakshmi, S., Shailajha, S., & Shanmugapriya, B. (2023). Bio-physical investigation of calcium silicate biomaterials by green synthesis-osseous tissue regeneration. *Journal of Materials Research*, 38(19), 4369–4384. DOI: 10.1557/s43578-023-01149-9

Samadi, A., Salati, M. A., Safari, A., Jouyandeh, M., Barani, M., Singh Chauhan, N. P., Golab, E. G., Zarrintaj, P., Kar, S., Seidi, F., Hejna, A., & Saeb, M. R. (2022). Comparative review of piezoelectric biomaterials approach for bone tissue engineering. *Journal of Biomaterials Science. Polymer Edition*, 33(12), 1555–1594. DOI: 10.1080/09205063.2022.2065409 PMID: 35604896

Samal, P. P., Singh, C. P., Tiwari, S., Shah, V., & Krishnamurty, S. (2024). Indazole-5-amine (AIA) as competing corrosion coating to Benzotriazole (BTAH) at the interface of Cu: A DFT and BOMD case study. *Computational & Theoretical Chemistry*, 1239, 114762. DOI: 10.1016/j.comptc.2024.114762

Samipour, S., Setoodeh, P., Rahimpour, E., & Rahimpour, M. R. (2024). Functional nanoporous membranes for drug delivery. In Basile, A., Lipnizki, F., Rahimpour, M. R., & Piemonte, V. (Eds.), *Current Trends and Future Developments on (Bio-) Membranes* (pp. 255–288). Elsevier., https://doi.org/https://doi.org/10.1016/B978-0-323-90258-8.00023-7 DOI: 10.1016/B978-0-323-90258-8.00023-7

Sanaei, Z., Bahlakeh, G., & Ramezanzadeh, B. (2017). Active corrosion protection of mild steel by an epoxy ester coating reinforced with hybrid organic/inorganic green inhibitive pigment. *Journal of Alloys and Compounds*, 728, 1289–1304. DOI: 10.1016/j.jallcom.2017.09.095

Sang, Q., Hao, S., Han, J., & Ding, Y. (2022). Dealloyed nanoporous materials for electrochemical energy conversion and storage. *EnergyChem*, 4(1), 100069. https://doi.org/https://doi.org/10.1016/j.enchem.2022.100069. DOI: 10.1016/j.enchem.2022.100069

Santoliquido, O., Bianchi, G., Eggenschwiler, P. D., & Ortona, A. (2017). Additive manufacturing of periodic ceramic substrates for automotive catalyst supports. *International Journal of Applied Ceramic Technology*, 14(6), 1164–1173. DOI: 10.1111/ijac.12745

Santos, A. (2018). *Electrochemically Engineering of Nanoporous Materials*. MDPI., DOI: 10.3390/books978-3-03897-269-3

Saoût, G. L., Simon, P., Fayon, F., Blin, A., & Vaills, Y. (2008). Raman and infrared structural investigation of (PbO)x(ZnO)(0.6−x)(P2O5)0.4 glasses. *Journal of Raman Spectroscopy : JRS*, 40(5), 522–526. DOI: 10.1002/jrs.2158

Sappati, K. K., & Bhadra, S. (2018). Piezoelectric Polymer and Paper Substrates: A Review. *Sensors (Basel)*, 18(11), 3605. DOI: 10.3390/s18113605 PMID: 30355961

Saravanapavan, P., & Hench, L. L. (2001). Low-temperature synthesis, structure, and bioactivity of gel-derived glasses in the binary CaO-SiO2 system. *Journal of Biomedical Materials Research*, 54(4), 608–618. DOI: 10.1002/1097-4636(20010315)54:4<608::AID-JBM180>3.0.CO;2-U PMID: 11426607

Saravanapavan, P., Jones, J. R., Pryce, R. S., & Hench, L. L. (2003). Bioactivity of gel–glass powders in the CaO-SiO_2 system: A comparison with ternary (CaO-P_2P_5-SiO_2) and quaternary glasses (SiO_2-CaO-P_2O_5-Na_2O). *Journal of Biomedical Materials Research. Part A*, 66A(1), 110–119. DOI: 10.1002/jbm.a.10532 PMID: 12833437

Sarkar, K. (2024). Research progress on biodegradable magnesium phosphate ceramics in orthopaedic applications. *Journal of Materials Chemistry. B*, 12(35), 8605–8615. DOI: 10.1039/D4TB01123F PMID: 39140212

Sarmast Sh, M., George, S., Dayang Radiah, A. B., Hoey, D., Abdullah, N., & Kamarudin, S. (2022). Synthesis of bioactive glass using cellulose nano fibre template. *Journal of the Mechanical Behavior of Biomedical Materials*, 130, 105174. DOI: 10.1016/j.jmbbm.2022.105174 PMID: 35344755

Sayagués, M. J., Otero, A., Santiago-Andrades, L., Poyato, R., Monzón, M., Paz, R., Gotor, F. J., & Moriche, R. (2024). Fine-grained BCZT piezoelectric ceramics by combining high-energy mechanochemical synthesis and hot-press sintering. *Journal of Alloys and Compounds*, 176453, 176453. Advance online publication. DOI: 10.1016/j.jallcom.2024.176453

Scavetta, E., Casagrande, A., Gualandi, I., & Tonelli, D. (2014). Analytical performances of Ni LDH films electrochemically deposited on Pt surfaces: Phenol and glucose detection. *Journal of Electroanalytical Chemistry (Lausanne, Switzerland)*, 722–723, 15–22. https://doi.org/https://doi.org/10.1016/j.jelechem.2014.03.018. DOI: 10.1016/j.jelechem.2014.03.018

Schaller, T., & Stebbins, J. F. (1998). The Structural Role of Lanthanum and Yttrium in Aluminosilicate Glasses: A27Al and17O MAS NMR Study. *The Journal of Physical Chemistry B*, 102(52), 10690–10697. DOI: 10.1021/jp982387m

Schofield, Z., Meloni, G. N., Tran, P., Zerfass, C., Sena, G., Hayashi, Y., Grant, M., Contera, S. A., Minteer, S. D., Kim, M., Prindle, A., Rocha, P., Djamgoz, M. B. A., Pilizota, T., Unwin, P. R., Asally, M., & Soyer, O. S. (2020). Bioelectrical understanding and engineering of cell biology. *Journal of the Royal Society, Interface*, 17(166), 20200013. DOI: 10.1098/rsif.2020.0013 PMID: 32429828

Schönherr, J., Buchheim, J. R., Scholz, P., & Adelhelm, P. (2018). Boehm Titration Revisited (Part II): A Comparison of Boehm Titration with Other Analytical Techniques on the Quantification of Oxygen-Containing Surface Groups for a Variety of Carbon Materials. C – Journal of Carbon Research, 4(2), 22. DOI: 10.3390/c4020022

Sedik, A., Lerari, D., Salci, A., Athmani, S., Bachari, K., Gecibesler, İ., & Solmaz, R. (2020). Dardagan Fruit extract as eco-friendly corrosion inhibitor for mild steel in 1 M HCl: Electrochemical and surface morphological studies. *Journal of the Taiwan Institute of Chemical Engineers*, 107, 189–200. DOI: 10.1016/j.jtice.2019.12.006

Selvam, S., & Yim, J.-H. (2023). Effective self-charge boosting sweat electrolyte textile supercapacitors array from bio-compatible polymer metal chelates. *Journal of Power Sources*, 556, 232511. DOI: 10.1016/j.jpowsour.2022.232511

Sengupta, S., Murmu, M., Mandal, S., Hirani, H., & Banerjee, P. (2021b). Competitive corrosion inhibition performance of alkyl/acyl substituted 2-(2-hydroxybenzylideneamino)phenol protecting mild steel used in adverse acidic medium: A dual approach analysis using FMOs/molecular dynamics simulation corroborated experimental findings. *Colloids and Surfaces. A, Physicochemical and Engineering Aspects*, 617, 126314. DOI: 10.1016/j.colsurfa.2021.126314

Senthilkumar, G., Kaliaraj, G. S., Vignesh, P., Vishwak, R. S., Joy, T. N., & Hemanandh, J. (2021). Hydroxyapatite–barium/strontium titanate composite coatings for better mechanical, corrosion and biological performance. *Materials Today: Proceedings*, 44, 3618–3621. DOI: 10.1016/j.matpr.2020.09.758

Seo, H. J., Cho, Y. E., Kim, T., Shin, H. I., & Kwun, I. S. (2010). Zinc may increase bone formation through stimulating cell proliferation, alkaline phosphatase activity and collagen synthesis in osteoblastic MC3T3-E1 cells. *Nutrition Research and Practice*, 4(5), 356–361. DOI: 10.4162/nrp.2010.4.5.356 PMID: 21103080

Seppänen-Kaijansinkko, R. (2019). Hard Tissue Engineering. In *Tissue Engineering in Oral and Maxillofacial Surgery*. Springer International Publishing., DOI: 10.1007/978-3-030-24517-7_7

Serafin, J., & Dziejarski, B. (2023). Activated carbons—preparation, characterization and their application in CO2 capture: A review. *Environmental Science and Pollution Research International*, 31(28), 40008–40062. DOI: 10.1007/s11356-023-28023-9 PMID: 37326723

Sergi, R., Bellucci, D., & Cannillo, V. (2020). A comprehensive review of bioactive glass coatings: State of the art, challenges and future perspectives. *Coatings*, 10(8), 757. DOI: 10.3390/coatings10080757

Serrato Ochoa, D., Nieto Aguilar, R., & Aguilera Méndez, A. (2015). Ingeniería de tejidos. Una nueva disciplina en medicina regenerativa. *Investigación y Ciencia. Universidad Autónoma de Aguascalientes*, 64(64), 61–69. DOI: 10.33064/iycuaa2015643597

Shamjitha, C., & Vargeese, A. A. (2023). Development of bimetallic spinel catalysts for low-temperature decomposition of ammonium dinitramide monopropellants. *Defence Technology*, 30, 47–54. DOI: 10.1016/j.dt.2023.05.007

Shamzhy, M., De, O., & Ramos, F. S. (2015). Tuning of acidic and catalytic properties of IWR zeolite by post-synthesis incorporation of three-valent elements. *Catalysis Today*, 243, 76–84. DOI: 10.1016/j.cattod.2014.06.041

Sharifianjazi, F., Moradi, M., Abouchenari, A., Pakseresht, A. H., Esmaeilkhanian, A., Shokouhimehr, M., & Shahedi Asl, M. (2020). Effects of Sr and Mg dopants on biological and mechanical properties of SiO2–CaO–P2O5 bioactive glass. *Ceramics International*, 46(14), 22674–22682. DOI: 10.1016/j.ceramint.2020.06.030

Sharifianjazi, F., Parvin, N., & Tahriri, M. (2017). Synthesis and characteristics of sol-gel bioactive SiO2-P2O5-CaO-Ag2O glasses. *Journal of Non-Crystalline Solids*, 476, 108–113. DOI: 10.1016/j.jnoncrysol.2017.09.035

Sharma, A. (2018). *Applications of Nanoporous Materials in Gas Separation and Storage*. https://digitalscholarship.unlv.edu/thesesdissertations/3326

Sharma, A. (2023). *Hydroxyapatite coating techniques for Titanium Dental Implants—an overview*. Qeios., DOI: 10.32388/2E6UHN

Sharma, A., Kokil, G. R., He, Y., Lowe, B., Salam, A., Altalhi, T. A., Ye, Q., & Kumeria, T. (2023). Inorganic/organic combination: Inorganic particles/polymer composites for tissue engineering applications. *Bioactive Materials*, 24, 535–550. DOI: 10.1016/j.bioactmat.2023.01.003 PMID: 36714332

Sharma, H. B., & Mansingh, A. (1998). Sol-gel processed barium titanate ceramics and thin films. *Journal of Materials Science*, 33(17), 4455–4459. DOI: 10.1023/A:1004576315328

Sharma, M., Yadav, S. S., Sharma, P., Yadav, L., Abedeen, M. Z., Kushwaha, H. S., & Gupta, R. (2023). An experimental and theoretical investigation of corrosion inhibitive behaviour of 4-amino antipyrine and its schiff's base (BHAP) on mild steel in 1 M HCl solution. *Inorganic Chemistry Communications*, 157, 111330. DOI: 10.1016/j.inoche.2023.111330

Sharma, V., Chowdhury, S., Bose, S., & Basu, B. (2022). Polydopamine Codoped $BaTiO_3$-Functionalized Polyvinylidene Fluoride Coating as a Piezo-Biomaterial Platform for an Enhanced Cellular Response and Bioactivity. *ACS Biomaterials Science & Engineering*, 8(1), 170–184. DOI: 10.1021/acsbiomaterials.1c00879 PMID: 34964600

Shearer, A., Montazerian, M., Sly, J. J., Hill, R. G., & Mauro, J. C. (2023). Trends and perspectives on the commercialization of bioactive glasses. *Acta Biomaterialia*, 160, 14–31. DOI: 10.1016/j.actbio.2023.02.020 PMID: 36804821

Shekhawat, D., Singh, A., Banerjee, M. K., Singh, T., & Patnaik, A. (2021). Bioceramic composites for orthopaedic applications: A comprehensive review of mechanical, biological, and microstructural properties. *Ceramics International*, 47(3), 3013–3030. DOI: 10.1016/j.ceramint.2020.09.214

Shelby, J. E., & Lord, C. E. (1990). Formation and Properties of Calcia-Calcium Fluoride-Alumina Glasses. *Journal of the American Ceramic Society*, 73(3), 750–752. DOI: 10.1111/j.1151-2916.1990.tb06586.x

Shen, C., Brozena, A. H., & Wang, Y. (2011). Double-walled carbon nanotubes: Challenges and opportunities. *Nanoscale*, 3(2), 503–518. DOI: 10.1039/C0NR00620C PMID: 21042608

Shi, G., & Kioupakis, E. (2015). Electronic and Optical Properties of Nanoporous Silicon for Solar-Cell Applications. *ACS Photonics*, 2(2), 208–215. DOI: 10.1021/ph5002999

Shi, L., Shi, Y., Zhuo, S., Zhang, C., Aldrees, Y., Aleid, S., & Wang, P. (2019). Multi-functional 3D honeycomb ceramic plate for clean water production by heterogeneous photo-Fenton reaction and solar-driven water evaporation. *Nano Energy*, 60, 222–230. DOI: 10.1016/j.nanoen.2019.03.039

Shirke, X., & Balivada, S. (2024). Lead Zirconate Titanate (PZT) Synthesis and Its Applications. I*nternational Journal of Health Technology and Innovation*, 3(02), 15–19. https://ijht.org.in/index.php/ijhti/article/view/135

Shi, S., Jiang, Y., Ren, H., Deng, S., Sun, J., Cheng, F., Jing, J., & Chen, Y. (2024). 3D-Printed Carbon-Based Conformal Electromagnetic Interference Shielding Module for Integrated Electronics. *Nano-Micro Letters*, 16(1), 85. DOI: 10.1007/s40820-023-01317-w PMID: 38214822

Silva, C. C., Pinheiro, A. G., Figueiró, S. D., Goes, J. C., Sasaki, J. M., Miranda, M. A. R., & Sombra, A. S. B. (2002). Piezoelectric properties of collagen-nanocrystalline hydroxyapatite composites. *Journal of Materials Science*, 37(10), 2061–2070. DOI: 10.1023/A:1015219800490

Singh, A., Ansari, K., Sharma, N. R., Singh, S., Singh, R., Bansal, A., Ali, I. H., Younas, M., Alanazi, A. K., & Lin, Y. (2023). Corrosion and bacterial growth mitigation in the desalination plant by imidazolium based ionic liquid: Experimental, surface and molecular docking analysis. *Journal of Environmental Chemical Engineering*, 11(2), 109313. DOI: 10.1016/j.jece.2023.109313

Singh, J., Bhunia, H., & Basu, S. (2019). Adsorption of CO2 on KOH activated carbon adsorbents: Effect of different mass ratios. *Journal of Environmental Management*, 250, 109457. DOI: 10.1016/j.jenvman.2019.109457 PMID: 31472376

Sinha, S., Elbaz-Alon, Y., & Avinoam, O. (2022). Ca^{2+} as a coordinator of skeletal muscle differentiation, fusion and contraction. *The FEBS Journal*, 289(21), 6531–6542. DOI: 10.1111/febs.16552 PMID: 35689496

Sirimanne, D. C. U., Kularatna, N., & Arawwawala, N. (2023). Electrical Performance of Current Commercial Supercapacitors and Their Future Applications. *Electronics (Basel)*, 12(11), 2465. DOI: 10.3390/electronics12112465

Sivakumar, P. R. (2021). Corrosion protection behaviour of some unsymmetrical oxadiazoles on mild steel surface in 1 m H2SO4 acid medium. *Journal of Bio- and Tribo-Corrosion*, 8(1), 17. Advance online publication. DOI: 10.1007/s40735-021-00614-4

Slowing, I., Viveroescoto, J., Wu, C., & Lin, V.SLOWING. (2008). Mesoporous silica nanoparticles as controlled release drug delivery and gene transfection carriers☆. *Advanced Drug Delivery Reviews*, 60(11), 1278–1288. DOI: 10.1016/j.addr.2008.03.012 PMID: 18514969

Sneddon, G., Greenaway, A., & Yiu, H. H. P. (2014). The Potential Applications of Nanoporous Materials for the Adsorption, Separation, and Catalytic Conversion of Carbon Dioxide. *Advanced Energy Materials*, 4(10), 1301873. Advance online publication. DOI: 10.1002/aenm.201301873

Sohrabi, M., Yekta, B. E., Rezaie, H., Naimi-Jamal, M. R., Kumar, A., Cochis, A., Miola, M., & Rimondini, L. (2021). The effect of magnesium on bioactivity, rheology and biology behaviors of injectable bioactive glass-gelatin-3-glycidyloxypropyl trimethoxysilane nanocomposite-paste for small bone defects repair. *Ceramics International*, 47(9), 12526–12536. DOI: 10.1016/j.ceramint.2021.01.110

Soleilhavoup, A., Delaye, J. M., Angeli, F., Caurant, D., & Charpentier, T. (2010). Contribution of first-principles calculations to multinuclear NMR analysis of borosilicate glasses. *Magnetic Resonance in Chemistry*, 48(S1), S159–S170. DOI: 10.1002/mrc.2673 PMID: 20818801

Solomon, M. M., Essien, K. E., Loto, R. T., & Ademosun, O. T. (2021). Synergistic corrosion inhibition of low carbon steel in HCl and H2SO4 media by 5-methyl-3-phenylisoxazole-4-carboxylic acid and iodide ions. *Journal of Adhesion Science and Technology*, 36(11), 1200–1226. DOI: 10.1080/01694243.2021.1962091

Solsana, F., & Méndez, J. P. (2003). *Water Disinfection, Pan American Center for Sanitary Engineering and Environmental Sciences*. Pan American Health Organization.

Soltani, N., Keshavarzi, B., Moore, F., Busquets, R., Nematollahi, M. J., Javid, R., & Gobert, S. (2022). Effect of land use on microplastic pollution in a major boundary waterway: The Arvand River. *The Science of the Total Environment*, 830, 154728. DOI: 10.1016/j.scitotenv.2022.154728 PMID: 35331773

Song, J., Zhao, G., Li, B., & Wang, J. (2017). Design optimization of PVDF-based piezoelectric energy harvesters. *Heliyon*, 3(9), e00377. DOI: 10.1016/j.heliyon.2017.e00377 PMID: 28948235

Song, Q., Zhu, M., & Mao, Y. (2021). Chemical vapor deposited polyelectrolyte coatings with osteoconductive and osteoinductive activities. *Surface and Coatings Technology*, 423, 127522. DOI: 10.1016/j.surfcoat.2021.127522

Souagh, A., Remissa, I., Atamanov, M., Alaoui, H. E., & Amrousse, R. (2022). Comparative study of the thermal decomposition of hydroxylammonium nitrate green energetic compound: Combination between experimental and DFT calculation. *International Journal of Energetic Materials and Chemical Propulsion*, 21(4), 31–38. DOI: 10.1615/IntJEnergeticMaterialsChemProp.2022044056

Spirandeli, B. R., Campos, T. M. B., Ribas, R. G., Thim, G. P., & Trichês, E. D. S. (2020). Evaluation of colloidal and polymeric routes in sol-gel synthesis of a bioactive glass-ceramic derived from 45S5 bioglass. *Ceramics International*, 46(12), 20264–20271. DOI: 10.1016/j.ceramint.2020.05.108

Sprio, S., Antoniac, I., Chevalier, J., Iafisco, M., Sandri, M., & Tampieri, A. (2023). Editorial: Recent advances in bioceramics for health. *Frontiers in Bioengineering and Biotechnology*, 11, 1264799. DOI: 10.3389/fbioe.2023.1264799 PMID: 37593328

Stamnitz, S., & Klimczak, A. (2021). Mesenchymal Stem Cells, Bioactive Factors, and Scaffolds in Bone Repair: From Research Perspectives to Clinical Practice. *Cells*, 10(8), 1925. DOI: 10.3390/cells10081925 PMID: 34440694

Stan, G., King, S. W., & Cook, R. F. (2009). Elastic modulus of low- *k* dielectric thin films measured by load-dependent contact-resonance atomic force microscopy. *Journal of Materials Research*, 24(9), 2960–2964. DOI: 10.1557/jmr.2009.0357

Stein, A. (2003). Advances in Microporous and Mesoporous Solids—Highlights of Recent Progress. *Advanced Materials*, 15(10), 763–775. DOI: 10.1002/adma.200300007

Stevensson, B., & Edén, M. (2013). Structural rationalization of the microhardness trends of rare-earth aluminosilicate glasses: Interplay between the RE3+ field-strength and the aluminum coordinations. *Journal of Non-Crystalline Solids*, 378, 163–167. DOI: 10.1016/j.jnoncrysol.2013.06.013

Stochero, N., De Moraes, E., Moreira, A., Fernandes, C., Innocentini, M., & De Oliveira, A. N. (2020). Ceramic shell foams produced by direct foaming and gelcasting of proteins: Permeability and microstructural characterization by X-ray microtomography. *Journal of the European Ceramic Society*, 40(12), 4224–4231. DOI: 10.1016/j.jeurceramsoc.2020.05.036

Studers, P., Jakovicka, D., Solska, J., Bladiko, U., & Radziņa, M. (2023). Mid-term clinical and radiographic outcomes after primary total hip replacement with fully hydroxyapatite-coated stem: A cross-sectional study. *Journal of Orthopaedics, Trauma and Rehabilitation*, 30(2), 241–247. DOI: 10.1177/22104917231171937

Sudaryanto, Y., Hartono, S. B., Irawaty, W., Hindarso, H., & Ismadji, S. (2005). High surface area activated carbon prepared from cassava peel by chemical activation. *Bioresource Technology*, 97(5), 734–739. DOI: 10.1016/j.biortech.2005.04.029 PMID: 15963718

Sudipta, S. M., & Murugavel, S. (2022). Biomineralization behavior of ternary mesoporous bioactive glasses stabilized through ethanol extraction process. *Journal of Non-Crystalline Solids*, 589, 121630. DOI: 10.1016/j.jnoncrysol.2022.121630

Suga, M., Asahina, S., Sakuda, Y., Kazumori, H., Nishiyama, H., Nokuo, T., Alfredsson, V., Kjellman, T., Stevens, S. M., Cho, H. S., Cho, M., Han, L., Che, S., Anderson, M. W., Schüth, F., Deng, H., Yaghi, O. M., Liu, Z., Jeong, H. Y., & Terasaki, O. (2014). Recent progress in scanning electron microscopy for the characterization of fine structural details of nano materials. *Progress in Solid State Chemistry*, 42(1), 1–21. https://doi.org/https://doi.org/10.1016/j.progsolidstchem.2014.02.001. DOI: 10.1016/j.progsolidstchem.2014.02.001

Su, K., Wang, C., Pu, Y., Wang, Y., Ma, P., Liu, L., Tian, X., Du, H., & Lang, J. (2024). Dilute aqueous hybrid electrolyte endows a high-voltage window for supercapacitors. *Journal of Alloys and Compounds*, 1002, 175354. DOI: 10.1016/j.jallcom.2024.175354

Sukenaga, S., Ogawa, M., Yanaba, Y., Ando, M., & Shibata, H. (2020). Viscosity of Na–Si–O–N–F Melts: Mixing Effect of Oxygen, Nitrogen, and Fluorine. *ISIJ International*, 60(12), 2794–2806. DOI: 10.2355/isijinternational.ISIJINT-2020-326

Sun, Y., Zhao, Z., Qiao, Q., Li, S., Yu, W., Guan, X., Schneider, A., Weir, M. D., Xu, H. H. K., Zhang, K., & Bai, Y. (2023). Injectable periodontal ligament stem cell-metformin-calcium phosphate scaffold for bone regeneration and vascularization in rats. *Dental materials: official publication of the Academy of Dental Materials, 39*(10), 872–885. https://doi.org/DOI: 10.1016/j.dental.2023.07.008

Sun, L., Ma, Y., Ding, Y., Mei, X., Liu, Y., & Feng, C. (2021). Removal of antibiotic resistance genes in secondary effluent by slow filtration-NF process. *Water Science and Technology*, 85(1), 152–165. DOI: 10.2166/wst.2021.607 PMID: 35050873

Sun, T. W., Yu, W. L., Zhu, Y. J., Yang, R. L., Shen, Y. Q., Chen, D. Y., He, Y. H., & Chen, F. (2017). Hydroxyapatite Nanowire@Magnesium Silicate Core-Shell Hierarchical Nanocomposite: Synthesis and Application in Bone Regeneration. *ACS Applied Materials & Interfaces*, 9(19), 16435–16447. DOI: 10.1021/acsami.7b03532 PMID: 28481082

Sun, Y., Chen, J., Li, A., Liu, F., & Zhang, Q. (2005). Adsorption of resorcinol and catechol from aqueous solution by aminated hypercrosslinked polymers. *Reactive & Functional Polymers*, 64(2), 63–73. DOI: 10.1016/j.reactfunctpolym.2005.03.004

Surmenev, R. A., Orlova, T., Chernozem, R. V., Ivanova, A. A., Bartasyte, A., Mathur, S., & Surmeneva, M. A. (2019). Hybrid lead-free polymer-based nanocomposites with improved piezoelectric response for biomedical energy-harvesting applications: A review. *Nano Energy*, 62, 475–506. DOI: 10.1016/j.nanoen.2019.04.090

Su, Y., Wang, X., Ye, Y., Xie, Y., Xu, Y., Jiang, Y., & Wang, C. (2024). Automation and machine learning augmented by large language models in a catalysis study. *Chemical Science (Cambridge)*, 15(31), 12200–12233. https://doi.org/https://doi.org/10.1039/d3sc07012c. DOI: 10.1039/D3SC07012C PMID: 39118602

Swain, S., Padhy, R. N., & Rautray, T. R. (2020). Electrically stimulated hydroxyapatite–barium titanate composites demonstrate immunocompatibility in vitro. *Journal of the Korean Ceramic Society*, 57(5), 495–502. DOI: 10.1007/s43207-020-00048-7

Świątkowski, A. (1999). Industrial carbon adsorbents. In Studies in surface science and catalysis (pp. 69–94). https://doi.org/DOI: 10.1016/S0167-2991(99)80549-7

Szauer, T., & Brandt, A. (1983). Equilibria in solutions of amines and fatty acids with relevance to the corrosion inhibition of iron. *Corrosion Science*, 23(12), 1247–1257. DOI: 10.1016/0010-938X(83)90075-6

Szu, S.-P., Klein, L. C., & Greenblatt, M. (1992). Effect of precursors on the structure of phosphosilicate gels: 29Si and 31P MAS-NMR study. *Journal of Non-Crystalline Solids*, 143, 21–30. DOI: 10.1016/S0022-3093(05)80548-4

Tabard, L., Garnier, V., Prud'homme, E., Courtial, E., Meille, S., Adrien, J., Jorand, Y., & Gremillard, L. (2021). Robocasting of highly porous ceramics scaffolds with hierarchized porosity. *Additive Manufacturing*, 38, 101776. DOI: 10.1016/j.addma.2020.101776

Tale, B. U., Nemade, K. R., & Tekade, P. V. (2024). Novel graphene based MnO2/polyaniline nanohybrid material for efficient supercapacitor application. *Journal of Porous Materials*, 31(6), 2053–2065. Advance online publication. DOI: 10.1007/s10934-024-01656-y

Tang, M., Liu, L., Gao, X., Zhang, Z., & Wang, Y. (2024). Collaborative Optimization of Curie Temperature and Piezoelectricity in Ternary $BiFeO_3$-$BiScO_3$-$PbTiO_3$. *ACS Applied Materials & Interfaces*, 16(31), 41185–41193. DOI: 10.1021/acsami.4c08300 PMID: 39069883

Tang, Y., Chen, L., Duan, Z., Zhao, K., & Wu, Z. (2020). Graphene/barium titanate/polymethyl methacrylate bio-piezoelectric composites for biomedical application. *Ceramics International*, 46(5), 6567–6574. DOI: 10.1016/j.ceramint.2019.11.142

Tanwer, S., & Shukla, S. K. (2021b). *Corrosion inhibition activity of Cefixime on mild steel surface in aqueous sulphuric acid.* Progress in Color, Colorants and Coatings., DOI: 10.30509/pccc.2021.166889.1133

Taurbekov, A., Fierro, V., Kuspanov, Z., Abdisattar, A., Atamanova, T., Kaidar, B., Mansurov, Z., & Atamanov, M. (2024). Nanocellulose and carbon nanotube composites: A universal solution for environmental and energy challenges. *Journal of Environmental Chemical Engineering*, 12(5), 113262. DOI: 10.1016/j.jece.2024.113262

Tebbji, K., Faska, N., Tounsi, A., Oudda, H., Benkaddour, M., & Hammouti, B. (2007). The effect of some lactones as inhibitors for the corrosion of mild steel in 1M HCl. *Materials Chemistry and Physics*, 106(2–3), 260–267. DOI: 10.1016/j.matchemphys.2007.05.046

Ten Elshof, J. E. (2022). Chemical solution deposition of oxide thin films. In *Epitaxial Growth of Complex Metal Oxides* (pp. 75–100). Woodhead Publishing., DOI: 10.1016/B978-0-08-102945-9.00012-5

Thomas, P., Arenberger, P., Bader, R., Bircher, A. J., Bruze, M., de Graaf, N., Hartmann, D., Johansen, J. D., Jowitz-Heinke, A., Krenn, V., Kurek, M., Odgaard, A., Rustemeyer, T., Summer, B., & Thyssen, J. P. (2024). A literature review and expert consensus statement on diagnostics in suspected metal implant allergy. *Journal of the European Academy of Dermatology and Venereology*, 38(8), 1471–1477. DOI: 10.1111/jdv.20026 PMID: 38606660

Thomas, S., Balakrishnan, P., & Sreekala, M. S. (2018). *Fundamental biomaterials: ceramics.* Woodhead Publishing., DOI: 10.1016/B978-0-08-102203-0.00001-9

Thombre, N. V. (2023). Oxidation in water and used water purification. In *Springer eBooks* (pp. 1–23). https://doi.org/DOI: 10.1007/978-3-319-66382-1_174-1

Thommes, M., Kaneko, K., Neimark, A. V., Olivier, J. P., Rodriguez-Reinoso, F., Rouquerol, J., & Sing, K. S. (2015). Physisorption of gases, with special reference to the evaluation of surface area and pore size distribution (IUPAC Technical Report). *Pure and Applied Chemistry*, 87(9–10), 1051–1069. DOI: 10.1515/pac-2014-1117

Thommes, M., & Schlumberger, C. (2021). Characterization of Nanoporous Materials. *Annual Review of Chemical and Biomolecular Engineering*, 12(1), 137–162. DOI: 10.1146/annurev-chembioeng-061720-081242 PMID: 33770464

Thuau, D., Abbas, M., Wantz, G., Hirsch, L., Dufour, I., & Ayela, C. (2016). Piezoelectric polymer gated OFET: Cutting-edge electro-mechanical transducer for organic MEMS-based sensors. *Scientific Reports*, 6(1), 38672. DOI: 10.1038/srep38672 PMID: 27924853

Tian, M., Wang, X. D., & Zhang, T. (2016). Hexaaluminates: A review of the structure, synthesis and catalytic performance. *Catalysis Science & Technology*, 6(7), 1984–2004. DOI: 10.1039/C5CY02077H

Troyan, J. E. (1953). Properties, Production, and Uses of Hydrazine. *Industrial & Engineering Chemistry*, 45(12), 2608–2612. DOI: 10.1021/ie50528a020

Tüysüz, H., Lehmann, C. W., Bongard, H., Tesche, B., Schmidt, R., & Schüth, F. (2008). Direct Imaging of Surface Topology and Pore System of Ordered Mesoporous Silica (MCM-41, SBA-15, and KIT-6) and Nanocast Metal Oxides by High Resolution Scanning Electron Microscopy. *Journal of the American Chemical Society*, 130(34), 11510–11517. DOI: 10.1021/ja803362s PMID: 18671351

Ubaidah Saidin, N., Kuan Ying, K., & Inn Khuan, N. (n.d.). *ELECTRODEPOSITION: PRINCIPLES, APPLICATIONS AND METHODS ELEKTRO-PEMENDAPAN: PRINSIP, APLIKASI DAN KAEDAH.*

Udayakumar, K. V., Gore, P. M., & Kandasubramanian, B. (2021). Foamed materials for oil-water separation. *Chemical Engineering Journal Advances*, 5, 100076. DOI: 10.1016/j.ceja.2020.100076

United Nations Statistics Division. (n.d.). — SDG indicators. https://unstats.un.org/sdgs/report/2022/Goal06/?_gl=1*2lpenp*_ga*MTU4NjM3MzgyNy4xNzI3Njc3OTgx*_ga_TK9BQL5X7Z*MTcyNzY3Nzk4MS4xLjEuMTcyNzY3Nzk5MS4wLjAuMA. (assessed online, 30/09/2024)

Urade, A. R., Lahiri, I., & Suresh, K. S. (2023). Graphene Properties, Synthesis and Applications: A Review. *JOM*, 75(3), 614–630. DOI: 10.1007/s11837-022-05505-8 PMID: 36267692

Vaills, Y., Luspin, Y., & Hauret, G. (2001). Annealing effects in SiO2–Na2O glasses investigated by Brillouin scattering. *Journal of Non-Crystalline Solids*, 286(3), 224–234. DOI: 10.1016/S0022-3093(01)00523-3

Valencia, J. A. (2000). *Teoría y práctica de la purificación del agua Tercera edición.* https://z-lib.io/book/13759703

Van Gestel, N. A. P., Geurts, J., Hulsen, D. J. W., Van Rietbergen, B., Hofmann, S., & Arts, J. J. (2015). Clinical Applications of S53P4 Bioactive Glass in Bone Healing and Osteomyelitic Treatment: A Literature Review. *BioMed Research International*, 2015, 1–12. DOI: 10.1155/2015/684826 PMID: 26504821

Van Voorhees, E. J., & Green, D. J. (1991). Failure behavior of Cellular-Core ceramic sandwich composites. *Journal of the American Ceramic Society*, 74(11), 2747–2752. DOI: 10.1111/j.1151-2916.1991.tb06838.x

Vandrovcova, M., Tolde, Z., Vanek, P., Nehasil, V., Doubková, M., Trávníčková, M., Drahokoupil, J., Buixaderas, E., Borodavka, F., Novakova, J., & Bacakova, L. (2021). Beta-Titanium alloy covered by ferroelectric coating–physicochemical properties and human osteoblast-like cell response. *Coatings*, 11(2), 210. DOI: 10.3390/coatings11020210

Vaněk, P., Kolská, Z., Luxbacher, T., García, J. A. L., Lehocký, M., Vandrovcová, M., Bačáková, L., & Petzelt, J. (2016). Electrical activity of ferroelectric biomaterials and its effects on the adhesion, growth and enzymatic activity of human osteoblast-like cells. *Journal of Physics. D, Applied Physics*, 49(17), 175403. DOI: 10.1088/0022-3727/49/17/175403

Vasquez-Sancho, F., Catalan, G., & Sort Viñas, J. (2018) *Flexoelectricity in biomaterials*. Universitat Autònoma de Barcelona. Accesed: https://dialnet.unirioja.es/servlet/tesis?codigo=229380

Vaughn, W. L., & Risbud, S. H. (1984). New fluoronitride glasses in zirconium-metal-F-N systems. *Journal of Materials Science Letters*, 3(2), 162–164. DOI: 10.1007/BF00723104

Venkateshwarlu, A., Singh, S., & Melnik, R. (2024). Piezoelectricity and flexoelectricity in biological cells: The role of cell structure and organelles. *arXiv preprint arXiv:2403.02050*. https://doi.org//arXiv.2403.02050DOI: 10.48550

Verma, C., Quraishi, M. A., Ebenso, E. E., & Hussain, C. M. (2021). Amines as corrosion inhibitors: A review. In Organic Corrosion Inhibitors: Synthesis, Characterization, Mechanism, and Applications (pp. 77-94). wiley. DOI: 10.1002/9781119794516.ch5

Verma, S., Daverey, A., & Sharma, A. (2017). Slow sand filtration for water and wastewater treatment–a review. *Environmental Technology Reviews*, 6(1), 47–58. DOI: 10.1080/21622515.2016.1278278

Verma, S., Padha, B., Singh, A., Khajuria, S., Sharma, A., Mahajan, P., Singh, B., & Arya, S. (2021). Sol-gel synthesized carbon nanoparticles as supercapacitor electrodes with ultralong cycling stability. *Fullerenes, Nanotubes, and Carbon Nanostructures*, 29(12), 1045–1052. DOI: 10.1080/1536383X.2021.1928645

Vestergaard, P., Jorgensen, N. R., Schwarz, P., & Mosekilde, L. (2007b). Effects of treatment with fluoride on bone mineral density and fracture risk - a meta-analysis. *Osteoporosis International*, 19(3), 257–268. DOI: 10.1007/s00198-007-0437-6 PMID: 17701094

Vikneshvaran, S., & Velmathi, S. (2019). Schiff Bases of 2,5-Thiophenedicarboxaldehyde as Corrosion Inhibitor for Stainless Steel under Acidic Medium: Experimental, Quantum Chemical and Surface Studies. *ChemistrySelect*, 4(1), 387–392. DOI: 10.1002/slct.201803235

Vlasea, M., Pilliar, R., & Toyserkani, E. (2015). Control of structural and mechanical properties in bioceramic bone substitutes via additive manufacturing layer stacking orientation. *Additive Manufacturing*, 6, 30–38. DOI: 10.1016/j.addma.2015.03.001

Vyazovkin, S., & Wight, C. A. (1997). Ammonium Dinitramide: Kinetics and Mechanism of Thermal Decomposition. *The Journal of Physical Chemistry A*, 101(31), 5653–5658. DOI: 10.1021/jp962547z

Walkley, B., Rees, G. J., Nicolas, R. S., Van Deventer, J. S. J., Hanna, J. V., & Provis, J. L. (2018). New Structural Model of Hydrous Sodium Aluminosilicate Gels and the Role of Charge-Balancing Extra-Framework Al. *The Journal of Physical Chemistry. C, Nanomaterials and Interfaces*, 122(10), 5673–5685. DOI: 10.1021/acs.jpcc.8b00259

Wang, C., Tao, Y., & Wang, S. (1982). Effect of nitrogen ion-implantation on silicate glasses. *Journal of Non-Crystalline Solids*, 52(1–3), 589–603. DOI: 10.1016/0022-3093(82)90336-2

Wang, D., Luo, H., Kou, R., Gil, M. P., Xiao, S., Golub, V. O., Yang, Z., Brinker, C. J., & Lu, Y. (2004). A General Route to Macroscopic Hierarchical 3D Nanowire Networks. *Angewandte Chemie International Edition*, 43(45), 6169–6173. https://doi.org/https://doi.org/10.1002/anie.200460535. DOI: 10.1002/anie.200460535 PMID: 15549745

Wang, H., Fan, H., & Zheng, J. (2003). Corrosion inhibition of mild steel in HCl solution by a mercapto-triazole compound. *Materials Chemistry and Physics*, 77(3), 655–661. DOI: 10.1016/S0254-0584(02)00123-2

Wang, J., Holt-Hindle, P., MacDonald, D., Thomas, D. F., & Chen, A. (2008). Synthesis and electrochemical study of Pt-based nanoporous materials. *Electrochimica Acta*, 53(23), 6944–6952. https://doi.org/https://doi.org/10.1016/j.electacta.2008.02.028. DOI: 10.1016/j.electacta.2008.02.028

Wang, M., Zhang, G., Li, W., & Wang, X. (2015). Microstructure and Properties of BaTiO3 Ferroelectric Films Prepared by DC Micro Arc Oxidation. *Bulletin of the Korean Chemical Society*, 36(4), 1178–1182. DOI: 10.1002/bkcs.10220

Wang, S., & Stebbins, J. F. (1999). Multiple-Quantum Magic-Angle Spinning 17O NMR Studies of Borate, Borosilicate, and Boroaluminate Glasses. *Journal of the American Ceramic Society*, 82(6), 1519–1528. DOI: 10.1111/j.1151-2916.1999.tb01950.x

Wang, T., Liu, Z., Li, P., Wei, H., Wei, K., & Chen, X. (2023). Lignin-derived carbon aerogels with high surface area for supercapacitor applications. *Chemical Engineering Journal*, 466, 143118. DOI: 10.1016/j.cej.2023.143118

Wang, X., Cheng, H., Ye, G., Fan, J., Yao, F., Wang, Y., Jiao, Y., Zhu, W., Huang, H., & Ye, D. (2022). Key factors and primary modification methods of activated carbon and their application in adsorption of carbon-based gases: A review. *Chemosphere*, 287, 131995. DOI: 10.1016/j.chemosphere.2021.131995 PMID: 34509016

Wang, Y., Sha, J., Zhu, S., Ma, L., He, C., Zhong, C., Hu, W., & Zhao, N. (2023). Data-driven design of carbon-based materials for high-performance flexible energy storage devices. *Journal of Power Sources*, 556, 232522. DOI: 10.1016/j.jpowsour.2022.232522

Water, U. States. E. P. Agency. O. of D. (1990). *Technologies for Upgrading Existing or Designing New Drinking Water Treatment Facilities*. US Environmental Protection Agency.

Wereszczak, A. A., & Anderson, C. E.Jr. (2014). Borofloat and Starphire Float Glasses: A Comparison. *International Journal of Applied Glass Science*, 5(4), 334–344. DOI: 10.1111/ijag.12095

Wheaton, B. R., & Clare, A. G. (2007). Evaluation of phase separation in glasses with the use of atomic force microscopy. *Journal of Non-Crystalline Solids*, 353(52–54), 4767–4778. DOI: 10.1016/j.jnoncrysol.2007.06.073

Wilkinson, J. L., Thornhill, I., Oldenkamp, R., Gachanja, A., & Busquets, R. (2023). Pharmaceuticals and Personal Care Products in the Aquatic Environment: How Can Regions at Risk be Identified in the Future? *Environmental Toxicology and Chemistry*, 43(3), 575–588. DOI: 10.1002/etc.5763 PMID: 37818878

Willatzen, M. (2024). *Piezoelectricity in Classical and Modern Systems*. IOP Publishing. DOI: 10.1088/978-0-7503-5557-5

Willinger, M.-G., Clavel, G., Di, W., & Pinna, N. (2009). A general soft-chemistry route to metal phosphate nanocrystals. *Journal of Industrial and Engineering Chemistry*, 15(6), 883–887. DOI: 10.1016/j.jiec.2009.09.017

Wingborg, N. (2006). Ammonium Dinitramide–Water: Interaction and Properties. *Journal of Chemical & Engineering Data*, 51(5), 1582–1586. DOI: 10.1021/je0600698

Woldeamanuel, M. M., Mohapatra, S., Senapati, S., Bastia, T. K., Panda, A. K., & Rath, P. (2024). Role of magnetic nanomaterials in environmental remediation. In *Nanostructure science and technology* (pp. 185–208). https://doi.org/DOI: 10.1007/978-3-031-44599-6_11

Wondraczek, L., Behrens, H., Yue, Y., Deubener, J., & Scherer, G. W. (2007). Relaxation and Glass Transition in an Isostatically Compressed Diopside Glass. *Journal of the American Ceramic Society*, 90(5), 1556–1561. DOI: 10.1111/j.1551-2916.2007.01566.x

Wondraczek, L., Krolikowski, S., & Behrens, H. (2010). Kinetics of pressure relaxation in a compressed alkali borosilicate glass. *Journal of Non-Crystalline Solids*, 356(35–36), 1859–1862. DOI: 10.1016/j.jnoncrysol.2010.06.009

Wood, R. J., & Lu, P. (2024). Coatings and surface modification of alloys for tribo-corrosion applications. *Coatings*, 14(1), 99. DOI: 10.3390/coatings14010099

Wu, J. (2024). *Piezoelectric Materials: From Fundamentals to Emerging Applications*. John Wiley & Sons. Accessed: https://www.wiley.com/

Wu, C., Wang, X., Zhou, X., Yang, T., & Zhang, T. (2015). Supported MnOx /SRO-Al_2O_3 high-cell density honeycomb ceramic monolith catalyst for high-concentration hydrogen peroxide decomposition. *International Journal of Energetic Materials and Chemical Propulsion*, 14(5), 421–436. DOI: 10.1615/IntJEnergeticMaterialsChemProp.2015011176

Wu, K.-J., Young, W.-B., & Young, C. (2023). Structural supercapacitors: A mini-review of their fabrication, mechanical & electrochemical properties. *Journal of Energy Storage*, 72, 108358. DOI: 10.1016/j.est.2023.108358

Wusirika, R. (1984). Problems Associated with the Melting of Oxynitride Glasses. *Journal of the American Ceramic Society*, 67(11). Advance online publication. DOI: 10.1111/j.1151-2916.1984.tb19492.x

Wu, Y., Sun, Y., Tong, Y., Liu, X., Zheng, J., Han, D., Li, H., & Niu, L. (2021). Recent advances in potassium-ion hybrid capacitors: Electrode materials, storage mechanisms and performance evaluation. *Energy Storage Materials*, 41, 108–132. DOI: 10.1016/j.ensm.2021.05.045

Wu, Y., Tang, L., Zhang, Q., Kong, F., & Bi, Y. (2021). A novel synthesis of monodispersed bioactive glass nanoparticles via ultrasonic-assisted surfactant-free microemulsion approach. *Materials Letters*, 285, 129053. DOI: 10.1016/j.matlet.2020.129053

Wu, Z.-P., Zhang, H., Chen, C., Li, G., & Han, Y. (2022). *Applications of in situ electron microscopy in oxygen electrocatalysis*. Microstructures., DOI: 10.20517/microstructures.2021.12

Wu, Z., Tang, T., Guo, H., Tang, S., Niu, Y., Zhang, J., Zhang, W., Ma, R., Su, J., Liu, C., & Wei, J. (2014). In vitro degradability, bioactivity and cell responses to mesoporous magnesium silicate for the induction of bone regeneration. *Colloids and Surfaces. B, Biointerfaces*, 120, 38–46. DOI: 10.1016/j.colsurfb.2014.04.010 PMID: 24905677

Wu, Z., Zheng, K., Zhang, J., Tang, T., Guo, H., Boccaccini, A. R., & Wei, J. (2016). Effects of magnesium silicate on the mechanical properties, biocompatibility, bioactivity, degradability, and osteogenesis of poly(butylene succinate)-based composite scaffolds for bone repair. *Journal of Materials Chemistry. B*, 4(48), 7974–7988. DOI: 10.1039/C6TB02429G PMID: 32263787

Xiao, B.-H., Xiao, K., Li, J.-X., Xiao, C.-F., Cao, S., & Liu, Z.-Q. (2024). Flexible electrochemical energy storage devices and related applications: Recent progress and challenges. *Chemical Science (Cambridge)*, 15(29), 11229–11266. DOI: 10.1039/D4SC02139H PMID: 39055032

Xiao, K., Liu, H., Li, Y., Yang, G., Wang, Y., & Yao, H. (2020). Excellent performance of porous carbon from urea-assisted hydrochar of orange peel for toluene and iodine adsorption. *Chemical Engineering Journal*, 382, 122997. DOI: 10.1016/j.cej.2019.122997

Xiao, X., Song, H., Lin, S., Zhou, Y., Zhan, X., Hu, Z., Zhang, Q., Sun, J., Yang, B., Li, T., Jiao, L., Zhou, J., Tang, J., & Gogotsi, Y. (2016). Scalable salt-templated synthesis of two-dimensional transition metal oxides. *Nature Communications*, 7(1), 11296. DOI: 10.1038/ncomms11296 PMID: 27103200

Xie, P., Yuan, W., Liu, X., Peng, Y., Yin, Y., Li, Y., & Wu, Z. (2021). Advanced carbon nanomaterials for state-of-the-art flexible supercapacitors. *Energy Storage Materials*, 36, 56–76. DOI: 10.1016/j.ensm.2020.12.011

Xin, X., Qi, C., Xu, L., Gao, Q., & Liu, X. (2022). Green synthesis of silver nanoparticles and their antibacterial effects. *Frontiers in Chemical Engineering*, 4, 941240. Advance online publication. DOI: 10.3389/fceng.2022.941240

Xiong, P., Tan, J., Lee, H., Ha, N., Lee, S. J., Yang, W., & Park, H. S. (2022). *Two-dimensional carbon-based heterostructures as bifunctional electrocatalysts for water splitting and metal–air batteries*. Nano Materials Science., DOI: 10.1016/j.nanoms.2022.10.001

Xiong, Y., & Cao, M. (2024). Application of surfactants in corrosion inhibition of metals. *Current Opinion in Colloid & Interface Science*, 73, 101830. DOI: 10.1016/j.cocis.2024.101830

Xue, L., Zhang, C., He, H., & Teraoka, Y. (2007). Catalytic decomposition of N2O over CeO2 promoted Co3O4 spinel catalyst. *Applied Catalysis B Environment and Energy*, 75(3–4), 167–174. https://doi.org/ DOI: 10.1016/j.apcatb.2007.04.013

Xu, H. (2001). Dielectric properties and ferroelectric behavior of poly (vinylidene fluoride-trifluoroethylene) 50/50 copolymer ultrathin films. *Journal of Applied Polymer Science*, 80(12), 2259–2266. DOI: 10.1002/app.1330

Xu, L., Campos, L. C., Li, J., Karu, K., & Ciric, L. (2021). Removal of antibiotics in sand, GAC, GAC sandwich and anthracite/sand biofiltration systems. *Chemosphere*, 275, 130004. DOI: 10.1016/j.chemosphere.2021.130004 PMID: 33640744

Xu, P., Li, X., Yu, H., & Xu, T. (2014). Advanced Nanoporous Materials for Micro-Gravimetric Sensing to Trace-Level Bio/Chemical Molecules. *Sensors (Basel)*, 14(10), 19023–19056. DOI: 10.3390/s141019023 PMID: 25313499

Xu, Q., Gao, X., Zhao, S., Liu, Y. N., Zhang, D., Zhou, K., Khanbareh, H., Chen, W., Zhang, Y., & Bowen, C. (2021). Construction of Bio-Piezoelectric Platforms: From Structures and Synthesis to Applications. *Advanced Materials*, 33(27), e2008452. DOI: 10.1002/adma.202008452 PMID: 34033180

Xu, T., Cheng, S., Jin, L., Zhang, K., & Zeng, T. (2020). High-temperature flexural strength of SiC ceramics prepared by additive manufacturing. *International Journal of Applied Ceramic Technology*, 17(2), 438–448. DOI: 10.1111/ijac.13454

Xu, W., Wang, T., Wang, H., Zhu, S., Liang, Y., Cui, Z., Yang, X., & Inoue, A. (2019). Free-standing amorphous nanoporous nickel cobalt phosphide prepared by electrochemically delloying process as a high performance energy storage electrode material. *Energy Storage Materials*, 17, 300–308. https://doi .org/https://doi.org/10.1016/j.ensm.2018.07.005. DOI: 10.1016/j.ensm.2018.07.005

Xu, Y., Fang, X., Xiong, J., & Zhang, Z. (2010). Hydrothermal transformation of titanate nanotubes into single-crystalline TiO2 nanomaterials with controlled phase composition and morphology. *Materials Research Bulletin*, 45(7), 799–804. DOI: 10.1016/j.materresbull.2010.03.016

Yadav, A. K., Chowdhury, A., & Srivastava, A. (2017). Interferometric investigation of methanol droplet combustion in varying oxygen environments under normal gravity. *International Journal of Heat and Mass Transfer*, 111, 871–883. DOI: 10.1016/j.ijheatmasstransfer.2017.03.125

Yadav, M., Behera, D., & Sharma, U. (2012). Nontoxic corrosion inhibitors for N80 steel in HCl. *Arabian Journal of Chemistry*, 9, S1487–S1495. DOI: 10.1016/j.arabjc.2012.03.011

Yadav, P., Gupta, S. M., & Sharma, S. K. (2022). A review on stabilization of carbon nanotube nanofluid. *Journal of Thermal Analysis and Calorimetry*, 147(12), 6537–6561. DOI: 10.1007/s10973-021-10999-6

Yang, F., Wang, X., Zhang, D., Yang, J., Luo, D., Xu, Z., Wei, J., Wang, J.-Q., Xu, Z., Peng, F., Li, X., Li, R., Li, Y., Li, M., Bai, X., Ding, F., & Li, Y. (2014). Chirality-specific growth of single-walled carbon nanotubes on solid alloy catalysts. *Nature*, 510(7506), 522–524. DOI: 10.1038/nature13434 PMID: 24965654

Yang, H., Xu, L., Li, Y., Liu, H., Wu, X., Zhou, P., Graham, N. J. D., & Yu, W. (2023). FexO/FeNC modified activated carbon packing media for biological slow filtration to enhance the removal of dissolved organic matter in reused water. *Journal of Hazardous Materials*, 457, 131736. DOI: 10.1016/j.jhazmat.2023.131736 PMID: 37295334

Yang, I., Jung, M., Kim, M., Choi, D., & Jung, J. C. (2021). Physical and chemical activation mechanisms of carbon materials based on the microdomain model. *Journal of Materials Chemistry. A, Materials for Energy and Sustainability*, 9(15), 9815–9825. DOI: 10.1039/D1TA00765C

Yang, J. W., Kwon, H. R., Seo, J. H., Ryu, S., & Jang, H. W. (2024). Nanoporous oxide electrodes for energy conversion and storage devices. *RSC Applied Interfaces*, 1(1), 11–42. https://doi.org/https://doi.org/10.1039/d3lf00094j. DOI: 10.1039/D3LF00094J

Yang, K., Fan, Q., Song, C., Zhang, Y., Sun, Y., Jiang, W., & Fu, P. (2023). Enhanced functional properties of porous carbon materials as high-performance electrode materials for supercapacitors. *Green Energy and Resources*, 1(3), 100030. DOI: 10.1016/j.gerr.2023.100030

Yang, R., Thakre, P., & Yang, V. (2005). Thermal Decomposition and Combustion of Ammonium Dinitramide [Review]. *Combustion, Explosion, and Shock Waves*, 41(6), 657–679. DOI: 10.1007/s10573-005-0079-y

Yang, S., Zavalij, P. Y., & Stanley Whittingham, M. (2001). Hydrothermal synthesis of lithium iron phosphate cathodes. *Electrochemistry Communications*, 3(9), 505–508. DOI: 10.1016/S1388-2481(01)00200-4

Yang, Z., Du, X., Ye, X., Qu, X., Duan, H., Xing, Y., Shao, L.-H., & Chen, C. (2021). The free-standing nanoporous palladium for hydrogen isotope storage. *Journal of Alloys and Compounds*, 854, 157062. https://doi.org/https://doi.org/10.1016/j.jallcom.2020.157062. DOI: 10.1016/j.jallcom.2020.157062

Yan, W., Hsiao, V. K. S., Zheng, Y. B., Shariff, Y. M., Gao, T., & Huang, T. J. (2009). Towards nanoporous polymer thin film-based drug delivery systems. *Thin Solid Films*, 517(5), 1794–1798. https://doi.org/https://doi.org/10.1016/j.tsf.2008.09.080. DOI: 10.1016/j.tsf.2008.09.080

Yan, Z., Luo, S., Li, Q., Wu, Z., & Liu, S. (2024). Recent Advances in Flexible Wearable Supercapacitors: Properties, Fabrication, and Applications. *Advancement of Science*, 11(8), 2302172. Advance online publication. DOI: 10.1002/advs.202302172 PMID: 37537662

Yao, J., Radin, S., Leboy, P. S., & Ducheyne, P. (2005). The effect of bioactive glass content on synthesis and bioactivity of composite poly (lactic-co-glycolic acid)/bioactive glass substrate for tissue engineering. *Biomaterials*, 26(14), 1935–1943. DOI: 10.1016/j.biomaterials.2004.06.027 PMID: 15576167

Yao, K., Chen, S., Guo, K., Tan, C. K. I., Mirshekarloo, M. S., & Tay, F. E. H. (2017). Lead-Free Piezoelectric Ceramic Coatings Fabricated by Thermal Spray Process. *IEEE Transactions on Ultrasonics, Ferroelectrics, and Frequency Control*, 64(11), 1758–1765. DOI: 10.1109/TUFFC.2017.2748154 PMID: 28880167

Yeetsorn, R., Tungkamani, S., & Maiket, Y. (2022). Fabrication of a ceramic foam catalyst using polymer foam scrap via the replica technique for dry reforming. *ACS Omega*, 7(5), 4202–4213. DOI: 10.1021/acsomega.1c05841 PMID: 35155913

Yeh, T. F., Lee, H. G., Chu, K. S., & Wang, C. B. (2004). Characterization and catalytic combustion of methane over hexaaluminates. *Materials Science and Engineering A*, 384(1–2), 324–330. DOI: 10.1016/S0921-5093(04)00835-4

Yergaziyeva, G., Kuspanov, Z., Mambetova, M., Khudaibergenov, N., Makayeva, N., Daulbayev, C. (2024). Advancements in catalytic, photocatalytic, and electrocatalytic CO2 conversion processes: Current trends and future outlook. Journal of CO2 Utilization, 80, 102682. https://doi.org/DOI: 10.1016/j.jcou.2024.102682

Yiannopoulos, Y. D., Chryssikos, G. D., & Kamitsos, E. I. (2001). Structure and properties of alkaline earth borate glasses. *Physics and Chemistry of Glasses*, 42(3), 164–172. https://helios-eie.ekt.gr/EIE/handle/10442/7036

Yi, H., Nakabayashi, K., Yoon, S., & Miyawaki, J. (2021). Pressurized physical activation: A simple production method for activated carbon with a highly developed pore structure. *Carbon*, 183, 735–742. DOI: 10.1016/j.carbon.2021.07.061

Yılmaz, E., Türk, S., Semerci, A. B., Kırkbınar, M., İbrahimoğlu, E., & Çalışkan, F. (2024). Bioactive apatite-wollastonite glass ceramics coating on metallic titanium for biomedical applications: effect of boron. *Journal of biological inorganic chemistry: JBIC: a publication of the Society of Biological Inorganic Chemistry, 29*(1), 75–85. https://doi.org/DOI: 10.1007/s00775-023-02032-y

Yin, Z., Tian, B., Zhu, Q., & Duan, C. (2019). Characterization and Application of PVDF and Its Copolymer Films Prepared by Spin-Coating and Langmuir-Blodgett Method. *Polymers*, 11(12), 2033. DOI: 10.3390/polym11122033 PMID: 31817985

Yoldas, B. E. (1975). Alumina gels that form porous transparent Al2O3. *Journal of Materials Science*, 10(11), 1856–1860. DOI: 10.1007/BF00754473

Yong, Y., Zhang, W., Hou, Q., Gao, R., Yuan, X., Hu, S., & Kuang, Y. (2022). Highly sensitive and selective gas sensors based on nanoporous CN monolayer for reusable detection of NO, H2S and NH3: A first-principles study. *Applied Surface Science*, 606, 154806. https://doi.org/https://doi.org/10.1016/j.apsusc.2022.154806. DOI: 10.1016/j.apsusc.2022.154806

Yoo, D., Kim, M., Oh, S. K., Hwang, S., Kim, S., Kim, W., Kwon, Y., Jo, Y., & Jeon, J. (2024). Synthesis of Hydroxylammonium Nitrate and Its Decomposition over Metal Oxide/Honeycomb Catalysts. *Catalysts*, 14(2), 116. DOI: 10.3390/catal14020116

Young-Rojanschi, C., & Madramootoo, C. (2014). Intermittent versus continuous operation of biosand filters. *Water Research*, 49(1), 1–10. DOI: 10.1016/j.watres.2013.11.011 PMID: 24316177

Youn, Y., Miller, J., Nwe, K., Hwang, K. J., Choi, C., Kim, Y., & Jin, S. (2019). Effects of Metal Dopings on CuCr2O4 Pigment for Use in Concentrated Solar Power Solar Selective Coatings. *ACS Applied Energy Materials*, 2(1), 882–888. DOI: 10.1021/acsaem.8b01976

Yuan, C., Xu, H. A., El-khodary, S., Ni, G., Esakkimuthu, S., Zhong, S., & Wang, S. (2024). Recent advances and challenges in biomass-derived carbon materials for supercapacitors: A review. *Fuel*, 362, 130795. DOI: 10.1016/j.fuel.2023.130795

Yuan, X., Wu, T., Lu, T., & Ye, J. (2023). Effects of Zinc and Strontium Doping on In Vitro Osteogenesis and Angiogenesis of Calcium Silicate/Calcium Phosphate Cement. *ACS Biomaterials Science & Engineering*, 9(10), 5761–5771. DOI: 10.1021/acsbiomaterials.3c00193 PMID: 37676927

Yurt, A., Balaban, A., Kandemir, S., Bereket, G., & Erk, B. (2004). Investigation on some Schiff bases as HCl corrosion inhibitors for carbon steel. *Materials Chemistry and Physics*, 85(2–3), 420–426. DOI: 10.1016/j.matchemphys.2004.01.033

Yusuf, K. O., Adio-Yusuf, S. I., & Obalowu, R. O. (2019). Development of a simplified slow sand filter for water purification. *Journal of Applied Science & Environmental Management*, 23(3), 389. DOI: 10.4314/jasem.v23i3.3

Yu, X., Wang, J., Huang, Z.-H., Shen, W., & Kang, F. (2013). Ordered mesoporous carbon nanospheres as electrode materials for high-performance supercapacitors. *Electrochemistry Communications*, 36, 66–70. DOI: 10.1016/j.elecom.2013.09.010

Zabukovec Logar, N., & Kaučič, V. (2006). She was visiting researcher in the Structural chemistry group at the University of Manchester, UK in 1995 and 1996. She was promoted to Assistant Professor in Chemistry at the University of Nova Gorica in 2004. She has worked in the Laboratory for Inorganic Chemistry and Technology at the National Institute of Chemistry in Ljubljana since. In *currently Acta Chim. Slov* (Vol. 53).

Zachariasen, W. H. (1932). THE ATOMIC ARRANGEMENT IN GLASS. *Journal of the American Chemical Society*, 54(10), 3841–3851. DOI: 10.1021/ja01349a006

Zanco, S. E., Joss, L., Hefti, M., Gazzani, M., & Mazzotti, M. (2017). Addressing the Criticalities for the Deployment of Adsorption-based CO2 Capture Processes. *Energy Procedia*, 114, 2497–2505. https://doi.org/https://doi.org/10.1016/j.egypro.2017.03.1407. DOI: 10.1016/j.egypro.2017.03.1407

Zearley, T. L., & Summers, R. S. (2012). Removal of trace organic micropollutants by drinking water biological filters. *Environmental Science & Technology*, 46(17), 9412–9419. DOI: 10.1021/es301428e PMID: 22881485

Zha, J., & Roggendorf, H. (1991). Sol–gel science, the physics and chemistry of sol–gel processing, Ed. by C. J. Brinker and G. W. Scherer, Academic Press, Boston 1990, xiv, 908 pp., bound—ISBN 0-12-134970-5. *Advanced Materials, 3*(10), 522–522. DOI: 10.1002/adma.19910031025

Zhang, C., Firestein, K. L., Fernando, J. F. S., Siriwardena, D., von Treifeldt, J. E., & Golberg, D. (2020). Recent Progress of In Situ Transmission Electron Microscopy for Energy Materials. *Advanced Materials*, 32(18), 1904094. Advance online publication. DOI: 10.1002/adma.201904094 PMID: 31566272

Zhang, C., Liu, B., Li, W., Liu, X., Wang, K., Deng, Y., Guo, Z., & Lv, Z. (2021). A well-designed honeycomb Co_3O_4@CdS photocatalyst derived from cobalt foam for high-efficiency visible-light H_2 evolution. *Journal of Materials Chemistry. A, Materials for Energy and Sustainability*, 9(19), 11665–11673. DOI: 10.1039/D0TA11433B

Zhang, C., Sui, J., Li, J., Tang, Y., & Cai, W. (2012). Efficient removal of heavy metal ions by thiol-functionalized superparamagnetic carbon nanotubes. *Chemical Engineering Journal*, 210, 45–52. DOI: 10.1016/j.cej.2012.08.062

Zhang, F., Nriagu, J. O., & Itoh, H. (2005). Mercury removal from water using activated carbons derived from organic sewage sludge. *Water Research*, 39(2–3), 389–395. DOI: 10.1016/j.watres.2004.09.027 PMID: 15644247

Zhang, L., Sadanandam, G., Liu, X., & Scurrell, M. S. (2017). Carbon surface modifications by plasma for catalyst support and electrode materials applications. *Topics in Catalysis*, 60(12–14), 823–830. DOI: 10.1007/s11244-017-0747-7

Zhang, Q., & Cao, G. (2011). Nanostructured photoelectrodes for dye-sensitized solar cells. *Nano Today*, 6(1), 91–109. DOI: 10.1016/j.nantod.2010.12.007

Zhang, W., Yin, J., Wang, C., Zhao, L., Jian, W., Lu, K., Lin, H., Qiu, X., & Alshareef, H. N. (2021). Lignin Derived Porous Carbons: Synthesis Methods and Supercapacitor Applications. *Small Methods*, 5(11), 2100896. Advance online publication. DOI: 10.1002/smtd.202100896 PMID: 34927974

Zhang, X., Zhang, K., Zhang, L., Wang, W., Li, Y., & He, R. (2022). Additive manufacturing of cellular ceramic structures: From structure to structure–function integration. *Materials & Design*, 215, 110470. DOI: 10.1016/j.matdes.2022.110470

Zhang, X., Zhang, Y., Su, Y., & Guan, S. (2024). Enhancing the corrosion inhibition performance of Mannich base on mild steel in lactic acid solution through synergistic effect of allicin: Experimental and theoretical study. *Journal of Molecular Structure*, 1304, 137658. DOI: 10.1016/j.molstruc.2024.137658

Zhao, X. S. (n.d.). *NANOPOROUS MATERIALS-AN OVERVIEW*.

Zhao, C., Ge, L., Zuo, M., Mai, L., Chen, S., Li, X., Li, Q., Wang, Y., & Xu, C. (2023). Study on the mechanical strength and iodine adsorption behavior of coal-based activated carbon based on orthogonal experiments. *Energy*, 282, 128450. DOI: 10.1016/j.energy.2023.128450

Zhao, D., Feng, J., Huo, Q., Melosh, N., Fredrickson, G. H., Chmelka, B. F., & Stucky, G. D. (1998). Triblock Copolymer Syntheses of Mesoporous Silica with Periodic 50 to 300 Angstrom Pores. *Science*, 279(5350), 548–552. DOI: 10.1126/science.279.5350.548 PMID: 9438845

Zhao, D., Xu, D., Wang, T., Yang, Z., Zhao, D., Xu, D., Wang, T., & Yang, Z. (2024). Nitrogen-Rich Nanoporous Carbon with MXene Composite for High-Performance Zn-ion Hybrid Capacitors. *Materials Today. Energy*, 101671, 101671. https://doi.org/https://doi.org/10.1016/j.mtener.2024.101671. DOI: 10.1016/j.mtener.2024.101671

Zhao, Z., Yan, K., Guan, Q., Guo, Q., & Zhao, C. (2024). Mechanism and physical activities in bone-skeletal muscle crosstalk. *Frontiers in Endocrinology*, 14, 1287972. DOI: 10.3389/fendo.2023.1287972 PMID: 38239981

Zheng, J., Ren, X., Song, Y., & Ge, X. (2009). Catalytic combustion of methane over iron- and manganese-substituted lanthanum hexaaluminates. *Reaction Kinetics and Catalysis Letters*, 97(1), 109–114. DOI: 10.1007/s11144-009-0013-5

Zheng, K., Yang, F., Wang, X., & Zhang, Z. (2014). Investigation of Self-Diffusion and Structure in Calcium Aluminosilicate Slags by Molecular Dynamics Simulation. *Materials Sciences and Applications*, 05(02), 73–80. DOI: 10.4236/msa.2014.52011

Zheng, Q., Potuzak, M., Mauro, J. C., Smedskjaer, M. M., Youngman, R. E., & Yue, Y. (2012). Composition–structure–property relationships in boroaluminosilicate glasses. *Journal of Non-Crystalline Solids*, 358(6–7), 993–1002. DOI: 10.1016/j.jnoncrysol.2012.01.030

Zhi, C., Bando, Y., Tang, C., & Golberg, D. (2005). Immobilization of proteins on boron nitride nanotubes. *Journal of the American Chemical Society*, 127(49), 17144–17145. DOI: 10.1021/ja055989+ PMID: 16332036

Zhong, J., & Greenspan, D. C. (2000). Processing and properties of sol-gel bioactive glasses. *Journal of Biomedical Materials Research*, 53(6), 694–701. DOI: 10.1002/1097-4636(2000)53:6<694::AID-JBM12>3.0.CO;2-6 PMID: 11074429

Zhou, T., Zhai, T., Ma, J., Wu, F., & Zhang, G. (2024). Sodium chloride-activated polymeric ferric sulfate-modified natural zeolite (Z-Na-Fe): Application of bio-slow filtration for the removal of ammonia and phosphorus from micro-polluted cellar water. *Journal of Water Process Engineering*, 61, 105367. DOI: 10.1016/j.jwpe.2024.105367

Zhou, X., Niu, C., Li, K., Lin, P., & Xu, K. (2024). Vitrification of lead–bismuth alloy nuclear waste into a glass waste form. *International Journal of Applied Glass Science*, 15(2), 139–147. DOI: 10.1111/ijag.16656

Zhou, Y., Niu, B., Wu, B., Luo, S., Fu, J., Zhao, Y., Quan, G., Pan, X., & Wu, C. (2020). A homogenous nanoporous pulmonary drug delivery system based on metal-organic frameworks with fine aerosolization performance and good compatibility. *Acta Pharmaceutica Sinica. B*, 10(12), 2404–2416. https://doi.org/https://doi.org/10.1016/j.apsb.2020.07.018. DOI: 10.1016/j.apsb.2020.07.018 PMID: 33354510

Zhou, Y., Zhang, Y., Tang, L., Long, B., & Zeng, G. (2019). Nanoporous Materials Based Sensors for Pollutant Detection. In Tang, L., Deng, Y., Wang, J., Wang, J., & Zeng, G. (Eds.), *Nanohybrid and Nanoporous Materials for Aquatic Pollution Control* (pp. 265–291). Elsevier., https://doi.org/https://doi.org/10.1016/B978-0-12-814154-0.00009-8 DOI: 10.1016/B978-0-12-814154-0.00009-8

Zhou, Z., Li, W., He, T., Qian, L., Tan, G., & Ning, C. (2016). Polarization of an electroactive functional film on titanium for inducing osteogenic differentiation. *Scientific Reports*, 6(1), 35512. DOI: 10.1038/srep35512 PMID: 27762318

Zhu, Y., Zhang, X., Chang, G., Deng, S., & Chan, H. F. (2024). Bioactive Glass in Tissue Regeneration: Unveiling Recent Advances in Regenerative Strategies and Applications. *Advanced materials (Deerfield Beach, Fla.)*, e2312964. https://doi.org/DOI: 10.1002/adma.202312964

Zhuang, Z., Chen, Y., Chen, K., Liu, Z., Guo, Z., & Huang, X. (2023). In-situ synthesis of ZnO onto nanoporous gold microelectrode for the electrochemical sensing of arsenic(III) in near-neutral conditions. *Sensors and Actuators. B, Chemical*, 378, 133184. https://doi.org/https://doi.org/10.1016/j.snb.2022.133184. DOI: 10.1016/j.snb.2022.133184

Zhu, B., Xu, Q., Bao, X., Yin, H., Qin, Y., & Shen, X. (2022). Highly selective CO_2 capture and photoreduction over porous carbon nitride foams/LDH monolith. *Chemical Engineering Journal*, 429, 132284. DOI: 10.1016/j.cej.2021.132284

Zhu, J., Yin, Z., Li, H., Tan, H., Chow, C. L., Zhang, H., Hng, H. H., Ma, J., & Yan, Q. (2011). Bottom-Up Preparation of Porous Metal-Oxide Ultrathin Sheets with Adjustable Composition/Phases and Their Applications. *Small*, 7(24), 3458–3464. DOI: 10.1002/smll.201101729 PMID: 22058077

Zhu, Q., Song, X., Chen, X., Li, D., Tang, X., Chen, J., & Yuan, Q. (2024). A high performance nanocellulose-PVDF based piezoelectric nanogenerator based on the highly active CNF@ ZnO via electrospinning technology. *Nano Energy*, 127, 109741. DOI: 10.1016/j.nanoen.2024.109741

Zobell, C. E. (1943). The Effect of Solid Surfaces upon Bacterial Activity. *Journal of Bacteriology*, 46(1), 39–56. DOI: 10.1128/jb.46.1.39-56.1943 PMID: 16560677

About the Contributors

Assia Mabrouk, PhD, is a luminary in the realm of materials physics, heralding from a distinguished academic background and a profound commitment to scientific inquiry. Her journey began with a steadfast pursuit of knowledge culminating in a Ph.D. in Materials Physics from the esteemed University of Orléans, France, in 2017. Following the completion of her doctoral studies, Assia embarked on a transformative journey at CEA Cadarache in France, where she honed her expertise in a cutting-edge project focused on the development of a potentiometric probe to measure oxygen in sodium. Her tenure at CEA Cadarache, spanning until 2019, was marked by a relentless pursuit of excellence and a deep-seated passion for pushing the boundaries of scientific innovation. Currently serving as a professor at Ibnou Zohr University in Morocco, Assia Mabrouk continues to leave an indelible mark on the landscape of materials science. Her primary research interests lie in the synthesis, structural characterization, and mechanical properties of glasses and ceramics, areas that hold immense promise for technological advancement and industrial applications. As a scholar, Assia's work represents the intersection of rigorous scientific inquiry and real-world impact. Through her groundbreaking research, she seeks to unravel the mysteries of material behavior and unlock the potential for transformative technological solutions. Her contributions not only advance our understanding of fundamental scientific principles but also hold the key to addressing pressing societal challenges. Beyond her remarkable achievements in academia, Assia Mabrouk is a beacon of inspiration for aspiring scientists and researchers, embodying the values of intellectual curiosity, perseverance, and innovation. Her unwavering commitment to pushing the boundaries of knowledge serves as a guiding light for future generations, inspiring them to dream big and pursue their passions with unwavering determination. In the ever-evolving landscape of materials science, Assia Mabrouk stands as a testament to the power of intellect, dedication, and the unyielding pursuit of excellence. Her journey serves as a reminder that through curiosity and perseverance, we have the ability to unlock the secrets of the universe and shape the course of human progress.

Ahmed Bachar's academic journey is marked by a profound dedication to the field of Materials Science. Graduating with distinction, he earned his professional master's degree in Materials Engineering from the esteemed University of Valenciennes in France in 2012. Fuelled by a passion for deeper understanding, Ahmed pursued further academic heights, culminating in the completion of his doctorate in Materials Science from the same institution. Since 2016, Ahmed has served as an esteemed accredited professor at Ibn Zohr University in Morocco, where he has imparted his wealth of knowledge and expertise to countless eager minds. His tenure at the university has been marked by a commitment to excellence in education, nurturing the next generation of aspiring scientists and engineers. At the core of Ahmed's scholarly pursuits lies a fascination with the synthesis, physicochemical properties, and intricate property-structure relationships inherent in ceramics and glassy compounds. His research endeavors delve into the fundamental principles governing these materials, seeking to unlock their full potential and push the boundaries of scientific understanding. Ahmed BACHAR's contributions to the field stand as a testament to his unwavering dedication and intellectual curiosity, enriching the academic landscape and paving the way for groundbreaking advancements in Materials Science.

Seitkhan Azat, PhD, Associate Professor, stands as a towering figure in the realm of materials science, chemistry, and nanotechnology. As the Head of the Laboratory of Engineering Profile at Satbayev University, his influence extends far beyond the confines of academia, shaping the forefront of research and innovation in multiple industries. With a relentless pursuit of knowledge and an insatiable curiosity, Seitkhan has carved out a distinguished career marked by groundbreaking discoveries and transformative contributions. His expertise lies in the intricate domain of nanotechnology, where he explores its multifaceted applications across diverse sectors including energy, chemistry, biomedicine, and environmental science. At the heart of Seitkhan's research endeavors lies a commitment to developing novel materials that redefine the boundaries of possibility. His pioneering work encompasses the synthesis of advanced materials, sorbents, and filters, all aimed at revolutionizing water purification techniques. Through his tireless dedication, Seitkhan has unlocked innovative solutions to address the pressing global challenge of water scarcity, offering hope for a more sustainable future. Beyond his pioneering work in water purification, Seitkhan's research extends to the synthesis and application of carbon materials, where he seeks to harness their unique properties for a myriad of purposes. From mitigating the impact of toxic gases to facilitating the separation and purification of liquids, his contributions have far-reaching implications across industries, driving progress and innovation on a global scale. A prolific author and esteemed scholar, Seitkhan's extensive body of published works and monographs serves as a testament to his intellectual prowess and enduring commitment to advancing scientific knowledge. His insights have not only expanded the frontiers of understanding but have also inspired countless researchers to push the boundaries of what is possible. As a visionary leader and mentor, Seitkhan Azat continues to inspire the next generation of scientists and engineers, instilling in them the values of curiosity, perseverance, and excellence. Through his unwavering dedication to scientific inquiry and innovation, he remains at the vanguard of transformative research, shaping the course of scientific discovery for generations to come.

Rachid Amrousse emerges as a pioneering figure in the realm of aerospace engineering, catalysis, and propulsion, with a career trajectory marked by relentless pursuit of scientific excellence and groundbreaking innovation. His journey began with humble origins, earning his high school certificate in experimental sciences in 2002, laying the foundation for a remarkable odyssey in academia and research. After obtaining his Bachelor's degree from El Jadida University in 2006, Prof. Rachid embarked on a transformative educational journey, setting his sights on the University of Poitiers, France. There, he earned his Master's degree in 2007, demonstrating exceptional aptitude and dedication to his field of study. The pinnacle of Prof. Rachid's academic journey was reached in 2010 when he attained his Ph.D. degree from the University of Poitiers, in collaboration with the prestigious French space agency (CNES) and Air Liquide Society. His doctoral dissertation, titled "The ignition of cold hydrogen-oxygen mixtures for satellite control reactions," showcased his expertise in the intricate mechanisms of space propulsion and combustion dynamics. In 2011, Prof. Rachid's career took an international turn as he joined the esteemed ranks of the Japan Aerospace Exploration Agency (JAXA) as a Research Associate, becoming the first foreign national to be recruited by JAXA a testament to his exceptional skills and reputation in the field. His research at JAXA focused on the development of high-energetic materials and the application of catalysis to space propulsion, encompassing a wide array of topics ranging from thermal decomposition to combustion mechanisms. A testament to his prowess as a researcher, Prof. Rachid secured competitive funding from the Japan Society for the Promotion of Science (JSPS) in 2016, enabling him to spearhead groundbreaking research initiatives aimed at developing green propellant thrusters for small satellite applications. This achievement underscored his status as a leading authority in the field of propulsion technology. In parallel to his illustrious career in Japan, Prof. Rachid made significant contributions to academia in his home country, Morocco, joining the University of Chouaib Doukkali (UCD) in El Jadida City in 2016. His rapid ascent within the academic ranks culminated in his promotion to the position of Associate Professor in 2020, further solidifying his status as a respected scholar and educator. Throughout his career, Prof. Rachid has garnered numerous accolades and awards for his groundbreaking research, receiving allowances and recognition from prestigious institutions such as the Poitou-Charentes Region, CNES, JAXA, and JSPS. His prolific output includes over 50 peer-reviewed papers and proceedings, along with an impressive roster of over 80 talks and presentations at conferences worldwide. Prof. Rachid's contributions extend beyond his research endeavors, as evidenced by his active involvement in professional societies such as the Japan Explosives Society, where he serves as an esteemed member. His commitment to advancing the frontiers of science and technology has earned him widespread recognition, including the esteemed National Medal of Merit from the King of Morocco, Mohammed VI, in 2016—a testament to his enduring impact and legacy in the field of aerospace engineering. Currently, Prof. Rachid Amrousse is at the forefront of knowledge dissemination as the co-editor of a forthcoming book titled "Recent Advancements in Green Propulsion" with Springer Nature, further cementing his status as a thought leader and influencer in the field. His journey serves as an inspiration to aspiring researchers worldwide, embodying the spirit of curiosity, perseverance, and innovation that drives scientific progress forward.

***

Tolganay Atamanova is a PhD researcher at the Institute of Combustion Problems. Her research interests include synthesizing nanomaterials and their applications, producing carbon fibers through electrospinning, obtaining supercapacitor electrodes, disposing of waste materials such as used tires and solid waste, and extraction cellulose from plant materials

Astrid L. Giraldo is Physical Engineer and Master in Science-Physics graduated from the Universidad Nacional de Colombia, campus Manizales. Doctor in Science- Materials science graduated from Cinvestav Campus Querétaro. Currently, she works as CONACYT commissioned to the Cinvestav unit Querétaro. Co-founder of the Mexican Laboratory of Thermal Spray - CENAPROT. Her research area is focused on multifunctional materials; especially in bioceramics: synthesis of raw material such hydroxyapatite from natural sources and bioglasses, hydroxyapatite-based composite materials to coating manufacture using different thermal spray technologies and their characterization. Additionally, she is working on the microencapsulation processes development for controlled drug release.

Stuart Hampshire is Emeritus Professor of Materials Science at the University of Limerick (UL). He was a founding member of the Materials and Surface Science Institute (MSSI) at the University and, as Associate Vice-president for Research, was coordinator of successful applications for funding totalling ~16 million euro from the Irish government and philanthropic institutions to establish the first phase of the MSSI, now the Bernal Institute, the leading research institute at UL. Professor Hampshire helped build its reputation as a major University for materials research on the international scene. In his research, he has investigated the relationships between chemistry, fabrication, structure and properties of silicon nitride and oxynitride ceramics and related materials which, because of their special properties, can be used in applications requiring high temperature resistance and wear resistance such as engine bearings and cutting tools. He has published over 200 papers in international peer-reviewed journals and is now recognized as a leading authority worldwide on oxynitride glasses, their properties and crystallization and is invited on a regular basis to give keynote lectures at international conferences. In 1989, Professor Hampshire was a Visiting Scientist for 3 months at the Government Industrial Research Institute, Nagoya, Japan where he was involved in collaborative research on Oxynitride Glasses. He was later involved in the Synergy Ceramics project, NEDO, Japan / Fine Ceramics Research Association, Tokyo. Professor Hampshire was elected Fellow of the American Ceramic Society in 2007 and received the International Bridge-building Award of the Engineering Ceramics Division of the Society in 2008. He was further honoured by the American Ceramic Society with a Symposium in his Honor (8th Intl. Symp. Adv. Processing and Manufacturing Technologies) as part of the 38th Intl. Conf. on Advanced Ceramics and Composites (ICACC), Daytona Beach, Florida (2014). In 2022, Professor Hampshire was conferred with the W. David Kingery Award which recognizes distinguished lifelong achievements involving multidisciplinary and global contributions to ceramic technology, science, education and art. Professor Hampshire received the prestigious Stuijts Award of the European Ceramic Society (ECerS) in 2007 which is presented each two years to a "ceramist noted for outstanding contributions in the field of ceramic science and the transfer to European industry". He was elected a Fellow of ECerS in 2013 and also serves as one of the two Irish members of the Council of ECerS. He has also served on the membership nomination committee of the World Academy of Ceramics. In 2009 Professor Hampshire was conferred with an Honorary Doctorate by the University of Limoges, France. He has been a regular visitor to Limoges and has been involved in various collaborative projects over the years. Professor Hampshire has been a Visiting Professor at the University of Rennes, France (1996, 1997), and at the University of Valenciennes et du Hainaut-Cambresis (now Université Polytechnique des Hauts-de-France), France (2010-2015), where he was involved in supervision and mentoring of researchers involved in special glasses for biomedical applications.

John Henao is currently professor at the National Council of Humanities, Science, and Technology (CONAHCYT) unit-CIATEQ A.C. in Querétaro, Mexico, and is member of the mexican system of researchers (SNI) level 1. He is materials engineer from the University of Antioquia, Colombia, and obtained his master degree in high performance materials engineering from the University of Limoges, France. He also obtained his PhD. in Engineering and Advanced Technologies from the University of Barcelona, Spain. During the last years, He has been working on the field of thermal spray, biomaterials, and materials for energy and corrosion applications.

Madhukumar K is a distinguished academician in Mechanical Engineering, specializing in Composites, holds a Ph.D. in the field and has dedicated over 13 years to academia. Currently serving as an Assistant Professor in the Mechanical Engineering Department at Sir M. Visvesvaraya Institute of Technology, this scholar's academic journey reflects a deep commitment to advancing knowledge in the discipline. With broad research interests spanning Material Science, Light Alloys, Composites, Artificial Aging, and Tool Design, the diverse expertise positions him as a well-rounded authority within Mechanical Engineering. In addition to teaching, substantial contributions have been made through various publications. As the author of the textbooks Beyond Boundaries: The World of Additive Manufacturing and Smart Materials and Applications, there is a profound understanding of these subjects. Furthermore, the scholarly impact is reinforced by the publication of five book chapters with respected international publishers. The academic and research contributions are highlighted by 18 research papers published in reputable international journals, and 10 papers presented at national and international conferences. Additionally, eight patents have been issued, underscoring practical contributions to the field. These accomplishments enrich both academic discourse and practical applications, solidifying a significant influence within the academic and research community.

Sampath Kumar L, a distinguished writer and educator, began his academic journey with a Bachelor's degree in Mechanical Engineering from SJMIT, Chitradurga, in 2008. He then earned a Master's degree in Engineering Analysis and Design at SDMCET, Dharwad, in 2011 and completed his Ph.D. in Nano Biomaterials with a focus on the Mechanical Stream from VTU Belagavi in 2023. He started his professional career as a Lecturer in the Mechanical Engineering Department at SIET, Tumakuru, from 2011 to 2013, and currently serves as an Assistant Professor in the Mechanical Engineering Department at Sir MVIT, Bengaluru, a role he has held since 2013. Dr. Kumar's research interests encompass Nano Biomaterials, Mechanical Characterization, and the green and chemical synthesis of nanomaterials. He has published 15 research papers in reputed journals and three patent publications. He authored the book "Beyond Boundaries: The World of Additive Manufacturing" (1st Edition), published by IIP Iterative International Publishers in 2024, with ISBN: 978-1-68576-470-8. Additionally, Dr. Kumar serves as a reviewer for Applied Nanoscience (APNA), a Springer Nature journal indexed in Scopus and listed in Q2 with an H-index of 80. His interdisciplinary approach and significant contributions to research and education make him an inspirational figure in academia, continuously advancing knowledge and innovation in his field.

Kiran Kumar M earned his B.E. in Mechanical Engineering from University BDT College of Engineering, followed by an M.Tech in Thermal Power Engineering and a Ph.D. in I.C. Engine Fuels and Combustion Systems. He is currently an Associate Professor in the Department of Mechanical Engineering at Sir Visvesvaraya Institute of Technology, Bengaluru, Karnataka, affiliated with Visvesvaraya Technological University (VTU). His research has been published in prestigious journals, including those indexed by SCOPUS, UGC, and other international databases. Innovative contributions to the field are highlighted by five national patents and one international patent. Additionally, he is the author of the textbook Advancements in Electric and Hybrid Vehicles. Active participation in international conferences, where numerous research articles have been presented, reflects a strong commitment to scientific collaboration. Practical engineering experience includes mentoring and leading teams in competitions such as the "Shell Eco-marathon" in the Asia Pacific region and designing a supercar for the "SAE Student F-1 Race Car" competition. With a wealth of experience in Mechanical Engineering, his diverse research interests and contributions to both academia and practical engineering projects underscore a significant dedication to the field.

Zulkhair Mansurov is a Doctor of Chemical Sciences. The area of scientific research is the study of the kinetics and mechanism of combustion of hydrocarbons and the structure of cold and sooty flames, the synthesis and study of nanocarbon materials for various functional purposes. He was the first to discover hydrogen atoms in cold hydrocarbon flames and established oscillatory regimes of temperature and concentration of radicals in cold flames of butane and hexane; a phenomenological model for explaining the oscillatory regime during oxidation is proposed.

Sasikumar N is a dedicated academician in the field of Physics, specialising in Solid State Physics. He has a master's and doctorate in physics, as well as an MBA in marketing. He has 22 years of academic experience and is currently the Assistant Professor of Physics at Sir M Visvesvaraya Institute of Technology in Bangalore, India. His research interests are on material science, nanomaterials, and computational physics.

Carlos Poblano is a metallurgical engineer graduated from Universidad Nacional Autonoma de Mexico where he also obtained a MSc degree in metallurgy. He obtained a PhD degree in materials science at the University of Sheffield, UK. Currently, he is director of the advanced materials department at CIATEQ A.C. His research lines include the fabrication of different metallic, ceramic,and composite materials by additive manufacturing and thermal spraying; he also works in powder metallurgy and corrosion science projects.

Sapargali I.O. Bachelor of Technical Sciences, majoring in "Engineering Physics and Materials Science" (KazNTU 2023.) In 2024, she entered the master's program at Tomsk Polytechnic University, majoring in "Materials Science and Technologies of New Materials."Currently he is a research fellow for scientific projects.

Azamat Taurbekov is a 3rd year postgraduate student of KazNU named after Al-Farabi, who is studying in the specialty "Nanotechnologies and Nanomaterials". Since 2017, he has been working as a researcher in the laboratory of functional nanomaterials of the Institute of Combustion Problems (Kazakhstan, Almaty). His research interests: extraction of nanocellulose from plant materials, synthesis of nanoporous carbon material by various methods and its application as electrodes for supercapacitors and for the extraction of noble metals.

Index

A

B

C

E

F

G

W

www.ingramcontent.com/pod-product-compliance
Ingram Content Group UK Ltd.
Pitfield, Milton Keynes, MK11 3LW, UK
UKHW012125171224
452514UK00011B/74